# Spiders of
# North America

## DEDICATION

This book is dedicated to the memory of Vincent D. Roth (1924-1997) who began work on this project over three decades ago. With the collaboration of many in the arachnological community, he compiled keys and information about spider genera found in North America and eventually produced three editions of his Spider Genera guide. We remain grateful for his tireless efforts and inspiration.

## AUTHORS

**Robert G. Bennett**
British Columbia Ministry of Forests
Saanichton, BC Canada
Robb.Bennett@gems6.gov.bc.ca

**Jason E. Bond**
Auburn University
Auburn, AL USA
jbond@mail.auburn.edu

**Allen R. Brady**
Hope College, Holland, MI USA
brady@hope.edu

**Donald J. Buckle**
Saskatoon, SK Canada
djbuckle@shaw.ca

**H. Don Cameron**
University of Michigan
Ann Arbor, MI USA
HDCamero@umich.edu

**James E. Carico**
Lynchburg College
Lynchburg, VA USA
deceased

**Jonathan A. Coddington**
National Museum of Natural History
Smithsonian Institution
Washington, D.C. USA
coddington@si.edu

**James C. Cokendolpher**
Museum of Texas Tech University
Lubbock, TX USA
Cokendolpher@aol.com

**Frederick A. Coyle**
Western Carolina University
Cullowhee, NC USA
coyle@email.wcu.edu

**Patrick R. Craig**
California Academy of Sciences
San Francisco, CA USA
amberid@inreach.com

**Sarah C. Crews**
University of California
Berkeley, CA USA
screws@nature.berkeley.edu

**Paula E. Cushing**
Denver Museum of Nature & Science
Denver, CO USA
Paula.Cushing@dmns.org

**Bruce Cutler**
University of Kansas
Lawrence, KS USA
bcutler@ku.edu

**Charles D. Dondale**
Canadian National Collection,
Ottawa, ON Canada
cjdondale@storm.ca

**Michael L. Draney**
University of Wisconsin
Green Bay, WI USA
draneym@uwgb.edu

**G.B. Edwards**
Florida State Collection of Arthropods Gainesville, FL USA
edwardg@doacs.state.fl.edu

**Marshal Hedin**
San Diego State University
San Diego, CA USA
mhedin@sciences.sdsu.edu

**Brent E. Hendrixson**
Millsaps College
Jackson, MS USA
hendrb@millsaps.edu

**Gustavo Hormiga**
George Washington University
Washington, DC USA
hormiga@gwu.edu

**Bernhard A. Huber**
Zoologisches Forschungsinstitut und
Museum Alexander Koenig
Bonn, Germany
b.huber.zfmk@Uni-Bonn.De

**Joel M. Ledford**
California Academy of Sciences
San Francisco, CA USA
JLedford@calacademy.org

**Herbert W. Levi**
Museum of Comparative Zoology
Harvard University
Cambridge, MA USA
levi@fas.harvard.edu
herblevi@mac.com

**Stephen E. Lew**
Oakland, CA USA
anothersteve lew@gmail.com

**Lara Lopardo**
Zoologisches Institut und Museum
Allgemeine & Systematische Zoologie
Ernst-Moritz-Arndt-Universität
Greifswald, Germany
laralopardo@gmail.com

**Daniel J. Mott**
Texas A & M International University
Laredo, TX USA
dmott@tamiu.edu

**Brent D. Opell**
Virginia Tech, Blacksburg, VA USA
bopell@vt.edu

**Pierre Paquin**
Shefford, QC, Canada
pierre.paquin@scienceinfuse.org

**Thomas R. Prentice**
University of California
Riverside, CA USA
prentice@citrus.ucr.edu

**David B. Richman**
New Mexico State University
Las Cruces, NM USA
rdavid@nmsu.edu

**Adalberto J. Santos**
Instituto Butantan, São Paulo, Brazil
oxyopes@yahoo.com

**Warren E. Savary**
California Academy of Sciences
San Francisco, CA USA
wsavary@yahoo.com

**Diana Silva Dávila**
California Academy of Sciences
San Francisco, CA USA
dsil_2000@yahoo.com

**Darrell Ubick**
California Academy of Sciences
San Francisco, CA USA
dubick@calacademy.org

## ILLUSTRATORS

**Nadine Dupérré**
American Museum of Natural History
New York, NY USA
nadineduperre@gmail.com

**Eric Parrish**
University of Colorado
Boulder, CO USA
Eric.Parrish@colorado.edu

**Marisa Sripracha**
Center for Conservation Biology
University of California
Riverside CA, USA

## ACKNOWLEDGMENTS

This book could not have been possible without the cooperative efforts of numerous individuals in the international arachnological community. We are particularly grateful for the organizational support of the American Arachnological Society (AAS) under whose auspices this project was completed. Very special thanks go to an anonymous arachnophile, whose generous financial donation made possible a much higher quality manual. Thanks must also be extended to Pat Craig, whose enthusiastic announcement of this budding project sparked the anonymous donation and to Barbara Roth who encouraged us to update Vince's book.

Thanks to all the taxonomists, arachnologists, and others who worked to make these identification keys as complete, accurate, and user-friendly as possible. Particular thanks go to the chapter authors and illustrators listed in the beginning of the manual as well to as the following (listed alphabetically): Ingi Agnarsson, Fernando Alvarez, Joe Beatty, Bill Bennett, Richard Bradley, Robert Breene, Chris Buddle, Dick Burrows, Karen Cangialosi, Allen Dean, Pete DeVries, Charles Griswold, Fran Haas, Norman Horner, Jack Kaspar, Jan Kempf, Jeremy Miller, Amanda Musser, Frank Pascoe, Marius Pfeiffer, Norman Platnick, Martín Ramírez, Barbara Roth, Jozef Slowik, Suzanne Ubick, Rick Vetter and Bea Vogel. Thanks also to the AAS marketing team: Jason Bond, Greta Binford, Chris Buddle, Gary Miller, Brent Opell, Linda Rayor, and Bob Suter. Thanks to all the scientists, teachers, students, and amateur arachnologists who field tested the keys and whose comments and suggestions greatly improved each chapter.

Thanks to Cor Vink, Charles Dondale and Scott Larcher for proofing the entire manual.

Appreciation is extended to the following institutions and curators for the loan of material used in preparing the keys and illustrations: American Museum of Natural History (Norman I. Platnick), California Academy of Sciences (Charles Griswold), Canadian National Collection (Charles D. Dondale), Denver Museum of Nature & Science (Paula E. Cushing), Florida State Collection of Arthropods (G.B. Edwards), Museum of Comparative Zoology (Gonzalo Giribet), Museum of Southwestern Biology (Sandra Brantley), National Museum of Natural History, Smithsonian Institution (Jonathan Coddington), University of California, Riverside (Richard Vetter) and the personal collections of Nadine Dupérré, Pierre Paquin, and Darrell Ubick.

A separate thanks is extended to those who assisted Mike Draney and Don Buckle in the monumental task of drafting the first ever comprehensive key to North American Linyphiidae. These chapter authors wish to acknowledge the following for comments and suggestions: Paula Cushing, Allen Dean, Charles Dondale, Gustavo Hormiga, Max Larivée, Steven Lew, Jeremy Miller, Amanda Musser, Thomas Prentice, David Richman, Petra Sierwald, Nina Sandlin, participants of Draney's December 2003 Linyphiid workshop at FMNH, and especially Nadine Dupérré and Pierre Paquin. Jeremy Miller provided some missing data on chaetotaxy.

Last, but certainly not least, we must acknowledge three references that provided the most important sources of information and inspiration. The *World Spider Catalog* (Platnick 2005) was indispensable in providing the current status of Nearctic spiders and the relevant references. *African Spiders* (Dippenaar-Schoeman & Jocqué 1997) was inspirational in its systematic organization of the family chapters and the illustrated glossary, both of which were used with modification in this manual. Finally, the *Guide d'identification des Araignées du Québec* (Paquin & Dupérré 2003) provided the format for the illustrated keys, numerous illustrations, and graphical concepts.

D. Ubick, P. Paquin, P. E. Cushing, and N. Dupérré

## TABLE OF CONTENTS

| | |
|---|---|
| **Authors** | iii |
| **Ackowledgements** | iv |
| **Introduction** | 1 |
| **Phylogeny and classification** | 18 |
| **Key to families** | 25 |

| | | | |
|---|---|---|---|
| Antrodiaetidae | 39 | Mimetidae | 171 |
| Atypidae | 41 | Miturgidae | 173 |
| Ctenizidae | 43 | Mysmenidae | 175 |
| Cyrtaucheniidae | 45 | Nesticidae | 178 |
| Dipluridae | 48 | Ochyroceratidae | 181 |
| Mecicobothriidae | 50 | Oecobiidae | 183 |
| Nemesiidae | 52 | Oonopidae | 185 |
| Theraphosidae | 54 | Oxyopidae | 189 |
| | | Philodromidae | 192 |
| Agelenidae | 56 | Pholcidae | 194 |
| Amaurobiidae | 60 | Pimoidae | 197 |
| Amphinectidae | 63 | Pisauridae | 199 |
| Anapidae | 64 | Plectreuridae | 201 |
| Anyphaenidae | 66 | Prodidomidae | 203 |
| Araneidae | 68 | Salticidae | 205 |
| Caponiidae | 75 | Scytodidae | 217 |
| Clubionidae | 77 | Segestriidae | 219 |
| Corinnidae | 79 | Selenopidae | 221 |
| Ctenidae | 83 | Sicariidae | 222 |
| Cybaeidae | 85 | Sparassidae | 224 |
| Deinopidae | 91 | Symphytognathidae | 226 |
| Desidae | 93 | Telemidae | 228 |
| Dictynidae | 95 | Tengellidae | 230 |
| Diguetidae | 102 | Tetragnathidae | 232 |
| Dysderidae | 103 | Theridiidae | 235 |
| Filistatidae | 104 | Theridiosomatidae | 244 |
| Gnaphosidae | 106 | Thomisidae | 246 |
| Hahniidae | 112 | Titanoecidae | 248 |
| Hersiliidae | 116 | Trechaleidae | 249 |
| Homalonychidae | 118 | Uloboridae | 250 |
| Hypochilidae | 120 | Zodariidae | 254 |
| Leptonetidae | 122 | Zoridae | 256 |
| Linyphiidae | 124 | Zorocratidae | 258 |
| Liocranidae | 162 | Zoropsidae | 259 |
| Lycosidae | 164 | | |

| | |
|---|---|
| **Glossary and pronunciation guide** | 260 |
| **Etymological dictionary** | 274 |
| **Pronunciation guide for taxa** | 331 |
| **Bibliography** | 334 |
| **Index** | 362 |

# Chapter 1
# INTRODUCTION

Paula E. Cushing

## History of the manual

This project had its formal genesis in 1972, during the first meeting of the American Arachnological Society in the Chiricahua Mountains, Arizona where Vince Roth suggested a collaborative effort to produce a catalog and identification keys to Nearctic spiders. Immediately after the meeting, he published the proposal (Roth 1972a) along with samples of completed chapters (Roth 1972b, c). In subsequent years many arachnologists contributed to the project, which was published as the *Handbook for Spider Identification* (Roth 1982). With advances in spider taxonomy, the book was completely revised and published under the new name, *Spider Genera of North America*, which appeared in two editions (Roth 1985, 1994).

Vince's inspiration for the project came from the works of two people, Willis J. Gertsch and B. J. Kaston. Gertsch revised and updated *The Spider Book* the most thorough textbook on arachnology at the time and the only compendium of North American taxa (Comstock 1912, revised in 1940). Kaston's *Spiders of Connecticut* (1948, 1981) was a complete taxonomic inventory of a small region, with keys and illustrations to the species level. However, it was Kaston's *How to Know the Spiders* (1978), comprising illustrated keys to much of the Nearctic fauna, which Vince wished to emulate. Vince's goal was to incorporate elements of all these works together with many revisionary studies by Gertsch, Levi, Platnick, and others into one volume. Unfortunately, Vince's untimely death in 1997 (Ubick 1999) curtailed further revision of the manual.

In 2001, the American Arachnological Society (AAS) determined that a new edition was necessary to reflect taxonomic changes that had occurred since 1994 and to include an expanded glossary and enhanced illustrations. The Spider Genera of North America Revision Team (SGNART) was formed. Paula E. Cushing was charged with coordinating the revision effort consistent with limited resources. Soon thereafter an anonymous donor offered sufficient funding to complete the project to a much higher standard than originally envisioned. Fittingly, this modest arachnophile was introduced to spiders by Darrell Ubick, who was himself mentored by Vince Roth.

The purpose of this manual is to provide a means of identifying spiders to the generic level while also incorporating a single source of bibliographic information to facilitate identification to species for spiders of the continental United States and Canada. The editors and authors of this manual consciously attempted to preserve the spirit of Vince's original work. However, this manual is so different from prior editions that SGNART decided that 2005 would mark the publication of the 1$^{st}$ edition of *Spiders of North America: an identification manual*.

A percentage of the proceeds from the sale of this manual will be directed to future revisions, thereby ensuring that it be periodically updated. The balance of the proceeds support the AAS. The authors of this manual encourage and welcome relevant comments. These can be e-mailed to Darrell Ubick at dubick@calacademy.org.

## Chapter organization

This manual comprises 73 chapters and an appendix. Chapter 1 is this Introduction. Chapter 2 provides an overview of our current understanding of spider phylogeny. Chapter 3 is a key to North American spider families. The subsequent chapters treat the mygalomorphs (Chapters 4 – 11) and araneomorphs (Chapters 12 – 71). The chapters within these two infraorders have been alphabetically arranged to facilitate locating families of interest. Each family chapter begins with its scientific name and a habitus of a representative of that family. Every chapter is laid out in the following format:

**Common name** – the accepted common name for the family if it has one (after Breene 2003).
**Similar families** – morphologically similar families.
**Diagnosis** – the morphological characteristics which distinguish that family from others.
**Characters** – morphological characters displayed by members of the family.
**Distribution** – where North American representatives of the family are found. This section generally does not include information about global distribution.
**Natural history** – behavioral and ecological characteristics of the family.
**Taxonomic history and notes** – a concise summary of current understanding of that family's taxonomy.
**Genera** – a list of the Nearctic genera found in the family, including the names of the scientists who first described those genera and the publication dates.
**Key to genera** – a dichotomous key to North American genera.

Chapter 72 is an illustrated glossary explaining directional and morphological terminology, common words used in a non-traditional sense, and terms familiar to biologists but not otherwise in common usage. Chapter 73 presents the etymology (derivation) of the names of spider genera.

A pronunciation guide for taxa, and index of taxonomic names and a bibliography are found at the end of the manual.

## Geographic Coverage —

Spider families and genera identified in this manual are found in North America north of Mexico. This region, hereafter referred to as North America, includes the contiguous United States, Canada, and Alaska. The geographic region and abbreviations of the states and provinces are illustrated below (Fig. 1.1). Although this guide is specific to North America north of Mexico, it should also be useful for identification of most of the spider genera found in northern Mexico.

## Geographic Abbreviations —

For consistency with the proposed series of publications on the fauna of North America, standard two-letter abbreviations are used for states, provinces and territories, and three-letter for the countries, United-States (USA), Canada (CAN) and Mexico (MEX). NA refers to North America.

**North America: NA**
**Mexico: MEX**

### Canada (CAN)
| | |
|---|---|
| Alberta | AB |
| British Columbia | BC |
| Labrador | LB |
| Manitoba | MB |
| New Brunswick | NB |
| Newfoundland, Island of | NF |
| Northwest Territories | NT |
| Nunavut | NU |
| Nova Scotia | NS |
| Ontario | ON |
| Prince Edward Island | PE |
| Québec | QC |
| Saskatchewan | SK |
| Yukon Territory | YT |

### United States (USA)
| | |
|---|---|
| Alabama | AL |
| Alaska | AK |
| Arizona | AZ |
| Arkansas | AR |
| California | CA |
| Colorado | CO |
| Connecticut | CT |
| Delaware | DE |
| Dist. of Columbia | DC |
| Florida | FL |
| Georgia | GA |
| Idaho | ID |
| Illinois | IL |
| Indiana | IN |
| Iowa | IA |
| Kansas | KS |
| Kentucky | KY |
| Louisiana | LA |
| Maine | ME |
| Maryland | MD |
| Massachusetts | MA |
| Michigan | MI |
| Minnesota | MN |
| Mississippi | MS |
| Missouri | MO |
| Montana | MT |
| Nebraska | NE |
| Nevada | NV |
| New Hampshire | NH |
| New Jersey | NJ |
| New Mexico | NM |
| New York | NY |
| North Carolina | NC |
| North Dakota | ND |
| Ohio | OH |
| Oklahoma | OK |
| Oregon | OR |
| Pennsylvania | PA |
| Rhode Island | RI |
| South Carolina | SC |
| South Dakota | SD |
| Tennessee | TN |
| Texas | TX |
| Utah | UT |
| Vermont | VT |
| Virginia | VA |
| Washington | WA |
| West Virginia | WV |
| Wisconsin | WI |
| Wyoming | WY |

**Fig. 1.1** North America, north of Mexico

## Taxonomic coverage

North America is presently home to 68 families, 569 genera, and approximately 3,700 species of spiders. This manual can be used to identify all families and nearly all genera. The number of species found in the region continually increases. A more detailed overview of spider taxonomy is given later in this chapter.

## Spider anatomy

Unlike insects, spiders and other arachnids have two major body parts rather than three (Figs. 1.2-1.3). The anterior body segment is the cephalothorax or prosoma. The dorsal (upper) part of the cephalothorax is called the carapace; the ventral (lower) part is the sternum. The posterior body part is termed the abdomen, or opisthosoma (Figs. 1.2-1.3).

All four pairs of legs are attached to the cephalothorax and are numbered I, II, III and IV from anterior to posterior. Each leg has seven segments: the coxa or basal segment, followed by the trochanter, femur, patella, tibia, metatarsus and tarsus (Fig. 1.4). The tarsus ends in either two or three claws depending on the family (Figs. 1.5, 1.28). Web-building spiders all have three claws (Fig. 1.5) and use the middle claw to grasp the silk strands. The arrangement of the spider's legs is often diagnostic for a particular kind of spider. For example, most spiders in the family Thomisidae have legs arranged laterigrade to the body, like those of a crab.

The front leg-like appendages are called the pedipalps, or palps. Pedipalps have only six segments: the coxa, trochanter, femur, patella, tibia and tarsus. In the male, the palpal tarsus is specialized to serve as a sperm transfer organ during copulation (see below). The coxae of the pedipalps are expanded laterally to form two plates, called the endites, which surround the opening to the mouth (Figs. 1.3, 1.9).

Several types of specialized setae (hairs) or macrosetae (spines) are found on the legs. These include long, fine sensory hairs called trichobothria, which originate in sockets with multiple nerve endings (Fig. 1.5). Trichobothria are extremely sensitive to air currents. Patterns of trichobothria are often diagnostic for certain families. Specialized spines or bristles can also be found on the legs of spiders and may be diagnostic for certain taxa. For example, spiders in the family Theridiidae are sometimes called comb-footed spiders because they have a row of serrated bristles on the ventral side of the tarsus of leg IV that is used for combing silk from the spinnerets (Fig. 1.6; a single serrated bristle is shown in the bottom part of this figure). Cribellate spiders (see later in the chapter for a description of cribellate versus ecribellate spiders) have a specialized comb called

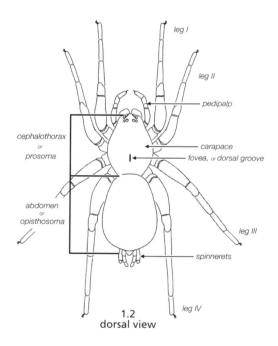

1.2 dorsal view

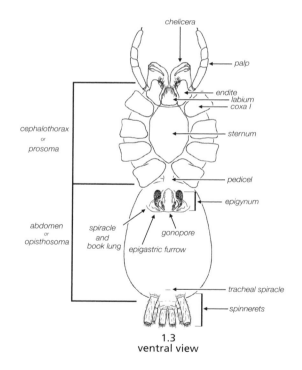

1.3 ventral view

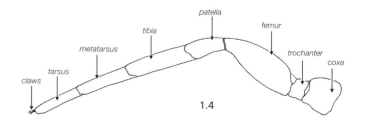

1.4

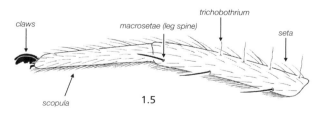

1.5

a calamistrum on the dorsal side of the metatarsus of leg IV used for combing silk from a flattened spinning plate, called the cribellum (Figs. 1.7-1.8).

The mouth is located between the base of the chelicerae and the labium. The coxae of the palps are expanded into plates called endites, and the labium is a plate situated between the endites (Figs. 1.3, 1.9). Anterior to the endites and labium are the jaws, or chelicerae, which terminate in hollow fangs. The base of each chelicera has a furrow on its distal (inner) surface in which the fang rests when not extended. In most spiders, teeth are found on one or both margins of the fang furrow (Fig. 1.9). The chelicerae are used to grasp and crush prey and are sometimes used in sexual and agonistic displays.

Most spiders have eight eyes arranged in two rows (Fig. 1.10), although some families are characterized by only six or even by only two eyes, and some cave dwelling species have lost the eyes altogether. Often the eye arrangement is diagnostic for a particular family (Fig. 1.11). A spider's eye is simple, unlike the compound eye of an insect, and is composed of a single outer lens, an inner cellular vitreous body, and a layer of visual and pigment cells which together make up the retina.

The silk-producing organs, called the spinnerets, are located near the posterior end of the abdomen. Most spiders have three pairs of spinnerets: the anterior lateral (ALS), posterior median (PMS), and posterior lateral (PLS) spinnerets (Fig. 1.8). Most mygalomorphs have only two pairs of functional spinnerets, having lost the anterior pair (Fig. 1.12). Posterior to the spinnerets is the anal tubercle that covers the anus through which waste is excreted (Fig. 1.8). In cribellate spiders, the cribellum (spinning plate) is located anterior to the spinnerets; this structure is replaced by a nonfunctional vestige called the colulus in most ecribellate spiders (Fig. 1.13). The spiracle, or opening to the respiratory tubes (tracheae), is located anterior to the colulus or cribellum (Figs. 1.3, 1.13). In some spiders the location of the tracheal spiracle is diagnostic. For example, in the Anyphaenidae, the tracheal spiracle is located approximately midway between the spinnerets and the epigastric furrow. The epigastric furrow is located in the anterior third of the ventral side of the abdomen (Fig. 1.3). Openings to the book lungs of the spider are found at the lateral extremes of the furrow (Fig. 1.3). Most spiders breathe by means of both book lungs and tracheae. However, some tiny spiders rely exclusively on tracheae while mygalomorphs have two pairs of book lungs and no tracheae (the respiratory systems of spiders are discussed later in this chapter).

The gonopore is an opening in the epigastric furrow leading to the internal testes of males and ovaries of females (Fig. 1.3). In male spiders, this is a simple opening with no associated sclerotized (hardened) external structures, although often with associated silk producing spigots

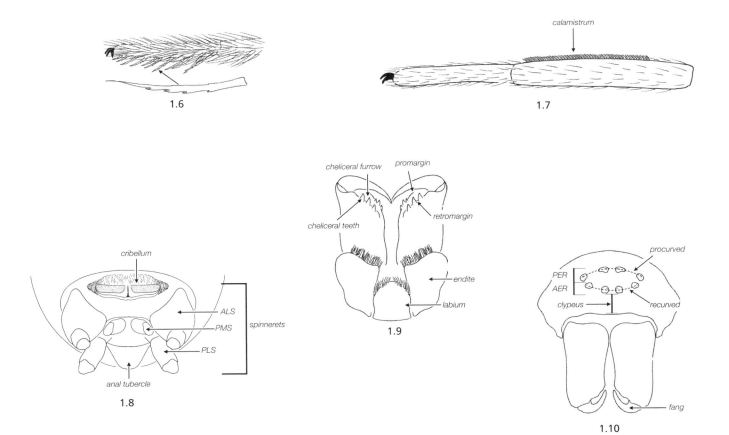

called epiandrous spigots. In females, the opening to the reproductive system is often more complex, with associated sclerotized structures that are collectively called the epigynum (discussed below). Females possess reproductive structures just internal to the gonopore that are collectively called the vulva. The vulva acts as a sperm storage organ and is usually composed of two spermathecae and associated ducts (Fig. 1.15).

Spiders in the infraorder Araneomorphae are divided into two groups according to their sex organs: the haplogynes, from Greek roots meaning "half woman" and the entelegynes, from Greek roots meaning "complete woman". In adult female mygalomorph, haplogyne, and some entelegyne (e.g., some Tetragnathidae) spiders, the gonopore serves as both a copulatory duct and a fertilization duct, being directly connected to the spermathecae. In other words, the male's semen enters and exits the spermathecae through the same duct. Eggs are individually fertilized during the process of oviposition. Haplogyne females usually have no complex sclerotized structures (epigyna) (Fig. 1.16) associated with their genital opening, and so may be mistaken for immature entelegyne spiders (Fig. 1.19). Entelegyne araneomorph females have usually a well-developed epigynum (Fig. 1.14). In these spiders, the genital opening is connected by copulatory ducts to the spermathecae; separate fertilization ducts lead from the spermathecae to the oviducts (Fig. 1.15). In entelegyne spiders, the male's semen enters the spermathecae through the copulatory ducts and is processed through side ducts called fertilization ducts.

An adult male spider's pedipalp is modified to serve as a sperm transfer organ. In mygalomorph and haplogyne males, the modifications to the palpal tarsus are usually very simple, consisting of a single extruded sclerotized bulb

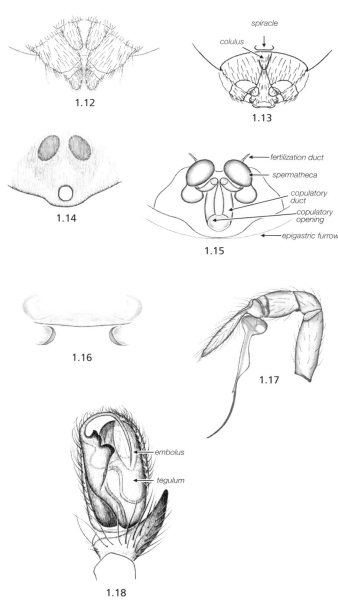

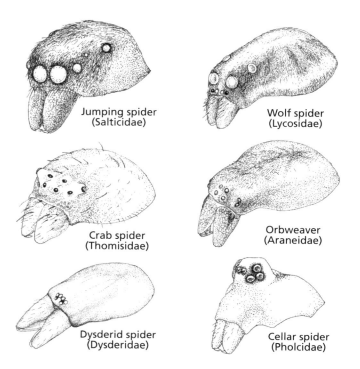

**Fig. 1.11** Comparison of 6 different eye arrangements

(Fig. 1.17). This bulb, the tip of which is called the embolus, is used to transfer sperm into the female's gonopore where it is stored in her spermathecae. In these spiders, the base of the embolus is attached ventrally to the palpal tarsus (Fig. 1.17). In entelegyne males, the structures associated with the pedipalp are much more complex (Fig. 1.18). Often the palpal tibia, femur, or patella has one or more apophyses, or sclerotized projections. The palpal tarsus is usually modified to form a spoon-shaped structure called the cymbium. The genital bulb rests in the cymbium. The bulb consists of several sclerites, or sclerotized structures, each with its own name. The largest sclerite is usually the tegulum. The tegulum gives rise to the embolus (Fig. 1.18). In some species, the embolus is a long curling tube; in others it is a straight, short tube. Other sclerites include the conductor, which supports the embolus while at rest, the median apophysis and the terminal apophysis.

All these structures associated with the male pedipalp and palpal tibia are important in species identification as are structures associated with the epigynum and the vulva.

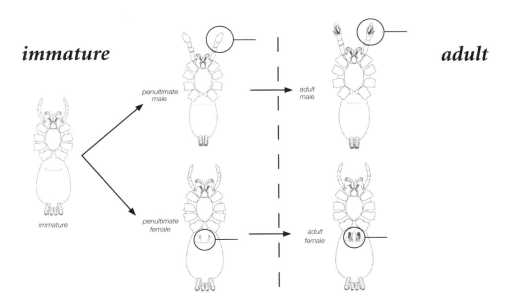

**Fig. 1.19** Differences between immature versus adult spiders

Because of the importance of genitalia in species identification, only adult specimens are generally identifiable to species. To distinguish between adults and immatures, one must examine the epigynum and the pedipalp. Immature spiders will have no sclerotized structures associated with the gonopore and will not have a well developed pedipalp with its associated sclerites (Fig. 1.19). Male spiders nearing the final molt to maturity have swollen pedipalps, but the sclerites will be absent. Because haplogyne female spiders do not have an epigynum, they are often difficult to distinguish from immatures, although the genital region on an adult female's abdomen is typically swollen or darker than that of an immature.

## SPECIMEN PREPARATION

When identifying specimens in the laboratory, place them in a Petri dish or watch glass immersed in alcohol. (See "How to Collect Spiders – field techniques" later in this chapter for a discussion of types of alcohol to use). Specimens should be immersed in alcohol from the moment of capture. A layer of fine sand, glass beads, silicon carbide (80 grit, lapidary grade, available from hobby shops), or black paraffin in the bottom of the dish will hold the specimen in place and will facilitate positioning of the specimen. A pair of forceps or fine tweezers should be used for manipulating the specimen. Care should be taken not to puncture the abdomen. A probe or a sewing needle mounted in a wooden handle is also useful. Many workers find sharpened insect pins held in pen-like vises to be excellent tools for manipulations and dissections.

Dissection is often necessary for examining the internal genitalia of females when identifying to species. With larger females, cut a three-sided flap around the epigynum, fold it back, and clean off adhering tissue with a tiny dissecting needle or a fine insect pin, the tip of which is sharpened with a whetstone. The sclerotized internal structures can clearly be seen. With tiny spiders, immerse the entire spider in a clearing agent such as clove oil, lactic acid, or methyl salicylate. Other techniques for dissecting and observing female vulvae are described in Millidge (1984b) and Griswold (1993).

To examine the internal genitalia in more detail, remove the epigynum with a pair of fine forceps and place it in a drop of the clearing agent in the depression of a concave microscope slide. For a clear view using incident light, clean the epigynum using enzymatic cleaner for contact lenses. This does not work well for specimens that have been preserved for a long time. Ten percent KOH can also be used; it both cleans and clears. Immerse specimens in a very hot solution of KOH for up to two minutes or soak overnight in a cold solution. The latter method gives better results. Some researchers use sodium hypochlorite (bleach) for the same purpose.

After examination, the dissected epigynum should be rinsed with alcohol and placed in a small genitalia vial, which is itself filled with alcohol. Plug the genitalia vial with cotton and place it in the larger vial containing the specimen of origin. Under no circumstances should the dissected epigynum be thrown away or kept separate from the specimen.

## HOW TO USE THE KEYS

This manual uses "dichotomous keys." A dichotomous key is arranged as a series of couplets. Each couplet describes sets of mutually exclusive characters. Select the character set best describing the specimen. The number at the end of the line indicates the next relevant couplet, finally leading to a probable family or genus identification for the specimen. Often family level identification can be carried out without the use of a microscope. A magnifying glass or loupe often reveals family-level diagnostic characters such as cheliceral position, eye arrangement, leg arrangement, and spinneret structure. Identification to genus and species level requires the use of a dissecting microscope for proper examination of the genitalic structures.

Proceeding through a dichotomous key often means using a series of best guesses. You may find it difficult at times to distinguish which set of characters in a couplet best describes the specimen and so may arrive at a couplet that does not describe your specimen at all; you may, in other words, arrive at a dead end. At this point, back-track in the key to the last clearly diagnostic couplet and follow the alternate route.

To identify a specimen, use the key to spider families in Chapter 3 then proceed to the appropriate chapter and key the spider to genus. Once the genus has been identified, the following information is provided below the couplet:

**Div**. (*Diversity*) the number of North American species in the genus.
**Dist**. (*Distribution*) the North American distribution of the species in the genus.
**Ref**. (*References*) papers that contain descriptions or other relevant information about the genus or species. (See the bibliography at the end of the manual for full citations.)
**Note**. additional taxonomic information.

Included in this information, you may see reference to an italicized species name. Species names are binomials, made up of a capitalized genus name followed by a lower case specific epithet. The species name is followed by the citation, which gives the name of the person who originally described the species and the date of the description. When the citation is in parentheses, this indicates that the species was originally described in another genus. For example, *Acanthepeira stellata* (WALCKENAER 1805) was originally described by Walckenaer as *Epeira stellata* and transferred to the genus *Acanthepeira* by later researchers. The International Code of Zoological Nomenclature (ICZN 1999) describes the rules used by zoologists for composing family, genus, and species names. In this manual, the name of the taxon author (i.e. the describer) is indicated by small, unitalicized capital letters. The author and date is a citation to a full reference.

## HOW TO COLLECT SPIDERS AND MAINTAIN COLLECTIONS
### Field techniques

Capturing spiders requires a quick hand and a ready container. Traps set up in the field or extraction traps set up in the lab can also be used. Film canisters are useful for field collecting, having tight fitting lids that allow for alcohol storage (semi-clear canisters are preferable to opaque canisters as you can then see the specimens inside). Most camera stores and film developers are only too glad to give away their leftover film canisters. Old pill vials can also be used. Clear plastic collecting vials can be purchased from a biological supply house (see "Worldwide web sources"). When collecting, carry three or four empty, dry vials of different sizes for live captures and one or two vials partly filled with alcohol for killing and preserving specimens. If you expect to find larger spiders, such as tarantulas, large wolf spiders, or large orb weaving spiders, bring one or two larger dry and alcohol vials as well.

Unlike insects, spiders are not pinned and dried, as they shrivel and fall apart when desiccated; rather, they are preserved in 70–80% alcohol, usually ethanol. It is best to avoid using denatured alcohol because some denaturing agents can damage the specimens (Simione 1995). However, methylated spirits (made up of 90% ethanol and about 10% methanol) can be used for specimen preservation and is readily available through laboratory supply companies. Methanol alone should be avoided (Simione 1995). Forty-five to 50% isopropyl alcohol (or propanol), available in drug stores as rubbing alcohol, can also be used; however, high percentages of propanol can shrivel specimens or make them brittle. Colored or scented rubbing alcohol and ethyl rubbing alcohol should be avoided.

Ethanol (95%) can be purchased in liquor stores; however, this alcohol is expensive and needs to be diluted to 75-80% with distilled water. Laboratory grade ethanol is difficult to obtain and often requires a permit for scientific use. Specimens may also be killed in hot water where they die instantly with their appendages conveniently extended for subsequent observation. These specimens must still be preserved in alcohol. Specimens to be used for molecular analyses involving DNA extraction and purification are best preserved in 95 – 100% ethanol (Prendini *et al.* 2002) provided they are stored at between -20°C to -80°C (Vink *et al.* 2005). A killing jar, as is used for collecting insects, can be used to kill spiders, as long as the spiders are transferred to alcohol soon after death. The advantage of this method is that the spider dies in a more relaxed state (i.e. the legs do not draw up close to the body).

A variety of nets and traps is used to capture spiders. Sweep nets, similar to butterfly nets but made of sturdy muslin instead of open mesh, are used for capturing spiders that spend their time on vegetation (Fig. 1.20). Beat sheets are used to collect spiders nesting and feeding in taller, sturdier vegetation. These beat sheets consist of squares of white cloth or canvas held open by sticks or dowel rods (Fig. 1.21). The beat sheet is placed beneath a tree branch or shrub while the vegetation is beaten with a stick, dislodging spiders onto the sheet. A small camel-hair paint brush can be used to coax spiders off the beat sheet and into collecting vials or an aspirator can be used for this purpose (see below).

Pitfall traps are good for capturing wandering spiders including web-building males out searching for mates. Pitfalls consist of cups or cans with preservative in the bottom -70% alcohol or propylene glycol, diluted with water to reduce surface tension, work well except the alcohol evaporates quickly so these traps must be checked frequently (Fig. 1.22). Ethylene glycol, or regular antifreeze, is toxic to mammals and should not be used. A pitfall trap can be constructed as follows: cut the top off a two-liter plastic soda bottle, place a plastic cup containing preservative in the bottom half, remove the bottle cap and invert the top half so it serves as a funnel. Use a trowel or a post hole digger to sink the entire trap into the ground so that the lip of the inverted bottle top is level with the soil. This is important as wandering spiders will tend to avoid perceived obstructions. Mark the spot with flagging tape, an upright stick, or a bright-colored rag tied to a nearby plant. Wandering spiders will topple into the opening and fall into the preservative. The funnel will prevent escape. Such a trap can be left in place for a long period. The inner cup is periodically checked for specimens, specimens removed and preserved in alcohol and the propylene glycol replenished.

Leaf litter spiders can be collected by sprinkling handfuls of leaf litter on a beat sheet or into a white enamel pan and gently agitating the sheet or pan, causing spiders to scurry from retreats. The white background aids in spotting the fugitives. Sifters can also be used to collect leaf litter spiders (Fig. 1.23). Leaf litter spiders can also be extracted with a Berlese, or Tullgren, funnel (Fig. 1.24). A Berlese funnel is a fairly large metal or plastic funnel with a metal screen set inside. The screen should have 1/2 cm sized mesh. A 25 watt light bulb is suspended over the funnel and a jar with preservative (propylene glycol or alcohol) is set below the funnel. A 5 – 10 cm thick layer of leaf litter is placed on the mesh. The warmth from the bulb gradually dries out the leaf litter driving the arthropods down until they finally fall through the mesh into the preservative. Depending on the moisture content of the leaf litter, it may take several days for the spiders and other arthropods to be driven into the

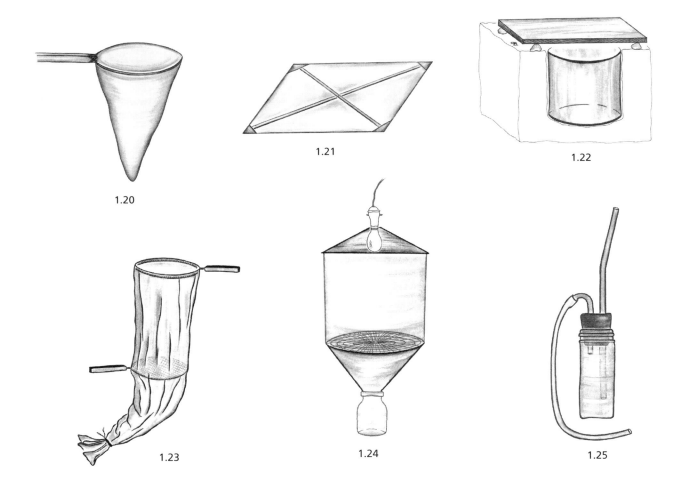

preservative. To avoid any risk of fire, never use more than a 25 watt bulb for the light source and make sure the light is suspended 15 – 20 cm above the leaf litter. Alternately, a mothball can be suspended over the leaf litter instead of a light bulb; the naphthalene vapor will drive out the arthropods.

Many collectors use aspirators, or pooters, to collect spiders (Fig. 1.25). An aspirator is made using two flexible tubes inserted into a two-holed rubber stopper. A small piece of nylon mesh is attached to the end of one of the tubes extending below the stopper. The rubber stopper is firmly inserted into a plastic or glass vial. By sucking on the end of the tube tipped with nylon mesh, a vacuum is created allowing you to suck a spider into the open tube. The spider then becomes trapped in the vial. A simpler type of aspirator is made using a flexible rubber tube attached to a small piece of glass or stiff plastic tubing. The glass tube has a piece of nylon mesh on the end inserted into the rubber tubing. Small spiders can be sucked into the glass tube and then blown into a collecting vial of alcohol or into a dry vial. To avoid sucking fungal spores or mold into your lungs, do not aspirate spiders from the ground or from rotting logs. Aspirators are useful for collecting spiders off a beat sheet.

Since many spiders are active only at night, night collecting often yields specimens not found during the day. For night surveys, it is essential to have a strong headlamp for spotting the eyeshine of wandering spiders, such as wolf spiders or fishing spiders, or to see the glint of light reflected off a spider's web. When using a headlamp, shine the light down several meters ahead of you. Move the beam slowly and look for tiny greenish lights on the ground or in the low vegetation and approach slowly, keeping the eyeshine in view, until you can see and capture the spider. Shining the light directly in front of you into the vegetation may reveal spider webs or the silken retreats of web-building spiders. Dusting vegetation using a sock filled with cornstarch or from a puffer will often reveal otherwise invisible webs. A puffer is a rubber bulb, partly filled with cornstarch that, when squeezed, releases a fine cloud of white powder.

### Rearing spiders

Because only mature spiders can be identified with accuracy to species, it is necessary to raise immature individuals in the lab. Useful information on rearing a variety of spiders in the laboratory can be found in Bruins (1999). All spiders are predatory and most are solitary, necessitating individual housing for specimens. Many types of containers can be used to rear spiders, from small plastic containers to larger terrariums. Burrowing spiders should be provided with a layer of soil mixed with sand that is at least twice as deep as the spider is long (Bruins 1999). Web-building spiders should be provided with silk attachment sites such as twigs, mesh, or bark and the container must be large enough to accommodate the web. Hiding places such as small rocks, bark, cardboard tubes or small ceramic plant pots should also be provided.

Spiders can live for extended periods of time without food but require access to a ready source of water. For small spiders, it is best to provide moistened toweling or cotton rather than an open container of water in which the spider could drown. Larger spiders can be provided water in a lid or small dish. A cotton plugged vial containing water can also be used; the spider will crawl inside the vial to imbibe the moisture from the saturated plug. Spiders will generally eat insects that are slightly smaller than themselves and require feeding once or twice a week. Crickets and fruit flies can be purchased at most pet supply stores. Field-collected insects from sites not sprayed with pesticides can also be used. Some prey taxa, such as ants, stink bugs, and June beetles are either distasteful or hazardous to spiders. Prey must be thrown into the web of a web-building spider to trigger feeding. Since spiders are most vulnerable during molting, uneaten insects should not be left in an immature spider's container or the prey may very well eat the spider before the new exoskeleton has hardened. Many spiders become lethargic and stop eating just prior to molting.

### Field observations and data collection

With practice, easily observed morphological characters such as eye arrangement (see Fig. 1.11) or leg posture permit field identification of live spiders to the family level. Habitat and behavior can also provide clues to a spider's identity. Jumping spiders (Salticidae), for example, are diurnally active, have enormous anterior median eyes (AME), are usually found on vegetation rather than on the ground, pounce on prey like cats when hunting, and are sometimes found inside thick silken retreats in curled leaves or under bark. A good field notebook will include detailed information about the spider's habitat, the location of the specimen in the habitat (e.g., on a bush, beneath bark, in a tree, inside a folded leaf, on the ground), and behavioral observations.

The minimum data to include with collected specimens are the collecting locality, the date collected, and the collector's name. Most arachnologists carry field notebooks for recording data, which are later transcribed onto specimen labels. The labels are placed in the alcohol with the spider (or in the dry vial with the spider if the spider is to be kept alive). Use white, unlined 100% cotton paper or Resistall© paper for labels. Write the information in waterproof India ink or Micron© or Pigma© permanent ink pen. Labels may also be written in pencil as the graphite is archival. Do not use your computer printer or a copy machine to make alcohol labels. The ink from laser printers and copy machines is burned onto the surface of the paper and can flake off in alcohol over time. The ink used for most ink jet printers is not of archival quality and may disintegrate in alcohol. Some newer deskjet printers, particularly those designed for archival photo printing, may produce adequate alcohol labels, but the company should be contacted first to determine if the ink will hold up when the label is immersed in alcohol. Dot matrix impact printers may be used for making labels when equipped with archivally-inked ribbons. Ballpoint pens are also inadequate for alcohol labels since the ink rapidly fades away. Labels should be small enough to fit easily into the vial. If all the data cannot be written on one label, divide it between two or more labels. The following information should be included on the labels.

**Locality data** — information about the exact location where the specimen was collected. Essential information is Country, State (two letter code) or Province, County or Parish, City (or distance and direction from nearest city or town); any other locale information that will help future collectors pinpoint the exact location. Latitude, longitude and elevation, as provided by a handheld Global Positioning System (GPS) are particularly helpful. Coordinates and elevation may be derived from these websites:

geonames.usgs/pls/gnis/web_query.gnis_web_query_fom
www.zipmgr.com/geocodeo.aspx
www.travelgis.com/geocode
www.ajmsoft.com/ac/geocode.php
www.geocode.com/modules.php?name=TestDrive_Eagle
www.topozone.com
geonames.nrcan.gc.ca

Search engines (such as yahoo.com or google.com) will find other sites that provide coordinates for collecting locales. Search for "free geocode."

**Ecological and habitat data** — Include the date on which the specimen was collected, the collector's name, collecting method, the type of habitat in which the specimen was found, and the time of day the collection was made. When writing the date, use Arabic numerals to indicate the day and year, but use letters or Roman numerals for the month (e.g., 12 Sept 2005 or 12.ix.2005) – this prevents future researchers from confusing the month and day. Use four digits for the year, as well-curated specimens could last for hundreds of years.

**Identification label** — A separate label should be included in the vial once a species determination has been made. Include the species name, the name of the scientist who originally described the species, the name of the person who keyed out the species (the determiner), the year the determination was made, and the number and gender of specimens in the vial. For example:

*Gnathonarium cambridgei* Schenkel
det. D. Buckle, 2004 3♂, 1♀, 1 imm

All labels must be placed inside the vials with the specimens. Labels attached to the outsides of the vials are too easily lost or damaged.

### Maintaining an alcohol collection

Different specimen storage techniques are used depending on available funds and space. Entomology supply companies, such as BioQuip (URL listed under "Worldwide web sources"), offer various sizes and types of storage vials for alcohol preserved specimens including tiny vials used for storing dissected genitalia. Genitalia (epigyna or palps) often must be dissected from the specimen in order to identify it to species. Specimens can be kept in glass vials with rubber, neoprene, or screw top caps with polyseal inserts. However, rubber hardens over time compromising the integrity of the seal; neoprene stoppers can cause leachates to form, discoloring the alcohol and the specimens; screw top lids without a polyseal (polypropylene) insert will not maintain a sufficient seal, and any type of screw top lid can loosen, allowing alcohol evaporation. Alternately, specimens can be kept in small straight-sided glass vials (shell vials) plugged with cotton and immersed in a larger container, such as a canning jar, filled with alcohol. No matter what storage method is used, the vials should be checked regularly to ensure that the alcohol has not evaporated and that the seals are still intact. If the alcohol in the vials is noticeably low or discolored, it should be replaced with fresh preservative. Exposure to light will fade specimens so collections should be kept in a closed cabinet or a dark room. Alcohol is flammable. Keep all heat sources away from the collection, and keep a well-charged fire extinguisher at hand. A catalog or a database of the specimen data should be kept in a separate location.

Extensive or important personal collections should be donated or bequeathed to a recognized museum or institute that maintains alcohol collections to ensure that the material will be widely available to researchers indefinitely. Holotypes and paratypes must be deposited in recognized research collections. In taxonomic research, the voucher specimen for a new species name is called the holotype specimen. The holotype is the single original specimen, male or female, used in the species description as the representative of the name of the new species. The formal species description is based upon characteristics displayed by the type specimen. Other specimens also used in the species description to capture morphological variations present in the species are called paratypes. Such specimens must be freely available to other researchers.

## Taxonomy of spiders

All organisms are arranged into taxonomic categories, which reflect their relationships to each other. The taxonomic categories from most to least inclusive are shown below. Subgroups may be assigned to each category as needed. The example illustrates the classification of one species of spider – the black widow.

**Kingdom** Animalia
 **Phylum** Arthropoda
  **Class** Arachnida
   **Order** Araneae
    **Infraorder** Araneomorphae
     **Family** Theridiidae
      **Genus** *Latrodectus*
       **Species** *Latrodectus mactans* (FABRICIUS 1775)

Spiders belong to the class Arachnida and the order Araneae. Other arachnids include scorpions (order Scorpiones), pseudoscorpions (order Pseudoscorpiones), camel spiders or wind scorpions (order Solifugae), vinegaroons or whip scorpions (order Uropygi), tailless whipscorpions or whip spiders (order Amblypygi), daddy longlegs or harvestmen (order Opiliones), mites and ticks (order Acari), as well as a few other minor orders. The small field guide, *Spiders and Their Kin* by Levi & Levi (2002) contains drawings and information about these orders of arachnids. A key to these orders can also be found in Kaston's *How to Know the Spiders* (1978). Other information is available from some of the websites listed near the end of this chapter.

The order Araneae is currently divided into the suborders Mesothelae and Opisthothelae (Coddington & Levi 1991). The suborder Mesothelae gets its name from the median position of the spinnerets on the venter of the abdomen. This group comprises a single family, Liphistiidae, found only in southeast Asia. Liphistiids are distinct from all other spiders because they retain the abdominal segmentation common to more ancestral arachnid orders such as scorpions, camel spiders, and vinegaroons. It is thought that the Mesothelae are the most evolutionarily primitive spiders. Spiders of the suborder Opisthothelae have spinnerets located at the posterior of the abdomen and are grouped into the infraorders Mygalomorphae and Araneomorphae. The two infraorders contain 110 families of spiders, divided into about 3,600 genera, which in turn comprise about 39,000 species (Platnick 2005). Norm Platnick's online *World Spider Catalog* (Platnick 2005) is the primary resource for species names and citations.

Mygalomorphae includes those spiders known in the New World as tarantulas and trapdoor spiders. Known elsewhere as bird spiders, mygalomorphs are distinguished from araneomorphs by the presence of two pairs of book lungs instead of one pair or no book lungs, and by chelicerae that move parallel to one another in the vertical plane (Fig. 1.26). Araneomorphae includes the majority of the world's spiders. Araneomorph spiders have chelicerae that move in opposition to each other in a pinching motion in the horizontal plane (Fig. 1.27) and usually possess either one pair of, or no book lungs. Most araneomorph spiders are divided further into the entelegynes and the haplogynes. As previously explained (page 6), entelegyne and haplogyne spiders are differentiated based upon the complexity of their genitalia.

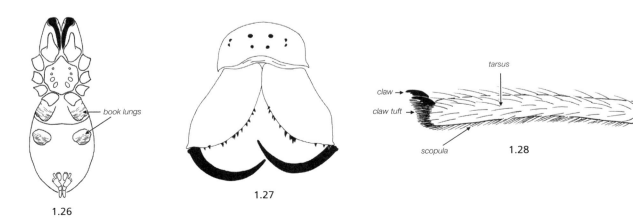

## Spider Biology
### Reproduction

The internal sexual organs, the testes or ovaries, are located in the abdomen. However, as with most arachnids (except Opiliones), male spiders have no direct intromittent organs associated with the internal genitalia and utilize indirect sperm transfer. Mature males use the embolus of their pedipalps as intromittent organs. Upon maturing, the male builds a small platform of silk called a sperm web, which may consist of only a few lines of silk. Onto this sperm web, the male deposits a droplet of sperm from his gonopore. He then inserts the embolus into this droplet of sperm and draws the sperm into the bulb of the pedipalp where it is stored. Mating requires the transfer of the sperm stored in the pedipalp into the female's gonopore.

Most spiders are solitary predators and opportunistic feeders, even upon conspecifics. A male spider must, therefore, approach a female with caution and must effectively communicate that he is a potential mate and not a meal. This is particularly important among web-building spiders because there is often dramatic dimorphism between males and females with males being considerably smaller than females. When a male web-building spider locates a female, probably via airborne pheromones, he plucks and taps at the web in a species-specific pattern communicating that he is a mate. If the female is mature and ready to mate, she will respond positively. If she is not ready to mate, she may attack the male.

Male wandering spiders, such as wolf spiders (Lycosidae) or jumping spiders (Salticidae), exhibit complex courtship behaviors after locating females, often via pheromones deposited in the female's silk (Barth 2002). Male jumping spiders often perform striking zig zag dances, incorporating species-specific waving of the legs and pedipalps. Wolf spiders and some jumping spiders may add a vibratory component by tapping their pedipalps or abdomen on the ground. Some male spiders have stridulatory structures on their bodies which are probably involved in courtship. Some males position the female for mating using their cheliceral spurs; others (e.g., some Thomisidae) spin a "bridal veil" around the female. When a male's courtship behaviors have been accepted by a female, he will approach her cautiously and may stroke her. Finally, he will insert his embolus into her genital opening to inject the sperm, which is then stored in the spermathecae. Some females will mate multiple times with the same or with different males. In some species, the male will deposit a mating plug into the female's genital opening in an attempt (often not effective) to prevent her from mating with other males.

### Hatching, development, and dispersal

A few weeks after copulation, the stored sperm is released from the female's spermathecae and egg fertilization occurs in the oviduct just prior to egg emergence from the gonopore. The number of eggs laid depends on the species. Very tiny females produce only 1 – 7 eggs per egg sac. Larger species can produce thousands of eggs (Foelix 1996). The female encloses her eggs in a sac made of a special tough silk. In many species, the female guards the egg sac against parasitoids and predators until hatching. For example, a wolf spider (Lycosidae) female attaches the silken egg sac to her spinnerets and carries it with her. When the young hatch, the female cuts a hole in the egg sac with her chelicerae and the spiderlings crawl up onto her abdomen where they hold on to specialized knobbed hairs (Fig. 1.29). The babies ride on their mother's back until the yolk supply left over from the egg stage is depleted and they must hunt for their own food. A jumping spider (Salticidae) female builds a silken retreat to house her egg sac and remains on guard until the eggs hatch. Female spiders in the family Pholcidae bind their eggs together loosely with silk and carry the egg sac in their chelicerae. Fishing spider (Pisauridae) females carry their egg sacs with the chelicerae until just prior to hatching. At that time, the female builds a nursery web and suspends the egg sac in it (Fig. 1.30). In other families, the female shows no guarding behavior after constructing her egg sac.

Usually, the eggs hatch within a few weeks. In temperate regions, some spiders overwinter in the egg stage, their development arrested until temperatures rise in the spring. In other species, the spiderlings overwinter by seeking protection in warmer microhabitats in cracks, crevices, or in leaf litter. Some spiders overwinter as subadults and mature in the spring, whereas others overwinter as adult females who lay eggs in the spring.

Many spiderlings disperse simply by walking away from their natal ground. However, a large number of araneomorph and some species of mygalomorph spiderlings often disperse by ballooning. In a few families of usually tiny spiders, even the adults will sometimes move to a new habitat by ballooning. These dispersing spiders crawl to the top of a blade of grass, a twig, a branch or some other structure

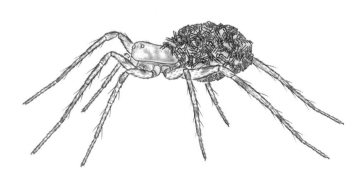

1.29

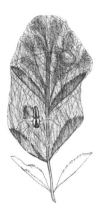

1.30

and release strands of silk. Air currents catch the silk, often called gossamer, and lift the spiders on an updraft (Decae 1986). Although some of these "flying spiders" may land only a meter or two away from the take-off point when their silk gets tangled in the branches of a nearby plant, others travel truly extraordinary distances. Some ballooning spiders have been found on ships far out at sea. Darwin noted spiders on gossamer threads caught on the rigging of the Beagle (Darwin 1909). Because of this ability to travel long distances by ballooning, spiders are often the first animals to establish themselves on distant oceanic islands. One of the most famous examples of this pioneer colonization is the appearance of a single small spider on the island of Krakatoa less than a year after a massive volcanic eruption obliterated all life from the island (Winchester 2003). Ballooning spiders were some of the first arthropods to re-colonize Mount St. Helens after its eruption in 1980 (Crawford 1985, Edwards *et al.* 1987).

Tiny spider species require about five molts to reach maturity whereas larger species may require 10 or more molts. Araneomorph spiders do not undergo any further molts after reaching maturity; mygalomorph females, however, continue to molt throughout their lives. During molting, all chitinous structures, including such internal structures as the trachea, the book lungs and, with mygalomorph females, the lining of the spermathecae, are shed. For adult mygalomorph females, this means that any sperm stored in the spermathecae is lost and they must mate again in order to reproduce (Foelix 1996).

Most male spiders, both araneomorphs and mygalomorphs, die shortly after maturation. In temperate regions, most species live only one or two years. Adults of many species die soon after the first hard frost. In tropical regions, lifespans can vary and there can be multiple generations of one species at any time. The longest-lived spiders are some mygalomorph species. Females have lived in the laboratory for over 20 years, and males have survived for up to 10 years (Jackman 1997.)

**Spinnerets and silk**

Silk production is characteristic of all spiders. Spider silk emerges from the spinnerets. Most extant spiders have three pairs of spinnerets: the ALS, PMS, and PLS (Figs. 1.8, 1.13). However, it is thought that ancestral spiders possessed four pairs; an anterior median in addition to the ALS, PMS and PLS. Extant Mesothelae, the most primitive of today's spiders, still have four spinneret pairs, although the anterior median pair is nonfunctional and vestigial. In comparison, many mygalomorphs have only two pairs of functional spinnerets, having lost the anterior pairs (Fig. 1.12).

Most araneomorph species have three pairs of functional spinnerets. The anterior median pair is represented by the cribellum or colulus (Figs. 1.8, 1.13). The colulus is considered homologous (having a common evolutionary origin) to the cribellum, which is present among some groups of both haplogyne and entelegyne araneomorph spiders. The cribellum is thought to be a plesiomorphic (ancestral) characteristic. Among the ecribellate spiders (those lacking a cribellum), the colulus may be the remnant of the lost cribellum and both are thought to be homologous to the anterior median spinnerets (Foelix 1996). The cribellum is covered with hundreds to thousands of individual silk spigots and, depending on the taxon, may be a single plate or may be divided into two parts. Cribellate spiders use the calamistrum (Fig. 1.7) to draw out and comb the silk. Cribellate silk is distinctively "fuzzy" and tends to have a bluish cast. The surface of each spinneret is covered with spigots, each connected to a silk gland in the abdomen of the spider. Spiders may possess up to six different types of silk glands, each secreting a unique kind of silk specific to a particular function (Foelix 1996).

For hundreds of years, poets and artists have depicted the spider web as a thing of beauty and wonder. Spiders are well known for their ability to construct silken nets for prey capture, but not all spiders build webs. Among the web-building groups, the web itself is often characteristic of a particular family of spider. For example, the majority

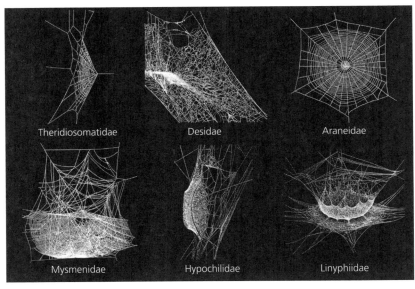

1.31

of spiders in the Araneidae (orbweavers), construct round webs with radial support strands attached from the center of the web to the outer edge like the spokes of a wheel. Orbweavers monitor their web either using a strand of silk attached to the web from a retreat in the vegetation or from the very center of the web. Most spiders in the families Uloboridae and Tetragnathidae also construct orb webs for prey capture; however, their orb webs are oriented on a different plane than the vertically positioned webs of the Araneidae. Spiders in the family Agelenidae are commonly called funnel weavers because they construct webs with a platform of silk stretching before a funnel-like retreat. The agelenid spider usually monitors its web from the funnel, rushing out to bite prey alighting on the platform of silk. Figure 1.31 illustrates several different types of webs characteristic of various araneomorph families.

**Prey capture and digestion**

Spiders usually immobilize their prey with venom, except for those in the family Uloboridae, which lack venom glands and are dependent on speedy wrapping of prey to subdue it. Other spiders may either first bite the prey, injecting venom through the hollow fangs, and then wrap it, or – in the case of large or dangerous prey like wasps – first wrap the prey before biting. The venom glands are located in the chelicerae or in the cephalothorax and are linked by ducts to the fangs.

Once prey has been envenomated, the spider pumps digestive enzymes from its mouth onto the bite area. These enzymes break down, or pre-digest, the body of the prey until the soft tissues of the body have been completely liquified. The spider has a specialized sucking stomach located in the dorsal portion of the cephalothorax and connected to the exoskeleton of the carapace by a series of muscular bands. The apodeme, or muscle attachment, of the dorsal band can be clearly seen on the carapace of most spiders as a dorsal groove or a dimple-like indentation (Fig. 1.2). When the stomach muscles contract, the stomach lumen expands; when they relax, it contracts. This pumping action enables the spider to regurgitate enzymes, draw food out of its prey, and direct the predigested food to the rest of its digestive system. Diverticula, extensions of the spider's digestive system, are present throughout the spider's body, sometimes extending even to the legs and eyes. This digestive network is an efficient means of nutritional transport and, combined with a low metabolic rate, allows spiders to go extended periods of time without eating. Many spiders have fringes of setae bordering the endites and labium (Fig. 1.9). These setae filter out particulate matter since spiders feed exclusively on predigested, liquefied food.

**Respiration and circulation**

Oxygen enters a spider's body through spiracles. For mygalomorph spiders, these spiracles lead to two pairs of book lungs (Fig. 1.26). Araneomorph spiders have either one pair of or no book lungs; it is thought that the posterior pair of book may have evolved into a pair of tubular tracheae. The spiracle openings into the book lungs can be opened or closed through muscular contractions (Strazny & Perry 1987). Air flows between thin plates of tissue called lamellae. In between these lamellae are evaginations of the body cavity filled with hemolymph, or blood. Oxygen diffuses through the lamellae into the hemolymph. Oxygen carried by the tracheal tubes also diffuses into the hemolymph (Schmitz & Perry 2001). The tracheal system varies in different araneomorph families. In some, the tubes branch out only into the abdomen; in others, the tubes branch all over the spider's body even into the cephalothorax. Some tiny spiders have completely lost the book lungs and rely on an extensive system of tracheal tubes to get their oxygen. Oxygen follows a gradient throughout the spider's body - as does carbon dioxide which diffuses into the tracheae or book lungs and is expelled from the body through the spiracles. The location of tracheal spiracles on a spider's abdomen is often diagnostic for a particular family or genus (e.g., Anyphaenidae).

The hemolymph of spiders contains an oxygen-carrying molecule called hemocyanin which functions very similarly to mammalian hemoglobin. However, hemoglobin contains iron atoms and is concentrated within blood cells, whereas spider hemocyanin has copper atoms and is not contained within blood cells. While the iron atoms give mammalian blood its red color when it becomes oxidized, the copper atoms give spiders' blood a blue-green cast. Hemocyanin is nearly as effective as hemoglobin in binding to oxygen molecules. The hemolymph of spiders also contains cells that are involved in blood clotting, wound healing, fighting off infections, and storing different types of compounds for transport throughout the body.

As with all arthropods, spiders have an open circulatory system; the hemolymph bathes the internal organs and tissues and is not contained in blood vessels. Nevertheless, a circulatory pathway does exist. A spider's tubular heart is located in the dorsal portion of the abdomen. As the heart muscle relaxes, holes in the sides of the heart, called ostia, open and oxygenated blood is drawn into the heart. Upon contraction, the ostia close, forcing blood anteriorly and posteriorly into a system of open-ended arteries. The oxygenated blood thus flows outward and, as it gives up the oxygen to the tissues, flows back towards the abdomen and the respiratory organs, following a gradient of decreasing pressure, before returning to the heart (Strazny & Perry 1987).

Spiders lack extensor muscles between the femur-patella and the tibia-metatarsal joints. Increasing hemolymph pressure into the legs is responsible for joint extension during locomotion (Foelix 1996). Increased hemolymph pressure also functions to force the sperm out of the male palp during copulation (Foelix 1996).

**Sensory systems**

Most spiders are nocturnal and so are not primarily dependent on sight. Tactile and chemical cues are probably more important. Web-building spiders, in particular, probably have poor vision by vertebrate standards (Land 1985, Yamashita 1985, Barth 2002). They respond to changes in light intensity or to changes in the polarization of incoming light but probably cannot see fully formed images of objects. However, cursorial hunters such as wolf spiders (Lycosidae), jumping spiders (Salticidae), and crab spiders (Thomisidae) are much more reliant on vision both to catch prey and to court mates (Williams & McIntyre 1980, Forster 1985, Land 1985). Some spiders may have limited color vision and it has been shown that the eyes of some spiders are sensitive to UV light and to polarized light (Blest 1985, Dacke et al. 2001a, b). Much more remains to be learned about spider eyes and vision (Barth 2002).

Hairs on a spider's body serve as chemosensors and as mechanoreceptors, responding to touch, vibrations, air currents, and positioning of the spider's joints. In some mygalomorph species the hairs on the abdomen, collectively known as urticating hairs, also serve for defense. These hairs are often barbed. When threatened, the spider "kicks" them off using a rear leg. Mygalomorph urticating hairs can lead to skin or mucosa irritation even when dead specimens or molts are handled or cages of pet tarantulas are cleaned. Some wandering spiders, or spiders that hunt without a web, such as most mygalomorphs and jumping spiders, have scopulae and claw tufts on the tarsi and metatarsi (Fig. 1.28). These are fine hair fringes that allow these spiders to climb and keep a foothold on smooth surfaces due to van der Waal's forces (Kesel *et al.* 2003). The scopula also facilitates prey capture (Rovner 1980a, b).

Leg segments of spiders exhibit long, fine hairs called trichobothria (Fig. 1.5). These hairs originate in sockets with multiple nerve endings at the base. Trichobothria are extremely sensitive to air currents – even those produced by the wings of an insect flying nearby. They probably serve in locating prey or enemies (Reissland & Görner 1985). Patterns of trichobothria are often diagnostic for certain families. Sensory structures called slit sense organs are located near the leg joints. These slits are also connected to nerves and are sensitive to airborne vibrations within certain frequencies, serve as gravity receptors, and respond to the spider's own movements (Barth 1985, 2002).

Open-ended hairs at the ends of the legs and pedipalps serve as taste receptors. Spiders can identify and avoid pungent or acidic substances upon contact, thereby probably enabling them to distinguish good prey from bad and allowing males to detect female pheromones deposited in silk. Spiders also respond to airborne odors. However, the structures or organs involved in olfaction are not known. Some evidence suggests that tarsal organs may be involved in chemoreception, including detection of airborne cues. Tarsal organs, also called Blumenthal's tarsal organs, are open pits on the tarsal segments of the legs and palpi. Each pit contains six or seven sensilla, each of which is connected to a dendrite. Strong evidence exists that the tarsal organ is involved in hygroreception and thermoreception and some data suggest it also is involved in chemoreception (Barth 2002). Emerit's glands, small pores with distinctive morphology, are found in a variety of spider families. Their function is unknown, but they may produce defensive secretions (Bennett 1989).

## MEDICALLY IMPORTANT SPIDERS
### General information and caution

Of the over 38,000 species of spiders described to date, only about 100 species worldwide have venom demonstrated to be detrimental to humans (Maretić 1987). Of these, only the widow spiders in the genus *Latrodectus* (Theridiidae) and the recluse spiders, including the brown recluse, in the genus *Loxosceles* (Sicariidae) have been definitively shown to have venom of medical importance in North America. It has been suggested that other species in this region also have venom of medical importance to humans, such as the hobo spider, *Tegenaria agrestis* (Agelenidae) (Akre & Myhre 1991) and the yellow sac spiders in the genus *Cheiracanthium* (Miturgidae) (Krinsky 1987). However, the evidence that bites from the hobo spider can cause medical problems in humans is circumstantial at best (Vetter & Isbister 2004). Evidence for lesions from bites of *Cheiracanthium* species is slightly better (Krinsky 1987). That said, since nearly all species of spiders have venom, the potential exists for some people to be allergic to components in the venom. Allergic responses to venoms of many different types of biting or stinging arthropods are possible. Spiders can control the amount of venom released so bites, even from widow spiders, can be asymptomatic if little or no venom is injected.

Whenever a person gets bitten or stung by any arthropod, it is critical to collect the offending insect or spider, or collect its remains after it has been killed, and bring it to the physician so the animal can be sent away for positive identification by an expert. Very few physicians have sufficient training in entomology or arachnology to accurately identify invertebrates. Many, if not most, necrotic lesions, sores, blisters, or other dermatological injuries that are diagnosed as spider bites without a spider having been seen at the site of the injury are likely due to other conditions. Just having spiders in the home is not valid justification for blaming the spiders for unusual wounds or sores. Many necrotic lesions or ulcerous sores diagnosed as brown recluse bites in areas of the U.S. where the recluse spiders are infrequent visitors or entirely absent are more likely caused by such conditions as diabetes (diabetic ulcers), *Staphylococcus* or *Streptococcus* bacterial infections, lymphoma, tick bites, herpes viruses, etc. (Vetter 2000b, Vetter *et al.* 2003). In those regions of the country outside the range of the recluse spiders, spiders are the least likely culprits for such ulcerous sores. Other more likely factors or conditions should first be ruled out before such a sore is diagnosed as a spider bite. Lymphoma, diabetic ulcers, bacterial infections, and tick bites can all be effectively treated if a correct diagnosis is made early on in treatment. If such conditions are initially misdiagnosed as spider bites, very serious medical complications could result.

Spiders, including widow spiders and recluse spiders, are generally timid creatures that are far more likely to try to escape when encountered than to attack and bite. Spiders typically only bite when seriously provoked – e.g., when a person inadvertently presses down on a hidden spider or reaches his/her hand into a spider's retreat.

## IMPORTANCE OF SPIDERS

Material engineering research is increasingly focusing on the biosynthesis of spider silk proteins. Spider silk, because of its strength and elasticity, is a potentially valuable substance if it can be biosynthesized and mass produced. Some success has already been seen in this field (Lazaris *et al.* 2002). Spider silk research is a growing field which may soon lead to lightweight bullet-proof vests, lightweight parachutes, and other super-strong material made from spider silk proteins.

Research is also very active in spider venom research. Since spider venom is very effective at killing insects, research is being done to develop pesticides derived from spider venoms (King *et al.* 2002). Medical uses of spider venom are also being explored since recent research indicates that venom components are useful in treating cardiac disorders (Bode *et al.* 2001).

Spiders are found in every terrestrial ecosystem except Antarctica and are significant components of many freshwater and marine ecosystems, functioning as major predators of aquatic insect populations in shoreline, estuarine, and riparian habitats. They are one of the top invertebrate predators regardless of habitat and comprise the seventh largest order of living organisms on earth in global species diversity (Coddington *et al.* 1990, 1991, Coddington & Levi 1991). Most are generalist predators and were not historically considered useful biocontrol agents against agricultural pests because they feed both on harmful and beneficial insects. However, recent studies have indicated that spiders can, in fact, help suppress pest populations (Riechert & Lockley 1984, Sunderland 1999, Hlivko & Rypstra 2003, Maloney *et al.* 2003).

## SOURCES OF INFORMATION
### Books
The following books, still widely available, provide information about spider biology and diversity. Other useful texts are cited at the International Society of Arachnology website. Full citations of the following texts can be found in the Bibliography:

*Guide d'identification des Araignées (Araneae) du Québec* (2003), by Pierre Paquin & Nadine Dupérré provides identification for all spider species of Québec, Canada (in French).

*An Introduction to the Spiders of South East Asia* (2000), by Frances and John Murphy provides descriptions of the spider genera found in SE Asia.

*Spiders of China* (1999) by Song, Zhu, and Chen provides illustrations of all Chinese species.

*Spiders of Panama* (1993) by Wolfgang Nentwig contains checklists and keys to the families and most genera found in Panama.

*Spinnen Mitteleuropas* (1991) by Stefan Heimer & Wolfgang Nentwig provides identification keys to the species of northern Europe (in German).

*African Spiders: An Identification Manual* (1997) by Ansie Dippenaar-Schoeman & Rudy Jocqué is a well-organized manual to the spiders of Africa.

*Spiders of Britain & Northern Europe* (1995) by Michael J. Roberts is a field guide to the spiders of Great Britain and northern Europe.

*Spiders of Japan in Colour* (1960) by Takeo Yaginuma is an illustrated guide to the spiders of Japan (in Japanese).

*Spiders of New Zealand and their Worldwide Kin* (1999) by Ray and Lyn Forster is a profusely illustrated guide to the common spiders of New Zealand.

*Biology of Spiders*, by Rainer Foelix (1996) provides an overview of many aspects of spider biology.

*A Spider's World*, by Friedrich G. Barth (2002) is an excellent source of information about the sensory systems and structures of spiders.

*Spiders in Ecological Webs* (1993) by David H. Wise provides a good summary of ecological research on spiders.

*Spiders and Their Kin. A Golden Guide®* (2002) by Herb & Lorna Levi, a small field guide to common spiders and other terrestrial non-insect arthropods found throughout the world.

*How to Know the Spiders* (1978) by B.J. Kaston: an illustrated key for identification to genus. The introductory material is particularly helpful. Allen Dean has produced a list updating the taxonomy in Kaston's guide that can be found at http://kaston.transy.edu/spiderlist/kaston78.htm.

### Worldwide web sources
The URLs listed below provide current information about spider taxonomy, biology, scientific societies, and biological supply houses. The reader is reminded that, although the internet contains a mind boggling amount of information about spiders, much of it is either worthless or downright incorrect and should always be checked. Information posted on the internet by professional arachnological organizations or by professional arachnologists can generally be trusted.

The following websites contain good information about spiders and other arachnids as well as links to other trustworthy sites:

### *World Spider Catalog*
research.amnh.org/entomology/spiders/catalog/index.html

This continuously updated online catalog provides a list of all 38,000+ described species of spiders and bibliographic citations for spider taxonomic literature. It is the basis for the taxonomy described in this identification manual. Readers are urged to use this catalog to determine current species names, current genus placements, and the names of species authors.

### American Arachnological Society (AAS) website
www.americanarachnology.org

The official website of the AAS. The AAS has among its 600+ members professional arachnologists, students, teachers, and enthusiastic amateurs. Regular membership costs $40 US, student membership costs $25 US annually (2005 rates). Online payments can be made through PayPal. Membership includes subscription to the Journal of Arachnology and the American Arachnological Society newsletter, reduced registration costs for annual conferences, and a discount on the purchase price of this identification manual!

### International Society of Arachnology (ISA) website
www.arachnology.org

The ISA website produces a catalog of arachnid-related research publications and registration information about the ISA conference held every three years. This website also serves as a clearinghouse for hundreds of other arachnid-related internet resources. Links are provided to sites containing information about every aspect of arachnid biology.

### Insect and Spider Collections of the World
hbs.bishopmuseum.org/codens/codens-r-us.html

Information about major museum and university collections of spiders.

### Spider Conservation Page
www.geocities.com/RainForest/9801/index.html

Promotes awareness of conservation concerns and issues related to this often overlooked group of organisms.

### Spiders of North America
kaston.transy.edi/spiderlist/index.html

Provides a list of current names for the spiders (Araneae) which occur in North America north of Mexico as well as common synonyms for species. Translation lists of genus names for Kaston's *How to Know the Spiders* and *Spiders of Connecticut* as well as Emerton's Common Spiders of the United States also can be found via this site.

### Tree of Life Web Project
tolweb.org/tree/phylogeny.html

Provides information about the diversity of organisms on Earth and their evolutionary history. Various phylogenies of groups of arachnids, including spiders, can be found at this site.

### BioQuip Products
www.bioquip.com

This biological supply company specializes in entomology-related equipment, supplies, and books. Their phone number is (310) 667-8800.

**Carolina Biological Supply Company**
www.carolina.com
　This company sells biology-related equipment and supplies. Their phone number is 1-800-334-5551.

**Edmund Scientific**
www.scientificsonline.com
　This company sells less expensive microscopes for home use. Their phone number is 1-800-728-6999.

Chapter 2

# PHYLOGENY AND CLASSIFICATION OF SPIDERS

Jonathan A. Coddington

## ARACHNIDA

Spiders are one of the eleven orders of the class Arachnida, which also includes groups such as harvestmen (Opiliones), ticks and mites (Acari), scorpions (Scorpiones), false scorpions (Pseudoscorpiones), windscorpions (Solifugae), and vinegaroons (Uropygi). All arachnid orders occur in North America. Arachnida today comprises approximately 640 families, 9000 genera, and 93,000 described species, but the current estimate is that untold hundreds of thousands of new mites, substantially fewer spiders, and several thousand species in the remaining orders, are still undescribed (Adis & Harvey 2000, reviewed in Coddington & Colwell 2001, Coddington et al. 2004). Acari (ticks and mites) are by far the most diverse, Araneae (spiders) second, and the remaining taxa orders of magnitude less diverse. Discounting secondarily freshwater and marine mites, and a few semi-aquatic or intertidal forms, all extant arachnid taxa are terrestrial. Arachnida evidently originated in a marine habitat (Dunlop & Selden 1998, Dunlop & Webster 1999), invaded land independently of other terrestrial arthropod groups such as myriapods, crustaceans, and hexapods (Labandeira 1999), and solved the problems of terrestrialization (desiccation, respiration, nitrogenous waste removal without loss of excess water, and reproduction) in different ways. Although the phylogeny of Arachnida is still controversial (Coddington et al. 2004), specialists agree that the closest relative of Araneae is a group of orders collectively known as Pedipalpi: Amblypygi, Schizomida, and Uropygi (Shultz 1990).

## PHYLOGENETIC THEORY AND METHOD

Systematics is the study and classification of the different kinds of organisms and the relationships among them. Good classifications are predictive: knowing one feature predicts many others. If one knows that an animal has spinnerets on the end of the abdomen, it will also have fangs and poison glands (lost in a few spiders), eight legs, two body regions, male palpi modified for sperm transfer, and it will spin silk: in short, it is a spider. All spiders share these features because they inherited them from a common ancestor, but today's spiders have evolved to differ among themselves. For example, the earliest spiders had fangs that worked in parallel (orthognath, like tarantulas and their allies), but later in spider evolution one lineage developed fangs that worked in opposition (labidognath, like the majority of spiders in North America). Much later within the labidognath lineage, some evolved the ability to coat silk lines with a viscid, semi-liquid glue, useful for entrapping and subduing prey. This nested pattern of branching lineages (phylogeny), results from evolutionary descent with modification (Fig. 2.1). The vast majority of similarities and differences among species are due to phylogeny. Jumping spiders (Salticidae) all have huge anterior median eyes because they are relatively closely related, and wolf spider (Lycosidae) eyes exhibit their characteristic eye pattern for the same reason. Phylogeny explains more biological patterns than any other scientific theory (e.g., ecology, physiology, ethology, etc.), and therefore classifications based on phylogeny will be maximally predictive. Besides huge front eyes, jumping spiders also share many other anatomical, behavioral, ecological, and physiological features. Most important for the field arachnologist they all jump, a useful bit of knowledge if you are trying to catch one. Taxonomic prediction works in reverse as well: that spider bouncing about erratically in the bushes is almost surely a salticid.

Another reason that scientists choose to base classification on phylogeny is that evolutionary history (like all history) is unique: strictly speaking, it only happened once. That means there is only one true reconstruction of evolutionary history and one true phylogeny: the existing classification is either correct, or it is not. In practice it can be complicated to reconstruct the true phylogeny of spiders and to know whether any given reconstruction (or classification) is "true." Indeed, scientists generally regard "truth" in this absolute sense as beyond their reach. Instead they strive to make their hypotheses as simple as possible, and as explanatory as possible. Simpler and more general hypotheses win. They win through comparison of predictions made by the hypothesis to factual observation. Scientific hypotheses (e.g., explanations, classifications, taxonomies, phylogenies) are constantly tested by discovery of new traits and new species. To the extent that the hypothesis is good, it accommodates and comfortably explains new data. If the new data do not fit the theoretical expectations, sooner or later a new hypothesis or a revised version of the old one takes its place. In biological classification, and phylogeny reconstruction in particular, scientists have developed a number of technical terms to describe the various ways that classifications or phylogenies do, or do not, correspond to fact (Fig. 2.1). Any group in a classification is said to be a taxon (plural taxa) or clade, and in theory corresponds to one common ancestral species and all of its descendants. Such clades are said to be monophyletic ("mono" = single, and "phylum" = race).

In the preceding examples, spiders (Araneae), labidognath spiders (now called Araneomorphae), sticky-silk spinners (Araneoidea), jumping spiders (Salticidae) and wolf spiders (Lycosidae) are all thought to be monophyletic groups, clades, and taxa. Each of these groups is distinguished by one or more uniquely evolved features or innovations. Such characters are said to be "derived," because they are transformations of a more primitive trait. Orthognath chelicerae is the original, primitive (plesiomorphic) condition for spiders, and labidognath chelicerae is the later, derived (apomorphic) condition. The only acceptable evidence for monophyletic groups are shared, derived characters, or synapomorphies ("syn" = shared, "apomorphy" = derived morphology) such as the evolution of viscid silk in Araneoidea (Fig. 2.1).

Sometimes systematists (scientists who infer phylogeny and use the results to classify organisms) make mistakes and group taxa based on primitive characters or plesiomorphies. Such groups, containing a common ancestor and some but not all of its descendants, are then termed paraphyletic. In Figure 2.1, the grouping "Orthognatha" is paraphyletic because it is based on a primitive character, orthognath or paraxial chelicerae, and because it includes the common ancestor of all spiders but excludes some descendants, i.e. the Labidognatha. Even worse, sometimes groups don't even include any common ancestor at all and are then termed polyphyletic ("Big Spiders," Lycosidae + Mesothelae in Fig. 2.1 would be polyphyletic). Polyphyletic groups are usually based on convergent features and paraphyletic groups on primitive features.

Classifications (and phylogenies) need not be strictly binary or dichotomous: in Figure 2.1 the three-way fork uniting Araneoidea, Lycosidae, and Salticidae intentionally doesn't indicate which is most closely related to which. If nodes are dichotomous, the two daughter lineages are often called sister taxa, or, informally, sisters.

In practice systematists infer phylogeny by compiling large tables or matrices of taxa and their traits or features. Traits may be anything presumed to be genetically determined and heritable, such as morphology, physiology, behavior, or, increasingly, DNA sequences. The ideal approach would encapsulate all comparative knowledge about the group in question. Based on evidence external to the analysis (or even an a priori assumption) one taxon in the analysis is specified to join at the root of the tree, and powerful computer algorithms are used to find the most plausible tree (or branching diagram, also termed a cladogram) that unites all taxa and best explains the data. Systematists adopt the initial null hypothesis that all similarities are due to phylogeny. The fit between the tree and the data decreases to the extent that one must suppose the "same" trait arose two or more times independently (convergent evolution) or was lost secondarily. An example of the former might be "big." Not all "big" spiders are each other's closest relatives (but some are). An example of the latter is the absence of true abdominal segmentation in all spiders. Spiders are arthropods and arthropods typically have segmented abdomens; spider relatives also have segmented abdomens. Rather than suppose that all arthropods with segmented abdomens gained the condition independently, and thus that spiders reflect the ancestral unsegmented arthropod, it becomes very much simpler

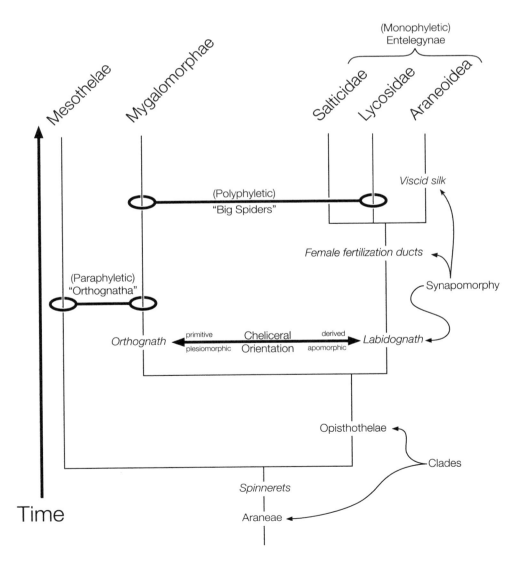

**Fig. 2.1** Taxa are in regular, characters in italic font. Only synapomorphies (shared, derived characters) are valid evidence of monophyletic groups (clades). Paraphyletic groups are usually based on plesiomorphies, polyphyletic groups on convergences.

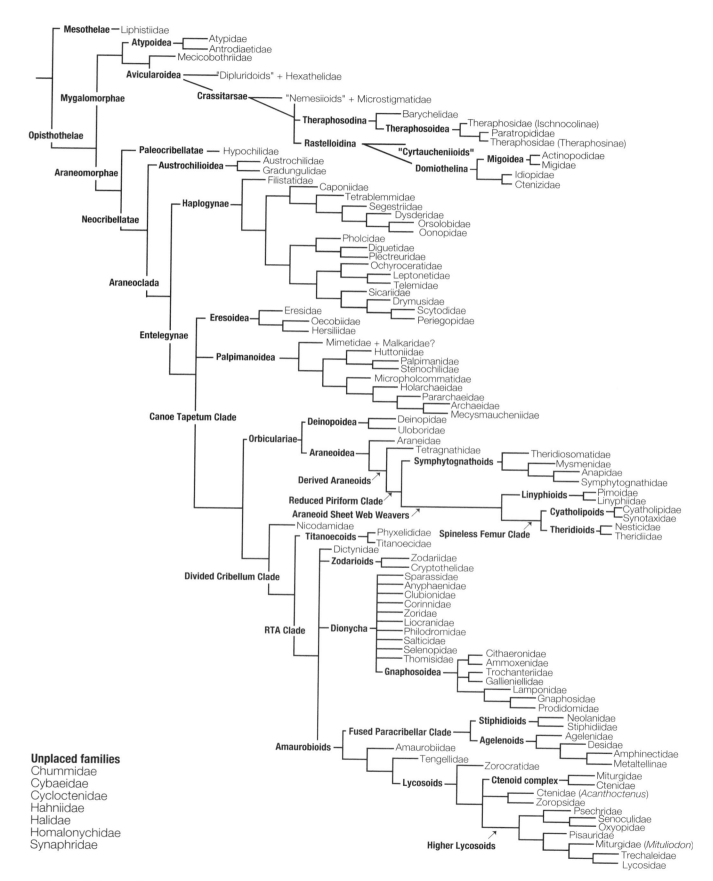

**Fig. 2.2** Phylogeny of Araneae

to suppose that the original arthropod was segmented and that it was spiders that changed to lack abdominal segmentation. The tree that requires the fewest hypotheses of convergent evolution and/or secondary loss is preferred as the current working hypothesis. Of course new taxa and characters can be added, so that in practice the preferred tree can be -and usually does -change at least a little bit with each new analysis. The best classifications are derived from phylogenetic analyses, but to date rather few groups of spiders have been analyzed phylogenetically.

## SPIDER PHYLOGENY

Spiders currently comprise 110 families, about 3,600 genera, and nearly 39,000 species (Platnick 2005). Paleontologists to date have described roughly 600 fossil species (Selden 1996, Dunlop & Selden 1998), but these have primarily been significant in dating lineages; thus far fossils have not seriously challenged or refuted inferences based on the recent fauna. Strong evidence supports spider monophyly: cheliceral venom glands, male pedipalpi modified for sperm transfer, abdominal spinnerets and silk glands, and lack of the trochanter-femur depressor muscle (Coddington & Levi 1991). Roughly 67 quantitative phylogenetic analyses of spiders at the generic level or above have been published to date, covering about 905 genera (about 25% of the known total), on the basis of approximately 3,200 morphological characters (summarized in Coddington & Colwell 2001 and Coddington et al. 2004). On the one hand, overlap and agreement among these studies is just sufficient to permit "stitching" the results together manually (Fig. 2.2); on the other, they are so sparse that many relationships in Fig. 2.2 are certain to change as more information accumulates and more taxa are studied. Figure 2.2 is not itself the result of a quantitative analysis but is, essentially, an amalgamation of individual cladograms published for particular lineages. Although there are several spider phylogenies above the species level based on DNA (Huber et al. 1993, Garb 1999, Gillespie et al. 1994, Piel & Nutt 1997, Hedin & Maddison 2001, Vink et al. 2002, Maddison & Hedin 2003a, b, Arnedo et al. 2004), this field is still in its infancy.

The basics of spider comparative morphology have been known for over a century, but the first explicitly phylogenetic treatment of spider classification was not published until the mid-1970's (Platnick & Gertsch 1976). This analysis resolved a long-standing debate by clearly showing a fundamental division between two suborders: the plesiomorphic mesotheles (the southeast Asian Liphistiidae with two genera and about 85 species) and the derived opisthotheles (everything else). Whereas mesotheles show substantial traces of segmentation, for example in the abdomen and nervous system, the opisthothele abdomen shows no segmentation (although color patterns in many spider species still reflect ancient segmentation patterns) and the ventral ganglia, or nerve centers, are fused. Opisthotheles include two major lineages: the baboon spiders (tarantulas) and their allies (Mygalomorphae, 15 families worldwide with about 300 genera and 2,500 species) and the so-called "true" spiders (Araneomorphae, 94 families worldwide with about 3,200 genera and 36,000 species) (Platnick 2005).

Mygalomorphs look much more like mesotheles than araneomorphs. They tend to be rather large, often hirsute animals with large, powerful chelicerae. Nearly all lead quite sedentary lives, usually in burrows, which they rarely leave, and they rely little on silk for prey capture. Some do fashion "trip-lines" from silk or debris, which effectively increase their sensory radius beyond the immediate area of the burrow entrance, and some diplurid species do spin elaborate sheet webs, but they lack one key innovation present in araneomorphs: the piriform silk essential to cement silk to silk, or silk to substrate. Whereas any araneomorph can dab a tiny dot of piriform cement to anchor its dragline and almost instantly trust its life to the bond, mygalomorphs must spin structures several centimeters across to anchor silk to substrate, and that is a long and laborious process. Without piriform silk, substantial innovations in web architecture are essentially impossible.

Mygalomorphs rarely balloon, and therefore their powers of dispersal are limited to walking. Usually, the juveniles do not walk far, and so mygalomorph populations are highly clumped: when you find one, its siblings and cousins are generally not far away. Mygalomorph known species diversity is barely 7% of araneomorph diversity, so the diversification rates of these two sister taxa clearly differ, but the reasons remain mysterious. The contrast in dispersal mechanisms may be a partial explanation, but it also may be that mygalomorph species are simply much more difficult to discriminate morphologically. Araneomorphs lead much more vagile lives (including dispersal by ballooning), and the group is much more diverse.

Within mygalomorphs, the atypoid tarantulas (Atypidae, Antrodiaetidae) seem to be the sister group of the remaining lineages (Raven 1985a, Goloboff 1993), although some evidence suggests including Mecicobothriidae in the atypoids. The sister group to the atypoids is the Avicularioidea, of which the basal Dipluridae may be a paraphyletic assemblage (Goloboff 1993). One of the larger problems in mygalomorph taxonomy worldwide concerns the paraphyletic Nemesiidae, currently 38 genera and 325 species (Goloboff 1995). The remaining mygalomorph families (represented in North America only by Ctenizidae, Cyrtaucheniidae, and Theraphosidae, Fig. 2.2) divide into two distinct groups: the theraphosodines – baboon spiders or true "tarantulas" and their allies (Pérez-Miles et al. 1996) and the typically trap-door dwelling rastelloidines (Bond & Opell 2002). Less work has gone into mygalomorph phylogeny, and because the diverse Dipluridae, Nemesiidae, and Cyrtaucheniidae seem to be paraphyletic, the number of mygalomorph families may increase substantially. For all their size and antiquity, mygalomorphs have remarkably uniform morphology. Because fewer reliable, distinctive features can be compared, mygalomorph phylogeny has been a difficult and frustrating subject. Perhaps molecular data will advance the subject.

Araneomorphs include over 90% of known spider species: they are derived in numerous ways and appear quite different from mesotheles or mygalomorphs. Mesotheles are the only spiders with an anterior median pair of distinct spinnerets and mygalomorphs have lost them completely. A complex, important synapomorphy of araneomorphs is the fusion and reduction of the anterior median spinnerets to a cribellum, a flat sclerotized plate that bears hundreds to thousands of silk spigots that produces very fine, dry, yet extremely adhesive, silk (cribellate silk). Spider dragline silk is justly famous because it is tougher than Kevlar (Craig 2003, Gosline et al. 1986, 1999, Scheibel 2004), but in many ways, cribellate silk is even more amazing. Its stickiness seems to be based on electron-electron interactions (van der Waals forces) between the silk and the surface to which it sticks (Hawthorn & Opell 2002). Insofar as other natural and man-made glues operate on gross chemical principles, this atomic-level universal glue mechanism also seems worthy of biotechnological attention.

Other animals use silk throughout their lives, but no other group of animals even comes close to the diversity, intricacy, and elegance of silk use by spiders (Eberhard 1990, Craig 2003). Systematists first became aware of this richness through comparative studies of behavior (Eberhard 1982, 1987b, 1990, Coddington 1986b, c), and only later by paying attention to the morphology that produced all that diversity. Although cytologists had been studying spider silk glands since the early 20th century (Kovoor 1987), it was not until the advent of the scanning electron microscope, and the cladistic reinterpretation of cytological data in the light of spigot diversity (Coddington 1989) that systematists began to plumb the immense variation in spinneret spigots and silks for phylogenetic research. Now spigots, silks, and silk use are some of the richest sources of comparative data in spiders (e.g., Platnick 1990a, Platnick et al. 1991, Eberhard & Pereira 1993, Griswold et al. 1999a).

Although many araneomorph lineages independently abandoned the sedentary web-spinning lifestyle to become vagabond hunters, the plesiomorphic foraging mode seems to be a web equipped with cribellate silk. Paleocribellatae contains just one family, Hypochilidae (two genera, 11 species). It is a famous North American taxon because it is sister to all remaining araneomorphs (Neocribellatae), and therefore retains a fair number of primitive traits (Platnick et al. 1991, Catley 1994). Within Neocribellatae, the monophyly of Haplogynae is weakly based on cheliceral (chelicerae fused with a lamina instead of teeth), palpal (tegulum and subtegulum fused rather than free), and spinneret characters (vestiges of spigots in former molts lacking) (Platnick et al. 1991, Ramírez 2000). The cribellate Filistatidae (three North American genera) is sister to the remaining haplogynes, and a quantitative phylogeny of the family has been published (Ramírez & Grismado 1997). All haplogynes except Filistatidae evidently lost the cribellum, but Pholcidae (Huber 2000), Diguetidae, Ochyroceratidae, and, debatably, Scytodidae and Segestriidae still build prey-catching webs. The cellar spiders (Pholcidae) are exceptional for their relatively elaborate, large webs. Some pholcid genera have independently invented viscid silk (Eberhard 1992). Some of the most common and ubiquitous synanthropic spider species are pholcids, so one should not assume that a taxon that branched off relatively early in evolution is necessarily primitive, poorly adapted, or non-competitive. The remaining haplogyne families live either in tubes (Dysderidae) or are vagabonds that tend to occur in leaf litter or other soil habitats and are not commonly encountered by the casual collector. Because they are still poorly known, many of the new spider species discovered each year tend to come from haplogyne families.

The araneomorph Entelegynae is supported by several important, yet poorly understood synapomorphies (Griswold et al. 1999a) in reproductive and spinning systems. Entelegynes have more complex reproductive systems in which the female genitalia (epigynum) has external copulatory openings. In all other spiders copulation takes place through the gonopore. This secondary set of openings provides a "flow-through" sperm management system: the male deposits sperm in the epigynum that connects to spermathecae that connect to the uterus. This has major implications for reproductive behavior and the relation between the sexes (Eberhard 2004). For one thing, the flow-through system means the first male to mate with a female usually sires most of the spiderlings (Austad 1984). The copulatory ducts that lead from the epigynum to the spermathecae are often extremely contorted in entelegyne females. This has led to hypotheses that such complexity actually make it more difficult for males to inseminate females (Eberhard 1985, 1996). For whatever reason, fertilization ducts in entelegyne spiders have been secondarily lost only five times and in mostly small groups: twice in distal palpimanoid families (this may be primary absence rather than secondary loss, Huber 2004), a small sub-clade of uloborids, some anapids, and a rather uniform, if speciose, sub-clade of tetragnathids (Hormiga et al. 1995). A second consistent but enigmatic entelegyne synapomorphy is cylindrical gland silk. These glands and spigots appear only in adult females, and it is thought that the silk is used only in egg sacs, but the specific contribution of cylindrical gland silk to egg sac function remains unknown.

Male entelegyne genitalia are also greatly modified. Plesiomorphic male spider genitalia are usually simple, tapering, pyriform bulbs. Pyriform bulbs lack apophyses or, if apophyses are present, they are small and quite simple (e.g., Fig. 2.3). Other parts of the male palp, such as the cymbium, patella, and tibia, are likewise unadorned. In contrast, entelegyne male genitalia can be bewilderingly complex. The bulb has two or three divisions (always subtegulum and tegulum, sometimes with an elaborate embolic division). The tegulum usually has two apophyses (conductor and median apophysis) in addition to the embolus. Any or all of these in entelegynes can be wonderfully complex, with knobs, levers, grooves, hooks, serrations, sinuous filaments, and spiraling parts (e.g., Figs. 2.4-2.5). Unraveling the homology of entelegyne male genitalia is a major problem (Coddington 1990). Entelegyne bulbs also work differently. The plesiomorphic bulb ejaculates via muscles that force the sperm out. Entelegyne bulbs lack those muscles and work hydraulically instead. The male pumps blood into the bulb to raise its internal pressure, which serves both to expand and "uncoil" its various parts. Glands empty their contents into the sperm duct and force the sperm out (Huber 2004). For sperm transfer to occur, this complicated structure must interact precisely with the correspondingly complex female genitalia (Huber 1994, 1995). In addition, the various parts of an entelegyne bulb are usually connected only by thin, flexible membranes that inflate like balloons during copulation. As a whole the bulb is so flexible that at least some of the male's complexity doubtless serves only to stabilize and orient his own genitalia during copulation.

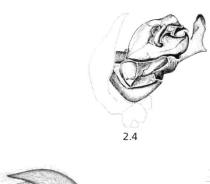

2.4

2.3

2.5

Female epigyna have corresponding ledges, pockets, ridges and protuberances externally, and often labyrinthine ductwork internally. One hypothesis is that the female complexity is essentially defensive, the result of antagonistic co-evolution (Chapman *et al.* 2003): females as a whole invest so much more in their offspring than do males (even in spiders), that they should choose mates carefully (Alexander *et al.* 1997). Female genitalic complexity may be a "challenge" to males such that only high-quality males succeed, or succeed much better than low-quality males. Males, in turn, have evolved complex and highly flexible genitalia, the better to win over choosy females. This explanation assumes, of course, that overall quality of males is tightly correlated with male copulatory prowess. In any event, the difficulty of attaching to and navigating female genitalia may give the female more time to assess the male and break off mating if she chooses.

Three small families, known as "eresoids," have thus far always clustered near the base of Entelegynae in phylogenetic analyses: Oecobiidae, Hersiliidae, and Eresidae (Coddington 1990, Griswold *et al.* 1999a). Only the first two occur in North America. Their phylogenetic relationships are controversial. Oecobiids (Glatz 1967) and hersiliids (Li *et al.* 2003) share a unique attack behavior: they are the only spiders known to run swiftly around the stationary prey encircling it with silk as they go. The behavior could also be convergently evolved, although in accordance with the null hypothesis of phylogenetics, until proven otherwise, we presume the similarity is explained by descent. Certainly no obvious features tie any of these families closely to any other entelegynes. Because they are entelegyne yet share no more derived apomorphies with other entelegyne clades, they seem to be the basal entelegyne group.

The Palpimanoidea is a controversial entelegyne group (10 families, 54 genera), of which only the pirate spiders (Mimetidae) occur in North America. For years mimetids were thought to be araneoids based on setal morphology, general appearance, and the overall complexity of their palps, but then were transferred to palpimanoids (Forster & Platnick 1984). Recent research, however, suggests that mimetids are araneoids after all, in which case Palpimanoidea is polyphyletic (Schütt 2000, 2003, Huber 2004).

Males can also have knobs or apophyses elsewhere on their palpi. About half of entelegyne species have one in particular, the "retrolateral tibial apophysis," that defines a clade of 39 entelegyne families (the "RTA clade": Sierwald 1990, Coddington & Levi 1991, Griswold 1993, Fig. 2.2). Huber (1994, 1995) found that the RTA usually, but not always, serves to anchor and orient the male bulb to the female genitalia prior to expansion of the hematodochae.

Orbiculariae is one of the largest entelegyne lineages. It consists of two superfamilies, Araneoidea (13 families, 1,000 genera), and Deinopoidea (2 families, 22 genera). The monophyly of Orbiculariae is controversial because the strongest apomorphies are all behavioral: both groups spin orbwebs (Coddington 1986c and references therein). Prior to strictly phylogenetic classification in spiders the two groups were thought to be only distantly related. The cribellate Deinopidae and Uloboridae were included in the "Cribellatae," and authors often commented on the detailed similarity of these orbs to those of the classical, ecribellate Araneidae (reviewed in Scharff & Coddington 1997). However, as noted above, the cribellum is primitive for Araneomorphae. On the one hand, araneoids or their ancestors must therefore have lost it and, on the other, groups based on plesiomorphies are false. When the old, polyphyletic Cribellatae collapsed (Lehtinen 1967, Forster 1967, 1970b), deinopids and uloborids had nowhere else to go, as it were, and so the form of the web (and the striking similarities in behavioral details) constituted a strong block of synapomorphies. But if orbweavers were monophyletic, the six araneoid families that spin sheet or cobwebs must have lost the orbweb. Against this view is the hypothesis that the orbweb is an unusually efficient and profitable design to catch prey. In general the more adaptive a feature is, the more likely it is to evolve independently; perhaps the araneoid and deinopoid web forms are convergent. This view argues that the orbweb is so superior a predation strategy that any spider lineage capable of it would have evolved it independently (and never lost it). Little evidence thus far suggests that orbwebs are drastically better than other web architectures (although they are widely regarded as better-looking!). Indeed, ecological evidence points the other way (Blackledge *et al.* 2003). Another difficulty for the monophyly hypothesis is that the deinopoid orb is cribellate (dry adhesive silk), and the araneoid orb uses viscid silk. The "missing link," it is argued, would have had neither. The obvious rejoinder is that perhaps they had both at one point, but one of the good effects of modern quantitative analysis is that people spend less time arguing about irresolvable issues, and more time seeking new evidence. The orb web diphyly argument particularly needs evidence that deinopoids share strong synapomorphies with some non-orb weaving group. Evidence against orbweaver monophyly is starting to appear from molecular evidence (Hausdorf 1999, Wu *et al.* 2002), but these studies are small, omit many important taxa, and do not confirm each other's results.

Araneoidea (ca. 11,000 species) is much larger than Deinopoidea (ca. 300 species). Only one deinopid species occurs in North America (in Florida and, possibly, Alabama). Araneoids are ecologically dominant species throughout the world but especially in north temperate areas such as North America, where Linyphiidae swamps any other spider family in both species diversity and sheer abundance. Current phylogenetic results (Hormiga 1994b, 2000, Griswold *et al.* 1998) indicate that Linyphiidae and five other families form the monophyletic "araneoid sheet weaver clade," which thus implies that within Araneoidea, the orb was lost only once (or transformed into a "sheet" web). Linyphiidae spin sheets as do Pimoidae. The classic cobwebs of Theridiidae and Nesticidae would then be derivations from a basic sheet, which, considering the web of black widows, *Steatoda*, and other apparently basal theridiid genera (Benjamin & Zschokke 2002, Agnarsson 2004, Arnedo *et al.* 2004, ), seems plausible. Araneoid sheet web weavers account for the bulk of araneoid species diversity (713 genera, 7,600 species worldwide). Perhaps sheet or cobwebs are not so bad after all (Griswold *et al.* 1998, Blackledge *et al.* 2003).

Although the most recent analysis suggests that the sister taxon of Orbiculariae is approximately all remaining entelegyne families (possibly including eresoids and palpimanoids), the evidence for this is quite weak for several reasons (Griswold *et al.* 1999a). First, non-orbicularian entelegyne families have received little phylogenetic research, so such overarching conclusions are premature. Second, the problem is intrinsically difficult. Resolving the entelegyne node requires an analysis that includes several representatives from all major entelegyne clades, including relevant enigmas such as Nicodamidae (Harvey 1995) and Zodariidae (Jocqué 1991a). That means a very large matrix and an even larger scope of characters. Such a matrix is not easily constructed, and will probably require collaboration of numerous specialists.

Although araneologists refer to large groups like the "amaurobioids" (Davies 1998a, 1999, Davies & Lambkin 2000, Wang 2002), "lycosoids" (wolf spiders, Griswold 1993), and "Dionycha" (two-clawed hunters, Platnick 1990a, 2000a, 2002), their monophyly is also tenuous at best. Amaurobioids (which currently contains lycosoids as a subgroup) are defined by a few small changes in spinneret spigot morphology only visible with the scanning electron microscope. Basal amaurobioid families present in North America are Desidae, Amaurobiidae, and Agelenidae (a mixture, by the way, of cribellate and ecribellate groups). The group is quite heterogeneous, including everything from hunters to elaborate web builders. Dictynidae may fall close to these families as well (e.g., Bond & Opell 1997).

Lycosoids were formerly thought to be defined by quite an unusual and convincing synapomorphy, the "grate-shaped" tapetum. The tapetum is a reflective layer within the eye that probably serves to increase sensitivity. In most lycosoids, the tapetal architecture is like a barbecue or street-drain grate -an arrangement of parallel bars and holes, whereas in other spiders the tapetum shows no particular pattern, or is in the form of a simple "canoe" (Canoe Tapetum Clade, Fig. 2.2). The reflection of the grate-shaped tapetum of the posterior median or lateral eyes can be visible at great distances in the field. At night it is common with a headlamp to see the green eye shine of a lycosoid 5, or even 15 meters away. In the latest analysis, however, the grate-shaped tapetum evolves twice (in stiphidiids and lycosoids *sensu stricto*). In North America, the lycosoid families are Ctenidae, Lycosidae, Miturgidae, Oxyopidae, Pisauridae, Trechaleidae, and Zoropsidae (introduced). Among these families a few genera still spin webs, but the majority have given up webs for a "vagabond" lifestyle.

The monophyly of "Dionycha" is equally tenuous. Dionycha is an old hypothesis -at one time many classifications divided entelegyne families into two-clawed (Dionycha) and three-clawed spiders (Trionycha). But three claws is primitive for a group even larger than all spiders, so a group defined by it would be paraphyletic. Two claws, however, is arguably a derived condition. This argument is weak (some spider groups placed elsewhere are two-clawed), and if strong evidence connected a dionychan family elsewhere, it would be preferred. As it happens, however, phylogenetically rigorous arguments link any dionychan group outside Dionycha, so the group stands solely on the simple fact of the two-clawed condition. One group within Dionycha, the Gnaphosoidea (7 families, Gnaphosidae and Prodidomidae in North America), does share an apomorphy with interesting functional implications (Platnick 1990a, 2000a, 2002): all have obliquely angled posterior median eyes with flat, rather than rounded lenses. As a flat lens cannot bend light, and there seems little reason otherwise to have eyes, the flat lens has always been a mystery. It turns out that at least the gnaphosid *Drassodes* uses these modified eyes to orient to polarized light, which in turn allows them to move about in the habitat and to return to the same spot (Dacke *et al.* 2001a, b).

The phylogenetic relationships of these non-orbicularian entelegynes, therefore, is poorly known. Because they were originally defined by plesiomorphies, many of the classical entelegyne families (most seriously Agelenidae, Amaurobiidae, Clubionidae, Ctenidae, and Pisauridae) were, and still probably are, paraphyletic. Dismembering these assemblages into monophyletic units has been difficult because the monophyly of their components or related families is also often doubtful (e.g., Amphinectidae, Corinnidae, Desidae, Liocranidae, Miturgidae, Tengellidae, Stiphidiidae, Titanoecidae). Corinnids, liocranids, zoropsids, and ctenids have been recently studied (Bosselaers & Jocqué 2002, Silva Dávila 2003), and neither analysis recovered a monophyletic Liocranidae, Corinnidae, or Ctenidae (indeed, rather the opposite). Bosselaers (2002) analyzed Zoropsidae, and it does seem to be monophyletic. The basic phylogenetic structure of Anyphaenidae was studied by Ramírez (1995a, 2003).

Therefore neither the RTA clade, nor the two-clawed hunting spider families (Dionycha) may be strictly monophyletic, although each presumably contains within it a large cluster of closely related lineages. Dionychan relationships are quite unknown, although some headway has been made in the vicinity of Gnaphosidae (Platnick 1990a, 2000a, 2002). In contrast, Lycosoidea was supposedly based on a clear apomorphy in eye structure, but recent results suggest that this feature evolved more than once, or, less likely, has been repeatedly lost (Griswold *et al.* 1998). The nominal families Liocranidae and Corinnidae are massively polyphyletic. The nodes surrounding Entelegynae will certainly change in the future.

In summary, most of the major clades in 20$^{th}$ century spider classifications were fundamentally flawed, which means books and overviews published prior to the last two decades have been superceded. Major lineages such as Trionycha, Cribellatae, Tetrapneumonae, Orthognatha, and Haplogynae (older definition) were all based on plesiomorphies. By the mid-1970's the higher-level classification of spiders had collapsed to the extent that catalogs of that era began to ignore higher classification (and still do). From that rubble has emerged the hypothesis presented in Figure 2.2, but events on the horizon suggest that its clarity may be shortlived. For one thing, a small cadre of workers from the mid 1980's on tried to cover as much ground as possible on the family-level, via studies that barely overlapped. Thus, the various studies underlying Figure 2.2, although quantitative, have not been tested by other workers, denser taxonomic sampling, or new sources of data. Figure 2.2 remains a first draft of spider phylogeny. As the latter processes proceed, our understanding of spider phylogeny will doubtless improve. The underlying observations are solid, and as more data accumulates, we can expect more stability in the results. In sum, phylogenetic understanding of spiders has advanced remarkably since the early 1980's. We are approaching a truly quantitative estimate of spider phylogeny at the family level, but phylogenies below that are going to require much denser taxonomic sampling and rigorous comparative study.

# Chapter 3

# KEY TO SPIDER FAMILIES OF NORTH AMERICA NORTH OF MEXICO

Darrell Ubick

L = length; W = width; > = greater than; < = less than; PT/C = patellar + tibial length of leg I / carapace length; size = body size from the anterior edge of the carapace to the end of the abdomen as viewed dorsally. Anterior tibia refers to tibia of legs I and II.

## INFRAORDERS OF ARANEAE

1  Chelicerae paraxial (fangs parallel, Fig. 3.1); with two pairs of book lungs (Fig. 3.1); AMS completely absent; with eight eyes; legs stout (PT/C < 2) .................................................. **MYGALOMORPHAE ... 2**

—  Chelicerae diaxial (fangs opposing each other or oblique Fig. 3.2); usually with at most one pair of book lungs (Fig. 3.2), if with two pairs of book lungs then with cribellum and slender legs (PT/C = 3-5); AMS represented by cribellum, colulus, or absent; with eight or fewer eyes; leg thickness variable .......................... **ARANEOMORPHAE ... 11**

## Families of MYGALOMORPHAE

2(1)  Abdomen with 1-3 tergites (Fig. 3.3); anal tubercle separated from spinnerets (Fig. 3.4) ......................... **3**

—  Abdomen without tergites (Fig. 3.5); anal tubercle adjacent to spinnerets (Fig. 3.6) ............................. **5**

3(2)  Endites long, 3/4 width of sternum (Fig. 3.7); labium and sternum fused (Fig. 3.7); thoracic furrow quadrangular or sub-oval (Fig. 3.8) .................................. **ATYPIDAE, p. 41**
**Dist.** e half of USA

—  Endites short, at most 1/2 width of sternum (Fig. 3.9); labium and sternum separated by a suture, groove, or at least a depression (Fig. 3.9); thoracic furrow longitudinal or rounded and pit-like ...................................................... **4**
**Dist.** widespread

4(3)  Distal segment of PLS slender, at least 5 times as long as basal width, tapering to a point, flexible, pseudosegmented (Fig. 3.10) ........................... **MECICOBOTHRIIDAE, p. 50**
**Dist.** BC-CA, AZ

—  Distal segment of PLS stouter, about 3 times as long as wide, neither flexible nor pseudosegmented (Fig. 3.11) ... .................................................. **ANTRODIAETIDAE, p. 39**
**Dist.** widespread

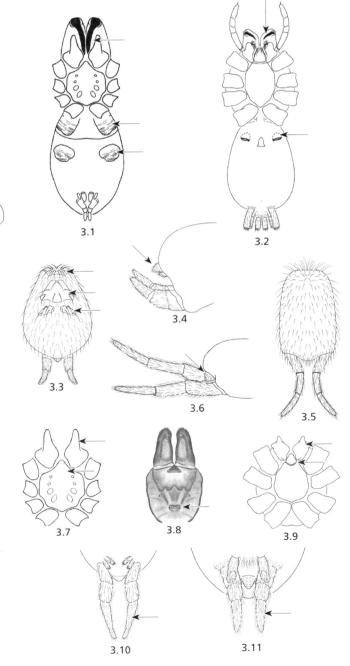

**5(2)** Tarsi with 2 claws and claw tufts (Fig. 3.12) ........................
............................................................. THERAPHOSIDAE, p. 54
**Dist.** sw USA, e to FL

— Tarsi with 3 claws, lacking claw tufts (Fig. 3.13) ............. **6**
**Dist.** widespread

**6(5)** PLS long (at least 1/2 carapace length) and slender (distal segment L > 2 X W) (Fig. 3.14) ........................................ **7**

— PLS short (at most 1/2 carapace length) and stout (distal segment L < 2 X W) (Fig. 3.15) ......................................... **8**

**7(6)** Endites with cuspules (Fig. 3.16); cheliceral inner surface with row of 4-7 short black rod-like setae (Fig. 3.17); thoracic furrow transverse (Fig. 3.18); PLS 1/2 to 2/3 carapace length (Fig. 3.18); size 16-23 mm; live in open burrows ...
............................................ NEMESIIDAE (*Calisoga*), p. 52

— Endites lacking cuspules; cheliceral inner surface without row of black rod-like setae; thoracic furrow pit-like or longitudinal (Fig. 3.19); PLS at least 3/4 carapace length (Fig. 3.19); size 2.5-17 mm; live in silken tubes or on sheet webs ........................................................ DIPLURIDAE, p. 48
**Dist.** BC-OR, AZ-TX, NC-TN

**8(6)** Abdomen posteriorly truncated, sclerotized, with longitudinal grooves (Fig. 3.20) ..................................................
............................. CTENIZIDAE, in part (*Cyclocosmia*), p. 43
**Dist.** se USA

— Abdomen unmodified, posteriorly rounded, not sclerotized, and without grooves ...................................................**9**

**9(8)** Females with a dense scopula on tarsus I (Fig. 3.21) and with cheliceral promargin toothed and retromargin with a row of short, stout, rounded tubercles (Fig. 3.22) .............
.............................. CYRTAUCHENIIDAE, in part (*Eucteniza*), p. 45
**Dist.** TX

— Males or females with tarsus I lacking scopula but with lateral spines (Fig. 3.23); if tarsus I with a weak scopula and lacking spines then the cheliceral retromargin without teeth (but may have a few denticles) ............................. **10**
**Dist.** widespread

**10(9)** Both cheliceral margins toothed (Fig. 3.24); thoracic furrow strongly procurved; anterior tarsi and metatarsi of female with lateral rows of short, thorn-like spines (Fig. 3.25) ...................................... CTENIZIDAE, *in part*, p. 43

— Only cheliceral promargin toothed (Fig. 3.26); thoracic furrow strongly procurved to straight; anterior tarsi and metatarsi of female with few spines which are usually long and slender (Fig. 3.27) ...................................................
............................................ CYRTAUCHENIIDAE, *in part*, p. 45

# Family key

**20(19)** Anal tubercle enlarged and fringed with long setae (Fig. 3.44); chelicerae free; entelegyne .......................... OECOBIIDAE (*Oecobius*), p. 183

— Anal tubercle not so modified (Fig. 3.45); chelicerae fused (Fig. 3.46); haplogyne ..................... FILISTATIDAE, p. 104

**21(19)** Anterior tibiae with at least 4 pairs of ventral spines (Fig. 3.47) ............................................................. **22**

— Anterior tibiae with fewer ventral spines ................... **23**

**22(21)** Anterior tibiae with 4-5 pairs of ventral spines, PER straight to weakly procurved (Fig. 3.48); tarsi II-IV with 2 claws; body concolorous .................................................. ZOROCRATIDAE (*Zorocrates*), p. 258
**Dist.** AZ-TX

— Anterior tibiae with 6-7 pairs of ventral spines; PER recurved (Fig. 3.49a); all tarsi with 2 claws; body patterned ..... ZOROPSIDAE (*Zoropsis*), p. 259
**Dist.** CA, introduced

— Anterior tibiae with 7-8 pairs of ventral spines; both PER and AER strongly recurved (Fig. 3.49b) ............................ CTENIDAE, in part (*Acanthoctenus*), p. 83
**Dist.** s USA, introduced

**23(21)** Calamistrum extends over more than half the length of metatarsus IV (Fig. 3.50) ................................................. **24**

— Calamistrum extends over no more than half the length of metatarsus IV (Fig. 3.51) ............................................... **26**

**24(23)** Femora II-IV with rows of long trichobothria (Fig. 3.52); metatarsus IV dorsally concave (Fig. 3.53) and with ventral rows of short spines extending to tip of tarsus (Fig. 3.54); cribellum entire (Fig. 3.55) ................................................ ULOBORIDAE, in part, p. 250

— Femora II-IV without rows of long trichobothria; metatarsus IV not so modified and without ventral row of short spines; cribellum divided or entire ................... **25**

**25(24)** Endites converging apically (Fig. 3.56); cribellum usually entire (Fig. 3.57); legs usually without spines ................................................ DICTYNIDAE, in part, p. 95

— Endites parallel (Fig. 3.58); cribellum divided (Fig. 3.59); legs with spines ...... TITANOECIDAE (*Titanoeca*), p. 248

**26(23)** Cheliceral margins with 5 to 7 stout teeth (Fig. 3.60); male palpus with embolus threadlike, enclosed in membranous conductor (Fig. 3.61) ..................................................... AMPHINECTIDAE (*Metaltella*), p. 63
**Dist.** CA, TX-FL

— Cheliceral margins usually with no more than 4 teeth (Fig. 3.62), if more, then teeth small and slender; male palpus with embolus variable ........................................ **27**
**Dist.** widespread

**27(26)** AME about 1.4 X larger than ALE (Fig. 3.63); male palpus with embolus long and sigmoid, enclosed in membranous conductor (Fig. 3.64) .................................................... DESIDAE, in part (*Badumna*), p. 93

— AME at most 1.2 X larger than ALE (Fig. 3.65); male palpus with embolus usually short and stout, if long, then arcuate, never enclosed in conductor (Fig. 3.66) ............ AMAUROBIIDAE, in part, p. 60

# Families of ARANEOMORPHAE
## key to sections

**11(1)** With cribellum (Fig. 3.28) and calamistrum (Fig. 3.29) ............................................. (Section 1) ... **14**

— Without cribellum (Fig. 3.30) and calamistrum ......... **12**

**12(11)** With fewer than 8 eyes .................. (Section 2) ... **28**
— With 8 eyes ..................................................... **13**

**13(12)** Tarsi 3-clawed, lacking scopulae and claw tufts (Fig. 3.31), mostly web builders. **Note** if legs are slender and relatively delicate, assume tarsi are 3-clawed ..................... (Section 3) .. **53**
— Tarsi 2-clawed, usually with scopulae and claw tufts (Fig. 3.32). **Note** if claw tufts are present, assume tarsi are 2-clawed ..................................... (Section 4) .. **83**

# Section 1
## Cribellate ARANEOMORPHAE

**Note** male cribellates have a degenerate cribellum and consequently a reduced calamistrum

**14(11)** With 2 pairs of book lungs (Fig. 3.33) ..................................... HYPOCHILIDAE (*Hypochilus*), p. 120
— With 1 pair of book lungs (Fig. 3.34) ............................ **15**

**15(14)** With fewer than 8 eyes ................................................ **16**
— With 8 eyes ..................................................................... **18**

**16(15)** With 4 eyes (Fig. 3.35); leg I greatly enlarged (Fig. 3.36) .. ............... ULOBORIDAE, in part (*Miagrammopes*), p. 250
**Dist.** s TX
— With 6 eyes; leg I not enlarged ........................ **17**
**Dist.** widespread

**17(16)** Cribellum entire (Fig. 3.37); male palpus lacking median apophysis (Fig. 3.38) ............................................................ .......................................... DICTYNIDAE, in part (*Lathys*), p. 95
**Dist.** widespread
— Cribellum divided (Fig. 3.39); male palpus with median apophysis (Fig. 3.40 ........................................................................ AMAUROBIIDAE, in part (*Cavernocymbium*, in part), p. 60
**Dist.** CA

**18(15)** PME huge, several times the diameter of remaining eyes (Fig. 3.41) ........................ DEINOPIDAE (*Deinopis*), p. 91
— PME not so enlarged ....................................................... **19**

**19(18)** Eyes clustered on central mound, occupying < 1/2 width of cephalic region (Fig. 3.42) ......................................... **20**
— Eyes spread across carapace, occupying > 1/2 width of cephalic region (Fig. 3.43) ............................................. **21**

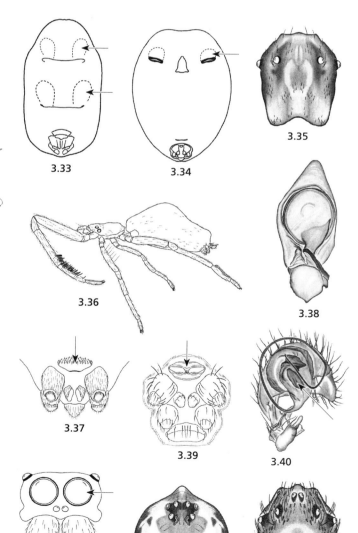

# Section 2
## Ecribellate ARANEOMORPHAE with < 8 eyes

**28(12)** Eyes completely lacking (but may have small eye spots) ........................................................................ **29**

— At least 2 eyes present .................................. **35**

## 0 eyes

**29(28)** Anterior tibiae with 2-3 pairs of ventral spines .......... **30**

— Anterior tibiae with few scattered ventral spines or none . ........................................................................ **31**

**30(29)** ALS contiguous, longer than PLS (Fig. 3.67) .................... ................................................ CYBAEIDAE, in part, p. 85

— ALS slightly separated, shorter than PLS (Fig. 3.68) ......... ................................................ DICTYNIDAE, in part, p. 95

**31(29)** Male palp with exposed bulb, female lacking epigynum ... ................................................ (Haplogynes) ... **32**

— Male palp with bulb enclosed by cymbium, female with epigynum ............................................ (Entelegynes) ... **33**

**32(31)** Abdomen with anterodorsal sclerotized ridge (Fig. 3.69); with a pair of tracheal spiracles between epigastric furrow and spinnerets (Fig. 3.70); colulus pentagonal, wider than ALS (Fig. 3.71) ...... TELEMIDAE (*Usofila*, in part) p. 228
**Dist.** CA

— Abdomen without anterodorsal sclerotized ridge; with one tracheal spiracle near spinnerets (Fig. 3.72); colulus not so modified ............ LEPTONETIDAE, in part, p. 122
**Dist.** TX, GA

**33(31)** Tarsus IV lacking ventral comb; cheliceral retromargin toothed (Fig. 3.73); chelicerae usually with stridulatory file on outer face (Fig. 3.74); leg break at patella-tibia joint (Fig. 3.75) ............................ LINYPHIIDAE, in part, p. 124

— Tarsus IV with ventral comb of serrated setae (Fig. 3.76); cheliceral retromargin edentate or with small denticles; chelicerae without stridulatory file; leg break at coxa-trochanter joint .................................................. **34**

**34(33)** Cheliceral retromargin edentate, lacking both teeth and denticles; male palp with inconspicuous paracymbium (Fig. 3.77); female palp with small claw ............................ ........................ THERIDIIDAE (*Thymoites*, in part) p. 235
**Dist.** s AZ

— Cheliceral retromargin with small denticles; male palp with large retrolateral paracymbium (Fig. 3.78); female palp with large claw ............ NESTICIDAE, in part, p. 178
**Dist.** CA, TX, Appalachian Mountains

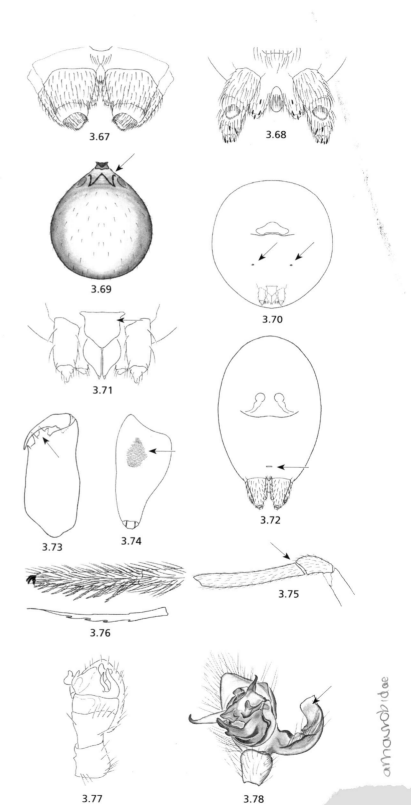

# Family key

## 2-4 eyes

**35(28)** Two eyes present (Fig. 3.79) .................................................
................................................ CAPONIIDAE, in part, p. 75

— More than two eyes present ......................................... **36**

**36(35)** Four eyes present ............................................................. **37**

— Six eyes present ............................................................ **38**

**37(36)** Two eyes pigmented; size 0.6 mm; litter spiders ...............
................ SYMPHYTOGNATHIDAE (*Anapistula*), p. 226
**Dist. FL**

— All eyes unpigmented; size 3.6 mm; cave spiders ...............
............................. NESTICIDAE (*Nesticus*, in part) p. 178
**Dist. TN**

## 6 eyes

**38(36)** Male palp with exposed bulb, tarsus not modified into cymbium (Fig. 3.80); female lacking epigynum (but may have some sclerotization at epigastric area) ........................
................................................... (Haplogynes) ... **39**

— Male palp with bulb enclosed by cymbium (Fig. 3.81), female with epigynum ........................ (Entelegynes) ... **49**

**39(38)** Chelicerae fused at base (Fig. 3.82) .............................. **40**

— Chelicerae not fused at base, can be moved apart (Fig. 3.83) .............................................................................. **43**

**40(39)** Eyes in two triads (Fig. 3.84) .........................................
................................................ PHOLCIDAE, in part, p. 194

— Eyes in three diads (Fig. 3.85) ...................................... **41**

**41(40)** Carapace strongly convex (Fig. 3.86) ............................
.......................................... SCYTODIDAE (*Scytodes*), p. 217

— Carapace flat ................................................................ **42**

**42(41)** PER strongly recurved; carapace pear-shaped (Fig. 3.87) .
.......................................... SICARIIDAE (*Loxosceles*), p. 222

— PER slightly recurved; carapace oval (Fig. 3.88) .............
.......................................... DIGUETIDAE (*Diguetia*), p. 102

**43(39)** Size > 5 mm; tracheal spiracles paired, conspicuous, located near book lung openings (Fig. 3.89) ...................... **44**

— Size < 5 mm; tracheal spiracles inconspicuous, if paired, then not near book lungs (Fig. 3.90) (except in Oonopiidae Fig. 3.91) ..................................................... **45**

**44(43)** Tarsi with two claws; leg III not anteriorly directed ..........
.......................................... DYSDERIDAE (*Dysdera*), p. 103

— Tarsi with three claws; leg III anteriorly directed (Fig. 3.92) .............................................. SEGESTRIIDAE, p. 219

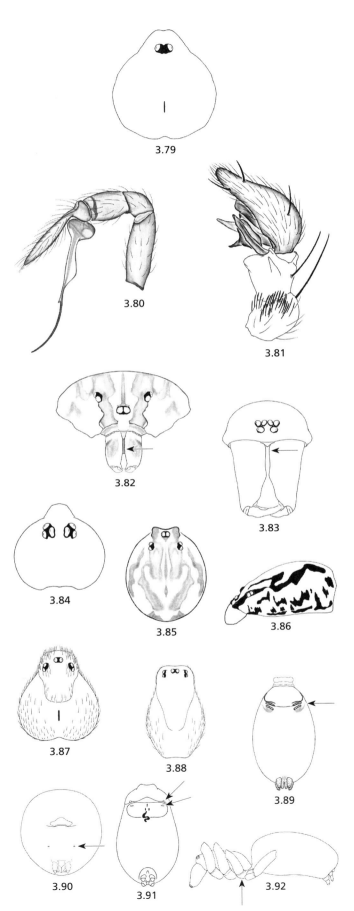

**45(43)** Abdomen with sclerotized ridge on anterior dorsal surface (Fig. 3.93) ............. TELEMIDAE (*Usofila*, in part), p. 228

— Abdomen without sclerotized ridge ............................ **46**

**46(45)** Abdomen with 1 or 2 scuta (Fig. 3.94) ..................................
............................................................ OONOPIDAE, in part, p. 185

— Abdomen without scuta ............................................. **47**

**47(46)** Legs relatively long (PT/C > 1.5); PME usually posteriorly displaced from LE (Fig. 3.95), if eyes contiguous then occupying less than 1/2 cephalon width (Fig. 3.96) ..........
........................................... LEPTONETIDAE, in part, p. 122

— Legs shorter (PT/C ca 1); eyes in transverse arrangement, if contiguous then occupying more than 1/2 cephalon width .................................................................. **48**

**48(47)** Eyes in compact group (Fig. 3.97) if in transverse row (Fig. 3.98) then femur IV enlarged (Fig. 3.99); colulus absent ..
............................................. OONOPIDAE, in part, p. 185
**Dist.** widespread

— Eyes in transverse row (Fig. 3.100); femur IV not enlarged; colulus large, broad (Fig. 3.101) .........................................
........................ OCHYROCERATIDAE (*Theotima*), p. 181
**Dist.** s FL

**49(38)** Abdomen with scutum (Fig. 3.102); size < 2 mm .............
............................... ANAPIDAE, in part, (*Comaroma*), p. 64
**Dist.** CA

— Abdomen without scutum; size variable ...................... **50**

**50(49)** Tibia I with 2 to 3 pairs of ventral spines ...................... **51**

— Tibia I with fewer ventral spines .................................. **52**

**51(50)** ALS contiguous, longer than PLS (Fig. 3.103); eyes small, widely separated (Fig. 3.104) ...............................................
........................... CYBAEIDAE, in part (*Cybaeozyga*), p. 85

— ALS slightly separated, shorter than PLS (Fig. 3.105); eyes larger, in two contiguous triads (Fig. 3.106) ......................
................................................ DICTYNIDAE, in part, p. 95

**52(50)** Tarsus IV with ventral comb of serrated bristles (Fig. 3.107); male palpus with large rigid basal paracymbium (Fig. 3.108) ........................... NESTICIDAE, in part, p. 178

— Tarsus IV without ventral comb; male palpus with smaller basal paracymbium flexibly attached by membrane (Fig. 3.109) .................................. LINYPHIIDAE, in part, p. 124

# Section 3
## Ecribellate 8-eyed ARANEOMORPHAE with 3 claws

**53(13)** Male palpus with bulb exposed, tarsus not modified into cymbium (Fig. 3.110); female without epigynum ............. ................................................... (HAPLOGYNES) ... **54**

— Male palpus with bulb partially enveloped by cymbium (Fig. 3.111); female with epigynum (Fig. 3.112) ............. ................................................... (ENTELEGYNES) ... **57**

## Haplogynes

**54(53)** Eyes contiguous with AME surrounded by the others (Fig. 3.113) ............... CAPONIIDAE, in part, *(Calponia)*, p. 75
**Dist.** CA

— Eyes not so arranged ....................................... **55**

**55(54)** Eyes in three groups with AME forming a diad and the others two triads (Fig. 3.114); chelicerae fused at base (Fig. 3.115) .................... PHOLCIDAE, in part, p. 194

— Eyes in two transverse rows; chelicerae fused or not ... **56**

**56(55)** Chelicerae fused at base (Fig. 3.116); endites converging apically (Fig. 3.117) ................ PLECTREURIDAE, p. 201

— Chelicerae not fused (Fig. 3.118); endites parallel to diverging (Fig. 3.119) ...... TETRAGNATHIDAE, in part, p. 232

## Entelegynes

**57(53)** Chelicerae fused at base (Fig. 3.115); eyes in three groups with AME forming a diad and the others two triads (Fig. 3.114); legs long and slender, PT/C > 1.6, with tarsi usually flexible (Fig. 3.120) ............. PHOLCIDAE, in part, p. 194
**Note** although pholcids **are** haplogynes, some have complex genitalia and may be mistaken for entelegynes

— Chelicerae not fused; eyes not so arranged; legs usually shorter with rigid tarsi .................................... **58**

**58(57)** Tarsi with at most a single trichobothrium (0-1) ........ **59**
— Tarsi with 2 or more trichobothria ........................... **69**

## Tarsi with 0-1 trichobothrium

**59(58)** PLS with long apical segment, about as long as abdomen (Fig. 3.121); eyes clustered on a mound at center of cephalon (Fig. 3.122) ........................... HERSILIIDAE p. 116
**Dist.** TX, FL

— PLS shorter; eye arrangement different ....................... **60**
**Dist.** widespread

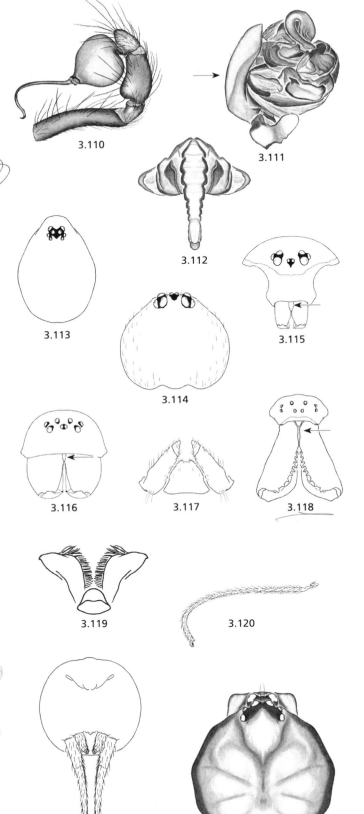

**60(59)** Anterior tibiae and metatarsi with prolateral row of curved spines in serrated series (Fig. 3.123); cheliceral armature of peg teeth (Fig. 3.124) ... MIMETIDAE, p. 171

— Anterior legs without such spines; chelicerae toothed or edentate but without peg teeth ...................................... **61**

**61(60)** Tarsus IV with ventral comb of serrated bristles (sometimes not distinct, lacking in *Argyrodes*) (Fig. 3.125); legs usually lacking spines; chelicerae usually with basal extension (Fig. 3.126); epigynum without scape .................. **62**

— Tarsus IV without ventral comb; legs with spines; chelicerae lacking basal extension; epigynum sometimes with scape ............................................................................................ **63**

**62(61)** Labium rebordered (Fig. 3.127); endites parallel (Fig. 3.127); male palpus with large basal paracymbium (Fig. 3.128) ........................ NESTICIDAE, in part, p. 178

— Labium not rebordered (Fig. 3.129); endites converging apically (Fig. 3.129); male palpus with paracymbium inconspicuous or represented by an apical or ectal notch (Fig. 3.130) ........................ THERIDIIDAE, in part, p. 235

**63(61)** Tarsi as long as or longer than metatarsi ................... **64**

— Tarsi shorter than metatarsi ........................................ **65**

**64(63)** Pedicel originating from opening on posterior slope of carapace (Fig. 3.131); male abdomen with scutum (Figs. 3.131-3.132); male metatarsus I without clasping spur; female without palpi ......................................
........................ ANAPIDAE, in part (*Gertschanapis*), p. 64

— Pedicel origin below edge of carapace (Fig. 3.133); male abdomen without scutum; male metatarsus I usually with clasping spine (Fig. 3.134); female with palpi ....................
........................................................ MYSMENIDAE, p. 175

**65(63)** Sternum with a pair of pits at labial margin (Fig. 3.135); tibia IV with long trichobothria (Fig. 3.136); legs stout; size < 2.5 mm ..............................................................
............. THERIDIOSOMATIDAE (*Theridiosoma*), p. 244

— Sternum without such pits; tibia IV without long trichobothria; legs variable; size usually larger ...................... **66**

**66(65)** Clypeus higher than 3 diameters of AME (Fig. 3.137); chelicerae usually with stridulatory file on lateral surface, lacking condyle (Fig. 3.138); legs weakly spined; tarsi cylindrical; autospasy at patella-tibia joint; sheet web builders ............................................................................ **67**

— Clypeus lower than 3 diameters of AME (Fig. 3.139); chelicerae without stridulatory file, usually with condyle; legs with strong spines; tarsi tapering distally; no autospasy at patella-tibia joint; orb web builders ....................... **68**

**67(66)** Male cymbium with integral retrolateral paracymbium (Fig. 3.140); cymbium usually with a dorsoectal spinose or cuspuliferous process (Fig. 3.141); female epigynum a stout, tongue-like projection with apical openings (Fig. 3.142). Size > 5 mm .............. PIMOIDAE (*Pimoa*), p. 197

— Male cymbium with retrolateral paracymbium attached by membrane (Figs. 3.143-3.144); cymbium usually lacking dorsoectal process; female epigynum variable, if tongue-like less than 5 mm. Size usually < 5 mm ............
............................................ LINYPHIIDAE, in part, p. 124

**68(66)** Endites square or rectangular (Fig. 3.145); epigynum usually 3 dimensional, usually with scape (flat in *Hyposinga*, *Mastophora*, and *Zygiella*) (Fig. 3.146); palpal tibia wider than long, cup-shaped with irregular distal margin (except in *Zygiella*); palpus with radix and median apophysis (Fig. 3.147); builders of vertical orb-webs .............
.................................................... ARANEIDAE, p. 68

— Endites elongate, widest at distal edge (Fig. 3.148); epigynum absent or a flat plate with at most a swelling (*Meta*, Fig. 3.149) or a large cone-shaped protuberance (*Plesiometa*); palpal tibia longer than wide (except *Nephila*) and conical (except *Meta*) (Fig. 3.150); palpus without radix or median apophysis (Fig. 3.150); most are builders of horizontal orb-webs (except *Nephila*) .............
.................................... TETRAGNATHIDAE, in part, p. 232

## Tarsi with 2 or more trichobothria

**69(58)** Tarsal and metatarsal trichobothria in dorsal row increasing in length distally (Fig. 3.151); at least anterior trochanters not notched .................................................. **70**

— Tarsal and metatarsal trichobothria in two irregular rows (Fig. 3.152); all trochanters with shallow to deep notches (Fig. 3.153) ............................................................... **80**

## Trichobothria in a single dorsal row

**70(69)** ALS enlarged, others reduced (Fig. 3.154); clypeus high (Fig. 3.155); endites strongly converging, lacking serrula (Fig. 3.156) ........................................ ZODARIIDAE, p. 254

— ALS at most only slightly larger than the others; endites not strongly converging; serrula present (Fig. 3.157) ........
............................................................................................. **71**

**71(70)** Colulus large and broad, occupying at least half the width of the spinning area (Fig. 3.158) ..................................... **72**

— Colulus width less than half of the spinning area (Fig. 3.159) ............................................................................ **73**

**72(71)** Distribution: FL ....................................................................
............................... DESIDAE, in part (*Paratheuma*), p. 93

— Distribution: s CA ..............................................................
............................... DICTYNIDAE, in part (*Saltonia*), p. 95

**73(71)** Spinnerets arranged in transverse row (Fig. 3.160); tracheal spiracle positioned well anterior of spinnerets (Fig. 3.160) .................. HAHNIIDAE, in part, p. 112
— Spinneret arrangement not so modified; tracheal spiracle near spinnerets ............................................................ **74**

**74(73)** Legs with plumose hairs (Fig. 3.161); both eye rows strongly procurved (Fig. 3.162), except in *Tegenaria*, which has distinctive sternum markings (Fig. 3.163) ........................
............................................................ AGELENIDAE, p. 56
— Legs lacking plumose hairs; at least anterior eye row straight (Fig. 3.164) ..................................................... **75**

**75(74)** ALS contiguous or nearly so, thicker and usually longer than PLS (Fig. 3.165) ............. CYBAEIDAE, in part, p. 85
— ALS distinctly separated, shorter than PLS (Fig. 3.166) ....
.................................................................................................. **76**

**76(75)** Anterior tibiae with 4 or more pairs of ventral spines (Fig. 3.167) ................... HAHNIIDAE, in part, p. 112
— Anterior tibiae with 3 or fewer pairs of ventral spines ......
.................................................................................................. **77**

**77(76)** Cheliceral retromargin with 2-5 equal sized teeth, no denticles ................................................................................. **78**
— Cheliceral retromargin with both teeth and denticles or with more than 5 teeth .................................................... **79**

**78(77)** Size > 5 mm ................... AMAUROBIIDAE, in part, p. 60
— Size < 5 mm ......................... HAHNIIDAE, in part, p. 112

**79(77)** Legs long and slender, PT/C > 1.25; male palp with apically produced cymbium (Fig. 3.168) ................................
..................... HAHNIIDAE, in part (*Calymmaria*), p. 112
— Legs shorter and stouter, PT/C < 1; male palp with unmodified cymbium ................ DICTYNIDAE, in part, p. 95

## Trichobothria scattered

**80(69)** Tarsi long and flexible (Fig. 3.169) ....................................
................................. TRECHALEIDAE (*Trechalea*), p. 249
— Tarsi not flexible ............................................................. **81**

**81(80)** Clypeus high; AME small, others larger, forming hexagon (PER procurved) (Fig. 3.170, frontal view); cheliceral margins with at most a single tooth; trochanters with shallow notches .............................................. OXYOPIDAE, p. 189
— Clypeus low; PER recurved; chelicerae strongly toothed; trochanters deeply notched ........................................... **82**

**82(81)** PER strongly recurved with PLE posterior to PME so that eyes appear as 3 rows (Fig. 3.171, frontal view); male palp lacking RTA (Fig. 3.172) .................... LYCOSIDAE, p. 164
— PER not as strongly recurved, PLE lateral to PME (Fig. 3.173, dorsal view); male palp with RTA (Fig. 3.174) ........
................................................................ PISAURIDAE, p. 199

# Section 4
## Ecribellate 8-eyed ARANEOMORPHAE with 2 claws

**83(13)** Legs laterigrade, extending laterally from the body and twisted so that the morphologically prolateral surface of the anterior legs is functionally dorsal (Fig. 3.175) ..... **84**

— Legs prograde: anterior pair directed forward, posterior backward (Fig. 3.176) .................................................. **88**

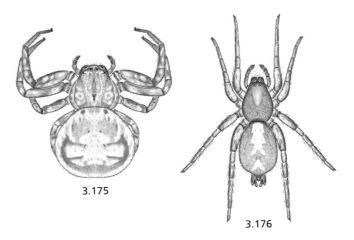

3.175

3.176

## Legs laterigrade

**84(83)** Carapace extremely flat; AER with 6 eyes (includes PME) (Fig. 3.177) .................. SELENOPIDAE (*Selenops*), p. 221

— Carapace not so flat; eye arrangement different .......... **85**

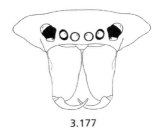

3.177

**85(84)** Anterior legs thicker and longer than posterior ones (Fig. 3.175); tarsi lacking scopulae; chelicerae lacking teeth ......
............................................................ THOMISIDAE, p. 246

— Anterior legs not so enlarged, (although leg II may be significantly longer than others); tarsi with scopulae; chelicerae usually with teeth or denticles ........................... **86**

**86(85)** Smaller spiders (size < 10 mm); cheliceral retromargin lacking teeth ............. PHILODROMIDAE, in part, p. 192

— Larger spiders (size usually > 10 mm); cheliceral retromargin with teeth or denticles ........................................ **87**

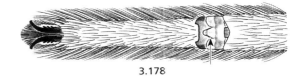

3.178

**87(86)** Metatarsi with dorsoapical trilobed membrane permitting hyperextension of tarsi (Fig. 3.178); tarsi with thick claw tufts; cheliceral retromargin with teeth; trochanters with deep notches ..................... SPARASSIDAE, in part, p. 224
**Dist.** sw USA and FL

— Metatarsi with dorsoapical trilobed membrane; tarsi with thick claw tufts; cheliceral retromargin with denticles (Fig. 3.179); trochanters not notched ..........................................
........... SPARASSIDAE, in part (*Pseudosparianthis*), p. 224
**Dist.** FL

— Metatarsi lacking dorsoapical trilobed membrane; tarsi lacking claw tufts, but tarsal scopulae apically produced and may give impression of tufts; cheliceral retromargin with teeth (Fig. 3.180); trochanters with shallow notches ....................... TENGELLIDAE, in part (*Lauricius*), p. 230
**Dist.** AZ-NM, CO

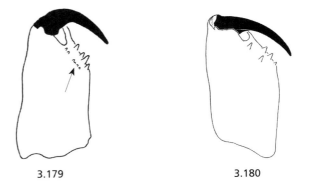

3.179    3.180

## Legs prograde

**88(83)** AME greatly enlarged, PER on lateral edge of carapace; carapace anteriorly truncate, face nearly vertical (Fig. 3.181) .................................................. SALTICIDAE, p. 205

— Eyes and carapace not so modified ............................... **89**

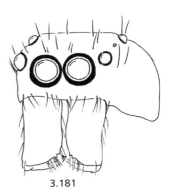

3.181

**89(88)** PER strongly recurved, eyes appear as 3 rows with anterior edge of PLE at or behind the posterior edge of PME (Fig. 3.184) .................................................................... **90**

— PER variable, but not strongly recurved ...................... **93**

**90(89)** Anterior tibiae with at least 5 pairs of strong ventral spines (Fig. 3.182) .................................................................... **91**

— Anterior tibiae with fewer ventral spines ...................... **92**

**91(90)** AER strongly recurved with ALE small and contiguous with PME and PLE (Fig. 3.183); size > 6 mm ................... CTENIDAE, p. 83

— AER straight to slightly recurved (Fig. 3.184); size < 6 mm ............................................... ZORIDAE (*Zora*), p. 256

**92(90)** Carapace narrow, longer than wide (Fig. 3.185); grass spiders .......... PHILODROMIDAE, in part (*Tibellus*), p. 192
**Dist.** widespread

— Carapace broad, as long as wide (Fig. 3.186); ground spiders ...... HOMALONYCHIDAE (*Homalonychus*), p. 118
**Dist.** s CA-AZ

**93(89)** Tracheal spiracle at middle of abdomen (Fig. 3.187); claw tufts of broad, lamelliform setae (Fig. 3.188) .................... ANYPHAENIDAE, p. 66

— Tracheal spiracle near spinnerets; claw tufts of normal setae ................................................................. **94**

**94(93)** Anterior spinnerets cylindrical and noticeably separated (may appear contiguous in *Micaria* and some *Orodrassus*) (Fig. 3.189); PME usually modified, either elliptical, oval or triangular (Fig. 3.190); cheliceral margins usually weakly toothed, with only denticles, or lacking armature, sometimes keeled or with only 2-3 teeth; endites usually with distinct oblique median depression (not conspicuous in *Callilepis*) (Fig. 3.191); claw tufts always present .... **95**

— ALS conical, contiguous at base (Fig. 3.192); PME usually round (if oval then anterior tibiae with 5-6 pairs of ventral spines); cheliceral margins strongly toothed; endites nearly parallel, often widened distally, usually lacking transverse median depression (if depressed, then anterior tibiae with 5-6 pairs of ventral spines); claw tufts present or absent ............................................................... **96**

**95(94)** PER straight (Fig. 3.190) or recurved, rarely procurved (*Scopoides*) (Fig. 3.193); ALS with fewer and only moderately elongated spigots (Fig. 3.194); tarsal claws and at least cheliceral promargin toothed .................................................. GNAPHOSIDAE, p. 106

— PER strongly procurved (Fig. 3.195); ALS with numerous greatly elongated spigots (Fig. 3.196); tarsal claws and cheliceral margins lacking teeth ............................................. PRODIDOMIDAE, p. 203

**96(94)** PLS distinctly two-segmented, distal segment conical (Fig. 3.197) ............................................. MITURGIDAE, p. 173

— PLS one-segmented or if two-segmented, distal segment rounded (Fig. 3.192) ...................................................... **97**

**97(96)** Legs lacking spines, but anterior tibiae with ventral cusps (Fig. 3.198) .................................................. CORINNIDAE, in part, (Trachelinae), p. 79
— Legs with spines ................................................. **98**

**98(97)** Anterior tibiae with more than 4 pairs of ventral spines .. **99**
— Anterior tibiae with fewer than 4 pairs of ventral spines .. **101**

**99(98)** Precoxal triangles present (Fig. 3.199); trochanters with at most shallow notches (Fig. 3.200) ....................................... CORINNIDAE, in part (Corinninae, Phrurolithinae) p. 79
— Precoxal triangles absent; trochanters deeply notched (Fig. 3.201) ................................................. **100**

**100(99)** Size greater than 5 mm; tarsi not pseudosegmented ......... ............................................. TENGELLIDAE, in part, p. 230
— Size less than 4 mm; tarsi pseudosegmented, posterior bent (Fig. 3.202) ................................................. ........................ LIOCRANIDAE, in part *(Apostenus)*, p. 162

**101(98)** Abdomen of adults with scutum [covering entire abdomen of male (Figs. 3.203, 3.204) and only anterior portion of female (Fig. 3.205)] ....................................... ................ CORINNIDAE, in part (Castianeirinae), p. 79
— Abdomen without scutum .......................................... **102**

**102(101)** Endites concave ectally (Fig. 3.206); precoxal triangles present (Fig. 3.207) ........................ CLUBIONIDAE, p. 77
— Endites straight or convex ectally (Fig. 3.208); precoxal triangles absent (but present in *Hesperocranum*, which has a distinctive dense brush of paired ventral bristles on anterior tibiae and metatarsi, Fig. 3.209) ........................... ............................................. LIOCRANIDAE, in part, p. 162

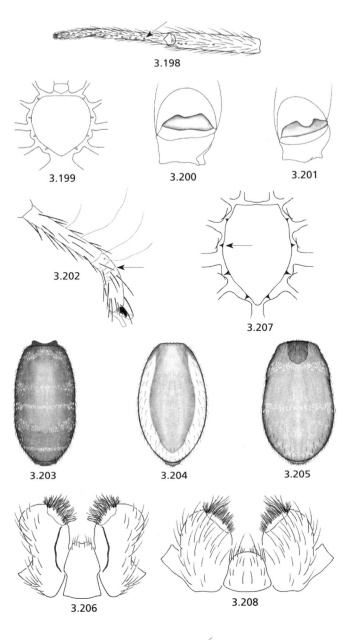

# Chapter 4
# ANTRODIAETIDAE
3 genera, 26 species

Frederick A. Coyle

## Common name —
Foldingdoor spiders or collardoor spiders (*Antrodiaetus*); turret spiders (*Atypoides*); trapdoor spiders (*Aliatypus*).

## Similar families —
Atypidae (p. 41), Ctenizidae (p. 43), Cyrtaucheniidae (p. 45), Mecicobothriidae (p. 50).

## Diagnosis —
Mygalomorph spiders with one or more patches of sclerotized cuticle anterodorsally on abdomen (tergites); endites small (shorter than 1/2 width of sternum); labium strongly inclined from plane of sternum; females with rastellum; distal segment of PLS digitiform. Live in burrows with a flexible collar, rigid turret, or trapdoor at the entrance.

## Characters —
**body size:** 6-25 mm.
**color:** tan to chestnut brown.
**carapace:** *pars cephalica* strongly elevated; thoracic furrow variable (longitudinal groove, deep pit, shallow rounded depression, or absent).
**sternum:** ovoid and longer than wide; 4 pairs of sigilla; anterior pair large, just behind labium, and sometimes indistinct; posterior pair larger than second and third pairs.
**eyes:** closely grouped on anterior median prominence; ALE the largest.
**chelicerae:** robust, especially so in females; males with dorsal process (cheliceral apophysis) absent, small, or elongate; females with well-developed rastellum; row of macroteeth on prolateral side of closed fang and microteeth (denticles) on retrolateral side of this row; sometimes macroteeth also on retrolateral side of closed fang.
**mouthparts:** endites small and without serrula or cuspules; labium strongly inclined from plane of sternum and without cuspules.
**legs:** female legs relatively stout with many spines; male legs more slender, with tibia and metatarsus of leg I sometimes (*Antrodiaetus*) with prominent spine patches or lobes specialized for clasping female; tarsi with 3 claws and no claw tufts or scopulae.
**abdomen:** males with 1-4 segmentally arranged sclerotized anterodorsal patches (sometimes fused), the second of which is most heavily sclerotized, tergite-like, and always present; females usually possess only this second patch.
**spinnerets:** 2-3 pairs; ALS 2-segmented and functional, reduced and unsegmented, or absent; PMS unsegmented, functional; PLS longest, 3-segmented and functional; distal article of PLS digitiform.
**respiratory system:** 2 pairs of book lungs.
**genitalia:** haplogyne; *female* with transverse row of 4 spermathecae, two opening into each side of the *bursa copulatrix*; *male* palpal organ with well-developed conductor divided into two parts; male pedipalp with swollen tibia and sometimes (*Aliatypus*) elongate patella.

## Distribution —
Except for two species of *Antrodiaetus* that live in Japan, this family is restricted to the continental USA (where it is widespread) and British Columbia.

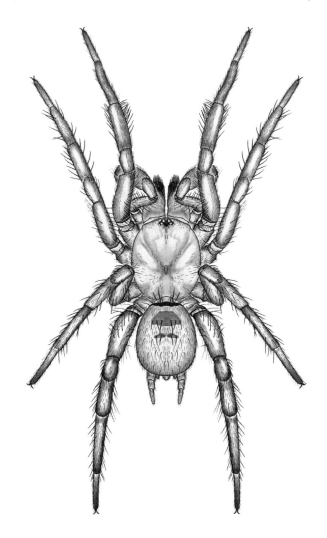

**Fig. 4.1** *Antrodiaetus unicolor* (Hentz 1842a), male

## Natural history —
Antrodiaetids excavate burrows that they line with silk, at least near the entrance. At the burrow entrance they construct a collapsible collar, rigid turret, or trapdoor composed of silk and particles of excavated soil and (sometimes) ground surface debris. When closed, collars and trapdoors are very difficult to locate because their external surfaces often closely match the surrounding substrate. When active, these spiders open their collars or trapdoors at night and wait just inside the entrance to strike at passing prey that the spider detects by sensing substrate vibrations. When mature, males abandon their burrows during rainy periods and wander in search of females. More natural his-

tory information is available in Coyle (1971, 1986) and Coyle & Icenogle (1994).

### Taxonomic history and notes —

Hendrixson & Bond (2005) showed that *Antrodiaetus unicolor* is a complex of species

### Genera —

*Aliatypus* SMITH 1908, *Antrodiaetus* AUSSERER 1871a, *Atypoides* O. PICKARD-CAMBRIDGE 1883

### Key to genera —
North America North of Mexico

1 — Thoracic furrow longitudinal (Fig. 4.1); 2 or 3 pairs of spinnerets; ALS absent (Fig. 4.2) or unsegmented with at most one spigot apically (Fig. 4.6); no cheliceral macroteeth on retrolateral side of closed fang (Fig. 4.3); collapsible collar or rigid turret at burrow entrance .................. **2**

— Thoracic furrow an irregular deep pit, shallow depression, or absent; 3 pairs of spinnerets (Fig. 4.4); ALS 2-segmented with at least several spigots clustered apically; female with short row of cheliceral macroteeth on retrolateral side of closed fang (Fig. 4.5); trapdoor at burrow entrance .......... ........................................................................... ***Aliatypus***
**Div.** 11 species — **Dist.** CA-AZ, s NV — **Ref.** Coyle 1975

2(1) Two pairs of spinnerets (Fig. 4.2); male without cheliceral apophysis, although some species have an anterodorsal swelling .................................................... ***Antrodiaetus***
**Div.** 12+ species — **Dist.** widespread — **Ref.** Coyle 1971

— Three pairs of spinnerets (Fig. 4.6) (ALS may be greatly reduced in some specimens of *Atypoides gertschi* COYLE 1968); male with large cheliceral apophysis (at least twice as long as wide) (Fig. 4.7) ............................. ***Atypoides***
**Div.** 3 species — **Dist.** s OR, n CA, s IL-e MO — **Ref.** Coyle 1968

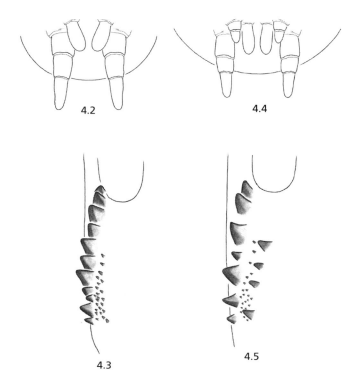

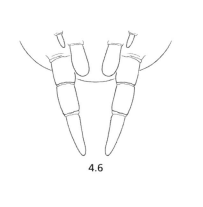

# Chapter 5
# ATYPIDAE
2 genera, 8 species

Frederick A. Coyle

**Common name —**
Purseweb spiders

**Similar families —**
Antrodiaetidae (p. 39), Ctenizidae (p. 43), Cyrtaucheniidae (p. 45), Mecicobothriidae (p. 50).

**Diagnosis —**
Mygalomorph spiders with one or more patches of sclerotized cuticle anterodorsally on abdomen (tergites); unusually long chelicerae (over 1/2 length of carapace) and long endites (longer than 1/2 width of sternum); no demarcation (seam, depression or change in slope) between labium and sternum; thoracic furrow quadrangular or suboval; 3 pairs of spinnerets, with PMS stout and possessing triangular tip. Live in burrows with a long silken tube extending out of the upper end of the burrow.

**Characters —**
**body size:** 8-27 mm.
**color:** yellow-brown to dark purple-black; legs of male *Sphodros rufipes* (LATREILLE 1829) and *Sphodros fitchi* GERTSCH & PLATNICK 1980 are bright orange-red.
**carapace:** *pars cephalica* strongly elevated; thoracic furrow quadrangular or suboval.
**sternum:** quadrate and nearly as wide as long; 4 pairs of sigilla, posterior pair largest and close to one another
**eyes:** closely grouped on anterior median prominence; ALE or AME the largest.
**chelicerae:** very large, elongate (over 1/2 length of carapace), and lacking a rastellum; straight row of sharp teeth prolateral to closed fang.
**mouthparts:** endites very long and without serrula or cuspules; labium fused to sternum (no seam, depression, or abrupt change in slope between it and sternum) and without cuspules.
**legs:** without many spines; tarsi with 3 claws and no claw tufts or scopulae.
**abdomen:** males with 1-4 segmentally arranged sclerotized dorsal patches (sometimes fused), the second of which is most heavily sclerotized, tergite-like, and always present; females usually possess only this second patch.
**spinnerets:** 3 pairs; ALS narrow and unsegmented; PMS wide, unsegmented, with obliquely triangular tips; PLS longest, usually 3-segmented (4-segmented in *Sphodros abbotii* WALCKENAER 1835); distal article of PLS digitiform.
**respiratory system:** 2 pairs of book lungs.
**genitalia:** haplogyne; **female** with transverse row of 4 spermathecae, two opening into each side of the *bursa copulatrix*; **male** palpal organ with well-developed undivided conductor; male pedipalp with swollen tibia.

**Distribution —**
The family Atypidae is found in the Nearctic and Palearctic regions as well as the Old World tropics. The two genera living in North America are restricted to the eastern half of the USA and southern Ontario.

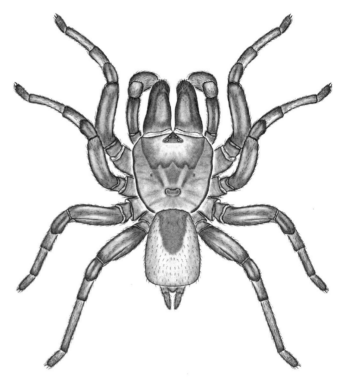

**Fig. 5.1** *Sphodros niger* (HENTZ 1842b), male

**Natural history —**
*Atypus* and *Sphodros* species excavate burrows, line them with silk and extend this lining above the soil surface to make a long narrow tube of silk (the "purse-web") that is closed off at its terminus and camouflaged with soil particles. Most species extend this tube more or less vertically up the base of a tree trunk or occasionally herbs, but *Sphodros niger* extends it horizontally along the ground surface, typically under leaf litter or rocks where it is often well hidden. Relying on vibrations transmitted by the silk, the spider detects arthropods that walk on the tube and captures these by striking through the tube with its very long fangs. More natural history information is available in Gertsch & Platnick (1980) and Coyle & Shear (1981).

**Genera —**
*Atypus* LATREILLE 1804b, *Sphodros* WALCKENAER 1835

## Key to genera —
## North America North of Mexico

1. Sternum of males with marginal ridges (Fig. 5.2); embolus straight and spine-shaped (Fig. 5.3); females with 2 heavily sclerotized, basal, epigynal plates, each bearing 2 short-stalked bulbous spermathecae (Fig. 5.4) ............. ***Atypus***
   <small>**Div.** 1 species: *Atypus snetsingeri* S<small>ARNO</small> 1973 — **Dist.** PA — **Ref.** Gertsch & Platnick 1980</small>

— Sternum of males lacking marginal ridges (Fig. 5.5); embolus slender, curved, and whip-like (Fig. 5.6); females lacking sclerotized epigynal plates; 4 long coiled tube-like spermathecae (Fig. 5.7) .................................... ***Sphodros***
   <small>**Div.** 7 species — **Dist.** e half of USA and ON — **Ref.** Gertsch & Platnick 1980</small>

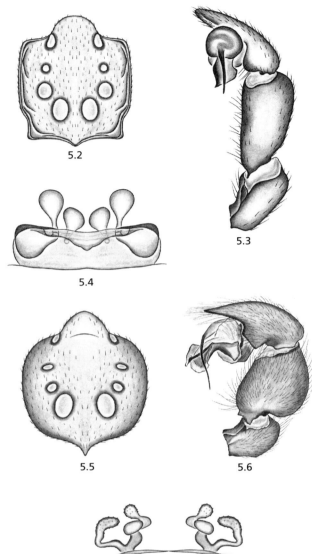

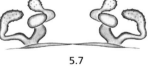

## Chapter 6
# CTENIZIDAE
4 genera, 14 species

Jason E. Bond
Brent E. Hendrixson

### Similar families —
Cyrtaucheniidae (p. 45).

### Diagnosis —
Mygalomorph spiders lacking scopulae (females); numerous robust lateral spines ("digging spines") on pedipalp and tarsus, metatarsus, and tibia of legs I and II; chelicerae with two distinct rows of teeth; a very deep transverse thoracic furrow that is procurved; lacking abdominal tergites. Chelicerae lack row of black rod-like setae. PLS short, distal-most article very short and dome shaped, lacking three prominent spigots on the anterior-most article; spigots are fused type. Live in burrows covered with a wafer-type (*Cyclocosmia*) or a thick, cork-type trapdoor (other genera).

### Characters —
**body size:** 10-> 30 mm.
**color:** tan, dark chestnut brown, and black (some species of *Ummidia*).
**carapace:** carapace with a few distinct spines, usually glabrous; transverse thoracic furrow typically very deep and procurved; *pars cephalica* elevated slightly.
**sternum:** longer than wide, wider posteriorly at coxae II and III; usually 3 pairs of sternal sigilla, posterior pair noticeably much larger than anterior pairs and located more towards the center of the sternum (sigilla reduced in some species).
**eyes:** eyes closely grouped on a low tubercle or widely separated (*Cyclocosmia*), subequal in diameter.
**chelicerae:** paraxial with rastellum.
**mouthparts:** palpal endites longer than wide, with cuspules, lacking serrula; labium typically short, with cuspules.
**legs:** female anterior legs (I, II) with numerous lateral spines on anterior articles; anterior legs of males – tibia, metatarsus, and tarsus of most species with prominent spine patches and lobes; posterior legs more stout with numerous heavy spines; females lack preening comb on metatarsus IV; superior tarsal claw of female with reduced number of teeth, many species with only a single, elongate basal tooth.
**abdomen:** abdominal coloration varies from tan, light brown, to black; most species lack distinct markings or coloration (in some species of *Ummidia*, males have distinct dorsal – anterior white patch); abdomen of *Cyclocosmia* very distinct, caudally truncate terminating in a very ornate sclerotized disk.
**spinnerets:** 2 pairs; PMS short, unsegmented and functional; PLS longer yet short relative to other mygalomorphs, 3 segments all with functional spigots; distal most segment is the shortest segment (~1/2 length of medial segment) with an almost triangular apex.
**respiratory system:** 2 pairs of book lungs.
**genitalia:** haplogyne; *female* with 2 simple spermathecae, arranged in a transverse row, consisting of a lightly sclerotized stalk and an enlarged apical bulb; *male* palpal organ with a relatively thin embolus that lacks a conductor; male pedipalp cymbium without spines.

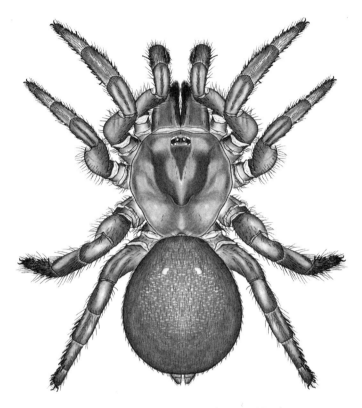

**Fig. 6.1** *Hebestatis theveneti* (SIMON 1891g),

### Distribution —
Widespread throughout United States (*Ummidia*), the Southeast (*Cyclocosmia*), and California (*Bothriocyrtum*, *Hebestatis*).

### Natural history —
Burrow structure: all known North American species of ctenizids are fossorial, living in silk-lined terrestrial burrows. Burrow linings are typically thick, consisting of a heavy layer of parchment-like silk and packed soil. Burrows are not branched. Both species of *Cyclocosmia* cover their burrow entrances with a thin, wafer-type trapdoor; members of the remaining genera cover their burrow entrances with a thick, silk and soil trapdoor, often referred to as a "cork door", that is often very cryptic. These trapdoors are typically used in prey capture. The burrow inhabitant waits beneath a slightly ajar trapdoor for unsuspecting passersby and upon detecting the prey item the spider lunges from the burrow, bites the prey, and then returns to the bottom of the burrow to feast. Although there is considerable variation in burrow depths across genera, *Hebestatis theveneti* (SIMON 1891g) builds very short, shallow burrows that often form small pockets on the sides of rocks and other

structures; *Cyclocosmia* typically build vertical burrows and are often found with their heavily-sclerotized abdomens facing upwards, probably to deter predators and perhaps to prevent burrow flooding. Habitat: members of this family occur in a wide variety of habitats ranging from the more mesic climates of the southeast to the xeric climates of the Californian deserts. *Bothriocyrtum* is widespread throughout southern California (chaparral, oak scrub, desert), whereas *Hebestatis* tends to be restricted to more mesic oak scrub habitats. *Ummidia* is likewise widespread with respect to habitat, found in the temperate more mesic climates of the southeast to the more xeric habitats of the southwest (e.g., Arizona and Texas). *Cyclocosmia* is restricted to mesic habitats in the southeast (Alabama, Florida, Georgia).

## Taxonomic history and notes —

With the exception of *Cyclocosmia*, revised by Gertsch & Platnick (1975), the taxonomic history of the North American Ctenizidae is sparse. The majority of literature, aside from original generic descriptions, consists of occasional redescriptions and descriptions of North American species, most of these by W. Gertsch. Gertsch (1935a) and Gertsch & Wallace (1936) redescribed the sole North American species representatives of *Hebestatis* and *Bothriocyrtum*, respectively. The most recent taxonomic work on North American *Ummidia* (described under the junior synonym *Pachylomerides* STRAND 1934) consists of descriptions of species from Texas by Gertsch & Mulaik (1940). There are numerous undescribed species of *Ummidia*

## Genera —

*Bothriocyrtum* SIMON 1891g, *Cyclocosmia* AUSSERER 1871a, *Hebestatis* SIMON 1903d , *Ummidia* THORELL 1875a

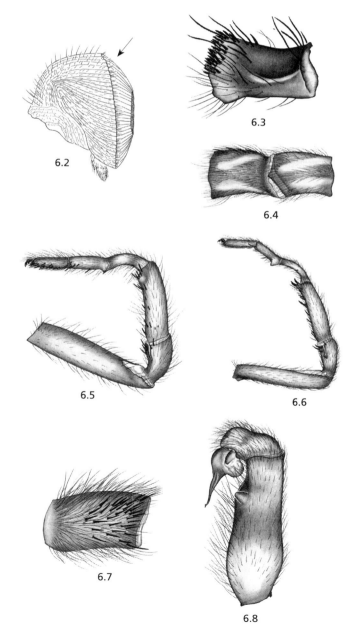

## Key to genera —
### North America North of Mexico

1  Abdomen caudally truncate terminating in a very ornate heavily sclerotized disk (Fig. 6.2) ................ **Cyclocosmia**
   **Div.** 2 species: *Cyclocosmia torreya* GERTSCH & PLATNICK 1975, *Cyclocosmia truncata* (HENTZ 1841b) — **Dist.** se USA — **Ref.** Gertsch & Platnick 1975

—  Abdomen of typical form, not truncate, or heavily sclerotized ...................................................................... **2**

2(1)  Dorsum of tibia III with a deep saddle-like depression (Fig. 6.3); *pars cephalica* elevated posterior to eye group; male carapace very rough with a granulate appearance; male metatarsus I unmodified ........................ **Ummidia**
   **Div.** ~50 species (10 described) — **Dist.** widespread e USA, sw USA to CO — **Refs.** Gertsch & Mulaik 1940, Roth 1993

—  Dorsum of tibia III flat or with only a shallow depression (Fig. 6.4); pars cephalica not elevated higher behind eye group; female labium anterior border with a linear row of cuspules; male carapace glabrous; male metatarsus I modified (Figs. 6.5-6.6) .................................................. **3**

3(2)  Tibia III unmodified, lacking a shallow depression; female tibia IV with a distinct cluster of prolateral, distal spines (Fig. 6.7); male palpal tibia with a distinct retrolateral distal knob (Fig. 6.8) and tibia leg I with cluster of spines (Fig. 6.5) .................................................. **Bothriocyrtum**
   **Div.** 1 species: *Bothriocyrtum californicum* O. PICKARD-CAMBRIDGE in MOGGRIDGE 1874 — **Dist.** s CA and sw NV — **Refs.** Gertsch & Wallace 1936

—  Tibia III of female only with a shallow dorsal depression (Fig. 6.4); tibia IV lacking a distinct cluster of numerous spines; male palpal tibia without a distal knob and tibia of leg I with two prominent spines borne on an apophysis (Fig. 6.6) .................................................. **Hebestatis**
   **Div.** 1 species: *Hebestatis theveneti* SIMON 1891g — **Dist.** c to s CA — **Refs.** Gertsch 1935a — **Note** the male is still undescribed

# Chapter 7
# CYRTAUCHENIIDAE
7 genera, 17 species

Jason E. Bond

## Similar families —
Ctenizidae (p. 43).

## Diagnosis —
Mygalomorph spiders with relatively dense scopulae on tarsi and metatarsi I and II; slender anterior legs with only a few dorsal spines; chelicerae with a single row of promarginal teeth and a small patch of retromarginal denticles (retromarginal denticles larger in *Eucteniza*); a transverse thoracic furrow that is straight, recurved, or procurved; lacking abdominal tergites. Chelicerae lack row of black rod-like setae. PLS of intermediate length with three prominent spigots visible on the anterior most article. Live in burrows with a short flexible, open turret (*Apomastus*) or a thin flexible trapdoor (all other species).

## Characters —
**body size:** 7-32 mm.
**color:** tan to dark chestnut brown.
**carapace:** carapace with a few distinct spines to very hirsute with spines; transverse thoracic furrow variable (straight, procurved, or slightly recurved).
**sternum:** shape of sternum highly variable, typically longer than wide, much wider posteriorly (ovoid to very slender in some species); 4 pairs of sternal sigilla, posterior pair typically much larger than anterior pairs.
**eyes:** eyes closely grouped on a low tubercule; subequal in diameter.
**chelicerae:** paraxial with rastellum.
**mouthparts:** palpal endites longer than wide, with cuspules, lacking serrula; labium typically short, with or without cuspules.
**legs:** female anterior legs (I, II) very slender with only a few dorsal spines; anterior legs of males usually more stout, tibia, metatarsus, and tarsus of most species with prominent spine patches and lobes; posterior legs more spinose and stout; females of some genera with a preening comb on metatarsus IV.
**abdomen:** abdominal coloration and markings across representatives of this family highly variable; light tan to dark chestnut brown; uniformly colored (*Apomastus, Eucteniza, Myrmekiaphila, Neoapachella*, some *Aptostichus*) or with distinct markings [broad dusky chevrons (*Promyrmekiaphila*) or distinct tiger-like stripes arranged in a chevron pattern (some *Aptostichus* Fig. 7.1)].
**spinnerets:** 2 pairs; PMS short, unsegmented and functional; PLS longer, 3 segments all with functional spigots; distal most segment is the shortest segment (~1/2 length of medial segment) with a broad rounded apex and 2 – 3 enlarged spigots (usually visible with the aid of a dissecting microscope).
**respiratory system:** 2 pairs of book lungs.
**genitalia:** haplogyne; *female* with 2 simple spermethecae, arranged in a transverse row, consisting of a lightly sclerotized stalk and an enlarged apical bulb; base of stalk sometimes enlarged such that it takes on the appearance of a smaller, reduced secondary bulb (some *Aptostichus* species); *male* palpal organ with a relatively thin embolus that lacks a conductor; male pedipalp cymbium with (*Aptostichus*) or without apical spines (all other genera).

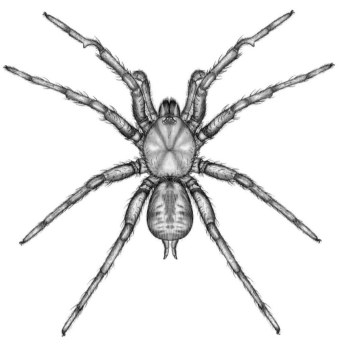

**Fig. 7.1** *Aptostichus atomarius* Simon 1891g

## Distribution —
Widespread throughout the Southeastern United States (*Myrmekiaphila*), southwestern USA and Mexico. The majority of species are endemic to California (*Aptostichus, Apomastus,* and *Promyrmekiaphila*).

## Natural history —
Burrow structure: all known North American species of cyrtaucheniids are fossorial, living in silk-lined terrestrial burrows. Burrow linings are typically thin, consisting of a moderate layer of white silk and packed soil. Burrows of many species are branched (*Aptostichus, Myrmekiaphila,* and *Promyrmekiaphila*); some species cover the underground branches with thin trapdoors (*Myrmekiaphila*). Most species cover their burrow entrances with a thin silk and soil trapdoor that is often very cryptic and is used in prey capture; however members of the genus *Apomastus* build a short flexible turret that is not covered with a door. Habitat: members of this family occur in a wide variety of habitats ranging from the more mesic climates of the southeast to the xeric climates of the Californian deserts. Activity: mature males of the southwestern species tend to abandon their burrows to go in search of females during the more rainy winter months. Females of most desert species (*Aptostichus*) excavate their burrows during the same time period. Excavation work results in the appearance of a small mound of soil at the burrow entrance.

# Cyrtaucheniidae

### Taxonomic history and notes —

The Cyrtaucheniidae is one of the more diverse families of mygalomorph spiders in North America. In addition to the documented diversity, there are over 30 undescribed species of *Aptostichus* known from California (Bond *in prep.*). Until recently, the North American Cyrtaucheniidae comprised numerous genera fated for synonymy. A number of these genera like *Astrosoga* and *Nemesoides* described by Chamberlin (1940b, 1919b) were long considered by Roth (1994) likely junior synonyms of *Eucteniza* and *Aptostichus*, respectively. However, the status of a number of other existing cyrtaucheniid generic names like *Entychides*, *Promyrmekiaphila*, and *Actinoxia* was not as straightforward. Although Gertsch & Wallace (1936) made a correct assessment of the limits of *Entychides*, their original work was confused by the subsequent work of Chamberlin (1937b) when he transferred *Entychides arizonicus* Gertsch & Wallace 1936 to *Actinoxia* Simon 1891g. These problems and others were formally resolved by the recent revision of the southwestern North American members of the Cyrtaucheniidae (subfamily Euctenizinae) by Bond & Opell (2002). Euctenizines were traditionally placed in the family Ctenizidae but removed from this family by Raven (1985a). However, phylogenetic analyses to date suggest that the Euctenizinae may deserve familial status (Goloboff 1993, Bond & Opell 2002). The recent revision of this group also discarded into synonymy a number of generic names (*Actinoxia*, *Nemesoides*, *Astrosoga*).

### Genera —

*Apomastus* Bond & Opell 2002, *Aptostichus* Simon 1891g, *Entychides* Simon 1888b, *Eucteniza* Ausserer 1875, *Myrmekiaphila* Atkinson 1886, *Neoapachella* Bond & Opell 2002, *Promyrmekiaphila* Schenkel 1950b

### Key to genera —
North America North of Mexico

| | | |
|---|---|---|
| **1** | Males | 2 |
| — | Females | 8 |
| **2(1)** | Palpal tibia with a retrolateral distal flange (Fig. 7.2) ........ ***Myrmekiaphila*** | |
| | Div. 4 species + ~5 undescribed — Dist. e USA — Ref. Gertsch & Wallace 1936 | |
| — | Palpal tibia without a retrolateral distal flange | 3 |
| **3(2)** | Tibia of leg I with a large mid-ventral megaspine (Fig. 7.3) | 4 |
| — | Tibia of leg I without a large mid-ventral megaspine (e.g., Fig. 7.5) | 5 |
| **4(3)** | Tibia of leg II with a large mid-ventral megaspine ........... ***Eucteniza*** | |
| | Div. 2 species: *Eucteniza rex* (Chamberlin 1940b), *Eucteniza stolida* (Gertsch & Mulaik 1940) — Dist. TX — Ref. Bond & Opell 2002 | |
| — | Tibia of leg II without a large mid-ventral megaspine ....... ***Neoapachella*** | |
| | Div. 1 species: *Neoapachella rothi* Bond & Opell 2002 + 2 undescribed — Dist. AZ, NM — Ref. Bond & Opell 2002 | |
| **5(2)** | Cymbium with dorsal apical spines (Fig. 7.4) ........ ***Aptostichus*** | |
| | Div. 5 species + ~25 undescribed — Dist. AZ, CA, NV — Ref. Bond & Opell 2002 | |
| — | Cymbium lacks dorsal spines | 6 |

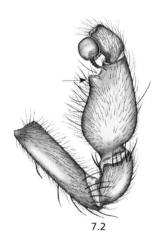

7.2

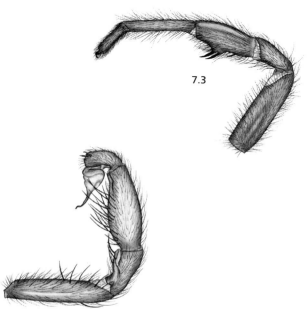

7.3

7.4

**6(5)** Spines on tibia of leg I borne on a low, distally placed retrolateral apophysis (Fig. 7.5) .................... ***Entychides***
        **Div.** 1 species: *Entychides arizonicus* Gertsch & Wallace 1936 — **Dist.** AZ, NM, TX — **Ref.** Bond & Opell 2002

— Spines on tibia of leg I not borne on such an apophysis ... ........................................................................................ **7**

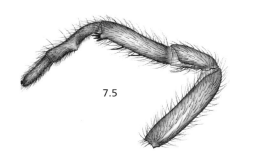

7.5

**7(6)** Thoracic groove straight or procurved, large patch of long thin spines and setae on ventral aspect of tibia I ............... ........................................................ ***Promyrmekiaphila***
        **Div.** 3 species — **Dist.** central CA — **Ref.** Bond & Opell 2002

— Thoracic groove recurved, no distinct spine patch on tibia I .............................................. ***Apomastus***
        **Div.** 2 species: *Apomastus schlingeri* Bond & Opell 2002 and *Apomastus kristenae* Bond 2004 — **Dist.** s CA — **Refs.** Bond & Opell 2002, Bond 2004

**8(1)** Thoracic groove straight or recurved ............................... **9**

— Thoracic groove procurved ............................................ **10**

**9(8)** Very distinct comb-like arrangement of setae on ventral aspect of tarsus IV (Fig. 7.6), lacks preening combs on metatarsus IV ........................................... ***Neoapachella***
        **Div.** 1 species: *Neoapachella rothi* Bond & Opell 2002 + 2 undescribed — **Dist.** AZ, NM — **Ref.** Bond & Opell 2002

— No distinct comb-like arrangement of setae on ventral aspect of tarsus IV, preening combs on metatarsus III and IV; burrow lacks a trapdoor ......................... ***Apomastus***
        **Div.** 2 species: *Apomastus schlingeri* Bond & Opell 2002, *Apomastus kristenae* Bond 2004 — **Dist.** s CA — **Refs.** Bond & Opell 2002, Bond 2004

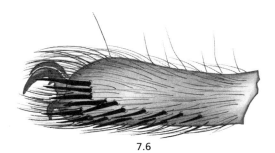

7.6

**10(8)** Dense spinule patch on prolateral aspect of patella IV (Fig. 7.7), posterior aspect of carapace lightly sclerotized ......... ........................................................................ ***Eucteniza***
        **Div.** 2 species: *Eucteniza rex* (Chamberlin 1940b), *Eucteniza stolida* (Gertsch & Mulaik 1940) — **Dist.** TX — **Ref.** Bond & Opell 2002

— No spinule patch on patella IV ...................................... **11**

**11(10)** Preening combs on metatarsus IV absent .... ***Entychides***
        **Div.** 1 species: *Entychides arizonicus* Gertsch & Wallace 1936 — **Dist.** AZ, NM, TX — **Ref.** Bond & Opell 2002

— Preening combs on metatarsus IV present ................. **12**

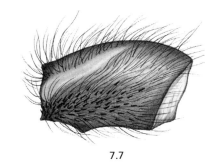

7.7

**12(11)** Rastellar spines borne on a very prominent mound that extends beyond the ventral surface of the fang, cuspules on palpal endite sparsely distributed from the endite midline posteriorly .............................. ***Myrmekiaphila***
        **Div.** 4 species + ~5 undescribed — **Dist.** e USA — **Ref.** Gertsch & Wallace 1936

— Rastellar spines not borne on a prominent mound, numerous endite cuspules distributed across entire face of endite or restricted to posterior most margin of endite .... ........................................................................................ **13**

**13(12)** Cuspules on palpal endites distributed across entire face of endite, wide dark bands of coloration on abdomen .......... ........................................................ ***Promyrmekiaphila***
        **Div.** 3 species — **Dist.** central CA — **Ref.** Bond & Opell 2002

— Endite cuspules concentrated posteriorly, distinct mottled band of abdominal coloration in most non-desert/sand dune species (others pale or without markings) ............... ........................................................................ ***Aptostichus***
        **Div.** 5 species + ~30 undescribed — **Dist.** AZ, CA, NV — **Ref.** Bond & Opell 2002

## Chapter 8
# DIPLURIDAE
2 genera, 5 species

Frederick A. Coyle

**Common name —**
Funnelweb spiders.

**Similar families —**
Mecicobothriidae (p. 50).

**Diagnosis —**
Mygalomorph spiders lacking abdominal tergites, rastellum, and claw tufts; endites small (shorter than 1/2 width of sternum) and lack cuspules; thoracic furrow longitudinal or a shallow to deep pit; four spinnerets, PLS at least 3/4 length of carapace, distal article of PLS digitiform. Live in ground webs of irregular flattened tubular retreats and sheets largely or wholly hidden under rocks, moss mats or other materials.

**Characters —**
**body size:** 3.5-17.0 mm.
**color:** pale tan to purple-brown.
**carapace:** *pars cephalica* may or may not be weakly elevated above *pars thoracica*; thoracic furrow a longitudinal groove or a shallow rounded depression or pit.
**sternum:** slightly to clearly longer than wide; 4 pairs of sigilla; anterior pair fused just behind labium, other three pairs small, subequal, round, marginal.
**eyes:** closely grouped on low anterior median prominence; ALE the largest.
**chelicerae:** less than 1/2 length of carapace; no rastellum; row of macroteeth on prolateral side of closed fang and microteeth (denticles) on retrolateral side of this row.
**mouthparts:** endites small, with serrula but no cuspules; labium moderately to strongly inclined from plane of sternum and without cuspules.
**legs:** female legs moderately spiny; midway on tibia I or II of male a prominent ventral apophysis armed with one or more spines used in clasping female; tarsi with 3 claws and no claw tufts or scopulae.
**abdomen:** no dorsal patches of sclerotized cuticle (tergites).
**spinnerets:** 2 pairs; PMS unsegmented; PLS nearly as long or longer than carapace, 3-segmented, distal article digitiform.
**respiratory system:** 2 pairs of book lungs.
**genitalia:** haplogyne; *female* with a spermathecal trunk opening into each side of the *bursa copulatrix*, trunk opens into one to many spermathecal bulbs; *male* pyriform palpal organ without conductor.

**Distribution —**
The family Dipluridae is worldwide but primarily tropical. The two North American genera live in the southwestern USA and Mexico (*Euagrus*) and in the southern Appalachians and the Pacific Northwest (*Microhexura*).

**Natural history —**
Both *Microhexura* and *Euagrus* build ground webs composed of flattened, often branching, tubular retreat passages connected to small irregular sheets. These webs are largely or completely hidden under objects or in soft

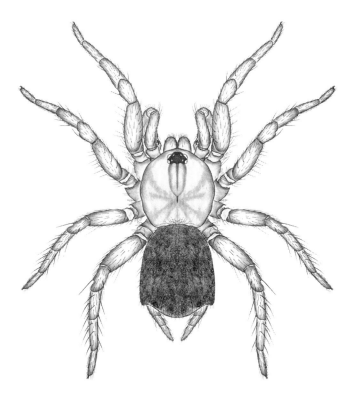

**Fig. 8.1** *Microhexura montivaga* CROSBY & BISHOP 1925a, male

porous organic substrates. *Euagrus* prefers riparian, woodland, and forest habitats, and while most of the web is typically under a rock, inconspicuous irregular funnels and sheets often extend out into surrounding leaf litter and cavities in the humus. *Microhexura montivaga* CROSBY & BISHOP 1925a, a federally listed endangered species, is restricted to remnant patches of fir forest on a few high peaks in the southern Appalachian Mountains, where it lives almost exclusively under moss mats on rock outcrops. *Microhexura idahoana* CHAMBERLIN & IVIE 1945a is widespread in conifer forests of the Pacific Northwest where its webs are often common in or under rotting logs and other organic debris. More natural history information is available in Coyle (1981, 1985a, 1988).

**Genera —**
*Euagrus* AUSSERER 1875, *Microhexura* CROSBY & BISHOP 1925a

**Key to genera —**
North America North of Mexico

1   Small spiders (2.5-6.0 mm long); thoracic furrow longitudinal (Fig. 8.1); tibia I of male with ventral median apophysis armed with one large spine (Fig. 8.2) .............. ........................................................... ***Microhexura***
    **Div.** 2 species: *Microhexura idahoana* CHAMBERLIN & IVIE 1945a, *Microhexura montivaga* CROSBY & BISHOP 1925a — **Dist.** NC, TN, WA, ID, OR, BC and MT — **Ref.** Coyle 1981

—   Medium to large spiders (6.5-17.0 mm long); thoracic furrow a shallow rounded depression or deeper pit; tibia II of male with ventral median apophysis with large spines (Fig. 8.3) ............................................................ ***Euagrus***
    **Div.** 3 species — **Dist.** TX, NM and AZ — **Ref.** Coyle 1988

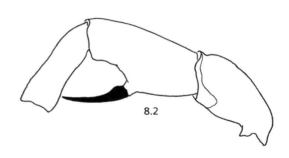

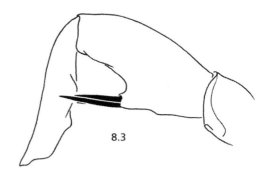

# Chapter 9
# MECICOBOTHRIIDAE
3 genera, 6 species

Frederick A. Coyle

### Similar families —
Antrodiaetidae (p. 39), Dipluridae (p. 48).

### Diagnosis —
Mygalomorph spiders with 2 patches of sclerotized cuticle anterodorsally on abdomen (these "tergites" may be fused); endites shorter than 1/2 width of sternum; labium distinct from sternum; thoracic furrow longitudinal; 2 or 3 pairs of spinnerets; distal segment of PMS long, tapered and pseudosegmented. Live in sheet web with complex of silken tubes usually hidden under objects on the ground.

### Characters —
**body size:** 3.5-18.0 mm.
**color:** whitish to light red-brown.
**carapace:** *pars cephalica* moderately elevated; thoracic furrow longitudinal.
**sternum:** ovoid and longer than wide; 4 pairs of sigilla, anterior pair large, others small, marginal, and may be inconspicuous.
**eyes:** closely grouped on low anterior median prominence; ALE the largest.
**chelicerae:** up to 1/2 length of carapace, laterally compressed, and lacking a rastellum; usually row of teeth prolateral to closed fang and microteeth (denticles) on retrolateral side of this row.
**mouthparts:** endites are moderately large rounded lobes with serrula and no cuspules; labium distinctly set off from sternum and without cuspules.
**legs:** not especially spinose; tarsi with 3 claws and no claw tufts or scopulae.
**abdomen:** 2 segmentally arranged patches of sclerotized cuticle anterodorsally (sometimes fused).
**spinnerets:** 3 pairs except in *Hexura rothi* GERTSCH & PLATNICK 1979 (only 2 pairs); ALS narrow and 2-segmented, except unsegmented or missing in *Hexura*; PMS as long or longer than ALS and unsegmented; PLS longest, usually 3-segmented (4-segmented in *Hexurella*); distal article of PLS (distal two articles in *Hexurella*) elongate, tapered and pseudosegmented.
**respiratory system:** 2 pairs of book lungs.
**genitalia:** haplogyne; *female* with transverse row of 4 spermathecae, two opening into each side of the *bursa copulatrix*; *male* palpal organ with well-developed undivided conductor; tibia of male pedipalp swollen.

### Distribution —
The family is restricted to the New World with one genus (*Mecicobothrium* HOLMBERG 1882) found in Argentina and Uruguay and the other three in North America.

### Natural history —
These spiders use their long flexible spinnerets to construct roughly horizontal sheets and tubes of silk which they use for retreats and for capturing prey. In the small-bodied taxa (*Hexura* and the especially tiny *Hexurella*), these webs are mostly hidden on or in the ground under rocks, wood, leaf litter and conifer duff. Webs of the larger-bodied *Megahexura fulva* (CHAMBERLIN 1919b) "to some extent resemble the webs of agelenids and occur in holes and crevices in the banks of ravines, usually under the shade of oak or conifer trees" (Gertsch & Platnick 1979).

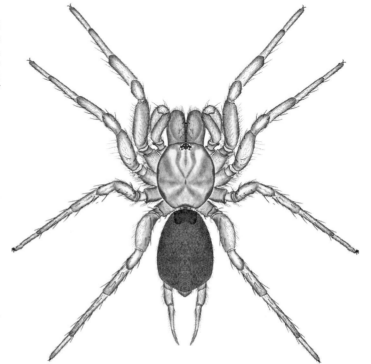

**Fig. 9.1** *Hexura picea* SIMON 1884m, male

### Genera —
*Hexura* SIMON 1884m, *Hexurella* GERTSCH & PLATNICK 1979, *Megahexura* KASTON 1972a

## Key to genera —
## North America North of Mexico

1     Abdomen with separate anterior and dorsal tergites (Fig. 9.2); ALS 2-segmented (Fig. 9.3) ........................ **2**

—     Anterior and dorsal tergites of abdomen fused (Fig. 9.1); ALS unsegmented or absent ................................ ***Hexura***
          **Div.** 2 species: *Hexura picea* SIMON 1884m, *Hexura rothi* GERTSCH & PLATNICK 1979 — **Dist.** OR, WA, BC — **Ref.** Gertsch & Platnick 1979

2(1)     PLS 3-segmented, distal article pseudosegmented (Fig. 9.3). Adults 10-18 mm long ........................ ***Megahexura***
          **Div.** 1 species: *Megahexura fulva* (CHAMBERLIN 1919b) — **Dist.** CA — **Ref.** Gertsch & Platnick 1979

—     PLS 4-segmented, distal two articles pseudosegmented (Fig. 9.4). Adults 2.5-5.0 mm long ................. ***Hexurella***
          **Div.** 3 species — **Dist.** AZ, CA — **Ref.** Gertsch & Platnick 1979

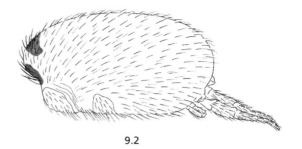

9.2

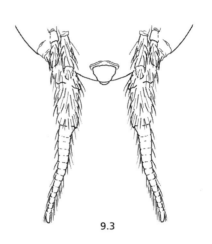

9.3

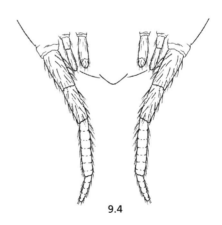

9.4

# Chapter 10
# NEMESIIDAE
1 genus, 5 species

Darrell Ubick
Joel M. Ledford

**Similar families —**
Cyrtaucheniidae (p. 45), Theraphosidae (p. 54).

**Diagnosis —**
*Calisoga* differs from other mygalomorphs in having a row of 3-11 black rod-like setae on the upper inner cheliceral faces (Fig. 10.2). Additionally these large hairy mygalomorphs resemble theraphosids, but lack claw tufts and have three tarsal claws.

**Characters —**
**body size:** 16-30 mm.
**color:** concolorous golden brown to dark gray, sometimes with faint chevron pattern on abdomen.
**carapace:** densely hairy, subrectangular, covered with fine recumbent hairs; cervical groove present, caput slightly elevated; fovea deep and transverse.
**sternum:** oval with scalloped margins, pointed posteriorly; with 3 pairs of marginal sigillae, posterior largest (Fig. 10.3).
**eyes:** eight, clustered on low central mound; AME round, others oval; AER procurved; PER recurved.
**chelicerae:** promargin (inner margin) with 8-13 teeth, retromargin with 4-20 denticles; apex with weakly defined rastellum consisting of several rows of short stout setae; mesobasal surface with row of 3-11 black rod-like setae (Fig. 10.2).
**mouthparts:** endites subrectangular, pointed mesoapically, with many mesobasal cuspules, with thick mesal scopula, serrula absent; labium oval, wider than long, with few apical cuspules.
**legs:** densely hairy, with numerous spines; tarsi and anterior metatarsi scopulate; tarsi with three claws; male tibia I with mesoapical process bearing 2 stout apical spines (=megaspine) (Fig. 10.4).
**abdomen:** densely hairy, oval, lacking tergites.
**spinnerets:** 4 spinnerets; PMS short, 1-segmented; PLS long, 1/2 to 2/3 length of carapace, with three subequal segments (Fig. 10.5).
**respiratory system:** 2 pairs of book lungs.
**genitalia:** haplogyne; *female* gonopore lacking external modification, internally with two longitudinal spermathecae, usually with lateral branches; *male* palpus with pyriform bulb and long embolus (Fig. 10.6).

**Distribution —**
*Calisoga* is known from northern and central California, with one record from western Nevada (Bentzien 1976).

**Natural history —**
*Calisoga* occurs in a variety of biomes, primarily oak grassland, riparian, and coniferous forests, from about sea level to 2300 m in elevation, and occasionally even in urban areas. These spiders live in burrows, or crevices in the ground, which they line with silk and where they wait for passing prey. Burrows are periodically reexcavated, and piles of small balls of soil wrapped in silk may be found

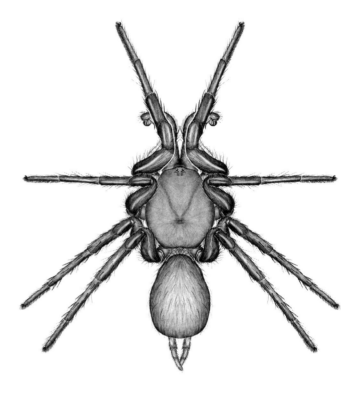

**Fig. 10.1** *Calisoga longitarsis* (Simon 1891f)

near the entrance (Ledford, pers. obs.). Mature males leave their retreats to search for mates and females are often flooded out of their burrows in the rainy season. Females place their egg sacs in the burrow (Bentzien 1976). These spiders often enter homes where they generate concern, given their large size and conspicuous threat display and aggressive behavior.

**Taxonomic history and notes —**
The taxonomic history of the Nearctic nemesiids is confusing as the species were originally placed in three different families. Simon (1891f) described two species from California, *Brachythele longitarsis* and *Brachythele theveneti*, which he placed in the family Dipluridae. Later, Chamberlin (1937b) described the genus *Calisoga* in Ctenizidae for his new species, *Calisoga sacra*, and with Ivie (1939b) erected *Hesperopholis* in Theraphosidae for *Hesperopholis centronethus*. Finally, a fifth species, *Brachythele anomala*, was described by Schenkel (1950b).

Raven (1985a) moved all these species to Nemesiidae and argued that they belong to *Calisoga* (with *Hesperopholis* as synonym) and not the European genus, *Brachythele*. But, as he neglected to transfer *Brachythele longitarsis* and *Brachythele anomala*, they are still catalogued in *Brachythele*, although clearly congeneric with *Calisoga*. Furthermore, Bentzien (1976) retained only the two Simon names and considered the others synonyms, adding that probably even *Calisoga theveneti* represents just the smallest individuals of *Calisoga longitarsis*. However, as Bentzien's thesis was never published, all five names are still catalogued as valid.

### Genus —

*Calisoga* CHAMBERLIN 1937b

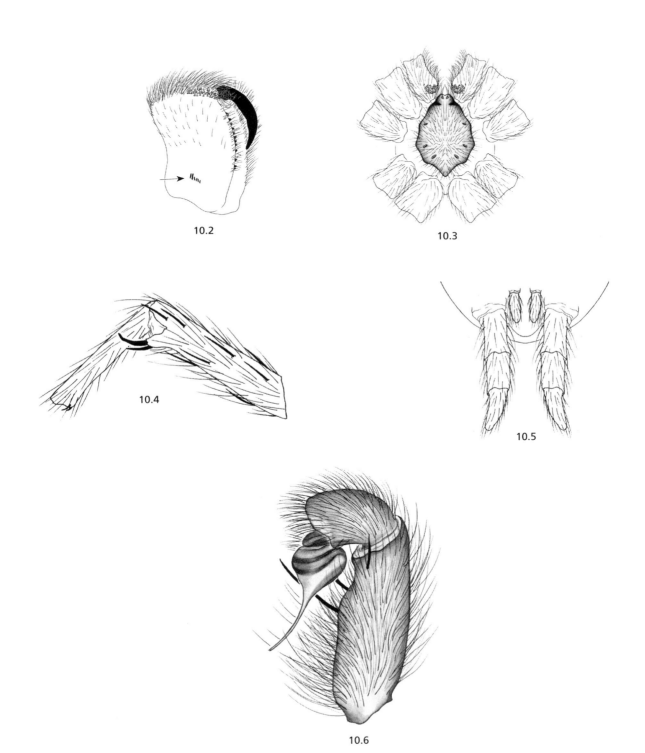

## Chapter 11
# THERAPHOSIDAE
2 genera, 55 species

Thomas R. Prentice

**Common name —**
Tarantulas, Baboon spiders.

**Similar families —**
Nemesiidae (*Calisoga* p. 52): mygalomorph spiders that look remarkably like tarantulas; however they are distinguishable by lack of claw tufts and urticating hairs and by possessing three tarsal claws; occur primarily in central and northern California. Although certain Lycosidae (p. 164) and Pisauridae (p. 199) achieve a size similar to some of the smaller tarantula species, they are not easily mistaken for tarantulas, having three tarsal claws, lacking claw tufts, and having much more widely spaced eyes (generally in three rows).

**Diagnosis —**
Mygalomorph spiders with claw tufts on all tarsi (Figs. 11.2-11.3); a distinctly produced lobe (Fig. 11.4) on moderately large endite; a patch of urticating hairs (visible at magnification of 36X) on the posterior abdominal dorsum (Fig. 11.5); leg I tibia of males with a distoventral bifid mating spur (Fig. 11.6).

**Characters —**
  **body size:** 9-75 mm.
  **color:** carapace light brown to black, abdomen generally clothed with blackish colored pubescence with much longer orange-buff setae.
  **carapace:** pubescent (hairy); cephalic region of carapace elevated; thoracic furrow a transverse to procurved pit.
  **sternum:** longer than wide; 3 pairs of sigilla, located laterally and increasing in maximum diameter from anterior to posterior.
  **eyes:** closely grouped on anterior marginal median prominence; AME and ALE generally subequal.
  **chelicerae:** robust, especially so in females, and without rastellum; row of macroteeth on prolateral side of closed fang and microteeth (denticles) on basal pro and retrolateral side of this row.
  **mouthparts:** endites large with anterior projection; labium fused to and dropping steeply from plane of sternum; cuspules present on basal portion of endite (anterobasal) and labium.
  **legs:** with lateral and ventral spines; male with distoventral bifid mating spur on tibia I.
  **abdomen:** clothed with short pubescence and much longer reddish-orange setae interspersed; urticating hair patch on posterior 1/2-3/4 of dorsal surface.
  **spinnerets:** 2 pairs; PMS unsegmented; PLS longer and 3-segmented, distal article digitiform.
  **respiratory system:** 2 pairs of book lungs.
  **genitalia:** haplogyne; *female* with one pair of weakly sclerotized spermathecae opening broadly into *bursa copulatrix*; *male* palpal organ pyriform, embolus slender (*Aphonopelma*, except *Aphonopelma anax* (CHAMBERLIN 1940a) or apically widened (*Brachypelma*).

**Distribution —**
*Aphonopelma* species occur west of the Mississippi River

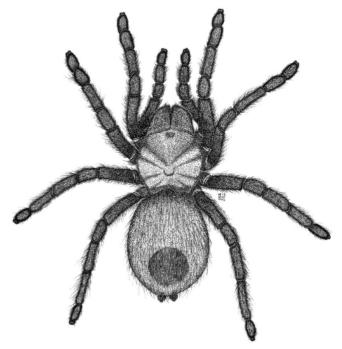

**Fig. 11.1** *Aphonopelma paloma* PRENTICE 1993

from the southern border of the US, north to the approximate midline through Missouri, Kansas, and Colorado and somewhat further north in Utah, Nevada, and California. *Brachypelma vagans* (AUSSERER 1875) has been introduced to Florida and appears to have become established.

**Natural history —**
All of our native tarantula species (genus *Aphonopelma*) and the introduced species, *Brachypelma vagans*, live in subterranean burrows. There is no door-like closure at the burrow entrance, but often the entrance perimeter is lined with silk. The opening is often covered during the day by an ephemeral, thin silk curtain and plugged with soil in the winter months and occasionally during the hot summer months in more arid areas. Tarantulas are nocturnal sit-and-wait predators, rarely found more than a few centimeters away from their burrow entrances. Females reared in captivity have matured at a minimum of seven years of age in three species, *Aphonopelma iodius* (CHAMBERLIN & IVIE 1939b), *Aphonopelma behlei* CHAMBERLIN 1940a, and *Aphonopelma chalcodes* CHAMBERLIN 1940a (pers. obs.); males of a single undescribed species from New Mexico are reported to mature at approximately two years of age (D. Richman, pers. comm.). Females are long-lived and continue to molt (often irregularly with increasing age) throughout the remainder of their lives. Several of our native species are

known to survive in captivity up to 30+ years but few under natural conditions probably survive for that long. Females continue to reside in burrows after maturity whereas mature males abandon their burrows in search of females during the respective breeding season and generally die shortly thereafter. Mating activity of fall breeding species is generally inhibited when nighttime temperatures are below 15°C (60°F) and daytime temperatures are above 26°C (80°F). Mating activity of summer breeding species, however, is almost exclusively carried out after sunset because of high daytime temperatures.

## Taxonomic history and notes —

*Phormictopus platus* CHAMBERLIN 1917 and *Brachypelma aureoceps* (CHAMBERLIN 1917), both described from the Tortugas, are not included in this publication because they are probably not resident in the US. No additional specimens of either species have been reported in the literature since the original descriptions, quite probably because the types were accidentally imported in boats delivering supplies from Mexico and Cuba to the inhabitants of the Tortugas.

*Brachypelma vagans* naturally occurs from Veracruz and the Yucatan Peninsula south along the Gulf Coast to Costa Rica but has apparently become established within the last few decades in St. Lucie Co., Florida (Edwards & Hibbard 1999).

A comprehensive revision of *Aphonopelma* and its close relatives is badly needed. Many of the names of the US tarantula fauna are undoubtedly synonyms, the result of much careless taxonomy. Males of all *Aphonopelma* species from the US are equipped with a slender, apically bent embolus (Fig. 11.7), except *Aphonopelma anax* (CHAMBERLIN 1940a) (broad and apically bent) and *Aphonopelma texense* (SIMON 1891h) (narrow but not apically bent), both from southern Texas. This slender bent form is quite distinct from the distally broadened, apically straight emboli of the type species of *Aphonopelma* [*Aphonopelma seemanni* (F.O. PICKARD-CAMBRIDGE, 1897a)] and the type species of *Dugesiella* POCOCK 1901c [*Aphonopelma crinitum* (POCOCK 1901c)]. Consequently, a revision of the genus will necessitate resurrection of a synonymized generic or subgeneric name to house those species in which males have slender, bent emboli.

## Genera —

GRAMMOSTOLINAE
  *Aphonopelma* POCOCK 1901c, *Brachypelma* SIMON 1891h

## Key to genera —
### North America North of Mexico

1   West of the Mississippi River; embolus of male slender (Fig. 11.7) (except *Aphonopelma anax* from southern Texas); no lanceolate plumose setae (enlarged setae) on any leg or palp segment; spermathecae capitate (Fig. 11.8) (form of spermathecae of *Aphonopelma anax* non-capitate, somewhat similar to that of *Brachypelma vagans*) ................................................ **Aphonopelma**
  **Div.** 54 species — **Dist.** widespread w of the Mississippi in the southern half of the US — **Refs.** Chamberlin & Ivie 1939b, Chamberlin 1940a, Smith 1995, Pérez-Miles *et al*. 1996, Prentice 1993, 1997

—   East of the Mississippi River in Florida; embolus distally spatulate and spooned out (Fig. 11.9); lanceolate plumose setae (both sexes) on the posterior surfaces of trochanter and femur of palp and anterior surface of trochanter and femur of leg I; spermathecae not capitate (Fig. 11.10) ..... ............................................................. **Brachypelma**
  **Div.** 1 species: *Brachypelma vagans* (AUSSERER 1875) — **Dist.** recently established in FL, a non-native species — **Refs.** Ausserer 1875, Locht *et al*. 1999, Edwards & Hibbard 1999

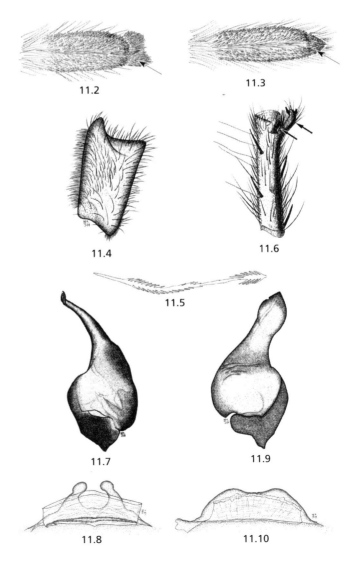

11.2   11.3   11.4   11.6   11.5   11.7   11.9   11.8   11.10

## Chapter 12
# AGELENIDAE
9 genera, 85 species

Robert G. Bennett
Darrell Ubick

### Common name —
Funnel-web weavers.

### Similar families —
Agelenids are similar to various RTA clade spiders especially amaurobiids with whom they have been classified in Amaurobioidea *sensu* Lehtinen (1967) and others. Agelenids are most likely to be confused with coelotine Amaurobiidae (p. 60) and the Lycosidae *Sosippus* (p. 164), less likely so with ecribellate Cicurininae (*Blabomma, Cicurina, Yorima,* p. 95) or larger Tricholathysinae (*Saltonia*) Dictynidae (p. 95) or larger Cryphoecinae Hahniidae (p. 112) (especially *Calymmaria*). The distinctive eye arrangements and plumose hairs of agelenids will serve to distinguish them from all of these spiders.

### Diagnosis —
Eight subequal eyes in two transverse rows, strongly procurved (8 genera – Fig. 12.6) except straight or slightly procurved in *Tegenaria* (Fig. 12.5). Tarsi with single row of trichobothria, increasing in length distally (Fig. 12.2). Plumose hairs on cephalothorax, abdomen, and legs (Fig. 12.3). Ecribellate with distal segment of posterior spinnerets 2/3 (Fig. 12.10) to twice (Fig. 12.9) length of basal segment. Web (Fig. 12.4) a more-or-less open sheet narrowing to a funnel-like retreat.

### Characters —
**body size:** approximately 4-18 mm.
**color:** variable shades of light to dark brown or gray, often with reddish highlights; abdomen and cephalothorax often with dark longitudinal banding, abdomen of larger specimens often with pale chevrons posteriorly.
**cephalothorax:** generally pyriform; fovea longitudinal, plumose hairs present.
**sternum:** (Figs. 12.7-12.8) slightly longer than wide, truncate anteriorly, bluntly acuminate posteriorly and extending between slightly separated hind coxae.
**eyes:** eight more or less subequal eyes in two transverse rows; rows strongly procurved in frontal view (Fig. 12.6) except in *Tegenaria* (where the rows are slightly procurved or nearly straight).
**chelicerae:** slightly to moderately geniculate with boss; promargin with 3-5 teeth; retromargin with 2-5 teeth + occasional denticles.
**mouthparts:** height of clypeus about 1.5 to 3 times diameter of AME (Figs. 12.5-12.6); labium slightly longer than wide; endites about twice as long as wide and slightly convergent.
**legs:** often lightly to strongly banded; trochanters III and IV slightly notched; macrosetae conspicuous, usually paired ventrally on tibiae and metatarsi; plumose hairs present (Fig. 12.3); tarsi with single row of 6-9 trichobothria increasing in length distally (Fig. 12.2); tarsi 3 clawed.
**abdomen:** plumose hairs present; pale dorsal chevrons and/or longitudinal dark marks often present.
**spinnerets:** (Figs. 12.1, 12.9-12.10) anterior spinnerets moderately separated; posterior spinnerets widely separated with distal segment distinct and slightly shorter

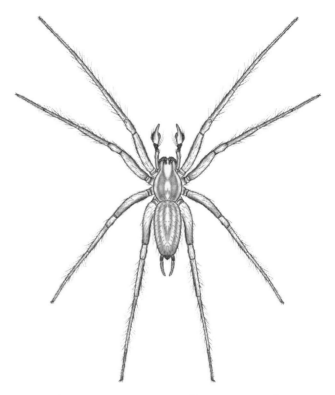

**Fig. 12.1** *Agelenopsis potteri* (Blackwall 1846a)

than to twice as long as basal segment; colulus divided, low, wider than long, and marked with two patches of setae.
**respiratory system:** paired anterior book lungs; spiracle small and just anterior of colulus; tracheae with 4 simple branches limited to abdomen.
**genitalia:** entelegyne; ***female*** epigynum variable (Figs. 12.13, 12.17, 12.21, 12.25, 12.29, 12.31); atrium conspicuous or not, divided or not, and with or without variously developed median septum, scape, and lateral spurs; spermathecae and associated ducts often highly complex and with various pore-bearing regions; ***male*** patellar apophysis absent; retrolateral tibial apophysis present and variably developed (Figs. 12.16, 12.19, 12.26, 12.28, 12.30); apex of cymbium more-or-less extended; tegular processes (true conductor) present (TA in Fig. 12.19) or absent; median apophysis and apophyses associated with embolus [e.g., fulcrum (F in Figs. 12.16, 12.28), conductor, terminal apophysis)] present and variably developed; embolus highly variable and ranging from simple, short spur (E in Fig. 12.30) to very large, elongate, and curved (E in Fig. 12.24).

### Distribution —
Agelenidae has a world-wide distribution, primarily in Holarctic, Neotropical, and Australian biogeographic realms. Some *Tegenaria* are cosmopolitan synanthropes. In North America, only *Agelenopsis* and one species of *Tege-*

*naria* are widespread; *Barronopsis* is found in the eastern USA; most other known North American agelenids occur in the west.

### Natural history —

Agelenids construct typical non-sticky funnel-webs with a flat open sheet-like area for prey capture and a more-or-less hidden (and often extensive) funnel-like component at one corner (Fig. 12.4) serving as a retreat for protection, mating, and egg sac production. Webs often have some amount of webbing extending above the sheet which may serve to knock down or otherwise disrupt the flight of airborne insects. Agelenids move about within the funnel and upon the dorsal surface of the sheet. Funnel-webs are built in a wide variety of habitats: on or around buildings, under various objects in natural or disturbed habitats, in grass, in or under bushes, within caves, etc. Unlike many web-building spiders, outside of their webs agelenids are agile and can move very quickly. Agelenids are primarily nocturnally active but may respond to and capture prey in daylight.

The majority of Nearctic agelenids reach maturity in the late summer and early fall. For some species, maturity may take more than one year. In some areas, some species may have two generations per year (M. Pfeiffer, pers. comm.). Mature males usually die by late fall; some mature females may overwinter and live through a second year of adulthood. Egg cases are usually produced in the late fall or early spring.

*Tegenaria* species are widely known because they are often large and synanthropic and one species has been (probably falsely) accused of being medically important.

### Taxonomic history and notes —

Historically (e.g., Simon 1898a), Agelenidae was a paraphyletic assemblage of spiders including genera now placed elsewhere in the RTA clade. Among recent workers, Lehtinen (1967) removed genera to families in his concept of Dictynoidea (Cybaeidae, Dictynidae, and Hahniidae) to create an Agelenidae composed of agelenines (Agelenidae as recognized here) plus coelotines (*Coras, Coelotes, Wadotes,* etc.). Wunderlich (1986) recognized a close relationship of coelotines with *Amaurobius* and proposed the synonymy of Agelenidae under Amaurobiidae. Platnick (2005) listed coelotines with Amaurobiidae but retained Agelenidae for agelenines (*Tegenaria, Agelenopsis,* etc.), a classification accepted by most arachnologists today. However, much controversy over the classification and taxonomy of various components of the current concept of Agelenidae remains. For instance, in-depth analysis of morphological characters (Griswold *et al.* 1999a) suggest agelenids are more closely related to amphinectids and desids rather than amaurobiids and initial analysis of recent molecular work (Spagna & Gillespie 2003) suggests that coelotines may indeed, be better placed in Agelenidae than in Amaurobiidae.

Gering (1953) and Lehtinen (1967) and, more recently Bennett (1992) and Stocks (1999) provided detailed analyses of agelenid genitalic characters.

Chamberlin & Ivie (1941a, 1942b), Roth (1954, 1968) and Roth & Brame (1972) provided most of the taxonomic context for the Nearctic Agelenidae. Chamberlin & Ivie (1941a) reviewed *Agelenopsis* (creating *Barronopsis* as a sub-genus of *Agelenopsis*) and described the new genera *Calilena, Ritalena,* and *Tortolena*. The following year Chamberlin & Ivie (1942b) revised *Hololena* and *Melpomene* and described *Novalena* and *Rualena*. Roth (1954) revised the sub-genus *Barronopsis* [subsequently elevated to genus by Lehtinen (1967)] and (Roth 1968) the Nearctic species of *Tegenaria*. Roth & Brame (1972) provided keys and descriptions for all Nearctic Agelenidae, including the genera now placed elsewhere, and synonymized *Ritalena* under *Melpomene*. Using a similar concept of Agelenidae, Roth & Brown (1986) catalogued all described Nearctic species. Subsequently Stocks (1999) revised *Barronopsis* although his work is currently (2005) unpublished.

With the exception of *Barronopsis*, all Nearctic agelenid genera are in need of revision. In particular, *Novalena* and *Melpomene* are poorly diagnosed and appear to be dumping grounds for otherwise unrelated agelenid species. In part because some species of *Tegenaria* are synanthropic and consequently have become widespread, the taxonomy and nomenclature of this genus is confused. Recent molecular work by Ayoub *et al.* (2005) suggests that taxonomic problems associated with *Agelenopsis* may be due to recent or ongoing speciation within this widespread genus.

### Genera —

AGELENINAE

*Agelenopsis* Giebel 1869a, *Barronopsis* Chamberlin & Ivie 1941a, *Calilena* Chamberlin & Ivie 1941a, *Hololena* Chamberlin & Gertsch 1929, *Melpomene* O. Pickard-Cambridge 1898a, *Novalena* Chamberlin & Ivie 1942b, *Rualena* Chamberlin & Ivie 1942b, *Tegenaria* Latreille 1804b, *Tortolena* Chamberlin & Ivie 1941a

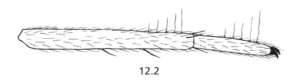

12.2

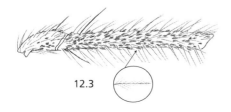

12.3

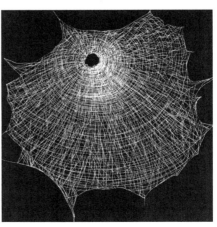

12.4

## Agelenidae

**Key to genera —**
North America North of Mexico

1. Eight eyes in two straight to slightly procurved rows in frontal view (Fig. 12.5); sternum usually with contrasting, variable pattern of pale median stripe and lateral spots (Figs. 12.7-12.8) .................................... ***Tegenaria***
   **Div.** 5 to 7 species – **Dist.** *Tegenaria domestica* (CLERCK 1757) widespread, *Tegenaria pagana* C.L. KOCH 1840c from CA-AL, others primarily western, from BC s to CA e to NM – **Ref.** Roth 1968 – **Note** both *Tegenaria duellica* SIMON 1875a and *Tegenaria saeva* BLACKWALL 1844 have been recorded from North America (Platnick 2005). Bennett (unpublished data) believes that all North American records of *Tegenaria saeva* are misidentifications of *Tegenaria duellica*. Paquin & Dupérré (2003) record *Tegenaria atrica* C.L. KOCH 1843c in Québec, but this species is not likely yet established in North America.

— Eight eyes in two strongly procurved rows in frontal view (Fig. 12.6); sternum not so patterned ............................... **2**

2(1). Distal segment of PLS ~ twice length of basal segment (Fig. 12.9) .................................................................. **3**

— Distal segment at most only slightly longer than basal (Fig. 12.10) ......................................................................... **5**

3(2). Embolus a conspicuous large open circle (Figs. 12.11-12.12). Epigynum with coupling cavity on posterior margin (arrows in Figs. 12.13-12.15) ........ ***Agelenopsis***
   **Div.** 13 species – **Dist.** widespread – **Ref.** Chamberlin & Ivie 1941a

— Embolus inconspicuous, not a large open circle (Figs. 12.16, 12.19). Epigynum lacking coupling cavity on posterior margin (Figs. 12.17-12.18, 12.20-12.21) ............... **4**

4(3). Male palp (Fig. 12.16) with RTA complex, distal and lateral components occupying most of retrolateral face of palpal tibia; embolus (E) short, simply curved; transparent fulcrum (F) present (may be very small) at base of embolus; genital bulb with median apophysis (MA) and conductor (CN) but lacking conspicuous tegular process (Fig. 12.16). Epigynum with slender, elongate scape projecting posteriorly from anterior median margin (SC in Figs. 12.17-12.18). 2 teeth on retromargin of cheliceral fang furrow .................................... ***Calilena***
   **Div.** 16 species – **Dist.** WA & ID s to CA & NM – **Refs.** Chamberlin & Ivie 1941a, 1942b

— Male palp (Fig. 12.19) with RTA simpler, confined to distal half of retrolateral face of palpal tibia; embolus (E) slender, compoundly curved; transparent fulcrum absent; tegular process(es) (TA) well developed (Fig. 12.19). Epigynum with anterior hood (H in Figs. 12.20-12.21) but lacking slender, elongate scape. 3-4 teeth on retromargin of cheliceral fang furrow ............................ ***Melpomene***
   **Div.** 1 species: *Melpomene rita* (CHAMBERLIN & IVIE 1941a) [Figs. 12.19-12.20] – **Dist.** s AZ – **Ref.** Chamberlin & Ivie 1941a, 1942b – **Note** the male of *Melpomene rita*, which has not been described, is illustrated here for the first time (Fig. 12.19). Roth (1985) mentioned an undescribed species from w TX and illustrated the epigynum. This species remains undescribed and unnamed but the epigynum is illustrated here again [Fig. 12.21]

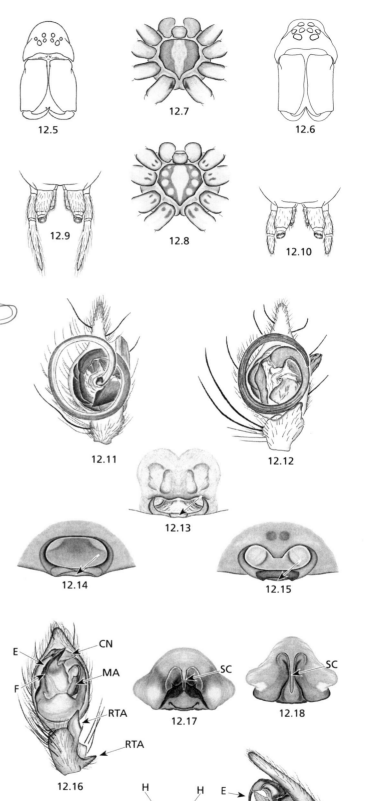

Spiders of North America — Agelenidae | 59

**5(2)** Embolus conspicuous, either tightly coiled basally (E in Fig. 12.22) or figure-8 shaped (E in Fig. 12.24). Epigynal atrium with deeply notched anterior margin (arrow in Fig. 12.23) or marked by paired ridges spiraling inwards to small copulatory openings (arrows in Fig. 12.25) .......... **6**

— Male with embolus inconspicuous and female with atrium otherwise ................................................................ **7**

**6(5)** Embolus in tight spiral coil basally (E in Fig. 12.22). Atrium with deeply notched anterior margin (arrow in Fig. 12.23) ........................................ ***Barronopsis***
    **Div.** 4 species (+ 1 described but unpublished) – **Dist.** MD s to FL & e TX – **Ref.** Chamberlin & Ivie 1941a, Stocks 1999

— Nearctic male apparently unknown (see Note) but embolus probably in form of transverse figure-8 (E in Fig. 12.24). Atrium marked by paired ridges spiraling inwards to small copulatory openings (arrows in Fig. 12.25) ..........
..................................................................................... ***Tortolena***
    **Div.** 1 species: *Tortolena dela* Chamberlin & Ivie 1941a – **Dist.** s TX – **Ref.** Chamberlin & Ivie 1941a – **Note** Figs. 12.24-12.25 illustrate the palp and epigynum of *Tortolena glaucopis* (F.O. Pickard-Cambridge 1902a) which is recorded from Mexico to Costa Rica, but is likely a senior synonym of *Tortolena dela*.

**7(5)** RTA confined to distal half of palpal tibia (Fig. 12.26). Epigynum with atrium undivided (or only partially divided) and with posterior margin bordered by a thickened ridge (arrow in Fig. 12.27). 2-3 teeth on retromargin of cheliceral fang furrow ........................................ ***Rualena***
    **Div.** 9 species – **Dist.** CA – **Ref.** Chamberlin & Ivie 1942b

— RTA occupying most of retrolateral face of palpal tibia (Figs. 12.28, 12.30). Epigynum otherwise (Figs. 12.29, 12.31). 2-4 teeth on retromargin of cheliceral fang furrow
..................................................................................... **8**

**8(7)** Male palp (Fig. 12.28) with RTA bipartite with basal component very large and distinctly separated from distal; embolus (E) sinuous with tip enclosed in transparent fulcrum (F) arising anteriorly from surface of tegulum. Epigynum large with atrium broad and divided longitudinally by more-or-less complete ridge and with pair of stout postero-lateral spurs (arrows in Fig. 12.29) (atrium of some *Rualena* is very similar-see couplet 7 for separation). 2 teeth on retromargin of cheliceral fang furrow ..... ***Hololena***
    **Div.** 30 species – **Dist.** WA & ID s to CA & NM – **Ref.** Chamberlin & Ivie 1942b – **Note** literature records of *Hololena* from "Canada" and Kodiak, Alaska originating in Roewer (1954b) and Banks (1900b) are likely inaccurate. No Canadian specimens are known (although *Hololena curta* McCook 1894 may occur in extreme sw BC) and the Kodiak records, if true, most likely were based on accidental introductions.

— Male palp (Fig. 12.30) with RTA not clearly bipartite, basal and distal components not distinctly separated; embolus (E) evenly curved, transparent fulcrum absent. Epigynum much smaller and lacking broad, medially ridged atrium and postero-lateral spurs (Fig. 12.31). 2-4 teeth on retromargin of cheliceral fang furrow ...................... ***Novalena***
    **Div.** 5 species – **Dist.** sw BC & MT s to CA & NM – **Ref.** Chamberlin & Ivie 1942b

| | |
|---|---|
| **CN** | conductor |
| **E** | embolus |
| **F** | fulcrum |
| **H** | hood |
| **MA** | median apophysis |
| **RTA** | retrolateral tibial apophysis |
| **SC** | scape |
| **TA** | tegular apophysis |

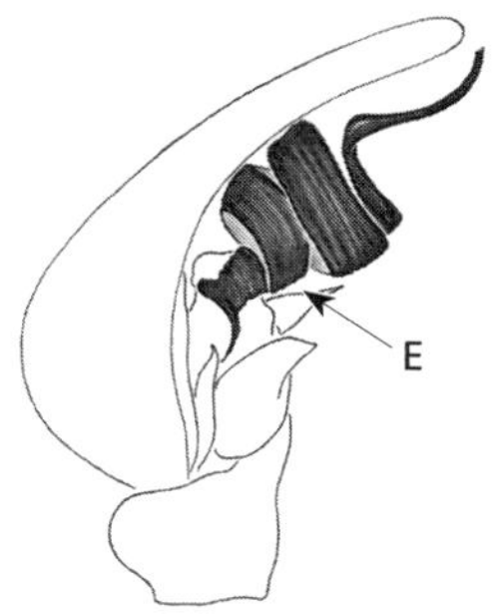

12.22

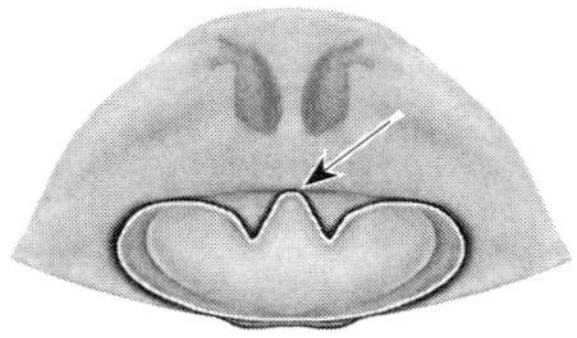

12.23

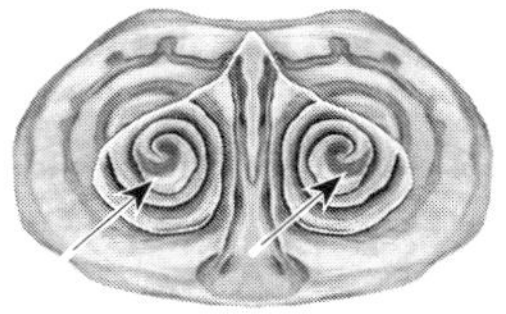

12.25

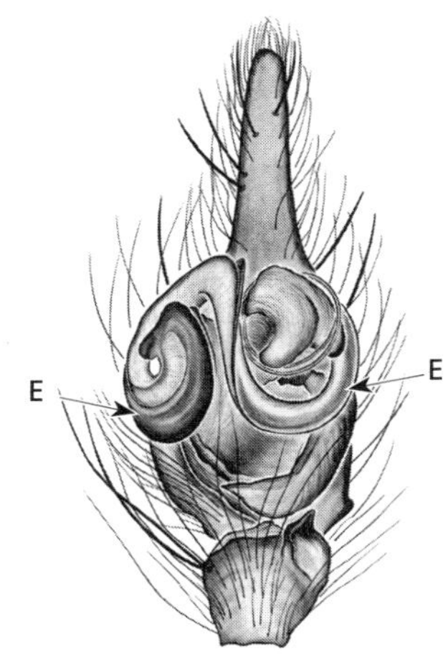

12.24

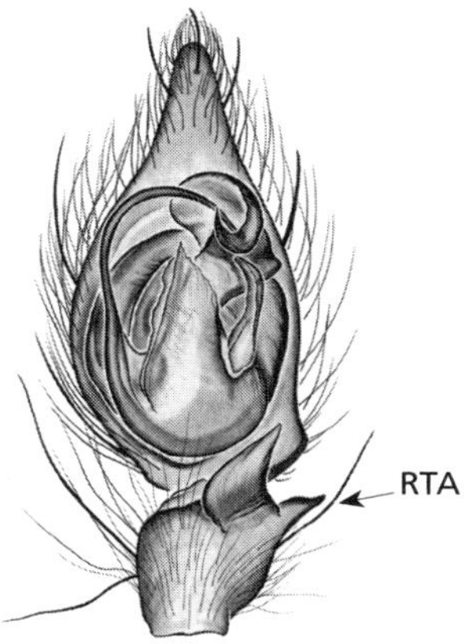

12.26

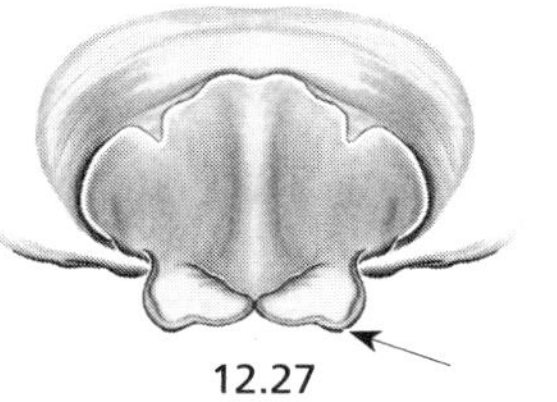

12.27

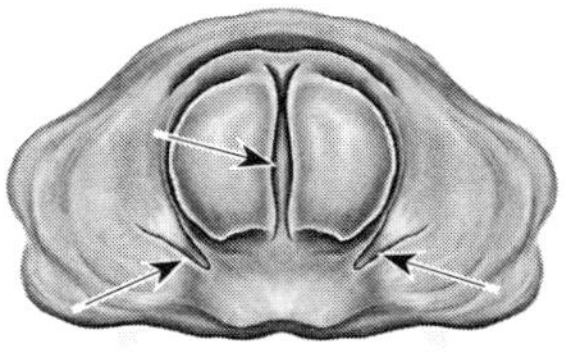

12.29

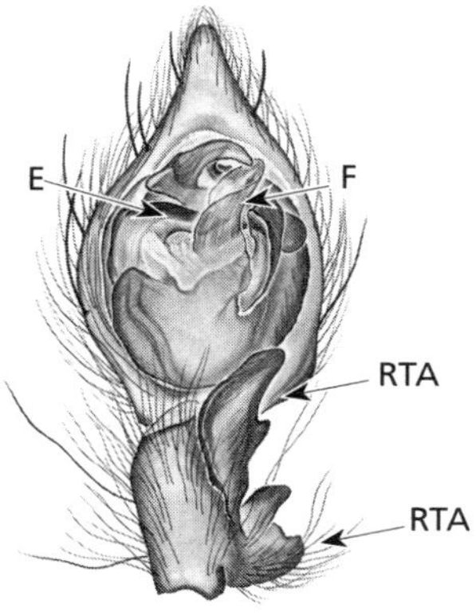

12.28

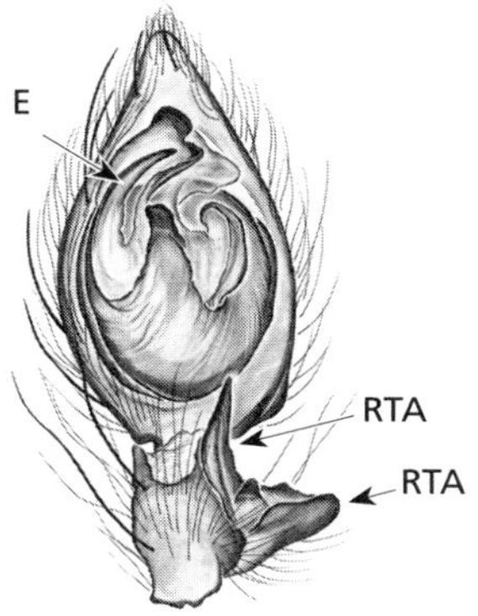

12.30

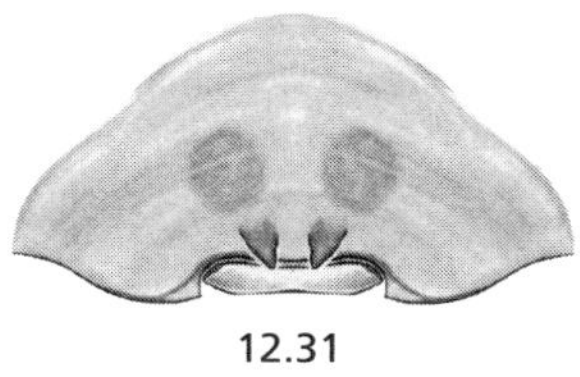

12.31

# Chapter 13
# AMAUROBIIDAE
10 genera, 97 species

Darrell Ubick

## Similar families —
Agelenidae (p. 56), Amphinectidae (p. 63), Titanoecidae (p. 246).

## Diagnosis —
Eight eyes in two transverse rows; AME often reduced in size, rarely absent. Tarsi and metatarsi with dorsal row of trichobothria in distally increasing lengths. Cribellate species with divided cribellum. Ecribellate species larger than 5 mm, with PLS longer than ALS, and lacking plumose hairs.

## Characters —
**body size:** 1.3-20.0 mm.
**color:** varies from yellowish brown to orange to grayish brown.
**carapace:** usually pyriform, round in males of some species; fovea longitudinal, absent in a few small species.
**sternum:** anteriorly truncate, posteriorly pointed or extended between coxae IV.
**eyes:** usually eight eyes in two transverse rows; AME may be reduced, vestigial, or rarely absent.
**chelicerae:** with boss, usually geniculate; margins toothed
**mouthparts:** labium subquadrate, basally notched, longer than wide; endites subquadrate to ectally rounded, two times longer than wide.
**legs:** lacking paired ventral spines; with dorsal trichobothria in distally increasing lengths; tarsi three clawed, with at least one dorsal trichobothrium.
**abdomen:** usually with chevron markings.
**spinnerets:** cribellum usually present, bipartite; ecribellate species with elongate PLS.
**respiratory system:** anterior book lungs and posterior spiracle near spinnerets; with 4 tracheal trunks, usually not branched.
**genitalia:** entelegyne; *female* with epigynum often divided into lobes and often with lateral teeth; *male* cymbium with or without trichobothria, retrolateral margin often modified as declivity, groove, or (rarely) pocket; median apophysis present, hook-like or plate-like; conductor usually reduced and hyaline, may be well developed, but never enclosing embolus.

## Distribution —
Widespread, about 2/3 of the species occur in California. Ecribellate coelotines (*Wadotes* and *Coras*) are restricted to the eastern Nearctic.

## Natural history —
Most amaurobiids occur in cryptozoic habitats on the forest floor, in and under decomposing logs, in leaf litter, and under rocks. Some (as *Callobius*) are found in the bark cavities of standing trees and others have been recorded from rocky grasslands, buildings, and caves. Most appear to be web builders that construct cribellate retreats with a radiating tangle. Ecribellate coelotines build sheet webs with one or more retreats (Bennett 1987, Wang 2002). Cribellate coelotines (*Zanomys* and relatives) construct small sheet webs from which they hang in an inverted position

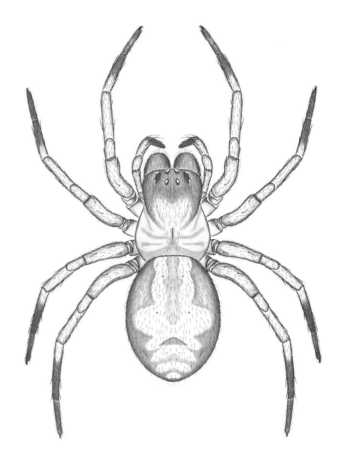

**Fig. 13.1** *Callobius bennetti* (BLACKWALL 1846a)

(Ubick 2005). Females guard their egg sacs, which are attached to their webs (Wang 2002).

## Taxonomic history and notes —
Most of the early studies were by Chamberlin. The cribellate species were initially placed in Dictynidae and largely included in the genus *Amaurobius*. Chamberlin & Ivie (1947a) revised the *bennetti* group of *Amaurobius* which Chamberlin later in the same year (1947) elevated to genus (*Callobius*), described two additional genera (*Pimus* and *Walmus*), and reviewed the known species; in the following year (1948b) he described *Zanomys*. Bishop & Crosby (1935a) described *Callioplus* which was later placed as a junior synonym of *Cybaeopsis* by Lehtinen (1967) and Yaginuma (1987a). Lehtinen (1967) described *Arctobius* and elevated *Titanoeca* and relatives to family status. Leech (1972) revised the Nearctic Amaurobiidae in which he included Titanoecinae (*Titanoeca* and relatives), Desinae (*Badumna*, as *Ixeuticus*), and Metaltellinae (*Metaltella*). The former two are now recognized as families (Gray 1983a) and the latter

has been transferred to Amphinectidae by Davies (1988a). The ecribellate genera were studied by Chamberlin (1925b), who described *Wadotes*, Muma, who revised *Coras* (1946) and *Wadotes* (1947), and Bennett (1987), who again revised *Wadotes*. These genera were placed in the subfamily Coelotinae of Agelenidae by F.O. Pickard-Cambridge (1893c) where they remained until Wunderlich (1986) transferred them to Amaurobiidae. Wang (2002) revised the Coelotinae at the genus level. Ubick (2005) described two genera considered related to *Zanomys* and representing cribellate Coelotinae.

**Genera —**
AMAUROBIINAE
   *Amaurobius* C.L. Koch 1837g, *Callobius* Chamberlin 1947, *Cybaeopsis* Strand 1907b, *Pimus* Chamberlin 1947
ARCTOBIINAE
   *Arctobius* Lehtinen 1967
COELOTINAE
   *Cavernocymbium* Ubick 2005, *Coras* Simon 1898a, *Parazanomys* Ubick 2005, *Wadotes* Chamberlin 1925b, *Zanomys* Chamberlin 1948b

**Key to genera —**
North America North of Mexico

1   Cribellum absent, PLS long (Fig. 13.2) .................... 2
—   Cribellum present, PLS short (Fig. 13.3) ................. 3

2(1) Cheliceral retromargin with 2 teeth (Fig. 13.4) .................
     ........................................................ **Wadotes**
     Div. 11 species — Dist. se CAN; MI s to GA — Ref. Bennett 1987
—   Cheliceral retromargin with 3-5 teeth (Fig. 13.5) ..............
     ........................................................... **Coras**
     Div. 15 species — Dist. se CAN; WI s to LA — Ref. Muma 1946

3(1) Legs lacking spines; chelicerae not geniculate; tiny species (< 3 mm) ........................................................ 4
—   Legs with spines; chelicerae geniculate (Fig. 13.6); larger species (except 1 species < 3 mm) ................... 6

4(3) Abdomen pale, concolorous; AME minute (< ¼ ALE) (Fig. 13.7) or absent (Fig. 13.8); male cymbium with deep retrolateral furrow (Fig. 13.9) ........... **Cavernocymbium**
     Div. 2 species *Cavernocymbium prentoglei* Ubick 2005 and *Cavernocymbium vetteri* Ubick 2005 — Dist. s CA — Ref. Ubick 2005
—   Abdomen with dark markings. AME small (> ¼ ALE); male cymbium with shallow cymbial furrow ................. 5

5(4) Male with retrolateral furrow extending the entire length of cymbium (Fig. 13.10); female with transversely coiled copulatory ducts (Fig. 13.11) .................... **Parazanomys**
     Div. 1 species: *Parazanomys thyasionnes* Ubick 2005 — Dist. CA: Tulare Co. — Ref. Ubick 2005
—   Male with retrolateral furrow restricted to the basal part of cymbium (Fig. 13.12); female lacking transversely coiled copulatory ducts (Fig. 13.13) ............................ **Zanomys**
     Div. 8 species — Dist. w USA, e to UT, CO and SD — Ref. Leech 1972

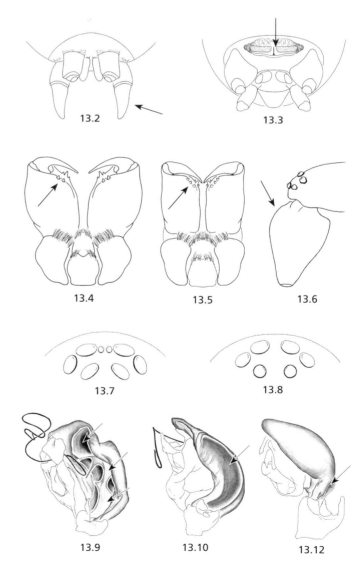

| | | |
|---|---|---|
| **6(3)** | AME equidistant from each other and ALE (Fig. 13.14, frontal view) .................................................. ***Arctobius*** | |
| | **Div.** 1 species: *Arctobius agelenoides* (EMERTON 1919a) — **Dist.** AK to w CAN — **Refs.** Lehtinen 1967, Leech 1972 | |
| — | AME closer to each other than to ALE (Fig. 13.15, frontal view) .................................................................. **7** | |
| | **Div.** 8 species — **Dist.** w USA, e to UT, CO and SD — **Ref.** Leech 1972 | |
| **7(6)** | Males ................................................................................... **8** | |
| — | Females ............................................................................ **11** | |
| **8(7)** | TA with simple tapering prongs (Figs. 13.16-13.18) ...... **9** | |
| — | TA with at least one prong bearing branches or teeth (Figs. 13.19-13.20) ...................................................................... **10** | |
| **9(8)** | TA with 2 prongs (Fig. 13.16) ................................ ***Pimus*** | |
| | **Div.** 10 species — **Dist.** OR to CA — **Ref.** Leech 1972 | |
| — | TA with 3 prongs; dorsomesal (= prolateral) prong usually at least as long as tibia (Fig. 13.17), short in 3 species (Fig. 13.18) ................................................................ ***Callobius*** | |
| | **Div.** 25 species — **Dist.** widespread — **Refs.** Leech 1972, Vetter & Prentice 1997 | |
| **10(8)** | TA with dorsal prong bearing 1-2 slender branches (Fig. 13.19) ............................................................ ***Cybaeopsis*** | |
| | **Div.** 9 species — **Dist.** AK, CAN, n USA, s to NC and TN — **Ref.** Leech 1972 — **Note** *Callioplus* BISHOP & CROSBY 1935a is a synonym | |
| — | TA with dorsal prong otherwise; usually stout, bearing short tooth-like branches (Fig. 13.20) ........ ***Amaurobius*** | |
| | **Div.** 15 species — **Dist.** Most in CA, few from AK to CAN to ne USA — **Ref.** Leech 1972 — **Note** *Walmus* CHAMBERLIN 1947 is a synonym | |
| **11(7)** | Epigynum with lateral lobes large, posteriorly convergent, containing spermathecae (Figs. 13.21-13.22); calamistrum delimited by a spine at either or at both ends ................ **12** | |
| — | Epigynum with lateral lobes small or absent, spermathecae under epigynal plate (Figs. 13.23-13.24); calamistrum not delimited by spine ........................................................ **13** | |
| **12(11)** | Epigynum with median lobe (Fig. 13.21) ........ ***Callobius*** | |
| | **Div.** 25 species — **Dist.** widespread — **Refs.** Leech 1972, Vetter & Prentice 1997 | |
| — | Epigynum lacking median lobe (Fig. 13.22) ....................... .................................................................... ***Cybaeopsis*** | |
| | **Div.** 9 species — **Dist.** AK, CAN, n USA, s to NC and TN — **Ref.** Leech 1972 — **Note** *Callioplus* BISHOP & CROSBY 1935a is a synonym | 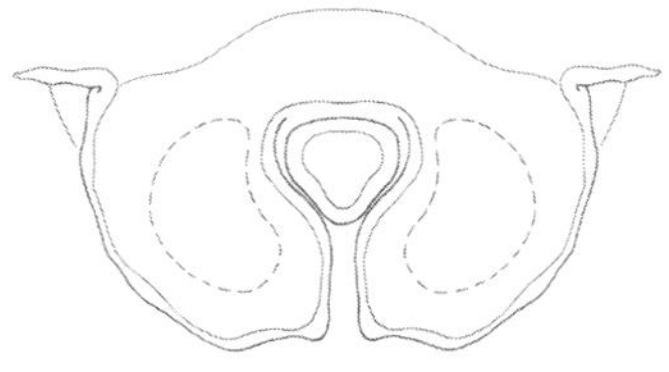 |
| **13(11)** | Epigynum with stabilizing pit at anterior edge (Fig. 13.23) ............................................................................. ***Pimus*** | |
| | **Div.** 10 species — **Dist.** OR to CA — **Ref.** Leech 1972 | |
| — | Epigynum lacking stabilizing pit (Fig. 13.24) ..................... ................................................................... ***Amaurobius*** | |
| | **Div.** 15 species — **Dist.** Most in CA, few from AK to CAN to ne USA — **Ref.** Leech 1972 — **Note** *Walmus* CHAMBERLIN 1947 is a synonym | |

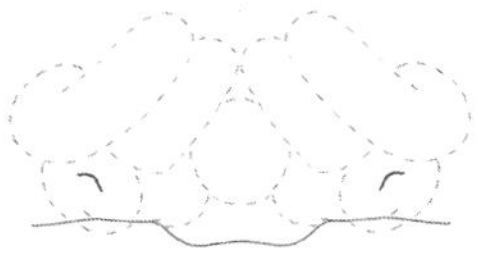

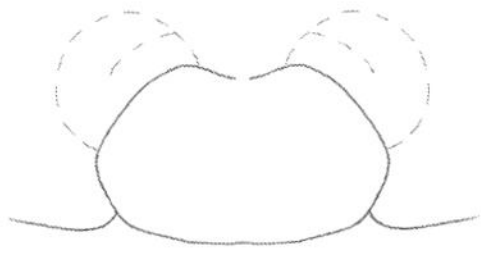

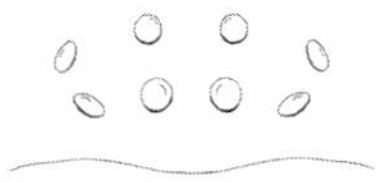

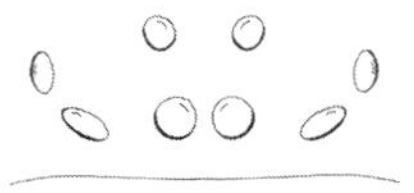

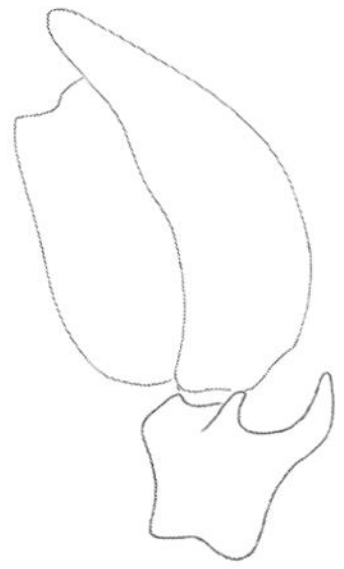

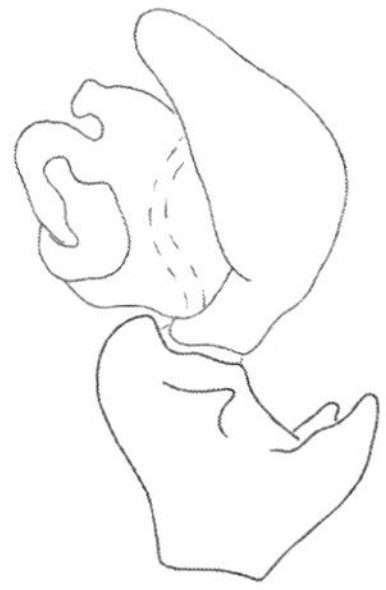

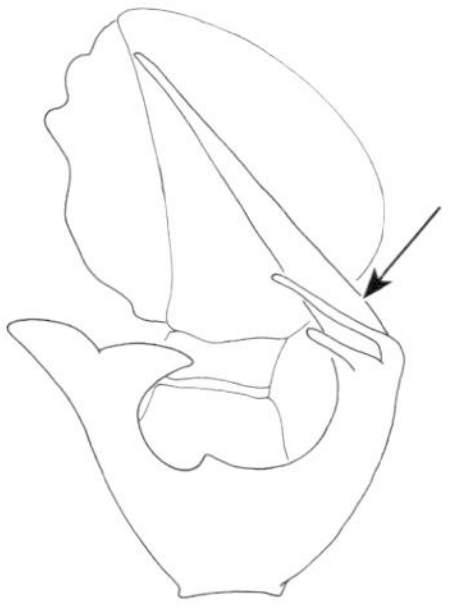

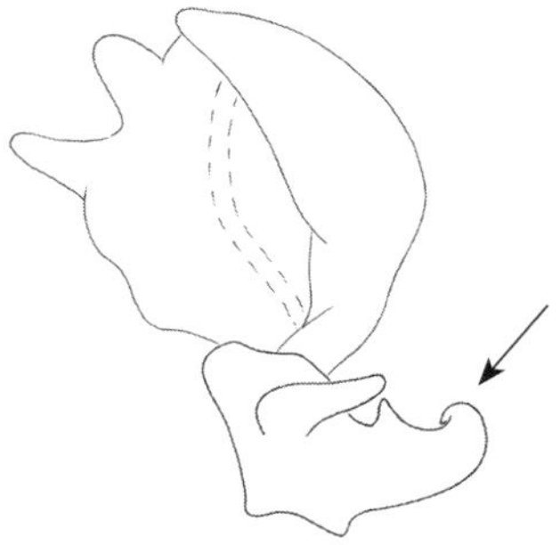

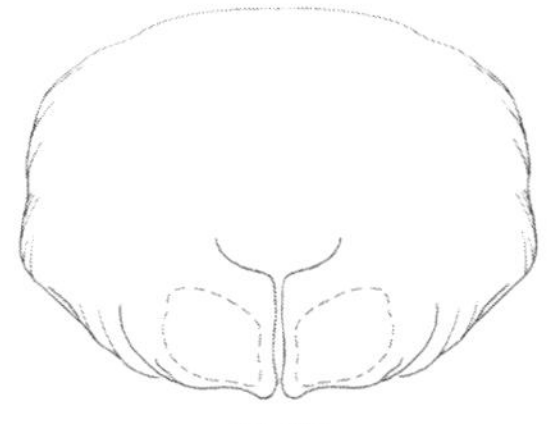

13.11   13.13

13.14   13.15

13.16   13.17   13.18

13.19   13.20

13.21   13.22

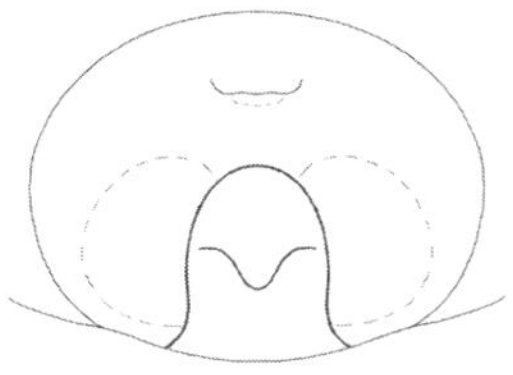

13.23 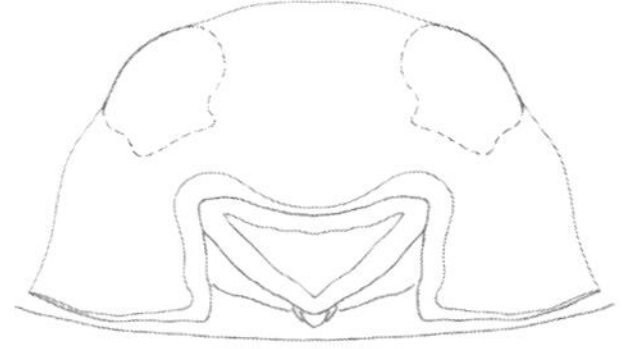   13.24

# Chapter 14
# AMPHINECTIDAE
1 genus, 1 species

Bruce Cutler

### Similar families—
Amaurobiidae (p. 60), Desidae (p. 93), Titanoecidae (p. 246), Zorocratidae (p. 258).

### Diagnosis —
In the field resembles a medium sized dark amaurobiid, this is the only cribellate spider in our region with 5 or more teeth on both the promargin and retromargin of the chelicerae (Fig. 14.2).

### Characters —
**body size:** medium size, 6-10 mm body length.
**color:** carapace orange to dark orange, dark orange to brown cephalic region; chelicerae dark brown; legs yellow to orange; abdomen gray to almost black, in well marked specimens pale median chevrons may be visible
**carapace:** pyriform, longer than wide.
**sternum:** heart shaped, anterior edge truncate.
**eyes:** eyes subequal in size; anterior eyerow straight, posterior eyerow procurved.
**chelicerae:** female chelicerae robust, geniculate, with a prominent boss; male chelicerae slender and tapering distally; both pro-and retrolateral margins with 5 or more teeth (Fig. 14.2).
**mouthparts:** labium and endites longer than wide, endites parallel.
**legs:** three-clawed; three pairs of ventral macrosetae on tibia and metatarsus; calamistrum in single row, in female occupies 40% or more of proximal length of metatarsus with a macroseta at each end, in male greatly reduced with a few thicker setae present.
**abdomen:** elongate oval.
**spinnerets:** divided cribellum; anterior spinnerets conical and contiguous.
**respiratory system:** two book lungs, tracheal spiracle near spinnerets.
**genitalia:** entelegyne; **female** epigynum with raised median smooth area, narrow lateral flaps pointed posteriorly and medially, a central posterior projection with paired lateral openings at the sides (Fig. 14.3); **male** palpus with tibial apophysis, and additional dorsal process on retrolateral edge; embolus threadlike, sheathed in membranous elongate conductor (Fig. 14.4).

### Distribution —
Our only species, the introduced *Metaltella simoni* (Keyserling 1878a), has long been established in the Gulf States from Florida-Texas, earliest USA record from Louisiana in 1944 (Leech 1971). More recently, it has become established in southern California (Vetter & Visscher 1994, Vetter 2000a) and found in Alberta (collected in a greenhouse) and North Carolina (Leech & Steiner 1992). The natural distribution appears to be Argentina, Brazil and Uruguay.

### Natural history —
*Metaltella* spins a tangled cribellate web under debris on the ground. Clearly synanthropic in most instances, in Florida and California it has moved into less disturbed habitat as well. Egg sacs are roughly globular and are suspended in the web.

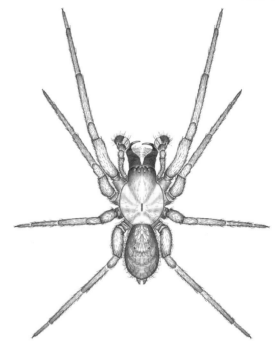

**Fig. 14.1** *Metaltella simoni* (Keyserling 1877a)

### Taxonomic history and notes —
Amphinectidae is an austral family with 33 genera and 175 species. It includes both cribellate and ecribellate taxa. Most taxa are from New Zealand followed by Australia and then temperate South America, including *Metaltella*. The family was established by Forster & Wilton (1973). *Metaltella* was transferred from Dictynidae to Amaurobiidae by Lehtinen (1967), and illustrated and keyed in Leech (1972). *Metaltella* then was transferred to Amphinectidae by Davies (1998a).

### Genus —
Metaltellinae
*Metaltella* Mello-Leitão 1931c

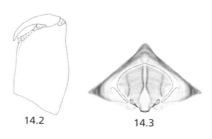
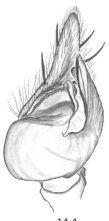

14.2  14.3  14.4

# Chapter 15
# ANAPIDAE
2 genera, 2 species

Jonathan A. Coddington

### Similar families —
Theridiidae (p. 235), Mysmenidae (p. 175), Symphytognathidae (p. 226), Theridiosomatidae (p. 244).

### Diagnosis —
*Gertschanapis* is monotypic and contains only *Gertschanapis shantzi* (GERTSCH 1960a). The females can be distinguished from all other North American spiders except *Anapistula* (Symphytognathidae) by the lack of palpi, and from *Anapistula* by the possession of eight, not four, eyes. The males are small (<2mm) and have an abdominal scutum. The lateral sides of the carapace and the dorsum of the abdomen are liberally provided with small pits or sclerotized dots. The species makes a small, horizontal web similar only to *Maymena* (Mysmenidae). *Comaroma* is a Palearctic anapid genus, but by the palp structure and presence of a $4^{th}$ tarsal comb, "*Comaroma*" *mendocino* (LEVI 1957d) is a theridiid, not an anapid. It is the only six-eyed North American theridiid under 2mm, and, like *Gertschanapis*, has an abdominal scutum.

### Characters —
**body size:** males ~ 1.0-1.5 mm; females 1.0-1.5 mm.
**color:** carapace and abdomen coriaceous to sclerotized, reddish, legs lighter.
**carapace:** highest at eye region (Figs. 15.1-15.2).
**sternum:** broadly rounded or nearly square, $4^{th}$ coxae widely separate.
**eyes:** six (AME absent) or eight, laterals contiguous (Figs. 15.1-15.2, 15.4).
**chelicerae:** promargin with three teeth, retromargin with denticles or bare.
**legs:** length order 1243.
**abdomen:** slightly flattened dorsally, males with scuta (Figs. 15.1-15.2, 15.5).
**spinnerets:** Six spinnerets small, fleshy colulus between and in front of anterior spinnerets.
**respiratory system:** book lungs absent in *Gertschanapis*, normal in "*Comaroma*" *mendocino*.
**genitalia:** entelegyne, *Gertschanapis shantzi* embolus long, spiraled (Fig. 15.3); "*Comaroma*" *mendocino* short and straight (Fig. 15.6).

### Distribution —
*Gertschanapis* thus far occurs in forests of California and Oregon. The range of "*Comaroma*" *mendocino* is restricted to California.

### Natural history —
*Gertschanapis shantzi* inhabits humid California and Oregon forests (coastal, oak, conifer), where it spins small, horizontal orb webs with out-of-plane radii (Fig. 15.7) in the interstices of litter, among fern fronds, and other similar spaces. Web construction has not been observed in *Gertschanapis*, but anapid web construction is generally like the mysmenid *Maymena* (Eberhard 1986, see Chapter 40:175). "*Comaroma*" *mendocino* is known from one ber-

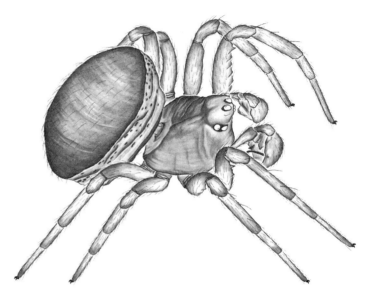

**Fig. 15.1** *Gertschanapis shantzi* (GERTSCH 1960a)

lese sample from a northern California mesic woodland. The biology is unknown.

### Taxonomic history and notes —
Gertsch (1960a) described *G. shantzi* in *Chasmocephalon* O. PICKARD-CAMBRIDGE 1889a, but Platnick and Forster removed it to its own genus in 1990. It is a typical anapid. A number of short, mainly descriptive papers on anapids have been published in recent years and the family needs a review at the generic level. Levi (1957d) described *C. mendocino* in a new monotypic genus, *Archerius*, but Oi (1960a) transferred it to the anapid genus *Comaroma*, where it has remained ever since. Anapidae is part of the symphytognathoid clade (Griswold *et al.* 1998), although the composition of the family and its monophyly have been questioned (Schütt 2003). Phylogenies of Theridiidae have been recently published, but they omit "*Comaroma*" *mendocino* (Arnedo *et al.* 2004, Agnarsson 2004)

### Genera —
*Comaroma* BERTKAU 1889, *Gertschanapis* PLATNICK & FORSTER 1990

# Anapidae

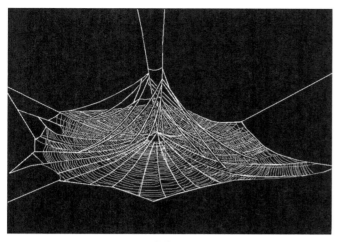

15.7

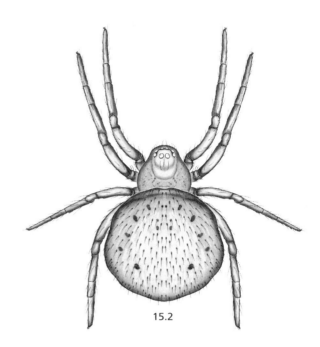

15.2

## Key to genera —
### North America North of Mexico

1   Six eyes (Fig. 15.4) .................................. ***"Comaroma"***
    **Div.** 1 species: *"Comaroma" mendocino* (LEVI 1957d) [Figs. 15.4-15.6] — **Dist.** CA — **Ref.** Levi 1957d — **Note** originally described in *Archerius* LEVI 1957d

—   Eight eyes (Fig. 15.2) ............................. ***Gertschanapis***
    **Div.** 1 species: *Gertschanapis schantzi* (GERSTSH 1960a) [Figs. 15.1-15.3] — **Dist.** CA and OR — **Ref.** Gertsch 1960a

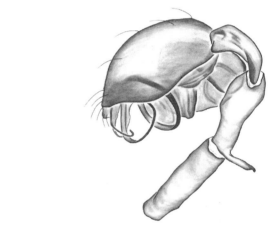

15.3

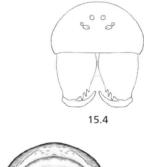

15.4

15.5

15.6

## Chapter 16
# ANYPHAENIDAE
6 genera, 37 species

David B. Richman
Darrell Ubick

### Similar families —
Clubionidae (p. 77), Liocranidae (*Agroeca*) (p. 160), Miturgidae (p. 173).

### Diagnosis —
Small to medium-sized ecribellate entelegyne spiders with two tarsal claws distinguished from other dionychans by the anteriorly displaced tracheal spiracle and lamelliform claw tufts.

### Characters —
**body size:** 3.0-8.5 mm.
**color:** pale brownish to yellowish or white, often with darker markings.
**carapace:** pyriform, fovea present.
**sternum:** ovoid; precoxal triangles present.
**eyes:** eight in two rows; about equal in size, small and nearly round.
**chelicerae:** medium to long, may be stout (especially in males); margins with teeth or denticles.
**mouthparts:** endites and labium longer than wide.
**legs:** prograde and moderately long; leg I longest; coxae of males may have spurs (in some *Anyphaena*); anterior tarsi and metatarsi scopulate; tarsi with two claws and lamelliform claw tufts (Fig. 16.2).
**abdomen:** oval, sometimes elongated.
**spinnerets:** ALS conical and nearly contiguous, colulus represented by setae.
**respiratory system:** one pair of book lungs; tracheal spiracle anteriorly displaced, midway between spinnerets and epigastric furrow or closer to epigastric furrow.
**genitalia:** entelegyne; *female* epigyna usually only partly sclerotized, often with anterior hood, copulatory openings small; *male* RTA variable and distinctive (absent in *Arachosia*), VTA and retrolateral patellar apophysis sometimes present; embolus short to long; median apophysis present or absent.

### Distribution —
*Anyphaena* and *Hibana* are widely distributed in the Nearctic. *Wulfila* and *Arachosia* are primarily eastern; *Pippuhana* and *Lupettiana* southern.

### Natural history —
Anyphaenids are wandering hunters and mostly nocturnal, although *Hibana* is diurnal. They are found on vegetation, in dead leaves, and under loose bark and rocks, especially in winter. In the eastern part of the range, most species are long-legged and occur in foliage, whereas in the West most are short-legged and occupy terrestrial microhabitats (Platnick 1974). They make little use of silk apart from building sac-like retreats and constructing egg sacs. At least one species, *Hibana velox* (BECKER 1879a), is known to eat insect eggs (Richman et al. 1983) and feed at extrafloral nectaries (Taylor & Foster 1996). Some anyphaenids, especially *Hibana incursa* (CHAMBERLIN 1919b), are believed to be important in the control of certain tree crop pests, such as aphids (Richman 2003).

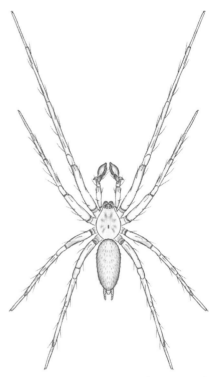

**Fig. 16.1** *Wulfila saltabundus* (HENTZ 1847)

### Taxonomic history and notes —
Although first recognized as a family by Bertkau (1878), most arachnologists followed Simon (1897a) by including these spiders as a subfamily of Clubionidae. Lehtinen (1967) argued that anyphaenids are distinct from clubionids, to which Platnick (1974) provided additional evidence. The Nearctic fauna was revised by Platnick (1974) and the Canadian species subsequently treated by Dondale & Redner (1982). In their continuing revisions of the Neotropical fauna, Brescovit transferred the Nearctic species of *Aysha* to *Hibana* (1991b) and those of *Teudis* to *Pippuhana* and *Lupettiana* (1997a), and Ramírez (2003) that of *Oxysoma* to *Arachosia*. A subfamily classification was proposed by Ramírez (1995a). Coddington & Levi (1991) associated this family with the Clubionidae and Salticidae, based on the absence of cylindrical glands in female spinnerets.

### Genera —
ANYPHAENINAE
*Anyphaena* SUNDEVALL 1833a, *Hibana* BRESCOVIT 1991b, *Lupettiana* BRESCOVIT 1997a, *Pippuhana* BRESCOVIT 1997a, *Wulfila* O. PICKARD-CAMBRIDGE 1895a
AMAUROBIOIDINAE
*Arachosia* O. PICKARD-CAMBRIDGE 1882c

## Key to genera —
North America North of Mexico

1. Tracheal spiracle distinctly closer to epigastric furrow than to spinnerets (Fig. 16.3); genitalia Figs. 16.4-16.5 ............ **Hibana**
   Div. 6 species — Dist. se CAN + e USA to s TX to CA — Refs. Platnick 1974, Brescovit 1991b — Note previously included in *Aysha* Keyserling 1891

— Tracheal spiracle about halfway between spinnerets and epigastric furrow (Fig. 16.6) ............................................. **2**

2(1) Legs long and thin, PT/C = 1.5-2.6; genitalia Figs. 16.7-16.8 ...................................................................... **Wulfila**
   Div. 7 species — Dist. se CAN + e USA to s TX to AZ — Ref. Platnick 1974

— Legs shorter and stouter, PT/C about 1.0 ................... **3**

3(2) Cheliceral retromargin with 2 teeth, no denticles; genitalia Figs. 16.9-16.10 .................................................. **Arachosia**
   Div. 1 species: *Arachosia cubana* (Banks 1909b) — Dist. e coast USA + se AZ — Refs. Platnick 1974, Ramírez 2003 — Note previously included in *Oxysoma* Nicolet in Gay 1849

— Cheliceral retromargin with 4-9 denticles ................... **4**

4(3) Carapace usually with 2 dark paramedian bands (Fig. 16.11); chelicerae not porrect; leg segments nearly concolorous; genitalia Figs. 16.12-16.13 .............. **Anyphaena**
   Div. 22 species — Dist. se + sw CAN, widespread in USA — Ref. Platnick 1974

— Carapace without dark paramedian bands; either chelicerae porrect or femora darker than other leg segments ... **5**

5(4) Anterior metatarsi with two pairs of ventral spines; chelicerae porrect; leg segments concolorous; genitalia as in Figs. 16.14-16.15 ........................................... **Lupettiana**
   Div. 1 species: *Lupettiana mordax* O. Pickard-Cambridge 1896a — Dist. MD and FL to CA — Refs. Platnick 1974, Brescovit 1997a — Note previously included in *Teudis* O. Pickard-Cambridge 1896a

— Anterior metatarsi with one pair of ventral spines; chelicerae not porrect; femora darker than other leg segments; genitalia Figs. 16.16-16.17 ............................ **Pippuhana**
   Div. 1 species: *Pippuhana calcar* (Bryant 1931) — Dist. FL and TX — Refs. Platnick 1974, Brescovit 1997a — Note previously included in *Teudis* O. Pickard-Cambridge 1896a

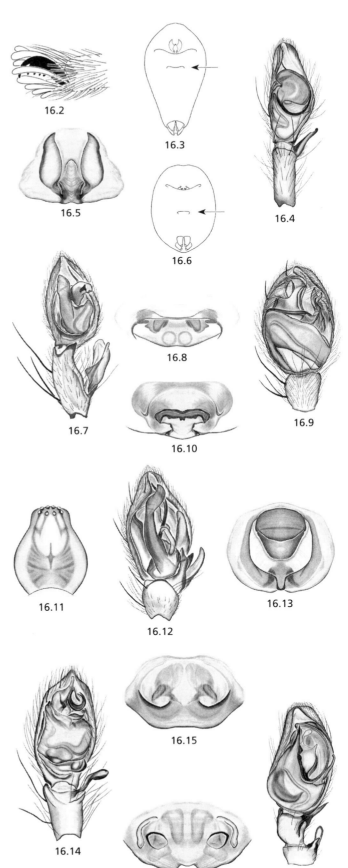

Chapter 17

# ARANEIDAE
31 genera, 161 species

Herbert W. Levi

**Common name** —
Orb weavers.

**Similar families** —
Theridiosomatidae (p. 244), Mysmenidae (p. 175), Tetragnathidae (p. 230).

**Diagnosis** —
Three tarsal claws. No cribellum. Eight eyes in two transverse rows with four ME clustered into a trapezoid at some distance from the two LE, which are adjacent to each other (Fig. 17.31). Height of clypeus usually less than two diameters of anterior median eyes (Figs. 17.11-17.12). Chelicerae with a proximal boss, and usually fewer than 5 strong teeth (Fig. 17.5). No trichobothria on tarsi.

With the exception of *Zygiella*, araneids differ from Tetragnathidae by the following: canoe-shaped tapetum of PME narrow, with rhabdoms in parallel rows toward the median (Levi 1973, diagram 1), and in having the male palpus with an irregular bowl-shaped tibia (Figs. 17.47, 17.51, 17.66). Labium wider than long with distal margin swollen, rebordered (center of Fig. 17.2) and endites squarish, only slightly longer than wide (Fig. 17.2). Males of all species differ in having the bulb with radix and median apophysis. The paracymbium (PC) is attached to the cymbium (Figs. 17.57, 17.73). The median apophysis (MA in Fig. 17.77) is to the right of the radix. Above is the semicircular embolus partly covered by the terminal apophysis (TA).

Most araneids make vertical orb webs to capture prey. Unlike many spiders, they chew their food.

**Characters** —
**body size:** 1.5-30.0 mm.
**color:** drab brown to colorful, greens and reds (which wash out in preserving fluid).
**carapace:** usually longer than wide with cephalic region set off and narrower (Fig. 17.6). Carapace sometimes swollen (Fig. 17.14); only *Mastophora* has projections and tubercles on the carapace (Fig. 17.10).
**sternum:** longer than wide, with pointed posterior tip (Fig. 17.58).
**eyes:** ME arranged in a trapezoid some distance from LE, which are adjacent. PME anterior to LE (Fig. 17.60) or PME posterior to LE in *Argiope*, *Gea* and *Mecynogea* (Fig. 17.9). Tapetum of PME narrow with rhabdoms in rows toward the median (Levi 1973, diagram 1). The median eyes of *Ocrepeira* and sometimes *Eustala* on slight hump with eyes facing laterally (Figs. 17.18-17.19).
**chelicerae:** with proximal boss and at most 4 or 5 teeth in two rows on distal margin. Fang relatively short and stout (Figs. 17.4-17.5).
**mouthparts:** labium wider than long with distal margin swollen, rebordered (center in Fig. 17.2, 17.58). Endites not much longer than wide (Figs. 17.2, 17.58); those of larger males usually with a tooth on the side toward the palpal femur.
**legs:** legs short, relatively stout. First leg longest, third shortest, second usually slightly longer than fourth (Fig.

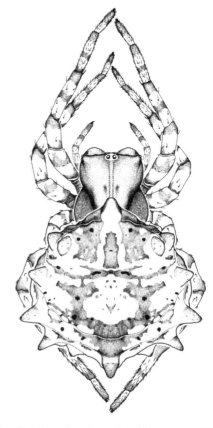

**Fig. 17.1** *Acanthepeira stellata* (WALCKENAER 1805)

17.11) except in *Micrathena*, which has fourth longest (Fig. 17.12). Legs of males often longer than those of females, the second tibia (rarely first) armed with many macrosetae; however macrosetae are variable within species and cannot be used as a diagnostic character for species (Carmichael 1973). Tarsi without trichobothria. First coxae of most larger-sized males have a hook on distal edge (Fig. 17.58), facing a slit on the second femur. Males of some genera have tubercles or macrosetae on fourth coxae.
**abdomen:** usually longer than wide, (except wider than long in *Gasteracantha* and *Mastophora*), often with spines or modified shapes (Figs. 17.16, 17.26, 17.28).
**spinnerets:** six spinnerets; small colulus between and in front of anterior spinnerets (Fig. 17.3). Spinnerets with aggregate glands producing viscid silk.
**respiratory system:** Araneids have two book lungs and an inconspicuous spiracle anterior to the colulus. The book lung covers of larger sized species have transverse supportive grooves, those of some *Micrathena* have fine

stridulating ridges.

**genitalia**: entelegyne; **female** epigynum often three dimensional and with a scape (Figs. 17.40, 17.45, 17.48, 17.74); **male** palpal bulb is rotated (Figs. 17.47, 17.54), with a conductor, embolus, and median apophysis. Median apophysis (arrow in Figs. 17.41, 17.51, MA in 17.73) has radix at its left (in the left palpus), often with terminal apophysis distally (TA in Figs. 17.47, 17.77). *Eriophora* has a paramedian apophysis (PM in Fig. 17.47). Paracymbium more or less U-shaped, attached to cymbium (PC in Fig. 17.73), modified into irregular shape in *Micrathena* and *Zygiella*, (arrow in Fig. 17.57). Palpal tibia usually shallow, saucer-shaped, asymmetrical (Figs. 17.47, 17.51, 17.65, 17.77). Palpal patella usually with one (arrow in Fig. 17.42) or two macrosetae (arrow in Figs. 17.64-17.65); a few small males with none (e.g., *Mastophora*). Larger sized males usually with tooth on endite and a coxal hook (arrow in Fig. 17.58).

## Distribution —
Araneidae is one of the largest families of spiders and is distributed worldwide. Some species are early invaders of isolated areas.

## Natural history —
Most orb weavers spin vertical orbs; vertical webs may be taken down in the morning and eaten by nocturnal weavers, remade in the evening; diurnal spiders remake the web in the morning for day use. *Mecynogea* constructs a fine-meshed horizontal web. *Mastophora* attracts male moths with scent and catches the moth by swinging a thread with a sticky ball attached at the approaching moth. Most larger orb weavers of North America mature in summer and fall, others mature in spring. They live one or two years. Dew covered webs are readily found by walking toward the sun in the morning. Orb weavers are collected by beating branches or sweeping vegetation with a net.

## Taxonomic history and notes —
Levi (1968-to present) and his students have revised and illustrated species of American Araneidae. Dondale *et al.* (2003) illustrate some species that were missed earlier.

The following groups can be recognized within Araneidae, and are used in the list of genera. These are informal groups, not established subfamilies.

*Zygiella* group: palpal bulb faces ventrally; radix indistinct; cylindrical palpal tibia.
*Argiope* group: PE row procurved; no terminal apophysis; large embolus.
*Araneus* group: usually two palpal patellar setae; no paramedian apophysis; large terminal apophysis; usually spines on median apophysis.
*Zilla* group: with a paramedian apophysis.
*Cyrtophora* group: make horizontal webs.
*Eustala* group: soft median apophysis often placed ventrally; scape of epigynum usually directed anteriorly.
*Mastophora* group: no orb web; hunt with a scent and sticky globule on thread.

An alternate classicifation is found in Scharff & Coddington (1997)

## Genera —
ZYGIELLA GROUP
  *Zygiella* F.O. Pickard-Cambridge 1902a
ARGIOPE GROUP
  *Argiope* Audouin 1826, *Gea* C.L. Koch 1843b
ARANEUS GROUP
  *Aculepeira* Chamberlin & Ivie 1942a, *Araniella* Chamberlin & Ivie 1942a, *Araneus* Clerck 1757, *Cercidia* Thorell 1869, *Colphepeira* Archer 1941b, *Hypsosinga* Ausserer 1871b, *Kaira* O. Pickard-Cambridge 1889d, *Larinia* Simon 1874a, *Larinioides* di Caporiacco 1934b, *Mangora* O. Pickard-Cambridge 1889d, *Metepeira* F.O. Pickard-Cambridge 1903a, *Neoscona* Simon 1864, *Singa* C.L. Koch 1836a
EUSTALA GROUP
  *Eustala* Simon 1895a, *Metazygia* F.O. Pickard-Cambridge 1904
ZILLA GROUP
  *Acacesia* Simon 1895a, *Acanthepeira* Marx *in* Howard 1883, *Allocyclosa* Levi 1999, *Cyclosa* Menge 1866, *Eriophora* Simon 1864, *Gasteracantha* Sundevall 1833b, *Micrathena* Sundevall 1833b, *Ocrepeira* Marx *in* Howard 1883, *Scoloderus* Simon 1887d, *Verrucosa* McCook 1888a, *Wagneriana* F.O. Pickard-Cambridge 1904
CYRTOPHORA GROUP
  *Cyrtophora* Simon 1864, *Mecynogea* Simon 1903d
MASTOPHORA GROUP
  *Mastophora* Holmberg 1876

## Key to genera —
### North America North of Mexico

**1**    PME posterior to LE in dorsal view (Figs. 17.6-17.9) .... **2**
—    PME in line with, or anterior to, LE (Figs. 17.12, 17.14, 17.60) ................................................................. **4**

**2(1)**    Abdomen with distinctive dorsal, white, longitudinal stripes and with dark w-shaped mark in middle (Fig. 17.6) ............................................................ ***Mecynogea***
        **Div.** 1 species: *Mecynogea lemniscata* (Walckenaer 1841) — **Dist.** se USA — **Ref.** Levi 1980a
—    Abdomen otherwise ....................................... **3**

17.2

17.3

17.4

17.5

17.6

**3(2)** Distance between PME less than distance from PLE (Figs. 17.7-17.8) ............................................................ ***Argiope***
**Div.** 5 species — **Dist.** widespread — **Ref.** Levi 1968a

— Distance between PE about equal (Fig. 17.9) ............ ***Gea***
**Div.** 1 species: *Gea heptagon* (Hentz 1850a) — **Dist.** widespread — **Ref.** Levi 1968a

**4(1)** Carapace tuberculate, posteriorly with two branched tubercles (Fig. 17.10) ................................. ***Mastophora***
**Div.** 15 species — **Dist.** e CAN, e sw USA, CA — **Ref.** Levi 2003

— Carapace smooth without dorsal tubercles (Figs. 17.11-17.14) .................................................................... **5**

**5(4)** Third tibia with cluster of feathery trichobothria (arrow, Fig. 17.11), carapace with eye region half width of thoracic region, thoracic region very high with median longitudinal line (Fig. 17.11) .......................................... ***Mangora***
**Div.** 7 species — **Dist.** e CAN, e sw USA, CA — **Ref.** Levi 1975b

— Third tibia without such trichobothria, carapace otherwise ..................................................................... **6**

**6(5)** Female abdomen with spines (Fig. 17.12), never with anterior median spine; fourth femur longer than first (arrow, Fig. 17.12); male abdomen flattened with almost trapezoid, rectangular dorsal plate with four almost 90° angles (Fig. 17.13) ....................................... ***Micrathena***
**Div.** 4 species — **Dist.** e CAN, e sw USA, CA — **Ref.** Levi 1978a

— Female abdomen without spines, or if with spines, first femur longer than fourth (Fig. 17.11); males never with rectangular, flattened abdomen ...................................... **7**

**7(6)** Abdomen with a notch at posterior end (Figs. 17.20-17.21) ............................................................................... **8**

— Abdomen without notch ................................................. **9**

**8(7)** LE separated (Fig. 17.20) ............................ ***Cyrtophora***
**Div.** 1 species: *Cyrtophora citricola* (Forskål 1775) — **Dist.** introduced tropical, not yet found in USA — **Ref.** Levi 1997

— LE touching (Fig. 17.21) .............................. ***Allocyclosa***
**Div.** 1 species: *Allocyclosa bifurca* (McCook 1887) — **Dist.** FL — **Ref.** Levi 1977a — **Note** species previously placed in *Cyclosa* in Levi 1977a

**9(7)** Abdomen vertical, attached to pedicel near its middle (Figs. 17.14-17.19) ........................................................ **10**

— Abdomen attached near anterior end (Figs. 17.21-17.22) ............................................................................................ **13**

**10(9)** Carapace swollen (Fig. 17.14) ....................... ***Scoloderus***
**Div.** 2 species: *Scoloderus nigriceps* (O. Pickard-Cambridge 1895a) and *Scoloderus tuberculifer* (O. Pickard-Cambridge 1889d) — **Dist.** se USA — **Ref.** Traw 1996

— Carapace low (Figs. 17.15-17.19) ................................ **11**

**11(10)** Abdomen subspherical (Fig. 17.15) .......... ***Colphepeira***
**Div.** 1 species: *Colphepeira catawba* (Banks 1911) — **Dist.** se USA, w to CO — **Ref.** Levi 1978a

— Abdomen longer than wide (or higher than wide) (Figs. 17.16-17.19) .................................................................. **12**

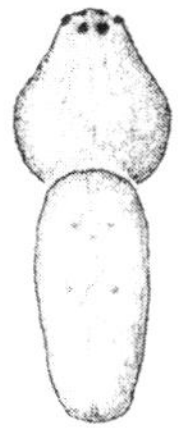
17.7

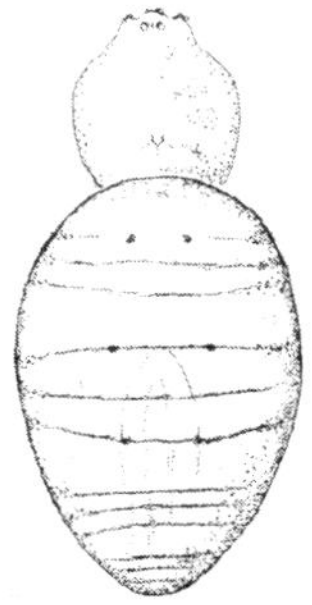
17.8

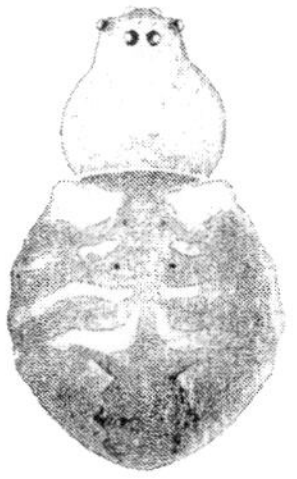
17.9

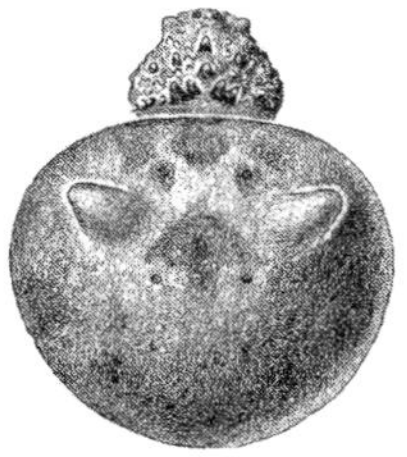
17.10

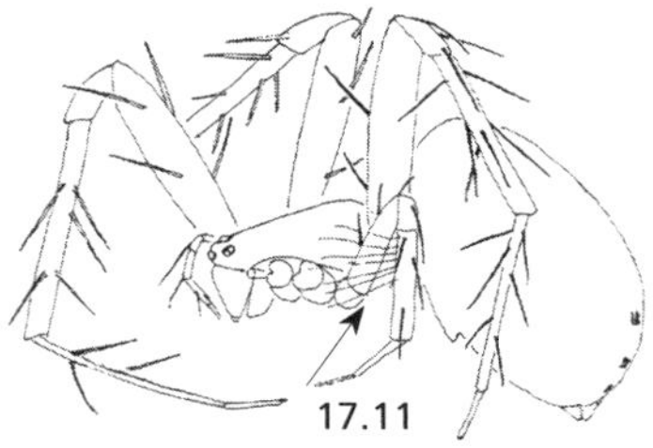
17.11

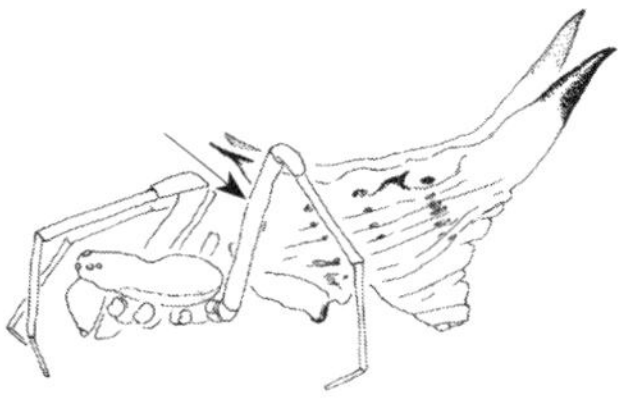
17.12

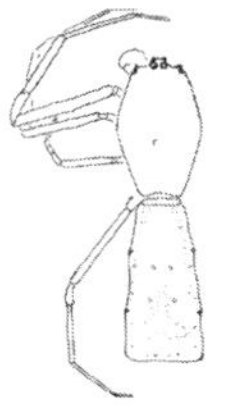
17.13

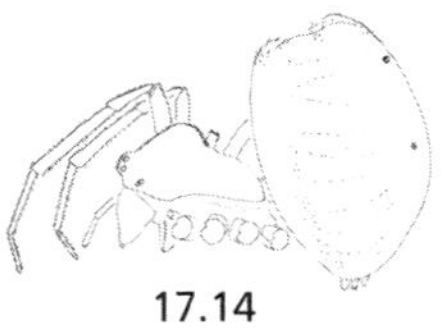
17.14

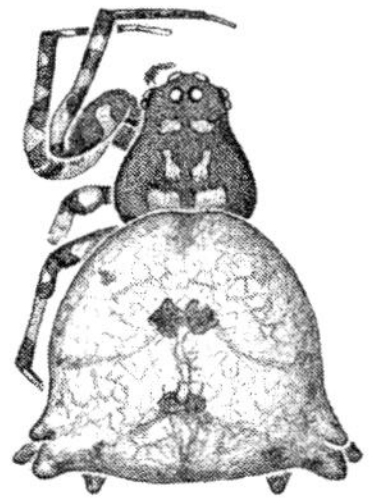
17.15

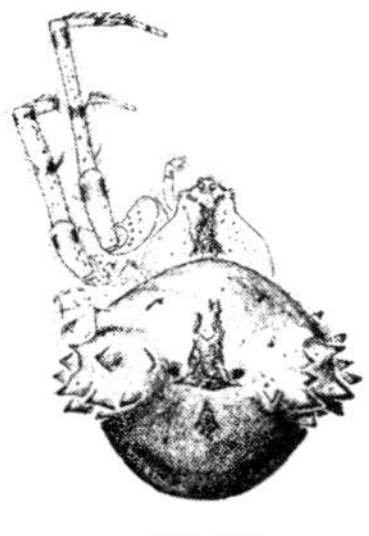
17.16

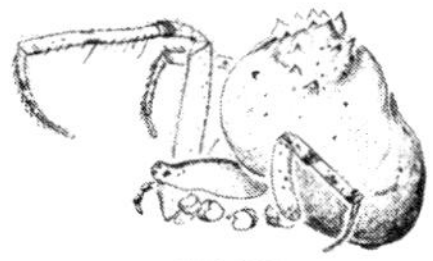
17.17

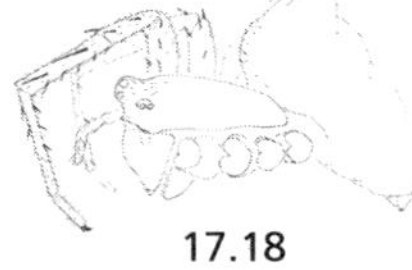
17.18

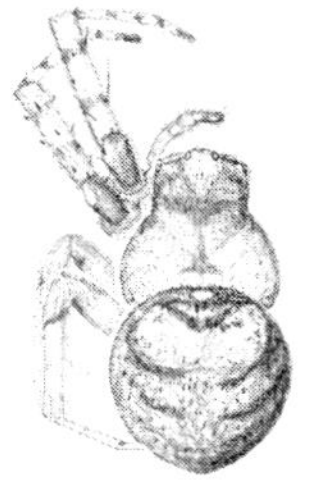
17.19

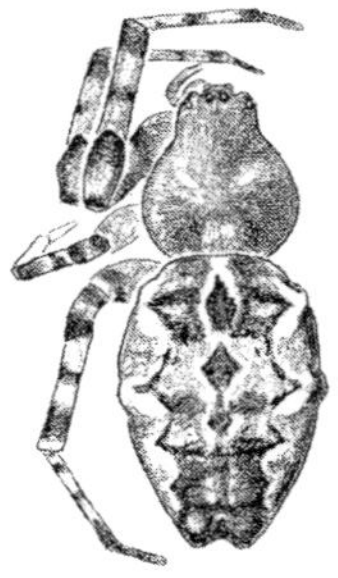
17.20

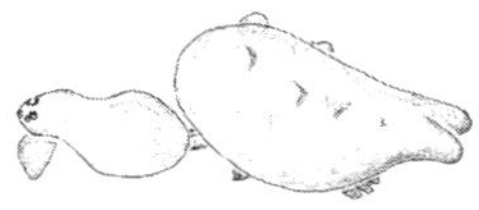
17.21

Spiders of North America — *Araneidae* | 71

**12(11)** Posterior median eyes facing laterally (Figs. 17.18-17.19); dark colored .................................................... ***Ocrepeira***
  **Div.** 3 species — **Dist.** e, sw USA — **Ref.** Levi 1976, 1993b — **Note** species previously placed in *Wixia* by Levi (1976)

— Posterior median eyes facing dorsolaterally (Figs. 17.16-17.17); light colored .................................................. ***Kaira***
  **Div.** 4 species — **Dist.** e, sw USA — **Ref.** Levi 1977b, 1993c

**13(9)** Abdomen with one to three posterior conical projections (Figs. 17.22-17.24); PME their diameter or less apart (Figs. 17.23-17.24) ...................................................... ***Cyclosa***
  **Div.** 4 species — **Dist.** widespread — **Ref.** Levi 1977a

— Abdomen otherwise; distance between PME variable ...... **14**

**14(13)** Abdomen orange, oval and with anterior median tubercle bearing a row of several macrosetae (Fig. 17.25) .................................... ***Cercidia***
  **Div.** 1 species: *Cercidia prominens* Westring 1851 — **Dist.** e CAN, e USA, introduced? — **Ref.** Levi 1975b

— Abdomen not orange, or if orange, not with anterior median tubercle .................................................... **15**

**15(14)** Abdomen with spines or tubercles on sides and/or posterior (Figs. 17.26, 17.28) or male as in Fig. 17.27 .......... **16**

— Abdomen lacks spines or tubercles, at most one pair of anterior humps; rarely anteriorly or posteriorly pointed .. **19**

**16(15)** Venter of abdomen with large, median bulge; female with 6 spines and abdomen much wider than long (Fig. 17.26); male abdomen without spines, as wide as long (Fig. 17.27) .................................................... ***Gasteracantha***
  **Div.** 1 species: *Gasteracantha cancriformis* (Linnaeus 1767) — **Dist.** se USA, s CA — **Ref.** Levi 1978a

— Abdomen without ventral bulge; female abdomen square to longer than wide rarely slightly wider than long (Fig. 17.28-17.29, 17.32) ....................................... **17**

**17(16)** Abdomen with anterior median spine or tubercle (Fig. 17.28) ...................................................... ***Acanthepeira***
  **Div.** 4 species — **Dist.** e CAN, e sw USA, CA — **Ref.** Levi 1976

— Abdomen without anterior median spine or tubercle (Figs. 17.29, 17.31-17.32) ....................................... **18**

**18(17)** Female abdomen square to slightly wider than long with white triangle (Fig. 17.29); male abdomen longer than wide (Fig. 17.31); male's first tibia with branch (Fig. 17.30) ...................................................... ***Verrucosa***
  **Div.** 1 species: *Verrucosa arenata* (Walckenaer 1841) — **Dist.** e USA — **Ref.** Levi 1976

— Female and male abdomen longer than wide (Fig. 17.32), male's first tibia unbranched ...................... ***Wagneriana***
  **Div.** 1 species: *Wagneriana tauricornis* (O. Pickard-Cambridge 1889d) — **Dist.** se USA — **Ref.** Levi 1976

**72** | *Araneidae*

**19(15)** Abdomen with distinct dorsal pattern of two lines, the outer one diamond-shaped (Fig. 17.33) .......... ***Acacesia***
  **Div.** 1 species: *Acacesia hamata* (Hentz 1847) — **Dist.** e USA — **Ref.** Levi 1976

— Abdomen without such marks ..................... **20**

**20(19)** Venter of abdomen with a median, light longitudinal band (Figs. 17.35, 17.37, 17.39, 17.43) .................... **21**

— Venter of abdomen without median light band or area (Figs. 17.48, 17.71) ....................... **24**

**21(20)** Abdomen elongated, often with median swelling on anterior (Fig. 17.36), ventral, white streak may contain a median dark line (Figs. 17.37) .............. ***Larinia***
  **Div.** 3 species — **Dist.** widespread — **Ref.** Levi 1975b

— Abdomen oval, short (Figs. 17.34, 17.38, 17.42) .......... **22**

**22(21)** Metatarsus and tarsus longer than patella and tibia (arrow, Fig. 17.34); abdomen with dorsal folium (Fig. 17.34) .......
  ............................................ ***Metepeira***
  **Div.** 11 species [Figs. 17.34-17.35] — **Dist.** widespread — **Refs.** Levi 1977b, Piel 2001

— Metatarsus and tarsus shorter than patella and tibia (Figs. 17.38, 17.44) .................... **23**

**23(22)** Epigynum scape directed anteriorly (Figs. 17.39-17.40); palpus with soft, whitish "vertical" median apophysis (arrow, Fig. 17.41) ................... ***Eustala**, in part*
  **Div.** 12 species [Figs. 17.38-17.42] — **Dist.** widespread — **Ref.** Levi 1977a

— Scape directed posteriorly (rarely broken off) (Fig. 17.45); median apophysis sclerotized with teeth and two flagellum-shaped projections (arrow, Fig. 17.46) ......................
  ............................................ ***Aculepeira***
  **Div.** 2 species: *Aculepeira packardii* Thorell 1875d and *Aculepeira carbonarioides* Keyserling 1892 [Figs. 17.43-17.46] — **Dist.** AK, CAN, w + ne USA — **Ref.** Levi 1977b

**24(20)** Abdomen subspherical; venter of abdomen with a discrete black trapezoid patch surrounded by white (Fig. 17.48); scape of epigynum with tip reaching more than halfway to spinnerets (Fig. 17.48); palpus with two patellar setae and paramedian apophysis (PM in Fig. 17.47) ..... ***Eriophora***
  **Div.** 2 species *Eriophora ravilla* (C.L. Koch 1844c) and *Eriophora edax* (Blackwall 1863a) — **Dist.** se + sw USA — **Ref.** Levi 1971b

— Venter of abdomen, epigynum and palpi otherwise ... **25**

**25(24)** Round abdomen light yellowish, pink or green with two or three pairs of discrete round black spots on rear (Figs. 17.63-17.64) .................... ***Araniella***
  **Div.** 2 species: *Araniella proxima* Kulczyński 1885a and *Araniella displicata* (Hentz 1847) [Figs. 17.62-17.64] — **Dist.** widespread — **Refs.** Levi 1974b, Dondale *et al.* 2003

— Abdomen otherwise, or if with spots, spots reddish ... **26**

**26(25)** Females ......................................... **27**
— Males ......................................... **34**

**27(26)** Epigynum flat, with depression, or with raised ridge or keel, without annuli (Figs. 17.49, 17.53, 17.56) .......... **28**

— Epigynum with scape (Figs. 17.59, 17.62, 17.67, 17.72, 17.74) (rarely scape is torn off) ..................... **30**

**28(27)** PME almost touching, their diameter or less apart (Fig. 17.50) ........................................ ***Metazygia***
  **Div.** 4 species [Figs. 17.49-17.51] — **Dist.** se USA — **Ref.** Levi 1977a

— PME their diameter or more apart (Figs. 17.52, 17.55) ........................................ **29**

**29(28)** Epigynum flat with depressions, usually with septum (Fig. 17.53); median eye area usually black (Fig. 17.52) ........................................ ***Hypsosinga***
  **Div.** 5 species [Figs. 17.52-17.54] — **Dist.** widespread — **Refs.** Levi 1972a, 1975c

— Epigynum without septum (Figs. 17.56) or with transverse ridge bearing a pair of posterior depressions (Figs. 17.56); median eye area light gray to reddish (Fig. 17.55) ........................................ ***Zygiella***
  **Div.** 5 species — **Dist.** widespread — **Refs.** Levi 1974a, Dondale et al. 2003

**30(27)** Scape projecting anteriorly (Figs. 17.39-17.40) ........................................ ***Eustala,*** *in part*
  **Div.** 13 species — **Dist.** s TX — **Ref.** Levi 1977a

— Scape projecting posteriorly (Figs. 17.45, 17.48) .......... **31**

**31(30)** Abdomen dorsally with three white bands separated by two longitudinal black bands (Fig. 17.60) and black cephalic region (Fig. 17.60) ..................... ***Singa***
  **Div.** 2 species: *Singa cyanea* (Worley 1928) and *Singa eugenii* Levi 1972a [Figs. 17.59-17.61] — **Dist.** e CAN, e USA — **Refs.** Levi 1972a, 1975c

— Abdomen and eye region colored otherwise ................ **32**

**32(31)** Scape smooth, rounded, with lip, not set off from base (Figs. 17.67-17.69) ..................... ***Neoscona***
  **Div.** 8 species [Figs. 17.65-17.69] — **Dist.** widespread — **Refs.** Berman & Levi 1971, Levi 1993a

— Scape usually annulate and set off from base (Figs. 17.62, 17.72, 17.74) ........................................ **33**

**33(32)** Abdomen oval, dorsoventrally flattened (Figs. 17.70-17.71); scape pointed (Fig. 17.72) or tipped by flat knob . ........................................ ***Larinioides***
  **Div.** 3 species — **Dist.** widespread — **Ref.** Levi 1974b — **Note** species previously placed in *Nuctenea* Simon 1864

— Abdomen ovoid to spherical (Fig. 17.75), variable often with anterior humps (Fig. 17.76); scape annulate, rounded at tip with distal border or shallow pocket (Fig. 17.74) .... ........................................ ***Araneus***
  **Div.** 51 species — **Dist.** widespread — **Refs.** Levi 1971a, 1973

**34(26)** Palpal patella with two macrosetae (arrows, Figs. 17.64-17.65), one may be shorter than the other .................. **35**

— Palpal patella with only one macroseta (arrows, Figs. 17.24, 17.42) or none .................................. **39**

**35(34)** Median apophysis split into two parallel, projecting branches (arrow in Fig. 17.73); abdomen oval (Figs. 17.70-17.71), dorsoventrally flattened .................... ***Larinioides***
   **Div.** 3 species — **Dist.** widespread — **Ref.** Levi 1974b — **Note** species previously placed in *Nuctenea* SIMON 1864

— Median apophysis otherwise ......................................... **36**

**36(35)** Median apophysis with one proximal, curved spur (Fig. 17.65) positioned in center of bulb, almost touching or overhanging cymbium (Fig. 17.65-17.66) ...... ***Neoscona***
   **Div.** 8 species [Figs. 17.65-17.69] — **Dist.** widespread — **Refs.** Berman & Levi 1971, Levi 1993a

— Median apophysis otherwise, usually not close to cymbium or center of bulb ................................................. **37**

**37(36)** Median eye region light, usually on setose carapace; abdomen setose; median apophysis always with one or several spines, or teeth (MA in Fig. 17.77) ...... ***Araneus***, *in part*
   **Div.** 51 species — **Dist.** widespread — **Refs.** Levi 1971a, 1973, 1981b

— Median eye region usually black on glossy carapace; glossy abdomen with two dorsal longitudinal black bands or a black frame and dark band on side (Fig. 17.60); median apophysis usually without spines (except in *Hypsosinga*) . ........................................................................................ **38**

**38(37)** Distal rim of first coxa with hook (arrow, Fig. 17.58); palpus as in Fig. 17.61 .............................. ***Singa***
   **Div.** 2 species: *Singa cyanea* (WORLEY 1928) and *Singa eugenii* LEVI 1972a [Figs. 17.59-17.61] — **Dist.** e CAN, e USA — **Refs.** Levi 1972a, 1975c

— First coxa without hook; palpus as in Fig. 17.54 ............. ................................................................................. ***Hypsosinga***
   **Div.** 5 species [Figs. 17.52-17.54] — **Dist.** e CAN, e USA — **Refs.** Levi 1972a, 1975c

**39(34)** Median apophysis in a vertical position, soft, cone to cylinder-shaped (arrow, Fig. 17.41) ............ ***Eustala***, *in part*
   **Div.** 13 species — **Dist.** widespread — **Ref.** Levi 1977a

— Median apophysis usually more horizontal and usually sclerotized, often with hooks and edges (Fig. 17.47, arrow, Figs. 17.46, 17.51) ......................................................... **40**

**40(39)** Paracymbium modified (arrow, Fig. 17.57); palpal tibia often cone-shaped, as long or longer than wide (Fig. 17.57) .............................................................. ***Zygiella***
   **Div.** 5 species — **Dist.** widespread — **Ref.** Levi 1974a, Dondale *et al.* 2003

— Paracymbium U-shaped or a simple hook or spur (Fig. 17.73); palpal tibia bowl-shaped, as wide as long or shorter with distal margin indented, asymmetrical (Figs. 17.61, 17.73, 17.77) ................................................... **41**

**41(40)** PME their diameter or less apart (as in ♀, Fig. 17.50); carapace and abdomen glossy, abdomen oval without humps (as ♀, Fig. 17.50), widest in middle, slightly flattened (as ♀, Fig. 17.50); median apophysis smooth, without spines (arrow, Fig. 17.51) ........................... ***Metazygia***
   **Div.** 4 species [Figs. 17.49-17.51] — **Dist.** w CAN, w USA — **Ref.** Levi 1977a

— PME separated by their diameter or more; carapace and abdomen not appearing glossy, abdomen subspherical, widest anteriorly, median apophysis always with pointed spur (Fig. 17.77) .................................. ***Araneus***, *in part*
   **Div.** 51 species, 3 key out here — **Dist.** widespread — **Ref.** Levi 1973

# Chapter 18
# CAPONIIDAE
3 genera, 9 species

Darrell Ubick

### Similar families —
Dysderidae (p. 103), Oonopidae (p. 185), Prodidomidae (p. 203).

### Diagnosis —
Caponiids are ecribellate haplogynes that differ from other spiders in lacking book lungs and having the PMS anteriorly displaced to form a transverse row with the ALS. Although primitively 8-eyed, most have only two eyes and, unlike other spiders showing eye loss, retain the AME.

### Characters —
**body size:** 3-6 mm.
**color:** prosoma and appendages orange to orange brown, abdomen grayish white, rarely with chevron markings (*Orthonops zebra* Platnick 1995b).
**carapace:** oval to pyriform, flattened; fovea indistinct to absent; clypeus sloping.
**sternum:** oval, with pointed extensions to and between coxae.
**eyes:** AME present, dark in color; other eyes usually absent, but present in *Calponia* where they are very pale.
**chelicerae:** vertical, free (not fused), outer margin with stridulatory ridges, inner margin with median lamina and membranous distal lobe; fang with swollen base.
**mouthparts:** labium triangular; endites strongly convergent, with serrula.
**legs:** lacking spines; tarsi with trichobothria and three claws; Nopinae have tarsi subsegmented and metatarsi with ventral keel and ventroapical lobe.
**abdomen:** elongate oval, densely hairy.
**spinnerets:** 6 spinnerets, PLS largest, PMS anteriorly displaced and forming transverse row with ALS; colulus absent (Fig. 18.2).
**respiratory system:** book lungs absent, two pairs of tracheal spiracles at epigastric furrow.
**genitalia:** haplogyne; *female* genitalia lacking external modifications; *male* palpus with enlarged tarsus; bulb large, spherical, and with short to long embolus, lacking additional sclerites.

### Distribution —
Caponiids occur in the Americas and southern Africa. In the Nearctic, *Calponia* has the narrowest range, being endemic to the central Coast Ranges of California. Both *Orthonops* and *Tarsonops* range from southwest US to Mexico, the former being best represented in southern California and the latter, according to Schick (unpublished), in Mexico.

### Natural history —
Caponiids are wandering hunters, typically found on the ground, under rocks and in leaf litter. Some are known only from woodrat (*Neotoma*) nests. Most of the Nearctic species live in arid desert or desert scrub (*Orthonops* and *Tarsonops*), others (*Calponia* some *Orthonops*) occur in oak savannah to closed canopy forests. Observations suggest that caponiids prefer to prey on other spiders (Platnick 1995b), although they are not be strictly araneophagous

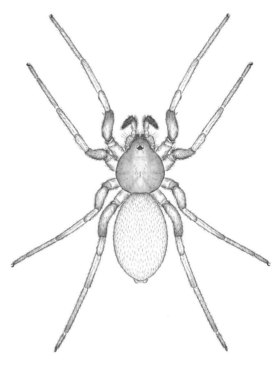

**Fig. 18.1** *Tarsonops systematicus* Chamberlin 1924b

as they readily eat termites in the laboratory (Dippenaar-Shoeman & Jocqué 1997). *Calponia* constructs a conical brown egg sac (D. Ubick, pers. obs.).

### Taxonomic history and notes —
Caponiidae has traditionally been placed in the Haplogynae, and was included in both the Dysderoidea (Simon 1892a) and Scytodoidea (Brignoli 1977a). At present, it is regarded as the most primitive member of the dysderoids (Platnick *et al.* 1991, Ramírez 2000). Two subfamilies are recognized, of which the Nopinae is monophyletic, having obvious modifications such as those on the anterior distal leg segments. It is restricted to the New World and represented in the Nearctic by *Orthonops* and *Tarsonops*. The subfamily Caponiinae is artificial, being supported only by primitive states, namely the absence of obvious modifications. Of its several included genera, two are primitive in retaining all eight eyes: *Caponia* from South Africa and *Calponia* from California.

The most important early work on Nearctic taxa was by Chamberlin (1924b), who described *Orthonops*, *Tarsonops*, and *Nopsides* from northwestern Mexico. In recent studies, Platnick described *Calponia* (1993a) and revised *Orthonops* (1995b).

### Genera —
"CAPONIINAE"
   *Calponia* Platnick 1993a
NOPINAE
   *Orthonops* Chamberlin 1924b, *Tarsonops* Chamberlin 1924b

# Caponiidae

**Key to genera —**
North America North of Mexico

1  Eight eyes (Fig. 18.3); tarsi not subsegmented; anterior metatarsi lacking ventral keel or lobe (Fig. 18.4); paired tarsal claws of anterior legs twice as large as those of posterior legs; embolus of male palpus sharply bent (Fig. 18.5) ................................................ **Calponia**
    **Div.** 1 species: *Calponia harrisonfordi* PLATNICK 1993a [Figs. 18.2-18.5] — **Dist.** central Coast Ranges of CA — **Ref.** Platnick 1993a

—  Two eyes (Fig. 18.6); tarsi subsegmented (Figs. 18.7-18.8); anterior metatarsi with ventral longitudinal keel and ventroapical lobe (Figs. 18.7-18.8); all paired tarsal claws subequal; embolus of male palpus straight to gradually bent (Figs. 18.10, 18.12) ....................................... **2**

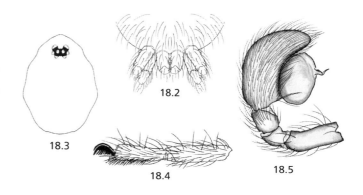

18.2  18.3  18.4  18.5

2(1)  Leg I long and slender (PT/C = 2); metatarsi subsegmented, tarsi often flexed (Fig. 18.9); embolus of male palpus short, less than 0.3 length of bulb (Fig. 18.10) ................................................................ **Tarsonops**
    **Div.** 1 species: *Tarsonops systematicus* CHAMBERLIN 1924b — **Dist.** s CA to AZ, TX — **Ref.** Chamberlin 1924b — **Note** this species, for which the male has not yet been described, has not formally been recorded from the US. Schick, in an unpublished manuscript, includes numerous records from sw US and regards the TX population as a different species.

—  Leg I stout (PT/C = 1), metatarsi entire (not subsegmented), tarsi stiff and straight (Fig. 18.11); embolus of male palpus long, more than 0.6 length of bulb (Fig. 18.12) ..... ......................................................................... **Orthonops**
    **Div.** 7 species — **Dist.** s CA to UT to TX — **Ref.** Platnick 1995b

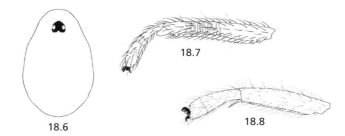

18.6  18.7  18.8

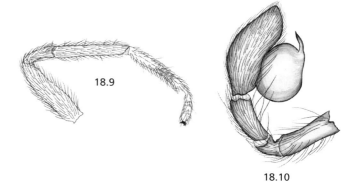

18.9  18.10

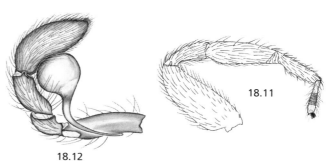

18.12  18.11

## Chapter 19
# CLUBIONIDAE
2 genera, 58 species and subspecies

David B. Richman
Darrell Ubick

**Common name —**
Sac spiders.

**Similar families —**
Anyphaenidae (p. 66), Corinnidae (p. 79), Liocranidae (p. 162), Miturgidae (*Cheiracanthium*) (p. 171).

**Diagnosis —**
Small to medium-sized ecribellate, entelegyne spiders with two tarsal claws distinguished from other dionychans by the combination of the following characters: legs prograde, eight subequal eyes in two rows, tracheal spiracle near spinnerets, claw tufts composed of slender setae, ALS conical and contiguous at base; PLS with distal segment small and rounded, anterior tibiae with fewer than 4 pairs of ventral spines, abdomen lacking scute, and endites concave ectally.

**Characters —**
**body size:** 3.0-10.0 mm.
**color:** usually pale brownish to yellowish; abdomen with brown heart mark, otherwise concolorous (*Clubiona*, Fig. 19.1) or with additional chevron markings (*Elaver*, Fig. 19.2).
**carapace:** longer than wide; cephalic region only slightly narrower than throracic; fovea present.
**sternum:** oval; precoxal triangles present.
**eyes:** eight in two rows, subequal in size, occupy entire cephalic width.
**chelicerae:** medium to long, may be stout (especially in males); margins toothed.
**mouthparts:** endites and labium longer than wide; endites apically scopulate, ectally concave (Fig. 19.3).
**legs:** prograde and moderately long; anterior tibiae with fewer than 4 pairs of ventral spines; tarsi with two claws and tufts composed of numerous slender setae (Fig. 19.4).
**abdomen:** flattened, longer than wide, scute absent.
**spinnerets:** ALS largest, usually conical and nearly contiguous; PLS with short, rounded apical segment; PMS cylindrical in male, enlarged in female.
**respiratory system:** one pair of book lungs; tracheal spiracle close to spinnerets.
**genitalia:** entelegyne; *female* epigyna usually weakly sclerotized; *male* palpus with RTA, simple in *Elaver*, large and variable in *Clubiona*; embolus short; median apophysis absent.

**Distribution —**
*Clubiona* is widespread, *Elaver* is found in e USA and extreme se Canada (Ontario and Nova Scotia) to Montana and Arizona.

**Natural history —**
Clubionids are wandering hunters and do not build snare webs but make sac-like silken retreats, often in rolled leaves, for molting and depositing egg sacs. They occur on foliage, beneath loose bark, in leaf litter, and under rocks (Edwards 1958).

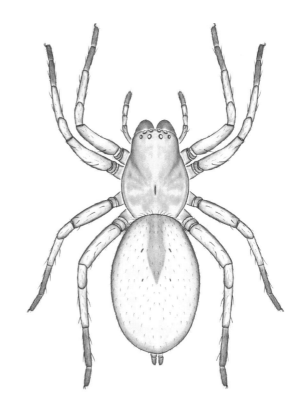

**Fig. 19.1** *Clubiona canadensis* EMERTON 1890

**Taxonomic history and notes —**
Historically, the family Clubionidae has been used as a repository for a wide variety of two clawed spiders. This broad family concept, as constructed by Simon in his "Histoire Naturelle" (1897a) included several subfamilies and groups that are presently given family status: Anyphaenidae, Corinnidae, Ctenidae, Liocranidae, Miturgidae, Sparassidae, and Zoridae (Petrunkevitch 1923, Lehtinen 1967, Platnick 1974). Additionally, several other genera previously included in Clubionidae have been transferred to different families: *Micaria* to Gnaphosidae and *Titiotus*, *Liocranoides*, and relatives to Tengellidae (Lehtinen 1967, Platnick 1999). At present, the family is restricted to those spiders believed to be closely related to *Clubiona*. Still unresolved is the placement of *Cheiracanthium*, which is considered a clubionid by Deeleman-Reinhold (2001) and apparently also by Raven & Stumkat (2003), and a miturgid by Platnick & Shadab (1989), Bonaldo & Brescovit (1992), and Ramírez *et al.* (1997). For the present publication, we follow Platnick (2005) in placing that genus in the Miturgidae. Nearctic clubionids were revised by Edwards (1958),

# Clubionidae

with additional species added by Roddy (1966) and Dondale & Redner (1976c) who also reviewed the Canadian fauna (1982). Brescovit *et al.* (1994) revalidated *Elaver* (for *Clubionoides*).

### Genera —
*Clubiona* LATREILLE 1804b, *Elaver* O. PICKARD-CAMBRIDGE 1898a.

### Key to genera —
North America North of Mexico

1   Trochanters II-IV not notched, or only IV with shallow notch (Fig. 19.5); femur I with one distal prolateral spine; abdomen typically concolorous, lacking chevrons or spots (Fig. 19.1); genitalia as in Figs. 19.6-19.9 ......... ***Clubiona***
    **Div**. 50 species and subspecies — **Dist**. widespread — **Refs**. Edwards 1958, Roddy 1966, Dondale & Redner 1976c, 1982, Paquin & Dupérré 2003

—   Trochanters II-IV notched (Fig. 19.10); femur I with two distal prolateral spines; abdomen usually with distinct chevrons or spots (Fig. 19.2); genitalia as in Figs. 19.11-19.12 .................................................................. ***Elaver***
    **Div**. 8 species — **Dist**. e USA + extreme se CAN to MT + AZ — **Refs**. Edwards 1958, Roddy 1966, Dondale & Redner 1982, Brescovit *et al.* 1994, Paquin & Dupérré 2003 — **Note** these species were previously included in *Clubionoides* EDWARDS 1958

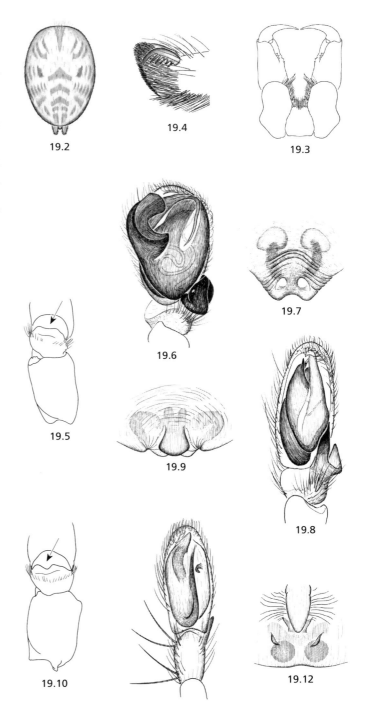

# Chapter 20
# CORINNIDAE
16 genera, 127 species

Darrell Ubick
David B. Richman

## Similar families —
Liocranidae (p. 162), Clubionidae (p. 77), Gnaphosidae (*Micaria*) (p. 105).

## Diagnosis —
Most corinnids may be distinguished from other 2-clawed, 8-eyed spiders in having precoxal triangles and more than 4 pairs of ventral spines on the anterior tibiae. Castianeirinae have fewer leg spines but are recognized by the presence of scutes in both sexes and their ant mimicry. Trachelinae lack typical leg spines, but have short cusps on the anterior legs, and usually have a bright reddish brown carapace.

## Characters —
**body size:** 1-15 mm.
**color:** mostly brown to orange brown; may be brightly colored in Castianeirinae; carapace reddish brown in Trachelinae and some Corinninae.
**carapace:** pyriform, anteriorly truncate to oval to elongate in ant mimics.
**sternum:** oval with precoxal triangles.
**eyes:** 8 in two rows, typically straight; PER recurved in some trachelines.
**chelicerae:** with toothed margins.
**mouthparts:** endites longer than wide, often with median constriction, with serrula; labium longer than wide.
**legs:** usually with several pairs of ventral spines on anterior tibiae, sometimes with fewer spines (Castianeirinae), or spines absent, represented by short cusps (Trachelinae); anterior tibiae and metatarsi rarely with short bristles (*Drassinella*); tarsi with two claws with or without tufts.
**abdomen:** oval; often with scute, especially in male.
**spinnerets:** 6, clustered; ALS conical and contiguous; PMS enlarged in female; PLS with short apical segment; colulus present.
**respiratory system:** one pair of book lungs; tracheal system restricted to abdomen with spiracle near spinnerets.
**genitalia:** entelegyne; *female* with epigynum weakly to strongly sclerotized, usually with two separate copulatory openings; *male* palpus with RTA, which is complex in Corinninae, absent in some Castianeirinae; bulb usually lacking median apophysis, with variable and often complex duct coiling.

## Distribution —
As a whole, corinnids are widespread in this region, although the major diversity is southern. The corinnine genera are restricted to the extreme south, from California to Florida. *Castianeira* is widespread, whereas the other castianeirines are limited to the east coast (*Myrmecotypus*, *Mazax*) or southern Texas (*Mazax*). *Drassinella* is widespread in the West and *Piabuna* in the Southwest, whereas *Scotinella* and *Phrurotimpus* are more widely distributed, as are the trachelines.

## Natural history —
Corinnids are largely ground dwellers, being commonly found beneath rocks and in leaf litter. *Trachelas* also occurs

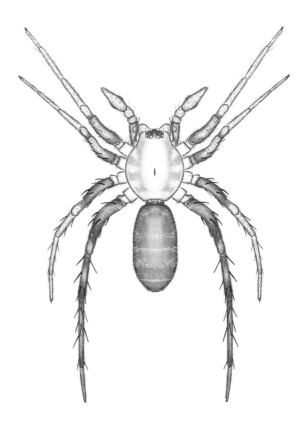

Fig. 20.1 *Castianeira longipalpa* (Hentz 1847)

on foliage and some species frequently enter houses, where they may be of minor medical concern (Platnick & Shadab 1974a). Ant mimicry is common in this family, especially in the Castianeirinae that are virtually all mimetic. Although the Nearctic *Castianeira* species are morphologically rather generalized mimics, those of *Mazax* and *Myrmecotypus*, which may have abdominal constrictions and expanded pedicels, bear a striking resemblance to ants. A number of species of *Castianeira* are brightly colored and so believed to mimic mutillids (Reiskind 1969). The Phrurolithinae also bear a generalized resemblance to ants and at least some species appear to be closely associated with them (Kaston 1981). The introduced corinnine, *Falconina gracilis* (Keyserling 1891), has been found in the nests of the imported fire ant, *Solenopsis invicta*, in Texas (J. Cokendolpher, pers. com.), but not in Florida (G.B. Edwards, pers. comm.). Although most corinnids are nocturnal, the ant mimics, as would be expected, tend to be diurnally active. As usual of dionychans, corinnids build saclike retreats. The egg sacs, which are typically flat discs, may be deposited in the retreats or attached to the substrate (Comstock 1940).

**Taxonomic history and notes —**

Simon (1897a) placed these spiders in his broadly defined Clubionidae. Corinninae included the groups Corinneae and Tracheleae, which correspond roughly to the current subfamilies of those names. Micariinae was defined largely on the basis of ant mimicry and shown by Lehtinen (1967) to contain both gnaphosoid and clubionoid taxa. Most of the latter genera were placed in a new grouping, Castianeirinae (Reiskind 1969). Liocraninae included the Phrurolithinae and *Drassinella*, which recent studies (Penniman 1985, Bosselaers & Jocqué 2002) place in Corinnidae.

The Nearctic corinnids, apart from the phrurolithines, are fairly well known. Castianeirinae was revised by Reiskind (1969), Trachelinae by Platnick & Shadab (1974a, 1974c), and Platnick & Ewing (1995), *Drassinella* by Platnick & Ubick 1989, and Corinninae by Bonaldo 2000.

**Genera —**

CASTIANEIRINAE
*Castianeira* Keyserling 1880a, *Mazax* O. Pickard-Cambridge 1898a, *Myrmecotypus* O. Pickard-Cambridge 1894a
CORINNINAE
*Creugas* Thorell 1878b, *Falconina* Brignoli 1985a, *Megalostrata* Karsch 1880c, *Septentrinna* Bonaldo 2000, *Xeropigo* O. Pickard-Cambridge 1882c
PHRUROLITHINAE
*Phrurolithus* C.L. Koch 1839e, *Phruronellus* Chamberlin 1921a, *Phrurotimpus* Chamberlin & Ivie 1935b, *Piabuna* Chamberlin & Ivie 1933a, *Scotinella* Banks 1911
TRACHELINAE
*Meriola* Banks 1895d, *Trachelas* L. Koch 1866a
INCERTAE SEDIS
*Drassinella* Banks 1904b

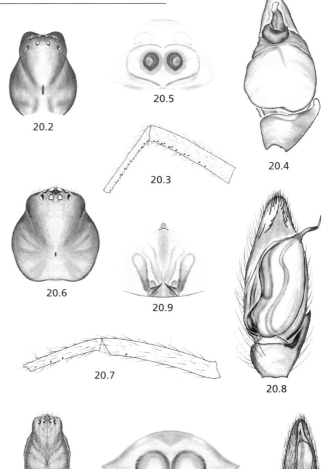

**Key to genera —**
**North America North of Mexico**

**Note** *Megalostrata unicolor* (Banks 1901a), described in *Chemmis* and recorded from CA, is not included in the key because its status is uncertain and it appears to be misplaced in *Megalostrata* (Platnick, pers com).

| | |
|---|---|
| **1** | Leg spines absent (although short cusps are present) ....... ................................................................ Trachelinae ... **2** |
| — | Leg spines present ............................................................. **3** |
| **2(1)** | PER strongly recurved (Fig. 20.2); leg cusps relatively short (Fig. 20.3); embolus a separate sclerite (Fig. 20.4); epigynum with coiled ducts (Fig. 20.5); total length 4-10 mm .................................................................. ***Trachelas*** |
| | **Div.** 8 species — **Dist.** widespread — **Ref.** Platnick & Shadab 1974a |
| — | PER almost straight (Fig. 20.6); leg cusps longer (Fig. 20.7); embolus integral with bulb (Fig. 20.8); epigynum lacking coiled ducts (Fig. 20.9); total length < 5 mm ........ .................................................................................. ***Meriola*** |
| | **Div.** 3 species — **Dist.** e US to UT, AZ to CA to WA — **Refs.** Platnick & Shadab 1974c, Platnick & Ewing 1995 — **Note** these species were previously included in *Trachelas*. *Meriola californica* (Banks 1904b) and *Meriola decepta* Banks 1895d are widespread native species, whereas *Meriola arcifera* (Simon 1886f) was introduced into CA from South America. |
| **3(1)** | Anterior tibiae with at most 3 pairs of ventral spines ....... ................................................................ Castianeirinae ... **4** |
| — | Anterior tibiae with 5-8 pairs of ventral spines .............. **6** |
| **4(3)** | Abdomen with distinct rugose petiole (Fig. 20.10); scutum with a pair of stout apical spines (Fig. 20.10); anterior eyes subequal or medians smaller; genitalia as in Figs. 20.11-20.12 ............................................................................. ***Mazax*** |
| | **Div.** 2 species: *Mazax kaspari* Cokendolpher 1978b and *Mazax pax* Reiskind 1969 — **Dist.** s TX; — **Refs.** Reiskind 1969, Cokendolpher 1978b — **Note** an undescribed species of *Mazax* has been discovered in e FL (G.B. Edwards, pers. comm). |
| — | Abdomen without rugose petiole; scutum usually with two pairs of slender apical setae; anterior eyes subequal or medians larger ................................................................. **5** |

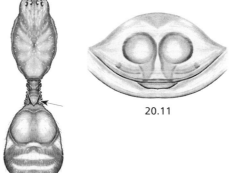
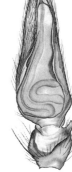

| | |
|---|---|
| **5(4)** | Fovea absent; PER recurved; abdomen strongly constricted (Fig. 20.13); genitalia as in Figs. 20.14-20.15 ................ ........................................................... ***Myrmecotypus*** |
| | **Div.** 1 species: *Myrmecotypus lineatus* (Emerton 1909) — **Dist.** MA-FL — **Ref.** Reiskind 1969 |
| — | Fovea present; PER slightly to moderately procurved; abdomen not constricted (Fig. 20.16); genitalia as in Figs. 20.17-20.18 ...................................... ***Castianeira*** |
| | **Div.** 26 species — **Dist.** widespread — **Ref.** Reiskind 1969 |
| **6(3)** | Body length < 5mm ......................................................... **7** |
| — | Body length > 5mm .............................. Corinninae ... **11** |
| **7(6)** | Femora with 2 dorsal spines; legs III and IV with ventral and lateral spines; genitalia as in Figs. 20.19-20.20 ............ ............................................................................. ***Drassinella*** |
| | **Div.** 6 species — **Dist.** w USA — **Ref.** Platnick & Ubick 1989 |
| — | Femora with fewer than 2 dorsal spines; legs III and IV without ventral or lateral spines (or with only 2 ventroapical spines on tibia IV) ..................... Phrurolithinae ... **8** |
| **8(7)** | Femora with 1 dorsal spine; carapace with paramedian dark markings (Fig. 20.21); tibia I usually dark; male palp with single apophysis originating dorsally on the tibia (Fig. 20.22); female epigynum large with curved copulatory ducts (Fig. 20.23) ............................ ***Phrurotimpus*** |
| | **Div.** 17 species and subspecies — **Dist.** widespread — **Refs.** Chamberlin & Ivie 1935b, Gertsch 1941b, Dondale & Redner 1982, Paquin & Dupérré 2003 — **Note** Roth (1994) recognizes 23 species. |
| — | Femora lacking dorsal spines; carapace lacking paramedian markings; tibia I not dark; male palpal tibia with apophysis double or bifid, if single, then originating ventrally on the tibia; epigynum otherwise ........................... **9** |
| **9(8)** | Body lacking markings, legs concolorous; AME large, round, dark and dorsally oriented (Fig. 20.24); male palpal tibia with two separate apophyses, dorsal and retrolateral (Fig. 20.25) (very small in *Piabuna brevispina* Chamberlin & Ivie 1935b); epigynum as in Fig 20.26; total length less than 2 mm ......................................................... ***Piabuna*** |
| | **Div.** 5 species — **Dist.** s CA-NM, UT — **Refs.** Chamberlin & Ivie 1933a, 1935b — **Note** Roth (1994) mentions an undescribed species from TX |
| — | Body and legs with at least some dark markings; AME otherwise; male palpal tibia with single apophysis which may be bifid; epigynum otherwise; total length usually greater than 2 mm ........................................................ **10** |

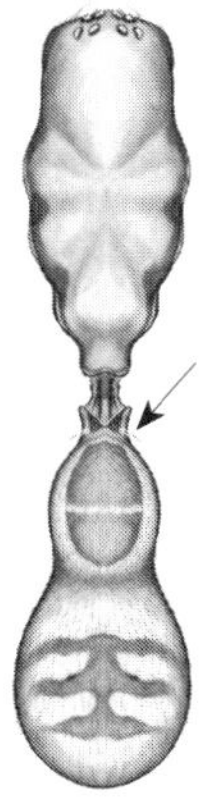

20.13
20.14 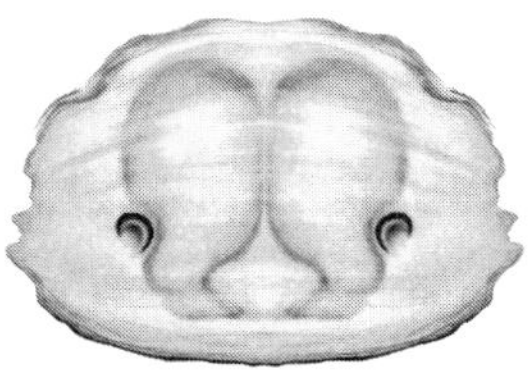
20.15 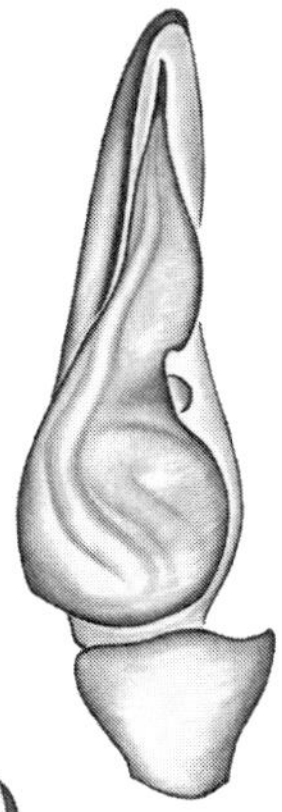
20.16 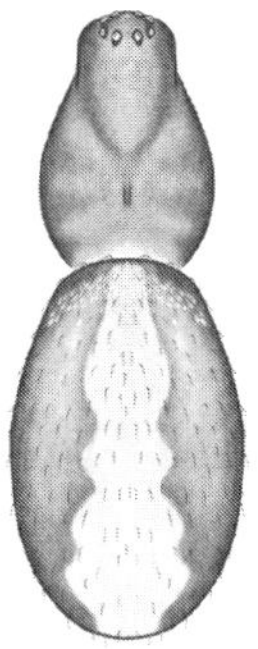
20.17 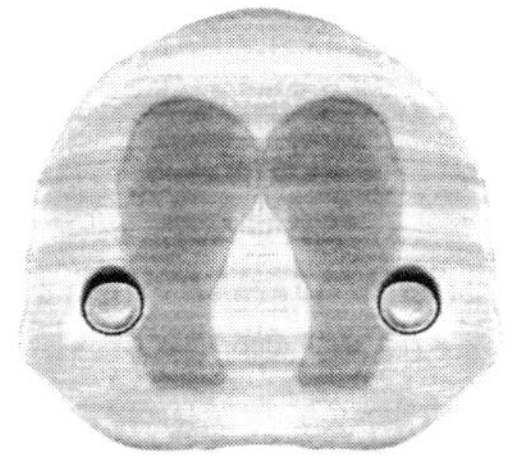
20.18 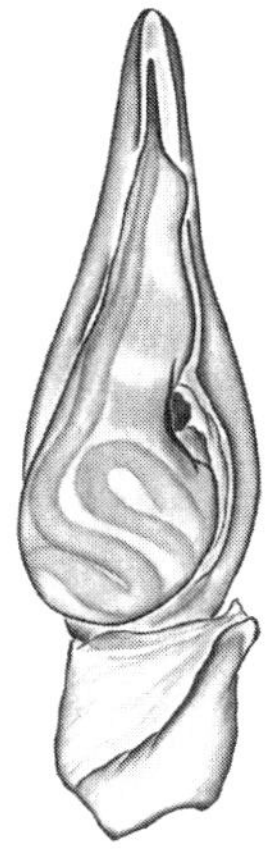
20.19 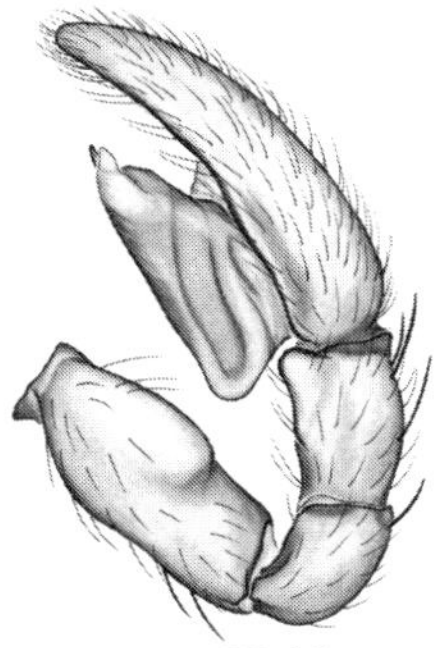
20.20 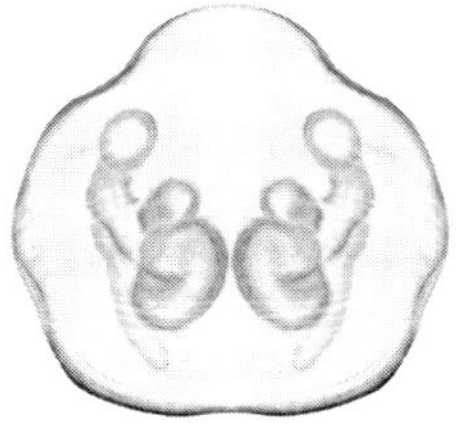
20.21 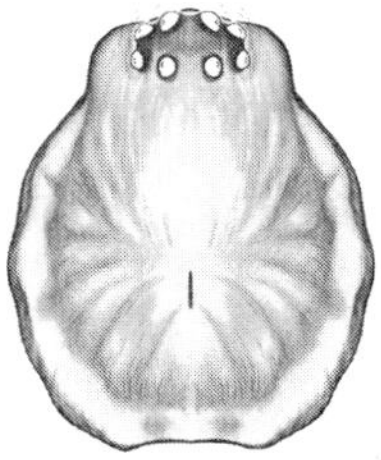
20.22 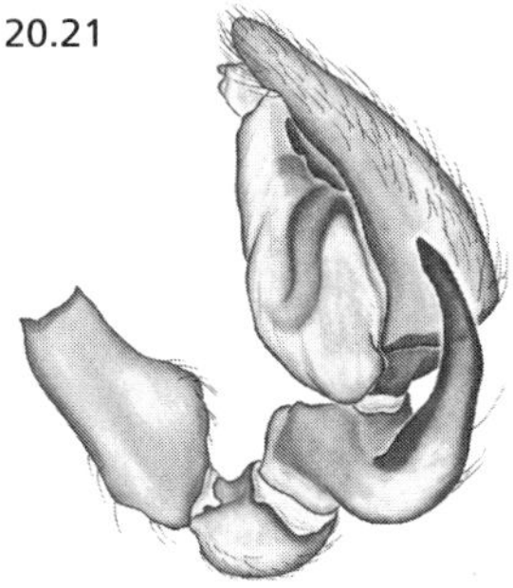
20.23 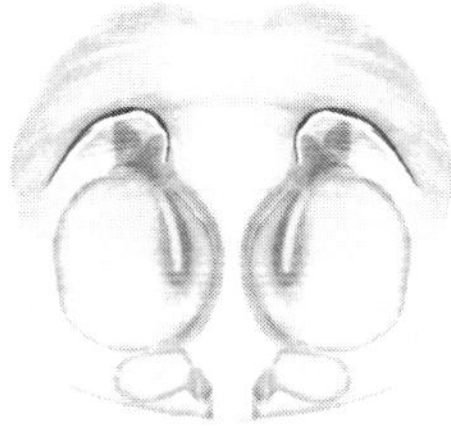
20.24 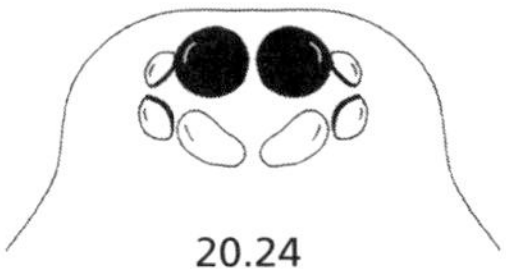
20.25 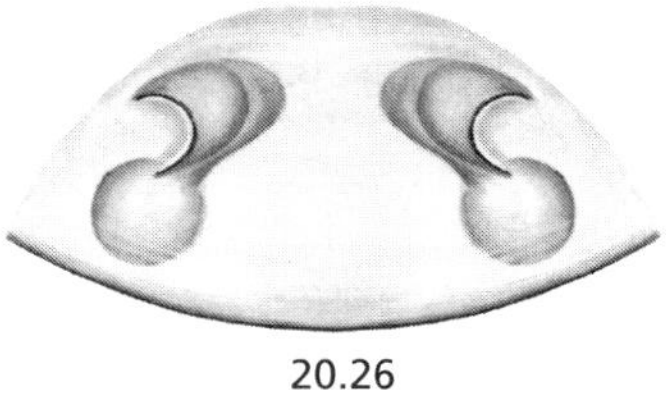
20.26

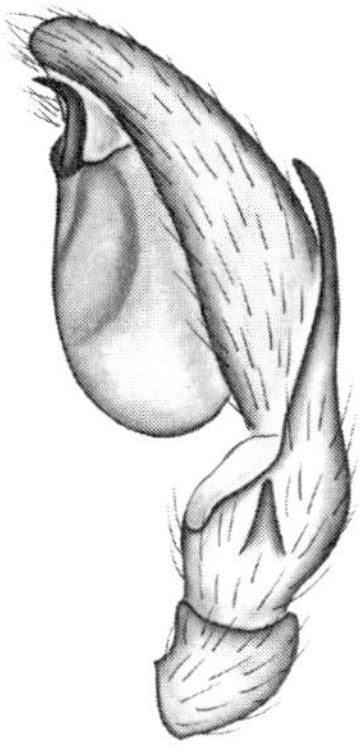

# 82 | Corinnidae

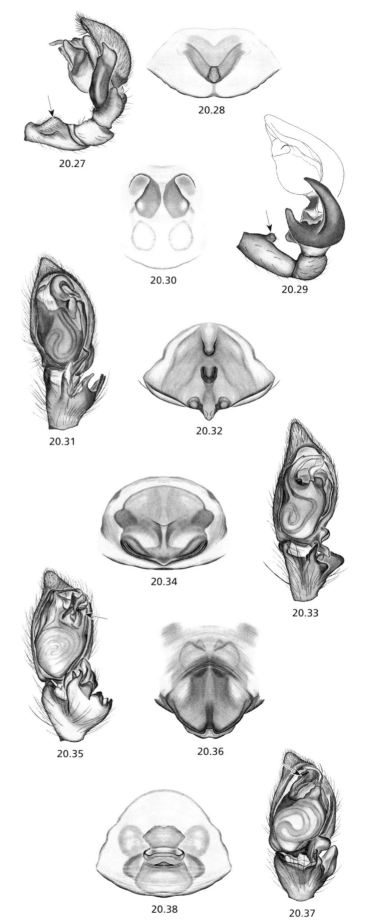

10(9) Femur I with one prolateral spine; male palp with RTA a single prong originating ventrally on the tibia, femoral apophysis distal and broadly attached (Fig. 20.27); female epigynum with single opening at posterior margin (Fig. 20.28) ........................................................ ***Phrurolithus***
  **Div.** 1 species: *Phrurolithus festivus* (C.L. KOCH *in* HERRICH-SHÄFFER 1835) — **Dist.** NY — **Refs.** Roberts 1985a, Grimm 1986 — **Note** *Phrurolithus festivus* was introduced from Europe into Onondaga Co., NY. The Nearctic species currently cataloged in *Phrurolithus* are believed to belong in *Scotinella* (Penniman 1985)

— Femur I with two prolateral spines; male palp with RTA bifid, femoral apophysis more basal and narrowly attached (knoblike) (Fig. 20.29); female epigynum with paired openings in median to anterior position (Fig. 20.30) .............
  ........................................................... ***Scotinella***
  **Div.** 50 species — **Dist.** widespread — **Refs.** Chamberlin & Gertsch 1930, Chamberlin & Ivie 1935b, Ivie & Barrows 1935, Gertsch 1935b, 1936a, 1941b,Muma 1945, Kaston 1981, Dondale & Redner 1982, Paquin & Dupérré 2003 — **Note** in addition to the 16 species currently catalogued in *Scotinella*, included here are also those listed in *Phruronellus* (5) and the Nearctic *Phrurolithus* (29), which key out here and which were considered to belong to *Scotinella* by Penniman (1985). Although this transfer of species appears in a dissertation and was never formally published, it is nonetheless accepted by various authors (Kaston 1981, Dondale & Redner 1982, Roth 1994). Bosselaers & Jocqué (2002), on the other hand, consider *Phruronellus* valid and not closely related to *Scotinella*

11(6) RTA of male palp with 4 sharp prongs (Fig. 20.31); epigynum with median projection on the posterior border (Fig. 20.32) ................................................ ***Xeropigo***
  **Div.** 1 species: *Xeropigo tridentiger* (O. PICKARD-CAMBRIDGE 1869d) — **Dist.** FL to Brazil — **Ref.** Bonaldo 2000

— RTA of male palp with fewer than 4 sharp prongs; epigynum without median projection .................................... **12**

12(11) Male palp with duct spiral restricted to the basal third of bulb (Fig. 20.33); epigynum with two small posterior copulatory openings or one large anterior opening, with the posterior border distinct (Fig. 20.34) .......... ***Creugas***
  **Div.** 2 species: *Creugas bajulus* (GERTSCH 1942) and *Creugas gulosus* THORELL 1878b — **Dist.** *Creugas bajulus* is a native species known from southern CA to Baja California, *Creugas gulosus* is cosmotropical and has been recorded from FL — **Ref.** Bonaldo 2000

— Male palp with duct spiral occupying the basal half of bulb (Figs. 20.35, 20.37); epigynum with one small copulatory opening or, when large, with anterior border distinct (Figs. 20.36, 20.38) ........................................................ **13**

13(12) Abdomen gray to black, with markings; male palp with median tegular process (Fig. 20.35); epigynum with inconspicuous copulatory opening covered by a fold in the epigynal plate (Fig. 20.36) ........................ ***Falconina***
  **Div.** 1 species: *Falconina gracilis* (KEYSERLING 1891) — **Dist.** TX-FL — **Ref.** Bonaldo 2000 — **Note** this South American species has been introduced to the Gulf states and seems to be associated with the imported fire ant, *Solenopsis invicta* in TX (J. Cokendolpher, pers. com.)

— Abdomen cream-colored, without markings; male palp with subapical tegular process (Fig. 20.37); epigynum with conspicuous copulatory opening (Fig. 20.38) ............
  ........................................................... ***Septentrinna***
  **Div.** 2 species: *Septentrinna bicalcarata* (SIMON 1896b) and *Septentrinna steckleri* (GERTSCH 1936a); — **Dist.** CA-TX — **Ref.** Bonaldo 2000

# Chapter 21
# CTENIDAE
4 genera, 8 species

Darrell Ubick
Diana Silva Dávila

## Similar families —
Lycosidae (p. 164), Pisauridae (p. 199), Tengellidae (p. 230), Zoridae (p. 256), Zorocratidae (p. 258), Zoropsidae (p. 259).

## Diagnosis —
Ctenids can be separated from all other spiders by the arrangement of eyes in three rows, with the small ALE positioned close to the PLE (Figs. 21.1-21.2). In addition, these 2-clawed lycosoid spiders have both cheliceral margins toothed, deeply notched trochanters, two to three rows of tarsal trichobothria, metatarsal-tarsal scopula, and claw tufts.

## Characters —
**body size:** 6-30 mm.
**color:** mottled brown to gray.
**carapace:** pyriform, with deep longitudinal fovea.
**sternum:** oval, posteriorly pointed; precoxal triangles absent.
**eyes:** eight in three rows; ALE close to PLE.
**chelicerae:** stout, geniculate, with toothed margins.
**mouthparts:** endites longer than wide; labium often longer than wide, occasionally as long as wide.
**legs:** generally stout, prograde; all trochanters deeply notched; anterior tibiae often with 5 pairs of ventral spines, sometimes with six or more ventral pairs; tarsi with 2 claws, light to dense scopula, and dense claw tufts.
**abdomen:** mottled grayish, typically with median pale stripe.
**spinnerets:** anterior spinnerets conical and contiguous.
**respiratory system:** one pair of book lungs; tracheal spiracle slightly anterior to spinnerets.
**genitalia:** entelegyne; *female* epigynum lightly to strongly sclerotized, with swollen median lobe (flat in *Anahita*); *male* palpus often with RTA, bulb with tegulum and subtegulum forming interlocking lobes, and median apophysis hook-like or spatulate.

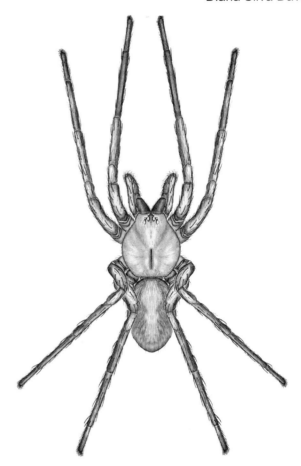

**Fig. 21.1** *Ctenus exlinae* Peck 1981

## Distribution —
Ctenids are primarily tropical spiders with few species ranging into the Nearctic. Native species occur in southeastern USA to Texas

## Natural history —
Ctenids are large nocturnal wandering hunters. Although some tropical species are arboreal and hunt on vegetation, all Nearctic species are strictly terrestrial. They may be collected at night using headlamps, under rocks and logs, and occasionally in caves. Egg sacs are either deposited on the substrate or carried with the chelicerae (Dippenaar-Schoeman & Jocqué 1997). Occasionally, some species of *Acanthoctenus*, *Cupiennius* Simon 1891a and *Phoneutria* Perty 1833 are introduced with tropical fruits.

## Taxonomic history and notes —
The family Ctenidae, as created by Keyserling (1877a), included both cribellate and ecribellate spiders and was considered closely related to Lycosidae and Oxyopidae. Simon (1897a), on the other hand, moved the cribellate *Acanthoctenus* to Zoropsidae in his distantly related group, Cribellatae, and placed the ecribellate ctenids as a subfamily of Clubionidae. As Lehtinen (1967) observed, many subsequent workers noted their general similarity to the lycosoids but, because they have only two claws, generally included them in a separate group, the Dionycha. A recent hypothesis (Silva Dávila 2003) indicates ctenids belong to a group of their own. Within the Ctenoidea, they are most closely related to a clade formed by Miturgidae *sensu stricto* and Zoridae *sensu stricto*. However, the limits of this family are still controversial, especially regarding the position of *Acanthoctenus*.

## Genera —
ACANTHOCTENINAE
*Acanthoctenus* Keyserling 1877a
HIGHER CTENIDS
*Anahita* Karsch 1879g, "*Ctenus*" Walckenaer 1805, "*Leptoctenus*" L. Koch 1878a

## Ctenidae

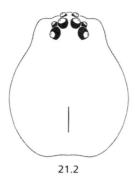

21.2

**Key to genera —**
North America North of Mexico

1   Cribellum and calamistrum present .... ***Acanthoctenus***
    **Div.** 2 species: *Acanthoctenus spiniger* KEYSERLING 1877a, *Acanthoctenus spinipes* KEYSERLING 1877a [Figs. 21.3-21.4] — **Dist.** CA, NY, WA — **Refs.** Lehtinen 1967, Silva Dávila 2003 — **Note** these Neotropical species have apparently become established in some urban areas

—   Cribellum and calamistrum absent ................................ 2

2(1) Small to medium sized spiders (6-9 mm); abdomen and legs yellow gray marked with numerous black spots; RTA absent (Fig. 21.5), instead with a retrodorsal modified spine; epigynum as in Fig. 21.6 ......................... ***Anahita***
    **Div.** 1 species: *Anahita punctulata* (HENTZ 1844) — **Dist.** se US to e TX — **Refs.** Peck 1981, Silva Dávila 2003

—   Medium to large sized spiders (9-30 mm); abdomen and legs lacking black dots; RTA present; epigynum otherwise
    .................................................................................... 3

3(2) Cheliceral retromargin with 3 teeth and often 1-2 denticles; RTA medial to basal (Fig. 21.7); epigynum as in Fig. 21.8 .................................................................. ***"Leptoctenus"***
    **Div.** 1 species: *Leptoctenus byrrhus* SIMON 1888b — **Dist.** s TX — **Ref.** Peck 1981 — **Note** this species and its relatives in northern MEX are misplaced in *Leptoctenus*, which is a genus apparently restricted to northeastern Australia (Raven, pers. comm).

—   Cheliceral retromargin with 4-5 teeth and 1-3 denticles; RTA apical (Fig. 21.9); epigynum as in Fig. 21.10 .............
    ........................................................................ ***"Ctenus"***
    **Div.** 4 species — **Dist.** se US to sw TX — **Refs.** Peck 1981, Sissom *et al.* 1999 — **Note** these species are not congeneric with the neotype of *Ctenus* (Brescovit pers. comm.)

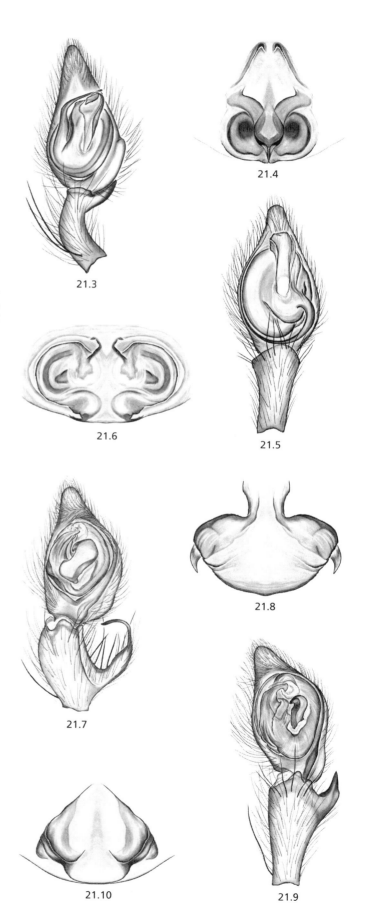

## Chapter 22
# CYBAEIDAE
8 genera, ~80 species

Robert G. Bennett

### Similar families —
Cybaeids are grouped with dictynids and hahniids (and other families not found in North America) in Dictynoidea. Cybaeids are similar to Agelenidae (p. 56), ecribellate Dictynidae (p. 95), Coelotinae (Amaurobiidae) (p. 59), and Cryphoecinae (Hahniidae) (p. 111) but are distinguished from them by spinneret characters. Because of their distinctive ventral tibial macrosetae and relatively small size, *Cybaeota*, *Cybaeina*, and new genera #1 and #2 may also be misidentified as 2-clawed Zoridae (p. 256) or Phrurolithinae (Corinnidae) (p. 78).

### Diagnosis —
ALS contiguous, slightly longer than PLS (Fig. 22.7); distal segment of PLS very short; tarsal trichobothria in a single dorsal row (Fig. 22.6); distal retrolateral patellar apophysis with peg setae present in males (PA in Figs. 22.9, 22.19-22.23, 22.45-22.47) except *Cybaeota*, *Cybaeozyga*, and new genus #3.

### Characters —
**body size:** small to moderately large, 1.4-14.0 mm.
**color:** carapace pale yellow to dark reddish brown, usually unmarked; abdomen pale and unmarked to dark gray with pale dorsal chevrons.
**cephalothorax:** generally pyriform, nearly glabrous, darkened around eyes; fovea longitudinal.
**sternum:** slightly longer than wide, truncate anteriorly, bluntly acuminate posteriorly and extending between slightly separated hind coxae.
**eyes:** usually eight in two transverse rows with AME slightly to very reduced and other eyes subequal (Figs. 22.2-22.4); AME slightly enlarged in *Cybaeus signifer* SIMON 1886d (Fig. 22.5) or absent in some *Cybaeota* and *Cybaeozyga*; all eyes very reduced or absent in some troglobitic *Cybaeus* and *Cybaeozyga*.
**chelicerae:** slightly to moderately geniculate with boss; margins toothed.
**mouthparts:** labium quadrate, wider than long; endites slightly longer than wide, slightly convergent.
**legs:** tarsi 3-clawed, lacking claw tufts and with 2-8 trichobothria in single dorsal row increasing in length distally (Fig. 22.6); macrosetae numerous and conspicuous, 2-5 complete linear pairs ventrally on tibia 1 (Figs. 22.13-22.14); trochanteral notches absent; integumentary Emerit's glands present in *Cybaeota*, especially dorsally on patellae and tibiae (EG in Figs. 22.15-22.16).
**abdomen:** ovoid, pale to dark gray, concolorous or with variable pale chevron-like markings dorsally.
**spinnerets:** (Fig. 22.7) 3 longitudinally arranged pairs, ALS contiguous and largest, PLS with very short terminal segment (appearing uni-segmented) and shorter and narrower than ALS; colulus vestigial, marked by 1 (*Cybaeota*, *Cybaeozyga*) to 10 (*Cybaeus*) pairs of setae.
**respiratory system:** paired anterior book lungs; spiracle small and just anterior of colulus setae; tracheae with two branches which usually extend into cephalothorax.
**genitalia:** entelegyne; *female* epigynum simple, lacking modifications; atrium inconspicuous, varying from

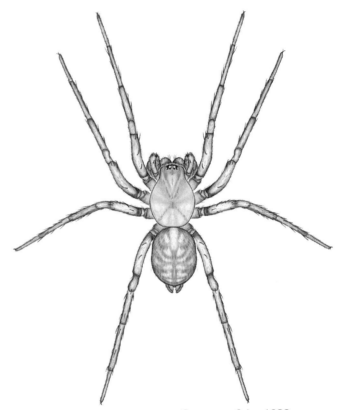

**Fig. 22.1** *Cybaeus eutypus* CHAMBERLIN & IVIE 1932

having only simple copulatory opening(s) (AT in Figs. 22.10, 22.18, 22.27-22.28, 22.39-22.40, 22.61, 22.64) to large concavities, single or paired (AT in Figs. 22.59, 22.62-22.63, 22.65-22.67); copulatory openings simple, single or paired and leading into variously developed copulatory ducts; spermathecae paired; head of each spermatheca marked by grouping of simple pores (PS in Figs. 22.27, 22.68, 22.75-22.77) and differentiated from copulatory ducts or not; single complex pore present medially or basally on each spermatheca (PC in Figs. 22.68, 22.75-22.77); spermathecal bases usually bulbous (BS in Figs. 22.18, 22.28, 22.39, 22.68-22.77); except undifferentiated from rest of spermathecal ducting in *Cybaeina* (BS in Fig. 22.27), and in new genera #1 (BS in Figs. 22.11, 22.40) and #4 (BS in Fig. 22.12); *male* variably developed distal retrolateral patellar apophysis with one to many short, stout, peg-like setae (PA in Figs. 22.9, 22.19-22.23, 22.45-22.47) present (except absent in *Cybaeota* and *Cybaeozyga*, vestigial or absent in new genus #3); longitudinally carinate RTA present and variably developed (Figs. 22.8 -22.9, 22.23-22.26, 22.48, 22.54-22.56); cymbium unmodified, containing relatively simple genital bulb; tegulum unmodified except for presence of short, bevelled, sclerotized ridge anteriorly (SR in Figs. 22.43, 22.54, 22.56) and, in new genus #3, a small posterior retrolateral protrusion (TP in Fig. 22.49); simple spiral embolus (E in Figs. 22.42-

22.44, 22.52, 22.54) coiled clockwise (left palpus, ventral view) and usually thin and more-or-less elongate (short and thick in *Cybaeota* (E in Fig. 22.55) and some *Cybaeus* (E in Fig. 22.56), occasionally modified apically in some *Cybaeozyga* (E in Fig. 22.52) and some new genus #3; median apophysis lacking; functional conductor a well developed and often complex tegular apophysis (TA in Figs. 22.42-22.44, 22.52, 22.54-22.56) with origin at base of embolus and supporting tip of embolus, form of tegular apophysis is usually species specific.

## Distribution —

Cybaeidae is primarily a Holarctic family with the great majority of species occurring in Japan and western North America. In North America, 1 species of *Cybaeota* and 3 of *Cybaeus* occur in forested uplands of the east; all others occur west of the western continental divide from far out on the Aleutian Islands (2 species) to the Mexico/USA border. *Cybaeus* is a Holarctic genus; the other genera considered here are apparent Nearctic endemics.

## Natural history —

Little has been published on the natural history of cybaeids. Typically they are forest floor litter inhabitants and are among the most commonly encountered spiders in the forests of western North America. However, recorded habitats range from coastal and riparian beach litter to alpine meadows (Bennett 1991). At least one species of *Cybaeota*, usually considered a rare spider, can be very abundant in moss on deciduous tree trunks within western coniferous forests (Bennett 1988). As well, a number of species live underground and several of these are true troglobites. One cybaeid, the well-known European *Argyroneta aquatica* (CLERCK 1757), is the world's only known entirely aquatic spider. Interestingly, in eastern North America where only 4 relatively uncommon cybaeids occur, the cybaeid "niche" is largely occupied by ecribellate coelotine amaurobiids, a grouping entirely absent from western North America.

Most cybaeids build small webs best described as a type of funnelweb and which probably function primarily as retreats. Ubick (pers. comm.) has described a *Cybaeus* funnelweb as a lacy, loosely woven squarish basket with an opening at each corner and a pair of long silk threads (signal lines?) extending from each opening. Some Japanese cave-dwelling cybaeids build specialized retreat constructs (Komatsu 1961a, 1968b) strikingly similar to those of the southern Appalachian dictynid *Cicurina bryantae* EXLINE 1936a. At least one species of *Cybaeus* is active on the bark of trees at night: *Cybaeus signifer* is commonly found (presumably hunting) on the heavily furrowed bark of old Douglas-fir trees on southern Vancouver Island (British Columbia) at night in cool, damp weather. It is interesting to note that *Cybaeus signifer* is the only Nearctic cybaeid with well developed AME – perhaps this may account for its activity. Ubick (pers. comm.) has observed similar behaviour in males and females of a large species of *Cybaeus* (perhaps also *Cybaeus signifer*) in the San Francisco Bay area (California).

The cybaeid life cycle is usually annual with newly matured adults appearing and mating in the late spring through to early fall, depending upon locality and species. Adult females may live for more than one year with some aged "matriarchs" still present when the next generation is maturing and mating (Bennett 1991, in press). Males do not survive long after maturing. Egg sacs usually are constructed shortly after mating with progeny hatching early the following spring.

## Taxonomic history and notes —

The world-wide cybaeid fauna currently comprises 10 genera with 139 catalogued species, fully 2/3 of which are described in the Holarctic genus *Cybaeus* (Platnick 2005). The Nearctic cybaeid fauna consists of 8 genera and about 80 species (these figures include 4 unpublished Nearctic new genera and 42 new species names contained in Bennett's (1991) dissertation (see below).

As with many of its "RTA clade" relatives, Cybaeidae has been difficult to diagnose and classify. The family still lacks a cohesive diagnosis and the monophyly of *Cybaeus* and several of the Palaearctic genera is questionable. Historically, *Cybaeus* and a variable number of other genera were classified as agelenids in Banks' (1892d) subfamily Cybaeinae (e.g., Simon 1898a, Roewer 1954a, Bonnet 1956a). Lehtinen (1967) relimited and reduced the subfamily to include only *Cybaeus* and its apparent closest relatives and transferred it from Agelenidae (Amaurobioidea) to Dictynidae (Dictynoidea). Subsequently, family status for the group within Dictynoidea was proposed (Forster 1970b) and various of the non-Nearctic genera covered in older versions of Cybaeinae (including *Argyroneta*) trickled back in (Platnick 2005).

*Cybaeus* was first proposed by L. Koch (1868) for two European species (one new and the other described previously in *Amaurobius*). Simon (1886d) and Banks (1892a) described the first four Nearctic species. A few more species were added by Barrows (1919b), Chamberlin (1919b), and Bishop & Crosby (1926) before Chamberlin & Ivie (1932, 1937, 1942a) described 22 new Nearctic species and one subspecies. Other workers (Exline 1935a, 1936c; Fox 1937c, Schenkel 1950b, Roth 1952b) added a few more species, Crawford (1988) elevated the subspecies to species status, and Roth (1956b) and Roth & Brown (1986) proposed some synonyms to bring the number of catalogued Nearctic species to 35. In his revision of the Nearctic species of *Cybaeus*, Bennett (1991) described 33 new species, transferred 1 species to new genus #3, proposed 10 new synonyms and rejected 4 older ones for a total of 61 species placed in 8 apparently monophyletic species groups. This revision remains largely unpublished although Bennett (in press) recently provided justification for the synonymy changes.

*Cybaeota* was erected by Chamberlin & Ivie (1933b) for the eastern North American species *Liocranum calcaratum* EMERTON 1911. Subsequently they described six more species from western North America (Chamberlin & Ivie 1937). Bennett (1988, 1989) revised the genus, proposing synonyms to reduce the species count to four, and described *Cybaeota*'s integumentary Emerit's glands. Also known in Leptonetidae, Telemidae and some Salticidae, Emerit's glands appear to be unique to *Cybaeota* among cybaeids.

*Cybaeina* was erected for *Cybaeus minutus* BANKS 1906b by Chamberlin & Ivie (1932). In later years three more species were described by Chamberlin & Ivie (1937, 1942a) and Roth (1952c). Bennett (1991) revised the genus, breaking it into two new genera (new genera # 1 & #2) and a reduced *Cybaeina* and described new species in each for total species counts of 5, 2, and 3 respectively. Bennett (1991) also described new genus #4 for a bizarre *Cybaeina*-like new species. As with his revision of *Cybaeus*, this work remains unpublished.

*Cybaeozyga* was described by Chamberlin & Ivie (1937) for a single new species and remains monotypic. However, several undescribed troglobitic species apparently exist in museum collections (Roth & Brame 1972).

Bennett (1991, 1992) described a new type of complex "dictynoid" pore in cybaeid spermathecae and discussed its morphology and distribution in dictynoid and other araneomorph spiders. Bennett (in press) described the ontogeny of female genitalic characters and other types of morphological variation in *Cybaeus*.

## Genera —

CYBAEINAE
*Cybaeina* CHAMBERLIN & IVIE 1932, *Cybaeota* CHAMBERLIN & IVIE 1933b, *Cybaeozyga* CHAMBERLIN & IVIE 1937, *Cybaeus* L. KOCH 1868, plus 4 unpublished manuscript names in Bennett (1991)

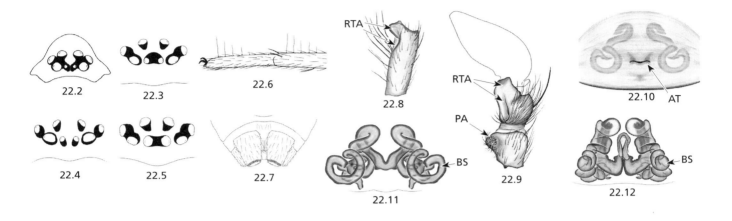

## Key to genera —
North America North of Mexico

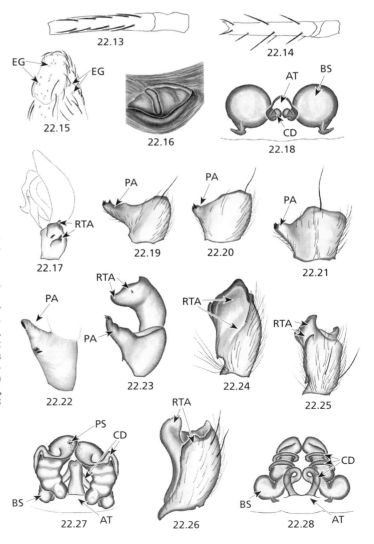

1   Four or five complete pairs of ventral tibia I macrosetae arranged in a smooth linear pattern (Fig. 22.13). w NA (one rarely collected *Cybaeota* species occurs in ne NA) .................................................................................. 2

—   Two or three complete pairs of macrosetae (Fig. 22.14). Widespread (e NA species of *Cybaeus* may have four or five pairs but not arranged in a smooth linear pattern) .... .................................................................................. 5

2(1)  Retromargin of cheliceral fang furrow with 2-5 small teeth; Emerit's glands present, especially dorsally on tibiae and patellae (best observed with SEM) (Figs. 22.15-22.16); retrolateral patellar apophysis and peg setae absent; RTA bipartite and with a single bristly seta separating distal medial components (Figs. 22.17, 22.55); internal female genitalia simple (Fig. 22.18) -connecting ducts (CD) very short, spermathecae bases (BS) large, circular; small spiders, usually <2.5 mm ..................... **Cybaeota**
      Div. 4 species — Dist. MB to NF and ne USA, sw AK to ID, UT & s CA
      — Ref. Bennett 1988

—   Retromargin of cheliceral fang furrow with >5 (usually 10-15) variably sized teeth; Emerit's glands absent; retrolateral patellar apophysis and peg setae present (PA in Figs. 22.19-22.23); RTA usually bipartite (Figs. 22.23-22.26) but not appearing as in *Cybaeota*; internal female genitalia more complex, a variable mass of coiled ducts with spermathecal bases (BS) undifferentiated (Figs. 22.11, 22.40) or somewhat enlarged (Figs. 22.28, 22.39); larger spiders, usually >2.8 mm .............................................. 3

3(2) Retrolateral patellar apophysis variable, very short (PA in Fig. 22.29) to about as long as width of patella (and retrolaterally directed) (PA in Fig. 22.30); medial component of RTA blade-like and terminating in a sharp point (Fig. 22.31) or ledge-like and terminating in a single macroseta (Fig. 22.32); copulatory ducts in cleared vulvae very lightly sclerotized, elongate, and encircling spermathecal stalks (CD in Figs. 22.33-22.34) ................................. **Cybaeina**
   **Div.** 3 species (2 described: *Cybaeina minuta* (BANKS 1906b), *Cybaeina confusa* CHAMBERLIN & IVIE 1942a, 1 unpublished ms description) — **Dist.** coastal BC to sw OR — **Refs**. Bennett 1991, Chamberlin & Ivie 1932 (*Cybaeina minuta*), Chamberlin & Ivie 1942a (*Cybaeina confusa*)

— Retrolateral patellar apophysis also variable, short (PA in Fig. 22.35) to up to as long as width of patella (and antero-dorsally directed) (PA in Figs. 22.36-22.37); medial component of RTA otherwise, may be an elongate ledge (Fig. 22.38) or reduced to a single small spine (Fig. 22.37); copulatory ducts in cleared vulvae very short, more heavily sclerotized, and not encircling spermathecal stalks (CD in Figs. 22.39-22.40) ........................................................ **4**

4(3) Patellar apophysis short, not as long as width of patella (PA in Fig. 22.35) or, if about as long as width of patella, then with a basal cluster of peg setae (arrow in Fig. 22.36) or a distinct prominence on leading edge (arrow in Fig. 22.41); distal end of tegular apophysis strongly developed (D in Figs. 22.42-22.43); spermathecal ducting long and convoluted but undifferentiated and of relatively constant diameter throughout (Figs. 22.11, 22.40) ..........................
........................................................... **New genus #1**
   **Div.** 4 species (1 described: *Cybaeina sequoia* ROTH 1952b, 3 unpublished ms descriptions) — **Dist.** sw OR, n CA — **Refs**. Bennett 1991, Roth 1952b

— Patellar apophysis elongate, as long as width of patella and lacking basal cluster of peg setae or distinct prominence on leading edge (PA Fig. 22.37); distal end of tegular apophysis reduced (D in Fig. 22.44); spermathecal ducts short and differentiated with spermathecal bases (BS) enlarged (Fig. 22.39) ............................. **New genus #2**
   **Div.** 2 species (1 described: *Cybaeina xantha* CHAMBERLIN & IVIE 1937 + 1 unpublished ms description) — **Dist.** extreme sw OR — **Refs**. Bennett 1991, Chamberlin & Ivie 1937

5(1) Males .................................................................. 6
— Females .............................................................. 9

6(5) Retrolateral patellar apophysis reduced or absent, no peg setae ................................................................. 7
— Retrolateral patellar apophysis and peg setae present (Figs. 22.45-22.47) ..................................................... 8

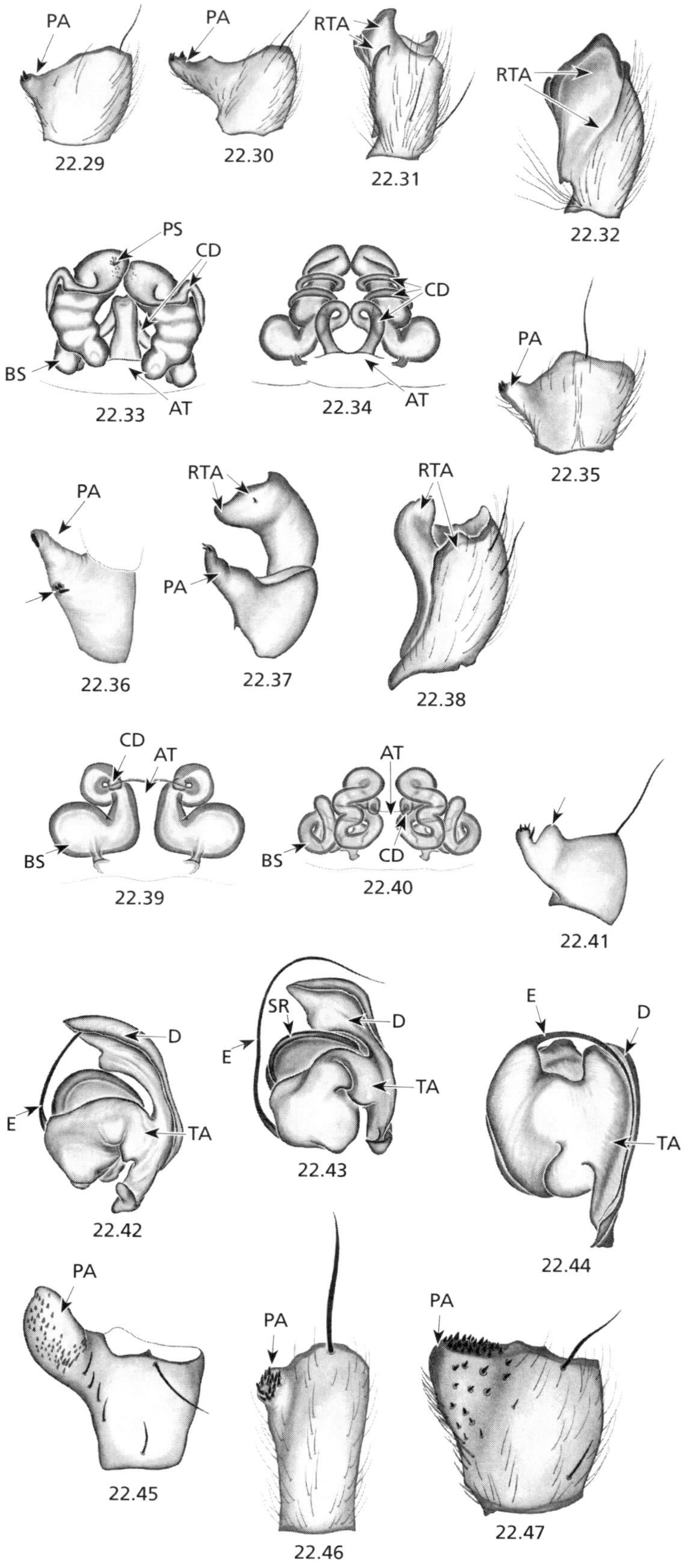

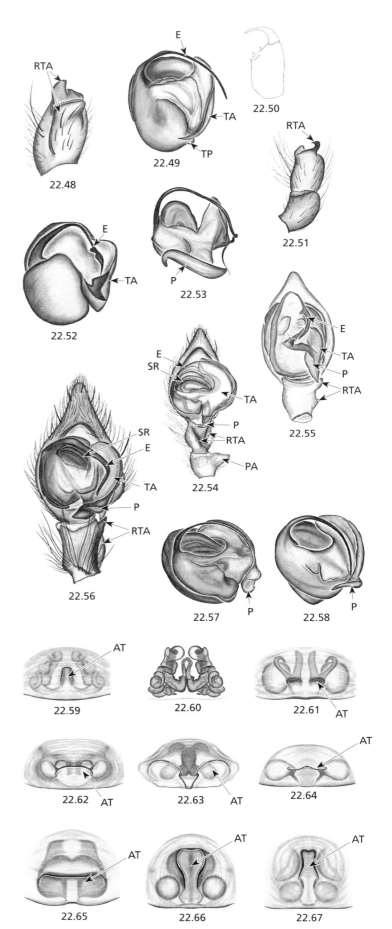

**7(6)** Promargin of cheliceral fang furrow with three relatively closely spaced teeth, retromargin with teeth and denticles. Retrolateral patellar apophysis reduced or absent; RTA with distal and medial components (Fig. 22.48); tegulum with proximal knob-like protrusion (TP in Fig. 22.49) .... ................................................................ **New genus #3**
  Div. 3 species (1 described: *Cybaeus perditus* CHAMBERLIN & IVIE 1932 + 2 unpublished ms descriptions) — **Dist.** nw WA, extreme sw OR and adjacent nw CA — **Refs.** Bennett 1991, Chamberlin & Ivie 1932

— Promargin of cheliceral fang furrow with three widely spaced teeth, retromargin with denticles only (Fig. 22.50). Patellar apophysis lacking; RTA with distal components only (Fig. 22.51); tegulum lacking knob-like protrusion (Fig. 22.52) .................................. ***Cybaeozyga***
  Div. 1 species: *Cybaeozyga heterops* CHAMBERLIN & IVIE 1937 — **Dist.** w OR & nw CA — **Refs.** Chamberlin & Ivie 1937, Roth & Brame 1972 — **Note** nw CA specimens are eyeless and apparently represent undescribed species

**8(6)** Tegular apophysis massively developed with proximal end flattened and plate-like (P in Fig. 22.53) ............................. ................................................................ **New genus #4**
  Div. 1 species (unpublished ms description) — **Dist.** nw CA (Mendocino Co.) — **Ref.** Bennett 1991

— Tegular apophysis well developed but never with flattened, plate-like proximal end (P in Figs. 22.54, 22.56-22.58) ..... ................................................................ ***Cybaeus***
  Div. 61 species (28 described, 33 unpublished ms descriptions) — **Dist.** w of western continental divide from western-most Aleutian Is. to s CA; e USA from OH and s NY to n GA and n AL — **Refs.** Bennett 1991, in press, Chamberlin & Ivie 1932, Roth 1952b

**9(5)** Epigynum with single, medially located, inverted U-shaped atrium (AT in Fig. 22.59); spermathecal ducting complex, convoluted and undifferentiated (Fig. 22.60); complex "dictynoid" pores apparently absent ................... ................................................................ **New genus #4**
  Div. 1 species (unpublished ms description) — **Dist.** nw CA (Mendocino Co.) — **Ref.** Bennett 1991

— Epignal atrium single or paired, variably shaped and located (AT in Figs. 22.61-22.67); spermathecal ducting variable but differentiated with conspicuously rounded or elongated spermathecal bases (BS in Figs. 22.68-22.75); complex spermathecal pores always present and usually conspicuous (PC in Figs. 22.68, 22.75-22.77) .............. **10**

**10(9)** Epigynum with posteriorly located paired atria (AT in Fig. 22.78); copulatory ducts elongate and well-defined (CD in Fig. 22.75); complex spermathecal pores everted (PC in Fig. 22.75) .................................. **New genus #3**
    **Div.** 3 species (1 described: *Cybaeus perditus* Chamberlin & Ivie 1932, 2 unpublished ms descriptions) — **Dist.** nw WA, extreme sw OR and adjacent nw CA — **Refs.** Bennett 1991, Chamberlin & Ivie 1932

— Without the above combination of characters ............ **11**

**11(9)** Epigynum with medially located paired atria (AT in Fig. 22.79); vulva very broad, usually at least twice as wide as long (Fig. 22.68) with copulatory ducts very short and inconspicuous and spermathecal components undifferentiated (CD, BS in Fig. 22.68) ..................... ***Cybaeozyga***
    **Div.** 1 species: *Cybaeozyga heterops* Chamberlin & Ivie 1937 — **Dist.** w OR & nw CA — **Refs.** Chamberlin & Ivie 1937, Roth & Brame 1972 — **Note** nw CA specimens are eyeless and apparently represent undescribed species

— Epignyal atrium single or paired, variably shaped and located (AT in Figs. 22.65-22.67, 22.80-22.81); vulva usually nearly as long as wide (Figs. 22.69-22.74) with copulatory ducts variably shaped but always conspicuous (CD in Figs. 22.69-22.74) and spermathecal components variable but always differentiated with variously shaped and arranged ducts and conspicuously rounded or elongated spermathecal bases (BS in Figs. 22.69-22.74) .......... ***Cybaeus***
    **Div.** 61 species (28 described, 33 unpublished ms descriptions) — **Dist.** w of western continental divide from western-most Aleutian Is. to s CA; e USA from OH and s NY to n GA and n AL — **Refs.** Bennett 1991, in press, Chamberlin & Ivie 1932, Roth 1952b

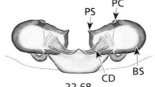
22.68

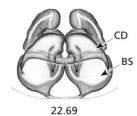
22.69

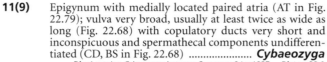
22.70     22.71

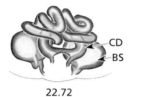
22.72

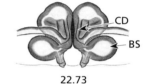
22.73

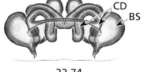
22.74

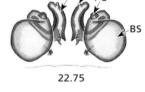
22.75

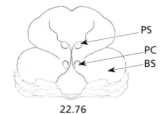
22.76

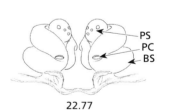
22.77

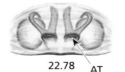
22.78

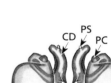
22.79

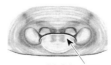
22.80

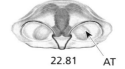
22.81

| | |
|---|---|
| AT | epigynal atrium |
| BS | base of spermatheca |
| CD | copulatory duct |
| D | distal end of TA |
| EG | Emerit's gland |
| P | proximal end of TA |
| PA | retrolateral patellar apophysis |
| PC | complex spermathecal pore |
| PS | simple spermathecal pore |
| RTA | retrolateral tibial apophysis |
| SR | sclerotized tegular ridge |
| TA | tegular apophysis (conductor) |
| TP | tegular protrusion |

# Chapter 23
# DEINOPIDAE
1 genus, 1 species

Jonathan A. Coddington

## Common name —
Ogre-faced spiders, net-casting spiders.

## Similar families —
None, although some Tetragnathidae (p. 232) also adopt similar resting postures along twigs.

## Diagnosis —
All instars of this cribellate, orbicularian family can be distinguished by the extremely large posterior median eyes (Fig. 23.2) or by the web architecture. In the field, *Deinopis* spin highly modified orbwebs placed slightly above or to the side of the substrate, and whose catching area is much smaller than the spider.

## Characters —
**body size:** males 10-14 mm; females 12-17 mm.
**color:** carapace tan, sparse black lateral mottling, abdomen with broad light median dorsal band and tan cardiac mark, faint posterior folium.
**carapace:** flat, less than half the length of the abdomen (Fig. 23.1).
**sternum:** broad white central region, gray borders.
**eyes:** eight, PME massive.
**chelicerae:** 6 pro-and retrolateral teeth.
**legs:** unusually long and thin, tarsi with three claws.
**abdomen:** fusiform, twice as long as carapace.
**spinnerets:** six spinnerets, entire cribellum in front of anterior spinnerets.
**respiratory system:** paired book lungs and median tracheal spiracle opening just anterior to spinnerets.
**genitalia:** entelegyne; *female* have an anchor-shaped epigynum with lateral copulatory slits (Fig. 23.3) leading to spiraled ducts; *male* have a simple, round tegulum with a centrally placed apophysis, around which the flat, thin, blade-like embolus tightly spirals (Fig. 23.4).

## Distribution —
Deinopidae is a pan-tropical family. *Deinopis spinosa* Marx 1889b also occurs in Jamaica but in North America is limited to extreme southeastern states.

## Natural history —
*Deinopis spinosa* spin modified orbwebs whose architecture is synapomorphic for the family (Figs. 23.5, 23.6). They are strictly nocturnal, and attack both flying and ambulatory prey. During the day the spiders are stick mimics and keep their legs stretched straight out in front and to the rear of their bodies while resting on twigs. Although deinopids are often considered among the rarest of spiders, *Deinopis spinosa* can be quite common where it occurs. Webs are placed at almost any height in the vegetation and have three radii on each side of the bilaterally symmetric web. The cribellate catching area is small relative to the span of the radii, which connect to one frame line on each side of the web, themselves attached to substrate. After sticky spiral construction the hub of the web is completely bitten out.

The spiders have two behaviorally stereotyped hunting styles, termed "forward" and "backward" strikes (Coddington & Sobrevila 1987). To hunt, the spider attaches its dragline to substrate above the midline of the web, and hangs head downward with its first four legs grasping the corners of the catching area. The rear legs reel in the dragline so that, if released, the web and spider swing forward and downward towards the substrate (usually a trunk or twig). During a forward strike at ambulatory prey the spider extends and spreads its first four legs so that the web is distorted into a large, planar sheet, with which they lunge at and enfold the prey. To elicit an attack, the prey must usually walk directly beneath or in front of the web. The forward motion of spider and web is due to the release of dragline slack through the fourth leg claws. For ambulatory prey, at least, *Deinopis spinosa* is primarily a sight hunter.

Their very large posterior median eyes are extremely sensitive with a very short focal length (the equivalent of a fish-eye lens) to better detect prey. In a closely related species the photoreceptors of the PME are destroyed by bright light (e.g., dawn) but synthesized anew at dusk (Blest & Land 1977, Blest et al. 1978a, b). Tapeta are absent.

During a "backward" strike at flying prey, the animal extends and stretches the web as in forward strike but instead pivots to sweep the air space in its rear. Each successful strike (and often unsuccessful strikes) involves the

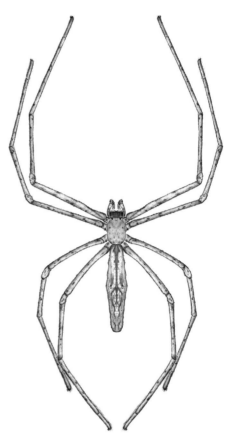

**Fig. 23.1** *Deinopis spinosa* Marx 1889b

destruction of the web, so to hunt again the animal must rebuild the web. Because their attack behavior and webs are so unusual, deinopid biology is unusually well-studied, of which the more recent literature is cited above, and see Laughlin *et al.* (1980), Peters (1992), Eberhard & Pereira (1993), Getty & Coyle (1996), and Opell (1997). Theuer (1954) is an unpublished master's thesis on the biology of *Deinopis spinosa* itself. Despite the odd form of the web, the actual motor patterns used during construction are fairly typical of orbweaving spiders (Coddington 1986c). Egg sacs are hard, brown spherical balls about the size of a pea. Tropical species bury their egg sacs in moist leaf litter, where presumably the hard casing rots in time to permit the spiderlings to escape (pers. obs.).

**Taxonomic history and notes —**

The first species of Deinopidae was described by MacLeay in 1839 from Cuba, who erected a new genus for this animal with such striking posterior median eyes. C.L. Koch (1850) raised the genus to family level. MacLeay did not explain his etymology, which led several subsequent authors to disagree with his orthography. They used the spelling "*Dinopis*" and considered the genus masculine (e.g., Marx 1889b, Bonnet 1956a), whereas the original spelling suggests it is feminine. The unjustified emendation went unnoticed until the mid-1980's. The family currently contains four genera: *Avella* O. Pickard-Cambridge 1877b, *Avellopsis* Purcell 1904, and *Menneus* Simon 1876f, in addition to *Deinopis*. None of the former have enlarged posterior median eyes, have not been revised, and may not be validly diagnosable. *Deinopis*, of course, is almost certainly monophyletic. Despite its rarity in the field, the family is known from Cretaceous amber (Penney 2003). Early authors usually assumed that Deinopidae and Uloboridae were closely related, which was confirmed by later phylogenetic analyses (Coddington 1990, Griswold *et al.* 1998).

**Genus —**

*Deinopis* MacLeay 1839

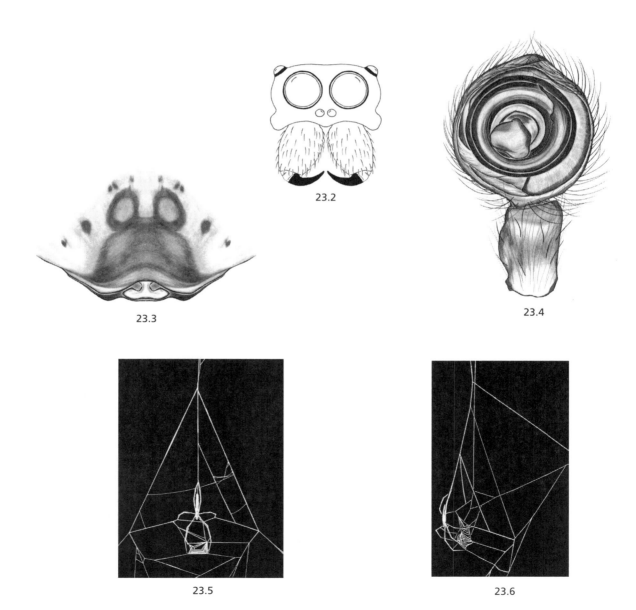

## Chapter 24
# DESIDAE
2 genera, 2 species

Darrell Ubick

### Similar families —
Amaurobiidae (p. 60), Dictynidae (e.g., *Saltonia*) (p. 94).

### Diagnosis —
Desidae is a heterogeneous family. The cribellate *Badumna* is distinguished from amaurobiids in having the AME distinctly larger than other eyes. The ecribellate *Paratheuma* is distinguished from other spiders by the combination of having a very broad colulus and enlarged, divergent chelicerae.

### Characters —
P = *Paratheuma*, B = *Badumna*

**body size:** 3-6mm (P) or 5-12mm (B).
**color:** orange to dark brown.
**carapace:** pyriform, cephalon > 1/2 carapace width; fovea logitudinal.
**sternum:** subtriangular, anteriorly truncate, posteriorly pointed.
**eyes:** eight eyes in two transverse rows; AME smaller (P) (Fig. 24.3) or larger (B) (Fig. 24.8) than others.
**chelicerae:** geniculate with boss; porrect and divergent (P) or vertical and more robust in female (B); promargin with 3-4 teeth, retromargin with 5-8 (P) or 2 (B) teeth.
**mouthparts:** labium about as long as wide, apically rounded, 2/3 length of endites; endites angular to ectally rounded.
**legs:** with spines weak (P) or strong (B); tarsi and metatarsi with dorsal trichobothrial row; tarsi with three claws.
**abdomen:** light to dark gray, with pale (P) or dark (B) markings.
**spinnerets:** cribellum present, bipartite (B) or absent with wide colulus (P); ALS contiguous (B) or widely separated (P).
**respiratory system:** anterior book lungs and posterior spiracle near spinnerets; with 4 tracheal trunks, usually branched.
**genitalia:** *entelegyne*; **female** epigynum lightly sclerotized with oval paired depressions and sclerotized ectal margins (P) or with central opening and a pair of lateral teeth (B); **male** palpus with RTA short and single (P) or multiple (B) prongs, bulb with embolus arched and received in large conductor (P) or sinuous and embedded in conductor (B).

### Distribution —
*Badumna* is known from coastal California; *Paratheuma* from Florida.

### Natural history —
*Badumna longiqua* is a native of Australia and was introduced to New Zealand and the Americas. In the Nearctic, the spider is known only from the coastal urban areas of California where it may be extremely abundant (pers. obs.). It is especially common on houses and associated structures, including shrubbery, where it constructs conspicuous cribellate webs. The web consists of a retreat with numerous sheets, radiating at various angles and made of a loose ladderlike mesh (Forster 1970b) (Fig. 24.2).

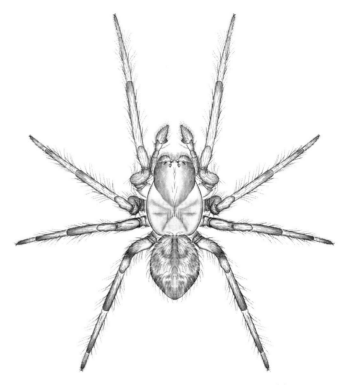

**Fig. 24.1** *Badumna longiqua* (C.L. Koch 1867b)

Spiders of the genus *Paratheuma* are wandering hunters and do not build webs. All species are from intertidal habitats. *Paratheuma insulana* (Banks 1902d) has been collected in coral rubble on beaches in the Florida Keys (Beatty & Berry 1988) and in oyster beds at low tide along the Gulf coast of Northern Florida (G.B. Edwards, pers. comm.). *Paratheuma interaesta* Roth & Brown 1975a, the only other Nearctic species (from Sonora, Mexico), lives along rocky shores and is associated with barnacles. The species is active nocturnally and builds retreats in abandoned barnacle shells (Roth & Brown 1975a).

### Taxonomic history and notes —
*Badumna longiqua* was first described from the Nearctic region by Gertsch, who believed it to be a native but distinct member of Amaurobiidae, and placed in a new genus and species, *Hesperauximus sternitzkii* Gertsch 1937. The species was subsequently recognized as being introduced and conspecific with *Ixeuticus martius* (Simon 1899b) (Marples 1959). Lehtinen (1967) transferred the species to *Badumna*, and placed *martius* as a synonym of *longiqua*. Although this synonymy was not accepted by many workers, for example Forster (1970b) and Leech (1972), it was eventually supported by Gray (1983a).

# Desidae

The species of *Paratheuma* have been placed in many different families. The genus was described by Bryant (1940) in the Gnaphosidae, and included only the type species from the Bermuda Islands, *Eutichurus insulanus* BANKS 1902d, which had been in Clubionidae. Many years later, Roth & Brown (1975a) described the genus *Corteza*, on the basis of a new species of intertidal spider from the Sea of Cortez, Mexico, and placed it into Desidae. Shortly thereafter, Platnick (1977c) moved *Paratheuma* to Desidae and showed it to be congeneric with *Corteza*. More recently, Beatty & Berry (1988, 1989) greatly expanded the size of the genus, indicating its greatest richness on south Pacific islands, and collected *Paratheuma insulana* from the Florida Keys.

### Genera —
*Badumna* THORELL 1890a, *Paratheuma* BRYANT 1940

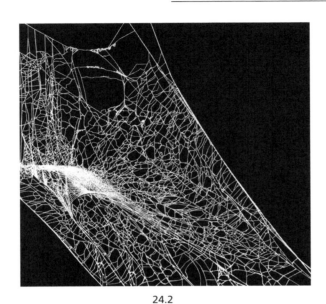

24.2

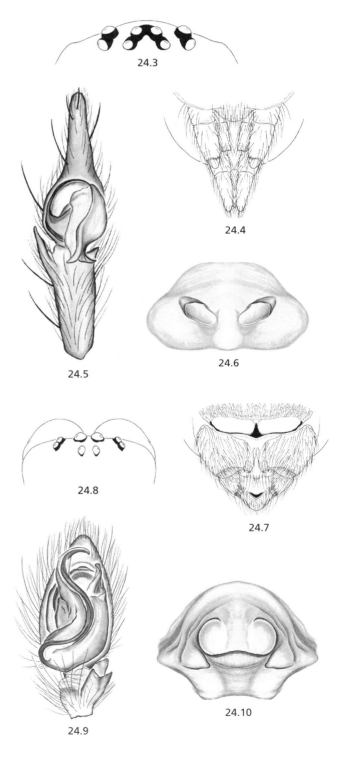

### Key to genera —
North America North of Mexico

1  Cribellum absent, but with broad colulus; ALS widely separated (Fig. 24.4); AME smaller than other eyes (Fig. 24.3); male palpus with arched embolus (Fig. 24.5); female epigynum with two depressions, lacking teeth (Fig. 24.6) .................................................. **Paratheuma**
   **Div.** 1 species: *Paratheuma insulana* (BANKS 1902d) — **Dist.** FL — **Refs.** Beatty & Berry, 1988 — **Note** a related species, *Paratheuma interaesta* (ROTH & BROWN 1975a), occurs in Sonora, Mexico.

—  Cribellum present, ALS contiguous (Fig. 24.7); AME larger than other eyes (Fig. 24.8); male palpus with sinuous embolus (Fig. 24.9); female epigynum with single depression and 2 lateral teeth (Fig. 24.10) ..... **Badumna**
   **Div.** 1 species: *Badumna longiqua* (L. KOCH 1867b) — **Dist.** CA — **Refs.** Leech 1972, Gray 1983a

## Chapter 25
# DICTYNIDAE
20 genera, ~ 290 species

Robert G. Bennett

### Similar families —
Dictynidae is grouped with Cybaeidae (p. 85) and Hahniidae (p. 112) (and other families not found in North America) in Dictynoidea. The cribellate dictynids (particularly the larger species) may be confused with cribellate Amaurobiidae (p. 60). Ecribellate dictynids may occasionally be misidentified as Agelenidae (p. 56) (with which they were formerly classified) but are more likely to be mistaken for Cybaeidae or, especially, Cryphoecinae (Hahniidae).

### Diagnosis—
Six or eight eyes in two transverse rows (Figs. 25.2, 25.31) or eyeless. Cribellate or ecribellate; ALS separated, sometimes widely; PLS as long as or longer than ALS, with distal segment of PLS $\leq 0.8$ length of proximal segment (Figs. 25.13-25.15). Male palpus lacking true median apophysis and with well developed tegular apophysis supporting embolus (TA, E in Figs. 25.25, 25.72). Female spermathecal ducting usually with pair of conspicuous complex pores (PC in Figs. 25.4, 25.23). Cribellate species small (most <4 mm, many <2mm), usually lacking leg macrosetae, 0-4 tarsal trichobothria (Fig. 25.32), calamistrum a single row of setae and nearly as long as metatarsus IV (Fig. 25.47), cribellum usually undivided (Fig. 25.15). Ecribellate species usually larger (most >4 mm), legs with conspicuous macrosetae, $\leq 3$ pairs of ventral tibial macrosetae, 3-6 tarsal trichobothria in single row increasing in length distally (Fig. 25.5).

### Characters —
**body size:** *cribellate* species small: 1.2-8 mm, most <4 mm; *ecribellate* species larger: 1.6-15 mm, most >4 mm.
**color:** various earth tones from dirty white to brown, reddish, or gray; carapace may be laterally banded; abdomen may be marked with chevrons posteriorly.
**carapace:** generally pyriform; fovea longitudinal
**sternum:** slightly longer than wide, truncate anteriorly, bluntly acuminate posteriorly and extending between slightly separated hind coxae.
**eyes:** usually eight in two transverse rows (Figs. 25.2-25.3, 25.33, 25.57, 25.59) with AME often reduced (Figs. 25.17, 25.34); some species lack AME (Figs. 25.16, 25.18, 25.31, 25.35); others (such as troglobites) lack eyes entirely.
**chelicerae:** slightly to moderately geniculate with boss, modified in males of *Dictyna* and close relatives (Figs. 25.33, 25.64); margins usually toothed (except retromargin in *Thallumetus*).
**mouthparts:** height of clypeus variable, narrow (Fig. 25.31) or wide (Figs. 25.33, 25.60) up to 4 times diameter of AME in some males of *Dictyna* and close relatives; labium quadrate, about as wide as long; endites slightly to moderately convergent.
**legs:** tarsi three clawed; *cribellate* species-macrosetae usually lacking, 0-4 tarsal trichobothria, (Fig. 25.32) calamistrum a single row of curved setae (usually nearly as long as metatarsus IV) (Fig. 25.47); *ecribellate* species-macrosetae conspicuous dorsally, laterally and/or ventrally, 3-6 tarsal trichobothria in single row and increasing in length distally (Fig. 25.5).

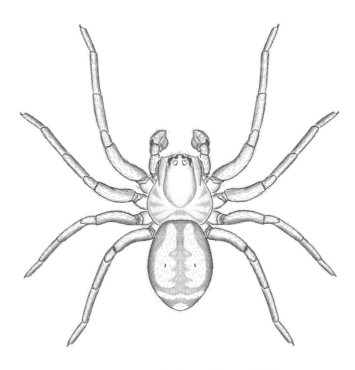

**Fig. 25.1** *Dictyna brevitarsus* Emerton 1915a

**abdomen:** ovoid, pale to dark gray, concolorous or with variable chevron-like markings dorsally.
**spinnerets:** ALS distinctly separated (Figs. 25.13, 25.15) widely in *Saltonia* (Fig. 25.14); PLS as long as or longer than ALS with distal segment very short or up to 0.8 length of basal segment; *cribellate* species-cribellum usually undivided or at least appearing so (Fig. 25.15), divided in some *Mallos*; *ecribellate* species-colulus vestigial and setose (Fig. 25.13) except low and broad in *Saltonia* (Fig. 25.14).
**respiratory system:** paired anterior book lungs; tracheal spiracle small and just anterior of cribellum or colulus (relatively large and somewhat further anterior in *Saltonia*); tracheae with two branches and usually extending into cephalothorax.
**genitalia:** entelegyne; *female* epigynum variable (Figs. 25.21-25.22, 25.29-25.30, 25.38-25.39, 25.42, 25.46, 25.48-25.49, 25.52, 25.54, 25.56, 25.62, 25.65-25.68); atrium (AT) conspicuous or not, divided or not, 'scape' or other adornments usually absent, except in *Tivyna* (SC in Fig. 25.68); copulatory ducts usually short and simple (CD in Figs. 25.4, 25.23, 25.29, 25.40); spermathecal ducts short and simple (SD in Figs. 25.4, 25.23, 25.40)

to long and complex (SD in Figs. 25.30, 25.43); bases of spermathecae usually simple, globular (SP in Figs. 25.4, 25.23, 25.40) to elongate ovoid (SP in Figs. 25.29, 25.43), occasionally doubled (SP in Fig. 25.30); simple vulval pores present (PS in Fig. 25.23), complex vulval pores present at least in ecribellate species (PC in Figs. 25.4, 25.23); **male** patella modified in some cribellate species (MP in Figs. 25.55, 25.61, 25.74-25.77) and in *Blabomma* (MP in Fig. 25.20); RTA present and variably developed (Figs. 25.6, 25.19, 25.26-25.28, 25.45, 25.55, 25.74-25.77), other tibial modifications present (A in Figs. 25.6, 25.70-25.72, 25.75) or absent; true median apophysis lacking; variously developed tegular apophysis ('true' conductor) present and (usually) supporting embolus (TA, E in Figs. 25.24-25.26, 25.37, 25.50, 25.69-25.73), a second tegular apophysis may be present (e.g., *Blabomma*, TA in Fig. 25.19), but usually absent; embolus highly variable and ranging from simple short spur (E in Fig. 25.24) to elongate curved (E in Figs. 25.10, 25.25-25.26, 25.37, 25.72-25.73), apically modified in *Emblyna* (Figs. 25.7-25.9, 25.71) and *Mexitlia* (Fig. 25.58), orientation always clockwise (left palp ventral view).

### Distribution —

Dictynidae has a world-wide distribution but the great majority of described species occur in the Northern Hemisphere, primarily the Holarctic region. Dictynids are widespread throughout North America with many endemic species in Texas (primarily troglobites) and the Pacific Coast region.

### Natural history —

Most dictynids have an annual life cycle; some species (e.g., some troglobites) may take two or more years to reach maturity. All ecribellate dictynids are found at or below ground level. A considerable number are specialized troglobites; most are forest inhabitants living deep in litter or under bark, logs, rocks and other objects. Webs are usually very reduced sheets and function primarily as a retreat. One species (*Cicurina bryantae* EXLINE 1936a) builds a highly specialized retreat (Bennett 1985) (Fig. 25.11). *Saltonia* is apparently restricted to salt flats and dry alkaline lake beds or similar areas. Cribellate dictynids occur in a much wider variety of habitats from deep litter to branch tips high in the canopy of old-growth conifers. Most dictynines are arboreal, making webs on foliage, flowers, branches, and dried stalks of plants. Webs are constructed of typical cribellate "hackled" silk in the form of an irregular mesh, sheet, or often a more-or-less symmetrical lattice (Fig. 25.12). *Mexitlia trivittata* (BANKS 1901b) is a well known communal species. Chamberlin & Gertsch (1958) provided some discussion of cribellate dictynid webs and behaviour. Bond & Opell (1997) discussed various aspects of the natural history of *Mallos* and *Mexitlia*.

### Taxonomic history and notes —

The current concept of Dictynidae joins cribellate and ecribellate taxa in three subfamilies: Dictyninae is entirely cribellate, Tricholathysinae and Cicurininae both contain a mix of cribellate and ecribellate genera. Historically, all cribellate dictynids were placed with a range of other cribellate groupings in a single paraphyletic (and entirely cribellate) family variously named Amaurobiidae (e.g., Thorell 1870b), Ciniflonidae (e.g., Emerton 1888a) or Dictynidae (e.g., Simon 1892a). Ecribellate dictynids usually resided in Agelenidae. Prior to Lehtinen's (1967) magnum opus, various workers (e.g., Gertsch 1946a) recognized Dictynidae as including only dictynines and cribellate cicurinines and tricholathysines. Lehtinen's (1967) radical relimitation of Dictynidae joined these elements with various "traditional" agelenids (including the ecribellate cicurinines and the enigmatic *Saltonia* as well as cybaeids and other groupings). Brignoli (1983c) first catalogued together the genera included in "modern" Dictynidae and, although the limits of Dictynidae remain unresolved (pending serious taxonomic revision and analysis), this provisional classification is widely accepted (see Platnick 2005). Several genera have had checkered histories: *Emblyna*, *Phantyna*, *Tivyna* were described by Chamberlin (1948b), synonymized under *Dictyna* by Chamberlin & Gertsch (1958), resurrected by Lehtinen (1967), but variably recognized or ignored subsequently depending upon personal preference. Bond & Opell (1997) confirmed the validity and analyzed the systematics of *Mallos* and *Mexitlia*. Paquin & Hedin (2004) recently analyzed the molecular systematics of eyed and eyeless *Cicurina* inhabiting Texas caves. The species of *Cicurina* occurring in caves near San Antonio (Texas), also were recently reviewed by Cokendolpher (2004a, b).

Much of the existing Nearctic dictynid literature is of limited usefulness for species identification and, with the exception of *Mallos* and *Mexitlia*, all genera are in need of revision. Building upon works by Chamberlin (1948b), Chamberlin & Ivie (1947b), Gertsch (1945, 1946a), and a smattering of earlier publications, Chamberlin & Gertsch (1958) revised the cribellate Nearctic dictynid fauna (with subsequent further revision of *Mallos* and *Mexitlia* by Bond & Opell in 1997). No such inclusive summary of the ecribellate Nearctic dictynid fauna exists except at the genus level (Roth 1994, Roth & Brame 1972). Important descriptive publications for ecribellate dictynid species were published by Chamberlin & Ivie (1937-*Blabomma*; 1940-*Cicurina*, 1942a-*Saltonia*), Gertsch (1992-troglobitic *Cicurina*), Roth (1956a-*Yorima*), and Roth & Brown (1975b-*Saltonia*). A large number of apparently undescribed species of *Blabomma* exist in museum collections. Roth & Brown (1986) listed and provided summaries of synonymy and distribution for all known ecribellate Nearctic cicurinines. Bennett (1992) detailed comparative morphology of the female genitalia of various exemplar dictynid species.

### Genera —

CICURININAE
  *Blabomma* CHAMBERLIN & IVIE 1937, *Brommella* TULLGREN 1948, *Cicurina* MENGE 1871, *Lathys* SIMON 1884o, *Yorima* CHAMBERLIN & IVIE 1942a

DICTYNINAE
  *Dictyna* SUNDEVALL 1833b, *Emblyna* CHAMBERLIN 1948b, *Mallos* O. PICKARD-CAMBRIDGE 1902a, *Mexitlia* LEHTINEN 1967, *Nigma* LEHTINEN 1967, *Phantyna* CHAMBERLIN 1948b, *Thallumetus* SIMON 1893b, *Tivyna* CHAMBERLIN 1948b

TRICHOLATHYSINAE
  *Arctella* HOLM 1945b, *Argenna* THORELL 1870b, *Argennina* GERTSCH & MULAIK 1936a, *Hackmania* LEHTINEN 1967, *Iviella* LEHTINEN 1967, *Saltonia* CHAMBERLIN & IVIE 1942a, *Tricholathys* CHAMBERLIN & IVIE 1935b

# Dictyniae

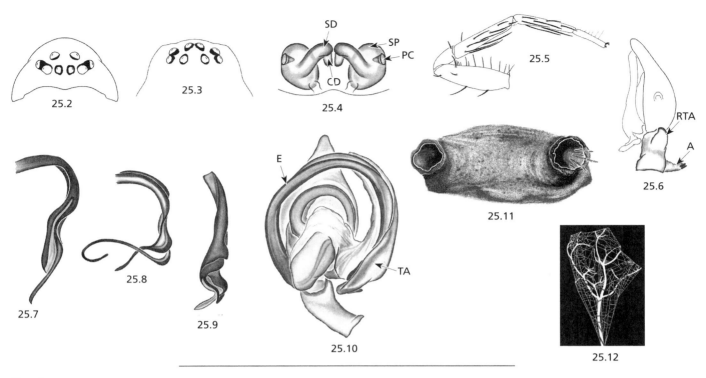

## Key to genera —
### North America North of Mexico

**1** Ecribellate (Figs. 25.13, 25.14) (CICURININAE and TRICHOLATHYSINAE, *in part*) .................................................. **2**

— Cribellate (Fig. 25.15) (cribellum or calamistrum may be difficult to discern in mature males) (CICURININAE and TRICHOLATHYSINAE *in part*, DICTYNINAE) .................................................. **5**

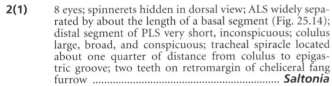

**2(1)** 8 eyes; spinnerets hidden in dorsal view; ALS widely separated by about the length of a basal segment (Fig. 25.14); distal segment of PLS very short, inconspicuous; colulus large, broad, and conspicuous; tracheal spiracle located about one quarter of distance from colulus to epigastric groove; two teeth on retromargin of cheliceral fang furrow .................................................. ***Saltonia***
  **Div.** 1 species: *Saltonia incerta* (BANKS 1898a) — **Dist.** extreme s CA (Mojave Desert and Imperial Valley) — **Refs.** Chamberlin & Ivie 1942a (male), Roth & Brown 1975b (female) — **Note** earlier thought to be extinct, this species has been collected from alkaline lake beds in the Mojave Desert and may be locally abundant there (D. Ubick and W. Savary, pers. com.)

— Usually 8 eyes (occasionally 6 or none); spinnerets usually visible in dorsal view; ALS separated by at most the diameter of a basal segment (Fig. 25.13); distal segment of PLS conspicuous, one third to nearly as long as basal segment; colulus small, inconspicuous, indicated by two groups of several setae; tracheal spiracle located just anterior of colulus, at least four teeth and/or "denticles" (minute teeth) on retromargin of cheliceral fang furrow ............. **3**

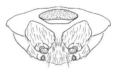

**3(2)** 6 eyes, AME absent; PER moderately recurved to nearly straight in dorsal view (Fig. 25.16); distal segment of PLS ½ to nearly as long as basal segment .................... ***Yorima***
  **Div.** 5 species — **Dist.** w CA from Sonoma Co. south — **Ref.** Roth 1956a — **Note** it can be difficult to separate this genus from six-eyed *Blabomma* and *Cicurina* (couplet 4). The latter are found only in TX. The former are best distinguished by details of the genitalia

— Usually 8 eyes (occasionally 6 or none), AME may be minute; PER moderately procurved to nearly straight in dorsal view (Fig. 25.17); distal segment of posterior spinnerets only about 1/3 as long as basal segment (Fig. 25.13) .................................................. **4**

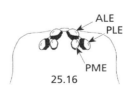

# Dictynidae

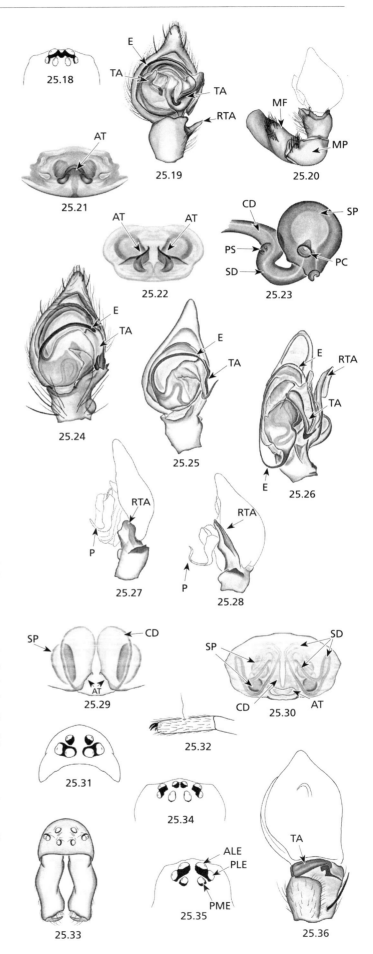

**4(3)** 8, 6, or no eyes, AME very minute when present (Figs. 25.17-25.18); male genital bulb with 2 tegular apophyses – one near base of embolus, second supporting embolus and with proximal tip variously developed but not resembling a "shepherd's crook" (TA, E in Fig. 25.19); RTA a simple distal spur; palpal patella and femur may be modified (MP, MF in Fig. 25.20); epigynum with single or paired atria usually located some distance anterior of epigastric groove (AT in Figs. 25.21-25.22); copulatory and spermathecal ducts simple, narrow and short; bases of spermathecae usually a single rounded pair (CD, SD, SP in Fig. 25.23) .................................................. ***Blabomma***
    **Div.** 10 species — **Dist.** s BC to s CA — **Ref.** Chamberlin & Ivie 1937 — **Note** a large number of undescribed species reportedly exist

— Eight, 6 or no eyes, AME usually present and about 2/3 diameter of ALE but may be larger or very minute; male genital bulb with single tegular apophysis supporting embolus (TA, E in Figs. 25.24-25.26) and with proximal tip often resembling a "shepherd's crook" (P in Figs. 25.27-25.28); RTA a conspicuous, often elaborate, flattened and cupped distal projection (Figs. 25.26-25.28); epigynum with single (rarely paired) atrium usually located close to epigastric groove (AT in Figs. 25.29, 25.30); copulatory ducts simple, short, may be widened (CD in Figs. 25.29-25.30); spermathecal ducts may be short and simple but often are very long and convoluted (SD in Fig. 25.30); bases of spermathecae elongate oval (SP in Fig. 25.29) or rounded, if rounded then usually doubled (SP in Fig. 25.30) ........................................................... ***Cicurina***
    **Div.** 115 species — **Dist.** widespread (6-eyed and blind species occur in caves in TX and AL) — **Refs.** Chamberlin & Ivie 1940, Gertsch 1992, Cokendolpher 2004a, b, Paquin & Hedin 2004 — **Note** many species in this genus are hard to identify because most descriptions are insufficiently detailed to ensure correct placement and a significant number of synonyms appear to be entrenched in the literature

**5(1)** Clypeus low, at best no wider than diameter of ALE (Figs. 25.2, 25.31); cephalic area low; tarsi with 1-4 dorsal trichobothria (Figs. 25.32); cribellum entire ............................. **6**

— Clypeus relatively high, greater than diameter of ALE, usually at least 1.5 x diameter of ALE (Figs. 25.33, 25.60, 25.64); cephalic area usually somewhat elevated; tarsi lacking dorsal trichobothria; cribellum usually entire (divided in some *Mallos*) ...................................................... **13**

**6(5)** 6 or 8 eyes, AME small (Fig. 25.34), tiny, or absent (Figs. 25.31, 25.35); in dorsal view PER strongly (Fig. 25.35) to moderately (Fig. 25.34) procurved; proximal end of tegular apophysis of male genital bulb tightly or loosely spiraled and situated dorsally over palpal tibia (TA in Fig. 25.36) ........................................................... ***Lathys***
    **Div.** 9 species — **Dist.** widespread — **Ref.** Chamberlin & Gertsch 1958

— 8 eyes, AME ≥2/3 diameter of PLE; PER straight or, at best, weakly procurved in dorsal view; proximal end of tegular apophysis variably shaped and situated ventrally ........... **7**

**7(6)** Promargin of cheliceral fang furrow with 4-5 teeth, retromargin with 3-4 ............................................................. **8**

— Cheliceral fang furrow teeth not exactly as above ......... **9**
    **Note** fang furrow teeth can be very difficult to discern, especially in small specimens. If in doubt, explore both halves of this couplet

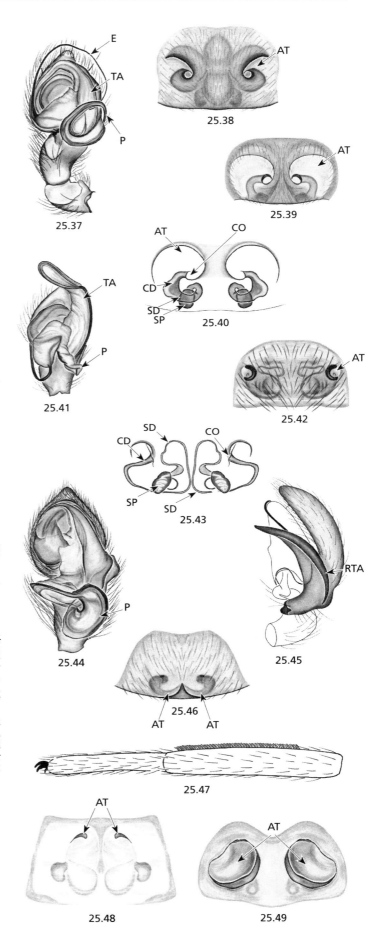

**8(7)** Tegular apophysis of male genital bulb weakly developed anteriorly, not extending beyond apex of cymbium and with proximal end a strongly developed loose coil (TA, P in Fig. 25.37); epigynal atria widely separated (AT in Fig. 25.38) or close-spaced (AT in Fig. 25.39), relatively conspicuous; copulatory and spermathecal ducts relatively broad and simple (CD, SD in Fig. 25.40) ............. **Tricholathys**
  **Div.** 9 species — **Dist.** w Nearctic, s BC to TX — **Ref.** Chamberlin & Gertsch 1958 — **Note** *Tricholathys* and *Iviella* are poorly diagnosed. They are best separated by comparing specimens with published figures

— Tegular apophysis of male genital bulb narrow, elongate, curved, extending anteriorly beyond apex of cymbium and with proximal end not a strongly developed loose coil (male of *Iviella reclusa* unknown) (TA, P in Fig. 25.41); epigynal atria wide-spaced, inconspicuous (AT in Fig. 25.42); copulatory and spermathecal ducts narrow and relatively complex (CD, SD in Fig. 25.43) ............. **Iviella**
  **Div.** 2 species: *Iviella ohioensis* (CHAMBERLIN & GERTSCH 1958) and *Iviella reclusa* (GERTSCH & IVIE 1936) — **Dist.** NF, ne USA w to NE, UT, & AZ — **Ref.** Chamberlin & Gertsch 1958 (as *Tricholathys*) — **Note** see note above

**9(7)** Promargin of cheliceral fang furrow with 5 teeth, retromargin with 2; proximal end of tegular apophysis very large, broad and spiralled (P in Fig. 25.44) ........ **Arctella**
  **Div.** 1 species: *Arctella lapponica* HOLM 1945b — **Dist.** AK, YT & w NT s to n BC — **Ref.** Holm 1945b — **Note** this species is catalogued as Palaearctic (Platnick 2005) but was reported from AK by Holm (1960b) and has subsequently been found in various localities in northern CAN.

— Promargin of cheliceral fang furrow with 3 teeth, retromargin with 0, 2, or 3; proximal end of tegular apophysis otherwise ........................................................................ **10**

**10(9)** Tibiae and metatarsi I, II, and IV with a few small macrosetae on dorsal, ventral and/or lateral surfaces ..................
............................................................................. **Argennina**
  **Div.** 1 species: *Argennina unica* GERTSCH & MULAIK 1936a — **Dist.** s TX — **Ref.** Chamberlin & Gertsch 1958 — **Note** known from only a single (female) specimen

— Macrosetae absent or, if present, only distally on tibia and metatarsus IV ................................................................ **11**

**11(10)** Tiny spiders, body length usually <2 mm; leg formula 1423; calamistrum primarily restricted to basal half of metatarsus IV; RTA conspicuous, curved, length ≥ palpal tibia (Fig. 25.45); paired atria located near posterior margin of epigynum (AT in Fig. 25.46) ....... **Brommella**
  **Div.** 3 species — **Dist.** w cordillera from MT s to e CA, NM, & n TX— **Ref.** Chamberlin & Gertsch 1958 (as *Pagomys*) — **Note** male of only one species described

— Larger spiders, body length usually ≥2 mm; leg formula 4123; calamistrum nearly as long as metatarsus IV (Fig. 25.47); RTA short, length ≤ width of palpal tibia; paired atria located near middle of epigynum or further anterior (AT in Figs. 25.48, 25.49) ............................................... **12**

**12(11)** Proximal end of tegular apophysis a prolaterally directed simple point (P in Fig. 25.50) or proximally directed anteriorly arched curve (P in Fig. 25.51); epigynal atria a pair of inconspicuous, oval or semi-circular, shallow depressions (AT in Fig. 25.52) .................................... **Argenna**
   **Div.** 2 species: *Argenna obesa* EMERTON 1911, *Argenna yakima* CHAMBERLIN & GERTSCH 1958 — **Dist.** widespread, se CAN, ne & w USA— **Ref.** Chamberlin & Gertsch 1958 — **Note** *Argenna* and *Hackmania* are best separated by comparison of specimens with published descriptions and illustrations

— Proximal end of tegular apophysis an elongate, proximo-retrolaterally directed point (P in Fig. 25.53); epigynal atria a pair of distinct, circular, deep openings (AT in Fig. 25.54) .................................... **Hackmania**
   **Div.** 2 species: *Hackmania prominula* (TULLGREN 1948) and *Hackmania saphes* (CHAMBERLIN & GERTSCH 1958) — **Dist.** AK to MB and s along w cordillera to NM — **Ref.** Chamberlin & Gertsch 1958 [as *Argenna lorna* CHAMBERLIN & GERTSCH 1958 (= *Hackmania prominula*) and *Argenna saphes*] — **Note** see note above. *Hackmania saphes* may also be a synonym of *Hackmania prominula* (Lehtinen 1967)

**13(5)** Retromargin of cheliceral fang furrow lacking teeth; male palpal femur, patella, and tibia greatly thickened (Fig. 25.55); epigynum elongate with tiny paired atria at anterior margin (AT in Fig. 25.56) .................. **Thallumetus**
   **Div.** 1 species: *Thallumetus pineus* (CHAMBERLIN & IVIE 1944) — **Dist.** TN & FL w to TX — **Ref.** Chamberlin & Gertsch 1958

— Retromargin of cheliceral fang furrow with 1-3 teeth; male palpal femur "normal"; patella usually "normal" but may be enlarged (*Tivyna*-MP in Fig. 25.77); tibia variously modified but not thickened; epigynum otherwise ...... **14**

**14(13)** Retromargin of cheliceral fang furrow with 2 or, rarely, 3 teeth ................................................................. **15**

— Retromargin of cheliceral fang furrow with 1 tooth ... **16**
   **Note** fang furrow teeth can be very difficult to discern, especially in small specimens. If in doubt, explore both halves of this couplet

**15(14)** Thoracic portion of carapace with pale marginal band (Fig. 25.57); embolus tip unbranched; cribellum divided in some species ...................................... **Mallos**
   **Div.** 7 species — **Dist.** w USA east to w MT and w TX — **Ref.** Bond & Opell 1997

— Thoracic portion of carapace without pale marginal band, usually uniformly brown; embolus tip bifurcate (Fig. 25.58); cribellum entire ...................................... **Mexitlia**
   **Div.** 1 species: *Mexitlia trivittata* (BANKS 1901b) — **Dist.** se ID & s WY south to NM, AZ & w CA — **Ref.** Bond & Opell 1997

**16(14)** Carapace with pale marginal blotches (Fig. 25.59); eyes small, widely separated, PME separated by ≥2 diameters (Figs. 25.59-25.60); male palpal patella with elongate retrolateral process (MP in Fig. 25.61); epigynal atria lacking posterolateral grooves or spatulate scape (AT in Fig. 25.62); sides of male chelicerae essentially parallel in frontal view (Fig. 25.60) ........................................ **Nigma**
   **Div.** 1 species: *Nigma linsdalei* (CHAMBERLIN & GERTSCH 1958) — **Dist.** s CA — **Ref.** Chamberlin & Gertsch 1958 (as *Heterodictyna*) — **Note** in life this spider is translucent green and often with a red or black spot on the abdomen

— Carapace without pale marginal marks; eyes larger and closer together, PME rarely separated by as much as 1.5 diameters (Figs. 25.63-25.64); male palpal patella without elongate retrolateral process (some *Phantyna* have a short retrolateral process, MP in Fig. 25.75); epigynal atria with more or less distinct posterolateral grooves (G in Figs. 25.65-25.67) or a spatulate scape (SC in Fig. 25.68); sides of male chelicerae bowed laterally in frontal view (Figs. 25.63-25.64) ......................................................... **17**
   **Note** the genera *Dictyna*, *Emblyna*, *Phantyna*, and *Tivyna* are not clearly diagnosed and can be very difficult to separate. Except for *Tivyna*, females are essentially indistinguishable at the genus level. Females (other than *Tivyna*), as well as males that cannot be keyed beyond this point, are best left identified as "*Dictyna*" sp. unless comparison can be made with species descriptive literature

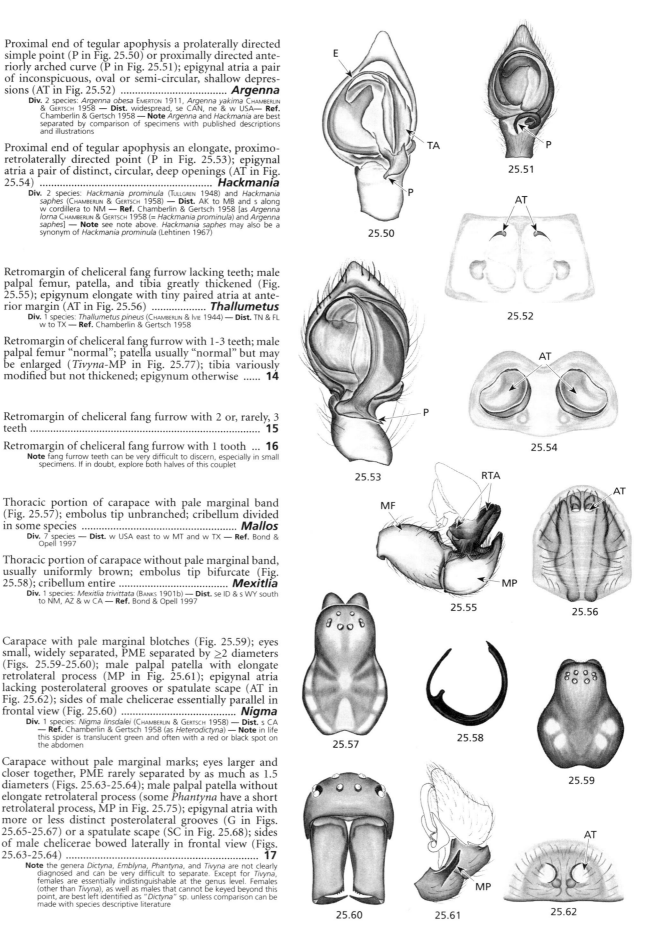

Spiders of North America — *Dictynidae* | **101**

**17(16)** Males .................................................................. **18**
— Females .............................................................. **21**

**18(17)** Embolus thick (E in Figs. 25.10, 25.69-25.71) and variously modified apically, often flattened, ribbed and/or twisted (E in Figs. 25.7-25.9, 25.71) ................. ***Emblyna***
**Div.** 63 species — **Dist.** widespread — **Ref.** Chamberlin & Gertsch 1958 (as *Emblyna* "section" of *Dictyna*)

— Embolus a simple, curved, thin structure, unmodified apically (E in Figs. 25.72-25.73) ..................................... **19**

**19(18)** Male palpal patella unmodified (Figs. 25.72-25.73) ..........
................................................................. ***Dictyna***
**Div.** 44 species — **Dist.** Widespread — **Ref.** Chamberlin & Gertsch 1958 (as *apacheca, brevitarsus, foliacea, longispina, major, personata, terrestris,* and *volucripes* "groups" and various unplaced species of *Dictyna* "section")

— Male palpal patella variously modified: greatly enlarged & bulbous, with dorsal or retrolateral projections, or globular with short, stout setae retrolaterally (MP in Figs. 25.74-25.77) .............................................................. **20**

**20(18)** Male palpal patella with variously developed retrolateral processes (MP in Fig. 25.74) or distinctly rounded and with short, stout setae retrolaterally (MP in Fig. 25.75) ....
................................................................. ***Phantyna***
**Div.** 9 species — **Dist.** widespread, s CAN s through USA — **Ref.** Chamberlin & Gertsch 1958 (as *bicornis, micro, rita,* and *varyna* "groups" of *Dictyna* "section")

— Male palpal patella greatly enlarged & bulbous, width ≥ width of cymbium (MP in Fig. 25.76) or moderately enlarged with distinctive dorsoapical acuminate projection (MP in Fig. 25.77) ............................. ***Tivyna***
**Div.** 4 species — **Dist.** s USA from UT & MD s to CA, TX & FL — **Ref.** Chamberlin & Gertsch 1958 (as *spathula* [sic] "group" of *Dictyna* "section")

**21(17)** Epigynum with more or less distinct posterolateral atrial grooves (G in Figs. 25.65-25.67) and lacking a spatulate scape ....................... ***Dictyna, Emblyna*** and ***Phantyna***
**Note** see genera details in couplet 18-20 above

— Epigynum with a spatulate scape (SC in Fig. 25.68) and lacking more or less distinct posterolateral atrial grooves .
................................................................. ***Tivyna***
**Div.** 4 species — **Dist.** s USA from UT & MD s to CA, TX & FL — **Ref.** Chamberlin & Gertsch 1958 (as *spathula* [sic] "group" of *Dictyna* "section")

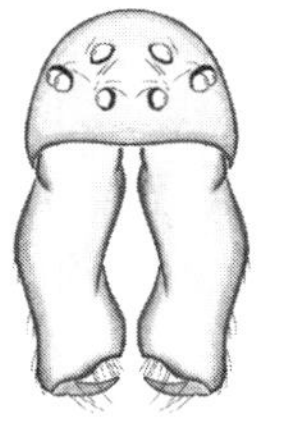

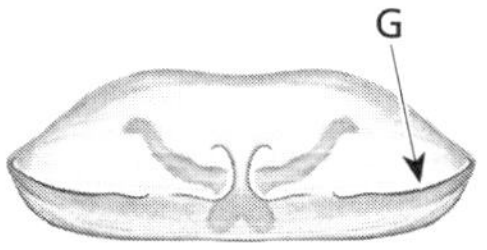

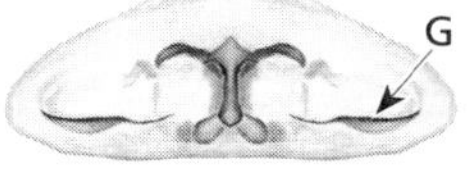

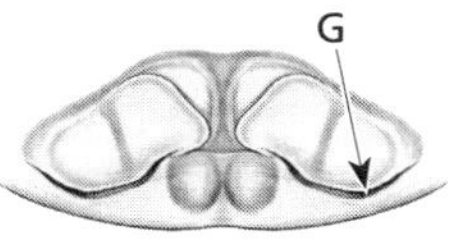

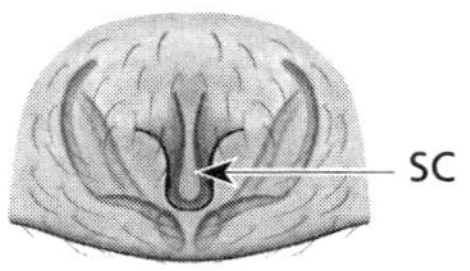

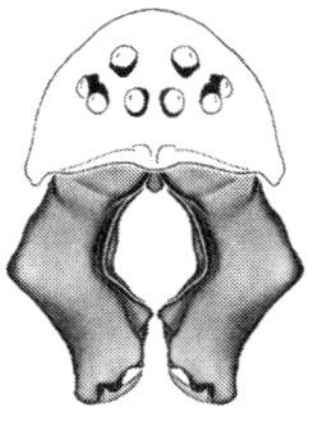

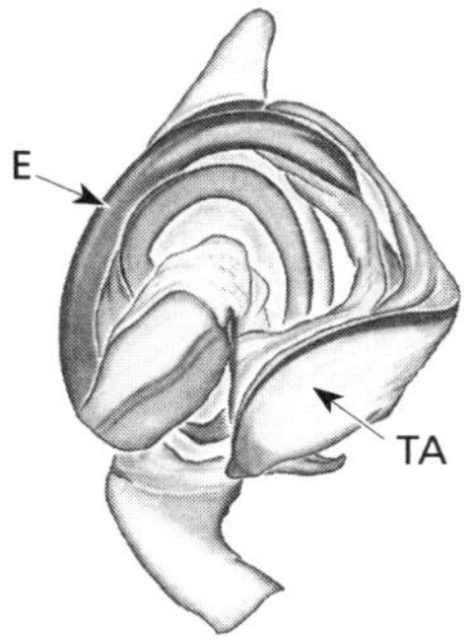

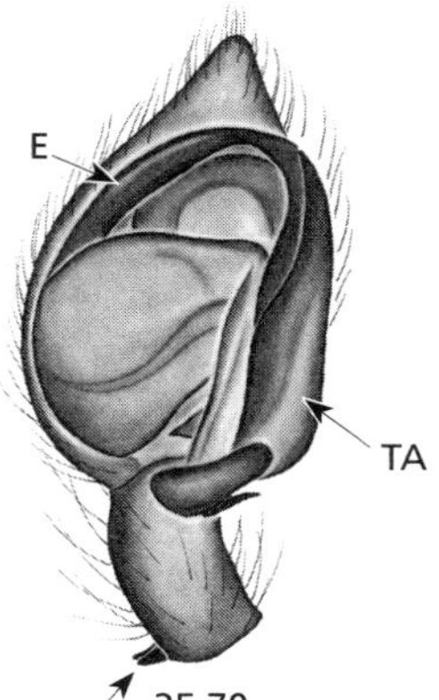

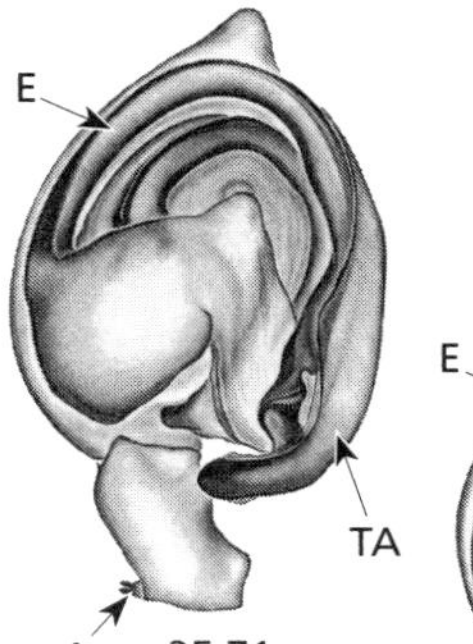

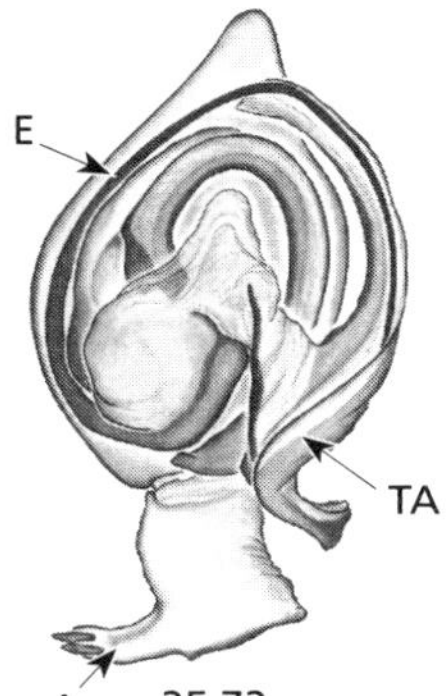

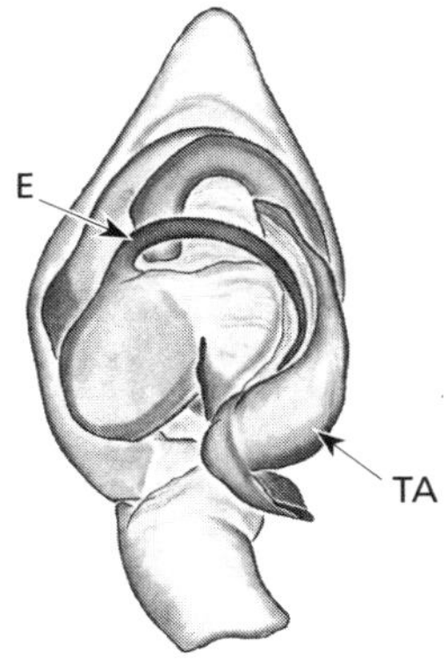

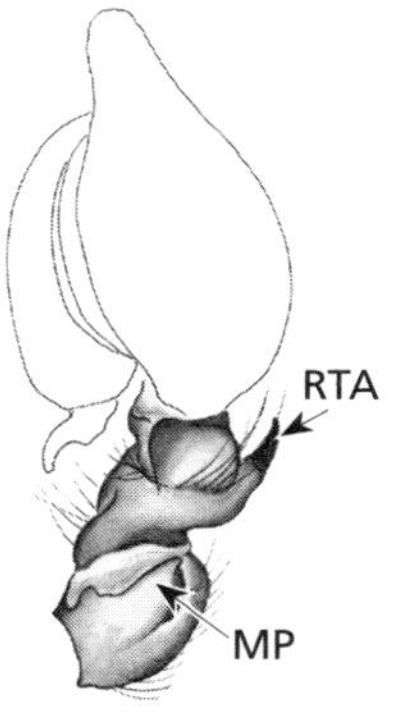

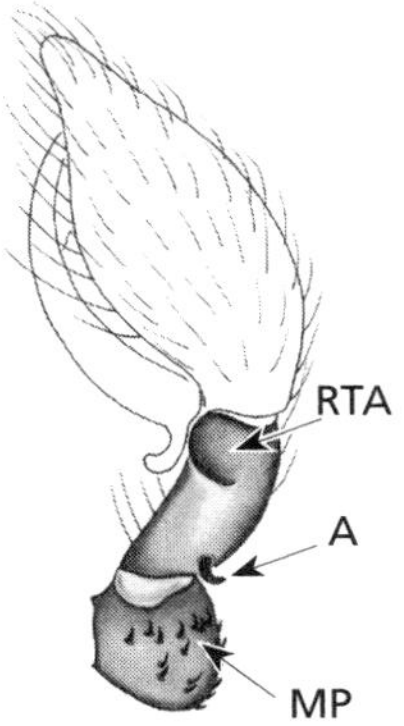

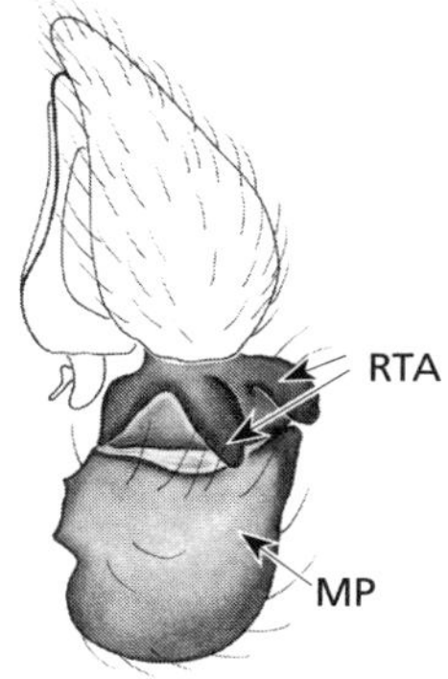

| A | non-RTA palpal tibial apophysis |
|---|---|
| AT | epigynal atrium |
| CD | copulatory duct |
| CO | copulatory opening |
| E | embolus |
| G | posterolateral groove |
| MF | modified palpal femur |
| MP | modified palpal patella |
| P | proximal end of TA |
| PC | complex spermathecal pore |
| PS | simple spermathecal pore |
| RTA | retrolateral tibial apophysis |
| SC | epigynal scape |
| SD | spermathecal duct |
| SP | spermatheca |
| TA | tegular apophysis |

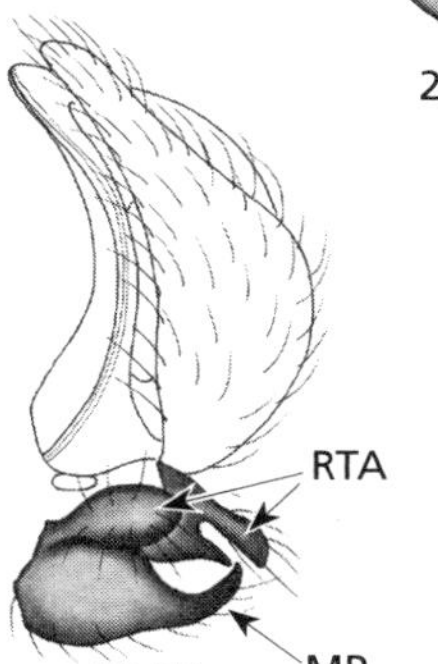

## Chapter 26
# DIGUETIDAE
1 genus, 7 species and subspecies

Darrell Ubick

### Similar families —
Plectreuridae (p. 201).

### Diagnosis —
Diguetids differ from most haplogynes in having basally fused chelicerae and 6 eyes arranged in 3 diads, from scytodids in having a flat carapace and from sicariids in having three tarsal claws.

### Characters —
**body size:** 5.0 – 10.0 mm.
**color:** prosoma and coxae brown, patterned with white setae; abdomen grayish white dorsally with pale folium or dark stripe, dark ventrally; legs orange brown with dark bands.
**carapace:** pyriform, flat; cephalic area broad; fovea present as shallow transverse groove.
**sternum:** narrow, with extensions between coxae.
**eyes:** six eyes in three diads, PME and ALE in straight row.
**chelicerae:** fused basally (connected by membrane), with median lamina produced into apical tooth; ectal margin with indistinct stridulatory file.
**mouthparts:** labium elongate, endites triangular and strongly convergent.
**legs:** long and thin, with few spines; tarsi with three claws; male leg I long, with tarsi subsegmented; female palp lacking claw.
**abdomen:** oval to spherical; rarely with caudal extension [*Diguetia albolineata* (O. PICKARD-CAMBRIDGE 1895a)].
**spinnerets:** six small and contiguous, PLS largest; colulus represented by small setose plate.
**respiratory system:** one pair of book lungs; tracheal system absent, but venter with groove anterior of spinnerets that can be mistaken for a tracheal spiracle (Fig. 26.2).
**genitalia:** haplogyne; *female* with gonopore region swollen and pigmented (Fig. 26.2); *male* palp with tibia swollen and tarsus bifurcate; bulb pyriform with spoon-shaped embolus and short tubular prong (Fig. 26.3).

### Distribution —
*Diguetia* is known from southwestern USA (Utah to Colorado s to California and Texas) to northern Mexico, and also from Argentina. The most widespread species is *Diguetia canities* (MCCOOK 1889), which has been divided into three subspecies of which two occur north of Mexico.

### Natural history —
*Diguetia* species are unusual among haplogynes, except pholcids, in that they build extensive snare webs. All species occur in arid desert scrub where they construct webs in low shrubbery, including cactus, and may be common among prickly pear (*Opuntia*). The web is a sheet with surrounding tangle and has a vertical tubular retreat. The retreat is conglomerated with plant debris and prey remains and also incorporates the egg sacs (Cazier & Mortenson 1962). The spiders move along the underside of the sheet and capture prey intercepted by the surrounding tangle web (Eberhard 1967). A study of the biology of *Diguetia imperiosa* GERTSCH

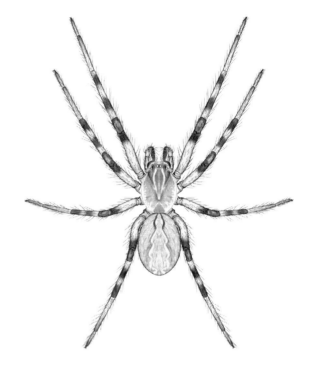

**Fig. 26.1** *Diguetia canities* (MCCOOK 1889)

& MULAIK 1940 concluded that the species probably has an annual life cycle (Bentzien 1973).

### Taxonomic history and notes —
The genus *Diguetia* was proposed by Simon (1895f) and included in Sicariidae, subfamily Periegopinae. F.O. Pickard-Cambridge (1899a) placed the genus in its own group, but it remained in Sicariidae until Gertsch (1949) elevated it to family and later (1958a) revised the North American species. Platnick (1989a) reinterpreted aspects of diguetid morphology and argued that *Diguetia* is derived relative to the South American diguetid genus, *Segestrioides* KEYSERLING 1883a. Recent analysis of haplogyne relationships (Platnick *et al.* 1991) places Diguetidae closest to Plectreuridae, also a strictly American family, in a clade that also includes Pholcidae.

### Genus —
*Diguetia* SIMON 1895f

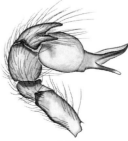

26.2

26.3

# Chapter 27
# DYSDERIDAE
1 genus, 1 species

Darrell Ubick

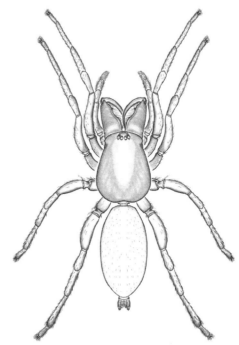

Fig. 27.1 *Dysdera crocata* C.L. Koch 1838e

## Similar families —
Caponiidae (p. 75), Corinnidae (Trachelinae) (p. 78), Segestriidae (p. 219).

## Diagnosis —
Dysderids differ from other haplogyne spiders by the combined presence of: six eyes with the PER forming slightly procurved row; chelicerae porrect, free (not basally fused), with toothed margins and long fangs; tarsi with two claws and tufts, and a pair of tracheal spiracles positioned near the epigastric furrow.

## Characters —
**body size:** 9.0-15.0 mm.
**color:** prosoma and appendages reddish orange, abdomen grayish white.
**carapace:** rectangular, slightly narrower in cephalic region, largely glabrous; fovea short, longitudinal (Fig. 27.1).
**sternum:** oval, with sclerotized extensions to and between coxae.
**eyes:** six eyes (AME absent) in cluster occupying < 1/3 cephalon width, PE in slightly procurved row.
**chelicerae:** large, porrect; promargin with setae, retromargin with four teeth; fangs long.
**mouthparts:** endites parallel, longer than wide, apically pointed, extending from mesal surfaces of palpal coxae; labium triangular, with apical notch, longer than wide; serrula absent.
**legs:** anterior legs lacking spines, posterior legs with spines on tibiae and metatarsi; tarsi with two claws and tufts.
**abdomen:** elongate, slightly flattened.
**spinnerets:** colulus represented by setal patch; ALS conical, contiguous, subequal to PLS.
**respiratory system:** one pair of book lungs; two tracheal spiracles near epigastric furrow (Fig. 27.2).
**genitalia:** haplogyne; *female* with gonopore region swollen (Fig. 27.2); *male* palpus with tarsus pointed, as long as tibia; bulb cylindrical, apically with broad embolus and accessory lobe (Fig. 27.3).

## Distribution —
Dysderids are native to the western Palearctic with most species being circum-Mediterranean. The synanthropic *Dysdera crocata*, however, has become nearly cosmopolitan and is widely distributed in the Nearctic.

## Natural history —
Dysderids are nocturnal wandering hunters. They are ground dwellers found in cryptozoic habitats, most commonly under rocks and logs, in both grasslands and forests. In the Nearctic, *Dysdera crocata* occurs primarily in urban areas and disturbed habitats. In the wild the preferred prey is terrestrial isopods, most of which are also introduced, whereas in captivity *Dysdera* may accept a variety of prey (Cooke 1965b). The spiders build retreats for molting and depositing eggs, which are loosely bound with silk and guarded by the female. In captivity, *Dysdera crocata* reaches maturity in about 18 months and then lives for an additional two to three years (Cooke 1965a).

## Taxonomic history and notes —
The ubiquitous *Dysdera crocata* has been well described numerous times (for example, Comstock 1940, Kaston 1948, Paquin & Dupérré 2003). The genus is placed in the Dysderinae, the largest and most derived of the three dysderid subfamilies (Deeleman-Reinhold & Deeleman 1988). Although Segestriidae has sometimes been included in Dysderidae (Simon 1892a, Beatty 1970), the two families are now regarded as distinct (Brignoli 1976g, Forster & Platnick 1985) and recent analysis of haplogyne relationships (Platnick et al. 1991) suggests that the two are equally closely related to each other as to the group comprising Oonopidae plus Orsolobidae.

## Genus —
DYSDERINAE
*Dysdera* LATREILLE 1804b

27.2     27.3

## Chapter 28
# FILISTATIDAE
3 genera, 7 species

Darrell Ubick

### Similar families —
Amaurobiidae (p. 60).

### Diagnosis —
Filistatids differ from all other cribellate spiders in having fused chelicerae. Additionally, the eyes are clustered on a central mound, the abdomen is rounded posteriorly and extends beyond the spinnerets; and the legs have autospasy at the tibia-patella joint.

### Characters —
**body size:** 1.5-18.0 mm.
**color:** orange to purplish brown to black, usually concolorous, sometimes with purplish chevron pattern on abdomen and banded legs.
**carapace:** oval, flat, with anterior projection; clypeus long, nearly horizontal; fovea deep to indistinct.
**sternum:** oval, with sigilla in larger species.
**eyes:** eight clustered on central mound.
**chelicerae:** small, fused basally, chelate with meso apical lamina.
**mouthparts:** endites convergent, labium fused to sternum.
**legs:** prograde, with numerous spines in larger species, tarsi with three claws, autospasy at tibia-patella joint.
**abdomen:** cylindrical to slightly flattened, posteriorly rounded and extended beyond spinnerets, covered with short dense hairs giving velvety appearance.
**spinnerets:** in ventral position, with divided cribellum, with ALS three-segmented.
**respiratory system:** with one pair of book lungs and tracheal openings positioned between spinnerets and epigastric furrow and represented by either a single wide tracheal spiracle or (in small species) two separate spiracles.
**genitalia:** haplogyne; *female* gonopore region not modified externally; *male* palp with elongated or incrassate segments, bulb simple, attenuated.

### Distribution —
The three genera in our region are all endemic to the New World. Of these, *Kukulcania* is widely distributed in the southern USA, with most species known from the West, and extends (synanthropically) to South America. *Filistatinella* is restricted to southwest USA and adjacent Mexico and *Filistatoides* ranges from the southwest USA to South America.

### Natural history —
Filistatids are sedentary spiders and, apart from wandering males, are typically hidden in crevices. Here they line their retreats with cribellate webbing, which extends out from the entrances into often extensive signal and capture lines. These spiders have been collected under rocks, under fallen bark and logs, in living and dead *Yucca*, occasionally in leaf litter, in pitfall traps, and frequently on buildings. They may take several years to mature and, probably unique among araneomorph spiders, the females may continue to molt after reaching maturity (Murphy & Murphy 2000). Also unusual among cribellates, except hypochilids,

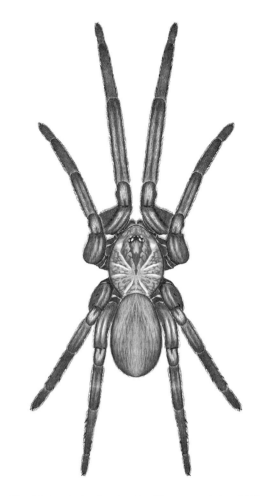

**Fig. 28.1** *Kukulcania arizonica* (Chamberlin & Ivie 1935b)

is that filistatids card cribellate silk through rapid vibrations of the fourth leg which rests on the third. In other cribellates, the fourth legs lock and rock slowly while carding. As in most spiders, male filistatids are smaller and have longer legs. However, in *Kukulcania* the dimorphism is even more extreme as the male is light brown in color, in contrast to the nearly black female.

### Taxonomic history and notes —
Although Filistatidae was a clearly recognized family since its inception, its species have generally been included in *Filistata* (Simon 1892a). Most of the Nearctic genera were recognized early: *Filistatoides* by F.O. Pickard-Cambridge (1899a) and *Filistatinella* by Gertsch & Ivie (1936). Major changes were introduced by Lehtinen (1967), who established a number of genera, including *Kukulcania* for the New World species previously included in *Filistata*.

Seven Nearctic species have been described, and about as many remain undescribed. The best known, and widest ranging, is *Kukulcania hibernalis* (Hentz 1842b). The western species of *Kukulcania* have been described by Chamberlin & Ivie (1935b, 1942a). The only other species described from the Nearctic is *Filistatinella crassipalpis* (Gertsch 1935a). *Filistatoides insignis* (O. Pickard-Cambridge 1896a) was originally described from Central America, but subsequently recorded from Texas (Comstock 1940).

Recent studies show that Filistatidae is the basalmost member of Haplogynae, being the only cribellate in that clade (Platnick *et al.* 1991). The family contains two subfamilies: the Filistatinae, with 3 genera of larger spiders, and the relatively derived Prithinae with several genera of small spiders, of which *Filistatinella* is basalmost (Gray 1995 and Ramírez & Grismado 1997).

### Genera —

FILISTATINAE
 *Kukulcania* Lehtinen 1967
PRITHINAE
 *Filistatinella* Gertsch & Ivie 1936, *Filistatoides* F.O. Pickard-Cambridge 1899a

### Key to genera —
North America North of Mexico

1  Size > 5 mm; thorax with deep fovea (Fig. 28.1); legs with numerous spines; ventral view of female abdomen (Fig. 28.2); palp (Fig. 28.3) .................. ***Kukulcania***
   **Div.** 6 species and subspecies — **Dist.** s USA to South America — **Refs.** Chamberlin & Ivie 1935b, Chamberlin & Ivie 1942a, Gray 1995, Ramírez & Grismado 1997 — **Note** species were previously included in *Filistata*

—  Size < 3 mm; thoracic groove indistinct; legs without spines ............................................................... 2

2  Color dark brown to grayish black, legs banded; carapace circular, about as wide as long (Fig. 28.4); male palpal tibia stout, femur much shorter than carapace (Fig. 28.5); size 1.5-3.5 mm ................................................ ***Filistatinella***
   **Div.** 1 species: *Filistatinella crassipalpis* (Gertsch 1935a) — **Dist.** central CA to s TX — **Ref.** Gertsch 1935a — **Note** this species was originally described (as a *Filistata*) from s TX. Roth (1994) mentions four undescribed species that range into central CA

—  Color orange brown, carapace with median dark stripe, abdomen with purplish chevrons, legs concolorous; carapace elongate, slightly longer than wide (Fig. 28.6); male palpal tibia slender, femur about as long as carapace (Fig. 28.7); size 2.5-4.0 mm ................................. ***Filistatoides***
   **Div.** 1 species: *Filistatoides insignis* (O. Pickard-Cambridge 1896a) — **Dist.** s CA to s TX — **Refs.** O. Pickard-Cambridge 1896a, F.O. Pickard-Cambridge 1899a, Comstock 1940, Ramírez & Grismado 1997 — **Note** although *Filistatoides* is currently known from a single species in North America, Roth (1994) mentions two undescribed species from the Nearctic and extends the known range to CA

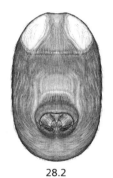

28.2

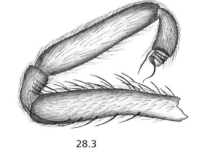

28.3

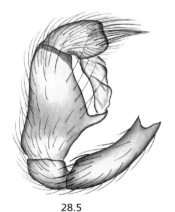

28.5

28.4

28.6

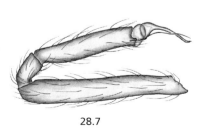

28.7

# Chapter 29
# GNAPHOSIDAE
24 genera, 255 species

Darrell Ubick

**Common name —**
Ground spiders; mouse spider [*Scotophaeus blackwallii* (THORELL 1871a)].

**Similar families —**
Prodidomidae (p. 203), Corinnidae (Phrurolithinae) (p. 78).

**Diagnosis —**
Gnaphosids may be distinguished from other 2-clawed entelegyne spiders, except prodidomids, by their enlarged and separated ALS, concave endites, and modified PME (flattened, oval, to irregular), and from prodidomids by having cylindrical (rather than flattened) ALS with relatively fewer and shorter spigots, PER straight to moderately procurved (strongly procurved in *Scopoides* and prodidomids), and pectinate tarsal claws (smooth in most prodidomids).

**Characters —**
**body size:** 2-17 mm.
**color:** usually brown to gray to black, abdomen sometimes with chevrons or white bands or stripes.
**carapace:** pyriform, low, with distinct fovea.
**sternum:** with precoxal triangles (except *Talanites*).
**eyes:** eight in two straight rows, PER rarely procurved; PME modified, oval to irregular in shape and flat.
**chelicerae:** usually not geniculate, margins with teeth, keels, lobes or carinae, rarely smooth.
**mouthparts:** endite with oblique medial depression, with serrula.
**legs:** prograde, moderately to lightly spined; tarsi with two pectinate claws, claw tufts, and often scopulae; posterior metatarsi of some genera with preening combs or brushes.
**abdomen:** elongate and somewhat flattened; often with scute in male.
**spinnerets:** ALS cylindrical and widely separated at base, composed of one segment, and with enlarged spigots.
**respiratory system:** with one pair of book lungs and with single tracheal spiracle near spinnerets.
**genitalia:** entelegyne; *female* epigynum at least moderately sclerotized, variable in structure; *male* palp with RTA, usually a simple prong, bulb with embolus, conductor, and usually with median apophysis.

**Distribution —**
Although most Nearctic gnaphosid genera are widely distributed, the majority of the species is concentrated in the western and southwestern regions. *Orodrassus*, *Parasyrisca*, and most *Gnaphosa* are primarily or exclusively northern and western, whereas *Synaphosus*, *Eilica*, *Camillina*, and *Gertschosa* are southern. *Litopyllus* is restricted to the eastern half of the area and *Scopoides* to the southwest.

**Natural history —**
Gnaphosids are primarily ground dwellers, only rarely found in arboreal habitats, and may be the most abundant spiders in drier and more open biomes (Dippenaar-Schoe-

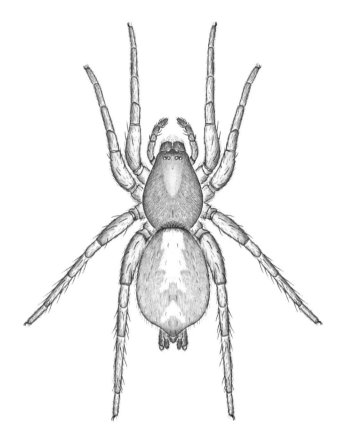

**Fig. 29.1** *Herpyllus ecclesiasticus* HENTZ 1832

man & Jocqué 1997). They are predominantly nocturnal and cryptozoic and are found in leaf litter, beneath rocks, and within and under decomposing wood. Synanthropic species, such as *Scotophaeus blackwallii* and *Urozelotes rusticus* (L. KOCH 1872b), occur in and around buildings. Being wandering hunters, gnaphosids use silk minimally, only for draglines, retreats, and egg sacs. The egg sacs may be constructed within a retreat and guarded by the female, or secured to the substrate and either guarded or abandoned. The egg sacs of at least some species are flat with a swollen center, like a fried egg, and made of a parchment-like silk. Gnaphosids are mostly generalized predators, utilizing a wide variety of prey, including other spiders. Species of *Micaria* and *Callilepis* have been observed to prey on ants (Heller 1976, Sauer & Wunderlich 1991). Ant association is more extreme in *Micaria*, which are convincing ant mimics and, unlike most gnaphosids, are diurnally active and occur in the company of various ant species (Platnick & Shadab 1988).

## Taxonomic history and notes —

As Gnaphosidae is defined by conspicuous derived characters, most Nearctic members have been traditionally clearly placed in the family. One exception is *Micaria* that, because of its ant mimicry and reduced and only moderately separated ALS, has previously been associated with the clubionoid myrmecomorphs-the castianeirines and phrurolithines-but recently clearly placed in Gnaphosidae (Wunderlich 1979b, Platnick & Shadab 1988). Also ambiguous has been the placement of the prodidomines which, based on the presence of morphologically intermediate taxa, were synonymized with the Gnaphosidae by Platnick & Shadab (1976c). After detailed study of spinneret morphology, the prodidomids have been restored to family status (Platnick 1990a). Interestingly, the gnaphosid subfamily Echeminae, represented here by *Scopoides*, shares some prodidomid characters: the procurved PER and anteriorly displaced ALS.

Most of the pioneering work on Nearctic gnaphosid taxonomy was by Chamberlin, who generated scores of poorly defined taxa. The entire fauna has been recently revised in a long series of papers by Platnick and collaborators and the genera have been clustered into provisional subfamilies, of which the "Drassodinae" were regarded as basal and probably not monophyletic (Platnick 1990a).

## Genera —

"DRASSODINAE"
　　*Drassodes* WESTRING 1851, *Haplodrassus* CHAMBERLIN 1922, *Orodrassus* CHAMBERLIN 1922, *Parasyrisca* SCHENKEL 1963, *Sosticus* CHAMBERLIN 1922, *Synaphosus* PLATNICK & SHADAB 1980a

ANAGRAPHIDINAE
　　*Talanites* SIMON 1893a
LARONIINAE
　　*Callilepis* WESTRING 1874, *Eilica* KEYSERLING 1891
GNAPHOSINAE
　　*Gnaphosa* LATREILLE 1804b
ZELOTINAE
　　*Camillina* BERLAND 1919b, *Drassyllus* CHAMBERLIN 1922, *Trachyzelotes* LOHMANDER 1944, *Urozelotes* MELLO-LEITÃO 1938a, *Zelotes* GISTEL 1848
HERPYLLINAE
　　*Cesonia* SIMON 1893a, *Gertschosa* PLATNICK & SHADAB 1981b, *Herpyllus* HENTZ 1832, *Litopyllus* CHAMBERLIN 1922, *Nodocion* CHAMBERLIN 1922, *Scotophaeus* SIMON 1893a, *Sergiolus* SIMON 1891c
MICARIINAE
　　*Micaria* WESTRING 1851
ECHEMINAE
　　*Scopoides* PLATNICK 1989b

## Distribution summaries —
**Note** *= introduced genera

**Worldwide** (*Drassodes, Micaria, \*Urozelotes, Zelotes*)
**Holarctic** (*Callilepis, Gnaphosa, Haplodrassus, Parasyrisca, Talanites*)
**Nearctic** (*Herpyllus, Orodrassus, Scopoides, Sosticus*)
**North America** (*Cesonia, Litopyllus, Nodocion, Sergiolus*)
**Nearctic** and **Neotropical** (*Drassyllus*)
**Neotropical** (*Gertschosa*)
**Neotropical** and **African** (*Camillina*)
**Gondwandan** to **South Nearctic** (*Eilica*)
**Palearctic** (*\*Scotophaeus, \*Synaphosus, \*Trachyzelotes*)

## Gnaphosidae

**Key to genera —**
North America North of Mexico

1  Body antlike (Fig. 29.2); abdomen with iridescent scales (Fig. 29.3); ALS small, often appearing only slightly separated (Fig. 29.4); genitalia, Figs. 29.5-29.6 .......... ***Micaria***
   **Div.** 41 species — **Dist.** widespread — **Refs.** Platnick & Shadab 1988, Platnick & Dondale 1992, Paquin & Dupérré 2003

—  Body not antlike; abdomen lacking iridescent scales; ALS large, well separated (Fig. 29.7) ......................... **2**

2(1)  Metatarsi III and IV with ventrodistal preening comb (Fig. 29.8a) .......................................................................... **3**

—  Metatarsi III and IV lacking ventrodistal preening comb, but may have a preening brush (Fig. 29.8b) ................... **7**

3(2)  Chelicera with cluster of stiff setae on anteromedian surface (Fig. 29.9); genitalia, Figs. 29.10-29.11 ....................... ............................................................................ ***Trachyzelotes***
   **Div.** 4 species — **Dist.** most in w USA, also IL, FL — **Ref.** Platnick & Murphy 1984 — **Note** all Nearctic species have been introduced from the Mediterranean region.

—  Chelicera without such cluster of setae ............................ **4**

4(3)  Genitalia as in Figs. 29.12-29.13 ................... ***Urozelotes***
   **Div.** 1 species; *Urozelotes rusticus* (L. Koch 1872b) — **Dist.** widespread in USA, ON — **Refs.** Platnick & Murphy 1984, Platnick & Dondale 1992 — **Note** this enigmatic species is a synanthrope with a worldwide distribution but whose origins are not known

—  Genitalia otherwise ......................................................... **5**

5(4)  PME usually round, rarely oval, barely larger than PLE (Fig. 29.14); PE nearly equidistant; PER straight, occasionally slightly procurved; cheliceral retromargin with 1 tooth and 1 denticle; male palpus with intercalary sclerite (Fig. 29.15); epigynum Figs. 29.16-29.17 ............ ***Zelotes***
   **Div.** 47 species — **Dist.** widespread — **Refs.** Platnick & Shadab 1983a, Platnick & Dondale 1992, Platnick & Murphy 1998, Platnick & Prentice 1999, Paquin & Dupérré 2003

—  PME usually oval, larger than PLE, closely spaced; PER slightly procurved (Fig. 29.18); cheliceral retromargin with 2-3 teeth; male palpus lacking intercalary sclerite .......... **6**

6(5)  Male with embolus stout (Fig. 29.19); female with median epigynal plate (Fig. 29.20); known from FL and AL .......... ............................................................................ ***Camillina***
   **Div.** 2 species — **Dist.** FL, AL — **Refs.** Platnick & Shadab 1982b — **Note** *Camillina elegans* (Bryant 1940) [Figs. 29.19-29.20] is native to FL and the Caribbean, *Camillina pulchra* (Keyserling 1891) recorded from FL-TX (A. Dean, pers. comm.), is apparently introduced from Brazil

—  Male with embolus long, whip-like (Fig. 29.21); epigynum variable, usually with paired depressions (Fig. 29.22); widespread ....................................................... ***Drassyllus***
   **Div.** 46 species — **Dist.** widespread — **Refs.** Platnick & Shadab 1982a, Platnick 1984d, Platnick & Corey 1989, Platnick & Dondale 1992, Paquin & Dupérré 2003

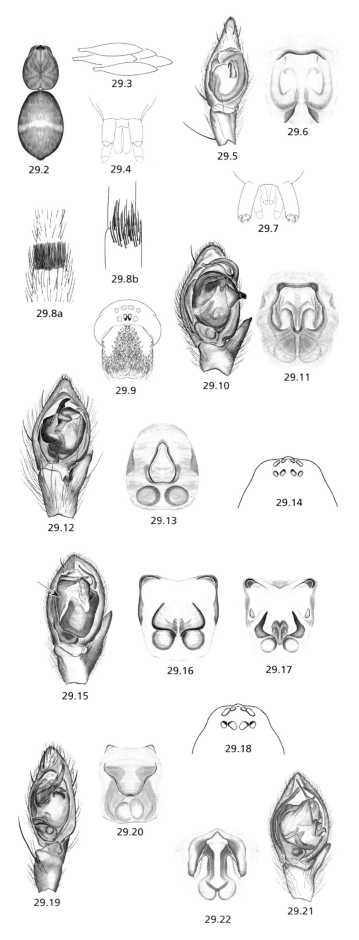

Spiders of North America — *Gnaphosidae*

7(2) Trochanters deeply notched (Fig. 29.23); male palpus with narrow cymbium, no wider than tibia (Fig. 29.24); epigynum as in Fig. 29.25 ......................... ***Drassodes***
**Div.** 7 species — **Dist.** widespread — **Refs.** Platnick & Shadab 1976a, Platnick & Dondale 1992, Paquin & Dupérré 2003

— Trochanters unnotched (Fig. 29.26) or with shallow notches (Fig. 29.27); male palpus with cymbium typically much wider than tibia .............................................. **8**

8(7) PER strongly procurved (Fig. 29.28); sw USA; genitalia, Figs. 29.29-29.30 ............................ ***Scopoides***
**Div.** 7 species — **Dist.** sw USA to Mexico — **Ref.** Platnick & Shadab 1976b — **Note** *Scopodes* Chamberlin 1922 is the original but unavailable name

— PER straight or slightly recurved, rarely slightly procurved; widespread .................................................... **9**

9(8) Cheliceral retromargin with serrated keel (Fig. 29.31) or lobes. (Fig. 29.32) ............................................. **10**
— Cheliceral retromargin lacking keel or lobes ................ **12**

10(9) Cheliceral retromargin with a serrated keel (Fig. 29.31); endites usually rounded laterally (Fig. 29.33); genitalia, Figs. 29.34-29.35 ....................... ***Gnaphosa***
**Div.** 20 species — **Dist.** widespread — **Refs.** Platnick & Shadab 1975a, Platnick & Dondale 1992, Paquin & Dupérré 2003

— Cheliceral retromargin with rounded lobe or lobes (Fig. 29.32); endites usually angular laterally (Fig. 29.36) ... **11**

11(10) Cheliceral retromargin with one rounded lobe (Fig. 29.32); sternum short, stout, about twice as wide as length of endites (Fig. 29.37); PME flattened, transverse (Fig. 29.38); genitalia, Figs. 29.39-29.40 .................. ***Callilepis***
**Div.** 7 species — **Dist.** widespread — **Refs.** Platnick 1975b, Platnick & Dondale 1992, Paquin & Dupérré 2003

— Cheliceral retromargin with 2-3 rounded lobes (Fig. 29.41); sternum narrow, elongate, about as wide as length of endites (Fig. 29.42); PME rectangular (Fig. 29.43); s USA ............................................................. ***Eilica***
**Div.** 1 species: *Eilica bicolor* Banks 1896e [Figs. 29.44-29.45] — **Dist.** s CA, s TX, FL — **Ref.** Platnick 1975c

12(9) Female epigynum with scape (Fig. 29.46) ........................
........................................................ ***Sosticus,*** in part
**Div.** 3 species, 2 key out here — **Dist.** widespread — **Refs.** Platnick & Shadab 1976b, Platnick & Dondale 1992, Paquin & Dupérré 2003

— Males, or female with epigynum lacking scape ............ **13**

13(12) Cheliceral retromargin with 2 or 3 teeth (Fig. 29.47) ........
........................................................................... **14**
— Cheliceral retromargin with at most one tooth (Figs. 29.48-29.49) ..................................................... **18**

14(13) Tibia IV lacking dorsal spines ........................ **15**
— Tibia IV with 2 dorsal spines ........................ **17**

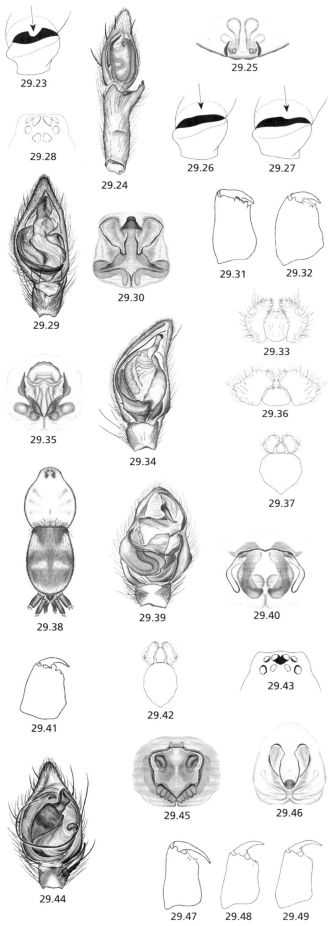

# Gnaphosidae

**15(14)** PME separated by no more than 1/2 their diameter (Fig. 29.50); tibia I lacking ventral spines; RTA flattened and dorsally positioned (Fig. 29.51); genitalia, Figs. 29.52-29.53 ............................................ ***Haplodrassus***
<small>**Div.** 9 species — **Dist.** widespread — **Refs.** Platnick & Shadab 1975b, Platnick & Dondale 1992, Paquin & Dupérré 2003</small>

— PME separated by at least their diameter (Fig. 29.54); tibia I usually with at least 1 ventral spine; RTA neither flattened nor dorsally positioned ............................................ **16**

**16(15)** Tibia I with 2 pairs of ventral spines; RTA with single prong (Fig. 29.55) ............................................ ***Parasyrisca***
<small>**Div.** 1 species: *Parasyrisca orites* (Chamberlin & Gertsch 1940) [Figs. 29.56-29.57] — **Dist.** BC, WA — **Refs.** Platnick & Shadab 1975b, Ovtsharenko et al. 1995, Platnick & Dondale 1992 — **Note** previously included in *Orodrassus*</small>

— Tibia I with 0-1 ventral spine; RTA with two prongs (Fig. 29.58); genitalia, Figs. 29.59-29.60 ............... ***Orodrassus***
<small>**Div.** 3 species — **Dist.** CAN, w USA — **Refs.** Platnick & Shadab 1975b, Ovtsharenko et al. 1995, Platnick & Dondale 1992, Paquin & Dupérré 2003</small>

**17(14)** Male RTA longer than tibia, and median apophysis in apical position (Fig. 29.61); palp ventral view (Fig. 29.62); female epigynum with scape (Fig. 29.46) ............................................
............................................ ***Sosticus*, in part**
<small>**Div.** 3 species, 2 key out here: *Sosticus californicus* Platnick & Shadab 1976b and *Sosticus insularis* (Banks 1895d) — **Dist.** CAN, e USA — **Refs.** Platnick & Shadab 1976b, Platnick & Dondale 1992, Paquin & Dupérré 2003</small>

— Male RTA shorter than tibia, and median apophysis in median position (Fig. 29.63); female epigynum without scape (Fig. 29.64) ............................................ ***Talanites***
<small>**Div.** 5 species — **Dist.** CAN, se USA — **Refs.** Platnick & Shadab 1976b, Platnick & Ovtsharenko 1991, Ubick & Moody 1995 — **Note** *Rachodrassus* Chamberlin 1922 is a synonym</small>

**18(13)** Abdomen with pale stripes or bands (Figs. 29.1; 29.65-29.66; 29.69; 29.71-29.73) ............................................ **19**

— Abdomen concolorous, rarely with faint chevron pattern posteriorly (some *Synaphosus*) ............................................ **22**

**19(18)** Cheliceral retromargin lacking denticles or teeth (Fig. 29.49); abdomen with transverse white markings (Fig. 29.65-29.66); genitalia Figs. 29.67-29.68 ......... ***Sergiolus***
<small>**Div.** 16 species — **Dist.** widespread — **Refs.** Platnick & Shadab 1981e, Paquin & Dupérré 2003 — **Note** species were previously included in the Palearctic genus, *Poecilochroa* Westring 1974</small>

— Cheliceral retromargin with one denticle (Fig. 29.48); abdominal markings variable ............................................ **20**

**20(19)** Abdomen with transverse white markings anteriorly (Fig. 29.69); s TX ............................................ ***Gertschosa***
<small>**Div.** 1 species: *Gertschosa amphiloga* (Chamberlin 1936b) [Fig. 29.70] — **Dist.** s TX — **Ref.** Platnick & Shadab 1981b — **Note** the male is not known</small>

— Abdomen with longitudinal white markings (Figs. 29.1; 29.71-29.73); widespread ............................................ **21**

Spiders of North America — *Gnaphosidae* | **111**

**21(20)** Abdomen with single median white stripe; carapace dark, concolorous (Fig. 29.1); PE equidistant ............................... ............................................................. ***Herpyllus***, *in part*
**Div.** 13 species, 2 key out here: *Herpyllus ecclesiasticus* Hentz 1832 [Fig. 29.1] and *Herpyllus propinquus* (Keyserling 1887b) — **Dist.** widespread — **Refs.** Platnick & Shadab 1977b, Platnick & Dondale 1992, Paquin & Dupérré 2003

— Abdomen with two or more white stripes (Fig. 29.71-29.73); carapace usually with white markings; PME closer to PLE than to each other; genitalia, Figs. 29.74-29.75 ...... ........................................................................... ***Cesonia***
**Div.** 10 species — **Dist.** widespread — **Refs.** Platnick & Shadab 1980b, Platnick & Dondale 1992, Paquin & Dupérré 2003

**22(18)** Trochanters not notched (Fig. 29.26) ............................. **23**

— At least posterior trochanters slightly notched (Fig. 29.27) ................................................................................................. **24**

**23(22)** Male palpus with embolus fully exposed and RTA bifid (Fig. 29.76); female epigynum with a scape (Fig. 29.46) ... ................................................. ***Sosticus***, *in part*
**Div.** 3 species, 1 keys out here: *Sosticus loricatus* (L. Koch 1866c) [Figs. 29.46; 29.76] — **Dist.** se CAN + ne USA to Rocky Mountains — **Refs.** Platnick & Shadab 1976b, Platnick & Dondale 1992, Paquin & Dupérré 2003 — **Note** *Sostogeus* Chamberlin & Gertsch 1940 is a synonym

— Male palpus with embolus partially enclosed in folded median apophysis and RTA single (Figs. 29.77-29.78); female with epigynum lacking a scape (Figs. 29.79-29.80) ..................................................................... ***Synaphosus***
**Div.** 2 species: *Synaphosus paludis* (Chamberlin & Gertsch 1940) [Figs. 29.78; 29.79] and *Synaphosus syntheticus* (Chamberlin 1924b) [Figs. 29.77; 29.80] — **Dist.** s USA to MEX — **Refs.** Platnick & Shadab 1980a, Ovtsharenko et al. 1994 — **Note** this genus is native to the Palearctic region. Ovtsharenko et al. 1994 mention that *Synaphosus paludis* is not congeneric and represents an undescribed genus

**24(22)** Cheliceral promargin toothed (Fig. 29.81); tibia IV with one dorsal spine; PME subequal to PLE, separated by their diameter ............................................................................. **25**

— Cheliceral promargin with carina (Fig. 29.82); tibia IV lacking dorsal spine; PME larger than PLE, usually separated by less than a diameter .................................................... **26**

**25(24)** Male palpus with embolus bearing enlarged base, median apophysis elongate, inconspicuous (Fig. 29.83); female with spermatheca lacking terminal bulbs (Fig. 29.84) ....... ............................................................. ***Herpyllus***, *in part*
**Div.** 13 species, 11 key out here — **Dist.** widespread — **Refs.** Platnick & Shadab 1977b, Platnick & Dondale 1992

— Male palpus with embolus sinuous, lacking enlarged base, median apophysis hooked (Fig. 29.85); female with spermatheca bearing terminal bulbs (Fig. 29.86) ....................... ..................................................................... ***Scotophaeus***
**Div.** 1 species: *Scotophaeus blackwalli* (Thorell 1871a) — **Dist.** w coast from CAN to MEX, Gulf states — **Refs.** Platnick & Shadab 1977b, Platnick & Dondale 1992 — **Note** the genus is native to the Palearctic region and this species is a cosmopolitan synanthrope

**26(24)** Cheliceral retromargin with single tooth (Fig. 29.87); PER procurved; male palpus with cymbium not excavated (Fig. 29.88); female with spermatheca clearly bipartite (Fig. 29.89) .................................................................. ***Litopyllus***
**Div.** 2 species *Litopyllus cubanus* (Bryant 1940) and *Litopyllus temporarius* Chamberlin 1922 — **Dist.** e USA — **Refs.** Platnick & Shadab 1980a, Platnick & Dondale 1992

— Cheliceral retromargin lacking teeth (Fig. 29.90); PER straight; male palpus with cymbium excavated to receive RTA (Fig. 29.91); female with spermathecae not conspicuously bipartite (Fig. 29.92) ............................ ***Nodocion***
**Div.** 6 species — **Dist.** widespread — **Refs.** Platnick & Shadab 1980a, Platnick & Dondale 1992

29.71  29.72  29.73  29.74  29.75

29.76  29.77  29.78

29.81  29.79  29.80

29.82  29.83  29.84

29.85  29.86  29.87  29.90

29.88  29.89  29.92  29.91

Chapter 30
# HAHNIIDAE
9 genera, 58 species

Robert G. Bennett

## Similar families —
Hahniidae is phylogenetically most closely related to Cybaeidae (p. 85) and Dictynidae (p. 95). The Nearctic hahniine taxa are unlikely to be confused with other spiders because of their distinctive spinnerets. Cryphoecinae may be misidentified as Agelenidae (p. 56), (with which they have been classified in the past), cybaeids, or dictynids.

## Diagnosis —
Hahniines (3 genera) with tracheal spiracle well in front of spinnerets, spinnerets in a transverse row (Figs. 30.10-30.12). Cryphoecines (6 genera) with spinnerets longitudinally arranged, ALS distinctly separated and shorter than PLS (Fig. 30.13), 8 eyes (AME may be reduced), tibia 1 with 4 or 5 pairs of ventral macrosetae (Fig. 30.21) *or* fewer pairs of ventral tibia 1 macrosetae and cheliceral retromargin with teeth but not denticles (Fig. 30.33) *or* fewer pairs of ventral tibia 1 macrosetae, cheliceral retromargin with teeth and denticles (TE and D in Fig. 30.29), long legs (length of patella/tibia I about 1.25 times as long as carapace), and tip of male palpal cymbium conspicuously constricted and elongated (C in Figs. 30.30, 30.31, 30.37).

## Characters —
**body size:** hahniines 1.2-5.1 mm; cryphoecines 2.0-7.8 mm.
**color:** variable shades of earth tones-pale to dark gray, brown, or reddish; often with darker radiating lines on cephalothorax and/or pale chevrons on abdomen.
**carapace:** generally pyriform; fovea longitudinal (somewhat reduced in hahniines); *Antistea* and *Neoantistea* with inconspicuous stridulatory "picks" on posterior margin (P in Fig. 30.16).
**sternum:** slightly longer than wide, broadly truncate anteriorly, posteriorly truncate (hahniines-Figs. 30.10-30.12) or bluntly acuminate (cryphoecines) and extending between separated hind coxae.
**eyes:** usually eight eyes in two transverse rows (in dorsal view PER procurved in hahniines (Fig. 30.14, 30.16, 30.17) but not so in cryphoecines (Fig. 30.23, 30.26)), AME reduced in *Hahnia* (Fig. 30.17) and *Ethobuella* (remaining eyes enlarged in *Ethobuella*-Fig. 30.23).
**chelicerae:** slightly to moderately geniculate, basal boss present or absent; margins toothed (Figs. 30.29, 30.33).
**mouthparts:** endites slightly convergent, about as long to twice as long as wide; labium variably subquadrate (Figs. 30.10-30.12).
**legs:** tarsi 3-clawed with 2-3 (occasionally 4) dorsal trichobothria; macrosetae very few to numerous, tibia and metatarsus I ventrally with conspicuous pairs of elongate macrosetae in *Dirksia* and *Ethobuella* (Fig. 30.21), trochanteral notches absent.
**abdomen:** ovoid, pale to dark earthtones, concolorous or with variable chevron-like markings dorsally; *Antistea* and *Neoantistea* with inconspicuous stridulatory "file" dorsolateral of petiole (F in Fig. 30.15).
**spinnerets:** ecribellate; hahniines-3 pairs arranged in single transverse row with PLS (outermost) and PMS (innermost) separated by AMS (Figs. 30.10-30.12), colulus

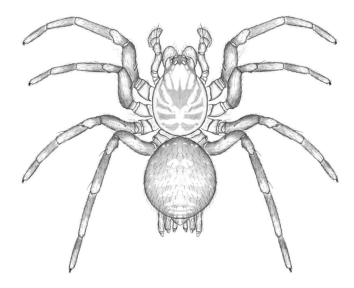

**Fig. 30.1** *Hahnia cinerea* EMERTON 1890

reduced or absent. Cryphoecines-3 pairs arranged in 2 "normal" staggered longitudinal rows (Fig. 30.13), colulus reduced and marked by only 2 setae (*Neocryphoeca*) to a broad swelling with up to 20 setae (*Calymmaria*).
**respiratory system:** paired anterior book lungs; spiracle broad and advanced a variable distance in front of colulus/PMS (hahniines-SPR in Figs. 30.10-30.12) or small and just anterior of colulus (cryphoecines); tracheae with two branches and extending into cephalothorax.
**genitalia:** entelegyne; HAHNIINES -*female* epigynum conspicuous with usually a pair of atria located centrally (AT in Fig. 30.4) or anteriorly (AT in Figs. 30.5-30.6); vulval ducting complex (Fig. 30.6) with or without bulbous spermathecal component, both simple and complex vulval pores present (PS, PC in Fig. 30.9); *male* palpal femur with retrolateral apophysis (*Antistea*-Fig. 30.18) or not; patella swollen (*Antistea*) or with variable basal retrolateral apophysis (PA in Figs. 30.2, 30.3 30.19); RTA conspicuously developed apically (Figs. 30.2, 30.3, 30.19); cymbium usually unmodified except for retrolateral groove; tegulum in some *Hahnia* with conspicuous basal row of setae (SE in Fig. 30.3), tegular apophyses absent or, at most, very reduced; embolus elongate, slender, unmodified, forming clockwise (left bulb, ventral view) complete (or nearly so) circle (E in Figs. 30.3, 30.20); CRYPHOECINES-*female* epigynum variable, usually inconspicuous with single or paired atrium variably located anteriorly, centrally, or posteriorly (AT in Figs. 30.32, 30.36, 30.38, 30.40); vulval ducting vari-

able, usually relatively simple with short ducts leading to bulbous spermathecal components, simple vulval pores present, complex pores usually present (PS, PC in Figs. 30.7-30.8); **male** palpal femur unmodified; patella unmodified or with small apical prolateral (*Calymmaria*-PA in Fig. 30.31) or retrolateral (*Willisus* PA in Fig. 30.35) lobe; RTA strongly developed and variable (with conspicuous apical projection in *Cryphoeca* Fig. 30.34); cymbium unmodified or, in *Calymmaria* and *Willisus*, with elongate abruptly narrowed tip (C in Figs. 30.30, 30.31, 30.37); 1 (TA in Fig. 30.41) or 2 (*Willisus*-TA in Fig. 30.37) variously developed tegular apophyses ("conductor"); embolus as in hahniines (*Cryphoeca* and *Neocryphoeca*-E in Figs. 30.39, 30.41) or short, stout, sinuous, and often slightly modified (*Calymmaria* and *Willisus* -E in Figs. 30.31, 30.37) to very short (*Ethobuella* and *Dirksia*-E in Figs. 30.25, 30.28).

## Distribution —

Hahniines have a world-wide distribution. *Hahnia* and *Neoantistea* are widespread in North America; *Antistea* is found in northeastern USA, adjacent parts of Canada and in British Columbia. Cryphoecines are Holarctic with the majority being endemic to the western Nearctic. One Nearctic species (*Cryphoeca montana* EMERTON 1909) is widely distributed across Canada and the adjacent northern USA. Except for two species [*Calymmaria persica* (Hentz 1847) and *Calymmaria virginica* Heiss & Draney 2004], all *Calymmaria* are found west of the continental divide from Alaska to California and Arizona.

## Natural history —

Opell & Beatty (1976) briefly reviewed the natural history of hahniine genera. Among Nearctic hahniines, only *Neoantistea* has been observed to build webs; these are small simple sheets without retreats built in moss or over small depressions in the ground (often these webs can be found in the footprints left by animals walking on soft ground). Other hahniines are litter inhabitants occurring in a wide variety of habitats. Collection records for some hahniines suggest the presence of adults in early spring and again in late summer. Hahniine egg cases apparently contain small numbers (<10) of relatively large eggs.

Roth & Brame (1972) summarized the natural history of cryphoecine genera. Cryphoecines are nocturnally active and primarily restricted to forested habitats; *Ethobuella* and *Dirksia* have been collected from vegetation but cryphoecines are usually ground-dwelling or litter-inhabiting spiders. Heiss & Draney (2004) described webs and feeding behavior of *Calymmaria*; Ubick (pers. comm.) has provided additional observations on the webs and eggs sacs of various species in this genus. Their webs are linyphiid-like sheets built in cavities and on the undersurface of various objects at or near ground level. Silk strands attached to the substrate below pull the sheet down in the middle to form a cone. Various anchoring and/or entanglement strands are attached to the upper surface of the web. The occupant travels linyphiid-like on the undersurface of the web on the outside of the cone. The egg sacs of *Calymmaria* are suspended from the undersurface of the web and are flattened, angular to somewhat rounded, and covered with a stiff coating of earth. The structure of the webs and egg sacs of other cryphoecine genera has not been recorded.

With the exception of *Calymmaria* and *Cryphoeca* (which are relatively commonly collected), cryphoecines are rarely encountered. However, as with many supposedly rare organisms, they may be abundant under certain conditions; Bennett (1988) found *Ethobuella* (as well as other "rare" RTA clade spiders) to be common on the south coast of British Columbia within moss mats on the standing trunks of riparian deciduous trees in old forests. *Neocryphoeca* and *Willisus* are known only from a very small number of localities (Roth 1970, 1981).

## Taxonomic history and notes —

As with most "RTA-clade" spider assemblages, Hahniidae has had a checkered past. Traditionally restricted to include only the distinctively spinnereted hahniines, the family has at times been placed within Agelenidae (e.g., Bonnet 1957a and included references) and, at higher levels, variously within Lycosoidea (e.g., Kaston 1948), Amaurobioidea (Lehtinen 1967), and currently Dictynoidea (Coddington & Levi 1991). Citing within-genus variability in the traditional hahniid spinneret pattern as well as genitalic similarities, Lehtinen (1967) first proposed including cryphoecine taxa (previously included in Agelenidae) in Hahniidae. Platnick (2005) and Coddington & Levi (1991) outline the current taxonomic make-up of Hahniidae and its higher level classification respectively. However, these remain controversial in the absence of in-depth phylogenetic analysis.

Opell & Beatty's (1976) revision of the Nearctic hahniines built upon a taxonomic foundation laid out by earlier workers including Chamberlin & Ivie (1942a) and especially Gertsch (1934c, 1946b). No inclusive revision of cryphoecines exists. Roth & Brame (1972) reviewed the Nearctic cryphoecine genera (except for *Willisus*); important papers describing Nearctic cryphoecine genera and species were published by Chamberlin & Ivie (1937-*Calymmaria* and *Ethobuella*, 1942a-*Calymmaria* and *Dirksia*) and Roth (1970-*Neocryphoeca*, 1981-*Willisus*, and 1988a-*Cryphoeca*). Roth & Brown (1986) catalogued all Nearctic cryphoecine taxa and literature; this work has subsequently been superseded by Platnick (2005). A revision of *Calymmaria* was the subject of Heiss's (1982) doctoral dissertation but his untimely death delayed publication. Fortunately, Draney recently completed the manuscript and this revision has now been published (Heiss & Draney 2004). Bennett (1992) described in detail the internal morphology of the female genitalia of *Neoantistea* and cryphoecines (exclusive of *Willisus*).

## Genera —

CRYPHOECINAE
  *Calymmaria* CHAMBERLIN & IVIE 1937, *Cryphoeca* THORELL 1870b, *Dirksia* CHAMBERLIN & IVIE 1942a, *Ethobuella* CHAMBERLIN & IVIE 1937, *Neocryphoeca* ROTH 1970, *Willisus* ROTH 1981
HAHNIINAE
  *Antistea* SIMON 1898a, *Hahnia* C.L. KOCH 1841a, *Neoantistea* GERTSCH 1934c

# Hahniidae

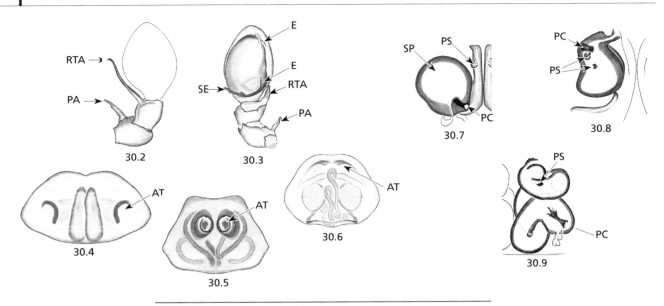

## Key to genera —
### North America North of Mexico

1. Spinnerets arranged in a single transverse row (Figs. 30.10-30.12) ................................. **HAHNIINAE** ... 2
— Spinnerets arranged in "normal" double longitudinal row (Fig. 30.13) ............................... **CRYPHOECINAE** ... 4

2(1). Tracheal spiracle (SPR) closer to spinnerets than to epigastric groove (Fig. 30.11); AME significantly reduced (Fig. 30.17); stridulatory apparatus lacking ................ **Hahnia**
   Div. 7 species — Dist. Widespread — Ref. Opell & Beatty 1976
— Tracheal spiracle (SPR) equidistant between epigastric furrow and spinnerets (Fig. 30.12) or closer to epigastric furrow (Fig. 30.10); AME not significantly reduced (Fig. 30.14, 30.16); abdominal/cephalothoracic pick (P) and file (F) stridulatory apparatus present (Figs. 30.15-30.16) ..... 3

3(2). Tracheal spiracle equidistant between epigastric furrow and spinnerets (Fig. 30.12); bases of PMS separated by ≤ ½ their diameter (Fig. 30.12); distal segment of lateral spinnerets shorter than median segment (Fig. 30.12); male palpal femur with retrolateral apophysis (Fig. 30.18), patella lacking apophysis ..................... **Antistea**
   Div. 1 species: *Antistea brunnea* (EMERTON 1909) — Dist. BC, se CAN, ne USA — Ref. Opell & Beatty 1976, Paquin & Dupérré 2003
— Tracheal spiracle closer to epigastric furrow (Fig. 30.10); bases of median spinnerets separated by about their diameter (Fig. 30.10); distal segment of lateral spinnerets about as long as median segment (Fig. 30.10); male palpal femur lacking apophysis, patella usually with proximal retrolateral spur (PA in Figs. 30.19) .......... **Neoantistea**
   Div. 11 species — Dist. Widespread — Ref. Opell & Beatty 1976

4(1). 4 (occasionally 1 distal macroseta lacking) or 5 complete pairs of ventral tibia I macrosetae arranged in a smooth linear pattern (Fig. 30.21) ................................. 5
— 1 or 2 complete pairs of ventral tibia I macrosetae (Fig. 30.22) (may be other, unpaired macrosetae present) or ventral tibia I macrosetae lacking entirely ...................... 6

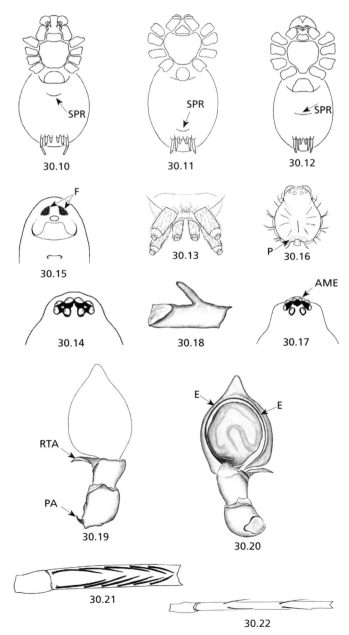

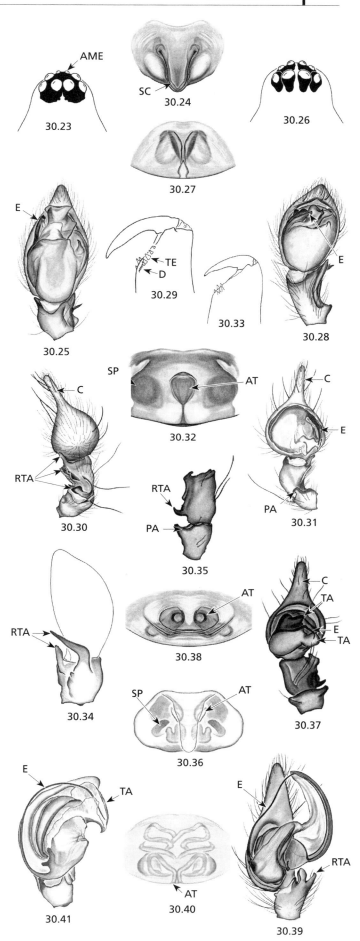

**5(4)** 4 pairs of ventral tibia I macrosetae (occasionally 1 distal macroseta missing); 3 pairs of ventral metatarsus I macrosetae; AME minute, other eyes equal and large (Fig. 30.23); epigynum with broad, posteriorly directed scape (SC in Fig. 30.24); palp as in Fig. 30.25 .................... **Ethobuella**
  **Div.** 2 species: *Ethobuella hespera* CHAMBERLIN & IVIE 1937 and *Ethobuella tuonops* CHAMBERLIN & IVIE 1937 — **Dist.** sw BC south to central CA — **Ref.** Chamberlin & Ivie 1937

— 5 pairs of ventral tibia I macrosetae (Fig. 30.21); 4 pairs of ventral metatarsus I macrosetae; AME larger than ALE, PE larger than AE (Fig. 30.26); epigynum lacking scape (Fig. 30.27); palp as in Fig. 30.28 ...................................... **Dirksia**
  **Div.** 1 species: *Dirksia cinctipes* (BANKS 1896e) — **Dist.** sw AK south to n coastal CA — **Ref.** Chamberlin & Ivie 1942a

**6(4)** Retromargin of cheliceral fang furrow with 3-4 teeth and 1 or more smaller "denticles" (TE and D in Fig. 30.29); usually only 1 complete pair of ventral tibia I macrosetae; RTA deeply excavated basally (Fig. 30.30); palpal bulb as in Fig. 30.31; vulval ducting simple with single pair of rounded spermathecae (SP in Figs. 30.7, 30.32); length 3.1-7.8 mm .................................................. **Calymmaria**
  **Div.** 31 species — **Dist.** s BC to s CA; IN and WV s to AL and GA — **Ref.** Heiss & Draney 2004

— Cheliceral retromargin with 3-5 equal-sized teeth only (Fig. 30.33); usually 2 complete pairs of ventral tibia I macrosetae (or lacking entirely); RTA otherwise (Figs. 30.34, 30.35, 30.39); vulval ducting otherwise, if simple then spermathecae variously shaped but not rounded (SP in Fig. 30.36); length <5 mm .......................... **7**

**7(6)** No ventral tibia I macrosetae; legs long, length of patella/tibia I 1.2-2 times length of carapace; legs most likely to break across patella; male palp (Figs. 30.35, 30.37), epigynum (Fig. 30.38); tiny spiders, length ≤2 mm ................. **Willisus**
  **Div.** 1 species: *Willisus gertschi* ROTH 1981 — **Dist.** s CA — **Ref.** Roth 1981

— 2 complete pairs of ventral tibia I macrosetae; legs shorter, patella/tibia I length about equal to that of carapace; legs most likely to break between coxa and trochanter; male palp (Figs. 30.34, 30.39, 30.41), epigynum (Figs. 30.36, 30.40); slightly larger spiders, length 2-4.5 mm ................... **8**

**8(7)** RTA with strongly developed pointed, dorso-apical extension (Fig. 30.34); epigynum with anterior pair of atria (AT in Fig. 30.36); colulus with about 10 setae; small spiders, length 2.5-4.5 mm, palp bulb as in Fig. 30.41 ..................................................................... **Cryphoeca**
  **Div.** 2 species: *Cryphoeca exlineae* ROTH 1988a and *Cryphoeca montana* EMERTON 1909 — **Dist.** widespread, AK to LB s to n USA (s to mid CA in west) — **Refs.** Roth & Brame 1972, Paquin & Dupérré 2003

— Males undescribed, RTA not as above, palpal bulb as in figure 30.39; epigynum with single or paired atria located close to epigastric groove (AT in Fig 30.40); colulus with only 2 setae; very small spiders, 2.0-2.3 mm ........................
.................................................. **Neocryphoeca**
  **Div.** 2 species: *Neocryphoeca beattyi* ROTH 1970 and *Neocryphoeca gertschi* ROTH 1970 — **Dist.** montane AZ — **Ref.** Roth 1970 — **Note** although Roth (1982) published a figure of the male palp (ventral view) of a *Neocryphoeca* sp. and another is published here, no males of this genus have yet been described

| | |
|---|---|
| AT | epignyal atrium |
| AME | anterior median eyes |
| C | tip of cymbium |
| D | cheliceral fang furrow denticles |
| E | embolus |
| F | abdominal stridulatory file |
| P | cephalothoracic stridulatory pick |
| PA | patellar apophysis |
| PC | complex spermathecal pore |
| PS | simple spermathecal pore |
| RTA | retrolateral tibial apophysis |
| SC | epigynal scape |
| SE | tegular setae |
| SP | spermatheca |
| SPR | spiracle |
| TA | tegular apophysis |
| TE | cheliceral fang furrow teeth |

# Chapter 31
# HERSILIIDAE
2 genera, 2 species

Bruce Cutler

### Similar families —
Oecobiidae (p. 183), Philodromidae (p. 192).

### Diagnosis —
Medium body size, body dorsoventrally flattened with ocular area raised; posterior spinnerets extremely long, as long or longer than abdomen; medial side of spinnerets with series of seta-like spigots; very long slender legs (Fig.31.1).

### Characters —
**body size:** 6-8 mm long (excluding spinnerets).
**color:** cephalothorax yellow to brown with thin black line around carapace except at posterior; legs yellow to brown; abdomen grayish brown with dark and pale marks.
**carapace:** flattened with ocular area raised; oval in female, as wide as long; longer than wide in male.
**sternum:** heart-shaped; anterior edge much wider than base of labium.
**eyes:** eight eyes in two strongly recurved rows; AME largest followed by PME in *Neotama*, PME largest followed by AME in *Yabisi*; in group at center anterior edge of carapace.
**chelicerae:** lacking boss; three promarginal teeth, series of 4+ denticles on retromargin.
**mouthparts:** labium rebordered; endites strongly convergent.
**legs:** prograde; very long and slender; three claws, median claw smooth; few macrosetae present; tarsus very short, lacking trichobothria; metatarsi as long or longer than corresponding femora; leg III shortest; flexible area (often curved) in distal part of metatarsi I, II and IV in *Neotama*.
**abdomen:** pyriform widest towards posterior; anal tubercle prominent.
**spinnerets:** colulus present; anterior spinnerets cylindrical, slightly tapering distally; posterior spinnerets extremely long, as long or longer than abdomen with a series of prominent seta-like spigots on mesal side.
**respiratory system:** one pair of book lungs, tracheal spiracle close to spinnerets.
**genitalia:** entelegyne; *female* epigyna with obvious paired narrow openings; spermathecae elongated; *male* palpus lacking tibial apophysis; with several macrosetae at and near tip of cymbium; bulb with median apophysis; embolus structure variable.

### Distribution —
*Neotama mexicana* (O. Pickard-Cambridge 1893b) in extreme southern Texas, *Yabisi habanensis* (Franganillo Balboa 1936b) in extreme southern Florida.

### Natural history —
Very little is known about the two Nearctic species. Label data indicate that specimens of *Neotama mexicana* are taken from woody vegetation. Some hersiliids are known to live on tree trunks where they rest on silken mats, frequently camouflaged by mosses and lichens. They capture approaching prey by rapidly circling and swathing them with silk. Other species live amidst stone fields where they build irregular webs under rocks. Camouflaged egg sacs are attached to tree limbs or under rocks (Dippenaar-Schoeman & Joqué 1997). Baehr & Baehr (1993) suggest that modified metatarsi may be indicative of an arboreal habitat. Based on this assumption, *Yabisi habanensis* which possesses unmodified metatarsi, may live under rocks and stones.

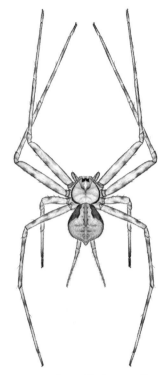

**Fig. 31.1** *Neotama mexicana* (O. Pickard-Cambridge 1893b)

### Taxonomic history and notes —
The family has long been recognized by its distinctive morphological features. Lehtinen (1967) placed it near Oecobiidae based on morphological similarities. Coddington & Levi (1991) put Hersiliidae in the Eresoidea which includes Eresidae and Oecobiidae. Both of our species were originally placed in *Tama*. Baehr & Baehr (1993) in their revision of southeast Asian hersiliids note that the Neotropical specimens they examined do not belong in *Tama* Simon 1882b and seemed close to their new genus *Neotama*. This was confirmed by Rheims & Brescovit (2004a) who placed *Tama mexicana* and numerous other New World specimens in *Neotama*, and erected several new genera. *Tama habanensis* was placed into one of these new genera, *Yabisi*.

### Genera —
*Neotama* Baehr & Baehr 1993, *Yabisi* Rheims & Brescovit 2004a

## Key to genera —
### North America North of Mexico

1  Metatarsi I, II, IV with flexible zone in distal portion (appears curved) (Fig. 31.2); male palpus with cymbium extending narrowly beyond bulb; coiled filiform embolus making a complete turn; middle of bulb with hammerhead median apophysis (arrow) (Fig. 31.3); epigynum with straight openings at posterior margin (Fig. 31.4); spermathecae cylindrical .................................. **Neotama**
**Div.** 1 species: *Neotama mexicana* (O. PICKARD-CAMBRIDGE 1893b) — **Dist.** extreme southern TX — **Ref.** Rheims & Brescovit 2004a

—  Metatarsi I, II, IV lacking a flexible zone in distal portion (appears straight) (Fig.31.5); male palpus with cymbium not narrowly extending beyond bulb; short, stout embolus tapering at tip; median apophysis not hammerhead shaped (Fig. 31.6); epigynum with curved anterior openings facing laterally near the midline, spermathecae kidney shaped ....................................................... **Yabisi**
**Div.** 1 species: *Yabisi habanensis* (FRANGANILLO BALBOA 1936b) — **Dist.** extreme southern FL — **Ref.** Rheims & Brescovit 2004a

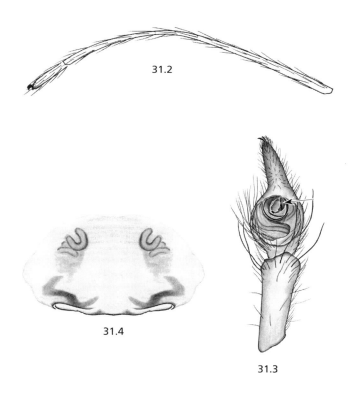

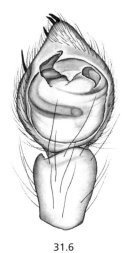

# Chapter 32
# HOMALONYCHIDAE
1 genus, 2 species

Sarah C. Crews

### Similar families —
Lycosidae (p. 164), Pisauridae (p. 199).

### Diagnosis —
Homalonychids can be distinguished from other spiders by the combination of strongly convergent endites, absence of a serrula and lack of teeth on the tarsal claws.

### Characters —
**body size:** 6.5-12.8mm.
**color:** brownish to reddish, usually resembling the color of the substrate, with variable darker markings on the dorsal surface: carapace darker in cephalic region, with a series of lateral spots, and with a w-shaped mark on posterior declivity; abdomen with cardiac mark and lateral stripes (Fig. 32.1). Legs with bands, especially prominent on the femora.
**carapace:** cephalic region narrow, distinct from thoracic; fovea deep and longitudinal; carapace rounded laterally and swollen dorsally, covered with modified erect setae; clypeus high; posterior declivity of carapace with w-shaped indentation connected to fovea (Fig. 32.1).
**sternum:** slightly longer than wide, truncated anteriorly, narrowly rounded between hind coxae (Fig. 32.2).
**eyes:** 8 eyes; secondary eyes with canoe shaped tapetum (Homann 1971); AER straight, PER strongly recurved; ALE smallest, posterior eyes nearly equal, largest.
**chelicerae:** paturon setose, small and straight with a distinct boss; no cheliceral armature; fang small, unserrated.
**mouthparts:** labium rhomboid (sides parallel and nearly equal, but narrowing apically) with truncate tip (Fig. 32.2); endites quadrate, strongly convergent, with scopula, lacking serrula.
**legs:** long, subequal in length, longer and more slender in male, tarsi with long, paired claws lacking teeth, median claw reduced to rounded lobe.
**abdomen:** longer than wide, pentagonal in shape, dorsoventrally flattened and covered with modified setae.
**spinnerets:** six contiguous; anterior lateral spinnerets stout; PLS slender; posterior median spinnerets smallest, difficult to see in male; colulus absent.
**respiratory system:** tracheal spiracle at base of spinnerets, minute; tracheae confined to abdomen (Petrunkevitch 1939a).
**genitalia:** entelegyne; *female* external epigynum with median lobe and two lateral lobes (Fig. 32.3: *Homalonychus theologus*, Fig. 32.4: *Homalonychus selenopoides* MARX 1891); *male* palp with process at base of cymbium; retrolateral tibial apophysis simple with several points along the top (Fig. 32.5); embolus distal, curving ectally; median apophysis basal; conductor formed from distal edge of tegulum (Fig. 32.6: *Homalonychus selenopoides*, Fig. 32.7: *Homalonychus theologus*). External epigyna of female and cymbial process and RTA of male highly variable (Roth 1984, pers. obs.).

### Distribution —
Southern California to central Arizona, including southern Nevada, south to northwest Mexico, including the Baja California Peninsula.

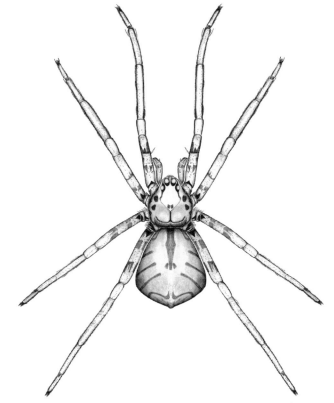

**Fig. 32.1** *Homalonychus theologus* CHAMBERLIN 1924b

### Natural history —
These spiders live in the desert and chaparral and have been found under rocks and plant debris on rocky hillsides and washes. They do not build webs and apparently do not make draglines (pers. obs.). Females presumably only use silk for egg sacs and males for wrapping the female during copulation. Adult males and juveniles have been collected wandering at night. The females and juveniles are cryptic in that they are usually found partially buried with particles of the substrate lodged between stiff setae on the carapace, abdomen and legs. In the field they have been observed eating other homalonychids and thysanurans. They are easily reared in the laboratory on crickets and fruitflies. The lifespan is certainly over two years. Adults have been collected throughout the year.

### Taxonomic history and notes —
The genus *Homalonychus* was described by Marx (1891) who associated it with the Sparassidae, particulary *Parahedrus* and *Selenops*, now placed in the Pisauridae and the Selenopidae, respectively. Later, Simon (1893a) included the genus in Zodariidae until Petrunkevitch (1923) ele-

vated it to family rank. A second species was described by Chamberlin (1924b) and the two were revised by Roth (1984). Patel & Reddy (1991) described a third species, *Homalonychus rhagavi*, from India. However, after examining the publication, Jocqué (pers. comm.) places this spider in the zodariid genus *Storenomorpha*.

The phylogenetic position of Homalonychidae is not fully resolved. Petrunkevitch (1933) and Bristowe (1938a) placed it close to the Gnaphosidae. di Caporicacco (1938f) and Gerhardt & Kästner (1938-1941) used Homalonychiformia for Homalonychidae and Cithaeronidae. Mello-Leitão (1941a) placed it as "posição incerta" under Ctenoidea. In 1967, Lehtinen stated it was clear that this family belongs to the Amaurobiides. He placed it in the superfamily Pisauroidea as its eye pattern, feathery hairs, notched trochanters and genitalia seemed similar to the Oxyopidae and Pisauridae. Roth (1984) did not agree with Lehtinen and thought the homalonychids shared a closer relationship to some zodariids, but preferred to keep the Homalonychidae as a separate family. Clearly, homalonychids share affinities with several groups, but their derived morphology and curious habits make their placement difficult at this time.

### Genus —
*Homalonychus* MARX 1891

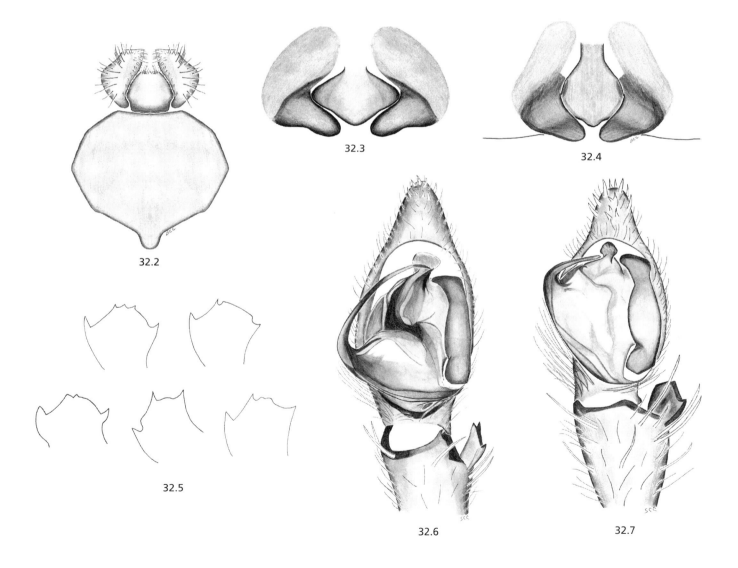

Chapter 33

# HYPOCHILIDAE
1 genus, 10 species

Pierre Paquin
Marshal Hedin

**Common name —**
Lampshade weavers.

**Similar families —**
Pholcidae (p. 194).

**Diagnosis —**
*Hypochilus* is a distinctive cribellate spider, easily recognized by the habitus (Fig. 33.1). The body is dorsoventrally flatened and of a fair size (10-20 mm). The long legs of males may reach up to 150 mm in some species. They are the only North American true spiders with 2 pairs of book lungs (Fig. 33.2). The chelicerae are of an intermediate state between the diaxial condition of Araneomorphae and paraxial of Mygalomorphae and Mesothelae, called semidiaxial by Catley (1994) (Fig. 33.3). Venom glands are restricted to chelicerae. *Hypochilus* spin distinctive webs that ressemble a lampshade (Fig. 33.4).

**Characters —**
- **body:** 10.0-20.0 mm, dorsoventrally flattened.
- **color:** yellowish-gray, often with dusky black patterns or lines; often iridescent.
- **carapace:** pear-shaped, longer than broad, the fovea is an indistinct longitudinal groove (Fig. 33.5).
- **sternum:** longer than broad.
- **eyes:** eight eyes in 2 rows, AER slightly procurved, PER moderately recurved; AME smaller (Fig. 33.5).
- **chelicerae:** semidiaxial (Fig. 33.3); chelicera stout, promargin with 5 stout teeth, retromargin with 1-2 small basal teeth .
- **mouthparts:** labium wider than long and fused to the sternum; endites divergent (Fig. 33.3).
- **legs:** prograde, long and thin without spines; males have longer legs than female; 3 well developed claws with paired claws pectinate and unpaired claw strongly curved; calamistrum biseriate, occupying one fouth of the metatarsus at base.
- **abdomen:** elongate and somewhat flattened; segmentation may be partially visible (Gertsch 1964b) (Fig. 33.5).
- **spinnerets:** the cribellum forms a thick lobe with large, suboval, undivided spinning plate; ALS large and bisegmented, MS short and one-segmented; PLS bisegmented (Fig. 33.2).
- **respiratory system:** two pair of book lungs; the posterior pair is situated midway between genital groove and spinnerets (Fig. 33.2).
- **genitalia:** haplogyne; *female* spermathecae consist of two distinctly separated seminal receptacles on each side (Fig. 33.6); *male* pedipalps are elongate, femora are long and cylindrical (Fig. 33.7); the bulb is terminal (Figs. 33.7-33.8); the shape of the apex of the palpal conductor is disgnotic for species (Fig. 33.8).

**Distribution —**
The distribution of *Hypochilus* is remarkable. Species are found in three distinct, disjunct geographic regions in North America: southern Appalachian Mountains (5 species), southern Rocky Mountains (2 species) and California mountains (3 species) (Catley 1994, Hedin 2001).

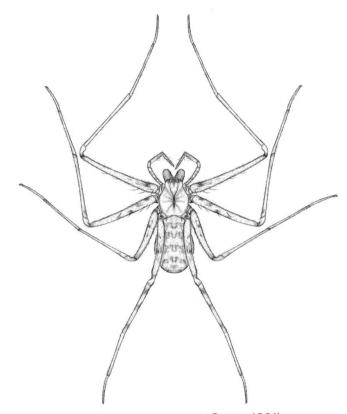

**Fig. 33.1** *Hypochilus bonneti* GERTSCH 1964b

**Natural history —**
*Hypochilus* is typically found on the bare surface of boulders, walls of overhanging cliffs often near streams and creeks, cave entrance walls, and sometimes in man-made habitat such as mines and culverts. They spin a distinctive web formed of heavy mesh of sticky, hackle-band threads, guyed by numerous dry lines to adjacent substrate. This structure leads to a more-or-less sheltered recess which takes the form of a deep lampshade (Gertsch 1958c). Most often, the spider hangs facing downward and has its long legs spread to touch the side of the mesh and presses its body close to the substrate. They can run quickly for a short distance if disturbed. *Hypochilus bonneti* GERTSCH 1964b spins suboval pouchlike egg sacs that they cover with detritus and hang on silk thread (Gertsch 1964b).

A few papers treat different aspects of *Hypochilus* biology: internal anatomy (Marples 1968); predatory behavior of *Hypochilus gertschi* HOFFMAN 1963 (Shear 1969); natural history of *Hypochilus pococki* PLATNICK in FORSTER et al. 1987 (Fergusson 1972); morphology of *Hypochilus thorelli* MARX 1888a (Foelix & Jung 1978); community ecology of *Hypochilus thorelli* (Riechert & Cady 1983, Hodge & Marshall 1996); life cycle of *Hypochilus pococki* (Coyle 1985b);

sperm morphology (Alberti & Coyle 1991); systematics of *Hypochilus coylei* PLATNICK *in* FORSTER *et al.* 1987 and *Hypochilus sheari* PLATNICK *in* FORSTER *et al.* 1987 (Huff & Coyle 1992); mating systems of *Hypochilus pococki* (Eberhard *et al.* 1993) and reproductive biology of *Hypochilus pococki* (Catley 1993). Catley (1994) provided data on the biology of *Hypochilus jemez* CATLEY 1994 and *Hypochilus bernardino* CATLEY 1994.

Based on molecular data, Hedin (2001) tested previous hypothesis on *Hypochilus* diversification and proposed a better understanding of the phylogeny and biogeography of the genus. Hedin (2001) and Hedin & Wood (2002) found that *Hypochilus* populations show an extraordinary degree of genetic divergence (at mitochondrial loci) that exceeds 10% between populations of a given species; in particular for *Hypochilus thorelli* which show a highly structured population structure (Hedin & Wood 2002).

Catley (1994) points out that species known only from a few populations may deserve conservation attention, probably as threatened or fragile species (including all California and Rocky mountain species).

### Taxonomic history and notes —

The genus was proposed by Marx 1888a and included one species: *Hypochilus thorelli*. *Hypochilus* was believed monotypic until Gertsch (1958c) described a second species, *Hypochilus petrunkevitchi* from California. Hoffman (1963) described *Hypochilus gertschi* from the Appalachian region and Gertsch (1964b) *Hypochilus bonneti* from Colorado. Platnick (in Forster *et al.* 1987) added three species from the Appalachians-*Hypochilus coylei*, *Hypochilus sheari* and *Hypochilus pococki*-and one species from California *Hypochilus kastoni* PLATNICK *in* FORSTER *et al.* 1987. Catley (1994) described 2 additional species: *Hypochilus bernardino* from California and *Hypochilus jemez* from New Mexico.

Simon (1889e) described *Hypochilus davidi* from China, but transferred it to *Ectatosticta* SIMON 1892a, the only other genus of Hypochilidae, leaving *Hypochilus* as a North American endemic.

The importance of the family has long been recognized for its phylogenetic position, as they are the sister group to all other araneomorph spiders (Platnick 1977g, Forster *et al.* 1987). Gertsch (1958c) referred to Hypochilidae as the aristocrats of Araneomorphae. Interestingly, most *Hypochilus* species were named after famous arachnologists.

### Genus —

*Hypochilus* MARX 1888a

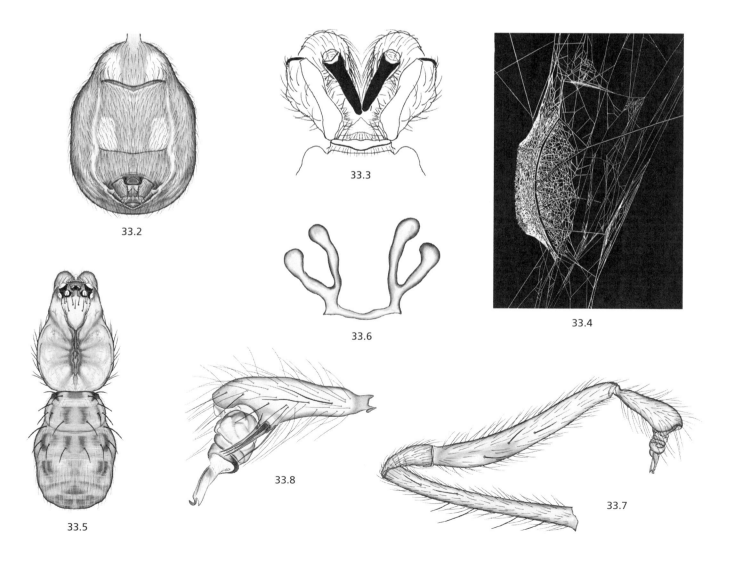

## Chapter 34
# LEPTONETIDAE
5 genera, 40 species

Joel M. Ledford
Darrell Ubick
James C. Cokendolpher

**Similar families —**

Telemidae (p. 228), Ochyroceratidae (p. 179), Pholcidae (p. 194).

**Diagnosis —**

Small (1.0-3.0mm), three-clawed, ecribellate, haplogyne spiders usually separated from other families by a distinctive 6-eyed pattern with the PME separated from the strongly recurved ALE and PLE (Fig. 34.7); rarely with eyes contiguous (*Archoleptoneta*) (Fig. 34.5), or blind (some *Neoleptoneta*); cephalothorax longer than wide, clypeus short, < 0.3 x length chelicerae; chelicerae unfused and not chelate, biserially dentate, promargin with teeth, retromargin with few small teeth and denticles; legs long and slender (PT/ C > 1.5), autospasy at the patella-tibia joint; abdomen globose, lacking anterodorsal sclerite, and with a single tracheal spiracle located near the spinnerets.

**Characters —**

**body size:** 1.0-3.0 mm.
**color:** pallid to creamy, may have purplish-brown maculations or chevrons on abdomen; in life appendages iridescent.
**cephalothorax:** longer than wide, pyriform, often with a faint longitudinal linear mark at the fovea.
**sternum:** oval.
**eyes:** six eyes (AME absent), often strongly convex (Fig. 34.7), PME usually posteriorly displaced, or all eyes contiguous and occupying < 1/ 2 width of cephalothorax (*Archoleptoneta*), or eyes absent in some cave species.
**chelicerae:** not fused basally, promargin with teeth, retromargin with few denticles.
**mouthparts:** labium short, wider than long; endites long, parallel-sided, often concave ectally; serrula present.
**legs:** long and slender (PT/C > 1.5), with or without scattered spines; integumentary glands on tibiae and patellae; autospasy at the patella-tibia joint; metatarsus III with a finely serrate comb (Fig. 34.2) tarsi with three claws; female palp with claw.
**abdomen:** globose, lacking anterodorsal sclerite.
**spinnerets:** in close cluster; colulus present as fleshy lobe; PMS with linear spigot arrangement.
**respiratory system:** one pair of book lungs; single tracheal spiracle located near spinnerets.
**genitalia:** haplogyne; *female* gonopore region lacking external modification, although may be slightly distended after insemination; atrium large and globular to triangular in shape (Fig. 34.3) with a pair of twisted spermathecae laterally (Fig. 34.3); *male* palpal femur short to long, 1.0-3.0 x cephalothorax length; tibia with a variety of modified setae, with or without a retrolateral tibial spine; tarsus with or without retroapical seta, bulb spherical to pyriform, expandable, with a variety of specialized apical processes (Figs. 34.6, 34.8-34.12).

**Distribution —**

*Archoleptoneta* is widespread in California and also known from isolated localities in Texas, Mexico (Chiapas), and Panama, giving it the widest distribution of any lep-

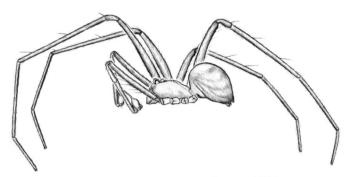

**Fig. 34.1** *Calileptoneta helferi* (Gertsch 1974)

tonetid genus. *Calileptoneta* is known only from California and Southern Oregon. *Neoleptoneta* occurs from eastern Arizona to Georgia, but is most richly represented in central Texas. *Appaleptoneta* and "*Leptoneta*" *sandra* Gertsch 1974 are found only in the southern Appalachian Mountains of Virginia to Tennessee, south to Georgia and Alabama.

**Natural history —**

Habitat: Leptonetids are cryptozoic spiders found in damp situations, under rocks, in log and leaf litter, and may be common in caves. Some species congregate at ideal habitats and careful collecting can reveal several specimens. Web architecture: Leptonetids build a small tangle or tightly woven sheet web beneath which they hang. Individuals drop from the web when disturbed and fold their legs close to the body to avoid detection.

Reproduction: Courtship and mating for *Calileptoneta* was recorded by Ledford (2004), and egg sac positioning and structure was observed in *Neoleptoneta* by Cokendolpher (2004c). Courtship for *Calileptoneta* consists of a slight plucking of the female's web by the male using his palpi. Both individuals remain inverted beneath the web during mating, and the male palpi rotate 180° before insertion. Males remain in the web with the female for several days. Egg sacs (Fig. 34.4) are covered in detritus and suspended on the outer margins of the web (Cokendolpher 2004c) or attached to the substrate (Roth 1994). Leptonetids are easily reared to maturity on a diet of pre-killed fruit flies (*Drosophila*), and females may be able to store sperm for extended periods (Cokendolpher 2004c).

**Taxonomic history and notes —**

Nearctic leptonetids have a controversial taxonomic history. Most Nearctic species were described by Gertsch (1974), who recognized only 2 genera for the family: *Archoleptoneta* for the few species with a clustered eye arrangement, and *Leptoneta* for all others. On the basis of genitalic differences, Brignoli (1977g) subdivided Gertsch's *Leptoneta* into several clusters and revalidated *Neoleptoneta*. Platnick (1986d) supported this division with characters of sensory organs and named *Appaleptoneta* and *Calilep-*

*toneta*. Additional species of *Neoleptoneta* were described by Brignoli (1979c), Cokendolpher & Reddell (2001b), and Cokendolpher (2004c). Ledford (2004) revised *Calileptoneta* and described additional species.

Leptonetids have long been associated with telemids and ochyroceratids, all tiny spiders and unusual among haplogynes in having neither the fused chelicerae characteristic of scytodoids nor the anteriorly displaced paired tracheal spiracles of dysderoids. All three were initially included in Leptonetidae (Simon 1892a), and the latter two eventually elevated to family status (Fage 1912, 1913). The relationship between these families was addressed by Brignoli (1979q) and Petrunkevitch (1933), who questioned their placement in the same suborder. Recent analyses of spinneret morphology (Platnick *et al.* 1991) and respiratory structures (Ramírez 2000) indicate that the three are indeed related, with Leptonetidae closest to Telemidae.

### Genera —
ARCHOLEPTONETINAE
  *Archoleptoneta* GERTSCH 1974
LEPTONETINAE
  *Appaleptoneta* PLATNICK 1986d, *Calileptoneta* PLATNICK 1986d, *Neoleptoneta* BRIGNOLI 1972i

34.4

### Key to genera —
North America North of Mexico

1    Eyes absent .................................. ***Neoleptoneta***, *in part*
     **Div.** 21 species of which 2 key out here, *Neoleptoneta anopica* (GERTSCH 1974) and *Neoleptoneta georgia* (GERTSCH 1974) — **Dist.** TX, GA — **Refs.** Gertsch 1974, Platnick 1986d — **Note** *Neoleptoneta microps* (GERTSCH 1974) has greatly reduced to absent eyes and may key here

—    Eyes present .............................................................. 2

2(1)    Eyes contiguous (Fig. 34.5); male palpal tarsus cylindrical (Fig. 34.6) .............................................. ***Archoleptoneta***
     **Div.** 2 species: *Archoleptoneta schusteri* GERTSCH 1974 and *Archoleptoneta garza* GERTSCH 1974 — **Dist.** CA, TX — **Refs.** Gertsch 1974, Platnick 1986d

—    Eyes with PME posteriorly displaced (Fig. 34.7); male palpal tarsus with dorsal transverse groove (Figs. 34.8-34.9) .................................................................... 3
     **Note:** It is not possible at this time to distinguish females of the following genera on the basis of external characters

3(2)    Dist s OR and CA; male palpal bulb with mesal lobe (Fig. 34.8) .................................................. ***Calileptoneta***
     **Div.** 9 species — **Dist.** s OR to CA — **Refs.** Gertsch 1974, Platnick 1986d, Ledford 2004

—    Dist AZ eastward; male palpal bulb lacking mesal lobe (Fig. 34.9) ............................................................... 4

4(3)    Male palpal femur with ventral row of spines (Fig. 34.10) ............................................................... ***"Leptoneta"***
     **Div.** 1 species: *"Leptoneta" sandra* GERTSCH 1974 — **Dist.** VA, WV — **Refs.** Gertsch 1974, Platnick 1986d — **Note** this species is misplaced in *Leptoneta* and may represent an undescribed genus (Brignoli 1977g, Platnick 1986d)

—    Male palpal femur lacking ventral row of spines ............ 5

5(4)    Male palpal tibia with RTA (Fig. 34.11) ............................... ***Neoleptoneta***
     **Div.** 21 species, of which 19 key out here — **Dist.** se AZ, TX to GA — **Refs.** Gertsch 1974, Brignoli 1979c, Platnick 1986d, Cokendolpher & Reddell 2001b

—    Male palpal tibia lacking RTA (Fig. 34.12) ............................... ***Appaleptoneta***
     **Div.** 7 species — **Dist.** Appalachian Mountains, KY s to SC and AL — **Refs.** Gertsch 1974, Platnick 1986d

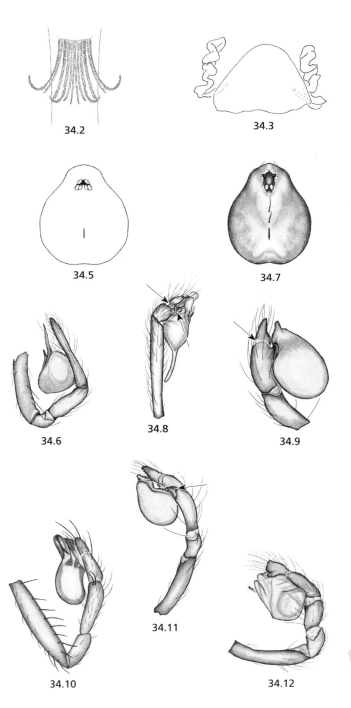

Chapter 35
# LINYPHIIDAE
157 genera, 952 species

Michael L. Draney
Donald J. Buckle

**Common name —**

Sheet-web spiders. Erigoninae are sometimes called dwarf spiders or money spiders.

**Similar families —**

Many small ground-dwelling spiders in other families have been mistaken for linyphiids, including Dictynidae (*Lathys*, p. 94), Mysmenidae (p. 175, *Microdipoena*), Theridiidae (p. 235, *Robertus*, *Thymoites*, *Pholcomma*, *Theonoe*), Tetragnathidae (p. 232, *Glenognatha*), Leptonetidae p. 121, Theridiosomatidae p. 242, and especially Pimoidae p. 195, which is the sister taxon to Linyphiidae (Hormiga 1994a, 2003).

**Diagnosis —**

Minute to small (<1–7 mm) three-clawed, ecribellate, entelegyne spiders belonging to the superfamily araneoidea. They may be distinguished from other araneoids, except Pimoidae, by the following combination of characters: male palp with discrete, intersegmental paracymbium (rarely lost); stridulatory file on the ectal side of the chelicerae (occasionally absent) and autospasy of the legs at the patella-tibia junction. Linyphiids and pimoids share cheliceral striae and patella-tibia autospasy. Linyphiids differ from pimoids in having a discrete paracymbium rather than one integrated with the cymbium, having the embolus part of a complex embolic division rather than attached directly to the tegulum, and by lacking a retrolateral cymbial sclerite attached to the cymbium via a membrane, which is a pimoid synapomorphy.

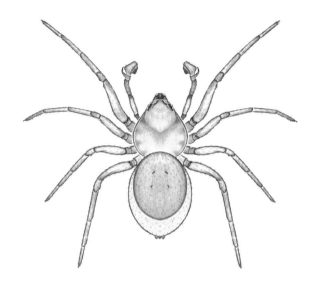

**Fig. 35.1** *Ceraticelus fissiceps* (O. Pickard-Cambridge 1874b)

**Characters —**

**body size:** from <1 to 7 mm; most are 1-4 mm long.
**color:** white, tan, brown to black. Occasionally shades of orange, red, and yellow. Some species (especially larger arboreal species) may have dorsal abdominal stripes, chevrons, or markings; or banded legs.
**carapace:** longer than wide; the carapace of many male linyphiids is modified in various ways. Eye region commonly elevated, sometimes exhibits bizarre forms. A pair of lateral pits may be present, opening near the posterior lateral eyes. Various clypeal protuberances or modifications, unusual setae or clusters of setae, or teeth around the carapace margin may be observed.
**sternum:** longer than wide, projecting behind coxae IV.
**eyes:** usually eight subequal eyes in two rows. Rarely reduced or absent (in cavernicolous species). Eyes of males with modified carapaces often in unusual configurations.
**chelicerae:** relatively robust for araneoid spiders, armed with teeth on both margins. Most with a stridulatory file on the lateral face. Some males have a mastidion or similar tooth-like apophysis on the anterior cheliceral face (e.g., *Zornella*).
**mouthparts:** endites are parallel, rarely armed with apophyses in males (e.g., some *Erigone* spp.).
**legs:** rather long and thin, with macrosetae. Autospasy occurs at the patella-tibia joint. All have a single metatarsal trichobothrium on legs I-III, and some also have one on leg IV. Setation on legs of males often somewhat reduced compared to females.
**abdomen:** usually ovoid to somewhat elongate. Some species have abdominal scuta (dorsal and sometimes ventral) of varying size.
**spinnerets:** colulus present, with setae. Spinnerets are of typical araneoid form.
**respiratory system:** one pair of book lungs, and a tracheal system opening onto a single spiracle anterior to the spinnerets, anteriorly displaced in some species (e.g., *Tennesseellum*). Most members of the subfamily Linyphiinae have unbranched tracheae confined to the abdomen while most members of the Erigoninae have branched tracheae extending into the cephalothorax.
**genitalia:** entelegyne; **female** epigynum highly variable; usually fairly simple, plate-like sclerotized structure with a dorsal and a ventral sclerite associated with paired copulatory openings, leading via complex ducts to spermathecae. These sclerotized structures are sometimes, particularly in the Linyphiinae, modified into a posteriorly directed scape; **male** palps nearly always have an intersegmental paracymbium (=flexibly attached). The embolus is part of a complex structure called the embolic division, attached by a stalk to the suprategulum. This and several other sclerites associated with the tegulum, subtegulum, and suprategulum produces structurally complex palp. Erigonines nearly always have a retrolateral tibial apophysis of varying size and shape, sometimes consisting of multiple projections.

### Distribution —

Worldwide with its greatest diversity in the north temperate regions. Well represented in North America.

### Natural history —

Linyphiids spin semi-permanent webs consisting of one or more horizontal sheets of silk often supported above and below by auxiliary support lines. The spider generally hangs upside down beneath the primary sheet, attacking prey from below. Most species live in leaf litter or on the ground surface, but some (especially larger species) make webs in vegetation. Both myrmecophiles and cavernicoles are known. Linyphiids generally feed on soft-bodied insects, especially springtails (Collembola) and flies (Diptera). They occupy almost all habitats but are most diverse and abundant in more mesic to hydric (i.e., wetlands) areas, and are prominent in northern peatlands. Most species are univoltine or annuals (one generation/year), but multivoltine (>1 generation/year) and merovoltine (>1 year/generation) strategies are common. Most species occur as adults for only a short period each year, but some multivoltine and merovoltine species are active as adults during most or all of the year. Most species overwinter as juveniles or adults, - some are even active under snow in winter - but a few overwinter as eggs.

### Taxonomic history and notes —

Linyphiidae is a large, poorly known, and taxonomically difficult family. Several genera and many species remain undescribed. While a number of North American genera have recently been revised, most have not. Their species are known from original descriptions, scattered through 150 years of literature, and from subsequent references under a variety of genus and species names.

Eugen von Keyserling, Octavius Pickard-Cambridge, and Nathan Banks were prolific early describers of North American species, each adding over 30 currently valid species. The first, and only, comprehensive survey of North American linyphiids was published by James Emerton in 1882. He subsequently produced a series of papers describing many new species. Cyrus Crosby and Sherman Bishop published a series of revisionary studies in the 1920s and 1930s. Wilton Ivie, often working with Ralph Chamberlin, described many western species and revised several genera before his untimely death in 1969. Between 1968 and 1981 Peter van Helsdingen, revised *Linyphia*, *Neriene*, *Microlinyphia*, *Stemonyphantes* and *Allomengea* on a worldwide basis and *Centromerus* and *Oreonetides* for North America. In the 1980s, Frank Millidge, revised many North American genera, including *Walckenaeria*, *Spirembolus* and *Eperigone*, and described over 100 species.

Buckle *et al*. (2001) provides an annotated catalog of North American taxa, and Hormiga *et al*. (2005) contains illustrations of all valid linyphiid genera worldwide. Paquin & Dupérré (2003) provide illustrations of all species known from Quebec. This is an invaluable reference for anyone working on linyphiids from northern and eastern North America. For a significant number of species it provides the first good illustrations or the first illustration of previously undocumented males or females.

The higher level classification of the linyphiids is not presently clear. While the subfamily Erigoninae is considered to be monophyletic (Hormiga 2000) the Linyphiinae, in its classical sense, is not (Millidge 1984b, Hormiga 2000). We evade this problem by considering everything not included in the Erigoninae to be a linyphiine. An additional problem is defining the limits of the Erigoninae. Typical linyphiines and erigonines can be distinguished by several characters. However, there are a dozen or so genera which possess some but not all of the defining characters, in various assortments. These genera have, over the years, been shuffled back and forth between the subfamilies by various authors. Here we follow Hormiga (2000) who considers most to be basal erigonines.

This key includes all linyphiid males and linyphiine females. Insufficient information is available at this time to permit construction of a key to erigonine females. There is no simple, straightforward way to distinguish between female linyphiines and erigonines. Almost all linyphiines have 2 dorsal spines on tibia IV, while many erigonines have only 1 or none. Many linyphiines have prolateral spines on tibia I and/or femur I while most erigonines do not. Many linyphiines have epigynal scapes, while most erigonines do not. Erigonines are small, while linyphiines may be small or medium sized; any linyphiid over 3.5 mm is likely to be a linyphiine. Species with patterned abdomens are common in the Linyphiinae, unusual in the Erigoninae. Females of the larger linyphiine genera have palpal claws while those of the erigonines and smaller linyphiines do not.

"*Clitolyna*" *electa* CROSBY 1905a and *Masikia* MILLIDGE 1984a are not included in the key because males are unknown. The male of *Pacifiphantes magnificus* (CHAMBERLIN & IVIE 1943) is unknown so *Pacifiphantes* is omitted from the male section of the key. *Styloctetor* keys out with *Ceratinopsis* because we were unable to find a way to consistently separate the two taxa. We key out two common western species which probably form an undescribed genus ("*Spirembolus*" *vasingtonus*, nomen dubium and "*Collinsia*" *wilburi* LEVI & LEVI 1955). We also include an undescribed western genus containing "*Spirembolus*" *approximatus* and other undescribed species, currently being examined and defined by J. Miller and T. Prentice, and another containing "*Linyphia*" *rita* GERTSCH 1951, "*Linyphia*" *catalina* GERTSCH 1951 and "*Linyphia*" *tauphora* GERTSCH 1951.

Generic treatment here differs in various small ways from the current World Catalog. *Meioneta* is considered to be a synonym of *Agyneta*, following Saaristo (1973b), *Collinsia* is treated as a synonym of *Halorates* (see Buckle *et al*. 2001) and *Nanavia monticola* CHAMBERLIN & IVIE 1933a, which was erroneously synonymized with *Leptorhoptrum robustum* (WESTRING 1851) by Eskov & Marusik (1994), has been restored to *Nanavia*. The recent deconstruction of *Lepthyphantes* into many small genera by Saaristo & Tanasevitch (1996b) and generated serious problems in the creation of this key. Only about half of the American *Lepthyphantes* have yet to be placed in the new genera, and some of the genera are distinguishable by characters possessed by one sex only. The only viable solution was to retain *Lepthyphantes* in the key in its old, broad sense. Students wishing to apply the Saaristo & Tanasevitch (1996b) names can follow an identification technique long used for American linyphiids -thumb through the literature, looking at illustrations, until the species, or a close relative is located, then check a catalog to determine its current placement.

### Definitions used in this key —

The following characters are useful for separating the vast number of species into smaller, more manageable groups. However, using these requires separating continuously varying characters into discrete groups, and is necessarily a somewhat subjective process. We provide the following precise definitions to minimize the subjectivity. Some questionable taxa also key out in more than one place, when the interpretation of these characters is inherently difficult.

Metatarsal trichobothria: A trichobothrium (Tm) is a long, very fine hair emerging from a pit (which looks like a distinct circle) on the dorsal surface of a leg. They often stand nearly perpendicular to the leg surface. Linyphiids have 1 trichobothrium on each metatarsus, but it may be absent from metatarsus IV. The relative position of TmI is sometimes useful as a character; we indicate this position as the decimal fraction of a/b, as shown in Fig. 35.2. More accurate measurements will be obtained if the entire metatarsus is in focus under the microscope. See Roberts (1987) for further explanations and figures of trichobothria.

Length of palpal tibia: The minimum distance parallel to the long axis of the tibia, between the point where the tibia and patella contact, and the point where the tibia contacts the "tarsus" (i.e., a palpal structure distal to the tibia, Figs. 35.3-35.5b).

Width of palpal tibia: The minimum diameter of the tibia, perpendicular to the long axis of the tibia, in the dorsal or ectal view (Figs. 35.4-35.5b).

Length of tibial apophysis: This is determined in dorsal or ectal view, whichever view the apophysis appears longest. In this view, the apophysis length is the straight line distance from the apical-most point to a point which appears to be midway along a line drawn across the "base" of the apophysis. This "base" is somewhat subjective, but is the point at which the apophysis begins to expand (Figs. 35.3-35.5b).

Normal versus modified cephalic region: The "normal" or unmodified linyphiid carapace is slightly raised or lobed in linyphiids (Figs. 35.6, 35.20, 35.120, 35.121), but a modified carapace is noticeably more raised (Fig. 35.57, 35.58) or raised in a distinctive way. The angle of the posterior declivity may be more acute than "normal" in lateral view, for example, producing a relatively sharp (but not particularly high) lobe (Fig. 35.165), or the entire carapace may be somewhat raised, or the posterior declivity of the cephalic region may be in an unusually posterior position, which is also considered modified (Fig. 35.203). Any noticeable protrusion or protuberance on the cephalic or clypeal region is also considered a modified carapace. Pits or excavations of an otherwise "normal" shaped carapace are not considered a cephalic modification, nor are any spines or projections of the lateral margins of the carapace (e.g., some *Erigone* species).

Note on linyphiid palps: Many characters (such as the size and shape of the tibial apophysis) are best viewed from a certain orientation, and the "best" orientation often varies between species. Removing a palp from the body allows easy examination from various orientations, and can facilitate identification in many cases. Note that the left palpus is generally illustrated throughout this work, but many of the revisions of American linyphiids illustrate right palpi.

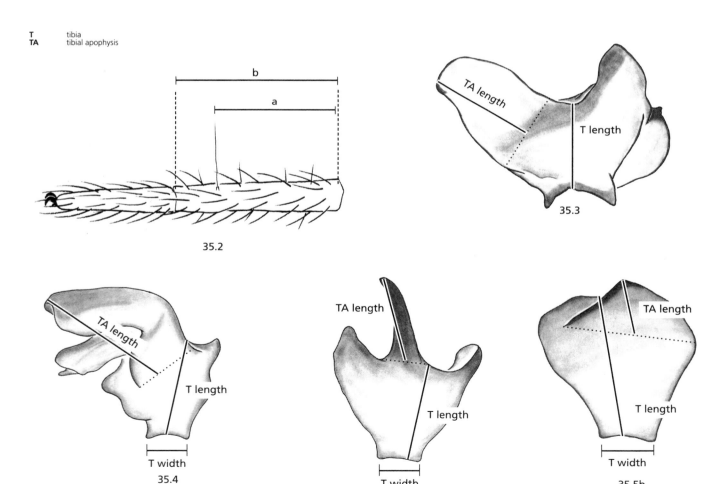

T   tibia
TA  tibial apophysis

35.2

35.3

35.4

35.5a

35.5b

**Genera —**

LINYPHIINAE
*Agyneta* HULL 1911a, *Allomengea* STRAND 1912b, *Anibontes* CHAMBERLIN 1924a, *Aphileta* HULL 1920, *Arcuphantes* CHAMBERLIN & IVIE 1943, *Bathyphantes* MENGE 1866, *Bolyphantes* C.L. KOCH 1837g, *Centromerita* F. DAHL 1912b, *Centromerus* F. DAHL 1886, *Drapetisca* MENGE 1866, *Estrandia* BLAUVELT 1936, *Florinda* O. PICKARD-CAMBRIDGE 1896a, *Frontinella* F.O. PICKARD-CAMBRIDGE 1902a, *Graphomoa* CHAMBERLIN 1924a, *Helophora* MENGE 1866, *Jalapyphantes* GERTSCH & DAVIS 1946, *Kaestneria* WIEHLE 1956, *Lepthyphantes* MENGE 1866 [here including *Agnyphantes* HULL 1932, *Improphantes* SAARISTO & TANASEVITCH 1996b, *Incestophantes* TANASEVITCH 1992, *Megalepthyphantes* WUNDERLICH 1994b, *Obscuriphantes* SAARISTO & TANASEVITCH 2000, *Tenuiphantes* SAARISTO & TANASEVITCH 1996b], *Linyphantes* CHAMBERLIN & IVIE 1942a, *Linyphia* LATREILLE 1804b, *Macrargus* F. DAHL 1886, *Maro* O. PICKARD-CAMBRIDGE 1906a, *Microlinyphia* GERHARDT 1928, *Microneta* MENGE 1869, *Neriene* BLACKWALL 1833c, *Oaphantes* CHAMBERLIN & IVIE 1943, *Oreonetides* STRAND 1901b, *Oreophantes* ESKOV 1984a, *Pacifiphantes* ESKOV & MARUSIK 1994, *Pityohyphantes* SIMON 1929, *Poeciloneta* KULCZYŃSKI in CHYZER & KULCZYŃSKI 1894, *Porrhomma* SIMON 1884a, *Saaristoa* MILLIDGE 1978b, *Stemonyphantes* MENGE 1866, *Tapinopa* WESTRING 1851, *Taranucnus* SIMON 1884a, *Tennesseellum* PETRUNKEVITCH 1925c, *Wubana* CHAMBERLIN 1919a

ERIGONINAE
*Anacornia* CHAMBERLIN & IVIE 1933a, *Annapolis* MILLIDGE 1984a, *Anthrobia* TELLKAMPF 1844, *Arcterigone* ESKOV & MARUSIK 1994, *Baryphyma* SIMON 1884r, *Blestia* MILLIDGE 1993a, *Carorita* DUFFEY & MERRETT 1964, *Caviphantes* OI 1960a, *Ceraticelus* SIMON 1884r, *Ceratinella* EMERTON 1882, *Ceratinops* BANKS 1905a, *Ceratinopsidis* BISHOP & CROSBY 1930, *Ceratinopsis* EMERTON 1882, *Cheniseo* BISHOP & CROSBY 1935c, *Cnephalocotes* SIMON 1884r, *Coloncus* CHAMBERLIN 1949, *Coreorgonal* BISHOP & CROSBY 1935c, *Dactylopisthes* SIMON 1884r, *Dicymbium* MENGE 1868, *Dietrichia* CROSBY & BISHOP 1933b, *Diplocentria* HULL 1911a, *Diplocephalus* BERTKAU in FÖRSTER & BERTKAU 1883, *Disembolus* CHAMBERLIN & IVIE 1933a, *Dismodicus* SIMON 1884r, *Entelecara* SIMON 1884r, *Epiceraticelus* CROSBY & BISHOP 1931, *Eridantes* CROSBY & BISHOP 1933b, *Erigone* AUDOUIN 1826, *Erigonella* F. DAHL 1901c, *Eskovia* MARUSIK & SAARISTO 1999, *Eulaira* CHAMBERLIN & IVIE 1933a, *Floricomus* CROSBY & BISHOP 1925b, *Glyphesis* SIMON 1926, *Gnathonargus* BISHOP & CROSBY 1935c, *Gnathonarium* KARSCH 1881a, *Gnathonaroides* BISHOP & CROSBY 1938, *Gonatium* MENGE 1868, *Goneatara* BISHOP & CROSBY 1935c, *Gongylidiellum* SIMON 1884r, *Grammonota* EMERTON 1882, *Halorates* HULL 1911a, *Hilaira* SIMON 1884a, *Horcotes* CROSBY & BISHOP 1933b, *Hybauchenidium* HOLM 1973, *Hypomma* F. DAHL 1886, *Hypselistes* SIMON 1894a, *Islandiana* BRAENDEGAARD 1932, *Ivielum* ESKOV 1988b, *Jacksonella* MILLIDGE 1951, *Lessertia* SMITH 1908, *Lophomma* MENGE 1868, *Masikia* MILLIDGE 1984a, *Maso* SIMON 1884r, *Masoncus* CHAMBERLIN 1949, *Masonetta* CHAMBERLIN & IVIE 1939a, *Mecynargus* KULCZYŃSKI in CHYZER & KULCZYŃSKI 1894, *Mermessus* O. PICKARD-CAMBRIDGE 1899a *Metopobactrus* SIMON 1884r, *Micrargus* F. DAHL 1886, *Microctenonyx* F. DAHL 1886, *Mythoplastoides* CROSBY & BISHOP 1933b, *Nanavia* CHAMBERLIN & IVIE 1933a, *Oedothorax* BERTKAU in FÖRSTER & BERTKAU 1883, *Origanates* CROSBY & BISHOP 1933b, *Ostearius* HULL 1911a, *Paracornicularia* CROSBY & BISHOP 1931, *Pelecopsidis* BISHOP & CROSBY 1935b, *Pelecopsis* SIMON 1864, *Perregrinus* TANASEVITCH 1992, *Perro* TANASEVITCH 1992, *Phanetta* KEYSERLING 1886b, *Pocadicnemis* SIMON 1884r, *Procerocymbium* ESKOV 1989b, *Satilatlas* KEYSERLING 1886b, *Savignia* BLACKWALL 1833a, *Sciastes* BISHOP & CROSBY 1938, *Scirites* BISHOP & CROSBY 1938, *Scironis* BISHOP & CROSBY 1938, *Scolopembolus* BISHOP & CROSBY 1938, *Scotinotylus* SIMON 1884r, *Scylaceus* BISHOP & CROSBY 1938, *Scyletria* BISHOP & CROSBY 1938, *Semljicola* STRAND 1906a, *Silometopoides* ESKOV 1990a, *Sisicottus* BISHOP & CROSBY 1938, *Sisicus* BISHOP & CROSBY 1938, *Sisis* BISHOP & CROSBY 1938, *Sisyrbe* BISHOP & CROSBY 1938, *Sitalcas* BISHOP & CROSBY 1938, *Smodix* BISHOP & CROSBY 1938, *Soucron* CROSBY & BISHOP 1936b, *Souessa* CROSBY & BISHOP 1936b, *Souessoula* CROSBY & BISHOP 1936b, *Sougambus* CROSBY & BISHOP 1936b, *Souidas* CROSBY & BISHOP 1936b, *Soulgas* CROSBY & BISHOP 1936b, *Sphecozone* O. PICKARD-CAMBRIDGE 1870d, *Spirembolus* CHAMBERLIN 1920a, *Styloctetor* SIMON 1884r, *Subbekasha* MILLIDGE 1984a, *Symmigma* CROSBY & BISHOP 1933b, *Tachygyna* CHAMBERLIN & IVIE 1939a, *Tapinocyba* SIMON 1884r, *Thaleria* TANASEVITCH 1984a, *Thyreosthenius* SIMON 1884r, *Tibioplus* CHAMBERLIN & IVIE 1947b, *Tiso* SIMON 1884r, *Tmeticus* MENGE 1868, *Traematosisis* BISHOP & CROSBY 1938, *Tunagyna* CHAMBERLIN & IVIE 1933a, *Tusukuru* ESKOV 1993, *Tutaibo* CHAMBERLIN 1916, *Typhochrestus* SIMON 1884r, *Vermontia* MILLIDGE 1984a, *Wabasso* MILLIDGE 1984a, *Walckenaeria* BLACKWALL 1833a, *Zornella* JACKSON 1932a, *Zygottus* CHAMBERLIN 1949

# Linyphiidae

**Key to genera —**
North America North of Mexico

| 1 | Male ..................................................... 2 |
| — | Female (Linyphiinae only) ............................ 197 |

2(1) With 2 dorsal spines on each tibia, or 1 dorsal spine on tibia IV and a prolateral spine on femur I; usually without a retrolateral apophysis on palpal tibia ................. 3

— With 1 or no dorsal spines on tibia IV and no prolateral spine on femur I; usually with a retrolateral apophysis on palpal tibia ................................................................. 51

## Linyphiinae and larger Erigoninae portion
(*Hilaira, Nanavia, Sisicus, Sougambus, Thaleria, Tibioplus*)

3(2) Prolateral spine present on tibia I ........................ 4
— No prolateral spine on tibia I ............................ 37

4(3) Some or all metatarsi with spines (check all legs) ........... 6
— No metatarsi with spines ........................................ 5

5(4) Femur I with prolateral spine ............................ 25
— Femur I without prolateral spine ........................ 28

6(4) Abdomen with conspicuous caudal tubercle (Fig. 35.6); body red (fading in alcohol) except for palps, eye region and caudal tubercle which are black; palp as shown (Fig. 35.7) ................................................... ***Florinda***
   **Div.** 1 species: *Florinda coccinea* (Hentz 1850a) — **Dist.** southeastern USA — **Ref.** Blauvelt 1936

— Abdomen without caudal tubercle; color brown or gray ................................................................. 7

7(6) Carapace pale with dark stripe bifurcating anteriorly (Figs. 35.8-35.9) ................................................ 8
— Carapace markings otherwise ............................ 9

8(7) Dorsal stripe on carapace narrow (Fig. 35.8); terminal apophysis of palp twisted in a thick, tight spiral (Fig. 35.10); palp without patellar apophysis .......................... ***Linyphia***
   **Div.** 1 introduced species: *Linyphia triangularis* (Clerck 1757) — **Dist.** ME — **Refs.** Helsdingen 1969, Jennings *et al*. 2003

— Dorsal stripe wider (Fig. 35.9); terminal apophysis not spiral (Fig. 35.11); palp with large patellar apophysis (Fig. 35.12) .................................................. ***Pityohyphantes***
   **Div.** 16 species — **Dist.** n and w NA — **Ref.** Buckle *et al*. 2001

9(7) Cymbium of palp with dorso-basal horn tipped with tuft of hairs (Figs. 35.13-35.14) ........................ ***Allomengea***
   **Div.** 3 species — **Dist.** n NA — **Ref.** Helsdingen 1974 — **Note** 2 of the species have metatarsal spines only on legs III & IV

— Cymbium otherwise .......................................... 10

10(9) Embolus forming an apically directed loop (Fig. 35.15); lamina large and shield-like ........ ***Bathyphantes***, in part
   **Div.** 26 species, 2 key out here: *Bathyphantes keeni* (Emerton 1917a) and *B.* ~~ski~~ 1916) — **Dist.** widespread — **Ref.** Ivie 1969
................................................................. 11

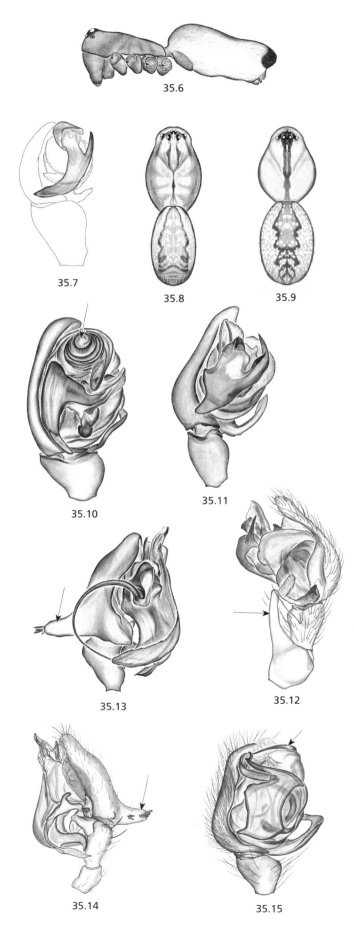

35.6

35.7  35.8  35.9

35.10  35.11

35.13  35.12

35.14  35.15

# Spiders of North America — Linyphiidae

**11(10)** Metatarsi with ventral spines ......................................... **12**

— Metatarsi without ventral spines ................................... **17**

**12(11)** Tibia I with 2-5 pairs of ventral spines (Fig. 35.16) ...... **13**

— Tibia I with one pair of ventral spines or 1 to 4 single ones (Fig. 35.17) ........................................................................... **16**

**13(12)** Terminal apophysis of palp twisted in a thick, tight spiral (Fig. 35.18) ................................................................. ***Neriene***
**Div.** 9 species — **Dist.** widespread — **Ref.** Helsdingen 1969 — **Note** some authors consider *Neriene* to be a synonym of *Linyphia*

— Terminal apophysis not spiral ........................................ **14**

**14(13)** Embolus long and threadlike, making loop beyond confines of cymbium (Fig. 35.19); carapace elongate in cephalic region; chelicerae elongate, angled posteriorly (Fig. 35.20) ............................................................... ***Microlinyphia***
**Div.** 4 species — **Dist.** n and w NA — **Ref.** Helsdingen 1970

— Embolus remaining within confines of cymbium (Figs. 35.21-35.22), carapace and chelicerae normal .............. **15**

**15(14)** Carapace pale with black median and marginal stripes; metatarsus I swollen (Fig. 35.23); palp as shown (Fig. 35.21) ................................................................. ***Stemonyphantes***
**Div.** 1 species: *Stemonyphantes blauveltae* Gertsch 1951 — **Dist.** n NA — **Ref.** Helsdingen 1968

— Carapace uniform brown; metatarsus I not swollen; palp as shown (Fig. 35.22) ...................... **Undescribed genus**
**Div.** *Linyphia catalina* Gertsch 1951, *Linyphia rita* Gertsch 1951 and perhaps *Linyphia tauphora* Chamberlin 1928a — **Dist.** sw USA — **Ref.** Gertsch 1951

**16(12)** Tibia I with pair of ventral spines near middle; palp with lamella of embolic division large and U-shaped (Fig. 35.24), Fickert's gland absent ......................... ***Helophora***
**Div.** 4 species — **Dist.** n and w NA — **Ref.** Buckle et al. 2001

— Tibia I with single ventral spine near base or 2-4 scattered ventral spines (Fig. 35.17); palp with lamella otherwise, Fickert's gland present (Fig. 35.25) ....................................
...................................................... ***Lepthyphantes,*** in part
**Div.** 29 species + several undescribed: a few of the larger species key out here — **Dist.** widespread — **Refs.** Zorsch 1937, Buckle et al. 2001

**17(11)** AME larger than other eyes; clypeus low, about height of AME; anterior eye row strongly recurved (Fig. 35.26); cymbium with long, angled basal apophysis (Figs. 35.27-35.28) ............................................................. ***Tapinopa***
**Div.** 2 species: *Tapinopa bilineata* Banks 1893a, *Tapinopa hentzi* Gertsch 1951 — **Dist.** eastern NA — **Refs.** Kaston 1981, Gertsch 1951

— AME same size or smaller than other eyes; clypeus height more than twice diameter of AME; anterior eye row nearly straight (Fig. 35.29); cymbium otherwise ....................... **18**

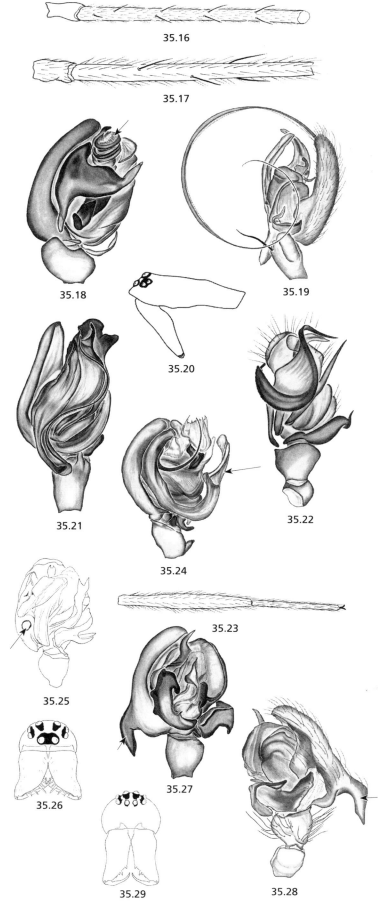

**18(17)** Prolateral face of chelicerae with 3 or 4 pairs of stout spines (Fig. 35.30), cheliceral teeth long; trichobothrium on leg IV present; palp as shown (Figs. 35.31-35.32) ........................................................................ ***Drapetisca***

**Div.** 2 species: *Drapetisca alteranda* Chamberlin 1909a, *Drapetisca oteroana* Gertsch 1951 — **Dist.** n NA, NM — **Refs.** Kaston 1981, Gertsch 1951

— Prolateral face of chelicerae without stout spines; cheliceral teeth relatively short; trichobothrium on leg IV absent ........................................................................ **19**

**19(18)** Tibia I with ventral spines ........................................ **20**

— Tibia I with no ventral spines ........................................ **23**

**20(19)** Tibia I with 3-5 pairs of ventral spines (Fig. 35.33); bulge on base of palpal tibia with brush of hairs (Fig. 35.34); palp as shown (Figs. 35.35-35.36); abdomen concolorous ........................................................................ ***Centromerita***

**Div.** 1 introduced species: *Centromerita bicolor* (Blackwall 1833c) — **Dist.** NF, NS, BC, WA — **Ref.** Roberts 1987

— Tibia with 1 or 2 pairs of spines or a few scattered ones; abdomen patterned or concolorous ............................ **21**

**21(20)** Palp with distinctive embolic division (Fig. 35.37); paracymbium with simple U-shape, distal arm very thin; abdomen with dorsal folium or uniformly black (Fig. 35.38) ........................................................ ***Estrandia***

**Div.** 1 species: *Estrandia grandaeva* (Keyserling 1886b) — **Dist.** n NA — **Ref.** Blauvelt 1936

— Palp and paracymbium otherwise ................................ **22**

**22(21)** Fang often stout and sinuous (Fig. 35.39); distal arm of paracymbium often with bifid tip (Fig. 35.40); Fickert's gland present (Fig. 35.41) ........................ ***Arcuphantes***

**Div.** 7 species — **Dist.** w NA — **Ref.** Chamberlin & Ivie 1943

— Fang normal; distal arm of paracymbium without bifid tip; Fickert's gland absent; palp as shown (Figs. 35.42-35.43) ........................................................ ***Oreophantes***

**Div.** 1 species: *Oreophantes recurvatus* (Emerton 1913a) — **Dist.** n NA — **Refs.** Helsdingen 1973b, 1981a

**23(19)** With short, stout legs (femur I no longer than carapace); basal segment of chelicera with anteriolateral row of short bristles (Fig. 35.44); body concolorous brown, palp as shown (Figs. 35.45-35.46) ........................ ***Centromerus***

**Div.** 9 species — **Dist.** e and n NA — **Ref.** Helsdingen 1973c

— With longer, thinner legs (femur I 1.1 to 1.5 times length of carapace); basal segment of chelicera without row of bristles; abdomen frequently patterned ........................ **24**

35.30

35.31

35.32

35.33

35.34

35.35

35.36

35.37

35.38

35.39

35.40

35.41

35.42

35.43

35.44

35.45

35.46

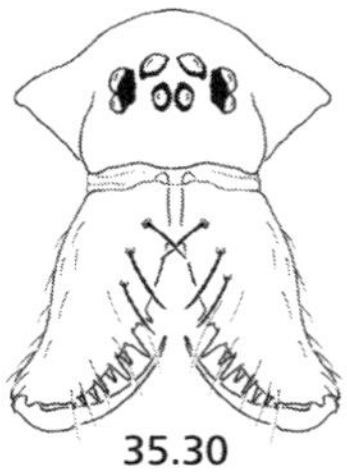

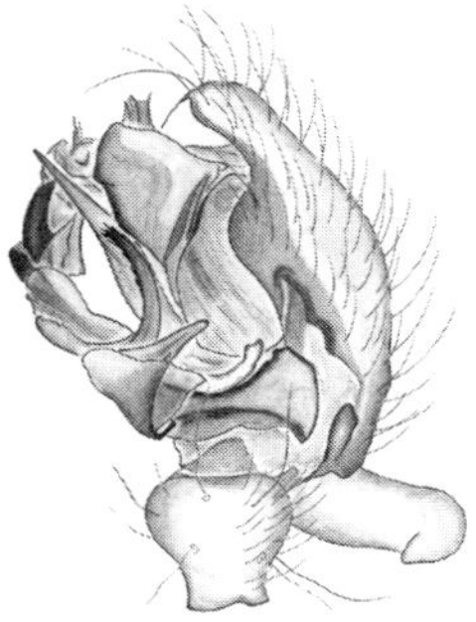

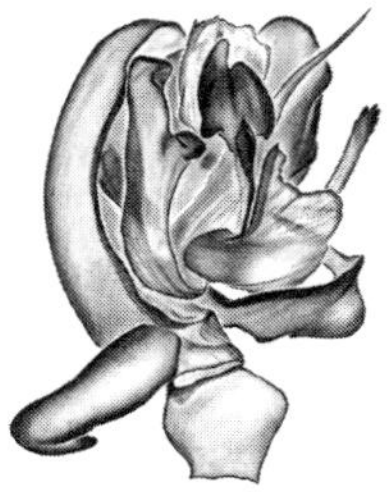

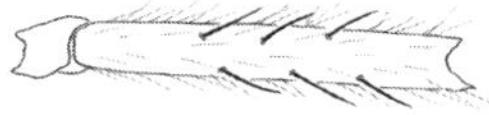

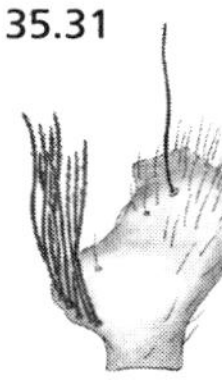

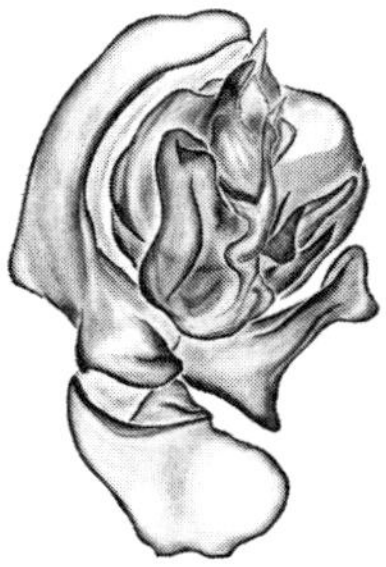

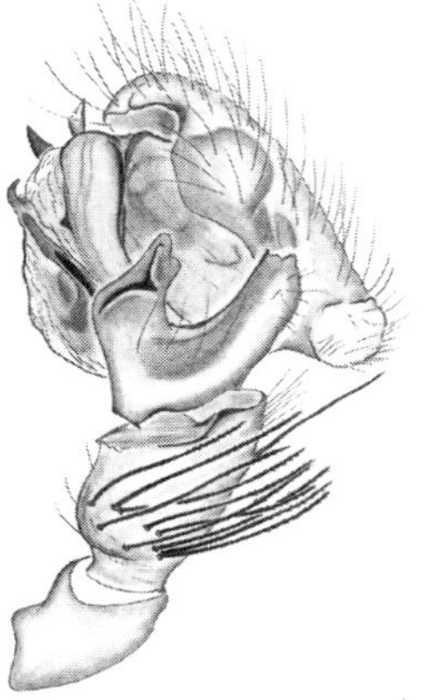

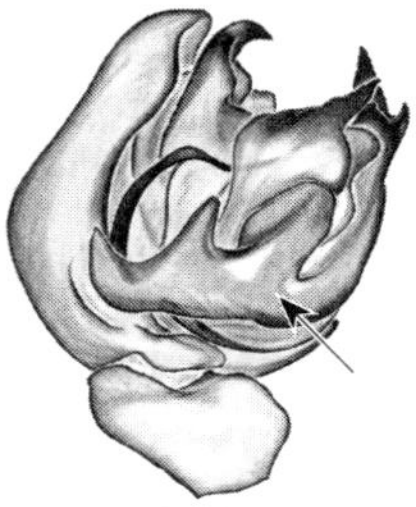

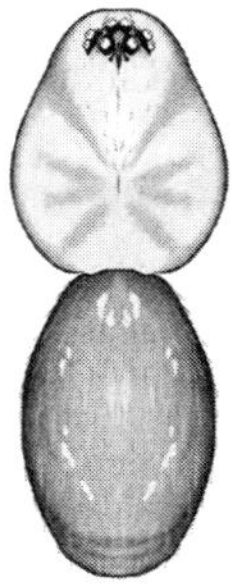

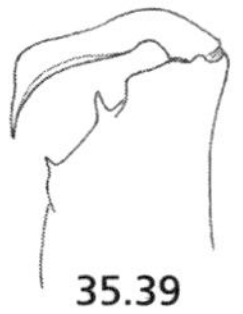

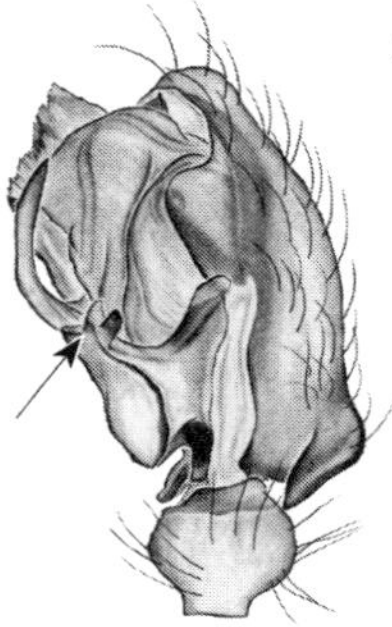

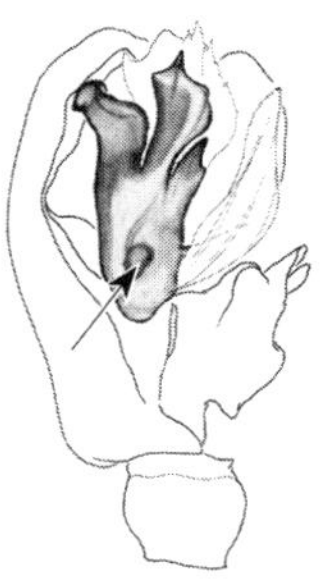

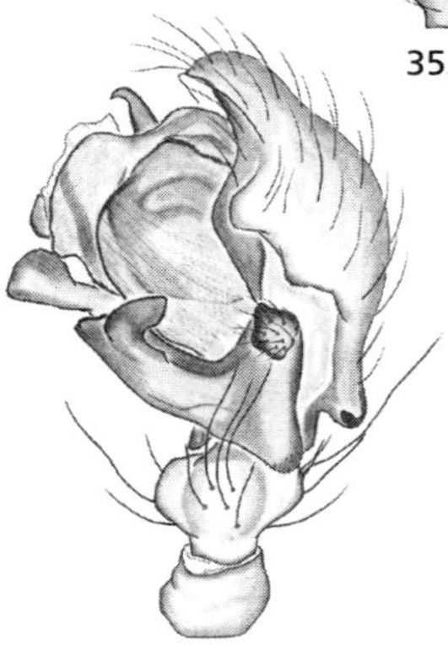

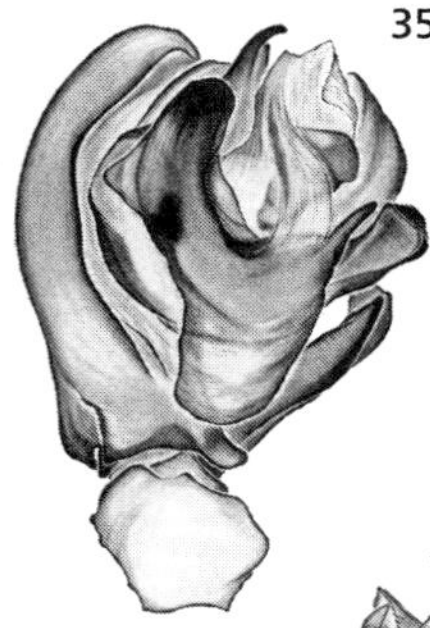

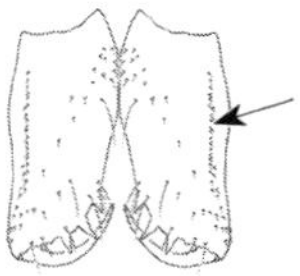

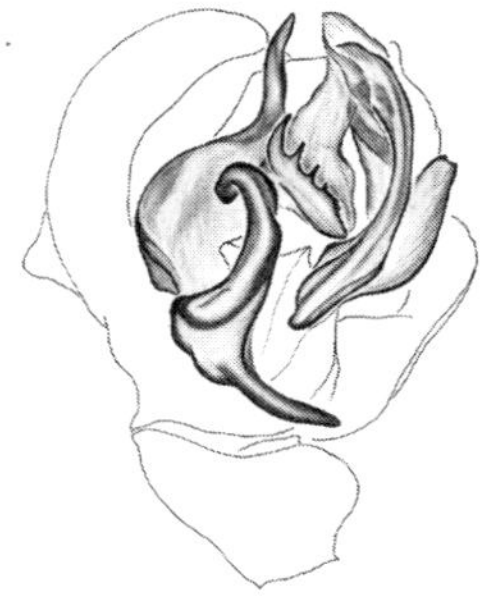

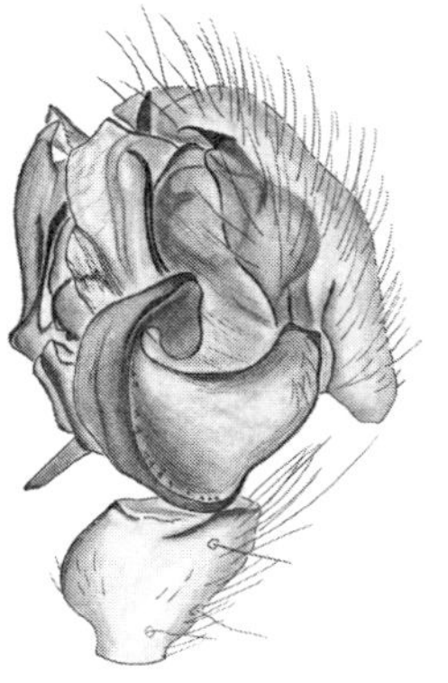

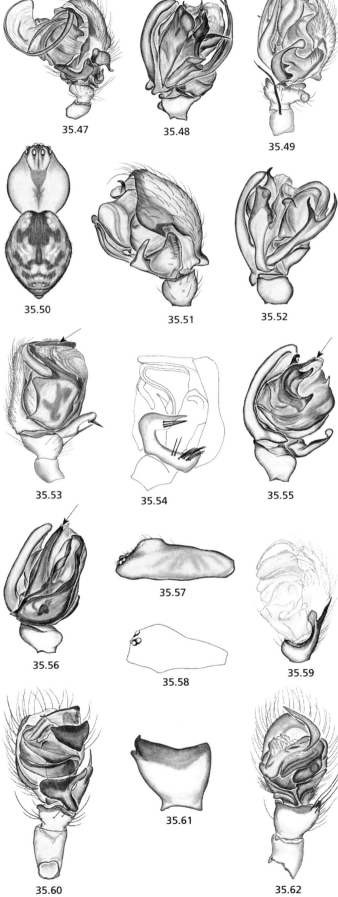

| | |
|---|---|
| **24(23)** | Palp with long curved embolus supported by a large, broad suprategular apophysis (Fig. 35.47); simple elbowed paracymbium; Fickert's gland absent .......... ***Taranucnus*** <br> **Div.** 1 species: *Taranucnus ornithes* (BARROWS 1940) — **Dist.** e NA from QC s to NC and TN — **Ref.** Helsdingen 1973b |
| — | Palp with embolus and suprategular apophysis otherwise (Fig. 35.48); paracymbium usually more complex, with teeth (Fig. 35.49); Fickert's gland present (Fig. 35.25) ...... ***Lepthyphantes,*** *in part* <br> **Div.** 29 species + several undescribed: most of the described species key out here — **Dist.** widespread — **Refs.** Zorsch 1937, Buckle *et al.* 2001 |
| **25(5)** | Carapace with medial dark shield-shaped mark (Fig. 35.50); legs long, femur I significantly longer than carapace; metatarsus IV with trichobothrium; palp as shown, similar to *Lepthyphantes* (Figs. 35.51-35.52) ...... ***Poeciloneta*** <br> **Div.** 10 species — **Dist.** n and w NA — **Ref.** Buckle *et al.* 2001 |
| — | Carapace without dark shield-shaped mark; legs shorter, femur I no longer than carapace; metatarsus IV without trichobothrium ............ **26** |
| **26(25)** | Embolus forming distal-facing coil (Figs. 35.53-35.54) .... ***Bathyphantes,*** *in part* <br> **Div.** 26 species, 24 key out here — **Dist.** widespread — **Ref.** Ivie 1969 |
| — | Embolus otherwise (Figs. 35.55-35.56) .......... **27** |
| **27(26)** | Embolus curved and coming to fine point, pointing ventrad (Fig. 35.55) ............ ***Porrhomma*** <br> **Div.** 9 species — **Dist.** widespread — **Ref.** Buckle *et al.* 2001 |
| — | Embolus straight and thick with blunt tip, pointing anteriad (Fig. 35.56) ............ ***Kaestneria*** <br> **Div.** 3 species — **Dist.** n NA — **Ref.** Ivie 1969 |
| **28(5)** | Carapace elevated behind eyes (Figs. 35.57-35.58) ....... **29** |
| — | Carapace not elevated behind eyes ................ **30** |
| **29(28)** | Trichobothria on metatarsus IV absent; palpal tibia with distinctive branched tibial apophysis (Figs. 35.59), palp as in Fig. 35.60 ................ ***Sougambus*** <br> **Div.** 2 species: *Sougambus bostoniensis* (EMERTON 1882) and "*Erigone*" *mentasta* CHAMBERLIN & IVIE 1947b — **Dist.** n NA — **Refs.** Crosby & Bishop 1936b, Hackman 1954 |
| — | Trichobothria on metatarsus IV present; tibial apophysis otherwise (Fig. 35.61), palp as in Fig. 35.62 ......... ***Hilaira*** <br> **Div.** 24 species — **Dist.** n NA — **Refs.** Buckle *et al.* 2001, Saaristo & Marusik 2004a — **Note** Saaristo & Marusik (2004a) placed 15 North American species, 10 of them new, in *Oreoneta* KULCZYŃSKI *in* CHYZER & KULCZYŃSKI 1894. They suggested that, of the remaining species, a few closely related to the type should be left in *Hilaira* and the remaining ones placed in new genera |

**30(28)** Palpal tibia with stout, thick, blunt tipped dorsal spine (Fig. 35.63); paracymbium a rectangular plate fused to cymbium (Fig. 35.64); palp with small cymbium; embolic division a flat plate, narrowed basally, and with a short, straight embolus extending distad (Fig. 35.65) ................................................................................... ***Frontinella***
   **Div.** 2 species — **Dist.** *Frontinella pyramitela* (WALCKENAER 1841), widespread and *Frontinella huachuca* GERTSCH & DAVIS 1946, AZ — **Refs.** Blauvelt 1936, Gertsch & Davis 1946 — **Note** *Frontinella pyramitela* is referred as *Frontinella communis* (HENTZ 1850a) in Platnick (2005)

— Palpal tibia without stout, thick, blunt tipped dorsal spine; paracymbium U-shaped sclerite, palp otherwise ......... **31**

**31(30)** Embolus an anteriad facing spiral (Fig. 35.66) ................................................................................... ***Linyphantes***
   **Div.** 19 species — **Dist.** Pacific northwest — **Ref.** Chamberlin & Ivie 1942a

— Embolus otherwise ......................................................... **32**

**32(31)** Carapace broad; abdomen dark with paired light spots (Fig. 35.67); prolateral spine on palpal tibial large and thick (Fig. 35.68) ......................................... ***Graphomoa***
   **Div.** 1 species: *Graphomoa theridioides* CHAMBERLIN 1924a — **Dist.** LA, TN — **Ref.** Chamberlin 1924a — **Note** according to Gustavo Hormiga (pers. comm.) *Graphomoa theridioides* is a synonym of *Pocobletus coroniger* SIMON 1894a and *Graphomoa* a synonym of *Pocobletus* SIMON 1894a

— Without the above combination of characters ............. **33**

**33(32)** Long threadlike embolus extending laterally far past the edge of the cymbium (Fig. 35.69); palpal tibia with ventral bump tipped with tuft of hairs (Fig. 35.70) ................................................................................... ***Jalapyphantes***
   **Div.** 1 species: *Jalapyphantes cuernavaca* GERTSCH & DAVIS 1946 — **Dist.** AZ — **Ref.** Gertsch & Davis 1946

— Without the above combination of characters ............. **34**

**34(33)** Trichobothrium present on tibia IV, palp as shown (Figs. 35.71-35.72) ........................................................ ***Microneta***
   **Div.** 1 species: *Microneta viaria* (BLACKWALL 1841) — **Dist.** n and w NA — **Refs.** Kaston 1981, Saaristo 1974a, Paquin & Dupérré 2003

— No trichobothrium present on tibia IV; palp otherwise (Fig. 35.73) .................................................................. **35**

**35(34)** Embolic division with radix (R), embolus (E) and lamella (L) not fused (Figs. 35.76-35.77) ......... ***Agyneta,*** *in part*
   **Div.** 45 species and perhaps as many more undescribed: a few western species key out here — **Dist.** widespread — **Refs.** Buckle *et al.* 2001, Paquin & Dupérré 2003 — **Note** some authors treat *Meioneta* HULL 1920 as a distinct subgenus or genus. *Agyneta sensu stricto* has a trichobothrium on metatarsus IV while *Meioneta* does not

— Embolic division of palp a compact unit with fused sclerites (Figs. 35.73, 35.74, 35.75) ............................ **36**

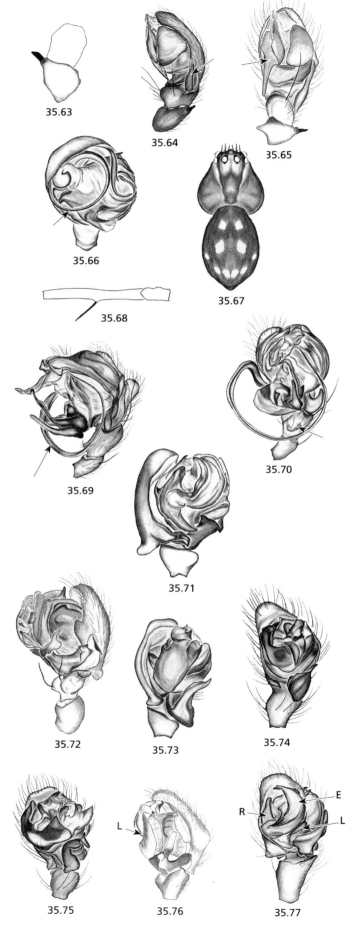

Spiders of North America — Linyphiidae | 133

**36(35)** Palp as shown (Figs. 35.74 -35.75, 35.78-35.79) ............... .................................................. **Oreonetides,** *in part*
    **Div.** 6 species + several undescribed: 3 described species key out here — **Dist.** n and w NA — **Ref.** Helsdingen 1981a

— Palp as shown (Figs. 35.73, 35.80) ................... **Saaristoa**
    **Div.** 1 species: *Saaristoa sammamish* (Levi & Levi 1955) — **Dist.** nw NA — **Ref.** Levi & Levi 1955

**37(3)** Chelicerae with mastidion (Figs. 35.81-35.82) ............. **38**
— Chelicera without mastidion ........................................ **42**

**38(37)** Palpal tibia with a long straight dorsal spur (Fig. 35.83) .. ........................................................ **Tibioplus**
    **Div.** 1 species: *Tibioplus nearcticus* Chamberlin & Ivie 1947b — **Dist** AK, YT — **Ref.** Chamberlin & Ivie 1947b

— Not so .................................................................... **39**

**39(38)** Abdomen red (fading to pale brown in alcohol) with black rump (Fig. 35.84); palp as shown (Figs. 35.85-35.86) ....... .................................................. **Ostearius**
    **Div.** 1 introduced species: *Ostearius melanopygius* (O. Pickard-Cambridge 1879d) — **Dist.** ne NA — **Refs.** Millidge 1985, Roberts 1987

— Abdomen brown or gray, usually concolorous; palp different ............................................................... **40**

**40(39)** Endite with single large setose tubercle (Fig. 35.87); legs relatively long, femur I length 1 1/3 times that of carapace; leg spines short and thin, barely distinguishable from setae; eyes small, especially AME; carapace higher than ocular area; palp as shown (Figs. 35.88-35.89) ................... .................................................. **Oaphantes**
    **Div.** 1 species: *Oaphantes pallidulus* (Banks 1904b) + 2 undescribed — **Dist.** CA to BC — **Ref.** Chamberlin & Ivie 1943

— Endite with several tiny setaceous tubercles or none; legs shorter, femur I no longer than carapace; leg spines of normal size; eyes larger; carapace no higher than ocular area ................................................................... **41**

**41(40)** Mastidion located anteriomesally on chelicera, above anterior teeth (Fig. 35.81); cephalothorax reddish, abdomen charcoal; epigastric plates not striate; palp as shown (Figs. 35.90-35.91) .......................... **Macrargus**
    **Div.** 1 species: *Macrargus multesimus* (O. Pickard-Cambridge 1875c) — **Dist.** n NA — **Ref.** Kaston 1981

— Mastidion located anteriolaterally on chelicera, just in front of stridulating ridges (Fig. 35.82); cephalothorax amber, abdomen brownish gray; epigastric plates striate (Fig. 35.92); palp as shown (Figs. 35.93-35.94) ................... .................................................. **Maro,** *in part*
    **Div.** 2 species, 1 keys out here: *Maro amplus* Dondale & Buckle 2001 — **Dist.** n NA — **Ref.** Dondale & Buckle 2001

**42(37)** Carapace elevated behind ocular area to a point which is adorned with one or more forward curving spines (Fig. 35.95); paracymbium with long basally extending branch (Fig. 35.96); palpal tibia modified (Fig. 35.96); palp as shown (Figs. 35.96-35.97) .................................. ***Wubana***
  **Div.** 7 species — **Dist.** n and w NA — **Refs.** Chamberlin & Ivie 1936b, Buckle *et al.* 2001

— Carapace not elevated; paracymbium of usual U-shape, palpal tibia unmodified .................................................. **43**

**43(42)** Clypeus bulging (Fig. 35.98); embolus long and straight, extending from base to apex of cymbium (Fig. 35.99) ...... ................................................................................ ***Thaleria***
  **Div.** 1 species: *Thaleria leechi* Eskov & Marusik 1992a — **Dist.** AK — **Ref.** Eskov & Marusik 1992a

— Clypeus normal, embolus otherwise .............................. **44**

**44(43)** Chelicerae with anterio-lateral row of denticles (Fig. 35.100); carapace narrowed anteriorly (Fig. 35.101); ant-mimic appearance ......................................... ***Anibontes***
  **Div.** 2 species: *Anibontes longipes* Chamberlin & Ivie 1944 and *Anibontes mimus* Chamberlin 1924a — **Dist.** se USA — **Refs.** Chamberlin 1924a, Chamberlin & Ivie 1944

— Chelicerae and carapace normal ................................... **45**

**45(44)** Palpal femur with ventral teeth (Fig. 35.102), spiracle forward by 1/4 to 1/3 distance between spinnerets and epigastric furrow (Fig. 35.103); palp is similar to that of *Agyneta* (Figs. 35.102, 35.104) .............. ***Tennesseellum***
  **Div.** 2 species — **Dist.** *Tennesseellum formicum* (Emerton 1882) widespread and an undescribed species from CA — **Ref.** Kaston 1981

— Palpal femur without ventral teeth, spiracle just in front of spinnerets (Fig. 35.105) .................................................. **46**

**46(45)** Tiny spiders (1.0-1.5 mm) with large, distinctive palps (Figs. 35.106-35.107) having complex embolic divisions .. ................................................................................. ***Sisicus***
  **Div.** 2 species: *Sisicus apertus* (Holm 1939b) and *Sisicus penifusifer* Bishop & Crosby 1938 — **Dist.** n NA — **Ref.** Paquin & Dupérré 2003

— Without the above combination of characters ............. **47**

**47(46)** Palp with distinctive paracymbium (Fig. 35.108) and simple embolic division (Fig. 35.109) .............. ***Aphileta***
  **Div.** 1 species: *Aphileta misera* (O. Pickard-Cambridge 1882b) — **Dist.** n NA — **Refs.** Bishop & Crosby 1938, Paquin & Dupérré 2003

— Without the above combination of characters ............. **48**

| | |
|---|---|
| **48(47)** | Palpal bulb and cymbium relatively small (Fig. 35.110); embolic division a simple plate (Fig. 35.111); palpal tibia one to two times length of cymbium; abdomen with dark heart mark and posterior chevrons .................... ***Nanavia***<br>**Div.** 1 species: *Nanavia monticola* CHAMBERLIN & IVIE 1933a + 1 undescribed — **Dist.** UT, CA — **Ref.** CHAMBERLIN & IVIE 1933a |
| — | Palpal bulb and cymbium larger (Figs. 35.112, 35.113); embolic division complex; palpal tibia shorter than cymbium; abdomen concolorous ......................................... **49** |
| **49(48)** | Palpal bulb with distinct arrangement of embolus and radix (Figs. 35.112, 35.114); abdomen concolorous or patterned ............................................. ***Agyneta***, *in part*<br>**Div.** 45 species and perhaps as many more undescribed: most key out here — **Dist.** widespread — **Refs.** Buckle *et al.* 2001, Paquin & Dupérré 2003 — **Note** some authors treat *Meioneta* HULL 1920 as a distinct subgenus or genus. *Agyneta sensu stricto* has a trichobothrium on metatarsus IV while *Meioneta* does not |
| — | Palp otherwise (Figs. 35.113, 35.114-35.117); abdomen concolorous .................................................................... **50** |
| **50(49)** | Body under 1.7 mm in length; stridulating ridges on epigastric plates (Fig. 35.118); palp as shown (Figs. 35.113, 35.115) ....................................... ***Maro***, *in part*<br>**Div.** 2 species, 1 keys out here: *Maro nearcticus* DONDALE & BUCKLE 2001 — **Dist.** n NA — **Ref.** Dondale & Buckle 2001 |
| — | Body usually over 1.7 mm in length; no stridulating ridges on epigastric plates; palp as shown (Figs. 35.116-35.117) . ....................................... ***Oreonetides***, *in part*<br>**Div.** 6 species + several undescribed: 3 key out here — **Dist.** n and w NA — **Ref.** Helsdingen 1981a |

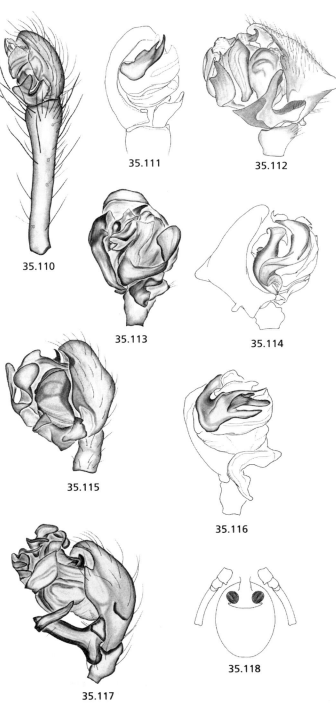

35.110, 35.111, 35.112, 35.113, 35.114, 35.115, 35.116, 35.117, 35.118

Linyphiidae

## Erigoninae portion, 107 genera

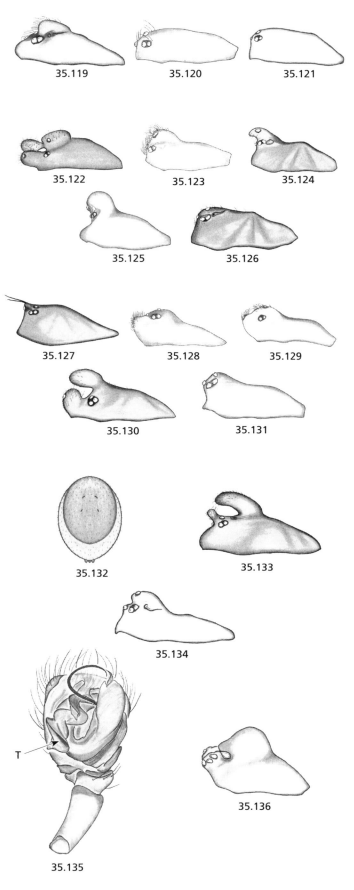

51(2) With a pair of distinct pits or excavations on the cephalothorax, usually just behind PLE (Fig. 35.119) ............ **52**

— Without distinct pits on the cephalothorax (Figs. 35.120-35.121) .................................................................. **53**

52(51) Cephalic region raised, or modified in the eye or clypeus region with one or more fairly distinct lobes, large setae or processes (Figs. 35.122-35.126) ...................... **54**

— Cephalic region "normal" (Figs. 35.120-35.121, although these illustrations lack cephalic pits) unmodified although the cephalic region is commonly slightly higher than the posterior portion of the cephalothorax, and some species have spinose carapace margins ........................ **93**

53(51) Cephalic region raised, or modified in the eye or clypeus region with one or more fairly distinct lobes, large setae or processes (Figs. 35.127-35.131) ...................... **98**

— Cephalic region "normal," unmodified (Figs. 35.120-35.121) although some species have spinose carapace margins .................................................................. **135**

### With pits and modified cephalic region, 41 genera

54(52) Abdomen with sclerotized scutum or shield (Fig. 35.132); reduced in some species to a small region at anteriormost portion of abdominal dorsum........................ ***Pelecopsis***
*Div.* 6 species — *Dist.* widespread except southwest USA — *Refs.* Buckle et al. 2001, Paquin & Dupérré 2003

— Without a scutum ............................................................ **55**

55(54) Cephalic region with both a clypeal and a distinct cephalic lobe or protuberance (Figs. 35.122, 35.133) ................ **56**

— Cephalic region with either a clypeal or one or two cephalic lobes, which may be quite small (Figs. 35.125-35.126, 35.134) ............................................................................. **57**

56(55) Cephalic lobe produced into a secondary protuberance that lies directly over the clypeal protuberance, giving the impression of stacked protuberances (Fig. 35.122); embolic division lacking obvious spiraled tailpiece ...............
.................................................................. ***Savignia***
*Div.* 1 species: *Savignia birostra* (CHAMBERLIN & IVIE 1947b) — *Dist.* Widespread except southwest USA — *Ref.* Chamberlin & Ivie 1947b

— Cephalic lobe without a secondary protuberance and not stacked over the clypeal protuberance (Fig. 35.133); embolic division with obvious spiraled tailpiece (at least half as long as tegulum, T in Fig. 35.135) ..............................
.......................................... ***Scotinotylus***, in part
*Div.* 34 species, 5 key out here — *Dist.* CAN, n and w USA — *Refs.* Millidge 1981a, Paquin & Dupérré 2003

57(55) With cephalothoracic lobe mostly posterior of eyes (not bearing eyes, Fig. 35.136) OR with pits (not just excavated area surrounding pit) distinctly larger than PME (Fig. 35.138) .............................................................................. **58**

— Lobe not mainly posterior to eyes (often bearing one or more pairs) AND with pits smaller than PME, although surrounding excavation may be larger than PME ........ **64**

**58(57)** Cephalothorax and legs orangish or reddish and abdomen black .................................................. ***Hypomma***
  **Div.** 3 species — **Dist.** CAN, AK, n USA — **Refs.** Crosby & Bishop 1933b, Chamberlin & Ivie 1947b, Paquin & Dupérré 2003

— Body coloration otherwise ............................................. **59**

**59(58)** Carapace with reticulate texture (like lines scratched with a needle), visible without highest magnification (visible at 30X); lobe mostly posterior of eyes and pits of normal size .................................................... ***Tapinocyba,*** *in part*
  **Note** 2 undescribed species from CA (T. Prentice, pers. comm.)

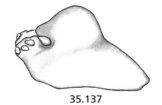

35.137

— Carapace smooth (except under very high magnification) ............................................................................................ **60**

**60(59)** With cephalic lobe mostly posterior of eyes and pits twice or more the diameter the PME (Fig. 35.137) ..................... ***Oedothorax***
  **Div.** 7 species — **Dist.** widespread — **Refs.** Buckle *et al.* 2001, Paquin & Dupérré 2003 — **Note** many North American *Oedothorax* probably do not belong in that genus

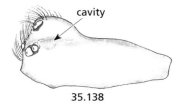

35.138

— Without either of the above characters ........................... **61**

**61(60)** Embolic division lacking spiraled tailpiece; cavity surrounding pit usually more than 3 times larger than diameter of PME (Fig. 35.138) ..................... ***Hybauchenidium***
  **Div.** 4 species — **Dist.** AK, CAN, northern USA — **Ref.** Buckle *et al.* 2001, Paquin & Dupérré 2003

— Embolic division with obvious spiraled tailpiece (at least half as long as tegulum); cavity surrounding pit less than 3 times the diameter of PME; usually with lobe mostly posterior of eyes ............................................................. **62**

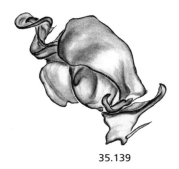

35.139

**62(61)** Tibial apophysis with thick basoectal macroseta (Fig. 35.139); tip of embolus retroverted or pointing more or less toward proximal end of palp (Fig. 35.139) ................. ***Disembolus,*** *in part*
  **Div.** 24 species, 9 key out here — **Dist.** widespread, more common in west — **Ref.** Millidge 1981b

— Tibial apophysis lacking thick basoectal macroseta; tip of embolus not retroverted, often terminates in wide coil at end of palp ................................................................. **63**

**63(62)** Tibal apophysis more than twice as long as length of tibia; paracymbium elongate ................. ***Spirembolus,*** *in part*
  **Div.** 43 species, 4 key out here — **Dist.** western NA— **Ref.** Millidge 1980b

— Tibial apophysis less than twice the length of the tibia; paracymbium of normal length, c-shaped ......................... ................................................................. ***Scotinotylus,*** *in part*
  **Div.** 34 species, 12 key out here — **Dist.** CAN, n and w USA — **Refs.** Millidge 1981a, Paquin & Dupérré 2003 — **Note** some members of an undescribed genus (see couplet 66) will key out here as well, including "*Spirembolus*" *approximatus* Chamberlin 1949 and four undescribed species

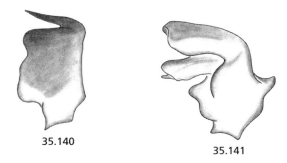

35.140           35.141

35.142

**64(57)** Apical portion of tibial apophysis pointing ectally (Figs. 35.140-35.142) ................................................................. **65**

— Tibial apophysis otherwise ............................................. **68**

**65(64)** Tibial apophysis with one apical portion, which points to the ectal side (Fig. 35.140) ........ ***Diplocephalus,*** *in part*
  **Div.** 4 species, 3 key out here — **Dist.** CAN, AK, northern USA — **Ref.** Buckle *et al.* 2001, Paquin & Dupérré 2003

— Tibial apophysis with two or more apical portions which point to the ectal side (Figs. 35.141-35.142) ................ **66**

Linyphiidae

66(65) Embolus coiled at end of palp; tailpiece spiraled ..............
............................................. **undescribed genus,** *in part*
**Div.** *"Ceratinopsis" palomara* CHAMBERLIN 1949 + undescribed species — **Dist.** CA and OR — **Ref.** Chamberlin 1949

— Without the above combination of characters ............. **67**

67(66) Embolus longer than cymbium, forms nearly a complete coil (Fig. 35.143) ................................ ***Horcotes***
**Div.** 3 species— **Dist.** widespread — **Refs.** Buckle *et al.* 2001, Paquin & Dupérré 2003

— Embolus shorter than cymbium, curves but does not form a coil (Fig. 35.144) ........................................ ***Entelecara***
**Div.** 3 species — **Dist.** CAN, AK, WA — **Refs.** Buckle *et al.* 2001, Paquin & Dupérré 2003

68(64) Tibial apophysis definitely longer than tibial length (Figs. 35.145, 35.148) ................................................................ **70**

— Tibial apophysis not longer than tibial length .............. **69**

69(68) Tibial apophysis definitely shorter than width of narrowest part of tibia (Figs. 35.146-35.147) ...................... **78**

— Tibial apophysis not shorter than width of narrowest part of tibia (Figs. 35.149-35. 150) ......................................... **86**

70(68) Tibial apophysis with conspicuous dorsoectal row of 3 or more long, thick macrosetae (Fig. 35.151) ..........................
............................................................ ***Glyphesis***
**Div.** 2 species: *Glyphesis idahoanus* (CHAMBERLIN 1949) and *Glyphesis scopulifer* (EMERTON 1882) — **Dist.** CAN and n USA — **Refs.** Buckle *et al.* 2001, Paquin & Dupérré 2003

— Tibial apophysis lacking conspicuous row of macrosetae . ................................................................................. **71**

71(70) Cephalic lobe bearing eyes terminates in a rather sharp, forward-pointing cone (Fig. 35.152) ........... ***Origanates***
**Div.** 1 species: *Origanates rostratus* (EMERTON 1882) — **Dist.** Eastern NA — **Ref.** Crosby & Bishop 1933b

— Cephalic lobe, if present, is rounded ............... **72**

72(71) Embolus very long, forming a large coil and a smaller terminal coil (Fig. 35.153); tibial apophysis clearly bifid with two long sharp points ............................ ***Traematosisis***
**Div.** 1 species: *Traematosisis bispinosus* (EMERTON 1911) — **Dist.** Eastern NA — **Ref.** Bishop & Crosby 1938

— Embolus not forming a large and small double coil; tibial apophysis otherwise ......................................... **73**

73(72) Tibial apophysis consists of one long process (Fig. 35.145) which terminates in 2-3 sharp points at extreme apical end (not visible in Fig. 35.145) ................................ ***Dietrichia***
**Div.** 1 species: *Dietrichia hesperia* Crosby & Bishop 1933b — **Dist.** CAN, CA — **Ref.** Crosby & Bishop 1933b

— Tibial apophysis otherwise ............................................... **74**

35.143

35.144

35.145

35.146

35.147

35.148

35.149

35.150

35.151

35.152

35.153

Spiders of North America — *Linyphiidae* **139**

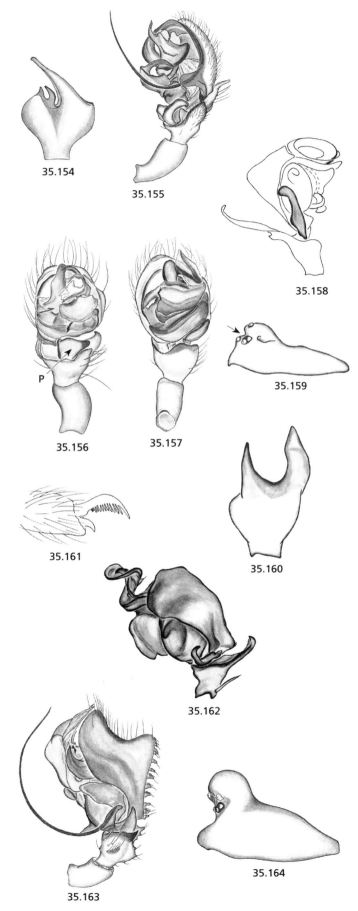

74(73) Tibial apophysis with 3 apical points (Fig. 35.154) OR embolus much longer than cymbium, making a spiral around entire palpal bulb, its long thin apical end held away from rest of palp (Fig. 35.155) ....................
................................................ **Baryphyma,** in part
Div. 4 species, 2 key out here: *Baryphyma gowerense* (LOCKET 1965) and *Baryphyma trifrons affine* (SCHENKEL 1930b) — Dist. CAN, n USA — Ref. Roberts 1987, Buckle *et al.* 2001, Paquin & Dupérré 2003

— Tibial apophysis with one or two apical points; embolus usually shorter than cymbium, apical end of embolus not as above ........................................................................ 75

75(74) Tailpiece of embolic division curved, but does not coil or spiral (Fig. 35.156-35.157); tibial apophysis terminates in one or two apical processes .............................................. 76

— Tailpiece of embolic division coils or spirals at least one turn; tibial apophysis terminates in one apical process (though an additional basal process may be present, Fig. 35.158 *Spirembolus*) ............................................................. 77

76(75) Cephalic region with small secondary lobe anterior to main lobe (Fig. 35.159), OR tibial apophysis with two separate apical processes, each longer than tibial width (Fig. 35.160); paracymbium normal, c-shaped; tarsal claws I-III strongly pectinate (Fig. 35.161) ....................
............................................. **Walckenaeria,** in part
Div. 68 species, 2 key out here: *Walckenaeria castanea* (EMERTON 1882), *Walckenaeria atrotibialis* (O. PICKARD-CAMBRIDGE 1878a), — Dist. widespread [*Walckenaeria atrotibialis*: CAN, AK, n USA — Refs. Millidge 1983, Paquin & Dupérré 2003

— With one cephalic lobe and tibial apophysis not as above; paracymbium reduced, triangular rather than c-shaped (P in Fig. 35.156); tarsal claws not pectinate ............................
.................................................. **Eridantes,** in part
Div. 2 species: *Eridantes erigonoides* (EMERTON 1882) and *Eridantes utibilis* CROSBY & BISHOP 1933b — Dist. eastern NA, AZ — Refs. Crosby & Bishop 1933b, Paquin & Dupérré 2003 — Note both species may key out here

77(75) Tibial apophysis with thick basoectal macroseta; tip of embolus retroverted or pointing more or less toward proximal end of palp (Fig. 35.162) ..................................
.................................................. **Disembolus,** in part
Div. 24 species, 11 key out here — Dist. widespread, more common in west — Ref. Millidge 1981b

— Tibial apophysis lacking thick basoectal macroseta; tip of embolus not retroverted, often terminates in wide coil at end of palp; sometimes with unusually long paracymbium (Fig. 35.158); sometimes with abdominal pattern of chevrons ............................................... **Spirembolus,** in part
Div. 43 species, 6 key out here — Dist. western NA — Ref. Millidge 1980b

78(69) Cymbium with row of stout tubercles across dorsum; embolus longer than cymbium, emerges toward proximal part of palp and hangs free well ventral of palp (Fig. 35.163) ............................................... **Cnephalocotes**
Div. 1 species: *Cnephalocotes obscurus* (BLACKWALL 1834b) — Dist. CAN — Refs. Roberts 1987, Paquin & Dupérré 2003

— Cymbium without row of stout dorsal tubercles; embolus of variable length, but not as above ............................... 79

79(78) Eyes in front of rather tall lobe (about as tall as basal width of lobe); pit located directly above PLE (Fig. 35.164); cephalothorax and legs orangish, abdomen black .............
............................................................ **Dismodicus**
Div. 4 species — Dist. CAN, n USA — Refs. Buckle *et al.* 2001, Paquin & Dupérré 2003

— Without the above combination of characters ............. 80

# Linyphiidae

**80(79)** Clypeus hairy (Fig. 35.165) and with a row of long spines on venter of tibia and metatarsus I and II (Fig. 35.166). Often found in wetlands .................. ***Satilatlas,*** in part
    **Div.** 6 species, several may key out here — **Dist.** CAN and n USA to CO — **Refs.** Millidge 1981c, Paquin & Dupérré 2003

— Without this combination of characters ...................... **81**

**81(80)** Embolus rather short, does not spiral around palp ........... **82**

— Embolus long, spirals around end of palp ................... **84**

**82(81)** Carapace and sternum smooth ...... ***Tapinocyba,*** in part
    **Div.** 15 species, 10 key out here — **Dist.** widespread — **Refs.** Buckle et al. 2001, Paquin & Dupérré 2003

— Carapace (and sometimes sternum) at least somewhat textured (rough, wrinkled, or reticulate, as visible even under lower magnification), at least along cervical grooves ........................................................................ **83**

**83(82)** Carapace and sternum often with rugose or reticulate texture (visible without highest magnification; embolus usually with a single terminal projection (except *C. crenatus*, which has undulating lateral margins of carapace, Fig. 35.167); TmI located at 0.40 or less ................................................................. ***Ceratinops,*** in part
    **Div.** 8 species, 6 key out here — **Dist.** widespread — **Refs.** Crosby & Bishop 1933b, Paquin & Dupérré 2003

— Carapace usually only textured along cervical grooves; embolus with two terminal projections (Fig. 35.168); TmI located at greater than 0.35, usually greater than 0.40 ...... ................................................................... ***Lophomma***
    **Div.** 6 species — **Dist.** eastern NA — **Ref.** Crosby & Bishop 1933b, Holm 1960b, Paquin & Dupérré 2003

**84(81)** Tibial apophysis broader apically than basally (Fig. 35.169) ............................................................ ***Tusukuru***
    **Div.** 1 species: *Tusukuru hartlandianus* (EMERTON 1913a) — **Dist.** eastern NA — **Ref.** Crosby & Bishop 1933b

— Tibial apophysis with one or more sharp points .......... **85**

**85(84)** Cymbium somewhat Y-shaped in dorsal view, much broader distally (Fig. 35.170); TmI located at 0.85-0.90; TmIV present ..................................................... ***Pocadicnemis***
    **Div.** 3 species — **Dist.** CAN and n USA to NM — **Ref.** Millidge 1976

— Cymbium of normal shape in dorsal view; TmI located at 0.30-0.42; TmIV absent ..................... ***Micrargus,*** in part
    **Div.** 2 species, 1 keys out here: *Micrargus longitarsus* (EMERTON 1882) — **Dist.** CAN, AK, and n USA — **Refs.** Crosby & Bishop 1933b, Paquin & Dupérré 2003 — **Note** Millidge (1977), without explicitly stating this to be a new combination, placed *Baryphyma longitarsum* in *Micrargus*, saying it was closely related to *Micrargus herbigradus*. Platnick (2005) retains it in *Baryphyma*

**86(69)** Clypeus hairy (Fig. 35.165) and with a row of long spines on venter of tibia and metatarsus I and II (Fig. 35.166). Often found in wetlands .................. ***Satilatlas,*** in part
    **Div.** 6 species, several may key out here — **Dist.** CAN and n USA to CO — **Refs.** Millidge 1981c, Paquin & Dupérré 2003

— Without this combination of characters ...................... **87**

35.165

35.166

35.167

35.168

35.169

35.170

**87(86)** Cephalic lobe hourglass shaped (constricted at sides, wider at top) when viewed from the front (Fig. 35.173); cephalothorax and legs reddish, abdomen black; tibial apophysis projects dorsally away from tibia; TmIV present; TmI at 0.87-0.84 ............................................ ***Hypselistes***
  **Div.** 4 species — **Dist.** CAN, AK, n USA — **Ref.** Buckle *et al.* 2001, Paquin & Dupérré 2003

— Cephalic lobe rarely so constricted; coloration otherwise; tibial apophysis does not project dorsally away from tibia (unless there are multiple apophyses: *Mythoplastoides*); TmIV absent; TmI at 0.38-0.70 ........................................ **88**

**88(87)** Embolic division long and relatively thin (>10 times longer than greatest width), and at least partly coiled or spiralled (Figs. 35.174-35.175) ........................................... **89**

— Embolic division rather thick (<10 times longer than greatest width), and does not coil or spiral, though it may be bent (Figs. 35.176-35.177) ............................................. **91**

**89(88)** Tibial apophysis consists essentially of a single process, although it may be bifid with two short processes apically (Fig. 35.178) ........................................ ***Eridantes,*** in part
  **Div.** 2 species: *Eridantes erigonoides* (EMERTON 1882) and *Eridantes utibilis* CROSBY & BISHOP 1933b — **Dist.** eastern NA, AZ — **Ref.** Crosby & Bishop 1933b, Paquin & Dupérré 2003 — **Note** both species may key out here

— Tibial apophysis consisting of more than one process joined near the base, or a single, distinctly bifid process (Fig. 35.179) ........................................................... **90**

**90(89)** Apical end of embolus thick and truncate (not sharply pointed) (Fig 35.174); end of embolus protrudes beyond cymbium somewhat (Fig 35.174) ................ ***Symmigma***
  **Div.** 1 species: *Symmigma minimum* (EMERTON 1923) — **Dist.** AK, northwestern NA — **Ref.** Crosby & Bishop 1933b

— Embolus thin and pointed at end, and does not protrude beyond cymbium (Fig. 35.175) ......... ***Mythoplastoides***
  **Div.** 2 species: *Mythoplastoides erectus* (EMERTON 1915b) and *Mythoplastoides exiguus* (BANKS 1892a) — **Dist.** widespread except southwestern USA — **Ref.** Crosby & Bishop 1933b

**91(88)** Cephalic lobe somewhat bifid longitudinally (Fig 35.180) ............................................................ ***Thyreosthenius***
  **Div.** 1 species: *Thyreosthenius parasiticus* (WESTRING 1851) — **Dist.** CAN, n USA — **Refs.** Crosby & Bishop 1933b as *Hormathion limnatum* CROSBY & BISHOP 1933b, Roberts 1987

— Cephalic lobe not bifid ................................. **92**

**92(91)** Embolus recurved, apical end points toward proximal portion of palp (Fig. 35.176) ................ ***Microctenonyx***
  **Div.** 1 species: *Microctenonyx subitaneus* (O. PICKARD-CAMBRIDGE 1875e) — **Dist.** introduced to MA — **Refs.** Crosby & Bishop 1933b, Roberts 1987

— Embolus not recurved, apical end usually points to distal portion of palp (Fig 35.177) .......... ***Tapinocyba,*** in part
  **Div.** 15 species, 5 key out here — **Dist.** widespread — **Refs.** Buckle *et al.* 2001, Paquin & Dupérré 2003

**With pits but unmodified cephalic region, 6 genera**

**93(52)** With prominent mastidion on cheliceral face; body length > 3 mm; TmIV present ........................................ ***Zornella***
  **Div.** 2 species: *Zornella cryptodon* (CHAMBERLIN 1920a) and *Zornella cultrigera* (L. KOCH 1879c) — **Dist.** AK, CAN, n USA — **Refs.** Buckle *et al.* 2001, Paquin & Dupérré 2003

— Without a cheliceral mastidion; body length < 3 mm; TmIV absent ................................................ **94**

35.173

35.174

35.175

35.176

35.177

35.178

35.179

35.180

**94(93)** Carapace and sternum somewhat rugose (visible without highest magnification); embolus short and spinelike ....... ................................................................ ***Ceratinops,*** in part
**Div.** 8 species, 2 key out here: *Ceratinops annulipes* (Banks 1892a) and *Ceratinops rugosus* (Emerton 1909) — **Dist.** widespread — **Ref.** Crosby & Bishop 1933b

— Carapace and sternum smooth; embolus variable ....... **95**

**95(94)** Embolus bifurcate terminally, shorter than width of cymbium, spinelike; often associated with ants ... ***Masoncus***
**Div.** 4 species — **Dist.** Widespread — **Refs.** Cushing 1995, Buckle et al. 2001

— Embolus not bifurcate terminally; embolus longer than width of cymbium; not associated with ants ................ **96**

**96(95)** Embolus recurved, points toward proximal portion of palp; palpal tibia with dorsoectal macroseta ..................... ................................................................ ***Disembolus,*** in part
**Div.** 24 species, 1 keys out here: *Disembolus alpha* (Chamberlin 1949) — **Dist.** CO — **Ref.** Millidge 1981b

— Embolus not recurved, palpal tibia lacking dorsoectal macroseta ..................................................................... **97**

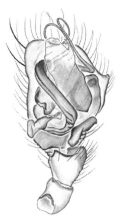

35.181

**97(96)** Embolus a large thick coil with a smaller apical coil (see Fig. 35.285 of *Sisis rotundus,* a similar species); tibial apophysis longer than length of tibia ................ ***Sisis,*** in part
**Div.** 2 species, 1 keys out here: *Sisis plesius* (Chamberlin 1949) — **Dist.** UT WY — **Ref.** Chamberlin 1949

— Embolus coils, forming two loops subequal in size (Fig. 35.181); tibial apophysis shorter than length of tibia ........ ................................................................ ***Micrargus,*** in part
**Div.** 2 species, 1 keys out here: *Micrargus longitarsus* (Emerton 1882) — **Dist.** n NA — **Ref.** Paquin & Dupérré 2003

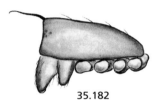

35.182

### Modified cephalic region, no pits, 33 genera

**98(53)** Most conspicuous carapace modification being one or two very large and thick forward-pointing macrosetae near apex of cephalic region (Figs. 35.182-35.184) ...... **99**

— Without one or two very large macrosetae near apex of cephalic region, although various smaller setae may be present ...................................................................... **101**

**99(98)** With one macroseta at apex of cephalic region (Fig. 35.182) .............................................. ***Islandiana,*** in part
**Div.** 14 species, 1 keys out here: *Islandiana unicornis* Ivie 1965 — **Dist.** TX — **Ref.** Ivie 1965

— With two macrosetae at apex of cephalic region (Figs. 35.183-35.184) .......................................................... **100**

35.183

**100(99)** Cephalic macrosetae nearly as long as width of eye area (Fig. 35.183); embolic division lacks obviously spiraled tailpiece .................................................. ***Paracornicularia***
**Div.** 1 species: *Paracornicularia bicapillata* Crosby & Bishop 1931 — **Dist.** MO, MS — **Ref.** Crosby & Bishop 1931

— Cephalic macrosetae much shorter than width of eye area (Fig. 35.184); embolic division with obviously spiraled tailpiece ......................................... ***Scotinotylus,*** in part
**Div.** 34 species, 1 keys out here: *Scotinotylus eutypus* (Chamberlin 1949) — **Dist.** BC, OR, WA — **Ref.** Millidge 1981a

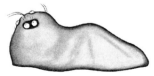

35.184

**101(98)** Abdomen with sclerotized scutum or shield (Fig. 35.185); reduced in some species to a small region at anteriormost portion of abdominal dorsum ..................................... **102**

— Abdomen lacking a scutum ......................... **104**

**102(101)** Embolic division with conspicuous tailpiece that is elongate (more than twice as long as wide), long (more than half as long as palp width), and oriented with its long axis parallel to the long axis of the palp (Fig. 35.186) .............. ***Ceraticelus,*** *in part*
**Div.** 41 species, 20 key out here — **Dist.** Widespread — **Refs.** Crosby & Bishop 1925c, Buckle *et al.* 2001, Paquin & Dupérré 2003

— Embolic division with tailpiece inconspicuous or otherwise than above ................................... **103**

**103(102)** Clypeus or clypeal projection covered with dense setae or with variously modified setae (Fig. 35.187) ...................... ***Floricomus***
**Div.** 14 species — **Dist.** widespread — **Refs.** Buckle *et al.* 2001, Paquin & Dupérré 2003

— Clypeal projection with sparse or no setae (Fig. 35.188); membrane accompanying embolus with a sharp point midway along its length; this point projects toward the end of the palp ............................ ***Pelecopsidis***
**Div.** 1 species: *Pelecopsidis frontalis* (Banks 1904a) — **Dist.** southeastern USA to CT — **Ref.** Bishop & Crosby 1935b

**104(101)** Palpal patella with ventral apical apophysis longer than width of segment (Fig. 35.189) ............. ***Erigone,*** *in part*
**Div.** 40 species, several may key out here — **Dist.** widespread — **Refs.** Buckle *et al.* 2001, Paquin & Dupérré 2003

— Palpal patella without ventral apical apophysis .......... **105**

**105(104)** With distinct knoblike or proboscis-like process (not bearing eyes) emerging from eye region (Fig. 35.190) ..... **106**

— Without such a process in the eye-region ................... **107**

**106(105)** Tibial apophysis longer than length of tibia; tarsal claws I-III strongly pectinate (Fig. 35.191) ................................... ***Walckenaeria,*** *in part*
**Div.** 68 species, 25 key out here — **Dist.** widespread — **Refs.** Millidge 1983, Paquin & Dupérré 2003

— Tibial apophysis a darkened knob, shorter than length of tibia; tarsal claws not pectinate .......... ***Micrargus,*** *in part*
**Div.** 2 species, 1 keys out here: *Micrargus pacificus* (Emerton 1923) — **Dist.** AK, BC, WA — **Ref.** Buckle *et al.* 2001 — **Note** the placement of this species in *Micrargus* is doubtful

**107(105)** With two cephalic lobes, both bearing eyes and many setae (Figs. 35.192-35.194) .................................. **108**

— One or two cephalic lobes, only one lobe bearing eyes ..... **110**

**108(107)** Posterior cephalic lobe much smaller than anterior lobe (Fig. 35.192); tibial apophysis emerges ectally and tip points mesally, without long ectally-bent terminal spine (Fig. 35.195) ............................................ ***Dactylopisthes***
**Div.** 1 species: *Dactylopisthes video* (Chamberlin & Ivie 1947b) — **Dist.** AK, YT — **Ref.** Chamberlin & Ivie 1947b

— Cephalic lobe subequal in size; tibial apophysis terminating in long spine bent abruptly to the ectal side ........ **109**

**109(108)** Clypeal and cephalic lobes indistinctly separated (Fig. 35.196); embolus a loose coil at end of palp; (Fig. 35.197) tibia as in Fig. 35.198) .................................. ***Dicymbium***
**Div.** 2 species: *Dicymbium elongatum* (Emerton 1882) and *Dicymbium nigrum* (Blackwall 1834a) — **Dist.** UT, WA — **Ref.** Buckle *et al.* 2001

— With two distinctly separated lobes (Fig. 35.199); embolus a bent spine rather than a coil .................................................. ***Diplocephalus,*** in part
**Div.** 4 species, 1 keys out here: *Diplocephalus cristatus* (Blackwall 1833a) — **Dist.** CAN n USA — **Refs.** Bishop & Crosby 1935c, Roberts 1987, Paquin & Dupérré 2003

**110(107)** With either a distinct projection in the clypeal area (Figs. 35.200-35.202) or with the cephalic/clypeal region divided into two lobes [though they may be closely appressed, as with the subocular sulci that separate the clypeus from the ocular area in *Blestia* (Fig. 35.203) Figs. 35.204-35.208] ... ................................................................................................ **111**

— With only one elevation or lobe in the cephalic region, or with the carapace broadly raised (Figs. 35.209-35.210) .... ................................................................................................ **121**

**111(110)** With distinct proboscis-like projection in the clypeal region (Figs. 35.200-35.202) ......................................... **112**

— With cephalic region divided into two lobes .............. **114**

**112(111)** Clypeal projection with a conspicuous sclerotized cap (Fig. 35.200) ..................................... ***Baryphyma,*** in part
**Div.** 4 species, 2 keys out here: *Baryphyma groenlandicum* (Holm 1967) and *Baryphyma kulczynskii* (Eskov 1979) — **Dist.** CAN, n USA. — **Refs.** Buckle *et al.* 2001, Paquin & Dupérré 2003

— Clypeal projection without sclerotized cap, although setae may be present ............................................................ **113**

**113(112)** Clypeal projection shorter than length of cheliceral bases, faintly resembling a human nose (Fig. 35.201); paracymbium of normal size ..................................... ***Perregrinus***
**Div.** 1 species: *Perregrinus deformis* (Tanasevitch 1982) — **Dist.** CAN — **Refs.** Buckle *et al.* 2001, Paquin & Dupérré 2003

— Clypeal projection longer than length of cheliceral bases, phallus-like (Fig. 35.202); paracymbium with a long and apically pointed distal arm ..................... ***Gnathonargus***
**Div.** 1 species: *Gnathonargus unicorn* (Banks 1892a) — **Dist.** NY — **Ref.** Bishop & Crosby 1935c

**114(111)** With the anterior cephalic lobe bearing eyes (Figs. 35.204-35.206) ............................................................................. **115**

— With the posterior cephalic lobe bearing one or more pairs of eyes (Figs. 35.207-35.208) ......................................... **119**

**115(114)** Embolic division with conspicuous tailpiece that is elongate (more than twice as long as wide) and long (more than half as long as palp width) (T in Figs. 35.211-35.212) ...................................................................... **116**

— Embolic division with tailpiece inconspicuous, absent or otherwise than above (Figs. 35.213-35.214) ............... **117**

Spiders of North America — *Linyphiidae* | 145

**116(115)** Tailpiece of embolic division twisted into a spiral (T in Fig. 35.215); posterior cephalic lobe usually quite tall and placed over the anterior cephalic lobe, with a considerable space between them (Fig. 35.216; except *Spirembolus mirus* Millidge 1980b with two small lobes); embolus coiled at end of palp ...................... ***Spirembolus,*** *in part*
  Div. 43 species, 3 key out here — Dist. Western NA — Ref. Millidge 1980b

— Posterior cephalic lobe comprised of a series of appressed longitudinal lobes (Fig. 35.217); tailpiece of embolic division not twisted into a spiral; embolus recurved (pointing toward proximal portion of palp) (Fig. 35.218) ................ ***Grammonota,*** *in part*
  Div. 24 species, 1 keys out here: *Grammonota gigas* (Banks 1896a) — Dist. Western NA — Refs. Bishop & Crosby 1933, Paquin & Dupérré 2003

**117(115)** Clypeus densely setose (Fig. 35.219); embolus rather short and not coiled .............................................. ***Goneatara***
  Div. 4 species — Dist. eastern NA — Refs. Bishop & Crosby 1935c, Barrows 1943, Paquin & Dupérré 2003

— Clypeus with at most a few setae; embolus long and curved or coiled ............................................................. **118**

**118(117)** Embolus forming more or less a complete coil, tip points toward apical end of palp (Fig. 35.220) ........................... ***Typhochrestus,*** *in part*
  Div. 3 species, 2 may key out here: *Typhochrestus latithorax* (Strand 1905) and *Typhochrestus pygmaeus* (Sørensen 1898) — Dist. CAN, AK, n USA — Refs. Holm 1960a, 1967, Paquin & Dupérré 2003

— Embolus recurved, tip points mesally (Fig. 35.221) .......... ***Souessa***
  Div. 1 species: *Souessa spinifera* (O. Pickard-Cambridge 1874b) — Dist. e NA — Refs. Crosby & Bishop 1936b, Paquin & Dupérré 2003

**119(114)** Posterior cephalic lobe bearing all eight eyes (Fig. 35.222); tibial apophysis shorter than width of tibia ......... ***Blestia***
  Div. 1 species: *Blestia sarcocuon* (Crosby & Bishop 1927) — Dist. NC, PA, VA — Ref. Millidge 1993a

— Posterior cephalic lobe bearing one or two pairs of eyes; tibial apophysis longer than width of tibia ................ **120**

**120(119)** Posterior cephalic lobe bearing AME and PME (Fig. 35.223); tibial apophysis with stout dorsoectal macroseta; tarsal claws not pectinate ........................... ***Coreorgonal***
  Div. 3 species — Dist. western USA, western CAN — Ref. Millidge 1981a

— Posterior cephalic lobe bearing PME (Fig. 35.224); tibial apophysis lacks a stout dorsoectal macroseta; tarsal claws I-III strongly pectinate (Fig. 35.225) ................................... ***Walckenaeria,*** *in part*
  Div. 68 species, about 13 species key out here (the *Walckenaeria tricornis* group) — Dist. widespread — Refs. Millidge 1983, Paquin & Dupérré 2003

**121(110)** Embolic division with conspicuous tailpiece that is elongate (more than twice as long as wide) and long (more than half as long as palp width) (T in Figs. 35.215, 35.226-35.227) ......................................................................... **122**

— Embolic division with tailpiece inconspicuous, absent, or otherwise than above (Figs. 35.228-35.230) ................ **127**

**122(121)** Embolus spinelike, much shorter than tailpiece; tibial apophysis bifid ................................................. ***Arcterigone***
  Div. 1 species: *Arcterigone pilifrons* (L. Koch 1879c) — Dist. AK, NT, YT — Refs. Leech & Ryan 1972, Buckle et al. 2001

— Embolus similar in length to much longer than, tailpiece; tibial apophysis variable ............................................. **123**

Linyphiidae

**123(122)** Tailpiece of embolic division twisted into a spiral (T in Fig. 35.231) ............................................................................ **124**

— Tailpiece of embolic division not twisted into a spiral (T in Figs. 35.232-35.233) .................................................... **125**

**124(123)** Tibial apophysis with thick basoectal macroseta; tip of embolus retroverted or pointing more or less toward proximal portion of palp ................ ***Disembolus,*** in part
**Div.** 24 species, 3 key out here — **Dist.** widespread, more common in west — **Ref.** Millidge 1981b

— Tibial apophysis lacking thick basoectal macroseta; tip of embolus not retroverted, often terminates in wide coil at proximal portion of palp ............. ***Spirembolus,*** in part
**Div.** 43 species, 20 key out here — **Dist.** western NA — **Ref.** Millidge 1980b

**125(123)** Tibial apophysis appears to be two appressed projections in dorsal view (Fig. 35.234) ...................... ***Epiceraticelus***
**Div.** 1 species: *Epiceraticelus fluvialis* Crosby & Bishop 1931 — **Dist.** NY, OH — **Ref.** Crosby & Bishop 1931

— Tibial apophysis otherwise ............................................ **126**

**126(125)** Tip of embolus initiates by an abrupt and acute bend (Fig. 35.232) ............................................ ***Ceratinopsis,*** in part
**Div.** 24 species, 3 key out here — **Dist.** widespread — **Ref.** Bishop & Crosby 1930, Buckle et al. 2001, Paquin & Dupérré 2003

— Tip of embolus without such a bend (Fig. 35.233) ............. ........................................................ ***Grammonota,*** in part
**Div.** 26 species, 8 key out here — **Dist.** widespread — **Refs.** Bishop & Crosby 1933, Dondale 1959, Buckle et al. 2001, Paquin & Dupérré 2003

**127(121)** Tibial apophysis definitely longer than length of rest of tibia (Fig. 35.235) ......................................................... **128**

— Tibial apophysis not longer than length of rest of tibia .... ............................................................................................. **130**

**128(127)** Palpal femur distally wider than palpal bulb, tipped disto-dorsally with a sharp spine and numerous tubercles (Fig. 35.236) ............................................................ ***Gonatium***
**Div.** 1 species: *Gonatium crassipalpum* Bryant 1933a — **Dist.** AK, CAN, eastern USA — **Refs.** Millidge 1981d, Paquin & Dupérré 2003

— Palpal femur normal, unarmed distally ...................... **129**

**129(128)** With two tibial apophyses (dorsal and ectal); embolus recurved, tip pointing toward proximal end of palp; cymbium hollowed out on ectal side; paracymbium lacking (Fig. 35.237) ................................................ ***Sphecozone***
**Div.** 1 species: *Sphecozone magnipalpis* Millidge 1993d — **Dist.** AZ — **Ref.** Millidge 1993d — **Note** this species may also key out at couplet 173

— With one tibial apophysis; embolus tip points toward apical end of palp; cymbium and paracymbium normal (Fig. 35.235) ............................................. ***Silometopoides***
**Div.** 1 species: *Silometopoides pampia* (Chamberlin 1949) — **Dist.** CAN — **Ref.** Chamberlin 1949 as *Minyriolus pampia*

**130(127)** Embolus threadlike, running a teardrop-shaped course across the palp (Fig. 35.238); eyes reduced in size; cephalic lobe sometimes densely clothed in short setae which are "parted" longitudinally (Fig. 35.239, *Anacornia microps*) . ................................................................................. ***Anacornia***
**Div.** 2 species: *Anacornia microps* Chamberlin & Ivie 1933a and *Anacornia proceps* Chamberlin 1949 — **Dist.** UT, WA — **Refs.** Chamberlin & Ivie 1933a, Chamberlin 1949 — **Note** both species may key out here

— Eyes normal in size, and without this combination of characters .............................................................................. **131**

**131(130)** Posterior declivity (PD) of cephalic lobe very steep, nearly to more than vertical; clypeus (C) setose (Fig. 35.240); embolus short .................................. **Cheniseo**
   **Div.** 4 species — **Dist.** AK, eastern NA — **Refs.** Bishop & Crosby 1935c, Paquin & Dupérré 2003

— Posterior declivity of cephalic lobe much shallower than vertical; clypeus with few or no setae; embolus variable ... .................................................................................. **132**

**132(131)** Carapace almost trapezoidal in lateral view, with overhanging clypeus, high peak, and long declivity (Fig. 35.241); cymbium with dark, setose projection dorsad of paracymbium (Fig. 35.242) .............................................. **Eskovia**
   **Div.** 1 species: *Eskovia exarmata* (Eskov 1989a) — **Dist.** YT — **Ref.** Eskov 1989a

— Carapace not trapezoidal in lateral view; cymbium lacking setose process ................................................................ **133**

**133(132)** Embolus long, whiplike, sinuous (Fig. 35.243); cephalic lobe peaks at level of PME (Fig. 35.244); TmIV present ... .................................................................. **Metopobactrus**
   **Div.** 1 species: *Metopobactrus prominulus* (O. Pickard-Cambridge 1872b) — **Dist.** CAN, n USA — **Refs.** Bishop & Crosby 1935c [as *Maso alticeps* Emerton 1909)], Roberts 1987, Paquin & Dupérré 2003]

— Embolus not sinuous; cephalic lobe peaks anterior or posterior to PME; TmIV variable ....................................... **134**

**134(133)** Cymbium with basal outgrowth that fits over tibial apophysis (Fig. 35.245); embolus long, curved; TmIV present ............................................... **Procerocymbium**
   **Div.** 1 species: *Procerocymbium dondalei* Marusik & Koponen 2001c — **Dist.** YT — **Ref.** Marusik & Koponen 2001c

— Cymbium normal; embolus short, thick, spinelike (Fig. 35.246); TmIV absent; sometimes with prominent stridulatory file on anterior of abdominal venter ........................ ................................................................ **Mecynargus**
   **Div.** 7 species — **Dist.** AK, CAN — **Refs.** Buckle et al. 2001, Paquin & Dupérré 2003 — **Note** *Smodix reticulata* (Emerton 1915b), which probably belongs in *Mecynargus*, should key out here

**No pits or cephalothoracic modification, 53 genera**

**135(53)** Eyes reduced in size or absent, usually cavernicolous ....... .................................................................................. **136**

— Eyes normal, habitat variable ....................... **139**

**136(135)** Eyes absent ..................................... **Anthrobia**
   **Div.** 1 species: *Anthrobia mammouthia* Tellkampf 1844 — **Dist.** AL, KY, TN, VA, WV — **Ref.** Buckle et al. 2001 — **Note** appeared as *Anthrobia monmouthia* in Tellkampf (1844), a *lapsus calami*. This genus includes two epigean species that will key out with *Sciastes*, and probably other troglobitic species in addition to *Anthrobia mammouthia* (J. Miller, pers. comm.)

— Eyes reduced in size but not absent ............. **137**

**137(136)** Paracymbium with 2-3 conspicuous, very long setae (Fig. 35.247) ............................................... **Islandiana,** in part
   **Div.** 14 species, 3 key out here — **Dist.** widespread — **Refs.** Ivie 1965, Eskov 1987d

— Paracymbium lacking very long setae ......................... **138**

**138(137)** Tibial apophysis with one subapical point (Fig. 35.248); paracymbium very thin (Fig. 35.248) ........ **Caviphantes**
   **Div.** 1 species: *Caviphantes saxetorum* (HULL 1916) — **Dist.** Palearctic, introduced to OR — **Ref.** Roberts 1987

— With multiple small tibial apophyses (Fig. 35.249); paracymbium normal ............................................. **Phanetta**
   **Div.** 1 species: *Phanetta subterranea* (EMERTON 1875) — **Dist.** e USA — **Ref.** Millidge 1984a

**139(135)** Abdomen with sclerotized scutum or shield (Fig. 35.250); reduced in some species to a small region at anteriormost portion of abdominal dorsum ....................... **140**

— Without abdominal scutum ............................ **141**

**140(139)** Cheliceral fang sinuous (Fig. 35.251) ............ **Ceratinella**
   **Div.** 14 species — **Dist.** widespread except for southwest USA — **Refs.** Crosby & Bishop 1925c, Chamberlin 1949, Paquin & Dupérré 2003

— Cheliceral fang not sinuous ............. **Ceraticelus,** *in part*
   **Div.** 41 species, 20 key out here — **Dist.** widespread — **Refs.** Crosby & Bishop 1925c, Buckle *et al.* 2001, Paquin & Dupérré 2003

**141(139)** With stridulating files on epigastric plates, activated by sharp retrolateral angle of coxae IV (Fig. 35.252) .............. ..................................................... **Semljicola**
   **Div.** 4 species — **Dist.** CAN, AK, CO — **Ref.** Saaristo & Eskov 1996

— Without stridulating ridge or retrolateral extensions of coxae IV ........................................................... **142**

**142(141)** Palpal patella with ventral apical apophysis (Fig. 35.253a-35.253b) ............................................. **143**

— Palpal patella without ventral apical apophysis .......... **148**

**143(142)** Patellar apophysis shorter to slightly longer than width of rest of patella (Fig. 35.254); chelicerae with tooth (mastidion) on face ................................................ **144**

— Patellar apophysis much longer than width of patella (Fig. 35.253b); chelicerae variable ............................ **147**

**144(143)** Cephalothorax orangish or reddish with black cephalic region ................................................. **Tmeticus**
   **Div.** 1 species: *Tmeticus ornatus* (EMERTON 1914a) — **Dist.** CAN, eastern USA — **Refs.** Bishop & Crosby 1935c, Paquin & Dupérré 2003

— Cephalothorax tan, yellowish, or brownish; cephalic region not black or only black in interocular area ...... **145**

**145(144)** Embolus longer than width of palp; tibial apophysis longer than tibial length ...................................... **Gnathonarium**
   **Div.** 2 species — **Dist.** AK, CAN — **Ref.** Buckle *et al.* 2001

— Embolic division without long embolus; tibial apophysis shorter than tibial length ................................. **146**

**146(145)** Body length less than 1.5 mm ................ **Erigone,** *in part*
   **Div.** 56 species, 1 keys out here: *Erigone praecursa* CHAMBERLIN & IVIE 1939a — **Dist.** NE — **Ref.** Chamberlin & Ivie 1939a

— Body length greater than 1.5 mm (usually greater than 2.0 mm) ................................................. **Mermessus,** *in part*
   **Div.** 35 species, 2 key out here: *Mermessus fradeorum* (BERLAND 1932a) and *Mermessus denticulatus* (BANKS 1898b) — **Dist.** NA except NW— **Refs.** Millidge 1987, Miller 2007 — **Note** senior synonym of *Eperigone* (Miller 2007).

35.248

35.249

35.250

35.251

35.252

35.253a

35.253b

35.254

**147(143)** Embolic division long, with multiple points, distinctively shaped (Fig. 35.255); TmI at 0.80 or greater ........................ ............................................................................. ***Ceratinopsidis***
   **Div.** 1 species: *Ceratinopsidis formosa* (BANKS 1892a) — **Dist.** e USA — **Refs.** Bishop & Crosby 1930

— Embolic division otherwise; TmI at 0.60 or less ................. ............................................................................. ***Erigone,*** in part
   **Div.** 56 species, 54 key out here — **Dist.** widespread — **Ref.** Crosby & Bishop 1928a, Buckle et al. 2001, Paquin & Dupérré 2003

**148(142)** Paracymbium with 2-3 conspicuous, very long setae (Fig. 35.256) ................................................ ***Islandiana,*** in part
   **Div.** 14 species, 10 key out here — **Dist.** widespread — **Refs.** Ivie 1965, Paquin & Dupérré 2003

— Paracymbium without very long setae ......................... **149**

**149(148)** Palpus with thick, nearly straight, conductor-like sclerite that is longer than width of palp. This sclerite extends well beyond the end of the palp (Fig. 35.257). Chelicerae robust, with a large tooth and hair tipped tubercles on its face ................................................ **undescribed genus**
   **Div.** 2 species — **Dist.** BC, MT, OR, WA — **Refs.** Chamberlin 1949, Levi & Levi 1955 — **Note** this udescribed genus includes "*Collinsia*" *wilburi* LEVI & LEVI 1955 and "*Spirembolus*" *vasingtonus* CHAMBERLIN 1949 a nomen nudum

— Palpus lacking such a sclerite; chelicerae rarely robust or toothed ............................................................... **150**

**150(149)** Embolic division with conspicuous tailpiece that is elongate (more than twice as long as wide), long (more than half as long as palp width), and oriented with its long axis parallel to the long axis of the palp (T in Figs. 35.258-35.260) ............................................................... **151**

— Embolic division with tailpiece inconspicuous, absent, or otherwise than above (Figs. 35.271, 35.282, 35.284, 35.288-35.289, 35.292, 35.304, 35.306, 35.309-35.310) ..... ................................................................................. **159**

**151(150)** Tailpiece distinctly spiraled (T in Fig. 35.259) ............ **152**

— Tailpiece not spiraled (T in Figs. 35.258, 35.260) ....... **153**

**152(151)** Embolic division with several short processes; dorsodistal margin of palpal tibia often serrated; cymbium often with raised dorsal knob (Fig. 35.262) ................... ***Tachygyna***
   **Div.** 15 species — **Dist.** Western NA — **Ref.** Millidge 1984a

— Embolus with a single point; palpal tibiae not serrated; tibial apophysis with a small secondary basal point; cymbium without raised dorsal knob ...................................... ....................................................... ***Spirembolus,*** in part
   **Div.** 43 species, 3 key out here — **Dist.** western NA — **Ref.** Millidge 1980b

**153(151)** Dorsum of metatarsus I with row of large curved spines (Fig. 35.263); TmI located at 0.50 to 0.60 ............ ***Scirites***
   **Div.** 1 species: *Scirites pectinatus* (EMERTON 1911) — **Dist.** CAN, n USA — **Ref.** Bishop & Crosby 1938

— Setation of metatarsus I otherwise; TmI variable ....... **154**

**154(153)** Metatarsus I with two thick curved prolateral spines distally (Fig. 35.264); tibial apophysis with two points .......... ................................................................................. ***Wabasso***
   **Div.** 2 species: *Wabasso cacuminatus* MILLIDGE 1984a and *Wabasso quaestio* (CHAMBERLIN 1949) — **Dist.** CAN, n USA — **Refs.** Millidge 1984a, Paquin & Dupérré 2003

— Metatarsus I lacking two thick curved prolateral spines distally; tibial apophysis variable ................................ **155**

35.255

35.256

35.257

35.258

35.259

35.260

35.262

35.263

35.264

**150** | *Linyphiidae*

**155(154)** Embolus looped back on itself at end of palp, forming virtually a complete loop (Fig. 35.265); tarsal claws I-III strongly pectinate (Fig. 35.266) .................................................. ................................................................. ***Walckenaeria***, in part
  **Div.** 68 species, 30 key out here (the *Walckenaeria acuminata* group) — **Dist.** widespread — **Refs.** Millidge 1983, Paquin & Dupérré 2003

— Embolus not looped, but only curved or bent; tarsal claws not pectinate ................................................................. **156**

**156(155)** Tibia with two apophyses, each longer than length of rest of tibia, one dorsal and one ectal (Fig. 35.267) ................. ........................................................................... ***Coloncus***
  **Div.** 5 species — **Dist.** CAN, n USA — **Refs.** Chamberlin & Ivie 1944, Chamberlin 1949, Paquin & Dupérré 2003

— Tibia with one apophysis, or with several short apophyses ................................................................................ **157**

**157(156)** Distal arm of paracymbium sharply pointed (Fig. 35.268); apical end of embolic division divided into a sharp point and a long curving free part (Fig. 35.269) .......... ***Tutaibo***
  **Div.** 1 species: *Tutaibo anglicanus* (Hentz 1850b) — **Dist.** southeastern USA — **Ref.** Bishop & Crosby 1930

— Distal arm of paracymbium not pointed; apical end of embolic division a long curving free part .................... **158**

**158(157)** Metatarsus IV with a trichobothrium ............................... .................................................................. ***Grammonota***, in part
  **Div.** 26 species, 18 key out here — **Dist.** widespread — **Refs.** Dondale 1959, Buckle et al. 2001, Paquin & Dupérré 2003

— Metatarsus IV lacking a trichobothrium ......................... ........................................... ***Ceratinopsis***, in part and ***Styloctetor***
  **Div.** 24 species, 21 key out here + 2 species: *Styloctetor purpurescens* (Keyserling 1886b) *and Styloctetor stativus* (Simon 1881g) — **Dist.** widespread — **Refs.** Bishop & Crosby 1930, Buckle et al. 2001, Paquin & Dupérré 2003

**159(150)** Tibial apophysis definitely longer than length of rest of tibia (Figs. 35.270-35.272) ............................................. **161**

— Tibial apophysis not longer than length of rest of tibia .... ................................................................................ **160**

**160(159)** Tibial apophysis definitely shorter than width of narrowest part of tibia, or lacking (Figs. 35.273-35.275) or as in Figure 35.376 ............................................................. **180**

— Tibial apophysis not shorter than width of narrowest part of tibia, but not as in Figure 35.376 .............................. **174**

**161(159)** Palpal patella and palpal femur both distinctively longer than cymbium (Fig. 35.270) ....................................... ***Tiso***
  **Div.** 2 species: *Tiso aestivus* (L. Koch 1872b) and *Tiso vagans* (Blackwall 1834b) — **Dist.** CAN, WA — **Refs.** Locket & Millidge 1953, Roberts 1987

— Cymbium as long or longer than palpal femur and/or patella ................................................................. **162**

**162(161)** Embolus less than half the length of cymbium .......... **163**

— Embolus longer than half the length of cymbium ..... **170**

**163(162)** Distal portion of tibial apophysis strongly bent or even recurved (Fig. 35.277) ................................................. **164**

— Distal portion of tibial apophysis not strongly bent or recurved ............................................................................. **167**

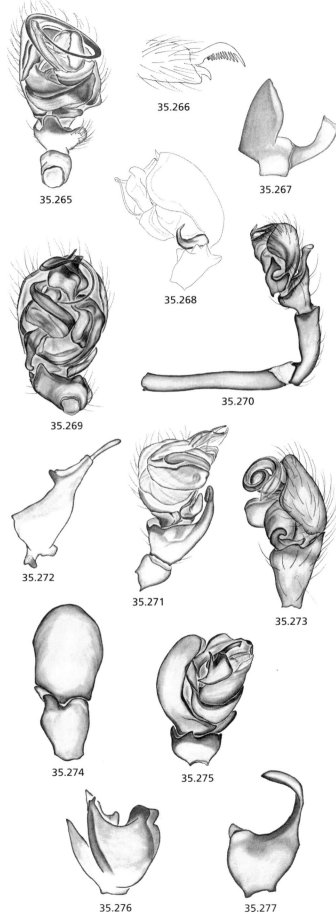

35.265  35.266  35.267  35.268  35.269  35.270  35.271  35.272  35.273  35.274  35.275  35.276  35.277

**164(163)** Cymbium somewhat hourglass-shaped in dorsal view; embolic division appears bifid (Fig. 35.278) ...... **Souidas**
   Div. 1 species: *Souidas tibialis* (EMERTON 1882) — Dist. NH, QC (eastern North America) — Refs. Crosby & Bishop 1936b, Paquin & Dupérré 2003

— Cymbium thicker in middle section in dorsal view; embolic division terminates in a single point ....................... **165**

**165(164)** Tibial apophysis bent so tip points ectally (Fig. 35.279); embolic division a sharp point .............................. **Ivielum**
   Div. 1 species: *Ivielum sibiricum* ESKOV 1988b — Dist. YT, Siberia — Ref. Eskov 1988b

— Tibial apophysis bent so tip points mesally ................. **166**

**166(165)** Embolic division with two processes, one rounded and one sharp and hooked (Fig. 35.280) ..... **Sciastes,** *in part*
   Div. 6 species, males are known for 5 of them, 1 species keys out here: *Sciastes hastatus* MILLIDGE 1984a — Dist. YT, Siberia — Ref. Millidge 1984a

— Embolic division with one process, which is blunt apically but bearing a short sharp spine; tibial apophysis as in Fig. 35.281 .................................................................... **Zygottus**
   Div. 2 species: *Zygottus corvallis* CHAMBERLIN 1949 and *Zygottus oregonus* CHAMBERLIN 1949 — Dist. BC, OR, WA — Ref. Chamberlin 1949

**167(163)** Embolus somewhat recurved, so tip points toward proximal end of palp (Fig. 35.282); palpal tibia with small ventral lobe; TmI located at 0.70 or greater ............. **Soulgas**
   Div. 1 species: *Soulgas corticarius* (EMERTON 1909) — Dist. eastern NA — Refs. Crosby & Bishop 1936b, Paquin & Dupérré 2003

— Embolus not recurved; palpal tibia lacking ventral lobe; TmI located at 0.60 or less ............................................ **168**

**168(167)** Tibial apophysis shape distinctive ("pistol shaped": Fig. 35.283); body uniformly pale in color; embolus long, curved ............................................................. **Scylaceus**
   Div. 1 species: *Scylaceus pallidus* (EMERTON 1882) — Dist. eastern NA — Refs. Bishop & Crosby 1938, Paquin & Dupérré 2003

— Tibial apophysis a plain point or lobe; embolus terminating in a short, spinelike process; color variable ......... **169**

**169(168)** Eye area black ................................... **Masonetta,** *in part*
   Div. 1 species: *Masonetta floridana* (IVIE & BARROWS 1935) — Dist. FL, GA — Ref. Ivie & Barrows 1935 — Note may also key out at couplet 179

— Eye area concolorous with rest of cephalothorax ............... ............................................................. **Sciastes,** *in part*
   Div. 6 species, males are known for 5 of them, 2 species key out here: *Sciastes dubius* (HACKMAN 1954) and *Sciastes extremus* HOLM 1967 — Dist. WA, UT, CAN — Refs. Millidge 1984a, Paquin & Dupérré 2003

**170(162)** Embolus very sinuous, nearly forming a figure-8 shape (Fig. 35.284); TmI located at 0.35 or less; tibial apophysis very long, thin, terminating in a sharp point ..................... ......................................................... **Gnathonaroides**
   Div. 1 species: *Gnathonaroides pedalis* (EMERTON 1923) — Dist. CAN and n USA — Refs. Bishop & Crosby 1938, Paquin & Dupérré 2003

— Embolus not forming a figure-8 shape; TmI usually located at less than 0.35; tibial apophysis otherwise ......... **171**

**171(170)** Embolus forms a wide proximal partial coil and terminates in a small partial coil or hook (Fig. 35.285); tibial apophysis with long terminal and short basal projections, neither sharply pointed ................................ **Sisis,** *in part*
   Div. 2 species, 1 keys out here: *Sisis rotundus* (EMERTON 1925a) — Dist. AK, CAN — Refs. Bishop & Crosby 1938, Paquin & Dupérré 2003

— Embolus coiled or not, but not forming large and small double coil .................................................................. **172**

35.278

35.279

35.280

35.281

35.282

35.283

35.284

35.285

**172(171)** Embolus sinuous, S-shaped (Fig. 35.286); palpal tibia with ventral lobe .................................................... ***Souessoula***
        **Div.** 1 species: *Souessoula parva* (BANKS 1899b) — **Dist.** eastern USA — **Ref.** Millidge 1984a

— Embolus not S-Shaped, but coiled or recurved .......... **173**

**173(172)** Paracymbium absent; with two separate tibial apophyses, dorsal and ectal; ectal margin of cymbium hollowed out; embolus recurved, so tip points toward proximal end of palp (Fig. 35.287) ............................ ***Sphecozone***, in part
        **Div.** 1 species: *Sphecozone magnipalpis* MILLIDGE 1993d — **Dist.** AZ — **Ref.** Millidge 1993d — **Note** specimens may also key out at couplet 129

— Paracymbium present; with one tibial apophysis with a single point; cymbium margins normal; embolus not recurved; embolus bifid ............. ***Typhochrestus***, in part
        **Div.** 3 species, 2 may key out here: *Typhochrestus uintanus* (CHAMBERLIN & IVIE 1939a) and *Typhochrestus pygmaeus* (SØRENSEN 1898) — **Dist.** CAN, AK, n USA — **Ref.** Holm 1967

**174(160)** Free part of embolic division longer than width of cymbium (Figs. 35.288-35.289) ............................................. **175**

— Free part of embolic division shorter than width of cymbium (Fig. 35.290) ............................................. **176**

**175(174)** Embolus makes complete coil around ventral face of palp (Fig. 35.288); TmI located at 0.30-0.45 ............... ***Scironis***
        **Div.** 2 species: *Scironis sima* CHAMBERLIN 1949 and *Scironis tarsalis* (EMERTON 1911) — **Dist.** CAN, AK, n USA — **Ref.** Buckle et al. 2001

— Embolus spirals at apical end of palp (Fig. 35.289); TmI beyond 0.45 ........................................ ***Sisicottus***, in part
        **Div.** 9 species — **Dist.** widespread — **Refs.** Miller 1999, Paquin & Dupérré 2003 — **Note** any of the species may key out here

**176(174)** Cymbium with basoectal groove that accepts the point of the tibial apophysis (Fig. 35.290) ............................. ***Perro***
        **Div.** 1 species: *Perro polaris* (ESKOV 1986b) — **Dist.** CAN, Siberia — **Ref.** Eskov 1986b

— Cymbium without such a groove ................................. **177**

**177(176)** Palp with large ventral conductor-like sclerite that points ventroapically in ectal view (the posterior tooth of the embolic division) (Fig. 35.291); fourth metatarsus often with a trichobothrium ........................ ***Halorates***, in part
        **Div.** 16 species, 6 key out here — **Dist.** widespread except southeastern USA — **Refs.** Crosby & Bishop 1928a, Buckle et al. 2001, Paquin & Dupérré 2003

— Palp lacking such a conductor-like process; fourth metatarsus always lacking trichobothrium .......................... **178**

**178(177)** Embolic division with multiple processes or points as long or longer than the body of the embolic division (Fig. 35.292); sometimes with abdominal pattern and/or teeth on cheliceral face ............................. ***Mermessus***, in part
        **Div.** 35 species, 20 key out here — **Dist.** widespread — **Refs.** Millidge 1987, Paquin & Dupérré 2003, Miller 2007 — **Note** senior synonym of *Eperigone* (Miller 2007).

— Embolic division terminates as a single point or process .. ............................................................................. **179**

**179(178)** Tibial apophysis with multiple processes or with deep notch producing appearance of two apophyses (Fig. 35.293); body uniformly pale; with tooth on cheliceral face ............................................. **Scyletria**, in part
   Div. 2 species, 1 key out here: Scyletria jona Bishop & Crosby 1938 — **Dist.** CAN, eastern USA — **Ref.** Bishop & Crosby 1938

— Tibial apophysis a single point; eye area dark; cheliceral face lacking teeth .............................. **Masonetta**, in part
   Div. 1 species: Masonetta floridana (Ivie & Barrows 1935) — **Dist.** GA, FL — **Ref.** Ivie & Barrows 1935 — **Note** may also key out at 169

**180(160)** Cymbium hourglass-shaped in dorsal view, narrowest in middle (Fig. 35.294); embolus long, making more than one complete coil .................................................. **Sisyrbe**
   Div. 1 species: Sisyrbe rustica (Banks 1892a) — **Dist.** NY — **Ref.** Bishop & Crosby 1938

— Cymbium not hourglass-shaped ................................. **181**

35.293

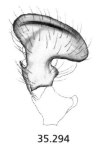

35.294

**181(180)** Tibia I somewhat swollen and curved (Fig. 35.295); with small mastidion on cheliceral face; median row of four setae emerging from small tubercles between the posterior eye row and the thoracic groove ........................ **Soucron**
   Div. 1 species: Soucron arenarium (Emerton 1925a) — **Dist.** CAN — **Ref.** Crosby & Bishop 1936b, Paquin & Dupérré 2003

— Without the above combination of characters ........... **182**

35.295

**182(181)** TmI located at 0.90 or greater; metatarsus IV usually with trichobothrium ...................................... **Maso**
   Div. 4 species — **Dist.** widespread — **Refs.** Buckle et al. 2001, Paquin & Dupérré 2003

— TmI located at 0.75 or less; metatarsus IV usually lacking trichobothrium ........................................... **183**

**183(182)** Free part of embolus clearly longer than cymbium length ................................................................. **184**

— Free part of embolus clearly shorter than cymbium length ................................................................. **186**

**184(183)** Embolus makes 1 ½ lateral loops (Fig. 35.296) ................
   ............................................................. **Lessertia**
   Div. 1 species: Lessertia dentichelis (Simon 1884a) — **Dist.** Palaearctic, introduced to BC — **Refs.** Locket & Millidge 1953, Roberts 1987

— Embolus sinuous, its path not coiling back on itself ......... ................................................................. **185**

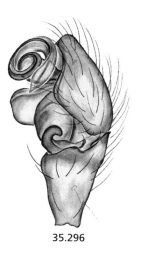

35.296

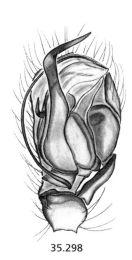

35.298

**185(184)** Embolus forms simple S-shape; with long, stout, bifid conductor-like sclerite (part of the embolic division, Fig. 35.298); cymbium narrowed basally in dorsal view, with ectal groove; paracymbium normal (Fig. 35.299) ............. ............................................................. **Tunagyna**
   Div. 1 species: Tunagyna debilis (Banks 1892a) — **Dist.** AK, CAN, n USA — **Refs.** Millidge 1984a, 1984b, Paquin & Dupérré 2003

— Embolus traces a complex, sinuous path (Fig. 35.300); lacking long, stout, bifid conductor-like sclerite; cymbium normal, without ectal groove; paracymbium very thin .... ............................................................. **Sitalcas**
   Div. 1 species: Sitalcas ruralis Bishop & Crosby 1938 — **Dist.** NY — **Ref.** Bishop & Crosby 1938

35.299

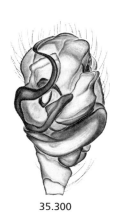

35.300

**186(183)** Cymbium with ectal groove which accepts ectal point of tibial apophysis; cymbium narrowed basally in dorsal view (Fig. 35.301); palpal tibia with rounded apical notch, giving the appearance of two tibial apophyses .................... ............................................................................ ***Scolopembolus***
**Div.** 1 species: *Scolopembolus littoralis* (EMERTON 1913a) — **Dist.** CT, NB — **Ref.** Bishop & Crosby 1938

— Cymbium lacking ectal groove or not narrowed basally; tibial apophysis usually otherwise .................. **187**

**187(186)** Cymbium of normal length, distinctly longer than wide .. ........................................................................................................... **188**

— Cymbium rather short, almost as wide as long ........... **191**

**188(187)** Embolic division with multiple apophyses .................. **189**

— Embolus terminates in a single point (although it may be bifid at apical end) .......................................................... **190**

**189(188)** Palp with large ventral conductor-like sclerite that points ventroapically in ectal view (posterior tooth of embolic division Fig. 35.302); never with abdominal pattern; rarely with teeth on cheliceral face ............... ***Halorates**, in part*
**Div.** 16 species, 10 key out here — **Dist.** widespread except southeastern USA — **Refs.** Crosby & Bishop 1928a, Buckle et al. 2001, Paquin & Dupérré 2003

— Palp lacking such a conductor-like process (Fig. 35.303); sometimes with dorsal abdominal pattern; sometimes teeth on face of chelicerae ................ ***Mermessus**, in part*
**Div.** 35 species, 15 key out here — **Dist.** widespread — **Refs.** Millidge 1987, Paquin & Dupérré 2003, Miller 2007— **Note** senior synonym of *Eperigone* (Miller 2007).

**190(188)** TmI located at 0.30-0.35; embolus and embolic membrane usually extend beyond end of cymbium (Fig. 35.304); embolus sometimes bifid apically ...................... ***Carorita***
**Div.** 2 species: *Carorita hiberna* (BARROWS 1945) and *Carorita limnaea* CROSBY & BISHOP 1927) — **Dist.** CAN, northern USA to CO — **Refs.** Zujko-Miller 1999, Paquin & Dupérré 2003

— TmI located at 0.48 or greater; embolus usually does not extend beyond end of cymbium; embolus with a single point, ending in a fine hook-like extension of the embolus, but not bifid apically (palp as in Fig. 35.305) ...................... ............................................................................ ***Sisicottus**, in part*
**Div.** 9 species — **Dist.** widespread except southeastern USA — **Refs.** Miller 1999, Paquin & Dupérré 2003— **Note** any of the species may key out here

**191(187)** Embolus and "conductor" (suprategular apophysis) point at each other ; embolic division with long but blunt-ended free part and sharp accessory spine (Fig. 35.306) ............. ............................................................................................ ***Vermontia***
**Div.** 1 species: *Vermontia thoracica* (EMERTON 1913a) — **Dist.** AK, CAN, northern USA — **Refs.** Millidge 1984a, Paquin & Dupérré 2003

— Palpal configuration and embolic division otherwise ........ ........................................................................................................... **192**

**192(191)** Lacking a palpal tibial apophysis; cymbium often with ectal lobe, or expanded/modified along ectal margin (Fig. 35.307) ............................................................ ***Eulaira***
**Div.** 15 species — **Dist.** widespread except for southeastern USA — **Refs.** Chamberlin & Ivie 1945b, Buckle et al. 2001

— With one or several apical palpal tibial apophyses; ectal margin of cymbium unmodified .................................... **193**

35.301

35.302

35.303

35.304

35.305

35.306

35.307

**193(192)** With several short tibial apophyses; size greater than 1.2 mm ................................................................. **194**

— Tibial apophysis a single point or process; size 1.2 mm or less ..................................................................... **196**

**194(193)** Palpal tibia laterally compressed and dark in color, the color contrasting with the palpal patella and femur (Fig. 35.308) ........................................... ***Scyletria,*** *in part*
>Div. 2 species, 1 keys out here: *Scyletria inflata* BISHOP & CROSBY 1938 — Dist. CAN, eastern USA — Refs. Bishop & Crosby 1938, Paquin & Dupérré 2003

— Palpal tibia concolorous with rest of palp, and not distinctly compressed laterally ........................................... **195**

**195(194)** Embolus spirals or curls around embolic membrane at end of palp (Fig. 35.309); distal arm of paracymbium curved and points toward proximal arm .......................... ***Diplocentria***
>Div. 4 species — Dist. CAN, n USA to CO — Refs. Buckle et al. 2001, Paquin & Dupérré 2003

— Embolic division rounded posteriorly and not spiralled; distal arm of paracymbium not curved ............................. ***Sciastes,*** *in part*
>Div. 6 species, males are known for 5 of them, 2 species key out here: "*Sciastes*" *acuminatus* (EMERTON 1913b) and *Sciastes truncatus* (EMERTON 1882) — Dist. CAN, n USA — Refs. Millidge 1984a, Paquin & Dupérré 2003 — Note "*Sciastes*" *acuminatus* probably belongs in *Anthrobia*. A new species of *Anthrobia* from eastern North America also keys out here (J. Miller, pers. comm.)

**196(193)** Embolic division with only one pointed process (Fig. 35.310) ............................................................. ***Annapolis***
>Div. 1 species: *Annapolis mossi* (MUMA 1945) — Dist. MD, NB, NS — Refs. Millidge 1984a, Paquin & Dupérré 2003

— Embolic division with more than one pointed process (Fig. 35.311) ............................................................. ***Jacksonella***
>Div. 1 species: *Jacksonella falconerii* (JACKSON 1908) — Dist. MN, palaearctic — Refs. Locket & Millidge 1953, Roberts 1987

35.308

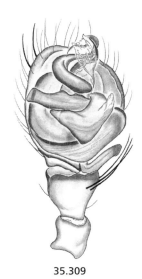

35.309

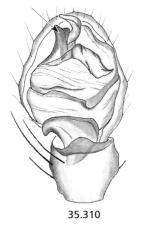

35.310

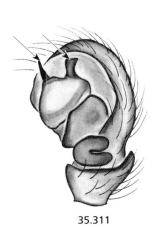

35.311

## Female Linyphiinae

**197(1)** Prolateral spine present on tibia I or femur I .............. **198**

— No prolateral spine on tibia I nor on femur I .............. **230**

**198(197)** Some or all metatarsi with spines (check all legs) ...... **199**

— No metatarsi with spines .................................. **218**

**199(198)** Abdomen with conspicuous caudal tubercle (Fig. 35.312); body red (fading to pale gray in alcohol) except for palps, eye region and caudal hump which are black (Fig. 35.312); epigynum as shown (Fig. 35.313) ...................... ***Florinda***
Div. 1 species: *Florinda coccinea* (Hentz 1850a) — Dist. southeastern USA — Ref. Blauvelt 1936

— Abdomen without tubercle; often patterned dorsally; coloration brown or gray ....................................... **200**

**200(199)** Carapace pale with median dark stripe bifurcating anteriorly; abdomen with pattern of narrow, dark chevrons on light background (Figs. 35.314-35.315) ........................ **201**

— Carapace and abdominal markings otherwise ............ **202**

**201(200)** Dorsal stripe on carapace narrow (Fig. 35.315); epigynum with triangular atrial opening and short dorsal scape (Fig. 35.316) ................................................. ***Linyphia***
Div. 1 introduced species: *Linyphia triangularis* (Clerck 1757) — Dist. ME — Refs. Helsdingen 1969, Jennings et al. 2003

— Dorsal stripe wider (Fig. 35.314); epigynum with scape usually extending well beyond epigastric furrow (Figs. 35.317-35.318), but short in *Pityohyphantes limitaneus* (Emerton 1915a) (Fig. 35.319), atrium in form of shallow lateral depressions .................................. ***Pityohyphantes***
Div. 16 species — Dist. n and w NA — Ref. Buckle et al. 2001

**202(200)** Epigynum with both dorsal and ventral scapes (Figs. 35.320-35.321) ............................. ***Bathyphantes,*** *in part*
Div. 26 species, 2 key out here: *Bathyphantes keeni* (Emerton 1917a) and *Bathyphantes reprobus* (Kulczyński 1916) — Dist. widespread — Ref. Ivie 1969

— Epigynum with dorsal scape only, or with no scape .......... **203**

**203(202)** Metatarsi with ventral spines ......................... **204**

— Metatarsi without ventral spines ................... **210**

**204(203)** Epigynum with an elongate scape extending well beyond epigastric groove ................................ **208**

— Epigynum otherwise ..................................... **205**

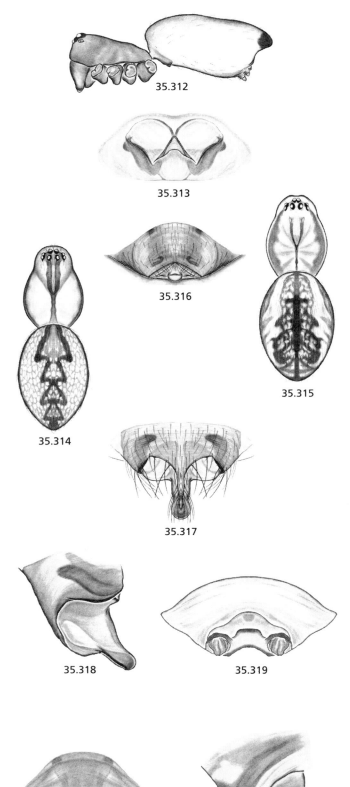

Spiders of North America — Linyphiidae

**205(204)** Epigynum with mesal atrium and small, short dorsal scape (Figs. 35.322-35.323) .......................................... **206**
— Epigynum with broad mesal plate (Figs. 35.324-35.325) .. ................................................................................. **207**

**206(205)** Atrial opening tiny (Fig. 35.322) ............ ***Microlinyphia***
  Div. 4 species — Dist. n and w NA — Ref. Helsdingen 1970
— Atrial opening large (Fig. 35.323) ...................... ***Neriene***
  Div. 9 species — Dist. widespread — Ref. Helsdingen 1969 — Note some authors consider Neriene to be a synonym of Linyphia

**207(205)** Mesal plate of epigynum much wider than long, its sides converging anteriorly (Fig. 35.324); carapace pale with black median and marginal bands ... ***Stemonyphantes***
  Div. 1 species: Stemonyphantes blauveltae Gertsch 1951 — Dist. n NA — Ref. Helsdingen 1968
— Mesal plate of epigynum with inverted "T" shape (Fig. 35.325); carapace uniform brown ....................................... ................................................................ **Undescribed genus**
  Div. Linyphia catalina Gertsch 1951, Linyphia rita Gertsch 1951 and perhaps Linyphia tauphora Chamberlin 1928a — Dist. sw USA — Ref. Gertsch 1951

**208(204)** Scape folded or otherwise modified at tip (Figs. 35.326-35.327) ........................................ ***Lepthyphantes***, in part
  Div. 29 species + several undescribed: a few of the larger species key out here — Dist. widespread — Refs. Zorsch 1937, Buckle et al. 2001
— Scape straight, unfolded (Figs. 35.329-35.331) ........... **209**

**209(208)** Scape relatively broad, with rounded tip (Fig. 35.328), sometimes with a medial notch (Figs. 35.329-35.330) ...... ................................................................................ ***Allomengea***
  Div. 3 species — Dist. n NA — Ref. Helsdingen 1974
— Scape long, with parallel sides and truncate tip (Fig. 35.331) .......................................................... ***Helophora***
  Div. 4 species — Dist. n and w NA — Ref. Buckle et al. 2001

**210(203)** AME larger than other eyes (Fig. 35.332); clypeus low, about height of AME; anterior eye row strongly recurved; cheliceral teeth long; epigynum as shown (Fig. 35.333) .... ................................................................................ ***Tapinopa***
  Div. 2 species: Tapinopa bilineata Banks 1893a, Tapinopa hentzi Gertsch 1951 — Dist. eastern NA — Refs. Kaston 1981, Gertsch 1951
— AME same size or smaller than other eyes (Fig. 35.334); clypeus height more than twice diameter of AME; anterior eye row nearly straight; cheliceral teeth short or long ....... ................................................................................ **211**

**211(210)** Chelicerae with 3-4 pairs of stout spines (Fig. 35.334); cheliceral teeth long; trichobothrium present on metatarsus of leg IV; epigynum as shown (Fig. 35.335) ................ ................................................................................ ***Drapetisca***
  Div. Div. 2 species: Drapetisca alteranda Chamberlin 1909a, Drapetisca oteroana Gertsch 1951 — Dist. n NA, NM — Refs. Kaston 1981, Gertsch 1951
— Chelicerae without stout spines; cheliceral teeth relatively short; no trichobothrium on metatarsus of leg IV ..... **212**

**212(211)** Tibia I with ventral spines ............................................. **213**
— Tibia I with no ventral spines ...................................... **216**

**213(212)** Tibia with 3-5 pairs of ventral spines (Fig. 35.336), epigynum as shown (Fig. 35.337); abdomen concolorous ......... ............................................................. ***Centromerita***
**Div.** 1 introduced species: *Centromerita bicolor* (Blackwall 1833c) — **Dist.** NF, NS, BC, WA — **Ref.** Roberts 1987, Paquin & Dupérré 2003

— Tibia with 1 or 2 pairs of spines or a few scattered ones; epigynum otherwise; abdomen patterned or concolorous ................................................................................ **214**

**214(213)** Abdomen with dorsal folium (Figs. 35.338-35.339); epigynum as shown (Fig. 35.340) .............................. ***Estrandia***
**Div.** 1 species: *Estrandia grandaeva* (Keyserling 1886b) — **Dist.** n NA — **Ref.** Blauvelt 1936, Paquin & Dupérré 2003

— Abdomen without folium, epigynum otherwise ........ **215**

**215(214)** Epigynum with short, broad, rigidly folded scape (Fig. 35.341) ............................................ ***Oreophantes***
**Div.** 1 species: *Oreophantes recurvatus* (Emerton 1913a) — **Dist.** n NA — **Refs.** Helsdingen 1973b, 1981a, Paquin & Dupérré 2003

— Epigynal scape long, tip abruptly narrowed, usually curved or folded (Figs. 35.342-35.343) ................. ***Arcuphantes***
**Div.** 7 species — **Dist.** w NA — **Ref.** Chamberlin & Ivie 1943

**216(212)** Legs short, femur I no longer than carapace; body concolorous brown; basal segment of chelicera with anteriolateral row of short bristles (Fig. 35.344); epigynum as shown (Figs. 35.345-35.346) .......................................... ***Centromerus***
**Div.** 9 species — **Dist.** e and n NA — **Ref.** Helsdingen 1973c

— Legs longer, femur I 1¼ to 2 times as long as carapace; abdomen frequently patterned; basal segment of chelicera without row of bristles .................................. **217**

**217(216)** Epigynum with short, straight scape, truncate posteriorly, with a medial notch (Fig. 35.347) ................ ***Taranucnus***
**Div.** 1 species: *Taranucnus ornithes* (Barrows 1940) — **Dist.** e NA from QC s to NC and TN — **Refs.** Helsdingen 1973b, Paquin & Dupérré 2003

— Epigynum with longer, narrower scape without medial notch, usually curved or folded (Fig. 35.348-35.349) ........ ................................................. ***Lepthyphantes,*** in part
**Div.** 29 species + several undescribed: most species key out here — **Dist.** widespread — **Refs.** Zorsch 1937, Buckle *et al.* 2001

**218(198)** Prolateral spine on femur I .......................................... **219**
— No prolateral spine on femur I ..................................... **223**

Spiders of North America — Linyphiidae | 159

**219(218)** Carapace with dark medial shield-shaped dark mark (Fig. 35.350); epigynal scape folded (Figs. 35.351-35.352); TmI 0.7 to 0.8; metatarsus IV with trichobothrium .................. ............................................................ ***Poeciloneta***
  **Div.** 10 species — **Dist.** n and w NA — **Ref.** Buckle *et al.* 2001

— Carapace without shield-shaped mark; scape straight or absent; TmI <0.5; metatarsus IV without trichobothrium .................................................................................................. **220**

**220(219)** Epigynum with both dorsal and ventral scapes (Figs. 35.353-35.357); ventral scape may be very short (Fig. 35.357), or long (Fig. 35.355) but always with terminal pit ................................................ ***Bathyphantes****, in part*
  **Div.** 26 species. 24 key out here — **Dist.** widespread — **Ref.** Ivie 1969

— Epigynum without ventral scape ................................ **221**

**221(220)** Dorsal scape relatively long, with terminal pit (Figs. 35.358-35.359) ................................................ ***Kaestneria***
  **Div.** 3 species — **Dist.** n NA — **Ref.** Ivie 1969

— Dorsal scape short (Fig. 35.360) or rudimentary (Fig. 35.361) ....................................................................... **222**

**222(221)** Carapace and abdomen concolorous; legs relatively short, with femur I about as long as carapace; epigynum as shown (Fig. 35.361) ........................................ ***Porrhomma***
  **Div.** 9 species — **Dist.** widespread — **Ref.** Buckle *et al.* 2001

— Carapace with dark cephalic region; abdomen patterned; legs relatively long, with femur I 1.5 times as long as carapace; epigynum as shown (Fig. 35.360) ................................ ............................................................ ***Pacifiphantes***
  **Div.** 1 species: *Pacifiphantes magnificus* (CHAMBERLIN & IVIE 1943) — **Dist.** Pacific northwest — **Refs.** Ivie 1969, Eskov & Marusik 1994 — **Note** the male of *Pacifiphantes magnificus* is unknown. It is unclear whether Eskov & Marusik (1994) were correct in placing it in *Pacifiphantes* or Ivie (1969) in placing it as a close relative of *Bathyphantes approximatus* (O. PICKARD-CAMBRIDGE 1871a).

**223(218)** Epigynum without scape ............................................. **224**

— Epigynum with scape ................................................... **225**

**224(223)** Epigynal plate ovoid, with anterior openings (Fig. 35.262); dark unicolorous carapace and distinctive abdominal pattern with lateral light bars (Fig. 35.363) ....... ***Frontinella***
  **Div.** 2 species — **Dist.** *Frontinella pyramitela* (WALCKENAER 1841), widespread and *Frontinella huachuca* GERTSCH & DAVIS 1946, AZ — **Refs.** Blauvelt 1936, Gertsch & Davis 1946 — **Note** *Frontinella pyramitela* is referred as *Frontinella communis* (HENTZ 1850a) in Platnick (2005)

— Epignal plate ovoid, with posterior openings (Fig. 35.364); carapace patterned; dorsum of abdomen dark with paired light spots (Fig. 35.365) ........... ***Graphomoa***
  **Div.** 1 species: *Graphomoa theridioides* CHAMBERLIN 1924a — **Dist.** LA, TN — **Ref.** Chamberlin 1924a — **Note** according to Gustavo Hormiga (pers. comm.) *Graphomoa theridioides* is a synonym of *Pocobletus coroniger* SIMON 1894a and *Graphomoa* a synonym of *Pocobletus* SIMON 1894a

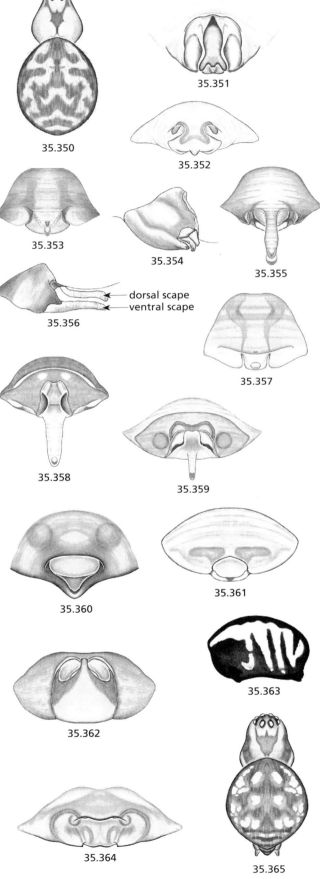

**225(223)** Epigynum as in Fig. 35.366 ...................... **Jalapyphantes**
　　　　**Div.** 1 species: *Jalapyphantes cuernavaca* Gertsch & Davis 1946 — **Dist.** AZ — **Ref.** Gertsch & Davis 1946

— Epigynum otherwise ...................................................... **226**

**226(225)** Abdomen patterned (Fig. 35.367); epigynal scape narrow, straight or slightly bent ventrally (Figs. 35.368-35.369) ....
................................................................................ **Linyphantes**
　　　　**Div.** 19 species — **Dist.** Pacific northwest — **Ref.** Chamberlin & Ivie 1942a

— Abdomen concolorous; epigynal scape otherwise ...... **227**

**227(226)** Epigynum scape rolled dorsally and with lateral lobes (Fig. 35.370-35.372) ............................................................. **228**

— Epigynum scape rigidly folded .................................... **229**

**228(227)** Trichobothrium present on metatarsus IV; epigynum as shown (Fig. 35.371) ......................................... **Microneta**
　　　　**Div.** 1 species: *Microneta viaria* (Blackwall 1841a) — **Dist.** n and w NA — **Refs.** Kaston 1981, Saaristo 1974a, Paquin & Dupérré 2003

— No trichobothrium on metatarsus IV; epigynum as shown (Figs. 35.372-35.373) ............................. **Agyneta,** *in part*
　　　　**Div.** 45 species and perhaps as many more undescribed, a few western species key out here — **Dist.** widespread — **Refs.** Buckle *et al.* 2001, Paquin & Dupérré 2003 — **Note** some authors treat *Meioneta* Hull 1920 as a distinct subgenus or genus. *Agyneta sensu stricto* has a trichobothrium on metatarsus IV while *Meioneta* does not

**229(227)** Basal part of scape concave distally, terminal part of folded scape does not extend beyond it (Fig. 35.374) ..................
................................................................................ **Saaristoa**
　　　　**Div.** 1 species: *Saaristoa sammamish* Levi & Levi 1955 — **Dist.** w NA — **Ref.** Levi & Levi 1955

— Basal part of scape convex distally, terminal part often extends beyond it (Fig. 35.375) .... **Oreonetides,** *in part*
　　　　**Div.** 6 species + several undescribed, 3 described species key out here — **Dist.** n and w NA — **Ref.** Helsdingen 1981a

**230(197)** Epigynum scape rolled dorsally (Figs. 35.370, 35.376-35.379) ............................................................................. **231**

— Scape not rolled ............................................................ **233**

**231(230)** Epigynal scape narrow, tightly rolled, without lateral lobes (Figs. 35.376, 35.378) ....................................... **Macrargus**
　　　　**Div.** 1 species: *Macrargus multesimus* (O. Pickard-Cambridge 1875c) — **Dist.** n NA — **Ref.** Kaston 1981, Paquin & Dupérré 2003

— Scape wider, more loosely rolled, with lateral lobes (Figs. 35.377, 35.379) ............................................................. **232**

**232(231)** Spiracle forward 1/4 to 1/3 of way to epigastric furrow (Fig. 35.380); epigynum very similar to that of *Agyneta* (Fig. 35.381) .................................... **Tennesseellum**
    **Div.** 2 species — **Dist.** *Tennesseellum formicum* (EMERTON 1882) widespread, and an undescribed species from CA — **Ref.** Kaston 1981, Paquin & Dupérré 2003

— Spiracle in normal position at base of spinnerets (Fig. 35.382); epigynum as shown (Figs. 35.383-35.384) .......... ................................................. ***Agyneta*, in part**
    **Div.** 45 species and perhaps as many more undescribed, most key out here — **Dist.** widespread — **Refs.** Buckle *et al.* 2001, Paquin & Dupérré 2003 — **Note** some authors treat *Meioneta* HULL 1920 as a distinct subgenus or genus. *Agyneta sensu stricto* has a trichobothrium on metatarsus IV while *Meioneta* does not. *Anibontes longipes* CHAMBERLIN & IVIE 1944 will key out here, and perhaps the unknown female of *Anibontes mimus* (CHAMBERLIN 1924a) as well

**233(230)** Anterior eyes approximately equally spaced; abdomen patterned (Fig. 35.385); epigynum with short, broad scape notched medially (Figs. 35.386-35.387) ............ **Wubana**
    **Div.** 7 species — **Dist.** n and w NA — **Refs.** Chamberlin & Ivie 1936b, Buckle *et al.* 2001, Paquin & Dupérré 2003

— AME much closer to each other than to ALE; abdomen concolorous; epigynum otherwise .............................. **234**

**234(233)** Legs relatively long and slim, with femur I length 1 1/3 times that of carapace; leg spines short and thin, barely distinguishable from setae; eyes small, especially AME (Fig. 35.388); epigynum long and straight with bluntly rounded tip (Fig. 35.389) .............................. **Oaphantes**
    **Div.** 1 species: *Oaphantes pallidulus* (BANKS 1904b) + 2 undescribed — **Dist.** CA to BC — **Ref.** Chamberlin & Ivie 1943

— Legs shorter, with femur I no longer than carapace; leg spines of normal size; eyes larger; epigynum with folded scape, tip with stretcher usually visible (Figs. 35.390-35.392) ........................................................ **235**

**235(234)** Body under 1.7 mm in length; stridulating ridges on epigastric plates (Fig. 35.393); epigynal scape short (Fig. 35.390) ................................................................. **Maro**
    **Div.** 2 species: *Maro nearcticus* DONDALE & BUCKLE 2001 and *Maro amplus* DONDALE & BUCKLE 2001 — **Dist.** n NA — **Refs.** Dondale & Buckle 2001, Paquin & Dupérré 2003

— Body usually over 1.7 mm in length; no stridulating ridges on epigastric plates; epigynal scape longer (Figs. 35.391-35.392) ............................................. ***Oreonetides*, in part**
    **Div.** 6 species + several undescribed, 3 described species key out here — **Dist.** n and w NA — **Ref.** Helsdingen 1981a

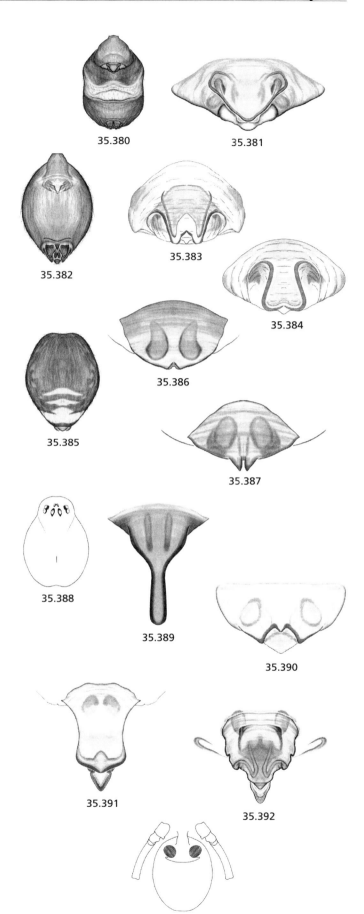

# Chapter 36
# LIOCRANIDAE
5 genera, 11 species

Darrell Ubick
David B. Richman

### Similar families —
Corinnidae (PHRUROLITHINAE) (p. 78.)

### Diagnosis —
These ecribellate, entelgyne, two clawed spiders may be distinguished from those of other families by the combination of: legs prograde; eyes eight, subequal, and in two straight or slightly curved rows; tracheal spiracle near spinnerets; anterior spinnerets conical and usually contiguous; PLS with short distal segment. *Hesperocranum* is unique among Nearctic spiders in having two dense rows of bristles ventrally on the anterior tibiae and metatarsi (Fig. 36.2). The other Nearctic Liocranidae differ from Clubionidae and Corinnidae in lacking precoxal triangles (*contra* Fig. 36.3).

### Characters —
**body size:** 2-10 mm.
**color:** orange to gray brown, often with black markings on carapace and chevron pattern on abdomen (Fig. 36.1).
**carapace:** pyriform.
**sternum:** oval with slight posterior extension between hind coxae; precoxal triangles present (*Hesperocranum*) or absent (other genera).
**eyes:** eight subequal eyes in two straight or slightly curved rows.
**chelicerae:** not geniculate, with toothed margins.
**mouthparts:** endites quadrate with rounded to straight ectal margins; labium slightly wider than long.
**legs:** anterior tibiae and metatarsi with ventral rows of bristles (*Hesperocranum*) or paired spines (others); tarsi entire (*Hesperocranum*) or at least posterior ones pseudosegmented (others); anterior tarsi with two rows of trichobothria; tarsi with two claws, lacking tufts (but may have tenent hairs, as *Apostenus*).
**abdomen:** oval elongate, slightly flattened.
**spinnerets:** ALS thicker than others, conical and contiguous to slightly separated (*Agroeca*); PMS enlarged in female.
**respiratory system:** one pair of book lungs, tracheal spiracle near spinnerets.
**genitalia:** entelegyne; *female* epigynum with median atrium or lobe; *male* palpus with single RTA, bulb with median apophysis and short embolus.

### Distribution —
The family is widespread in the Nearctic region. *Agroeca* is largely a northern genus but extends into California and the southeast has an isolated species in Georgia. The genus is best represented in northeastern US, where it is sympatric with *Liocranoeca*. *Neoanagraphis* is found from the Mojave Desert region of California to Texas. *Apostenus* is known from the mountains of southern California and *Hesperocranum* from the mountains of southern Oregon to the southern Sierra Nevada of California. The latter three genera are allopatric in California, but at least *Hesperocranum* has been collected with *Agroeca*. *Hesperocranum* and *Neoanagraphis* are Nearctic, the other genera being Holarctic.

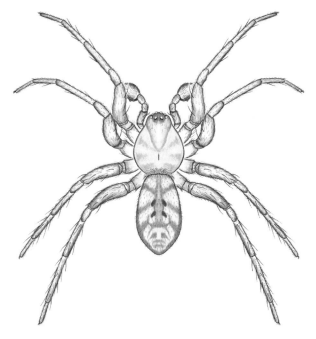

**Fig. 36.1** *Agroeca ornata* BANKS 1892a

### Natural history —
Liocranids are ground spiders and occur under rocks and in leaf litter. Most are forest species; *Apostenus* associated with oaks and *Hesperocranum* largely with conifers. *Neoanagraphis* is from arid regions and has been collected on the desert floor, in animal burrows, and pitfall traps. The spiders are wandering hunters and construct sac-like retreats. The egg sacs of some *Agroeca* are suspended from a stalk and covered with dirt (Roberts 1995). In captivity, liocranids are easily reared on flies and springtails.

### Taxonomic history and notes —
These spiders were traditionally included in Clubionidae. Simon (1897a) placed them in his group Liocraneae, one of four in the subfamily Liocraninae. Lehtinen (1967), in dismembering the Clubionidae, elevated three of the groups to family and placed the fourth, Phrurolithinae, in the Gnaphosidae and closely related to *Micaria*. This association proved to be erroneous (Platnick & Shadab 1988) and the phrurolithines were moved back to Liocranidae (Platnick 1989b). Recent studies of Dionycha relationships have shown that the phrurolithines actually belong to the Corinnidae (Penniman 1985, Bosselaers & Jocqué 2002). However, of the five Nearctic genera currently remaining in Liocranidae, only *Hesperocranum* is closely related to the type genus, *Liocranum* L. KOCH 1866a. As the remaining four genera appear to be more closely related to Corinnidae (Bosselaers & Jocqué 2002), even the currently residual Liocranidae is likely to undergo further reduction.

For the Nearctic fauna, *Agroeca* was revised by Kaston (1938b), an additional species described by Chamberlin & Ivie (1944), the Canadian species redescribed by Dondale & Redner (1982) and Paquin & Dupérré (2003), and *Agroeca emertoni* (Kaston 1938b) transferred to *Liocranoeca* (Wunderlich 1999a). *Neoanagraphis* species were described by Gertsch & Mulaik (1936a) and Gertsch (1941b) and revised by Vetter (2001). *Hesperocranum* was described by Ubick & Platnick (1991) and the first Nearctic species of *Apostenus* by Ubick & Vetter (2005).

### Genera —

LIOCRANINAE
 *Hesperocranum* Ubick & Platnick 1991
"AGROECA COMPLEX"
 *Agroeca* Westring 1861, *Apostenus* Westring 1851, *Liocranoeca* Wunderlich 1999a, *Neoanagraphis* Gertsch & Mulaik 1936a

### Key to genera —
North America North of Mexico

**1** Anterior tibiae with two dense rows of ventral bristles (bristles shorter than segment diameter) (Fig. 36.2); precoxal triangles present (Fig. 36.3); genitalia as in Figs. 36.4-36.5 ............................................... **Hesperocranum**
 **Div.** 1 species: *Hesperocranum rothi* Ubick & Platnick 1991 — **Dist.** s OR-c CA — **Ref.** Ubick & Platnick 1991

— Anterior tibiae with ventral spines (spines longer than segment diameter); precoxal triangles absent ................. **2**

**2(1)** Anterior tibiae with 5 pairs of ventral spines; genitalia as in Figs. 36.6-36.7 ............................................ **Apostenus**
 **Div.** 1 species: *Apostenus californicus* Ubick & Vetter 2005 — **Dist.** s CA — **Ref.** Ubick & Vetter 2005

— Anterior tibiae with 1-3 pairs of ventral spines .............. **3**

**3(2)** Tarsal claws of posterior legs long, longer than tarsal width, and with few small teeth confined to base (Fig. 36.8); genitalia as in Figs. 36.9-36.10 ..................................
 ............................................... **Neoanagraphis**
 **Div.** 2 species: *Neoanagraphis chamberlini* Gertsch & Mulaik 1936a and *Neoanagraphis pearcei* Gertsch 1941b — **Dist.** s CA to TX — **Ref.** Vetter 2001

— Tarsal claws of posterior legs unmodified: shorter than tarsal width and with teeth widespread ........................... **4**

**4(3)** Anterior metatarsi with 2 pairs of ventral spines and patella IV without retrolateral spine; genitalia as in Figs. 36.11-36.12 ................................................. **Liocranoeca**
 **Div.** 1 species: *Liocranoeca emertoni* (Kaston 1938b) — **Dist.** New England — **Ref.** Kaston 1938b, Wunderlich 1999a — **Note** this species was previously included in *Agroeca*

— Anterior metatarsi with 3 pairs of ventral spines, or if with only 2 pairs then patella IV with retrolateral spine; genitalia as in Figs. 36.13-36.14 .................................... **Agroeca**
 **Div.** 6 species — **Dist.** widespread — **Refs.** Kaston 1938b, Chamberlin & Ivie 1944, Dondale & Redner 1982, Paquin & Dupérré 2003

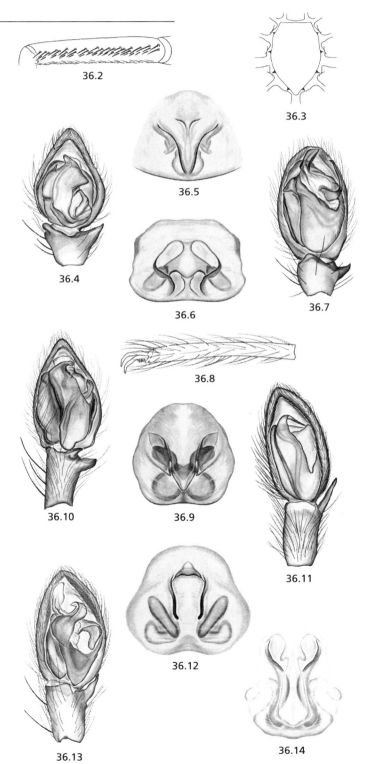

## Chapter 37
# LYCOSIDAE
16 genera, 238 species

Charles D. Dondale

**Common name —**
Wolf spiders.

**Similar families —**
Ctenidae (p. 83), Pisauridae (p. 199), Trechaleidae (p. 247), Zoridae (p. 256), Zoropsidae (p. 259).

**Diagnosis —**
Small to large, three-clawed ecribellate mainly hunting spiders with a high carapace and distinctive eye arrangement (Figs. 37.1, 37.2). The four anterior eyes small, not greatly different in size, and arranged in a nearly straight row across front of carapace; the four posterior eyes situated dorsally, much larger than the anterior eyes and forming a quadrangle which is usually wider posteriorly than anteriorly (Fig. 37.2). Palpus of male lacking tibial apophyses, often with stridulatory organ involving tip of tibia and base of cymbium. Egg cocoon carried by female's spinnerets, and newly hatched young transported on female's back for several days before dispersing.

**Characters —**
**size:** 2.2-35.0 mm.
**color:** drably marked with off-white, black, yellow, or red on a brown or blackish background; carapace often with two or more distinctive dark longitudinal stripes.
**carapace:** longer than wide, narrowed toward front, rather high (usually highest between dorsal groove and posterior lateral eyes), usually covered with short recumbent setae.
**eyes:** as in diagnosis.
**chelicerae:** long, robust, bare or hairy, often with prominent boss, with 3 prolateral and 2-4 (usually 3) retrolateral teeth, lacking sexual modifications in males.
**legs:** long, strong, often with dense scopulae; tarsi with 3 claws; trochanters with notch.
**abdomen:** longer than wide, tapered posteriorly, highest anteriorly, covered with short recumbent setae.
**genitalia:** entelegyne; *female* epigynum usually with deep atrium and distinct median septum, sometimes with hood which may be single or double; *male* palpus lacking tibial apophyses; embolus usually long and slender, arising with terminal apophysis at distal end of genital bulb; median apophysis minute to large, often with spur, hook, or sharp retrolateral teeth.

**Distribution —**
Widespread from the Arctic to the Subtropics.

**Natural history —**
Most members of this family use vision in detecting prey, and may use both stalking and ambushing in the capture. Among the North American representatives, only *Sosippus* spp. spin webs. With regard to habitat, some are found only or mainly in wet terrain such as bogs, fens, or swamps, whereas others seem to prefer open grasslands, deciduous forests, or deserts. *Geolycosa* spp. are deep burrowers in sandy soil (Marshall 1995, 1999, Marshall *et al.* 2000). A distinctive behavior of wolf spiders is the carrying of the

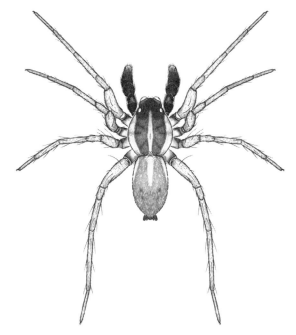

**Fig. 37.1** *Pardosa hyperborea* (Thorell 1872a)

egg sac by the female's spinnerets and the subsequent transport of the young for a time on the female's back (Rovner *et al.* 1973). Considerable research has gone into the courtship and mating of these spiders (Rovner 1968, 1974, 1975, Buckle 1972, Stratton & Lowrie 1984, Stratton 1985, Stratton & Uetz 1986, Hebets *et al.* 1996, Stratton *et al.* 1996). Life histories have been studied by Eason & Whitcomb (1965), Eason (1969), Dondale (1977), McQueen (1978), Buddle (2000), and Pickavance (2001), among others. Predatory behavior in the laboratory was investigated by Hardman & Turnbull (1974, 1980), and by Holmberg & Turnbull (1982).

**Taxonomic history and notes —**
Taxonomic work on the North American Lycosidae had its beginning through the interest of the Europeans C.A. Walckenaer, Eugen Keyserling, and T. Thorell. The Americans N.M. Hentz, J.H. Emerton, Nathan Banks, and T.H. Montgomery, Jr. then took up the task. Significant numbers of species were made known in the mid-1900s by Willis J. Gertsch, R.V. Chamberlin, and Wilton Ivie. Most of the genera were later defined by H.K. Wallace (*Geolycosa*), Wallace and H. Exline (*Pirata*), B.R. Vogel (*Pardosa*), A.R. Brady (*Trochosa, Gladicosa*), C.D. Dondale and J.H. Redner (*Schizocosa, Alopecosa, Arctosa, Allocosa, Pardosa*), and T. Kronestedt (*Pardosa*). The world catalogs by P. Bonnet, C.F. Roewer, and N.I. Platnick were, and are, major contributors to progress. Dondale (1986) attempted to show the relationships among the various genera, and Griswold (1993) the relationship of the family Lycosidae to other families of Lycosoidea.

As shown by Zyuzin & Logunov (2000), the genus *Lycosa* Latreille 1804b is properly restricted to a group of large burrowing wolf spiders in the Mediterranean region. Consequently, this genus is not represented in North America, and the species listed from this continent by Platnick (2005) require revision and placement.

Platnick's (2005) listing of *Paratrochosina insolita* (L. Koch 1879b) from a North American locality is probably based on a misidentification of specimens of the highly variable *Alopecosa hirtipes* (Kulczyński 1907a), and therefore the genus *Paratrochosina* Roewer 1960d does not appear in the key.

The genus *Rabidosa* Brady & McKinley 1994 has been found in the testing of the key to manifest no unique character by which specimens can be distinguished from those of other genera, and does not appear in the key.

The genus *Acantholycosa* Dahl 1908 is based mainly on the possession of supernumerary pairs of macrosetae under tibiae I and II. Similar macrosetae are found in specimens of some species of *Pardosa*, *Trabeops*, and *Pirata*. This casts doubt on the validity of the genus *Acantholycosa*. The external genitalia of the single North American species placed in this genus, *Acantholycosa solituda* Levi & Levi 1951, are of the type found throughout the genus *Pardosa*, and hence *Acantholycosa solituda* is keyed as a species of *Pardosa*. The North American species of *Hogna* need revising, of necessity in relation to the type species, *Hogna radiata* (Latreille 1817). Some lycosid species in North America are still poorly known or misplaced. Future research should eliminate these problems.

### Genera —
SOSIPPINAE
   *Sosippus* Simon 1888b
VENONIINAE
   *Pirata* Sundevall 1833b, *Trabeops* Roewer 1959b, *Trebacosa* Dondale & Redner 1981b
ALLOCOSINAE
   *Allocosa* Banks 1900c
PARDOSINAE
   "*Acantholycosa*" Dahl 1908, *Pardosa* C.L. Koch 1847a
LYCOSINAE-LYCOSA GROUP
   *Alopecosa* Simon 1885g, *Arctosa* C.L. Koch 1847a, "*Lycosa*" Latreille 1804b, *Melocosa* Gertsch 1937, "*Paratrochosina*" Roewer 1960d, *Varacosa* Chamberlin & Ivie 1942a
LYCOSINAE-TROCHOSA GROUP
   *Geolycosa* Montgomery 1904, *Gladicosa* Brady 1987, *Hesperocosa* Gertsch & Wallace 1937, *Hogna* Simon 1885g, "*Rabidosa*" Roewer 1960d, *Schizocosa* Chamberlin 1904a, *Trochosa* C.L. Koch 1847a

### Key to genera —
North America North of Mexico

**1**  Carapace in lateral view ascending steeply in height from dorsal groove to posterior lateral eyes (Fig. 37.3). Spider living in deep burrow in sandy soil ............... **Geolycosa**
   Div. 20 species — Dist. widespread in USA and s CAN, but best represented in se USA — Refs. Wallace 1942a, McCrone 1963, Dondale & Redner 1990

—  Carapace not ascending steeply between dorsal groove and posterior lateral eyes. Spider active on ground or in shallow retreat, or on web ..................................................... **2**

**2(1)**  Carapace with V-shaped mark (Fig. 37.4) .........................
   ........................................................... ***Pirata*, in part**
   Div. 28 species — Dist. widespread — Refs. Wallace & Exline 1978, Dondale & Redner 1981a, 1990

—  Carapace lacking V-shaped mark ..................................... **3**

**3(2)**  Tibiae I and II with more than 3 pairs of ventral macrosetae ............................................................................. **4**

—  Tibiae I and II with only 3 pairs of ventral macrosetae ..... ........................................................................... **7**

**4(3)**  Tibiae I and II with 4 pairs of long ventral macrosetae ..... ............................................................................. **5**

—  Tibiae I and II with 5-7 pairs of long ventral macrosetae .. ............................................................................. **6**

**5(4)**  Palpus of male with median apophysis in form of large crescent (Fig. 37.5). Epigynum of female with median septum flared anterolaterad (Fig. 37.6). Total length <4 mm. Spider restricted to se CAN and e USA ...................... ........................................................................... **Trabeops**
   Div. 1 species: *Trabeops aurantiacus* (Emerton 1885) — Dist. se CAN, e USA — Refs. Kaston 1948 (as *Trabea*), Dondale & Redner 1990

—  Palpus of male with median apophysis not in form of large crescent. Epigynum of female with median septum not flared anterolaterad. Total length >5 mm. Spider restricted to AK, YT ............................................ ***Pardosa*, in part**
   Div. 1 species: *Pardosa beringiana* Dondale & Redner 1987— Dist. AK, YT — Refs. Dondale & Redner 1987, 1990

37.2

37.4

37.3

37.5

37.6

**6(4)** Carapace and legs black or dark brown. Spider restricted to high elevations in the Rocky and Sierra Nevada Mountains. Tibiae I and II covered with short dense white pubescence .............................................. ***Pardosa,*** in part
    **Div.** 1 species: *Pardosa solituda* Levi & Levi 1951— **Dist.** high elevations BC-CA — **Refs.** Levi & Levi 1951, Dondale & Redner 1987, 1990, Kronestedt & Marusik 2002 (as *Acantholycosa*).

— Carapace and legs amber in color. Spider restricted to lower elevations. Tibiae I and II lacking short dense white pubescence ................................................ ***Pirata,*** in part
    **Div.** 1 species: *Pirata spiniger* (Simon 1898a) — **Dist.** e USA — **Ref.** Wallace & Exline 1978.

**7(3)** Carapace, abdomen, and legs I and II totally black ...........
...................................................................... ***Pirata,*** in part
    **Div.** 1 species: *Pirata bryantae* Kurata 1944 — **Dist.** boreal forest zone, AK-NF — **Refs.** Wallace & Exline 1978, Dondale & Redner 1990.

— Carapace, abdomen, and legs I and II not totally black, usually with pattern of pale spots and bands .................. **8**

**8(7)** Palpus of male with large angular median apophysis (Fig. 37.7). Epigynum of female with depressed triangular median septum (Fig. 37.8) .............................. ***Trebacosa***
    **Div.** 1 species: *Trebacosa marxi* (Stone 1890) — **Dist.** IL to NS, s to AR and FL — **Refs.** Dondale & Redner 1981b, 1990.

— Genitalia not as in Figs. 37.7 and 37.8 ............................ **9**

**9(8)** Palpus of male with embolus and terminal apophysis arched retrolaterally at tips (Fig. 37.9). Epigynum of female with posterior part of median septum rectangular and pearl-like (Fig. 37.10) .............. ***Gladicosa***
    **Div.** 5 species — **Dist.** se CAN, e USA w to SD, CO, TX — **Ref.** Brady 1987.

— Palpus of male with embolus and terminal apophysis not arched retrolaterally at tips (Figs. 37.11-37.13, 37.18, 37.21, 37.23). Epigynum of female with posterior part of median septum, if septum present, neither rectangular nor pearl-like (Figs. 37.14-37.17, 37.19, 37.22) ........... **10**

**10(9)** Palpus of male with embolus abruptly curved or angled toward distal end of genital bulb (Fig. 37.11), rarely a minute point. Epigynum of female with transverse piece of median septum excavated (Fig. 37.15) or truncated (Fig. 37.16) ..................................... ***Schizocosa***
    **Div.** 27 species — **Dist.** widespread — **Refs.** Dondale & Redner 1978a, 1990, Uetz & Dondale 1979, Stratton 1991.

— Palpus of male with embolus smoothly curved or not visible (Figs. 37.12, 37.18, 37.24, 37.26). Epigynum of female with transverse piece of median septum, if septum present, not excavated or truncated ............... **11**

**11(10)** Palpus of male with median apophysis possessing 2 distinct processes of different lengths (Fig. 37.12). Epigynum of female with smooth surface, lacking atrium and median septum (Fig. 37.17), often covered with fine dense setae .. .................................................................. ***Allocosa***
    **Div.** 18 species — **Dist.** widespread — **Ref.** Dondale & Redner 1983b — **Note** females of some *Arctosa* spp. possess shallow indistinct depressions that can be mistaken for an atrium.

— Palpus of male with median apophysis possessing short point(s) or none (Figs. 37.18, 37.21, 37.23-37.24, 37.26, 37.40, 37.41). Epigynum of female with well-defined atrium and median septum (Figs. 37.19, 37.22, 37.27-37.32, 37.42, 37.44) ........................................ **12**

37.7

37.8

37.10

37.9

37.11

37.12

37.13

37.14

**12(11)** Palpus of male with tiny median apophysis (Fig. 37.18). Epigynum of female with median septum broadly flask-shaped (Fig. 37.19). Spider restricted to mountains above tree line .................................................. ***Melocosa***
   **Div.** 1 species: *Melocosa fumosa* (EMERTON 1894) — **Dist.** AK to MT — **Refs.** Leech 1969b, Dondale & Redner 1990

— Palpus of male with well-developed median apophysis (Figs. 37.24-37.26). Epigynum of female lacking broadly flask-shaped median septum. Spider usually not restricted to mountains above tree line ............................. **13**

**13(12)** Male .................................................................... **14**
— Female ................................................................ **22**

**14(13)** Terminal apophysis sickle-like (Fig. 37.45) ............... **15**
— Terminal apophysis not sickle-like, or not visible in ventral view ..................................................................... **17**

**15(14)** Carapace with paired dark longitudinal streaks in pale area anterior to dorsal groove (Fig. 37.20) ....... ***Trochosa***
   **Div.** 2 species: *Trochosa terricola* THORELL 1856 and the introduced *Trochosa ruricola* (DE GEER 1778b) — **Dist.** widespread — **Refs.** Brady 1980 (part), Dondale & Redner 1990, Prentice 2001 — **Note** the carapace markings may be faded in poorly preserved specimens. Also, some specimens of *Hogna frondicola* (EMERTON 1885) key here; males of that species differ in possessing darkened leg coxae, sternum, and venter. Males of *Trochosa terricola* possess a curled embolus, those of *Trochosa ruricola* do not

— Carapace lacking paired dark longitudinal streaks in pale area anterior to dorsal groove .......................... **16**

**16(15)** Palpus as in Fig. 37.21. Total length 6.0 mm or less. Range restricted to southwestern USA ................ ***Hesperocosa***
   **Div.** 1 species: *Hesperocosa unica* (GERTSCH & WALLACE 1935) — **Dist.** AZ to TX — **Refs.** Gertsch & Wallace 1935 (as *Schizocosa*), 1937

— Palpus not as in Fig. 37.21. Total length usually more than 7.0 mm. Range not restricted to southwestern USA ......... ................................................................................. ***Hogna***
   **Div.** approximately 25 species — **Dist.** widespread — **Refs.** Gertsch & Wallace 1935 (as *Lycosa*), 1937 (part, as *Lycosa*), Wallace 1942b (as *Lycosa*), Dondale & Redner 1990 (part), Reiskind & Cushing 1996 (as *Lycosa*) — **Note** some species that key here are currently catalogued as "*Lycosa*" or "*Rabidosa*"

**17(14)** Conductor extending finger-like beyond margin of genital bulb (Fig. 37.23). Anterior lateral eyes situated on small tubercules. Spider living on funnel web ........... ***Sosippus***
   **Div.** 6 species — **Dist.** CA to FL — **Refs.** Brady 1962, Sierwald 2000

— Conductor not extending finger-like beyond margin of genital bulb. Anterior lateral eyes not situated on small tubercules. Spider not living on funnel web ................ **18**

**18(17)** Embolus with slender sinuous piece on flat truncate base (Fig. 37.13) .................................................. ***Varacosa***
   **Div.** 4 species — **Dist.** widespread — **Refs.** Brady 1980 (as *Trochosa*), Jiménez & Dondale 1988, Dondale & Redner 1990

— Embolus lacking slender sinuous piece on flat truncate base ..................................................................... **19**

37.15

37.16

37.17

37.18

37.19

37.20

37.21

37.22

37.23

**19(18)** Terminal apophysis plump, subdivided by a groove (Fig. 37.24). Carapace often bare and mottled .................................................................................... ***Arctosa,*** *in part*
        **Div.** 12 species — **Dist.** widespread — **Refs.** Dondale & Redner 1983a, 1990

— Terminal apophysis not plump, not subdivided by groove. Carapace usually covered with setae, not mottled. Note: males of *Pardosa moesta* BANKS 1892a may key here; they are distinguished by numerous minute lateral tubercles on the chelicerae ................................................................ **20**

**20(19)** Median apophysis like a large pointed tooth (Fig. 37.25) .. .................................................................................... ***Arctosa,*** *in part*
        **Div.** *Arctosa raptor* (KULCZYŃSKI 1885a) — **Dist.** AK to NF, s to ME — **Refs.** Dondale & Redner 1983a, 1990 — **Note** males of *Pardosa palustris* (LINNAEUS 1758) may key here; they differ in possessing a swollen pale tarsus I, which is fringed with pale setae

— Median apophysis not like a large pointed tooth ......... **21**

**21(20)** Median apophysis flat, oriented transversely, with minute pointed tooth at tip (Fig. 37.26). Cheliceral retromargin usually with 2 teeth ........................................ ***Alopecosa***
        **Div.** 6 species — **Dist.** widespread — **Refs.** Dondale & Redner 1979, 1990 — **Note** some males of *Pardosa glacialis* (THORELL 1872b) may key here; they lack the pointed tooth on the median apophysis and possess a broad sword-shaped embolus. Specimens of «*Paratrochosina insolita*» (L. KOCH 1879b) may key here

— Median apophysis round in cross section, with longitudinal piece sometimes oriented obliquely, usually with spur at or near base (Figs. 37.39-37.41); spur may be longer than longitudinal part. Cheliceral retromargin usually with 3 teeth [*Pardosa tesquorum* (ODENWALL 1901) is an exception]. Carapace usually vertical at sides (Fig. 37.1), with eyes of posterior row situated at carapace margins ... .................................................................................... ***Pardosa,*** *in part*
        **Div.** 69 species — **Dist.** widespread — **Refs.** Barnes 1959, Vogel 1964, 1970a, Lowrie & Dondale 1981, Kronestedt 1975a, 1981, 1986a, 1988, Dondale & Redner 1984, 1986, 1987, 1990, Dondale 1999

**22(13)** Anterior lateral eyes situated on small tubercles. Spider living on funnel web ............................................ ***Sosippus***
        **Div.** 6 species — **Dist.** CA to FL — **Refs.** Brady 1962, Sierwald 2000

— Anterior lateral eyes not situated on tubercles. Spider not living on web .................................................................... **23**

**23(22)** Median septum with ends of transverse piece strongly curved anteriorly (Fig. 37.27). Carapace often with paired dark longitudinal streaks in pale area anterior to dorsal groove (Fig. 37.20) ............................................. ***Varacosa***
        **Div.** 4 species — **Dist.** widespread — **Refs.** Brady 1980 (as *Trochosa*), Jiménez & Dondale 1988, Dondale & Redner 1990

— Median septum with ends of posterior transverse piece not curved anteriorly. Carapace rarely with paired dark longitudinal streaks in pale area anterior to dorsal groove .......................................................................................... **24**

**24(23)** Carapace with paired dark longitudinal streaks in pale area anterior to dorsal groove (Fig. 37.20) ....... ***Trochosa***
        **Div.** 2 species: *Trochosa terricola* THORELL 1856 and the introduced *Trochosa ruricola* (DE GEER 1778b) — **Dist.** widespread — **Refs.** Brady 1980 (part), Dondale & Redner 1990, Prentice 2001 — **Note** some specimens of *Hogna frondicola* (EMERTON 1885) may key here; they possess dark leg coxae, sternum, and venter

— Carapace lacking paired dark longitudinal streaks in pale area anterior to dorsal groove (but see note on *Hogna frondicola,* above) ......................................................... **25**

37.24

37.25

37.26

37.27

37.28

37.29

37.30

37.31

37.32

Spiders of North America — Lycosidae

**25(24)** Epigynal hood single (ventral view, Figs. 37.28-37.30) ..... 26

— Epigynal hood double (Figs. 37.22, 37.31), or hood lacking (Figs. 37.42-37.44) ........................................................ 27

**26(25)** Epigynal hood rectangular (Fig. 37.28) Cheliceral retromargin usually with 2 teeth. Carapace sloped at sides anteriorly ........................................... **Alopecosa,** in part
   **Div.** 1 species: *Alopecosa pulverulenta* (CLERCK 1757) — **Dist.** AK (1 doubtful record from the Aleutian Islands) — **Refs.** Dondale & Redner 1979, 1990

— Epigynal hood triangular (Fig. 37.29) or broadly crescent-shaped (Fig. 37.30). Cheliceral retromargin usually with 3 teeth. Carapace usually vertical at sides anteriorly with eyes of posterior row situated at carapace margins ........... ............................................................... **Pardosa,** in part
   **Div.** *Pardosa sternalis* group (6 species), *Pardosa distincta* group (6 species), *Pardosa nigra* group (8 species), *Pardosa milvina* group (9 species), *Pardosa lapponica* group (2 species), *Pardosa lapidicina* group (6 species), *Pardosa falcifera* group ( 2 species) — **Dist.** widespread — **Refs.** Barnes 1959, Vogel 1964, 1970a, Lowrie & Dondale 1981, Dondale & Redner 1984, 1986, 1990

**27(25)** Epigynal hood double (Figs. 37.22, 37.31, 37.32) ........ 29

— Epigynal hood lacking (Figs. 37.42-37.44) ................... 28

**28(27)** Copulatory tubes tortuous, with at least two anterior-posterior arms (Fig. 37.37) .......................... **Pardosa,** in part
   **Div.** *Pardosa coloradensis* group (3 species) — **Dist.** s AK to MB, s to CA, AZ, NM — **Refs.** Dondale & Redner 1986, 1990

— Copulatory tubes with a single anterior-posterior arm (Figs. 37.33, 37.38) ........................................................ 33

**29(27)** Copulatory openings well separated, semicircular in shape (Fig. 37.31) ................................................ **Arctosa,** in part
   **Div.** *Arctosa perita* (LATREILLE 1799b) — **Dist.** s BC (introduction from Eurasia) — **Refs.** Dondale & Redner 1983a, 1990

— Copulatory openings not well separated or semicircular in shape .................................................................. 30

**30(29)** Median septum shaped like an inverted T, sometimes with longitudinal piece expanded at sides (Figs. 37.22, 37.32). Note: females of *Pardosa algens* (KULCZYŃSKI 1908b) and of *Pardosa moesta* BANKS 1892a may key here, but they possess a vertical carapace at the sides anteriorly where the eyes of the posterior row are situated at the carapace margins ........................................................................ 31

— Median septum variously shaped but not like an inverted T .................................................................................. 32

**31(30)** Anterior lateral eyes larger than anterior median eyes. Total length < 7 mm. Range restricted to southwestern USA. Epigynum as in Fig. 37.22 ............... **Hesperocosa**
   **Div.** 1 species: *Hesperocosa unica* (GERTSCH & WALLACE 1935) — **Dist.** AZ to TX — **Refs.** Gertsch & Wallace 1935 (as *Schizocosa*), 1937

— Anterior lateral eyes smaller than or about equal to anterior median eyes. Total length usually > 7 mm (often much longer). Range not restricted to southwestern U.S.A. Epigynum often otherwise, e.g., Fig. 37.32 .......... **Hogna**
   **Div.** approximately 25 species — **Dist.** widespread — **Refs.** Gertsch & Wallace 1935, as *Lycosa*), 1937 (part, as *Lycosa*), Wallace 1942b (as *Lycosa*), Dondale & Redner 1990 (part), Reiskind & Cushing 1996 (as *Lycosa*) — **Note** some species that key here are currently catalogued as "*Lycosa*" or "*Rabidosa*"

37.33

37.34

37.35

37.36

37.37

37.38

37.39

37.40

37.41

37.42

**32(30)** Copulatory tubes each with 2 large dark swellings (Fig. 37.33). Cheliceral retromargin usually with 2 teeth. Carapace sloped at sides anteriorly .......... ***Alopecosa,** in part*
  **Div.** 2 species: *Alopecosa aculeata* (Clerck 1757) *Alopecosa kochii* (Keyserling 1877a) — **Dist.** widespread — **Refs.** Dondale & Redner 1979, 1990

— Copulatory tubes each with a single dark swelling or none (Figs. 37.34-37.36). Cheliceral retromargin usually with 3 teeth. Carapace vertical or nearly so at sides anteriorly with the eyes of the posterior row situated at the carapace margins .................................................. ***Pardosa,** in part*
  **Div.** *Pardosa tesquorum* group (1 species), *Pardosa atrata* group (1 species), *Pardosa monticola* group (1 species), *Pardosa modica* group (16 species), *Pardosa moesta* group (1 species), *Pardosa xerampelina* group (4 species), *Pardosa saltuaria* group (2 species) — **Dist.** widespread — **Refs.** Kronestedt 1975a, 1981, 1986a, 1988, Dondale & Redner 1986, 1987, 1990, Dondale 1999

**33(28)** Copulatory tubes with 1 dark swelling (Fig. 37.38), which may have 1 or more small secondary lobes that do not constitute a second swelling. Cheliceral retromargin usually with 3 teeth. Carapace often bare and mottled ..... ................................................................ ***Arctosa,** in part*
  **Div.** 11 species — **Dist.** widespread — **Refs.** Dondale & Redner 1983a, 1990

— Copulatory tubes with 2 dark swellings approximately equal in size (Fig. 37.33). Cheliceral retromargin usually with 2 teeth. Carapace covered with setae (often densely), not mottled ........................................ ***Alopecosa,** in part*
  **Div.** 4 species — **Dist.** widespread — **Refs.** Dondale & Redner 1979, 1990 — **Note** specimens of «*Paratrochosina insolita*» (L. Koch 1879b) may key here

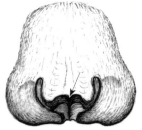
37.43

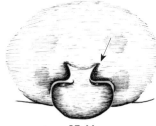
37.44

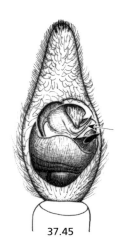
37.45

# Chapter 38
# MIMETIDAE
3 genera, 18 species

Stephen E. Lew
Daniel J. Mott

### Common name —
Pirate spiders, cannibal spiders

### Similar families —
In North America, mimetids are most likely to be mistaken for orbicularians such as Araneidae (p. 68) or Theridiidae (p. 235).

### Diagnosis —
Mimetids can be distinguished from all other entelegyne, ecribellate, three clawed spiders by the characteristic prolateral spination on the first two tibiae and metatarsi (Fig. 38.2).

### Characters —
**body size**: 2.5mm-7.0 mm
**color**: various
**carapace**: oval to pyriform, with or without distinct patterns, head region not distinct.
**sternum**: quadrate anteriorly, narrowing after coxae III, with acuminate posterior projection extending between coxae IV only in some *Ero*.
**eyes**: eight eyes in two rows, lateral eyes contiguous and well separated from median eyes.
**chelicerae**: elongate (*Mimetus*) to medium (*Ero*), the fang promargin lined with peg teeth (Fig. 38.3).
**legs**: I and II very long, III and IV shorter by comparison, tibiae and metatarsi I and II bearing characteristic spination pattern (Fig. 38.2).
**abdomen**: subspherical to ovate, sometimes with paired projections.
**spinnerets**: three pairs, conical and well developed. Colulus distinct.
**respiratory system**: 1 pair book lungs, 4-branched tracheal system with the spiracle near the spinnerets.
**genitalia**: entelegyne; *female* epigynum usually covered by a broad, flat, sclerotized plate narrowing posteriad, openings variable; *male* palpus complex with many sclerites. The cymbium is well-developed in the North American genera, always with a curved paracymbium, a proximal cup-shaped extension, and often with sculpting and/or dentition on the cymbial margin (Fig. 38.4). The palpus is rotated so that the morphologically ventral bulb is functionally ectal.

### Distribution —
Mimetid spiders are found on all continents except Antarctica. In North America, they are absent only from arctic regions.

### Natural history —
Mimetids are spider-predators, and share a highly adapted suite of predatory behaviors. Mimetids typically stalk orbicularians (e.g., Araneidae, p. 68, Theridiidae, p. 235, and related families) on their own webs, attacking the prey spider on its leg. Mimetids will opportunistically eat insects, but their venom is specialized to kill spiders rapidly. They have been observed eating the egg sacs of their host/prey spiders.

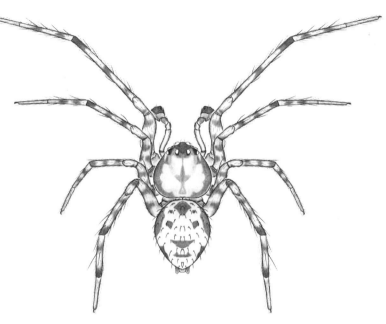

**Fig. 38.1** *Ero canionis* CHAMBERLIN & IVIE 1935b

### Taxonomic history and notes —
Worldwide diversity: 12 genera and 154 species. There are 10 undescribed species in North America. The genus *Mimetus* was described in the Epeiridae (=Araneidae) by Hentz (1832). The group was given family status by Simon (1890a), who had previously (Simon 1881a) placed the mimetid genera in the Theridionidae (=Theridiidae). Simon (1895a) noted that the peg teeth are a character shared with the Archaeidae, and also noted similarities to Theridiidae and Nesticidae. Mimetids were generally considered to be related to orbicularians until Forster & Platnick (1984) proposed a placement in the Palpimanoidea (which includes the archaeids). This placement was accepted by Coddington & Levi (1991) and others, but rejected by Schütt (2000) and others. The generic placement of *Reo eutypus* (CHAMBERLIN & IVIE 1935b) by Brignoli (1979k) is also disputed (Mott 1989), and the species probably belongs in *Mimetus*.

### Genera —
*Ero* C.L. KOCH *in* HERRICH-SHÄFFER 1836b, *Mimetus* HENTZ 1832, *Reo* BRIGNOLI 1979k

# Mimetidae

**Key to genera —**
North America North of Mexico

1  Chelicerae without anteromesal macrosetae, clypeus as high or higher than length of MOQ (Fig. 38.5) ....... **Ero**
   **Div.** 4 species — **Dist.** widespread — **Refs.** Chamberlin & Ivie 1935b, Archer 1941a, Griswold 1977, Shear 1981, Paquin & Dupérré 2003

—  Chelicerae with anteromesal macroseta, clypeus less than one half length of MOQ (Fig. 38.6) ............................. **2**

2(1)  Body dark and occurring in CA to NV; epigynum with very small lateral lobes (Fig. 38.7); palpus with sickle-shaped terminal apophysis AND terminal cymbial projection (Fig. 38.8) .......................................... **Reo**
   **Div.** 1 species: *Reo eutypus* (Chamberlin & Ivie 1935b) — **Dist.** CA and NV to Baja (MEX) — **Refs.** Chamberlin & Ivie 1935b, Brignoli 1979k

—  Body pale or, if dark, occurring east of NM (*Mimetus syllepsicus* Hentz 1832); epigynum without lateral lobes (Fig. 38.10); palpus without sickle-shaped terminal apophysis (Fig. 38.11), if terminal apophysis sickle-shaped, then cymbium lacking apical projection (Fig. 38.9, *Mimetus nelsoni* Archer 1950) .......................... **Mimetus**
   **Div.** 14 species + 10 undescribed, mostly in SW states — **Dist.** widespread — **Refs.** Chamberlin 1923, Chamberlin & Ivie 1935b, Gertsch & Mulaik 1940, Archer 1941a, 1950, Mott 1989, Paquin & Dupérré 2003

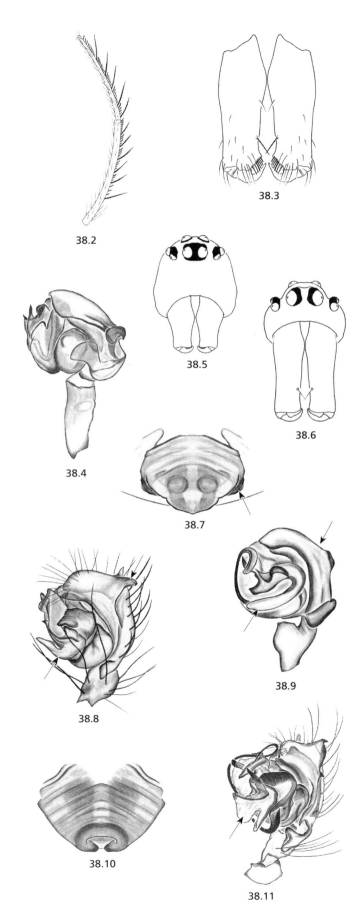

# Chapter 39
# MITURGIDAE
4 genera, 12 species

Darrell Ubick
David B. Richman

**Similar families —**
Clubionidae (p. 77), Lycosidae (p. 164).

**Diagnosis —**
Small to medium-sized (5-18 mm), ecribellate, entelegyne, eight-eyed spiders distinguished from other North American species in having two tarsal claws and claw tufts and the PLS with apical segment conical to elongate.

**Characters —**
**body size:** 5.0-18.0 mm.
**color:** yellowish to brown.
**carapace:** longer than wide.
**sternum:** oval, longer than wide; precoxal triangles present (*Cheiracanthium* and *Strotarchus*) or absent (*Syspira* and *Teminius*).
**eyes:** eight in two rows; PER usually straight (recurved in *Syspira*).
**chelicerae:** usually stout (especially in males), margins with teeth.
**mouthparts:** endites long (twice the width), labium longer than wide.
**legs:** prograde; tarsi with 2 claws, tufts large to weakly developed.
**abdomen:** ovoid to elongate, with or without markings.
**spinnerets:** ALS conical and nearly contiguous; PLS with distal segment conical to elongate.
**respiratory system:** one pair of book lungs; tracheal spiracle slightly anterior to spinnerets.
**genitalia:** entelegyne; *female* epigynum typically with small median plate; *male* palpus with RTA, cymbium modified in Eutichurinae (basal projection in *Cheiracanthium*, apical elongation in *Strotarchus*), bulb with median apophysis in Miturginae.

**Distribution —**
*Cheiracanthium* is widespread, *Strotarchus piscatorius* (Hentz 1847) occurs along the eastern coast, and all others are southern.

**Natural history —**
Miturgids are wandering hunters and live in forests, scrub, and rocky deserts. Most are ground dwellers and found under rocks and plant detritus. *Cheiracanthium*, however, which has large claw tufts, is largely to exclusively arboreal. It is also the best known genus, being frequently associated with humans. The life cycle of *Cheiracanthium inclusum* (Hentz 1847) was studied by Peck & Whitcomb (1970). This species may be common in certain crops and is of agricultural interest as it preys largely on caterpillars and flies and can recognize insect eggs as food (Peck & Whitcomb 1970, Buschman et al. 1977, Richman et al. 1980). *Cheiracanthium* is also common in houses, where it constructs white sac-like retreats, which may vary in form depending on their intended use: resting, molting, mating, breeding, or hibernating (Dippenaar-Schoeman & Jocqué 1997). The genus is also of some medical significance as the bite of *Cheiracanthium* may be painful although not serious (Foelix 1996).

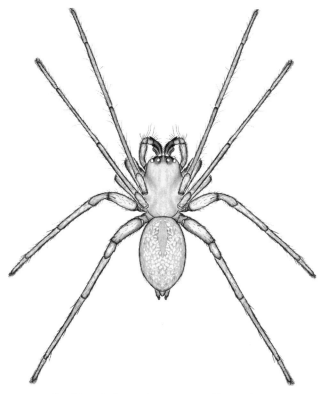

**Fig. 39.1** *Cheiracanthium inclusum* Hentz 1847

**Taxonomic history and notes —**
These spiders were originally placed in Clubionidae, as group Miturgeae of the subfamily Liocraninae (Simon 1897a). The group was elevated to family status by Lehtinen (1967) who recognized several subfamilies, most of which have been themselves elevated to family or transferred elsewhere, to where only the Miturginae and Eutichurinae currently remain. The composition of the subfamilies, and especially the placement of the Eutichurinae is hotly debated. Although most workers consider them miturgids (Platnick & Shadab 1989, Bonaldo & Brescovit 1992, Ramírez et al. 1997, Dippenaar-Schoeman & Jocqué 1997) others argue for returning them to Clubionidae (Deeleman-Reinhold 2001, Raven & Stumkat 2003). Placed by Lehtinen (1967) in the Amaurobioidea, the miturgids are now generally included somewhere in the Lycosoidea (Griswold 1993, Silva Dávila 2003).

The Nearctic species of *Cheiracanthium* and *Strotarchus* were revised by Edwards (1958); those of *Teminius* by Platnick & Shadab (1989). Species of *Syspira* were described by Chamberlin (1924b) and studied by Olmstead (1975).

# Miturgidae

**Genera —**
MITURGINAE
*Syspira* SIMON 1895e, *Teminius* KEYSERLING 1887b
EUTICHURINAE
*Cheiracanthium* C.L. KOCH 1839a, *Strotarchus* SIMON 1888b

## Key to genera —
### North America North of Mexico

1. Carapace with dark paramedian markings (Fig. 39.2); PER recurved .................................................. ***Syspira***
   **Div.** 4 species [Figs. 39.2-39.5] — **Dist.** s CA-w TX — **Refs.** Chamberlin 1924b, Simon 1897a — **Note** Olmstead (1975), in an unpublished thesis, recognizes three species, including a new one, and synonymizes and transfers some of the nominal species

— Carapace concolorous; PER straight to slightly procurved .................................................................... 2

2(1). Legs relatively short and stout, PT/C less than 1.3; anterior tarsi and metatarsi strongly scopulate (scopular setae longer than tarsal diameter) (Fig. 39.6); precoxal triangles absent .................................................. ***Teminius***
   **Div.** 2 species — **Dist.** *Temenius affinis* BANKS 1897 [Figs. 39.6-39.9]: OK-NM-s TX, LA; *Temenius insularis* (LUCAS 1857a): FL — **Ref.** Platnick & Shadab 1989 — **Note** these species were previously included in the African genus *Syrisca* SIMON 1885d

— Legs relatively long and slender, PT/C more than 1.4; tarsi and metatarsi weakly scopulate (scopular setae much shorter than tarsal diameter) (Figs. 39.10, 39.14); precoxal triangles present .................................................................... 3

3(2). Fovea present; claw tufts weakly developed, smaller than claws (Fig. 39.10); PLS with distal segment subequal to basal; male cymbium apically produced (Figs. 39.11-39.12) .................................................. ***Strotarchus***
   **Div.** 2 species — **Dist.** *Strotarchus piscatorius* (HENTZ 1847) [Figs. 39.10-39.13]: MA to FL; *Strotarchus planeticus* EDWARDS 1958: TX to AZ — **Ref.** Edwards 1958 — **Note** Roth 1994 mentions an additional two undescribed species.

— Fovea absent; claw tufts strongly developed, larger than claws (Fig. 39.14); PLS with distal segment much shorter than basal; male cymbium basally produced (Figs. 39.15-39.16) .................................................. ***Cheiracanthium***
   **Div.** 2 species: *Cheiracanthium mildei* L. KOCH 1864 [Figs. 39.14-39.17] *Cheiracanthium inclusum* (HENTZ 1847) — **Dist.** widespread in USA and s CAN — **Refs.** Edwards 1958, Dondale & Redner 1982 — **Note** *Chiracanthium* is an unjustified emendation of the original spelling.

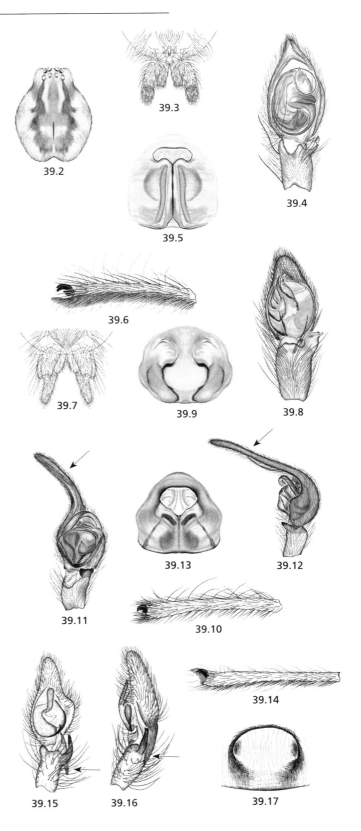

## Chapter 40
# MYSMENIDAE
5 genera, 6 species

Lara Lopardo
Jonathan A. Coddington

### Similar families —
Anapidae (p. 64), Symphytognathidae (p. 226), Theridiidae (p. 235), Theridiosomatidae (p. 244).

### Diagnosis —
This orbicularian family can be distinguished by the ventral, subapical, sclerotized spot on the femur of at least leg I on both sexes (Fig. 40.8); a prolateral clasping spine on male metatarsus or tibia I (or both) (Figs. 40.3, 40.5, 40.7); apically twisted cymbium and long embolus on male palps (Griswold *et al.* 1998, Schütt 2003); spinnerets near the genital opening; and elevated carapace.

### Characters —
**body size:** males 0.5-1.4 mm; females 0.8-2.0 mm.
**color:** abdomen dark (orange in *Maymena*) with yellowish dots or other distinct pattern (Figs. 40.1, 40.4, 40.6). Legs usually with stripes of contrasting colors, uniform in *Maymena* and *Parogulnius hypsigaster* ARCHER 1953. Sternum usually yellowish, dark in *Mysmenopsis* and *Trogloneta paradoxa* GERTSCH 1960a, and carapace orange (except *Mysmenopsis* dark; and *Trogloneta paradoxa* with two darker spots). The abdomen in *Trogloneta paradoxa* has a white pattern posteriorly (Fig. 40.4).
**carapace:** carapace short and high (as long as wide and high); remarkably high in males of *Microdipoena* and *Trogloneta*, with the eyes grouped in a tubercle, and the clypeus concave (Figs. 40.4, 40.6).
**sternum:** sternum convex in lateral view in most mysmenid species. Sternum usually truncated posteriorly, pointed in *Maymena*.
**eyes:** eight eyes in two rows; eyes white except AME.
**chelicerae:** cheliceral basal segment not tapering. Tiny denticles on cheliceral furrow (difficult to see).
**legs:** prograde, tarsi with three claws. Males have one prolateral clasping spine on mid metatarsus I, and/or one or two on distal tibia I (Figs. 40.3, 40.5, 40.7).
**abdomen:** globose, except in *Trogloneta paradoxa*, which is pointed posteriorly (Fig. 40.4).
**spinnerets:** six spinnerets with small, fleshy colulus.
**respiratory system:** diverse but poorly studied in the family: *Microdipoena* with two vestigial book lungs and two independent posterior tracheal spiracles; *Calodipoena* and *Mysmenopsis* with two anterior spiracles open to tracheae joined by a transverse duct and two posterior tracheal spiracles connected transversely in an atrium (Forster 1959); *Maymena* with two anterior book lungs and one posterior tracheal spiracle (Gertsch 1960a); the respiratory system of *Trogloneta* is unknown.
**genitalia:** entelegyne; *female* epigynum weakly sclerotized in *Microdipoena guttata* BANKS 1895d, *Parogulnius hypsigaster* and *Calodipoena incredula* GERTSCH & DAVIS 1936; a sclerotized plate is present in *Trogloneta paradoxa*, *Mysmenopsis cymbia* (LEVI 1956a), and *Maymena ambita* (BARROWS 1940)(Fig. 40.11). A finger-like scape extends posteriorly in *Calodipoena incredula* and *Parogulnius hypsigaster* (Fig. 40.12); *male* palp is complex, embolus long and usually coiled (Fig. 40.9), can be distally twisted as in *Maymena ambita* and *Microdipoena guttata* (Figs.

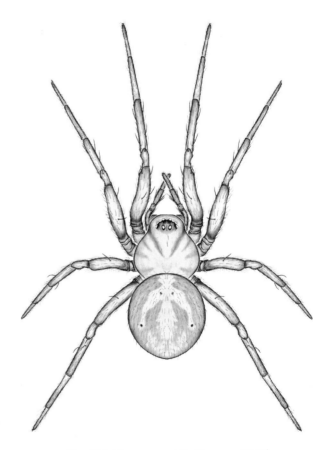

**Fig. 40.1** *Maymena ambita* (BARROWS 1940)

40.2, 40.10, respectively); cymbium with lobes or apophyses related to the embolus. Conductor and median apophysis absent (Coddington 1990).

### Distribution —
Mysmenids have been poorly collected. *Maymena ambita* has been reported only from Southeastern USA, although the genus also occurs in Central and Northern South America. *Microdipoena guttata* is widespread throughout Eastern USA, and Central and South America, but the rest of *Microdipoena* species occur mainly in Africa. *Calodipoena incredula* occurs in Texas and Florida, the Bahamas, and Cuba, and the genus has been reported from Central America and Africa. *Mysmenopsis cymbia* occurs only in Florida, other *Mysmenopsis* species are Neotropical. *Trogloneta* (including *Parogulnius*) is widespread in the Northern Hemisphere, *Trogloneta paradoxa* occurs in west (California, Utah, and Oregon) and western Canadian Columbia; R. Bennett, pers. comm.), and *Parogulnius hypsigaster* has been reported only from Alabama.

## Natural history —

Mysmenids mainly occur in leaf litter and other cryptic places in very humid habitats. North American web-spinning species seem to prefer the interstices of leaf litter and like habitats, i.e. small cavities perhaps 5-10cm in diameter (depending on the size of the spider) created by the top layer of leaves. The cavities usually open to the air on one side, and the webs presumably trap prey entering or leaving such cavities in the leaf litter or detritus. *Trogloneta paradoxa* has been reported from ground detritus, fir needles, and *Neotoma* (bushytail woodrat) midden (Gertsch 1960a). *Maymena ambita* was also collected from pine needles (Barrows 1940), broadleaf forest leaf litter and caves (Coddington, pers. obs.). *Microdipoena guttata* was found under dead leaves in dry woods (Banks 1895d, Levi 1956a), and at edges of swamps (Bishop & Crosby 1926). *Calodipoena incredula* has been collected in the soil (Levi 1956a). Many species of *Mysmenopsis* are kleptoparasites on the webs of other spiders (Platnick & Shadab 1978a), but the natural history of *Mysmenopsis cymbia* is still unknown. Mysmenid web architecture is known for a few species of *Maymena*, *Mysmena*, and *Microdipoena guttata*. Kleptoparasitic species are not known to build webs of their own and some have lost the ability to spin the sticky silk characteristic of orbweaving spiders. *Mysmena* species build highly derived three-dimensional orb webs, with a proliferation of out-of-plane radii that result in a characteristic spherical web (Fig. 40.13, Forster 1959). The out-of-plane radii result from the failure of the animal to remove radial lines constructed during exploration of the web site. Orbweavers normally retain only the exploratory lines that fall roughly in a plane. Standard frame and radius construction result in further eccentric radii, but sticky spiral construction is very irregular, resulting in a rind or layer of sticky silk spanning just the perimeter of the web. Radii near the hub are bare of sticky spiral attachments, so that if the outer sticky layer is perforated, the spherically symmetric radial lines are revealed (Fig. 40.14). The web of *Microdipoena guttata* is similar to that of *Mysmena* species (Fig. 40.15). In contrast, *Maymena ambita* and Central American *Maymena* webs are mainly planar but the hub is distorted upwards by one to several out-of-plane radial lines that attach to substrate above the web (Fig. 40.16). During sticky spiral construction, *Maymena* occasionally exit on these eccentric radii to make the next attachment, which results in sporadic sticky silk segments rising vertically out of the plane of the web. Presumably these eccentric sticky elements allow the web to catch more horizontally flying prey. The webs also contain "accessory radii" that are spun after completion of the sticky spiral but before the hub is remodeled. Accessory radii are easily distinguished because the sticky spiral contacts but does not "kink" at the junction with accessory radii (because there is no attachment), whereas the sticky spiral does kink when crossing true radii constructed during the exploration of the web site, or during frame or radius construction (Coddington, pers. obs.). Anapids make similar webs, not only in structure but also in the behavior used to construct the webs (Coddington 1986b, Eberhard 1987a, Griswold et al. 1998).

## Taxonomic history and notes —

Mysmenidae was first described as the group Mysmeneae within the Theridiidae (Simon 1922, 1926), and comprised only three genera: *Mysmena* Simon 1894a, *Cepheia* Simon 1894a and *Trogloneta* Simon 1922. The group was diagnosed mainly on the absence of female palpal claws, the voluminous male bulb with long embolus, the globular or conical abdomen with sparse, rather long hairs, and the inferior tarsal claws as long as the superior claws. Petrunkevitch (1928b) united Simon's Mysmeneae and Theonoeae in the subfamily Mysmeninae, thus adding six genera to the original group: *Epecthina* Simon 1895a, *Epecthinula* Simon 1903d, *Iardinis* Simon 1899a, *Mysmenopsis* Simon 1897d, *Synaphris* Simon 1894a, and *Theonoe* Simon 1881a. Forster (1959) transferred *Mysmena* (and consequently the entire subfamily) from the Theridiidae to the Symphytognathidae. Unaware of Forster's change, Gertsch (1960a) independently transferred Mysmeninae from the Theridiidae to the Symphytognathidae, and proposed new diagnostic features for the mysmenines, none of which are considered reliable by contemporary workers (e.g., size of the spiders from small to minute, with or without book lungs, pedipalps present and of normal size in females, lack of comb in the hind tarsi, and metatarsi longer or of equal size than tarsi). In 1977, Forster and Platnick concluded that the mysmenids were sufficiently distinct from the Symphytognathidae to warrant family rank. A year later, Platnick & Shadab (1978a) provided a diagnosis of the Mysmenidae, although based only on genera from the New World. Wunderlich (1986) (*contra* Forster & Platnick 1977) defended the monophyly of Symphytognathidae *sensu* Forster (1959), and split the mysmenids into two subfamilies: Mysmeninae and Synaphrinae. Two higher level phylogenetic analyses have included mysmenids. In an analysis of orbicularian relationships, Griswold et al. (1998) included two mysmenid genera: *Mysmena* and *Maymena*. Schütt (2003) included three mysmenid genera (*Trogloneta*, *Microdipoena* and *Cepheia*) in her study of symphytognathid relationships, and raised Synaphrinae to family level. Marusik & Lehtinen (2003) independently raised Synaphrinae to family rank based on morphology. Mysmenidae currently comprises 91 described species in 22 genera (Platnick 2005).

The monotypic genus *Parogulnius* Archer 1953 is troublesome. It was first described in Theridiosomatidae, but Gertsch (1960a) synonymized it with the mysmenid *Trogloneta*. Brignoli (1970b) rejected the synonymy, thus restoring it to theridiosomatids. Coddington (1986a) stated that it certainly was not a theridiosomatid, but left it *incertae sedis*. *Parogulnius hypsigaster* exhibits some diagnostic features of Mysmenidae and many of *Trogloneta*. We agree with Gertsch's synonymy and so include it here, keeping in mind that its placement is still dubious. In the following it will key out to *Trogloneta*.

## Genera —

*Calodipoena* Gertsch & Davis 1936, *Maymena* Gertsch 1960a, *Microdipoena* Banks 1895d, *Mysmenopsis* Simon 1897d, *Trogloneta* Simon 1922

# Spiders of North America — Mysmenidae | 177

40.13

40.14

40.15

40.16

## Key to genera —
### North America North of Mexico

1. AME equal to or larger than ALE (Figs. 40.3, 40.7), abdomen globose .................................................. **2**
— AME smaller than ALE (Fig. 40.5), one ventral sclerotized spot on femur I in both sexes, and high concave clypeus (Fig. 40.4) and long prolateral clasping spine on distal end of middle third of metatarsus I in males (Fig. 40.5) .......... ................................................................. ***Trogloneta***
   **Div.** 2 species — **Dist.** *Trogloneta paradoxa* GERTSCH 1960a: CA, UT and OR, n to BC and *Paroguinius hypsigaster* ARCHER 1953: AL — **Refs.** Archer 1953, Gertsch 1960a, Brignoli 1970b, 1980j — **Note** see Taxonomic history and notes for the generic placement of *Paroguinius hypsigaster*

2(1). One ventral sclerotized spot on femora I and II at least in females (Fig. 40.8) (difficult to see in *Mysmenopsis*), female genitalia not extended posteriorly, embolus surrounding tegulum in ventral view (Figs. 40.2, 40.10) ..... **3**
— One ventral sclerotized spot on femur I in both sexes, female genitalia with finger-like posteriorly extended scape (as in Fig. 40.12), male with prolateral clasping spine on distal end of middle third of metatarsus I and embolus spirally coiled from center of tegulum in ventral view (Fig. 40.9) .................................................................. ***Calodipoena***
   **Div.** 1 species: *Calodipoena incredula* GERTSCH & DAVIS 1936 — **Dist.** TX and FL — **Refs.** Gertsch & Davis 1936, Brignoli 1980j, Baert 1984a

3(2). Prolateral distal clasping spine on tibia I in males present, metatarsal clasping spine can be absent; chelicerae strong, longer than height of clypeus; embolus twisted distally ... ........................................................................................ **4**
— Prolateral clasping spine on tibia I in males absent, only on metatarsus I; female chelicerae small, shorter than height of clypeus .................................................. ***Mysmenopsis***
   **Div.** 1 species: *Mysmenopsis cymbia* (LEVI 1956a) — **Dist.** known only from FL — **Refs.** Gertsch 1960a, Platnick & Shadab 1978a, Brignoli 1980j

4(3). One very strong distal prolateral clasping spine on tibia I, arising from a spur (Fig. 40.3); metatarsal clasping spine absent; both sexes with one ventral sclerotized spot on femora I and II (the second lighter, Fig. 40.8) .................. ................................................................. ***Maymena***
   **Div.** 1 species: *Maymena ambita* (BARROWS 1940) — **Dist.** KY to AL, GA and VA — **Refs.** Gertsch 1960a, Brignoli 1980j
— Two distal prolateral clasping spines on tibia I and one metatarsal clasping spine basally on middle third (Fig. 40.7); females with one ventral sclerotized spot on femora I and II (equally sclerotized) and males with one weakly sclerotized femoral spot on femur I only; clypeus very high and concave in males (Fig. 40.6) .... ***Microdipoena***
   **Div.** 1 species: *Microdipoena guttata* BANKS 1895d — **Dist.** e USA, MA, NY, NJ, MO e to NC, se to FL — **Ref.** Brignoli 1980j

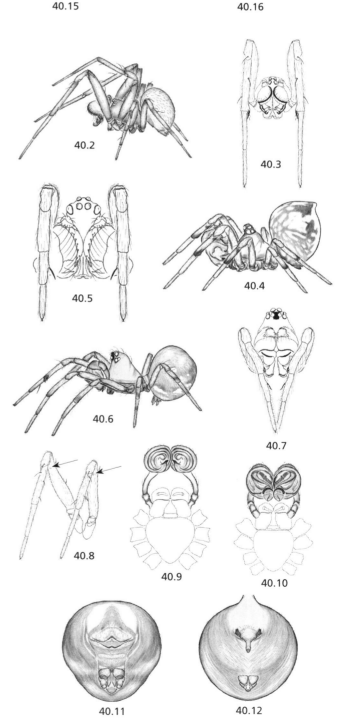

Chapter 41

# NESTICIDAE
3 genera, 38 species

Pierre Paquin
Marshal Hedin

**Similar families** —
Linyphiidae (p. 124), Theridiidae (p. 235).

**Diagnosis** —
Nesticidae are small to medium sized ecribellate spiders. They are distinguished by the combination of a rebordered labium (Fig. 41.4), the presence of a ventral serrated comb on tarsus IV (Fig. 41.2), and a male palpus with a rigid basal paracymbium (Figs. 41.7, 41.11, 41.14).

**Characters** —
**body size:** 1.8-10.0 mm.
**color:** abdomen grayish to dirty white; paler species decorated with darker spots or stripes and darker species may show lighter maculations; troglomorphic species typically paler; legs light orange to yellowish.
**carapace:** longer than broad, rounded on its side (Fig. 41.1); clypeus high; fovea shallow and round.
**sternum:** subtriangular, slightly longer than broad (Fig. 41.4)
**eyes:** 8 eyes in 2 rows, AME typically smaller; 6 eyes, AME absent; both AME and lateral eyes greatly reduced or absent.
**chelicerae:** stout, subparallel and slightly narrowed at apex; promargin with 3 teeth (1 long, 2 short in *Gaucelmus*); retromargin bearing 6-12 small teeth; male *Gaucelmus* with conspicuous bulbous enlargement at base of fang (Fig. 41.5).
**mouthparts:** labium rebordered, twice as long as broad; endites subparallel (Fig. 41.4).
**legs:** 3-clawed, long, usually 1432 rarely 1243; without true spines but bearing long dorsal setae on tibia and patella; sensory setae are present on tarsi and metatarsi; tarsi IV with a ventral comb of curved bristles (Fig. 41.2); female palpus with long claw.
**abdomen:** globular (Figs. 41.1, 41.3).
**spinnerets:** six spinnerets and colulus; anterior lateral and posterior lateral spinnerets subconical and two-segmented; posterior median spinnerets one-segmented, smaller, hidden in spinneret cluster.
**respiratory system:** two book lungs; tracheal spiracle opening in front of spinnerets.
**genitalia:** entelegyne; **female** epigynum variable, either protruding or inconspicuous, with a median septum (*Nesticus* Figs. 41.16-41.17, *Eidmannella* Fig. 41.12) or inconspicuous (*Gaucelmus* Fig. 41.8); spermathecae usually visible through cuticule; **male** palpal tibia without apophysis. Cymbium bearing a well developed paracymbium (Figs. 41.7, 41.11, 41.14); conductor tapered distally; embolus long and slender (Figs. 41.6, 41.10, 41.13).

**Distribution** —
Temperate Nesticidae show marked affinities for caves or cave-like (cool, moist, dark) habitats near the surface, and this largely influences their distribution. Several *Eidmannella* species are Texas cave endemics, and *Eidmannella pallida* (EMERTON 1875) is widespread throughout USA, Mexico, Central America and the West Indies. The northernmost records of this species are from s Ontario (Gertsch

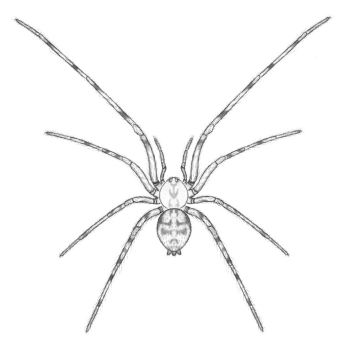

**Fig. 41.1** *Eidmannella pallida* (EMERTON 1875)

1984) and Québec (Paquin *et al.* 2001). *Gaucelmus augustinus* KEYSERLING 1884a is found in Florida and Gulf states to Texas, throughout Mexico to Panama, and in the West Indies. The North American *Nesticus* are found in five distinct regions (Gertsch 1984), three of which are Nearctic. The northeastern fauna only includes *Nesticus cellulanus* (CLERCK 1757), an introduced and synanthropic species. Its distribution is restricted to the East Coast and the northernmost records are from Québec (Paquin & Dupérré 2003). The Appalachian region is the richest with more than 26 cave-dwelling and surface species. The western fauna includes three species, *Nesticus silvestrii* FAGE 1929c found both in caves and in favorable surface habitats in California to s British Columbia, whereas the two other species are known only from a few California caves.

**Natural history** —
Nesticidae are troglophiles and some species show clear morphological evidence for an obligate cave-dwelling existence (e.g., eyeless, pale coloration, relatively long-legged, and typically large in size). Surface species occur in deep litter, under stones in boulder fields, and in crevices of rocky cliffs near creeks and streams. *Gaucelmus augustinus* has been recorded from cellars but is also a troglophile (Cokendolpher & Reddell 2001a). *Eidmannella pallida* is cosmopolitan (Lehtinen & Saaristo 1980), while *Nesticus cellulanus* is introduced and synanthropic (Gertsch 1949, 1984). Nesticids build webs similar to those of Theridiidae, with a tangled retreat suspended above sticky verti-

cal lines, and typical single vertical threads that split near the end. The webs usually span small spaces in appropriate microhabitats. Egg sacs of *Nesticus* are carried by the female with their chelicerae or spinnerets. Collections are seemingly biased towards adult females and juveniles, probably because of the more cryptic wandering lifestyle of adult males and probably shorter lives.

Reeves (2000) documented the distribution of the troglobite *Nesticus barrowsi* GERTSCH 1984 in the caves of the Great Smoky Mountains National Park and Hedin & Dellinger (2005) discuss the precarious status of a few Appalachian *Nesticus* species. Cokendolpher & Reddell (2001a) comment on similar status for *Eidmannella* spp. in Texas caves.

## Taxonomic history and notes —

The Nesticeae were proposed by Simon (1894a) in the subfamily Tetragnathinae of the family Argiopidae, and elevated to family by Dahl (1926). Gertsch (1971a) removed *Gaucelmus* from the synonymy of *Nesticus*, and later (1984) confirmed the occurrence of three genera in North America. Lehtinen & Saaristo (1980) proposed a worldwide revision of the genera.

Three species were described by early arachnologists: *Nesticus cellulanus*, *Eidmannella pallida*, and *Gaucelmus angustinus*, and only three *Nesticus* were described in the following century [Petrunkevitch (1925b), Chamberlin (1933b), and Bishop (1950)]. In 1984, Gertsch published an important revision of the family in which he described the bulk of our fauna: 5 new *Eidmannella* and 23 new *Nesticus* species. One cave and one surface species were described by Coyle & McGarity (1991) from the Appalachian region. Recent collecting efforts in caves and surface habitats of the southern Appalachians has resulted in the discovery of previously undescribed males and additional new species (Hedin & Dellinger 2005). Important collections from Texas caves led to the discovery of *Eidmannella tuckeri* COKENDOLPHER & REDDELL 2001a and provide a better understanding of the distribution of *Eidmannella* spp. and *Gaucelmus angustinus* (Cokendolpher & Reddell 2001a).

Nesticidae are proposed as the sister family of Theridiidae (Coddington 1986b). Characters that support this phylogenetic placement are: the use of viscid silk to wrap prey, the presence of a ventral comb on tarsus IV, and their similar webs. Hedin (1997a, b) provides data on speciation mechanisms and phylogenetics of Appalachian *Nesticus*.

## Genera —

*Eidmannella* ROEWER 1935, *Gaucelmus* KEYSERLING 1884a, *Nesticus* THORELL 1869.

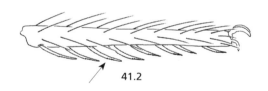

41.2

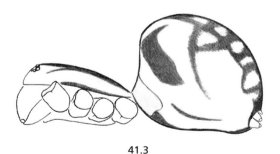

41.3

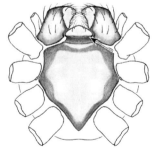

41.4

# Nesticidae

**Key to genera —**
North America North of Mexico

1. AME 80% of the diameter of ALE (Fig. 41.5); fang of male greatly enlarged at base (Fig. 41.5); genitalia as in Figs. 41.6-41.8 .................................................. **Gaucelmus**
   **Div.** 1 species: *Gaucelmus angustinus* Keyserling 1884a — **Dist.** Gulf states, FL to TX, MEX — **Refs.** Gertsch 1984, Cokendolpher & Reddell 2001a

— AME, when present, about 33% of the diameter of ALE (Fig. 41.9); fang of male barely enlarged at base; genitalia otherwise ........................................................... 2

2(1). Male palpus with short, curved paracymbium (Figs. 41.10-41.11); epigynum with a small inverse t-shaped septal element (Fig. 41.12) ......................... **Eidmannella**
   **Div.** 7 species — **Dist.** TX, except for *Eidmannella pallida* (Emerton 1875) [Figs., 41.1, 41.10-41.12] which is widespread — **Refs.** Gertsch 1984, Cokendolpher & Reddell 2001a — **Note** the degree of troglomorphy and the coloration pattern of *Eidmannella pallida* (Fig. 41.1) is variable.

— Male palpus with large paracymbium (Figs. 14.13 -14.15) that branches (Fig. 41.14) except for *Nesticus cellulanus* (Fig. 41.15); epigynum prominent occupying most of the genital groove (Figs. 41.16-41. 17) ..................... **Nesticus**
   **Div.** 30 + several undescribed species — **Dist.** the Holarctic *Nesticus cellulanus* (Clerck 1757) [Figs. 41.15, 41.17] is introduced on the ne Coast: NY, MA, ME, NS and QC; Appalachian group: 26 species; Californian group: 3 species n CA, w OR, w WA and s BC — **Refs.** Gertsch 1984, Coyle & McGarity 1991, Hedin 1997a, Hedin & Dellinger 2005.

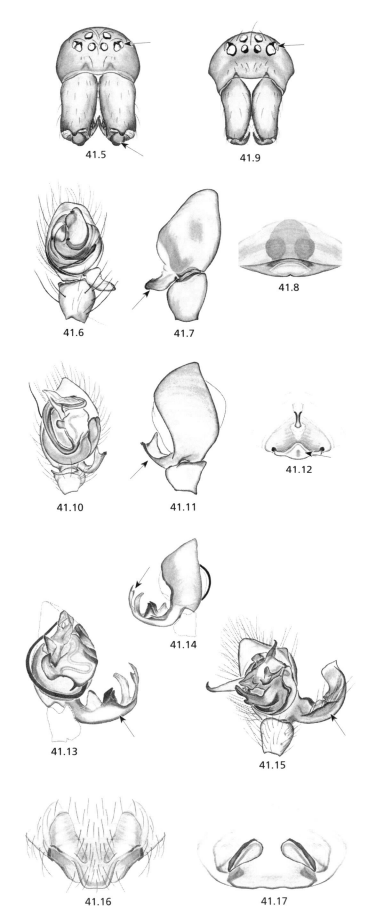

# Chapter 42
# OCHYROCERATIDAE
1(?) genus, 2 species

Pierre Paquin
Darrell Ubick

## Similar families —
Leptonetidae (p. 122), Telemidae (p. 228).

## Diagnosis —
Ochyroceratidae are small ecribellate haplogyne spiders that are distinguished by the combination of tarsi with three claws on an onychium (Fig. 42.2), chelicerae with a lamina (Fig. 42.3), 6 eyes with ALE and PME forming a straight row, colulus large and fleshy, and an abdomen lacking anterior sclerotized ridge. The genital openings of females of some species can extend laterally on the abdomen (Fig. 42.4).

## Characters —
**body size**: 1.0-2.0 mm.
**color**: cephalothorax dark brown, occasionally patterned, abdomen sometimes with purplish to greenish mottling.
**carapace**: as long as wide, clypeus almost vertical.
**sternum**: apically truncate.
**eyes**: six eyes (AME absent) in three diads, PME anteriorly displaced forming a straight row with ALE.
**chelicerae**: short, not fused, bearing a lamina (Fig. 42.3); promargin 6-7 teeth, retromargin 2 teeth.
**mouthparts**: tip of labium excised; endites converging.
**legs**: 3-clawed, short and spineless; tarsi with 3 claws on an onychium (Fig. 42.2); autospasy between coxa and trochanter.
**abdomen**: globular to elongate.
**spinnerets**: anterior spinnerets long with a short distal segment; median spinnerets with a single spigot; colulus longer than wide (Fig. 42.5).
**respiratory system**: book lungs absent, replaced by tracheae; posterior tracheal spiracle halfway between epigastric furrow and spinnerets.
**genitalia**: haplogyne; *female* often with external sclerotized copulatory ducts, spermathecae paired and close together; *male* palpus with tibia often enlarged and with apophysis; bulb typically large and spherical with a terminal embolus, sometimes with accessory processes.

## Distribution —
Ochyroceratidae are cosmotropical and known in NA from s Florida to South America. *Theotima minutissima* (PETRUNKEVITCH 1929b) has a transpacific distribution, being found in both Asian and American tropics (Shear 1976, Reddell 1981, Dumitrescu & Georgescu 1992, Deeleman-Reinhold 1995d).

## Natural history —
Very little is known about North American Ochyroceratidae as most of our knowledge comes from Africa (de Barros Machado 1951) and Indonesia (Deeleman-Reinhold 1995d). Although some ochyroceratids are associated with caves, the majority of the species occur in forest litter where they spin small irregular webs. All Nearctic records are from deep litter samples (Shear 1976). In the tropics, ochyroceratids are one of the ecological counterparts of the north temperate Linyphiidae (Deeleman-Reinhold 1995d).

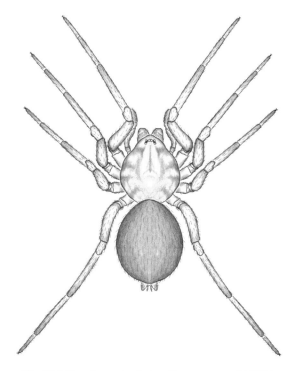

Fig. 42.1 *Theotima minutissima* (PETRUNKEVITCH 1929b)

*Theotima minutissima* is found in leaf litter typically at lower elevations. Females carry their eggs-usually a bundle of 4 to 6 eggs-in their chelicerae until they hatch (see Edwards *et al.* 2003). Recent studies demonstrate that this species reproduces parthenogenetically (Edwards *et al.* 2003), based on laboratory rearings and the absence of males in field samples. This supports earlier hypotheses about the occurrence of parthenogenesis in the family (de Barros Machado 1964, Deeleman-Reinhold 1995d). However, Bryant (1940) described an apparent male as *Theotima minutissima*, which was doubted by Shear (1976) and rejected by Saaristo (1998).

## Taxonomic history and notes —
Simon (1893a) included *Theotima* and *Ochyrocera* SIMON 1891c in the Leptonetidae until Fage (1912) erected the family Ochyroceratidae.

Fage & de Barros Machado (1951) pointed out that the family falls in two distinct morphological groups. Deeleman-Reinhold (1995d) proposed the Psilodercinae for one and Theotiminae for the "new subfamily" of Lehtinen (1986). The genus *Ochyrocera* cannot be placed in any of these two groups and constitutes its own subfamily: the Ochyroceratinae (Deeleman-Reinhold 1995d).

Shear (1976) reported two ochyroceratids from Florida but did not fully resolve their identities. He suspected that one is *Theotima minutissima*, and the other either *Theotima* or *Speocera* BERLAND 1914, two genera with species having external copulatory openings, as he observed.

Interestingly, a sample from the US Virgin Islands examined included 11 females of *Theotima minutissima* (Fig. 42.6) and a male, illustrated here (Figs. 42.7-42.8). The parthenogenetic reproduction known for *Theotima minutissima* does not preclude the existence of males, which may occur in extremely low abundance or be present only in certain conditions.

In short, a careful study is required to establish the identity of the species reported by Shear (1976) and to reassess the status of the male illustrated by Bryant and here.

### Genera —
THEOTIMINAE
  *Theotima* SIMON 1893a

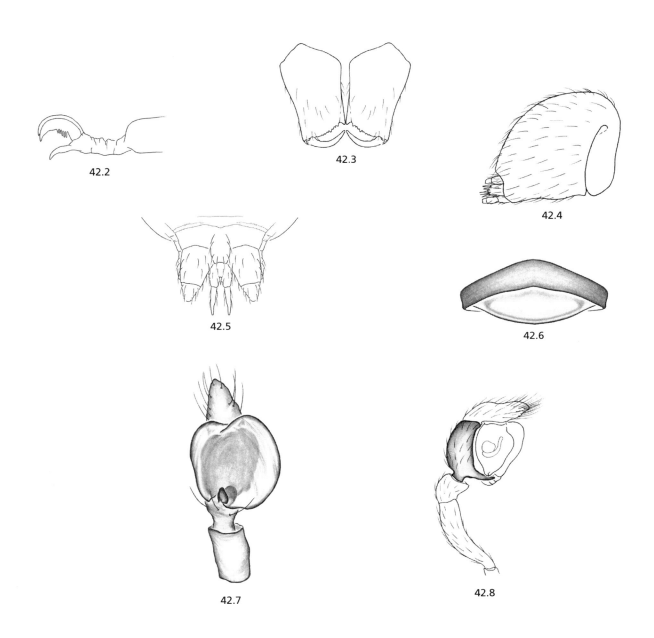

# Chapter 43
# OECOBIIDAE
2 genera, 8 species

Patrick R. Craig
Warren E. Savary
Darrell Ubick

### Similar families —
Oecobiids are unique and not likely confused with any other spiders, although they superficially resemble *Euryopis* (Theridiidae, p. 233).

### Diagnosis —
Oecobiids are distinguished from all other spiders in having a large anal tubercle fringed with long hairs (Fig. 43.2) and a subcircular carapace with eyes arranged in a central cluster. All North American species are cribellate.

### Characters —
**body size:** 0.90-3.2 mm.
**color:** brown to gray, rarely yellowish white; carapace usually with 3 pairs of submarginal maculations, sometimes concolorous brown, rarely pale; eyes ringed with black.
**carapace:** subcircular to heart-shaped; wider than long; fovea longitudinal, may be indistinct or absent.
**sternum:** heart-shaped, wider than long, posteriorly pointed, separating coxae IV; with marginal fringe of spatulate setae in males.
**eyes:** eight, arranged in a compact group near center of carapace; AME and PLE subcircular, others oval to triangular.
**chelicerae:** short, without condyle, lacking teeth; fangs short.
**mouthparts:** labium free, wider than long; endites coverging, almost touching, without scopula.
**legs:** subequal in length, curved away from body, covered with plumose hairs but with few or no spines; tarsi lacking trichobothria, metatarsi with one or two; metatarsus IV with calamistrum biseriate, extending at least 2/3 length of segment; tarsi with three claws; autospasy at the patella-tibia joint.
**abdomen:** oval to round, flattened; anal tubercle large, two-segmented, with subapical fringe of setae.
**spinnerets:** cribellum divided; ALS and PMS, contiguous; PLS two-segmented, distal segment long and curved.
**respiratory system:** one pair of book lungs; tracheal spiracle inconspicuous, near spinnerets.
**genitalia:** entelegyne; *female* epigynum diverse, usually with scape of variable thickness and bearing copulatory openings, rarely platelike and with anterior notch (*Platoecobius*); *male* palpus with tibia lacking apophyses; bulb with embolus of variable shape and with three additional tegular apophyses.

### Distribution—
*Oecobius cellariorum* (DUGÈS 1836a) and *Oecobius navus* BLACKWALL 1859b are widespread in the US and se Canada; *Oecobius interpellator* SHEAR 1970 has been recorded only from MA; other species of *Oecobius* are southern. *Platoecobius* occurs from South Carolina to Florida, with an apparently undescribed species from s California.

### Natural history —
*Oecobius* species are found on and under rocks and are especially common around human habitation, even in the case of native species. They typically lie concealed within

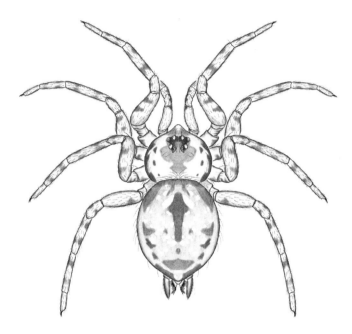

**Fig. 43.1** *Oecobius navus* BLACKWALL 1859b

small sheet webs placed over cavities. The webs are quite distinctive and consist of two horizontal sheets. The spider is positioned on the lower sheet and covered by the upper, which has radiating signal lines and gives the web a star-shaped appearance. Potential prey touching these lines flushes the spider out of its retreat. *Oecobius* captures prey by swathing it with silk while rapidly running around it in circles. The swathing band is combed out from the enlarged PLS by the specialized setae on the anal tubercle (Dippenaar-Schoeman & Jocqué 1997). *Platoecobius floridanus* (BANKS 1896e), on the other hand, is not known to construct a web and has been collected under pine bark scales (Chamberlin & Ivie 1935c).

The main prey of *Oecobius* appears to be ants. These spiders often form dense aggregations and some may be communal. Mating takes place in special webs constructed by the male adjacent to the web of the female. Females have a small clutch size (egg sacs contain three to ten eggs) but don't provide much maternal care (Shear 1970).

### Taxonomic history and notes —
The family Oecobiidae is so distinctive that it has been universally accepted since established by Blackwall (1862a). Its relationship to the genus *Uroctea* DUFOUR 1820c, however, has generated much controversy. As the latter lacks a cribellum, Simon (1892a) placed it in a separate family, Urocteidae, and considered it phylogenetically distant. This interpretation has been followed into modern times (Levi et al. 1968, Shear 1970, Kaston 1972a). However, given the striking similarity between the two taxa, they were combined into a single family (Millot 1931, Chamberlin & Ivie

1935c) or at least regarded as closely related (Petrunkevitch 1958). Thus, the oecobiids became the key evidence to challenge the validity of Simon's Cribellatae, which was eventually thoroughly dismantled by Lehtinen (1967). Currently, Oecobiidae includes both of the above taxa as well as recently discovered cribellate and ecribellate genera from southern Africa (Kullmann & Zimmermann 1976a, Lamoral 1981), which has the highest diversity of these spiders. Recent analysis places Oecobiidae near the ecribellate Hersiliidae and the largely cribellate and Old World Eresidae in the superfamily Eresoidea, which is the basalmost of all entelegyne araneomorph spiders (Griswold *et al.* 1999a).

Of its three subfamilies, only Oecobiinae occurs in the New World. Lehtinen recognized several genera on the basis of differences in genitalia and eye arrangement. Shear (1970), who revised the Nearctic species, synonymized most of these with *Oecobius*.

### Genera —
OECOBIINAE
    *Oecobius* Lucas 1846, *Platoecobius* Chamberlin & Ivie 1935c.

### Key to genera —
North America North of Mexico

1. Carapace with sides rounded (Fig. 43.3); legs long, tibia I about 6-7 X longer than wide; calamistrum extending 2/3 length of metatarsus IV; PLE largest eyes, PME separated by 1-2 diameters (Fig. 43.3), except for *Oecobius putus* O. Pickard-Cambridge 1876b (Fig. 43.4) .............. **Oecobius**
    Div. 7 species — Dist. widespread — Ref. Shear 1970 — Note *Oecobius navus* Blackwall 1859b [Figs. 43.5-43.6] (previously referred to as *Oecobius annulipes* Lucas 1846) and *Oecobius interpellator* Shear 1970 are introduced to the Nearctic region (Shear 1970)

— Carapace with sides sinuous (Fig. 43.7); legs shorter, tibia I about 4 X longer than wide; calamistrum extending entire length of metatarsus IV; AME largest, PME separated by 4 or more diameters (Fig. 43.7) .................... **Platoecobius**
    Div. 1 species: *Platoecobius floridanus* (Banks 1896e) [Figs. 43.8-43.9] — Dist. SC to FL — Ref. Shear 1970 — Note Roth (1994) mentions one undescribed species from CA

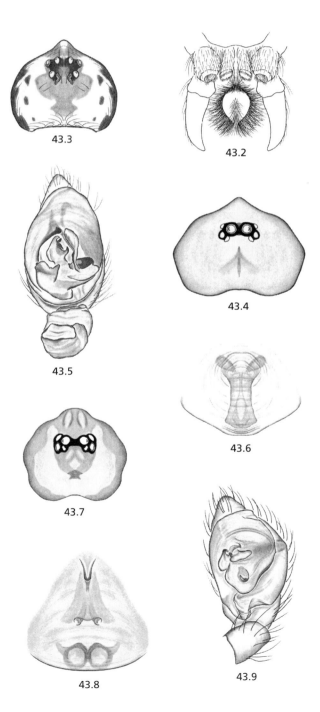

## Chapter 44
# OONOPIDAE
11 genera, 27 species

Darrell Ubick

### Similar families —
Linyphiidae (p. 124), Ochyroceratidae (p. 179), Telemidae (p. 228).

### Diagnosis —
These small yellowish to orange spiders can be distinguished from other haplogynes in having six eyes which are usually large and contiguous, an abdomen usually covered by scutes, tracheal spiracles which are paired and anteriorly positioned, chelicerae free and lacking teeth, tarsi with onychium bearing two claws, and the female palpi lacking a claw.

### Characters —
**body size:** 1.0-3.0 mm.
**color:** yellowish white to yellowish brown to orange.
**carapace:** pyriform, slightly to very strongly (*Orchestina*) convex, often microsculptured with tubercles or striae.
**sternum:** oval, may expand laterally to enclose coxal bases.
**eyes:** six eyes in cluster, PME usually between PLE, rarely (*Orchestina*) between ALE (eyes may be reduced or absent in some species outside of our region).
**chelicerae:** not fused, lacking teeth.
**mouthparts:** endites apically narrowed.
**legs:** relatively short; lacking spines, with weak spines, or with strong spines paired ventrally on anterior tibiae and metatarsi; tarsi with two claws on onychium; female lacking palpal claw.
**abdomen:** usually without markings, sometimes with purplish streaks; with or without scutes.
**spinnerets:** ALS small, conical, contiguous; colulus absent or represented by two setae.
**respiratory system:** anterior book lungs and 2 tracheal spiracle near epigastric furrow.
**genitalia:** haplogyne; ***female*** with gonopore region marked by sclerites or transparent with vulva visible; ***male*** with simple exposed bulb, sometimes fused to tarsus; embolus with or without accessory structures.

### Distribution —
Widespread, most species are known from Florida and the sw USA.

### Natural history —
Most oonopids occur in leaf litter and similar situations (as wood rat nests), under rocks, or interstitially; some enter homes, especially in the northern parts of the range. Movement is by short jerky motions, both forward and backward; some species can hop (*Orchestina*). Oonopids do not construct webs but do make silken molting chambers. Some species are known only from females and are thought to be parthenogenetic, as *Heteroonops spinimanus* (Simon 1891c) (Saaristo 2001a).

### Taxonomic history and notes —
The Nearctic oonopids have received only limited treatment. The most comprehensive study to date is the review of the Florida species by Chickering (1969c). Prior to that, the western fauna was treated in a series of species descrip-

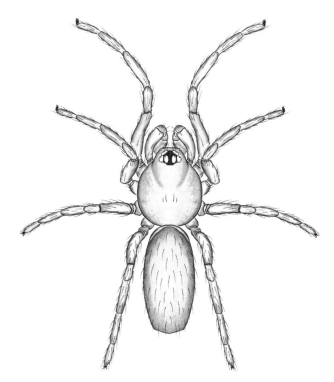

**Fig. 44.1** *Scaphiella hespera* (Chamberlin 1924b)

tions by Chamberlin (1924b), Chamberlin & Ivie (1935b, 1942a), Gertsch (1936a), and Gertsch & Davis (1936). The few known species from the ne USA are covered by Petrunkevitch (1920, 1929b) and Kaston (1948). Much more detailed descriptions of some pantropical species, including some which occur in Florida, are presented in recent studies by Dumitrescu & Georgescu (1983) and especially Saaristo (2001a).

This poorly studied family is badly in need of a revision. There are several known undescribed species in collections and, given the small size and cryptic habits of these spiders, many more will turn up. Apart from this, a number of the known species have been obviously misplaced in the catch-all genera *Opopaea* and *Oonops* and will need to be transferred to other, perhaps new, genera. Additionally, the placement of several species, known only from females, must remain tentative until males are described.

### Genera —
ORCHESTINA COMPLEX
*Orchestina* Simon 1882c
OONOPINAE
*Heteroonops* Dalmas 1916, *Oonops* Templeton 1835, "*Oonops*" Templeton 1835, *Stenoonops* Simon 1891c, *Tapinesthis* Simon 1914a
GAMASOMORPHINAE
*Gamasomorpha* Karsch 1881c, *Ischnothyreus* Simon 1893a, *Opopaea* Simon 1891c, "*Opopaea*" Simon 1891c, *Pelicinus* Simon 1891c, *Scaphiella* Simon 1891c, *Triaeris* Simon 1891c

## Oonopidae

**Key to genera —**
North America North of Mexico

| | | |
|---|---|---|
| 1 | Abdomen lacking dorsal or lateral scutes .................. | 2 |
| — | Abdomen with dorsal (Fig. 44.2) or lateral scutes (Fig. 44.3) ............................................................. | 7 |
| 2(1) | Femur IV enlarged (Fig. 44.4); PME located between ALE (Fig. 44.5) .................................... ***Orchestina*** | |
| | **Div.** 5 species — **Dist.** widespread — **Refs.** Petrunkevitch 1920, Chamberlin & Ivie 1935b, 1942a, Chickering 1969c | |
| — | Femur IV not enlarged; PME located between PLE (Figs. 44.9-44.11) ......................................................... | 3 |
| 3(2) | Legs without spines; abdomen with a narrow but well defined ventral scute posterior of epigastric furrow (Fig. 44.6) ............................................................ ***Stenoonops*** | |
| | **Div.** 1 species: *Stenoonops minutus* Chamberlin & Ivie 1935b — **Dist.** FL — **Refs.** Chamberlin & Ivie 1935b, 1942a, Chickering 1969c | |
| — | Legs with at least weak spines; abdomen lacking well defined ventral scute ..................................... | 4 |
| 4(3) | Palpal patella with prolateral swelling bearing robust spines; large spines also on palpal tibia and femur (Fig. 44.7) ......................................................... ***Heteroonops*** | |
| | **Div.** 1 species: *Heteroonops spinimanus* (Simon 1891c) — **Dist.** FL — **Refs.** Chickering 1969c, Dumitrescu & Georgescu 1983, Saaristo 2001a — **Note** known only from female | |
| — | Palpus without patellar swelling or enlarged spines ....... | 5 |
| 5(4) | Tibia I with 4 pairs of ventral spines; male palpal bulb separated from tarsus (Fig. 44.8) .................. ***Oonops*** | |
| | **Div.** at least 2 undescribed species — **Dist.** CA — **Ref.** Roberts 1995 | |
| — | Tibia I without ventral spines; male palpal bulb variable .. ...................................................................... | 6 |
| 6(5) | ALE separated by one diameter (Fig. 44.9); male palpus with bulb separated from tarsus (Fig. 44.10) ............ ................................................................. ***Tapinesthis*** | |
| | **Div.** 1 species: *Tapinesthis inermis* (Simon 1882c) — **Dist.** MA, BC — **Refs.** Kaston 1948, Kraus 1967a — **Note** This species, introduced from Europe, was initially recorded by Kaston (1948) on the basis of a single specimen and has recently turned up in BC (R. Bennett). | |
| — | ALE separated by less than one diameter (Fig. 44.11); male palpus with bulb fused to tarsus (Fig. 44.12) .................... ................................................................. ***"Oonops"*** | |
| | **Div.** 6 species — **Dist.** FL-TX, 4 species, CA, 2 undescribed species — **Refs.** Gertsch 1936a, Chamberlin & Ivie 1935b, Chickering 1969c — **Note** These "*Oonops*" are misplaced and not congeneric with the type species, *Oonops pulcher* Templeton 1835. The latter differs in having stronger leg spination, with paired ventral spines on Tibia I, and a male palpal bulb separated from the tarsus. | |

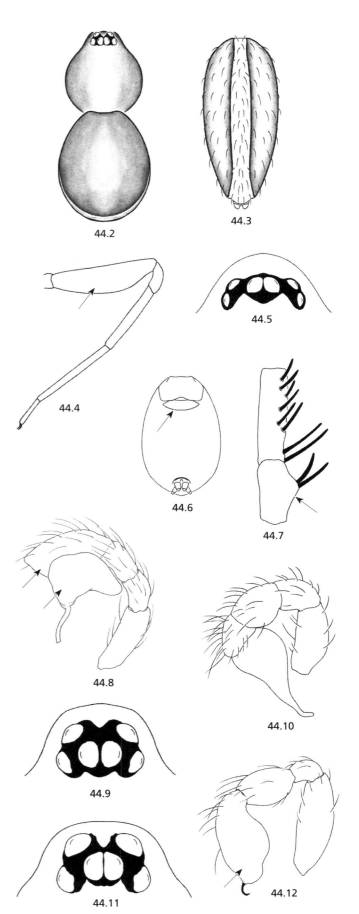

# Spiders of North America — Oonopidae

**7(1)** Abdomen with lateral scutes (Fig. 44.13) ............................ ........................................................... (female) ***Scaphiella***
        **Div.** 3 species — **Dist.** CA-TX — **Refs.** Chamberlin 1924b, Gertsch & Davis 1936

— Abdomen with dorsal scute (Fig. 44.14) ......................... **8**

**8(7)** Scute covering virtually entire abdominal length ........... **9**

— Scute covering less than 3/4 abdominal length ............ **17**

**9(8)** Males .......................................................................... **10**

— Females ...................................................................... **14**

**10(9)** Bulb separate from tarsus (Fig. 44.15) ......................... **11**

— Bulb fused to tarsus (Figs. 44.16-44.18) ...................... **12**

**11(10)** Legs short, Leg I PT/C <1; palpus as in Fig. 44.15 ............. ................................................................... ***Gamasomorpha***
        **Div.** 1 species: *Gamasomorpha lutzi* (Petrunkevitch 1929b) — **Dist.** FL — **Ref.** Chickering 1969c

— Legs long, Leg I PT/C at least 1 ......................... ***Pelicinus***
        **Div.** 1 species: *Pelicinus vernalis* (Bryant 1945b) — **Dist.** FL — **Note** known only from female; male presumed to key here

**12(10)** Palpal patella greatly enlarged, larger than femur (Fig. 44.16) ............................................................. ***Opopaea***
        **Div.** 7 nominal species of which only 2 (*Opopaea deserticola* Simon 1891c and *Opopaea bandina* Chickering 1969c) unambiguously belong to this genus, 4 are only known from females and based on the original descriptions cannot be placed with certainty, and 1 (*Opopaea calona* Chickering 1969c) clearly belongs elsewhere — **Dist.** CA-FL — **Refs.** Chickering 1969c, Saaristo 2001a

— Palpal patella not enlarged ........................................... **13**

**13(12)** Embolus a thin spine, lacking conductor (Fig. 44.17)........ ................................................................... ***Scaphiella***
        **Div.** 3 species — **Dist.** CA-TX — **Refs.** Chamberlin 1924b, Gertsch & Davis 1936

— Embolus thicker, with robust conductor (Fig. 44.18) ........ ........................................................................ ***"Opopaea"***
        **Div.** 1 species: "*Opopaea*" *calona* Chickering 1969c — **Dist.** FL

**14(9)** Ventral scute clearly shorter than dorsal .......... ***Pelicinus***
        **Div.** 1 species: *Pelicinus vernalis* (Bryant 1945b) — **Dist.** FL — **Note** known only from female

— Ventral and dorsal scutes subequal ............................... **15**

**15(14)** Epigastric region lacking 2 short longitudinal lines .......... ........................................................................ ***"Opopaea"***
        **Div.** 1 species: "*Opopaea*" *calona* Chickering 1969c — **Dist.** FL

— Epigastric region with 2 short longitudinal lines (Fig. 44.19) ............................................................................ **16**

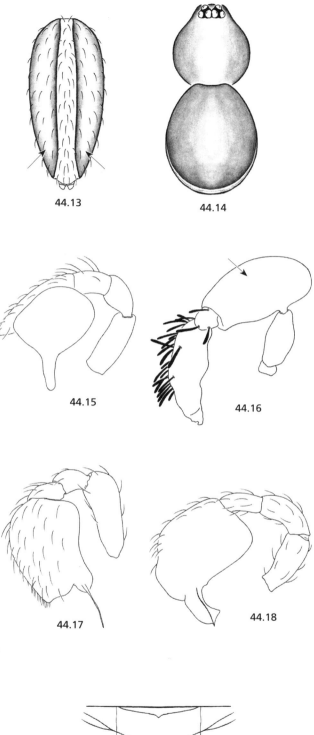

44.13    44.14

44.15    44.16

44.17    44.18

44.19

**16(15)** Lateral slope of carapace with fine longitudinal striae (Fig. 44.20) ................................................................ ***Opopaea***
> **Div.** 7 nominal species of which only 2 (*Opopaea deserticola* Simon 1891c and *Opopaea bandina* Chickering 1969c) unambiguously belong to this genus, 4 are only known from females and based on the original descriptions can not be placed with certainty, and 1 (*Opopaea calona* Chickering 1969c) clearly belongs elsewhere — **Dist.** CA-FL — **Refs.** Chickering 1969c, Saaristo 2001a

— Lateral slope of carapace lacking longitudinal striae ......... ................................................................ ***Gamasomorpha***
> **Div.** 1 species: *Gamasomorpha lutzi* (Petrunkevitch 1929b) — **Dist.** FL — **Ref.** Chickering 1969c

**17(8)** ALE nearly contiguous (Fig. 44.21) ........ ***Ischnothyreus***
> **Div.** 1 species: *Ischnothyreus peltifer* (Simon 1891c) — **Dist.** FL — **Refs.** Chickering 1969c, Dumitrescu & Georgescu 1983, Saaristo 2001a

— ALE separated by at least one radius ............................ **18**

**18(17)** Ventral scute covering less than half of abdomen (Fig. 44.22) ................................................................ ***Triaeris***
> **Div.** 1 species: *Triaeris stenaspis* Simon 1891c — **Dist.** FL — **Refs.** Bristowe 1948, Chickering 1968e — **Note** known only from female

— Ventral scute covering more than half of abdomen .......... ................................................................ ***Pelicinus***
> **Div.** 1 species: *Pelecinus vernalis* (Bryant 1945b) — **Dist.** FL — **Note** known only from female

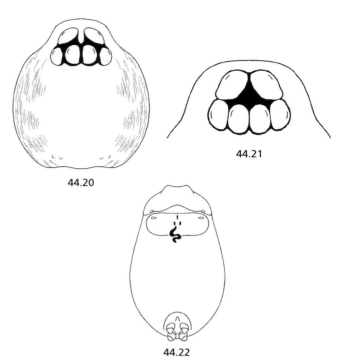

44.20

44.21

44.22

## Chapter 45
# OXYOPIDAE
3 genera, 18 species

Allen R. Brady
Adalberto J. Santos

### Common name —
Lynx spiders.

### Similar families —
Oxyopidae are distinctive but could be confused in the field with Salticidae (p. 205) because they both occur in the same general habitats and exhibit the same saltatory behavior.

### Diagnosis —
Oxyopids are entelegyne spiders with three tarsal claws and eight eyes. They can be distinguished by their hexagonal eye arrangement (Figs. 45.8-45.11), legs with prominent, erect spines (Figs. 45.1, 45.16, 45.18), and long tapering abdomen (Figs. 45.1, 45.12-45.15).

### Characters —
**body size:** *Oxyopes* and *Hamataliwa* 3.2-8.2 mm; *Peucetia* 8.5-21.5 mm.
**color:** in *Oxyopes* colors range from white to yellow background with contrasting darker brown and black stripes (Figs. 45.1, 45.8, 45.12). Flattened scale-like hairs contribute to the darker colors and produce metallic iridescent sheens ranging from purple to gold in some species (Townsend & Felgenhauer 2001). *Hamataliwa* is cryptic gray and brown in appearance (Figs. 45.9, 45.10, 45.13, 45.14); *Peucetia* green (Figs. 45.11, 45.15). Tufts of white cuticular scales occur in the eye region of all species.
**carapace:** high and convex with a distinct frontal vertical slope (face); more gradual in *Hamataliwa*.
**sternum:** heart or shield-shaped, tapering behind to a thin projection between the posterior coxae (Fig. 45.17).
**eyes:** AER recurved and PER procurved, producing a hexagonal arrangement (Figs. 45.8-45.15). AME much the smallest.
**chelicerae:** basal segment long and tapering at distal end with the fangs short. *Oxyopes* and *Hamataliwa* with two teeth on anterior and one tooth on posterior margin of the fang furrow. *Peucetia* species have an anterior distal projection on the internal margin of the basal segment, one tooth on anterior margin of fang furrow and no teeth on posterior margin.
**legs:** prograde and relatively long compared to body length with conspicuous long and stout spines on all legs (Figs. 45.1, 45.16, 45.18); tarsi with three claws.
**abdomen:** elongate, usually widest immediately behind the base and tapering narrowly posteriorly (Figs. 45.12, 45.13, 45.15).
**spinnerets:** cylindrical anterior and posterior pairs with a relatively large basal segment and an abbreviated, almost ring-like apical segment; median pair hidden by others.
**respiratory system:** one pair of book lung openings and a tracheal spiracle at base of spinnerets.
**genitalia:** entelegyne; *female* epigynum well-sclerotized, with a median excavated atrium (Figs. 45.2, 45.3, 45.22), or a large scape-like anteriorly directed projection (Figs. 45.4-45.5); *male* tibial apophysis conspicuous, present in *Oxyopes* and *Hamataliwa* (Figs. 45.6, 45.7, 45.19, 45.20). *Peucetia* with a conspicuous paracymbium (Fig.

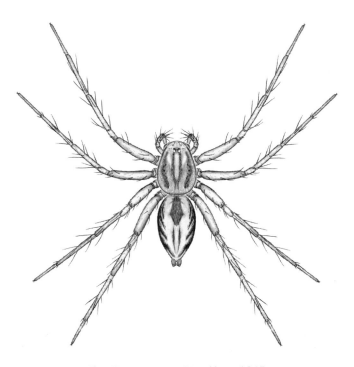

**Fig. 45.1** *Oxyopes salticus* Hentz 1845

45.21). In *Oxyopes* and *Hamataliwa*, the conductor is turned ventrally, almost touching the apex of the median apophysis (Figs. 45.6-45.7).

### Distribution —
*Oxyopes scalaris* Hentz 1845 occurs throughout the USA and southern Canada. *Oxyopes salticus* is the most widespread species of the family occurring from southern Canada to northern Argentina, frequently associated with disturbed habitats or agroecosystems (Young & Lockley 1985). The remaining species of *Oxyopes* and *Hamataliwa* are found in the southern tier of states, with a few species reaching northern California and Virginia in the coastal regions and southern Kansas centrally. North American species of *Peucetia* occur from southern USA to northern South America (Brady 1964, Santos & Brescovit 2003). The three genera recorded in North America occur all over the world with a primarily tropical and subtropical distribution (Platnick 2005).

### Natural history —
Most species of *Oxyopes* occur in tall grass and herbaceous vegetation. *Oxyopes scalaris* and *Hamataliwa* occur most often on woody shrubs and trees. *Peucetia* prefers tall grass and especially woody shrubs such as wild buckwheat (*Erigonium fasiculatum*) and dog fennel (*Eupatorium*

*capillifolium*). *Peucetia viridans* HENTZ 1832 and *Oxyopes salticus* have been reported as common inhabitants of agroecosystems, and as potential agents for biological control of insect pests (Young & Lockley 1985, Nyffeler *et al.* 1992). Oxyopids are basically "sit and wait" predators, but have been reported to stalk their prey like cats. There are no leg scopulae present, but the long spines on the front legs function as a net or basket to aid in prey capture. In addition, their characteristic darting movements and a tendency to repeatedly jump when disturbed provide useful features for identification in the field. Almost nothing is known about reproductive behavior, except that *Peucetia* males attach a resinous mating plug (together with the apex of the paracymbium in *Peucetia viridans*) into the female epigynum after copulation. Egg cases are hung within a mesh of silk in *Peucetia*, and fixed over leaves or twigs in *Oxyopes* and *Hamataliwa*.

### Taxonomic history and notes —

Oxyopidae is clearly defined on the basis of several diagnostic characters and has been recognized as a distinct family since its proposal by Thorell (1870b). This family has traditionally been placed in the superfamily Lycosoidea, supported by the presence of a grate-shaped tapetum on the posterior median eyes (Homann 1971). In a recent phylogenetic analysis of lycosoid spider families, Griswold (1993) hypothesized Senoculidae and Psechridae, respectively from the neotropics and Southeast Asia, as the successive sister groups of Oxyopidae. This study also suggested Stiphidiidae, from Australia and New Zealand, as closely related to lynx spiders, but this was later refuted by Griswold *et al.* (1999a), who showed this family was closer to the Agelenidae and Desidae than to lycosoids. The Oxyopidae-Senoculidae clade is supported by the undifferentiated median and lateral sectors of the female epigynum and the presence of dorsal spines on the female first tibia, and both families share with Psechridae large spermathecal heads and the tarsal tricobothria organized in a row (Griswold 1993). Recently, Silva Dávila (2003) corroborated Griswold's hypothesis, with the clade Oxyopidae + Senoculidae + Psechridae being close to a clade uniting Lycosidae, Pisauridae and Trechaleidae. There are no studies regarding the internal phylogeny of the family, except for a speculative hypothesis proposed by Rovner (1980b), which suggests the neotropical web-weaving *Tapinillus* SIMON 1898a as the most primitive lynx spider genus. Rovner suggested that oxyopids diverged from an ancestral theridiid or linyphiid-like web-weaving ancestor, which is not congruent with the currently accepted hypothesis of relationships for lynx spiders.

### Genera —

*Hamataliwa* KEYSERLING 1887b, *Oxyopes* LATREILLE 1804b *Peucetia* THORELL 1869

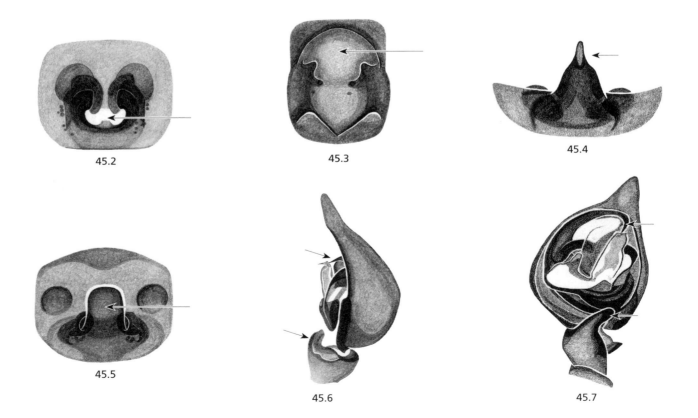

45.2

45.3

45.4

45.5

45.6

45.7

# Key to genera —
## North America North of Mexico

**1** Posterior cheliceral margin without teeth; ALE row obviously wider than PME row; posterior eye row (PME and PLE) only slightly procurved (Figs. 45.11; 45.15); living specimens bright green in color; size 8.5-21.5 mm . 
............................................................ ***Peucetia***
      **Div.** 2 species — **Dist.** *Peucetia viridans* (HENTZ 1832) in southern USA, *Peucetia longipalpis* F.O. PICKARD-CAMBRIDGE 1902a from CA to TX — **Ref.** Brady 1964

— Posterior cheliceral margin with a single tooth; ALE row subequal to PME row (Figs. 45.8-45.9; 45.12-45.13) or PME row much wider than ALE row (Figs. 45.10; 45.12); posterior eye row (PME and PLE) strongly procurved (Figs. 45.8-45.10; 45.12-45.14). Living specimens not green in color, size 3.2-8.2 mm ........................................ **2**

**2(1)** Leg IV robust, clearly longer than leg III; distance between PME subequal to distance between PME and PLE ............
............................................................ ***Oxyopes***
      **Div.** 13 species — **Dist.** widespread — **Refs.** Brady 1964, 1969

— Leg IV small, subequal to or shorter than leg III; distance between PME much greater than distance from PME to PLE (Figs. 45.10; 45.12; except in *H. helia*, Fig. 45.13) ......
............................................................ ***Hamataliwa***
      **Div.** 3 species — **Dist.** *Hamataliwa grisea* KEYSERLING 1887b southern USA, *Hamataliwa helia* (CHAMBERLIN 1929) from FL to TX, *Hamataliwa unca* BRADY 1964 in southern TX — **Ref.** Brady 1964

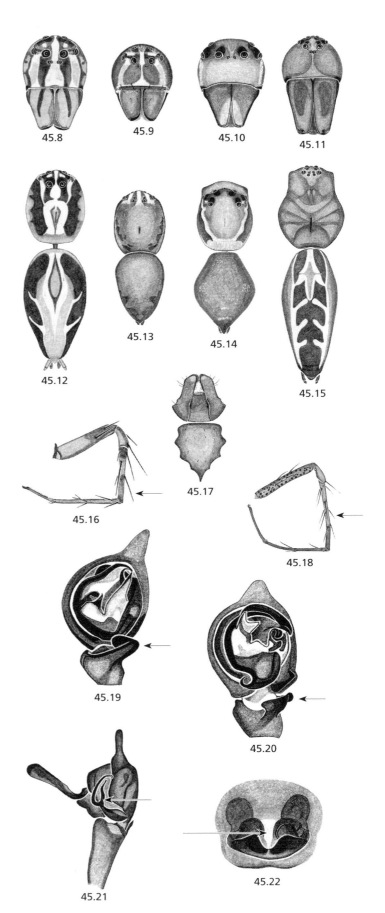

Chapter 46

# PHILODROMIDAE

5 genera, 96 species

Charles D. Dondale

### Common name —
Running crab spiders.

### Similar families —
Selenopidae (p. 221), Sparassidae (p. 224) Thomisidae (p. 246).

### Diagnosis —
Two-clawed, flat-bodied, somewhat crab-like hunting spiders with soft, recumbent body setae and with leg II longer (sometimes much longer) than the subequal legs I, III, and IV (Fig. 46.1).

### Characters —
**body size:** total length 3.0-8.2 mm.
**color:** carapace dull yellow, brown, red, or gray, usually with pale median band; abdomen often with pattern of dark heart mark and chevrons, sometimes with dark midstripe its full length.
**carapace:** low, as wide as long or somewhat longer than wide, covered with soft, recumbent setae.
**eyes:** eight, in two transverse recurved rows; usually uniform in size; eye tubercles small or lacking.
**chelicerae:** short, slender; retromargin lacking teeth.
**legs:** laterigrade, 2-clawed, long, slender, usually with scopulae and claw tufts.
**abdomen:** elongate-oval, rather flat, widest at or posterior to middle, covered with soft recumbent setae.
**genitalia:** entelegyne; ***female*** epigynum usually with flat median septum; copulatory openings situated at sides of median septum; ***male*** palp with retrolateral tibial apophysis (RTA) and usually with ventral tibial apophysis (VTA); tegulum flat or somewhat convex, with single open loop of seminal duct visible through integument; embolus short to long, slender, often arched around tip of tegulum.

### Distribution —
Widespread, with concentrations of species in the east, the southwest, and the North American Cordillera. Several species are transcontinental, and of these a proportion extend into Siberia or even to Europe. *Thanatus vulgaris* SIMON 1870b is a world traveler; specimens have been found in stores, restaurants, railroad stations, and ships, and as pests in commercial cricket cultures.

### Natural history —
Philodromids lead an active predatory life on the slippery surfaces of plant stems and leaves. Some are found only in coniferous trees, others only in deciduous. A few are desert inhabitants. Life histories have been studied by Dondale (1961) and Putman (1967), and predatory behavior by Haynes & Sisojevic (1966). Dondale (1964, 1967) used courtship behavior to distinguish some closely related species.

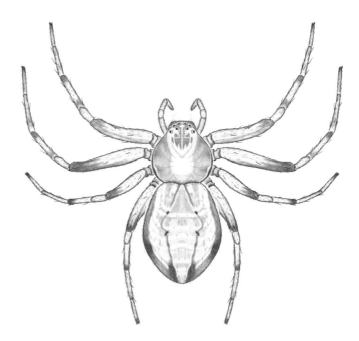

**Fig. 46.1** *Philodromus cespitum* (WALCKENAER 1802)

### Taxonomic history and notes —
Most of these spiders are readily identified with the use of existing keys. Some of the wide ranging species may prove, upon examination with newer methods, to comprise complexes of valid species. The Philodromidae were formerly treated as a subfamily of Thomisidae, but Homann (1975) showed the Philodromidae to be independent of Thomisidae. The actual relationship of the Philodromidae to other spider families is thought to lie with the family Sparassidae.

### Genera —
*Apollophanes* O. PICKARD-CAMBRIDGE 1898a, *Ebo* KEYSERLING 1884b, *Philodromus* WALCKENAER 1826, *Thanatus* C.L. KOCH 1837g, *Tibellus* SIMON 1875a

# Key to genera —
## North America North of Mexico

**1** Posterior median eyes (PME) distinctly closer to posterior lateral eyes (PLE) than to each other (Figs. 46.2-46.3) .......................................................................................... **2**

— PME approximately as far from PLE as from each other (Fig. 46.5) or closer to each other than to PLE (Figs. 46.6, 46.7) ................................................................ **3**

**2(1)** Leg II approximately twice as long as leg I (Fig. 46.4). Anterior median eyes (AME) distinctly larger than anterior lateral eyes (ALE) (Fig. 46.2) .............. ***Ebo***
   **Div.** 22 species — **Dist.** contiguous USA and southernmost CAN — **Refs.** Sauer & Platnick 1972, Schick 1965

— Leg II less than twice as long as leg I (Fig. 46.1). AME approximately equal in diameter to ALE (Fig. 46.3) ......... .............................................................. ***Philodromus***
   **Div.** 55 species — **Dist.** widespread — **Refs.** Schick 1965, Dondale & Redner 1968, 1969, 1975b, 1976a, 1978b

**3(1)** Body with dark midstripe its full length (Fig. 46.8). Abdomen elongate (Fig. 46.8). PME distinctly closer to each other than to PLE (Fig. 46.6) .............................. ***Tibellus***
   **Div.** 7 species — **Dist.** widespread — **Refs.** Schick 1965, Dondale & Redner 1978b, Marusik 1989a

— Body lacking dark midstripe its full length. Abdomen less elongate. PME approximately as far from each other as from PLE (Figs 46.5; 46.7) ............................................... **4**

**4(3)** Carapace distinctly longer than wide. Basitarsus I with 1 or more prolateral macrosetae. Palpus of male with tibia distinctly narrower than tegulum (Fig. 46.9). Epigynum lacking flat lips at sides of median septum (Fig. 46.10) .... .................................................................. ***Apollophanes***
   **Div.** 4 species — **Dist.** western and southwestern parts of the continent — **Refs.** Dondale & Redner 1975c

— Carapace only a little longer than wide. Basitarsus I usually lacking prolateral macrosetae (exceptions: male *Thanatus vulgaris* and both sexes of *Thanatus bungei* (KULCZYŃSKI 1908b). Palpus of male with tibia nearly as wide as tegulum (Fig. 46.11). Epigynum with flat lips at sides of median septum (Fig. 46.12) ............................ ***Thanatus***
   **Div.** 8 species — **Dist.** widespread — **Refs.** Dondale *et al.* 1964, Dondale & Redner 1978b, Marusik 1989a

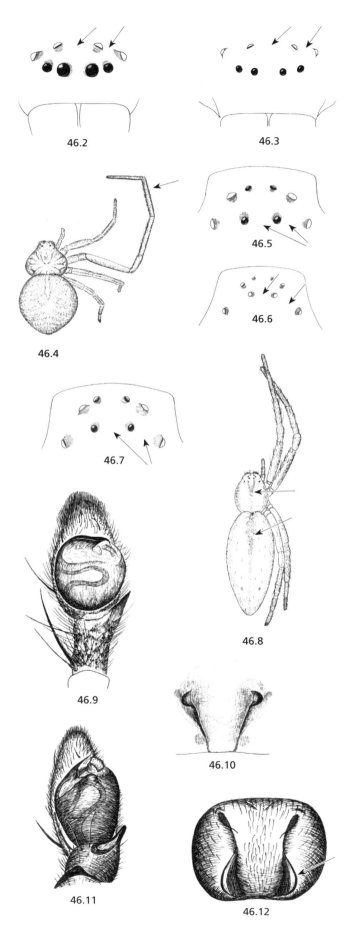

# Chapter 47
# PHOLCIDAE
12 genera, 34 species

Bernhard A. Huber

**Common name —**
Cellar spiders, daddylongleg spiders.

**Similar families —**
Hypochilidae (p. 120).

**Diagnosis —**
Small to medium (1-10 mm), three-clawed, ecribellate, haplogyne spiders, with six or eight eyes; prosoma about as long as wide, clypeus about as high as chelicerae; male palp with prominent retrolateral paracymbium ('procursus'), male chelicerae usually with sexual modifications; tarsi usually pseudosegmented.

**Characters —**
**body size:** 1.0-10.0 mm.
**color:** usually ochre-yellow to light brown, often with brown or black markings.
**carapace:** about as long as wide, with or without median indentation.
**eyes:** ALE and posterior eyes forming triads; AME present or absent.
**chelicerae:** chelate, with lamina opposing fang; male chelicerae usually sexually modified.
**mouthparts:** labium fused to sternum.
**legs:** often very long, tarsi usually pseudosegmented, with three claws.
**abdomen:** cylindrical to globular.
**spinnerets:** ALS large and cylindrical, PMS small and partly hidden by flattened PLS.
**respiratory system:** one pair of book lungs; tracheal spiracle often seems present, but tracheae are absent.
**genitalia:** haplogyne; *female* usually with sclerotized plate (flat or sculptured) covering internal genitalia; *male* with large palps, paracymbium (procursus) often large and complex.

**Distribution —**
Widespread, highest diversity in southwestern USA, almost absent from Canada.

**Natural history —**
Indigenous pholcids are found mainly in dark spaces, under ground objects, in leaf and plant detritus, in soil openings and caves. Six of the twelve genera are represented by introduced species only, and these occupy corners and dark spaces in and around buildings. Long-legged species perform rapid vibrations when disturbed, presumably to avoid predation. Females carry the egg sacs in their chelicerae. Web structure varies from domed sheets with a network of irregular threads above (e.g., *Holocnemus*, *Artema*), to tangled spacewebs without any apparent design (e.g., *Pholcus*), to a few lines close to the substrate (e.g., *Pholcophora*). Short-legged species can run quickly.

**Taxonomic history and notes —**
The taxonomy of Nearctic pholcids is mainly based on scattered species descriptions accumulated over the last 160 years. This is especially true for the two most diverse genera

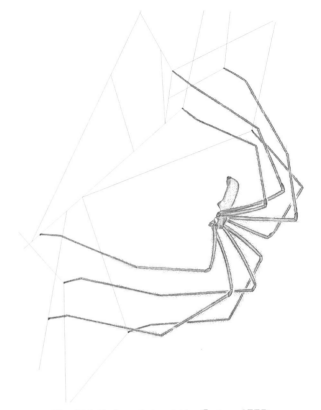

**Fig. 47.1** *Pholcus phalangioides* (Fuesllin 1775)

in the region, *Psilochorus* and *Physocyclus*, both of which have been severely neglected for about a half century and are in urgent need of revision.

**Genera —**
PHOLCINAE
  *Micropholcus* Deeleman-Reinhold & Prinsen 1987, *Pholcus* Walckenaer 1805, *Spermophora* Hentz 1841a
NEW WORLD CLADE
  *Modisimus* Simon 1893d, *Psilochorus* Simon 1893a
NINETINAE (possibly not monophyletic)
  *Chisosa* Huber 2000, *Pholcophora* Banks 1896e
"HOLOCNEMINAE" (very probably not monophyletic)
  *Artema* Walckenaer 1837, *Crossopriza* Simon 1893a, *Holocnemus* Simon 1873a, *Physocyclus* Simon 1893a, *Smeringopus* Simon 1890a

## Key to genera —
### North America North of Mexico

\* *Metagonia* is not included because the only record from the USA was one specimen in a banana bunch from Mexico (Gertsch 1986: 41).

**1**     Short-legged: tibia I < body length; retrolateral trichobothria on tibiae of walking legs very distal (at about 60% of tibia length); tibia I length/diameter: 10-25 .................... **2**

—     Long-legged: tibia I ≥ body length; retrolateral trichobothria on tibiae of walking legs proximal (at < 40% of tibia length); tibia I length/diameter: >30 ................ **3**

**2(1)**     Procursus massive and complex (Fig. 47.2); male sternum without anterior humps; carapace without median indentation: tiny spider (about 1.5 mm body length) ................
................................................................. ***Chisosa***
        **Div.** 1 species: *Chisosa diluta* (GERTSCH & MULAIK 1940) — **Dist.** only TX, Big Bend National Park — **Refs.** Gertsch & Mulaik 1940, Gertsch 1982b, Huber 2000 — **Note** previously in *Pholcophora*, transferred by Huber (2000)

—     Procursus simple; male sternum with pair of anterior humps (Fig. 47.3); carapace with shallow median groove.
................................................................. ***Pholcophora***
        **Div.** 2 species: *Pholcophora americana* BANKS 1896e (widespread) and *Pholcophora texana* GERTSCH 1935a (only TX) — **Dist.** w USA and southwestern CAN — **Refs.** Banks 1896e, Chamberlin & Ivie 1935b, Gertsch 1935a, 1982b, Gertsch & Mulaik 1940, Huber 2000 — **Note** a third species, *Pholcophora obscura* CHAMBERLIN & IVIE 1935b, was synonymized under *Pholcophora americana* by Gertsch (1982b)

**3(1)**     Carapace without median invagination; palpal trochanter with retrolateral projection; male chelicerae with lateral projections proximally (Fig. 47.4) ..................... **4**

—     Carapace with median groove .......................... **6**

—     Carapace with pit, i.e. rounded indentation behind ocular area ................................................................. **8**

**4(3)**     Six eyes (AME missing), small pale spider .........................
................................................................. ***Spermophora***
        **Div.** 1 species: *Spermophora senoculata* (DUGÈS 1836a) — **Dist.** widespread — **Refs.** Hentz 1841a, Emerton 1882, Petrunkevitch 1910, Gertsch & Mulaik 1940, Kaston 1948, 1977, Schenkel 1950b, Huber 2002 — **Note** introduced Old World species; mainly in houses; *Spermophora meridionalis* HENTZ 1841a synonymized by Yaginuma (1974a)

—     Eight eyes, AME small but well developed ...................... **5**

**5(4)**     Abdomen oval to globular, procursus with distinctive dorsal sclerite (Fig. 47.5), female genitalia with distinctive median structure shining through cuticle (Fig. 47.6) .......
................................................................. ***Micropholcus***
        **Div.** 1 species: *Microphorus fauroti* (SIMON 1887e) — **Dist.** Southern USA — **Refs.** Gertsch & Mulaik 1940, Roth 1985, Deeleman-Reinhold & Prinsen 1987 — **Note** introduced Old World species; *Pholcus unicolor* PETRUNKEVITCH 1929b synonymized by Roth (1985)

—     Abdomen elongated ............................................ ***Pholcus***
        **Div.** 3 species: *Pholcus phalangioides* (FUESSLIN 1775), *Pholcus muralicola* MAUGHAN & FITCH 1976, and *Pholcus manueli* GERTSCH 1937, plus several undescribed — **Dist.** *Pholcus phalangioides* widespread, *Pholcus muralicola* in KS, other species mainly eastern USA endemics — **Refs.** Gertsch 1937, Kaston 1948, Maughan & Fitch 1976, Huber 2000, Senglet 2001

**6(3)**     All eyes together on high median turret (Fig. 47.7); male femora densely covered with short vertical hairs ...............
................................................................. ***Modisimus***
        **Div.** 2 species: *Modisimus texanus* BANKS 1906b and *Modisimus culicinus* (SIMON 1893d), plus several undescribed — **Dist.** *Modisimus culicinus* widely distributed, others restricted to southwestern USA and FL — **Refs.** Banks 1906b, Gertsch & Mulaik 1940, Gertsch & Peck 1992, Huber 1997a, 1998e — **Note** includes *Hedypsilus* (synonymized by Huber 1997a)

—     Ocular area moderately elevated; very few short vertical hairs on male femora ....................................................... **7**

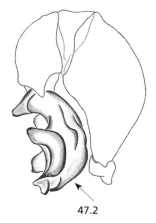

47.2

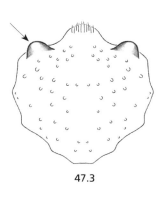

47.3

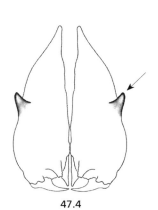

47.4

47.5

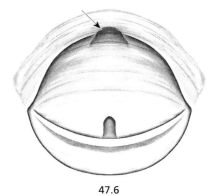

47.6

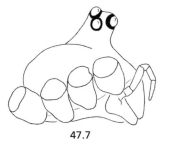

47.7

## Pholcidae

**7(6)** Body size usually < 4 mm, male palpal femur slender with pointed apophysis ventrally (Fig. 47.8), procursus slender and simple (Fig. 47.8); male chelicerae without stridulatory files ........................................................ ***Psilochorus***
  **Div.** 17 nominal species, plus several undescribed — **Dist.** w USA, e to OK, TX, only *Psilochorus pullulus* (Hentz 1850b) in the southeast — **Refs.** Hentz 1850b, Thorell 1877c, Keyserling 1887b, Chamberlin 1919a, b, Banks 1921, Chamberlin & Gertsch 1928, Banks *et al.* 1932, Gertsch 1935a, Gertsch & Ivie 1936, Gertsch & Mulaik 1940, Chamberlin & Ivie 1942a, Schenkel 1950b, Agnew *et al.* 1985

— Body size usually > 4 mm, male palpal femur enlarged (Fig. 47.9) without pointed apophysis ventrally, procursus massive and more complex (Fig. 47.9), male chelicerae with stridulatory files .................................... ***Physocyclus***
  **Div.** 3 species + several undescribed — **Dist.** western USA, e to TX — **Refs.** Chamberlin 1921b, Chamberlin & Gertsch 1929, Gertsch 1935a, Gertsch & Mulaik 1940, Brignoli 1979c

**8(3)** Abdomen globular; very large pholcid with massive black modification on male chelicerae (Fig. 47.10), female genital plate flat with pair of dark areas (Fig. 47.11) ................ ........................................................................ ***Artema***
  **Div.** 1 species: *Artema atlanta* Walckenaer 1837 — **Dist.** western USA — **Refs.** Gertsch 1935a, Brignoli 1981c — **Note** introduced Old World species

— Abdomen pointed dorso-posteriorly; male chelicerae with two pairs of distinctive apophyses (Fig. 47.12); female genital plate with distinctive median sclerotized area (Fig. 47.13) ........................................................ ***Crossopriza***
  **Div.** 1 species: *Crossopriza lyoni* (Blackwall 1867a) — **Dist.** southwestern and central USA — **Refs.** Roth 1985, Edwards 1993c, Huber *et al.* 1999, Guarisco & Cutler 2003 — **Note** introduced Old World species

— Abdomen cylindrical, not pointed dorso-posteriorly .... **9**

**9(8)** Thoracic pit shallow, female sternum without posterior projection, female palp not widened, no pigment specks on legs ...................................................... ***Smeringopus***
  **Div.** 1 species: *Smeringopus pallidus* (Blackwall 1858a) — **Dist.** southwestern and southern USA — **Refs.** Roth 1985, Edwards 1993c — **Note** introduced Old World species

— Thoracic pit very deep, female sternum with posterior projection (Fig. 47.14), female palp distally widened (Fig. 47.15), pigment specks on legs .................. ***Holocnemus***
  **Div.** 1 species: *Holocnemus pluchei* (Scopoli 1763) — **Dist.** southwestern USA — **Refs.** Roth 1985, 1994, Porter & Jakob 1990 — **Note** introduced Old World species

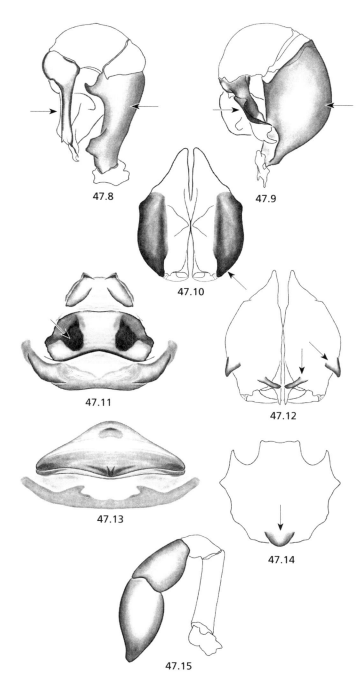

47.8  47.9  47.10  47.11  47.12  47.13  47.14  47.15

# Chapter 48
# PIMOIDAE
1 genus, 13 species

Gustavo Hormiga
Stephen E. Lew

### Similar families —
Pimoids are the sister group of the large family Linyphiidae (p. 124). The somatic morphology of *Pimoa* is similar to that of some linyphiid species (Fig. 48.1). Some species of *Meta* and closely related genera (Tetragnathidae, p. 230) are also similar to *Pimoa* in somatic morphology, but they can be easily distinguished by the web architecture (orb webs in the former genera). Larger species of the hahniid genus *Calymmaria* (p. 111) may cause confusion in the field as they have annulated legs, a superficially similar morphology, prefer the same habitats, and are common in much of *Pimoa*'s North American range.

### Diagnosis —
American species of *Pimoa* can be distinguished from other three-clawed, ecribellate, entelegyne spiders by the following combination of characters: autospasy at the patella-tibia junction, stridulatory striae on the ectal side of the chelicerae (absent in some species), male palp with retrolateral integral paracymbium, a retrolateral cymbial sclerite articulated by means of membrane, and a dorsoectal cymbial process with cuspules and an alveolar sclerite (Figs. 48.2-48.4). Conductor and median apophysis are present in most species. The embolus is continuous with the tegulum (the typical linyphiid embolic division is absent) and has an elongated filiform or lamelliform embolic process (Fig. 48.3). The epigynum protrudes more than its width, has a dorsal to lateral fold or groove with the copulatory openings at the distal end (Figs. 48.5-48.6). The fertilization ducts are anteriorly oriented (Figs. 48.7-48.8).

### Characters —
**body size:** 5.0-12.0 mm.
**color:** brownish to grayish, with light marks (often chevron-like), usually with two light bands ventrally; legs often annulated.
**carapace:** pyriform, longer than wide, thoracic furrow a conspicuous ovate pit.
**sternum:** longer than wide, projecting behind coxae IV, usually darkly pigmented.
**eyes:** typical araneoid arrangement, lateral eyes juxtaposed, eyes usually surrounded by pigment.
**chelicerae:** large, with ectal stridulatory striae, three prolateral and one to four retrolateral teeth.
**mouthparts:** labium free, wider than long.
**legs:** long (particularly in adult males) and often annulated, setose with many macrosetae (two dorsal on femur IV), tarsi with 3 claws; leg formula 1243 (except in *Pimoa laurae* HORMIGA 1994a and *Pimoa edenticulata* HORMIGA 1994a). Autospasy at patella-tibia.
**abdomen:** oval.
**spinnerets:** typical araneoid (orbweavers and relatives) general appearance. Colulus large and fleshy, with setae. PMS and/or PLS aciniform spigot fields reduced to one or no spigots. Peripheral cylindrical spigot in PLS with an enlarged base (relative to the other cylindrical spigot). Functional PLS triplet present.
**respiratory system:** one pair of book lungs and simple, unbranched median tracheae (haplotracheate) with the

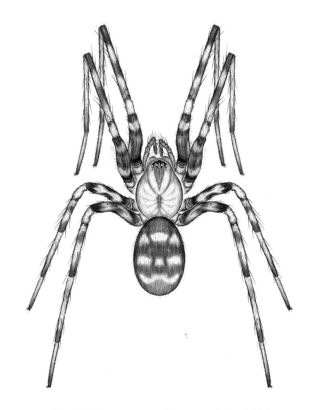

**Fig. 48.1** *Pimoa curvata* CHAMBERLIN & IVIE 1943

spiracle just anterior to colulus.
**genitalia:** entelegyne; *female* epigynum protruding more than its width, with dorsal and ventral plates which have a groove at their margins (epigynal fold; Figs. 48.5-48.6). Epigynal fold distally bears the copulatory opening. Copulatory ducts varying in length and degree of sinuosity. Spermathecae usually spherical, with short and lightly sclerotized fertilization ducts running anteriorly (Figs. 48.7-48.8); *male* pedipalpal tibia with a rounded dorsal process, with 2-4 prolateral and 2-5 retrolateral trichobothria. Cymbium with alveolus in an eccentric position, close to the prolateral margin, provided with an alveolar sclerite (a small plate located on the ventral side of the cymbium, anteroectal to the distal margin of the alveolus; this sclerite appears as a dark sclerotized plate that lies between the distal end of the alveolus and the attachment of the pimoid cymbial sclerite). Paracymbium continuous with the retrolateral cymbial margin (Figs. 48.2, 48.4). Cymbium wide, with a relatively large retrolateral cymbial sclerite ("pimoid cymbial sclerite") attached to the cymbium by means of a membrane (Figs. 48.2, 48.4). Dorsum of cymbium with projection bearing

dark denticles or cuspules in variable number (absent in *Pimoa edenticulata*). Tegulum large, more or less globular and bearing a membranous conductor and a small hook-shaped apophysis (median apophysis; absent in some species). Embolus long and filiform, curved following the margin of the tegulum. Embolus paralleled by a long process ("pimoid embolic process") sharing a common base with the embolus. The morphology of the pimoid cymbial sclerite, the denticulate cymbial process and the embolic process is species-specific.

### Distribution —

Pacific coast between the 35th and 60th parallels, including the Sierra Nevada and Cascades, Idaho and western Montana (Bitterroot Range).

### Natural history —

Pimoids build large sheet webs close to the ground (in some species up to one square meter in surface) in sheltered microhabitats such as fallen or hollow tree trunks and under overhanging cornices in banks and road cuts. Sometimes these spiders can be found in human made structures, such as sheds and outhouses. Several species can be found in caves. Pimoids are nocturnal, hunt on their webs at night and spend the daylight hours hiding in a retreat on the web's margin. Webs are often maintained for long periods of time and can show obvious signs of repair and senescence. Like linyphiids, pimoids hang from the undersurface of their webs.

### Taxonomic history and notes —

The taxonomic history of pimoids, summarized by Hormiga (1994a, 2003), is rife with generic and familial transfers, with some species having been classified at some point in the families Linyphiidae (p. 124) and Tetragnathidae (p. 232). These taxonomic changes are a consequence of the unusual combination of primitive and derived characters of the male palp together with a number of somatic features shared by several araneoid families (Hormiga 1994a, 2003). Wunderlich (1986) erected the linyphiid subfamily Pimoinae to include the genera *Pimoa* and *Louisfagea* BRIGNOLI 1971d (the latter is now a junior synonym of the former) and Hormiga (1993) elevated the subfamily to family rank. Hormiga (1994a) analyzed the phylogenetic relationships of *Pimoa* species and monographed the family. Griswold *et al.* (1999b) have described a new species from China. More recently, Hormiga (2003) has described a new genus (*Weintrauboa* from Japan and adjacent islands), reassessed the synapomorphies supporting Pimoidae and the "linyphioid" (Pimoidae + Linyphiidae) monophyly and redefined the diagnosis of pimoids and linyphiids.

As currently defined, Pimoidae includes two genera and 24 described species. The family has a relictual distribution, with 13 species in western North America, two in southern Europe and nine in Asia (the Himalayas and adjacent areas and Japan). A key to the species of *Pimoa* is found in Hormiga (1994a).

### Genus—

*Pimoa* CHAMBERLIN & IVIE 1943

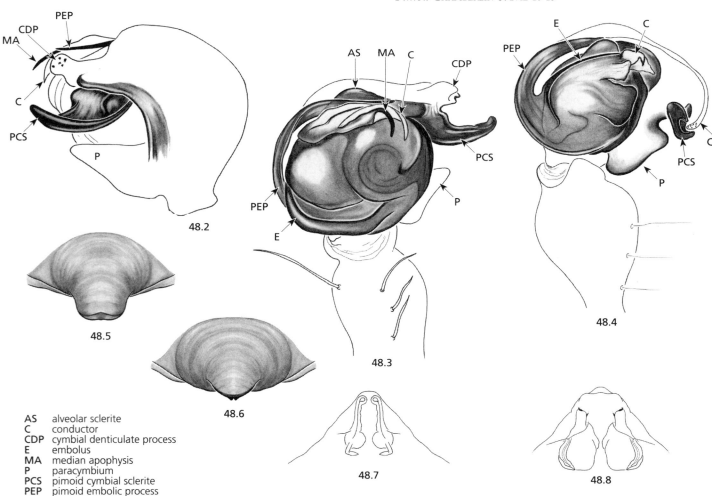

| AS | alveolar sclerite |
| C | conductor |
| CDP | cymbial denticulate process |
| E | embolus |
| MA | median apophysis |
| P | paracymbium |
| PCS | pimoid cymbial sclerite |
| PEP | pimoid embolic process |

## Chapter 49
# PISAURIDAE
3 genera, 14 species

James E. Carico

### Common name —
Nursery web spiders.

### Similar families —
Lycosidae (p. 164), Trechaleidae (p. 249).

### Diagnosis —
This family can be distinguished from other North American lycosoid families by the combination of: a retrolateral tibial apophysis, non-flexible tarsi, moderately recurved PE row, legs prograde, construction of a nursery web, globular form of the egg sac, and manner of carrying the egg sac under the prosoma held by the chelicerae while also attached to the spinnerets.

### Characters —
**body size**: 0.5-3.7 cm.
**color**: ground color varies from light tan to dark reddish brown to greenish; light submarginal lines or bands often present on carapace; variable, including light longitudinal bands and/or pairs of light spots on abdomen, dark chevrons, mottling.
**carapace**: widest posteriorly with the cephalic area distinct and occasionally elevated.
**sternum**: truncate anteriorly and pointed posteriorly.
**eyes**: 4 AE subequal eyes in straight or moderately recurved row; larger 4 PE equal in a recurved row.
**chelicerae**: boss lacking; fang furrow margins toothed.
**legs**: prograde; I, II, IV subequal or IV somewhat longer; III always shortest; pairs of macrosetae on venter of tarsi, metatarsi, and tibia; trichobothria on dorsum of tarsi, metatarsi and tibia; *Dolomedes* species also have trichobothria on dorsum of femora.
**abdomen**: ovoid, moderately high.
**spinnerets**: three pairs, unmodified.
**respiratory system**: book lung spiracles at anterior third; tracheal spiracles near spinnerets.
**genitalia**: entelegyne; **female** epigynum composed of three distinct elevations in most genera [*Tinus peregrinus* (BISHOP 1924b) excepted], lateral ones cover genital atria; **male** retrolateral tibial apophysis prominent; median apophysis, tegulum conspicuous on the ventral surface of a complex genital bulb.

### Distribution —
Several species are common East of the 100th parallel; only one extends to the west of this line in the northwestern USA and Canada. Only two species have distributions restricted to West of this line in the southwestern USA (*Dolomedes gertschi* CARICO 1973 and *Tinus peregrinus*).

### Natural history —
All pisaurid species carry the spherical egg sac in the chelicerae and by a thread from the spinnerets. They also make a nursery web into which they suspend the egg sac where the spiderlings will hatch and spend at least the first instar. The female will defend the nursery from a perch nearby. Most species of *Dolomedes* are semi-aquatic and are located near bodies of water where they rest on vegetation, rocks, or

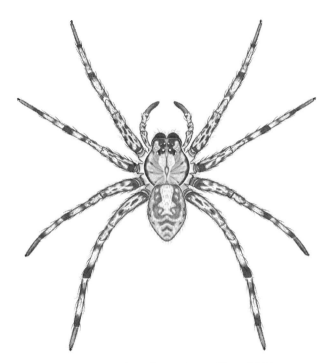

**Fig. 49.1** *Dolomedes tenebrosus* HENTZ 1844

debris at the margin of the water (Carico 1973) while hunting aquatic insect prey. Some species are known to catch small fish and therefore have earned the common name of "fishing spiders". *Pisaurina* are found in herbaceous vegetation and low shrubs typically in ecotonal areas beween grasslands and woods [*Pisaurina mira* (WALCKENAER 1837) and *Pisaurina brevipes* (EMERTON 1911)] or in vegetation at margins of streams and ponds [*Pisaurina dubia* (HENTZ 1847) and *Pisaurina undulata* (KEYSERLING 1887b), Carico (1976)]. Juveniles of *Pisaurina mira* are known to make silken retreats (Carico 1985). Mating pairs of *Pisaurina mira* copulate suspended from a dragline after the female is bound by a veil of the male's silk (Bruce & Carico 1988). *Tinus peregrinus* is found at the margin of streams (Carico 1976) similar to the habit of *Dolomedes*.

### Taxonomic history and notes —
The family is distributed worldwide (Platnick 2005) and is generally regarded as polyphyletic (Sierwald 1990). The family name "Dolomedidae" was proposed (Lehtinen 1967) for *Dolomedes* and other, mostly unrelated genera, but is not widely accepted. All North American genera were revised by Carico (1972a, 1973, 1976). Structure of genitalia was analyzed by Sierwald (1989, 1990). *Trechalea*, formerly regarded as a member of Pisauridae, is now placed into the related family, Trechaleidae (see chapter 66).

### Genera —
*Dolomedes* LATREILLE 1804b, *Pisaurina* SIMON 1898a, *Tinus* F.O. PICKARD-CAMBRIDGE 1901a

## Key to genera —
### North America North of Mexico

1. Transverse suture of lorum indistinct and not v-notched; 3 teeth on cheliceral retromargin; femora lacking trichobothria ............................................................................. **2**
   **Dist.** widespread

— Transverse suture of lorum distinct and v-notched (Fig. 49.2); 4 teeth on chelicera retromargin; trichobothria present on dorsal surface of femora .................. ***Dolomedes***
   **Div.** 9 species [Figs. 49.3; 49.6-49.7] — **Dist.** widespread primarily east of 100th parallel in aquatic and mesic forest habitats; restricted locally in the west — **Ref.** Carico 1973

2(1) Embolus a wide half to full circle (Fig. 49.4); conductor distal, transverse; broad median lobe on epigynum (Fig. 49.8) ................................................................ ***Pisaurina***
   **Div.** 4 species [Figs. 49.4; 49.8-49.9] — **Dist.** widespread in USA and s CAN e 100th parallel — **Ref.** Carico 1972a

— Embolus coiled 3-5 times (Fig. 49.5), conductor lateral and directed distally; only two elevations, laterally, on epigynum (Fig. 49.10) .......................................... ***Tinus***
   **Div.** 1 species: *Tinus peregrinus* (Bishop 1924b) [Figs. 49.5; 49.10-49.11] — **Dist.** limited to w USA in s CA-s TX, (?) MO — **Ref.** Carico 1976

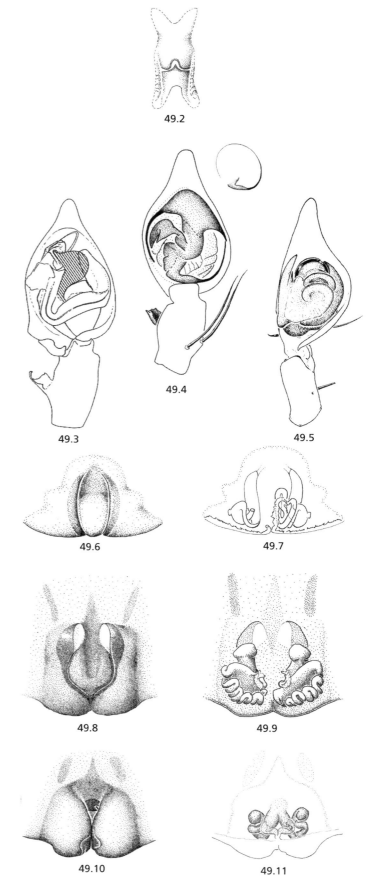

# Chapter 50
# PLECTREURIDAE
2 genera, 16 species and subspecies

Darrell Ubick

## Similar families —
Diguetidae (p. 102).

## Diagnosis —
Plecteurids differ from most other haplogynes in having eight eyes, and from those with eight eyes, in lacking a cribellum (present in Filistatidae), having fused chelicerae (free in *Calponia*, Caponiidae), and having relatively short and stocky legs (typically long and slender in Pholcidae. They are further separated from pholcids in having the eyes arranged in two straight rows (ALE and PE as triads in pholcids).

## Characters —
**body size:** 5.0 – 12.0 mm.
**color:** prosoma and appendages orange to chestnut brown to black; abdomen gray, often with pale cardiac mark (Fig. 50.1).
**carapace:** pyriform, cephalic area broad; fovea present, longitudinal.
**sternum:** suboval, anteriorly truncate.
**eyes:** eight eyes in two straight rows; AME somewhat smaller than others; LE on low mound.
**chelicerae:** fused basally (connected by membrane), with median lamina produced into apical tooth; ectal margin with indistinct stridulatory file.
**mouthparts:** labium triangular, endites strongly convergent, prolonged basoventrally into cone.
**legs:** with spines; tarsi with three claws; male tibia unmodified (*Kibramoa*) or with retrolateral clasping spur (*Plectreurys*); male tarsi subsegmented; female palp lacking claw; juvenile male palp elongated and sinuous.
**abdomen:** oval to spherical.
**spinnerets:** 6 small contiguous spinnerets, ALS with distal segment prolonged into hook, PLS lacking spigots but median pit enclosing row of setae; colulus represented by two setae.
**respiratory system:** 1 pair of book lungs; tracheal system absent, with small vestigial spiracle near spinnerets.
**genitalia:** haplogyne; ***female*** with gonopore region swollen, somewhat sclerotized, and pigmented; ***male*** palp with small tarsus and spherical bulb; embolus sinuous, threadlike to swollen; conductor absent.

## Distribution —
Plectreurids are known only from southwestern USA to Mexico and the Caribbean. *Kibramoa* ranges from central California and Nevada to Arizona and Baja California and *Plectreurys* extends the distribution n to Idaho and w to sw Texas. The majority of species are from southern California and Baja California.

## Natural history —
Plectreurids are sedentary hunters most abundant in deserts and other arid regions, but ranging also into mixed broadleaf and coniferous forests. They are ground spiders and build tangle webs with retreats under rocks, fallen bark, and other ground litter. Males leave their retreats upon maturity and may be seen wandering at night. Both sexes may be collected in pitfall traps. Females construct lacy cocoon-like retreats for guarding the egg sac. In captivity, plectreurids are easily reared on a diet of flies and may live for several years.

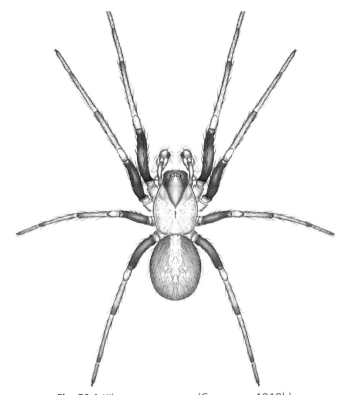

**Fig. 50.1** *Kibramoa suprenans* (CHAMBERLIN 1919b)

## Taxonomic history and notes —
These spiders were first recognized by Simon (1892a), who considered them a subfamily of his broad Sicariidae. Although shortly thereafter elevated to family rank (Banks 1898b), that placement was not always accepted. Chamberlin (1924b), who added new species from Baja California and described *Kibramoa*, placed them in the Scytodidae. Gertsch (1949) argued that, on the basis of their full eye complement, they are a distinct family, which he later revised (1958d). Plectreurids have not been studied subsequently and are in need of a revision.

Because of their fused chelicerae, plectreurids have been traditionally associated with the scytodoids and filistatids. Gertsch (1958d) placed them intermediate between these groups and, because of their eight eyes, considered them relatively primitive to the scytodoids (from which he excluded Pholcidae). Recent analysis, based on spinneret morphology, groups Plectreuridae with the 6-eyed Diguetidae, which together cluster with Pholcidae (Platnick et al. 1991).

## Genera —
*Kibramoa* CHAMBERLIN 1924b, *Plectreurys* SIMON 1893d.

## Key to genera —
### North America North of Mexico

**1**    Femur I straight and slender (barely thicker than femur II), much longer than carapace, with dorsal spines extending nearly to apex (Fig. 50.2); tibia I of male lacking clasping spur (Fig. 50.2); male palpal bulb with slender embolus (Fig. 50.3); female gonopore region (Fig. 50.4) ...
................................................................ ***Kibramoa***
         **Div.** 7 species and subspecies — **Dist.** central CA – AZ, NV — **Ref.** Gertsch 1958d — **Note** *Kibramoa suprenans* CHAMBERLIN 1924b is divided into two subspecies

—    Femur I curved and robust (clearly thicker than femur II), typically shorter than carapace, with dorsal spines absent or restricted to base (Fig. 50.5); tibia I of male with clasping spur (megaspine) (Fig. 50.5); male palpal bulb with embolus slender (*tristis* group) or broad (*castanea* group) (Fig. 50.6); female gonopore region (Fig. 50.7) ...........
................................................................ ***Plectreurys***
         **Div.** 9 species — **Dist.** sw USA e to NM and n to ID, most species in s CA – AZ — **Ref.** Gertsch 1958d — **Note** undescribed species are known in collections

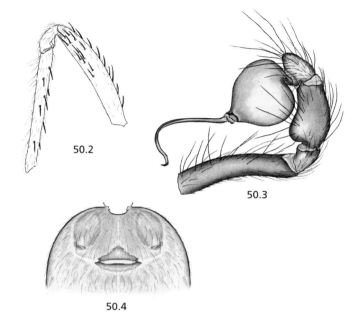

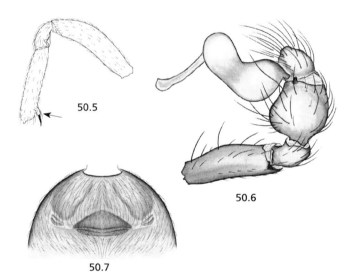

# Chapter 51
# PRODIDOMIDAE
2 genera, 2 species

Darrell Ubick

**Common name —**
Ground spiders.

**Similar families —**
Caponiidae (p. 75), Gnaphosidae (p. 106).

**Diagnosis —**
Prodidomids may be distinguished from other 2-clawed entelegyne spiders, except gnaphosids, by their enlarged ALS, concave endites, and modified PME (oval to triangular), and from gnaphosids by having flattened (rather than cylindrical) ALS with numerous long spigots, PER strongly procurved (straight to slightly procurved in gnaphosids, except *Scopoides*), and usually smooth tarsal claws (pectinate in gnaphosids and basal prodidomids).

**Characters —**
**body size:** 1.7-5.5 mm.
**color:** yellow to orange brown, abdomen pale gray to brown, sometimes with a pink tinge.
**carapace:** oval, low; fovea indistinct or absent.
**sternum:** oval, with precoxal triangles.
**eyes:** eight in two rows, PER strongly procurved; PME oval to triangular; AME dark, others silvery.
**chelicerae:** geniculate or not, margins lacking teeth (in Nearctic species).
**mouthparts:** endites triangular, with slight oblique medial depression, with serrula.
**legs:** prograde, weakly spined, tarsi with two smooth (usually) claws and claw tufts.
**abdomen:** elongate and somewhat flattened.
**spinnerets:** ALS enlarged, flattened and slightly separated, composed of one segment, apically with long spigots (pyriform) having greatly elongated bases.
**respiratory system:** with one pair of book lungs and single tracheal spiracle near spinnerets.
**genitalia:** entelegyne; *female* epigynum weakly sclerotized; *male* palp with RTA, consisting of a simple or complex prong, bulb simple.

**Distribution—**
Prodidomids are primarily tropical and south temperate in distribution and the two species that occur north of Mexico are from the extreme south of the region.

**Natural history—**
These spiders are relatively rare and little is known of their habits. They are nocturnal ground dwellers that seem to prefer relatively arid regions, where they occur under rocks and in ground litter, and occasionally even turn up in homes (Vetter 1996).

**Taxonomic history and notes—**
Prodidomidae has been recognized as a family since its inception by Simon (1893a) and considered closely related to Gnaphosidae, but clearly separated on the basis of eye and spinneret morphology. During the course of revisionary study of these spiders by Platnick & Shadab (1976c), several morphologically intermediate taxa were discovered

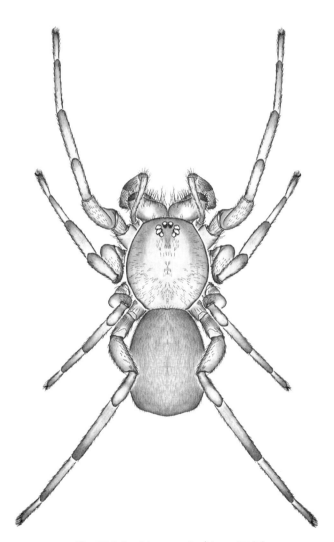

**Fig. 51.1** *Prodidomus rufus* (Hentz 1847)

and the two families synonymized. More recently, Platnick (1990a) examined spinneret microstructure in some detail and found evidence for revalidating Prodidomidae, which differ in the form of the spigots on the ALS.

**Genera —**
PRODIDOMINAE
*Prodidomus* Hentz 1847, *Neozimiris* Simon 1903a

# Prodidomidae

**Key to genera —**
North America North of Mexico

1  Chelicerae enlarged and divergent (Fig. 51.1); abdomen with pinkish tinge; PLS robust and curved, wider than ALS in lateral view (Figs. 51.2-51.3); male palp with RTA bifurcate and embolus sinuate (Fig. 51.4); female with epigynum having median septum (Fig. 51.5); size 3.0-5.5 mm .................................................... **Prodidomus**
   **Div.** 1 species: *Prodidomus rufus* (Hentz 1847) — **Dist.** s CA, TX to AL — **Refs.** Bryant 1935, Bryant 1949, Cooke 1964, Platnick 1976f — **Note** this species is believed to be a synanthrope, probably introduced into the New World (Platnick, pers. com.), and may occasionally turn up in northern regions, as the isolated record from New York City (Kaston 1981: 934)

—  Chelicerae normal (Fig. 51.6); abdomen grayish white; PLS not so enlarged, narrower than ALS in lateral view (Figs. 51.7-51.8); male palp with RTA a single prong and embolus straight (Fig. 51.9); female with epigynum lacking median septum (Fig. 51.10); size 1.7-4.0 mm ........... ............................................................. **Neozimiris**
   **Div.** 1 species: *Neozimiris pubescens* (Banks 1898b) — **Dist.** s CA to w AZ to Baja California — **Refs.** Cooke 1964, Platnick & Shadab 1976c

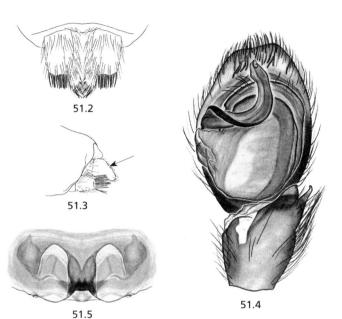

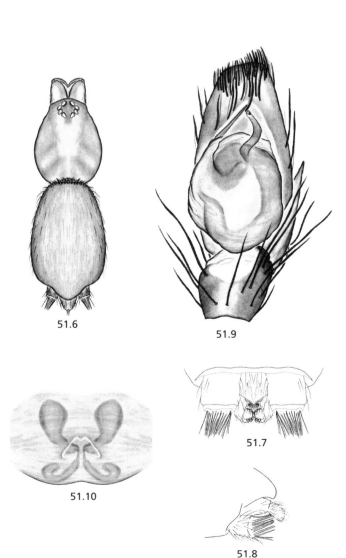

# Chapter 52
# SALTICIDAE
63 genera, 315+ species

David B. Richman
G.B. Edwards
Bruce Cutler

### Common name —
Jumping spiders.

### Similar families —
Salticidae are quite distinctive but could be confused with Corinnidae (p. 79), Oxyopidae (p. 189), and Thomisidae (p. 246).

### Diagnosis —
Tiny to large (1-22 mm) spiders. Ecribellate, entelegyne spiders with two tarsal claws and eight eyes with greatly enlarged AME. Both AER and PER recurved [according to Homann (1971), the PER would be procurved]; PER curved greatly, forming two eye rows, therefore typically with eyes in three rows with formula 4-2-2. A few genera with AER also greatly recurved forming a fourth eye row (2-2-2-2). PME and PLE near sides of carapace. Face truncate.

### Characters —
**body size:** 1-22 mm.
**color:** extremely variable, including brown, gray, black, etc., often with patterns of brilliant scales of red, yellow, metallic or iridescent green, blue, copper, gold, or silver, especially in males.
**carapace:** longer than wide (in ant-like species very elongated) to as wide as long in beetle-like species.
**sternum:** variable, from elongate to oval to nearly round, sometimes with narrow anterior end.
**eyes:** eight in three to four rows; AME especially enlarged, but ALE and PLE also large; PME small and sometimes difficult to see.
**chelicerae:** small to long, may be very stout, projecting and heavily toothed (especially in males).
**mouthparts:** endites and usually labium longer than wide; endites with well developed scopulae; labium variable in shape, may be narrowed anteriorly.
**legs:** prograde and moderately long to short, with first or fourth leg longest; tarsi with two claws, usually with claw tufts.
**abdomen:** quite variable, from elongate in some *Marpissa* and ant-like species to near round in some beetle–like species.
**spinnerets:** short and sometimes not obvious; anterior and posterior about same length.
**respiratory system:** one pair of book lungs; tracheal spiracle close to spinnerets.
**genitalia:** entelegyne; *female* epigyna variable, usually well sclerotized; *male* retrolateral tibial apophysis present. Although typically considered to have simple genitalia, these structures are as complex in salticids as other members of the Dionycha. This structural diversity has yet to be fully described.

### Distribution —
Salticidae is the largest family of spiders, presently with 5026 species in 550 genera worldwide (Platnick 2005). They are found abundantly on every continent except Antarctica, and occur in a broad range of microhabitats from under leaf litter to the top of the forest canopy. Tropical areas of the world are the most biodiverse. The United States and Canada combined have over 315 species in 63 genera (Richman & Cutler 1978, Platnick 2005, and unpublished sources). Although modest when compared to tropical faunas, this is a considerable number, and the task of identification can be daunting. Mexico, much of which is part of the Nearctic, probably has an even more complex salticid fauna than the United States and Canada, but only about 200 species are known from the country (Hoffman 1976, Richman & Cutler 1988). Although Mexico is not part of the geographic area covered by this book, the key should serve well for the most northern Mexican states.

### Natural history —
Apparent mimicry (e.g., of ants, beetles) and crypsis (e.g., resemblance to bark, grass, or stone) are abundantly represented in both body form and color pattern. Some species are gaudily colored, others have specially evolved decorations on various body parts. Males are particularly ornate, and utilize decorations in complex courtship behaviors. Some species (such as in some *Phidippus* and many, if not

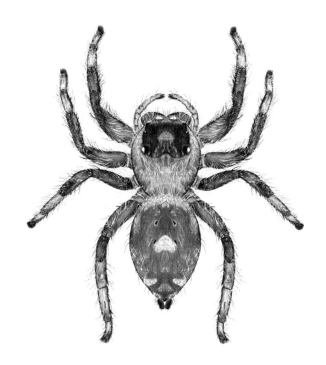

**Fig. 52.1** *Phidippus regius* C.L. Koch 1846d, female, gray phase

all, *Habronattus*) have stridulatory mechanisms for communication, which are often used in courtship. To mate, a male mounts over the female's anterior dorsum and turns her abdomen to apply the closest palp to the epigynum, then moves to the opposite side to engage the other palp. Ritualized male agonistic behavior exists in many genera and may to some extent resemble courtship. This may occasionally end in sparring contests using the chelicerae if the spiders are of similar size, although the smaller male usually decamps if there is a definite size disparity. Maternal care of the eggs seems to be the general rule, with females usually closely associated with both eggs and newly eclosed spiderlings until the spiderlings molt to the first free-living instar and disperse. The female may lay her eggs in a distinct sac (most salticids) or lay only a thin sheet of silk over them (e.g., *Lyssomanes*). Predatory behavior is strongly reminiscent of cats, with some species showing remarkable abilities to locate prey and develop prey capture tactics to optimize efficiency and avoid injury. Richman & Jackson (1992) reviewed the behavior of these fascinating spiders. Most species are probably univoltine, with adults present only during a few months of the year, as in *Phidippus johnsoni* PECKHAM & PECKHAM 1883 and *Lyssomanes viridis* (WALCKENAER 1837). Instars in these species apparently vary from 5 to 11 (Jackson 1978, Richman & Whitcomb 1981).

**Taxonomic history and notes —**

Salticidae includes such a large number of genera and species that its taxonomic history is extremely complicated. Hentz (1832, 1845, 1846), Walckenaer (1837), C.L. Koch (1846d), and Keyserling (1885) made early significant contributions to descriptions of North American Salticidae. George and Elizabeth Peckham (1888, 1909) and F.O. Pickard-Cambridge (1901a) laid the groundwork for modern studies on the Salticidae in North America with their revisionary work. Emerton (1891), Gertsch (1936a), Chamberlin & Ivie (1944), Kaston (1948), Lowrie & Gertsch (1955), and Gertsch & Riechert (1976) include various descriptions of species. Barnes (1955, 1958) revised the Marpissinae, including *Maevia* and *Paramaevia*. Cokendolpher & Horner (1978) revised *Poultonella*. Cutler (1981b, 1988a, 1990) revised *Paradamoetas*, *Synageles* and *Chalcoscirtus*; and (1982) recorded *Pseudeuophrys erratica* (WALCKENAER 1826) for the first time in the United States (as *Euophrys*). Richman (1978) wrote the first complete key to salticid genera of the entire region after Peckham & Peckham (1909) and Richman & Cutler (1978) published a list of the known species for the United States and Canada. Cutler (1979) and Richman (1979, 1980) published corrections to the key and list. Richman (1981) revised *Habrocestum* (now *Naphrys* and *Chinattus*), while also assigning several species to *Tylogonus* (now *Mexigonus*) and later (1989) revised the genus *Hentzia*. Richman & Vetter (2004) reviewed the US species of the genus *Thiodina*. Edwards (1979a) reviewed pantropical jumping spiders occurring in Florida; (1980) further updated the key by Richman and list by Richman & Cutler; (1999a) revised *Attidops*; (1999b) reassigned *Corythalia canosa* (WALCKENAER 1837) to *Anasaitis*. Edwards also (2003a) reviewed the Nearctic Euophryinae, in the process removing the species of *Habrocestum* and *Tylogonus* to the new genera *Naphrys* and *Mexigonus* (while moving one species of *Habrocestum* to *Chinattus*); (2003b) described a new species of *Neonella*; (2004) revised the genus *Phidippus*, while resurrecting the genus *Paraphidippus*. Hill (1979) described the new genus *Platycryptus*, to which he assigned the species *Metacyrba undata* (DE GEER 1778b) and *Metacyrba californica* (PECKHAM & PECKHAM 1888). Edwards (in press) transferred *Metacyrba arizonensis* BARNES 1958 to *Platycryptus*. Maddison (1978) transferred *Bianor aemulus* (GERTSCH 1934d) from *Sassacus*; this species was subsequently placed in a new genus *Sibianor* by Logunov (2001a). Maddison (1986) also distinguished between *Eris militaris* (HENTZ 1845) and *Eris flava* (PECKHAM & PECKHAM 1888); (1987) redescribed *Marchena*; and (1996) revised *Pelegrina* and *Metaphidippus*, while describing or redescribing and defining the genera *Phanias*, *Terralonus*, *Ghelna*, *Bagheera* and *Messua*. Marusik & Logunov (1998) reviewed the *Evarcha falcata* group, including *Evarcha hoyi* (PECKHAM & PECKHAM 1883) and described *Evarcha proszynskii*. Logunov & Cutler (1999) revised *Paramarpissa*, assigning *Pseudicius piraticus* (PECKHAM & PECKHAM 1888) to this genus. Platnick (1984a) reviewed *Cheliferoides*. Prószyński (1968c, 1971b, 1973a, 1980) revised several species groups of *Sitticus*. Griswold (1987a) revised the large genus *Habronattus*. Recent molecular analyses have resulted in proposed phylogenies for the genus *Habronattus* (Maddison & Hedin 2003a), the subfamily Dendryphantinae (Hedin & Maddison 2001), and a major portion of the family (Maddison & Hedin 2003b).

**Genera —**
**AMYCOIDA**
  HYETUSSINAE
    *Bredana* GERTSCH 1936a, *Cyllodania* SIMON 1902d
  SITTICINAE
    *Sitticus* SIMON 1901a
  SYNEMOSYNINAE
    *Sarinda* PECKHAM & PECKHAM 1892, *Synemosyna* HENTZ 1846
  THIODININAE
    *Thiodina* SIMON 1900b
**MARPISSOIDA**
  DENDRYPHANTINAE
    *Agassa* SIMON 1901a, *Bagheera* PECKHAM & PECKHAM 1896, "*Beata*" PECKHAM & PECKHAM 1895, *Bellota* PECKHAM & PECKHAM 1892, "*Cheliferoides*" F.O. PICKARD-CAMBRIDGE 1901a, *Dendryphantes* C.L. KOCH 1837b, *Eris* C.L. KOCH 1846d, *Ghelna* MADDISON 1996, *Hentzia* MARX in HOWARD 1883, *Messua* PECKHAM & PECKHAM 1896, "*Metaphidippus*" F.O. PICKARD-CAMBRIDGE 1901a, *Paradamoetas* PECKHAM & PECKHAM 1885b, *Paraphidippus* F.O. PICKARD-CAMBRIDGE 1901a, *Pelegrina* FRANGANILLO BALBOA 1930, *Phanias* F.O. PICKARD-CAMBRIDGE 1901a, *Phidippus* C.L. KOCH 1846d, *Poultonella* PECKHAM & PECKHAM 1909, "*Pseudicius*" SIMON 1885e, *Rhetenor* SIMON 1902d, *Sassacus* PECKHAM & PECKHAM 1895, *Terralonus* MADDISON 1996, *Tutelina* SIMON 1901a, *Zygoballus* PECKHAM & PECKHAM 1885b
  MARPISSINAE
    *Maevia* C.L. KOCH 1846d, *Marpissa* C.L. KOCH 1846d, *Metacyrba* F.O. PICKARD-CAMBRIDGE 1901a, *Paramaevia* BARNES 1955, *Platycryptus* HILL 1979
  MARPISSOIDA INCERTAE SEDIS
    *Admestina* PECKHAM & PECKHAM 1888, *Attidops* BANKS 1905a, *Cheliferoides* F.O. PICKARD-CAMBRIDGE 1901a
  SYNAGELINAE
    *Peckhamia* SIMON 1901a, *Synageles* SIMON 1876a
**PLEXIPPOIDA**
  PELLENINAE
    *Habronattus* F.O. PICKARD-CAMBRIDGE 1901a, *Pellenes* SIMON 1876a, *Sibianor* LOGUNOV 2001a
  PLEXIPPINAE
    *Evarcha* SIMON 1902a, *Plexippus* C.L. KOCH 1846d
**UNPLACED SUBFAMILIES**
  AELURILLINAE
    *Phlegra* SIMON 1876a

EUOPHRYINAE
> *Anasaitis* BRYANT 1950, *Chalcoscirtus* BERTKAU 1880c, *Corythalia* C.L. KOCH 1850, *Euophrys* C.L. KOCH in HERRICH-SCHÄFFER 1834, *Mexigonus* EDWARDS 2003a, *Naphrys* EDWARDS 2003a, *Pseudeuophrys* F. DAHL 1912b, *Talavera* PECKHAM & PECKHAM 1909

HELIOPHANINAE
> *Marchena* PECKHAM & PECKHAM 1909, *Menemerus* SIMON 1868b

LYSSOMANINAE
> *Lyssomanes* HENTZ 1845

MYRMARACHNINAE
> *Myrmarachne* MACLEAY 1839

SALTICINAE
> *Salticus* LATREILLE 1804b

INCERTAE SEDIS
> *Chinattus* LOGUNOV 1999a, *Hasarius* SIMON 1871a, *Neon* SIMON 1876a, *Neonella* GERTSCH 1936a, *Paramarpissa* F.O. PICKARD-CAMBRIDGE 1901a

## Characters unique to certain genera —

1. Eyes in 4 rows, spiders translucent green: *Lyssomanes*.
2. Ant-like species with female palpi swollen distally or flattened, male chelicerae projecting forward: *Sarinda*, *Myrmarachne*.
3. Two pair of ventral setae on tibia I with bulbous bases: *Thiodina*.
4. Tibia I about as wide as long, referred to as pseudoscorpion-like by some authors: *Admestina*, *Bellota*, *Cheliferoides*, and *Poultonella*; *Habronattus oregonensis* (PECKHAM & PECKHAM 1888) appears to have wide tibiae because of "hair" fringes.
5. Tibia I with 4 pairs of ventral macrosetae: *Maevia*, *Paramaevia* and *Marpissa*.
6. Femur I with a prolateral row of 3-6 seta-bearing tubercles: *Marchena*.
7. Long spatulate scales on the ventral retromargin of femur and patella I and below PME, especially in females: *Hentzia*.
8. Abdomen hard, sclerotized, shiny purple or green: *Agassa* and *Sassacus*.
9. Spiders with more or less extensive iridescent or metallic scales: *Agassa*, *Messua*, *Bagheera*, *Paradamoetas*, *Sassacus*, *Salticus peckhamae* (COCKERELL 1897), most *Paraphidippus*, females of *Eris*, and females of *Tutelina*.
10. Ant-like: *Bellota*, *Synemosyna*, *Sarinda*, *Myrmarachne formicaria* (DE GEER 1778b), *Paradamoetas*, *Peckhamia*, *Synageles*, *Tutelina formicaria* (EMERTON 1891) (and others in this genus to some extent), and an undescribed genus from Arizona.
11. Small (1.8-3.0 mm), sometimes lightly pigmented species found in leaf litter: *Neon* (carapace often dark), *Neonella* (one dark species from Florida), *Talavera* and *Chalcoscirtus*.
12. The largest, usually 8-14 mm long (up to 22 mm), stout jumping spiders, with dorsal hair tufts behind the PME (at least in females) and/or iridescent chelicerae (usually green, but varies throughout colors of visible spectrum): *Phidippus*.
13. A few genera are restricted to various geographic locales, such as South Texas: *Bellota*, *Bredana*, *Rhetenor*; Arizona: *Cyllodania*; West Coast: *Marchena*; or Florida: "*Beata*".

# Salticidae

**Key to genera —**
## North America North of Mexico
Modified after Richman (1978)

Edwards & Hill (1978) photographically illustrated many examples of the genera. Genera listed in quotation marks are almost certainly incorrect, but have not been officially changed

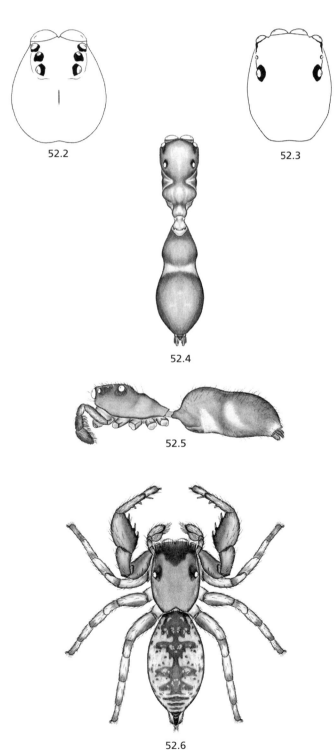

52.2

52.3

52.4

52.5

52.6

1   ALE well behind AME so that eyes appear in 4 rows (Fig. 52.2); living spiders bright translucent green with spindly legs, often found on the undersides of tree leaves ............ ........................................................... ***Lyssomanes***
    **Div.** 1 species: *Lyssomanes viridis* (WALCKENAER 1837) — **Dist.** se USA to c TX, s into MEX — **Ref.** Galiano 1980a

—   Eyes appear in three rows (in order ALE-AME, PME, PLE) (Fig. 52.3), not translucent green when alive .................. **2**

2(1) Ant-like (abdomen constricted, Figs. 52.4-52.5) or pseudoscorpion-like (tibia swollen, larger in diameter than the metatarsus or patella, Fig. 52.6), especially when alive, but resemblance is noticeable in most, even when dead ....... **3**

—   Not ant-like or pseudoscorpion-like, even when alive (*Paradamoetas* is not especially ant-like when dead and appears twice in the key) ................................................ **11**

3(2) Tibia I almost as wide as long (Fig. 52.6) ......................... **4**

—   Tibia I more than twice as long as wide ........................... **5**

4(3) Labium wider than long; male with spiral embolus; pseudoscorpion-like (Fig. 52.6) ...................... ***Cheliferoides***
    **Div.** 1 species: *Cheliferoides segmentatus* F.O. PICKARD-CAMBRIDGE 1901a — **Dist.** AZ, TX s into MEX — **Ref.** Platnick 1984a

—   Labium longer than wide; male with curved or straight embolus; ant-like ................................................. ***Bellota***
    **Div.** 2 species: *Bellota micans* PECKHAM & PECKHAM 1909, *Bellota wheeleri* PECKHAM & PECKHAM 1909 — **Dist.** TX — **Ref.** Galiano 1972a — **Note** an undescribed genus from AZ keys out to *Bellota*. This species appears to mimic tiny ants on oak trees; it is black and about 2-3 mm long

5(3) Carapace entirely flat, without constrictions or transverse lateral white bands; tibia I about 40% as wide as long, noticeably wider than patella ............... ***"Cheliferoides"***
    **Div.** 1 species: *"Cheliferoides" longimanus* GERTSCH 1936a — **Dist.** FL and TX — **Ref.** Gertsch 1936a — **Note** not a true *Cheliferoides*

—   Carapace at most flattened in cephalic area, with lateral constrictions or white bands imitating constrictions; tibia I not noticeably thickened or wider than patella ............ **6**

6(5) Posterior portion of carapace narrow, with parallel lateral margins, adding to apparent length of pedicel (Fig. 52.4). Very slender species resembling ants of the genus *Pseudomyrmex* ............................................. ***Synemosyna***
    **Div.** 2 species — **Dist.** e USA w to KS and TX [*Synemosyna formica* HENTZ 1846, a dark species], s FL [*Synemosyna petrunkevitchi* (CHAPIN 1922), a yellow species] — **Ref.** Galiano 1966a

—   Posterior portion of carapace without parallel lateral margins; resembling other ant genera ....................... **7**

**7(6)** Male chelicerae projecting forward greater than ½ length of carapace; female palpal tarsus dorsoventrally flattened, bent downward distally ............................ **Myrmarachne**
   **Div.** 1 species: *Myrmarachne formicaria* (De Geer 1778b) — **Dist.** OH — **Ref.** Bradley 2003 — **Note** this species recently discovered in NA is introduced from Europe

— Male chelicerae ½ length of carapace or less; female palpal tarsus never flattened and bent ......................... **8**

**8(7)** Distinct declivity and transverse groove behind PLE (*S. cutleri* lacks groove), both male and female with enlarged palpi (Fig. 52.5). Possible mimics of ponerine and formicine ants ............................................ **Sarinda**
   **Div.** 2 species — **Dist.** e USA, AZ [*Sarinda hentzi* (Banks 1913)], CA-AZ [*Sarinda cutleri* (Richman 1965)] — **Refs.** Galiano 1965b, 1969a

— Neither distinct declivity nor transverse groove; female with slender palpi ............................................. **9**

**9(8)** Cheliceral retromargin with simple tooth (Fig. 52.7) ....... ................................................................. **Paradamoetas**
   **Div.** 2 species — **Dist.** s TX [*Paradamoetas formicinus* Peckham & Peckham 1885b], Great Lakes region [*Paradamoetas fontanus* (Levi 1951)] — **Ref.** Cutler 1981b

— Cheliceral retromargin with compound tooth (Fig. 52.8) ..................................................................... **10**

**10(9)** Tibia I with 3 pairs of macrosetae; embolus forming spiral or at least making complete 360° turn (Fig. 52.9); epigynum with anterior rims sclerotized in form of two arcs (Fig. 52.10); eye region occupying half of carapace (Fig. 52.11). Possible mimics of arboreal *Crematogaster* and small formicine ants ........................................ **Peckhamia**
   **Div.** 4 species — **Dist.** widespread — **Ref.** Peckham & Peckham 1909

— Tibia I with 2 pairs of macrosetae (except *Synageles mexicanus* Cutler 1988a has 3 pairs); embolus a simple straight rod, curved arc or short spike (Fig. 52.12); epigynum lacking anterior sclerotized arcs; eye region occupying more than half of carapace. Possible mimics of small dolichoderine, small formicine and small *Crematogaster* ants ........................................... **Synageles**
   **Div.** 7 species (1 introduced) — **Dist.** widespread — **Refs.** Cutler 1988a, Hutchinson & Limoges 1998

**11(2)** Tibia I with 2 pairs of ventral bulbous setae (Fig. 52.13) .. ............................................................................. **Thiodina**
   **Div.** 3 species — **Dist.** e USA to TX and west to CA, OR, s into MEX — **Ref.** Richman & Vetter 2004

— Tibia I lacking bulbous setae ......................................... **12**

**12(11)** Femur I with a row of setae-bearing tubercles (Fig. 52.14) opposite a file of tubercles on the carapace ... **Marchena**
   **Div.** 1 species: *Marchena minuta* (Peckham & Peckham 1888) — **Dist.** West Coast — **Ref.** Maddison 1987

— Femur I and carapace lacking opposing tubercles ....... **13**

**13(12)** With three retromarginal teeth. Male with embolus broad, partially membranous, and arising from prolateral base of tegulum. Tiny, 2 mm ........................................ **Cyllodania**
   **Div.** 1 apparently undescribed species — **Dist.** AZ

— With less than 3 retromarginal teeth ............................. **14**

**14(13)** Tibia I without macrosetae [macrosetae difficult to see in male *Euophrys monadnock* Emerton 1891, which has three pairs on the ventral tibia; if spider is black, femora III and IV are red, orange or yellow, see *Euophrys*, couplet 57] .... ................................................................................... **15**

— Tibia I with at least one macroseta .............................. **17**

52.7

52.8

52.9

52.10

52.11

52.12

52.13

52.14

**15(14)** Tibia I almost as wide as long .................. ***Admestina***
  **Div.** 3 species — **Dist.** e USA w to SD, TX — **Refs.** Galiano 1989a, Piel 1992

— Tibia I much longer than wide ........................ **16**

**16(15)** Metatarsus I with at least a pair of ventral macrosetae ......
  ................................................................. ***Bredana***
  **Div.** 2 species: *Bredana complicata* Gertsch 1936a, *Bredana alternata* Gertsch 1936a — **Dist.** TX — **Ref.** Gertsch 1936a

— Metatarsus I without macrosetae ..................... ***Salticus***
  **Div.** 4 described plus several undescribed species from sw and Pacific coast — **Dist.** widespread in n USA, s CAN [*Salticus scenicus* (Clerck 1757), probably introduced from Europe], other 3 species from sw USA — **Ref.** Peckham & Peckham 1909

**17(14)** Tibia I with 4 pairs of ventral macrosetae ................... **18**

— Tibia I with fewer than 4 pairs of ventral macrosetae (if no ventral macrosetae, at least one lateral macroseta) ..... **20**

**18(17)** Carapace height no more than 40-60% of its greatest width. Male palp with proximal part of tegulum conical, projecting ventrally (Fig. 52.15). Epigynum with 2 distinct posterior copulatory openings .................... ***Marpissa***
  **Div.** 10 species — **Dist.** e USA to CA, s into MEX — **Ref.** Barnes 1955

— Carapace height at least 60-70% of its greatest width. Male palp with tegulum flat, not conical or projecting ventrally. Epigynum with single median anterior copulatory opening ................................................................. **19**

**19(18)** Embolus thread-like (Fig. 52.16), epigynal copulatory opening tiny ................................................ ***Maevia***
  **Div.** 3 species + 2 undescribed — **Dist.** e USA — **Refs.** Barnes 1955, 1958

— Embolus broader (Fig. 52.17), elongated tip sclerotized on each side with a membranous center (Fig. 52.18); epigynal copulatory opening enlarged ................... ***Paramaevia***
  **Div.** 3 species — **Dist.** AZ to FL, n to NC, s into MEX — **Refs.** Barnes 1955, 1958 — **Note** included in *Maevia* by Platnick 2005

**20(17)** Tibia I with no ventral macrosetae, one prolateral macroseta. Elongate body; long, thick leg I ....... ***Paramarpissa***
  **Div.** 3 species — **Dist.** CA to TX, s into MEX — **Ref.** Logunov & Cutler 1999

— Tibia I with one or more ventral macrosetae ............... **21**

**21(20)** Tibia I with one ventral macroseta .......................... **22**

— Tibia I with at least two ventral macrosetae ................ **24**

**22(21)** Tibia I as wide as long. Body covered with white or orange scales ................................................ ***Poultonella***
  **Div.** 2 species — **Dist.** NY, VA, IA, KS, OK, TX to AZ [*Poultonella alboimmaculata* (Peckham & Peckham 1883)], TX (*Poultonella nuecesensis* Cokendolpher & Horner 1978) — **Ref.** Cokendolpher & Horner 1978

— Tibia I longer than wide ........................................ **23**

**23(22)** Carapace almost as wide as long, not flattened; PME twice as far from PLE as from ALE (Fig. 52.19); small, metallic green or purple, somewhat resembling flea beetles ..........
  ................................................................. ***Agassa***
  **Div.** 1 species: *Agassa cyanea* (Hentz 1846) — **Dist.** e USA w to WI, NM — **Ref.** Peckham & Peckham 1909 — **Note** similar undescribed species in MEX

— Carapace much longer than wide, flattened; PME about halfway between PLE and ALE; neither metallic nor beetle-like ....................................................... ***Attidops***
  **Div.** 4 species — **Dist.** s-central CAN, e USA, e MEX — **Ref.** Kaston 1948 (as *Ballus*), Edwards 1999a

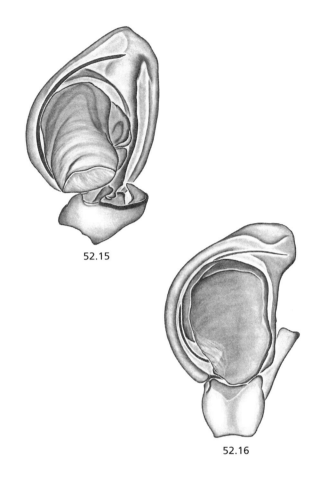

52.15

52.16

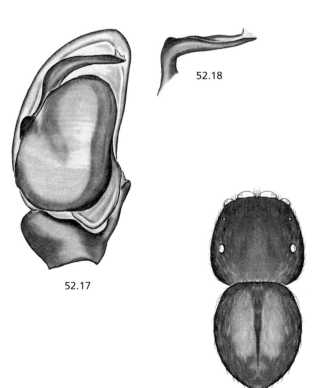

52.18

52.17

52.19

**24(21)** Cheliceral retromargin lacking teeth ............................ **25**
— Cheliceral retromargin with at least one tooth ............ **26**

**25(24)** Light colored scales on carapace, often forming a pattern. Often with narrow central white stripe, many with butterfly-like marking on abdomen ................................ ***Sitticus***
  **Div.** 13 species — **Dist.** — widespread — **Refs.** Prószyński 1968c, 1971b, 1973a, 1980

— No scales on carapace, most setae dark except on anterior edge of carapace ........................................ ***Chalcoscirtus***
  **Div.** 3 species — **Dist.** widespread — **Refs.** Cutler 1990, Edwards 2003a

**26(24)** Patella-tibia III as long or longer than patella-tibia IV ... **27**
— Patella-tibia III distinctly shorter than patella-tibia IV ... **35**

**27(26)** OQ as wide as or wider anteriorly than posteriorly .... **28**
— OQ wider posteriorly than anteriorly ............................ **33**

**28(27)** Teeth of cheliceral retromargin compound (bicuspid) ..... ................................................................................ ***Hasarius***
  **Div.** 1 species: *Hasarius adansoni* (AUDOUIN 1826) — **Dist.** introduced into MA, NY — **Ref.** Kaston 1948

— Teeth of cheliceral retromargin simple .......................... **29**

**29(28)** OQ distinctly wider anteriorly than posteriorly ........... **30**
— OQ as wide posteriorly as anteriorly .............................. **31**

**30(29)** Embolus a moderately short, 3-dimensional spiral with broad conical projection on basal embolar disk (Fig. 52.20); epigynum with two large translucent atria more or less encircled by rim-like grooves that posteriorly intersect medial copulatory openings (Fig. 52.21) .......... ***Naphrys***
  **Div.** 4 species — **Dist.** se CAN, e half USA to e NM, s into MEX — **Refs.** Richman 1981 (as *Habrocestum*), Edwards 2003a

— Embolus a flat attenuate projection from a broader base (Fig. 52.22); epigynum without large atria, copulatory openings lateral (Fig.52. 23) ........................... ***Chinattus***
  **Div.** 1 species: *Chinattus parvulus* (BANKS 1895e) — **Dist.** se CAN, e half of USA to n FL — **Refs.** Richman 1981 (as *Habrocestum parvulum*), Edwards 2003a

**31(29)** Embolus a short arc arising at acute angle, tegulum as wide as long (Fig. 52.24); epignal plate longer than wide with minute median raised arc. Male with dorsal black and white stripes extending onto face ............ ***Plexippus***
  **Div.** 1 species: *Plexippus paykullii* (AUDOUIN 1826) — **Dist.** introduced to southern USA, widespread in MEX — **Ref.** Edwards 1979a

— Embolus a distal spiral or short rod, not at acute angle, tegulum longer than wide (Figs. 52.25-52.26); epignal plate as wide as or wider than long (Figs. 52.27-52.28) .... ................................................................................ **32**

**32(31)** Embolus an extended spiral (Fig. 52.25), epigynum with strongly spiral atrial rims (Fig. 52.27). Legs of males dark metallic blue, any to all legs heavily fringed with black setae dorsally and ventrally ............................. ***Corythalia***
  **Div.** 2 species: *Corythalia conspecta* (PECKHAM & PECKHAM 1896) and *Corythalia opima* (PECKHAM & PECKHAM 1885b) — **Dist.** AZ — **Ref.** Peckham & Peckham 1909 — **Note** possibly neither occurs n of Mexico (Edwards 2003a)

— Embolus a bent, twisted rod (Fig. 52.26), epigynum without atrial rims, atria very faint or not discernable (Fig. 52.28). Male legs neither metallic blue nor fringed .. ................................................................................ ***Anasaitis***
  **Div.** 1 species: *Anasaitis canosa* (WALCKENAER 1837) — **Dist.** TX to FL, n to SC — **Ref.** Edwards 1999b

52.20
52.21
52.23
52.22
52.25
52.24
52.26
52.27
52.28

## Salticidae

**33(27)** Male tegulum with small, prolateral, proximal, conical projection (Fig. 52.29); epigynum with lateral pockets (Fig. 52.30) .................................................. ***Evarcha***
  **Div.** 2 species — **Dist.** CAN, n USA [*Evarcha hoyi* (Peckham & Peckham 1883)], w CAN, mountains of w USA [*Evarcha proszynskii* Marusik & Logunov 1998] — **Ref.** Marusik & Logunov 1998

— Male tegulum angulate or rounded, but never with prolateral conical projection; epigynum lacking lateral pockets .................................................................................. **34**

**34(33)** Males with embolus long, slender, tip exposed in transverse cymbial groove; tegular apophysis shorter, usually elbowed at base, arising basally to almost laterally (Figs. 52.31-52.32); with elaborate courtship ornaments on legs, palpi, and/or carapace. Female epigynum with copulatory openings hidden at side of crescent-shaped sclerotized bars lateral to or anterior to median hood-like pocket (Figs. 52.33-52.34) ........................ ***Habronattus***
  **Div.** 66 species and several undescribed — **Dist.** widespread — **Ref.** Griswold 1987a

— Males with embolus short, hidden by tegular apophysis; latter never elbowed, arising laterally to distally (Fig. 52.35); distal cymbial groove not transverse (except *Pellenes crandalli* Lowrie & Gertsch 1955 from AZ); lacking conspicuous courtship ornaments. Female epigynum with copulatory openings on surface posterior to pocket; sclerotized bars absent (Fig. 52.36) .......... ***Pellenes***
  **Div.** 11 species — **Dist.** widespread — **Ref.** Lowrie & Gertsch 1955

**35(26)** Tibia I with 3 ventral macrosetae ................................. **36**

— Tibia I with more than 3 ventral macrosetae (in *Sibianor* there may be 3 or 4) ..................................................... **37**

**36(35)** Leg IV longest, followed by leg III. Beetle-like ................... ................................................................... ***Sassacus*, in part**
  **Div.** 2 of 3 species key out here — **Dist.** s CAN throughout much of USA s into MEX — **Ref.** Peckham & Peckham 1909 — **Note** *Sassacus barbipes* (Peckham & Peckham 1888) is excluded from the key because it does not appear to actually be found north of southern Sonora in MEX

— Leg I longest, followed by leg IV ........... ***Phanias*, in part**
  **Div.** 8 species (see note) — **Dist.** sw CAN s to CA to TX, CO — **Ref.** Maddison 1996 — **Note** this genus is poorly known, number of species that key out here is uncertain

**37(35)** OQ occupying at least ½ of carapace ........................... **38**

— OQ occupying distinctly less than ½ of carapace ........ **45**

**38(37)** PME equidistant between ALE and PLE ...................... **39**

— PME closer to ALE than PLE .......................................... **42**

**39(38)** OQ greater than ½ of carapace length (Fig. 52.37). Tiny, pale to brown species found commonly in leaf litter ........ ............................................................................... ***Neon***
  **Div.** 6 species — **Dist.** widespread — **Ref.** Gertsch & Ivie 1955

— OQ occupying ½ of carapace length ............................ **40**

**40(39)** Length at least 3 mm; legs slender, often with lineate markings ................................................... ***Paradamoetas***
  **Div.** 2 species — **Dist.** s TX [*Paradamoetas formicinus* Peckham & Peckham 1885b] and Great Lakes region [*Paradamoetas fontanus* (Levi 1951)] — **Ref.** Cutler 1981b

— Length 1.0-2.6 mm; legs stout, lacking lineate markings .. ........................................................................................ **41**

52.29
52.30
52.31
52.32
52.33
52.34
52.35
52.36
52.37

**41(40)** Labium wider than long; body dull brownish, hairy (males gray in life). Tiny salticids found in leaf litter and on rocks in the north and high mountains in the west ... ***Talavera***
  Div. 1 species: *Talavera minuta* (Banks 1895e) [Figs. 52.38-52.39] — Dist. CAN (all but 4 prov.), n USA, KS, TX, NM, UT, CA — Ref. Kaston 1948, Paquin & Dupérré 2003

— Labium longer than wide; body shiny, yellow to orange, or dark brown with iridescent sheen. In leaf litter ................................................................ ***Neonella***
  Div. 2 species — Dist. se USA from NC to TX [*Neonella vinnula* Gertsch 1936a], s FL [*Neonella camillae* Edwards 2003b] — Refs. Gertsch 1936a, Galiano 1988b, Edwards 2003b

**42(38)** Patella-tibia I much enlarged and flattened, especially in the male ............................................................... ***Sibianor***
  Div. 1 species: *Sibianor aemulus* (Gertsch 1934d) [Figs. 52.40-52.41] — Dist. QC, ON, AB, ME, MN, WI — Refs. Gertsch 1934d (as *Sassacus*), Maddison 1978 (as *Bianor*), Logunov 2001a, Paquin & Dupérré 2003

— Patella-tibia I not enlarged and flattened ..................... **43**

**43(42)** Carapace rounded; male lacking cheliceral projections .... ............................................................... ***Sassacus***, in part
  Div. 3 species, 1 keys out here — Dist. s CAN throughout much of USA s into MEX — Ref. Peckham & Peckham 1909 — Note *Sassacus barbipes* (Peckham & Peckham 1888) is excluded from the key because it does not appear to actually be found north of southern Sonora in MEX. Individuals of *Sassacus papenhoei* Peckham & Peckham 1895 will key out here

— Carapace angulate, nearly vertical behind PLE (Fig. 52.42); male with cheliceral projections (Fig. 52.43) ............... **44**

**44(43)** Labium as long as or longer than wide; profile of carapace angulate and box-like ..................................... ***Zygoballus***
  Div. 3 species — Dist. e USA, AZ, s into MEX — Ref. Peckham & Peckham 1909

— Labium wider than long; carapace extremely angulate and box-like .......................................................... ***Rhetenor***
  Div. 1 species: *Rhetenor texanus* Gertsch 1936a — Dist. TX, s into MEX — Ref. Gertsch 1936a

**45(37)** PME closer to ALE than to PLE (measured between eye lenses) ......................................................... **46**

— PME closer to PLE than to ALE or equidistant from ALE and PLE ..................................................... **50**

**46(45)** Tooth of cheliceral retromargin simple (Fig. 52.44) .... **47**

— Tooth of cheliceral retromargin bicuspid (Fig. 52.45) ...... .......................................................... ***"Beata"***
  Div. 1 species: *"Beata" wickhami* (Peckham & Peckham 1894) — Dist. s FL — Refs. Richman 1989, Maddison 1996 — Note not a true *Beata*

**47(46)** Carapace lateral margins parallel (flat) ..... ***"Pseudicius"***
  Div. 1 species: *"Pseudicius" siticulosus* (Peckham & Peckham 1909) — Dist. CA to NM, KS — Ref. Peckham & Peckham 1909 — Note not a true *Pseudicius*, nor a Heliophaninae (see list of genera)

— Carapace lateral margins convex ............................... **48**

52.38

52.39

52.40

52.41

52.42

52.43

52.44

52.45

**48(47)** Carapace with sides strongly curved (Figs. 52.1, 52.46); PME twice as far from PLE as ALE (less in smaller species); dorsal and lateral hair tufts present behind and below PME (dorsal tufts in all females, but lacking in many species in males); male chelicerae projecting neither anteriorly nor laterally. Most with metallic chelicerae (usually green, varies from red to violet, rarely black). The largest of our salticids, ~ 7-22 mm .............................. **Phidippus**
    Div. 47 species — Dist. widespread — Ref. Edwards 2004

— Carapace with sides not so strongly curved; PME less than twice as far from PLE as ALE; without tufts of hair in eye region; male often with chelicerae projecting either anteriorly or laterally, not metallic. Mostly medium in size ..... **49**

52.46

**49(48)** Iridescent green, copper, and/or black dorsally, with red, orange, and/or white abdominal markings, often with broad white submarginal carapace bands in males, moderately large, ~ 6-15 mm ........................... **Paraphidippus**
    Div. 3 species — Dist. e USA to AZ, s into MEX — Refs. Kaston 1973 (as *Eris*), Edwards 2004

— Dark brown with white markings usually on both carapace and abdomen, abdomen of females lighter brown with several pairs of small white spots and bronze iridescence, <10 mm in body length ..................................... **Eris**
    Div. 4 species — Dist. e & n USA, CAN — Refs. Kaston 1973, Maddison 1986

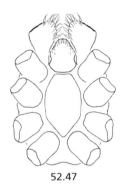

52.47

**50(45)** Front of sternum narrower than base of labium (Fig. 52.47) ............................................................................. **51**

— Front of sternum as wide as base of labium, or wider ....... ............................................................................. **54**

**51(50)** Leg I as slender as legs II-IV (except one species with enlarged tibia I); labium wider than long; legs with longitudinal lineate markings; most males with dorsal setal tufts. Females covered with iridescent scales, often green . .................................................................. **Tutelina**
    Div. 4 species + 4 undescribed — Dist. CAN, e USA w to NM, WA, OR and CA — Ref. Kaston 1948

— Leg I much thicker than legs II-IV; labium longer than wide; legs without lineate markings; males without dorsal setal tufts ...................................................................... **52**

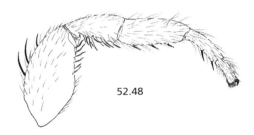

52.48

**52(51)** At least legs II-IV white and translucent; patch of long spatulate setae on ventral retromargin of patella and femur I (Fig. 52.48) and below PME (Fig. 52.49); male usually [except *Hentzia mitrata* (Hentz 1946)] with elongate chelicerae projecting anteriorly (Fig. 52.50) ............. .................................................................. **Hentzia**
    Div. 5 species — Dist. NS, e half of USA, AZ, s into MEX — Ref. Richman 1989

— Leg II-IV pigmented; no long spatulate setae on patella and femur I and below PME; chelicerae of male never elongate ...................................................................... **53**

**53(52)** Body dark red-brown to black; legs mostly black with distal segments yellow to red, or legs mostly red; dorsal abdomen with two pale thin stripes or rows of dash-like spots. Under rocks or bark ........................... **Metacyrba**
    Div. 3 species — Dist. widespread s of 40° N — Refs. Barnes 1958, Edwards in press

— Body not dark, mostly covered with gray scales; abdominal pattern broadly foliate with central chevrons. Under or on bark, buildings ........................ **Platycryptus**
    Div. 3 species — Dist. widespread — Refs. Barnes 1958 (as *Metacyrba*), Hill 1979, Edwards in press

52.49  52.50

**54(50)** PME closer to PLE than to ALE (measured between eyes); carapace somewhat flattened to slightly convex .......... **55**

— PME equidistant from ALE and PLE; carapace usually noticeably convex ...................... **58**

**55(54)** Mid-dorsal white stripe on abdomen in both sexes; male with clypeus aqua in color ................ ***Phlegra***
  **Div.** 1 species: *Phlegra hentzi* (MARX 1890) — **Dist.** CAN ?, USA w to MN, s to TX & FL — **Ref.** Kaston 1948 [as *Phlegra fasciata* (HAHN 1826)]

— No mid-dorsal white stripe on abdomen in either sex ...... ................................... **56**

**56(55)** Height of carapace at 3rd eye row about 1/3 length, flattened; male with dorsal black and white stripes (middle stripe black), female gray with yellow clypeus; 7-10 mm .. ................................... ***Menemerus***
  **Div.** 1 species: *Menemerus bivittatus* (DUFOUR 1831) — **Dist.** introduced, s USA, widespread in MEX — **Refs.** Barnes 1958, Edwards 1979a

— Height of carapace at 3rd eye row about 1/2 length; <5 mm ..................................... **57**

**57(56)** Males black with carapace shiny, females dark brown, not noticeably hairy; embolus a slender two-dimensional spiral; epigynal spermathecae oval, duct directed inside central openings, then curving outward to openings ........ ................................... ***Euophrys***
  **Div.** 1 species: *Euophrys monadnock* EMERTON 1891 — **Dist.** s CAN, n USA — **Refs.** Kaston 1948, Paquin & Dupérré 2003, Edwards 2003a

— Both sexes brown, somewhat hairy in appearance; embolus a flat, nearly straight blade, abruptly narrowed at tip; epigynal spermathecae elongate and bent, ducts directed straight toward lateral openings .......... ***Pseudeuophrys***
  **Div.** 1 species: *Pseudeuophrys erratica* (WALCKENAER 1826) — **Dist.** introduced, NJ, NY — **Refs.** Cutler 1982 (as *Euophrys*), Edwards 2003a

**58(54)** Embolus of male a ventral bowl-like spiral with narrow angulate gap between basal embolar disk and free apical portion (Fig. 52.51); epigynum with strong circular atrial rims continued across each atrium as a curved or sinuate sclerotized groove (Fig. 52.52) .................... ***Mexigonus***
  **Div.** 3 species — **Dist.** CA to NM — **Refs.** Richman 1981 (as *Tylogonus*), Edwards 2003a

— Embolus and epigynum variable but not as above ...... **59**

**59(58)** Embolus hook-shaped (Fig. 52.53), epigynum without flaps, openings widely separated, ducts prominent (Fig. 52.54); both sexes slightly elongated, with moderate cover of green to gold iridescent scales ......... ***Sassacus***, in part
  **Div.** 3 species, one keys out here: *Sassacus vitis* (COCKERELL 1894b) — **Dist.** sw CAN throughout much of w and central USA, s FL (introduced) s into MEX — **Ref.** Peckham & Peckham 1909 (as *Icius vitis*)

— Embolus not hook-shaped; epigynum with flaps (Fig. 52.55) or if lacking flaps, without posterior transverse ducts .......................... **60**

**60(59)** Both sexes iridescent green or bronze in color or with an iridescent green sheen; males with elongated chelicerae; embolus forming a prolateral spiral or at least rotated toward and initially curved toward the prolateral side (Figs. 52.56-52.57); epigyna without flaps; sw USA to MEX ............................. **61**

— Body not iridescent or bronze in color (rarely on abdomen), usually dark with white markings (except *Phanias* pale yellow in color, especially females); males with "normal" chelicerae; embolus not obviously spiralled; epigynal flaps variable; widespread ............................ **62**

52.51

52.52

52.53

52.54

52.55

52.56

52.57

**61(60)** Bright iridescent green in color (like a larger, more slender *Sassacus*), males with long, divergent chelicerae (Fig. 52.58); epignal openings widely separated (Fig. 52.59) ... ***Messua***
  **Div.** 2 species — **Dist.** AZ, s TX, s into MEX: *Messua limbata* (Banks 1898b), s TX, s into MEX: *Messua felix* (Peckham & Peckham 1901a) — **Refs.** Maddison 1996, Prószyński 2003 — **Note** Platnick (2005) refers to the last species as *Metaphidippus felix*, but Prószyński (2003) correctly places it in *Messua*

— Brown to bronze in color; male with long, parallel chelicerae (Fig. 52.60) and large spiral embolus on prolateral side of palpus; epignal openings in spiral atria (Fig. 52.61) ... ***Bagheera***
  **Div.** 1 species: *Bagheera prosper* (Peckham & Peckham 1901a) — **Dist.** TX, s into MEX — **Ref.** Maddison 1996

**62(60)** Embolus short (Fig. 52.62) and twisted in some species; both sexes with ventral macrosetae on tibia I and metatarsus I nearly as long as tarsus I (Fig. 52.63), femoral macrosetae reduced in size ... ***Ghelna***
  **Div.** 4 species — **Dist.** e CAN, e USA — **Refs.** Kaston 1973, Maddison 1996

— Embolus longer; macrosetae of both sexes normal in size. ... **63**

**63(62)** Embolus slender, sinuate, with large basal prong curving retrolaterally toward cymbium (Fig. 52.64); epignal flaps absent but large atria present (Fig. 52.65) ... ***Dendryphantes***
  **Div.** 1 species: *Dendryphantes nigromaculatus* (Keyserling 1885) — **Dist.** CO, UT, ID, NM, NH, ME, QC — **Refs.** Kaston 1973 (as *Eris*), Maddison 1996, Paquin & Dupérré 2003

— Embolus straight or slightly curved toward the prolateral side at tip, if sinuate then lacking basal prong; epignal flaps present ... **64**

**64(63)** Embolus tip narrow, lacking distal rami (Fig. 52.66) but often with shorter basal retrolateral flange that may be extended as a short prong; males usually with large white scale patch on chelicerae; some females yellow but not elongate, epignal flaps descend posteriorly into openings (Fig. 52.67) ... **"*Metaphidippus*" *manni* group**
  **Div.** 6 species — **Dist.** sw USA, s into MEX — **Ref.** Maddison 1996

— Embolus moderately broad, or if narrow, lacking basal projections; epignal flaps complete posteriorly ... **65**

**65(64)** Embolus tip usually wider and with two rami (or tip broad with single slender ramus on retrolateral side) (Fig. 52.68); males of most species without light patches of scales on chelicerae; epignal flaps strongly convex laterally, longitudinally oriented, more or less parallel ... ***Pelegrina***
  **Div.** 26 species — **Dist.** widespread — **Ref.** Maddison 1996

— Embolus long, sometimes sinuate, without long rami or basal prong (tip may be forked); epignal flaps usually transverse or convex medially ... **66**

**66(65)** Male palpus with tegular ledge expanded to cover the tegular shoulder (Fig. 52.69); arboreal, body usually pale (often yellow), may be slightly elongate ... ***Phanias***, in part
  **Div.** 8 species (see note) — **Dist.** sw CAN s to CA to TX, CO — **Ref.** Maddison 1996 — **Note** this genus is poorly known, number of species that key out here is uncertain

— Male palpus without tegular ledge expanded, tegular shoulder visible (Fig. 52.70); terrestrial, body dark in color ... ***Terralonus***
  **Div.** 7 species — **Dist.** w USA — **Ref.** Maddison 1996

Chapter 53
# SCYTODIDAE
1 genus, 7 species

Darrell Ubick

**Common name** —
Spitting spiders.

**Similar families** —
Pholcidae (p. 194), Sicariidae (p. 222).

**Diagnosis** —
Scytodids are easily differentiated from other haplogyne spiders in having a convex carapace and six eyes arranged in three diads that form a strongly recurved row.

**Characters** —
**body size:** 3.5-10.0 mm.
**color:** yellowish to dark brown with patterned dorsum.
**carapace:** convex, highest in thoracic region; fovea absent (Fig. 53.2).
**sternum:** oval, with extensions to coxae.
**eyes:** six eyes (AME absent), arranged in three diads forming strongly recurved row.
**chelicerae:** fused basally (connected by membrane), with median lamina produced into apical tooth.
**mouthparts:** labium quadrate, fused to sternum; endites strongly convergent.
**legs:** short to long, tarsi with three claws, female palp with claw.
**abdomen:** oval to spherical.
**spinnerets:** 6 small contiguous spinnerets with reduced spigots; colulus large, twice as long as wide.
**respiratory system:** one pair of book lungs; tracheal spiracle anteriorly displaced from spinnerets.
**genitalia:** haplogyne; *female* with sclerotized modifications posterior to epigastric furrow, consisting of circular foveae with mesal ridges (Fig. 53.3); *male* palpal tarsus apically attenuated with bulb attached basally; bulb spherical with embolus a simple prong or with additional apical projections (Fig. 53.4).

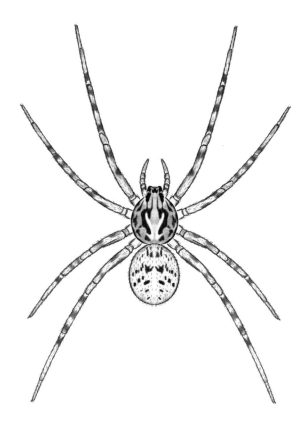

**Fig. 53.1** *Scytodes thoracica* (LATREILLE 1802a)

**Distribution** —
As currently understood, *Scytodes* is a diverse worldwide genus. In the Nearctic, apart from the synanthropic *Scytodes thoracica* (LATREILLE 1802b), which occurs in eastern USA north to se Canada, all species are strictly southern ranging from CA to FL. Two of these, *Scytodes fusca* WALCKENAER 1837 and *Scytodes longipes* LUCAS 1844a, are pantropical and probably synanthropic. Another synanthrope, *Scytodes globula* NICOLET *in* GAY 1849, has recently been recorded from Florida (Guarisco 2003). The remaining species, most of which have been recorded from Texas (Vogel 1970b), are probably Nearctic endemics.

**Natural history** —
Scytodids are unique among spiders in capturing prey by spitting strands of glue from the fangs; hence the common name. The glue is produced in an enlarged posterior lobe of the venom gland that occupies much of the prosoma and accounts for the convex carapace of the spider. Prey is bound to the substrate and immobilized, from a distance of up to a few cm, by fine strands of glue with venom that are sprayed extremely rapidly (140 ms). Not surprisingly, this unusual behavior has been much studied (Foelix 1996, containing additional references).

Many scytodids are synanthropic and occur in and around human habitation. The genus includes short-legged wandering hunters, which live on the ground, under rocks, and in leaf litter. Other species are long-legged and build pholcid-like webs and may be communal. Females lack cylindrical spigots on the spinnerets and so do not construct a typical egg sac, but bind the eggs with a few strands of silk and carry them beneath the body.

**Taxonomic history and notes** —
Nearctic species were described and the genus reviewed by Gertsch (1935a) and Gertsch & Mulaik (1940) but, as *Scytodes* was never revised, the names currently recorded are not a good indication of the actual diversity. Of the eight species listed by Vogel (1970b) for Texas, the record

of *Scytodes intricata* BANKS 1909a appears to be a misidentification of an undescribed species (Brown 1973) and similarly that of *Scytodes championi* F.O. PICKARD-CAMBRIDGE 1899a, which is believed to occur only from Central America to northern South America (Brescovit & Rheims 2001). Two undescribed species are known from Florida (G.B. Edwards pers. comm.). Additional descriptions of Nearctic species are provided by Kaston (1948), Brignoli (1976a), Valerio (1981), Brescovit & Rheims (2000), and Paquin & Dupérré (2003).

*Scytodes* is diverse in both genital and somatic morphology, as well as the degree of web building, and has been divided into a number of species groups (Brignoli 1976a, Valerio 1981). Recent workers, as for example Saaristo (1997c), have begun recognizing additional genera in the family although this has, so far, not affected the Nearctic fauna. Analysis of the relationships of haplogynes places Scytodidae as closest to the Neotropical and Afrotropical Drymusidae, which together form the sister group of Sicariidae (Platnick *et al.* 1991).

**Genus —**

*Scytodes* LATREILLE 1804b.

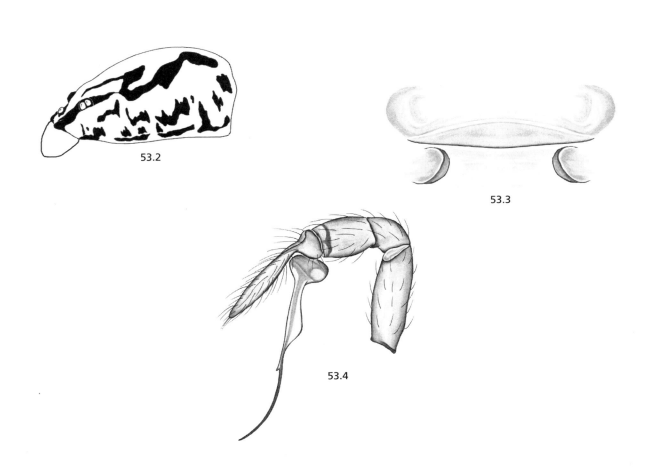

# Chapter 54
# SEGESTRIIDAE
2 genera, 7 species

Darrell Ubick

### Similar families —
Dysderidae (p. 103)

### Diagnosis —
Segestriids differ from other haplogyne spiders in having leg III anteriorly directed. They are further differentiated in having the chelicerae free (not basally fused) and with toothed margins, six eyes with the PER forming a straight to slightly recurved row, tarsi with three claws, and a pair of posterior tracheal spiracles positioned near the epigastric furrow.

### Characters —
**body size:** 4.0-16.0 mm.
**color:** prosoma and appendages brown, abdomen purplish brown with variable degrees of yellowish markings.
**carapace:** rectangular, anterior margin only slightly narrower than posterior.
**sternum:** oval, with precoxal triangles.
**eyes:** six eyes (AME absent); PER in straight to slightly procurved row.
**chelicerae:** promargin with three teeth, retromargin with one or two teeth.
**mouthparts:** endites parallel, longer than wide, rounded ectally, originating from mesal margin of palpal coxae; labium longer than wide.
**legs:** tibiae and metatarsi I-II with paired ventral spines (larger in *Ariadna*); metatarsi and tarsi I-III scopulate (denser in *Segestria*); metatarsus I curved and with lateral apophyses in most males of *Ariadna*; metatarsus IV with ventroapical preening comb; tarsi with three claws.
**abdomen:** elongated, cylindrical.
**spinnerets:** colulus represented by setal patch; ALS conical, contiguous; PS smaller (Fig. 54.2).
**respiratory system:** one pair of book lungs; two tracheal spiracles near epigastric furrow (Fig. 54.2).
**genitalia:** haplogyne; *female* gonopore region swollen with some sclerotization on anterior margin (Fig. 54.2); *male* palpus with tarsus short to long, bulb spherical to pyriform with spinelike embolus.

### Distribution —
*Ariadna* has a worldwide distribution and occurs throughout most of the Americas. In the US, the most widespread species is *Ariadna bicolor* (HENTZ 1842b), ranging from Maine to central Florida and west to southern California. The other three species all have restricted distributions: southern Florida (*Ariadna arthuri* PETRUNKEVITCH 1926b), Arizona (*Ariadna pilifera* O. PICKARD-CAMBRIDGE, 1898a), and the southern coast of California (*Ariadna fidicina* CHAMBERLIN 1924b). *Segestria* is primarily Holarctic and in North America is known only from the Pacific coast. The single widespread species, *Segestria pacifica* BANKS 1891c, extends from Baja California north to Canada and two isolated ones, *Segestria bella* CHAMBERLIN & IVIE 1935b and *Segestria cruzana* CHAMBERLIN & IVIE 1935b are known from central California and Santa Cruz Island, respectively. An undetermined species of *Segestria* has recently turned up in New Mexico.

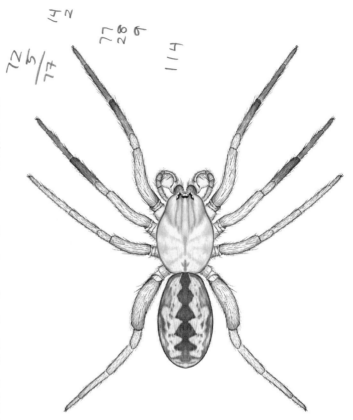

**Fig. 54.1** *Segestria pacifica* BANKS 1891c

### Natural history —
Segestriids are nocturnal sedentary hunters that live in crevices. They may be abundant in areas of favorable habitats, such as in rock walls, rock talus, and under loose tree bark (especially *Eucalyptus*). Here they build tubular retreats with elaborate webbing at the entrances, which are equipped with signal lines similar to that of another haplogyne, *Kukulcania* (Filistatidae). The signal lines are monitored by the spider's legs, including the anteriorly directed leg III, and indicate the position and size of approaching prey. This allows the spider to rapidly capture prey, which is then pulled into the retreat for feeding (Comstock 1940). Segestriids do well in captivity, accepting a variety of insect prey, and can live for few to several years (D. Ubick, pers. obs.).

### Taxonomic history and notes —
Segestriids are similar to and have long been associated with the dysderids. Simon (1892a) included both in the Dysderidae, a placement accepted by Beatty (1970). Petrunkevitch (1933), however, argued that the two deserve familial status, to which Brignoli (1976g) and Forster & Platnick (1985) gave additional evidence. In the recent analysis of haplogyne relationships (Platnick *et al.* 1991) not only are both families recognized but their relationship within the dysderoids is unresolved (i.e., they are consid-

ered equally closely related to each other as well as to the group comprising Oonopidae plus Orsolobidae).

The Nearctic segestriids have received little taxonomic attention. *Segestria* is very poorly known and since the original (and not very useful) description of *Segestria pacifica* BANKS 1891c, only Chamberlin & Ivie (1935b) have added any species. *Ariadna* has been recently revised by Beatty (1970), who recognized four Nearctic species based on differences in somatic and male genital morphology. The female genitalia of these species have not been described, although studies on the Mediterranean species (Brignoli 1976g) indicate they are taxonomically quite useful. In light of this, additional studies of the genus would be welcome, especially to resolve the validity of *Citharoceps* CHAMBERLIN 1924b, which differs from other *Ariadna*, where it currently resides, in possessing prominent stridulatory files on the carapace.

### Genera —

*Ariadna* AUDOUIN 1826, *Segestria* LATREILLE 1804b

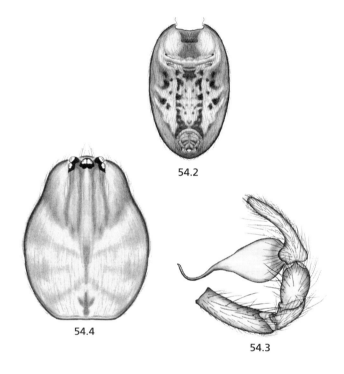

### Key to genera —
### North America North of Mexico

1   Abdomen with median dark stripe broken into chevrons and with lateral and ventral spots (Figs. 54.1-54.2); legs banded, relatively long (TP>C); cheliceral retromargin with two teeth; male palpus with tarsus long and apically attenuated (Fig. 54.3); male metatarsus I straight and lacking lateral processes; carapace Fig. 54.4 ......... ***Segestria***
      **Div.** 3 species — **Dist.** Pacific coast, from CAN to MEX, NM — **Refs.** Chamberlin & Ivie 1935b, Crawford 1988 — **Note** genus is badly in need of revision

—   Abdomen usually uniformly dark, sometimes with transverse pale bands; legs concolorous, relatively short (TP<C); cheliceral retromargin with one tooth; male palpus with tarsus short and apically notched (Fig. 54.5) male metatarsus I bent and with lateral processes (Fig. 54.6) ................................................................ ***Ariadna***
      **Div.** 4 species — **Dist.** from ME to central CA and south — **Ref.** Beatty 1970 — **Note** *Ariadna fidicina* (CHAMBERLIN 1924b), described only from the female, was originally placed in a separate genus (*Citharoceps* CHAMBERLIN 1924b) because it differs from other *Ariadna* in having prominent stridulatory files on the cephalon (Fig. 54.7)

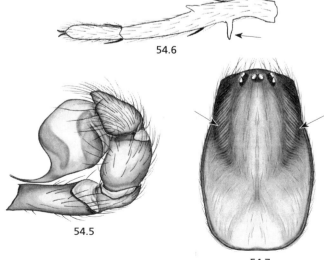

## Chapter 55
# SELENOPIDAE
1 genus, 5 species

Sarah C. Crews

### Similar families —
Philodromidae (p. 192), Sparassidae (p. 224), Thomisidae (p. 246).

### Diagnosis —
Selenopids can be distinguished from other spiders by the combination of their extremely dorsoventrally flattened habitus (Fig. 55.1), AER with 6 eyes (Fig. 55.2), the notched posterior projection of the sternum (Fig. 55.3) and smooth tarsal claws.

### Characters —
**body size:** 7.5-13.0 mm.
**color:** body yellowish to brownish to grayish, mottled with darker markings; legs with bands on the femora, tibiae and metatarsi.
**carapace:** dorsoventrally flattened with distinct cephalic region; thoracic region laterally convex; fovea longitudinal with radiating grooves (Fig. 55.1).
**sternum:** circular to ovoid, with notched posterior projection extending between coxae IV (Fig. 55.3).
**eyes:** 8 eyes outlined in black; 6 in the anterior row and 2 in the posterior.
**chelicerae:** geniculate, robust; paturon setose with distinct lateral condyles. Fang large, cheliceral margins toothed.
**mouthparts:** labium as wide as or wider than long, rounded anteriorly, truncate posteriorly; endites ectally convex, subparallel with terminal scopulae (Fig. 55.3).
**legs:** laterigrade, long; tarsi with two claws and scopulae; trichobothria present on all leg segments.
**abdomen:** longer than wide, oval in shape, extremely dorsoventrally flattened, slightly truncate posteriorly; mottled.
**spinnerets:** six spinnerets, anterior pair adjacent; colulus absent.
**respiratory system:** with 1 pair of book lungs and single tracheal spiracle near spinnerets.
**genitalia:** entelegyne; *female* epigynum (Fig. 55.5) with spermathecal openings located at the caudal end of median guide or in a single median pit (Muma 1953); *male* palpus (Fig. 55.4) with retrolateral tibial apophysis large and complex; embolus long, extending at least 1/3 – 1/2 of the distance around the cymbium; median apophysis with a single branch; conductor with spine-or spur-like apex (Muma 1953).

### Distribution —
Southern California east to Texas, south to Mexico, including the Baja California Peninsula; Florida and most Caribbean islands. Occasionally introduced elsewhere, but not established.

### Natural history —
These spiders live in habitats varying from dry desert and chaparral to mesic tropical areas. They are typically found under rocks and bark or in houses and occasionally on logs, debris on the ground and between the bases of leaves of tropical plants (Muma 1953). Their extreme flatness allows them to fit in very narrow spaces. This, coupled with their

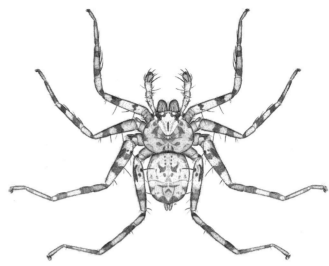

**Fig. 55.1** *Selenops nesophilus* CHAMBERLIN 1924b

speed, can make them difficult to collect. They are nocturnal and do not build webs. They are easily reared in the laboratory on crickets and fruitflies. Adults have been collected throughout the year.

### Taxonomic history and notes —
The first systematic study of the genus was completed by Walckenaer (1837) who recognized three groups based on the chelicerae, labium and leg lengths. However, Simon (1880a) could not corroborate the characters used by Walckenaer, and separated Old and New World species based on eye size. In 1905, O. Pickard-Cambridge distinguished species in the Western Hemisphere using eye size and position and genitalic characters. Petrunkevitch (1925a, 1930b) decided leg proportions were of group and specific importance. In 1953, Muma established six subgeneric groups in North and Central America and the Caribbean based on leg lengths, eye size and position and genitalic characters.

### Genus —
*Selenops* LATREILLE 1819

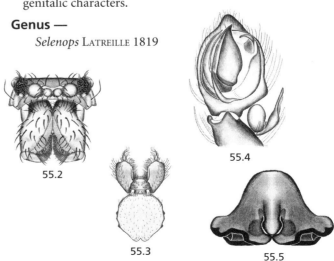

55.2

55.3

55.4

55.5

## Chapter 56
# SICARIIDAE
1 genus, 13 species

Darrell Ubick

### Common name —
Brown spiders, violin spiders, brown recluse spiders (*Loxosceles reclusa* GERTSCH & MULAIK 1940)

### Similar families —
Scytodidae (p. 217).

### Diagnosis —
Sicariids may be distinguished from most other haplogyne spiders in having basally fused chelicerae and six eyes arranged in three diads forming a strongly recurved row; and from scytodids in having a relatively flat carapace and tarsi with only two claws.

### Characters —
**body size:** 5.0 – 13.0 mm.
**color:** yellowish to grayish brown with cephalic area often darker.
**carapace:** flat, pyriform, longer than wide; fovea longitudinal (Fig. 56.2).
**sternum:** subcircular, with extensions to coxae.
**eyes:** six eyes (AME absent), arranged in three diads forming strongly recurved row (Figs. 56.2-56.3).
**chelicerae:** fused basally (connected by membrane), with median lamina produced into apical tooth; ectal margin with stridulatory file.
**mouthparts:** labium elongate, flexibly attached to sternum; endites strongly convergent.
**legs:** slender (PT/C about 1.5), tarsi with two claws; female palp lacking claw.
**abdomen:** flattened oval, hairy.
**spinnerets:** colulus large, pointed, about twice as long as wide; ALS largest, separated by width of colulus; PS contiguous.
**respiratory system:** one pair of book lungs; tracheal system with single spiracle near spinnerets.
**genitalia:** haplogyne; *female* lacking external modification; *male* palp with tibia swollen, tarsus short; bulb spherical, embolus a simple prong (Fig. 56.4).

### Distribution —
The genus *Loxosceles* has a worldwide distribution. Most of the native Nearctic species are restricted to the southwest, from southern California to southern Texas. *Loxosceles reclusa* GERTSCH & MULAIK 1940 has a wide natural distribution along the Mississippi Basin and adjacent parts of central USA, but has been spread synanthropically to isolated localities beyond. *Loxosceles rufescens* (DUFOUR 1820c) was introduced from Europe and is now established over much of the southern USA, and recently in New York. *Loxosceles laeta* (NICOLET *in* GAY 1849) is a South American species now known from a few localities on both coasts, including southwestern Canada (Gertsch & Ennik 1983).

### Natural history —
*Loxosceles* occurs in a variety of habitats, from deserts to mesic forests, and is commonly synanthropic. The spiders are typically hidden in crevices that they line with silk and extend beyond the retreat. The silk is sticky, readily ensnar-

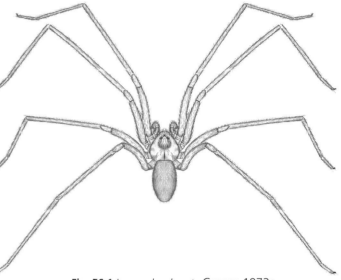

**Fig. 56.1** *Loxosceles deserta* GERTSCH 1973a

ing prey, and somewhat resembles cribellate silk although structurally different (Stern & Kullmann 1981). Studies on the biology of *Loxosceles reclusa* indicate that the species is a generalist feeder and has an annual life cycle but may live up to three years in captivity (Hite *et al.* 1966). Observations on the biology of *Loxosceles deserta* GERTSCH 1973a (as *Loxosceles unicolor* KEYSERLING 1887b) were presented by Ennik (1971) and those of *Loxosceles arizonica* GERTSCH & MULAIK 1940 by Richman (1973).

*Loxosceles* has received much notoriety because of its potent venom and has been associated with numerous cases of envenomation, both real and imagined (Vetter *et al.* 2003). The venom is cytotoxic to humans and produces lesions and persistent sores, although rarely death (6 cases recorded up to 1977, Gertsch & Ennik 1983).

### Taxonomic history and notes —
The Nearctic species of *Loxosceles* have been revised in a series of papers by Gertsch [1940 (with Mulaik), 1958b, 1973a, and 1983 (with Ennik)].

Because of its cheliceral modifications, *Loxosceles* has long been clearly associated with the scytodoid haplogynes, although its actual family designation has varied considerably. Simon (1893a) included it, as one of several subfamilies, in his broadly interpreted Sicariidae. Gertsch (1949) elevated Loxoscelinae, as well as most of the others, to family. Brignoli (1976a) considered *Loxosceles* closely related to *Scytodes* and kin and included them all in Scytodidae. Recent analysis of haplogyne relationships

(Platnick *et al.* 1991) indicates that *Loxosceles* is actually most closely related to *Sicarius* WALCKENAER 1847a. As the two families they represent are both monotypic, they were merged, and given that the older name has priority, *Loxosceles* is once again in Sicariidae.

**Genus —**

*Loxosceles* HEINECKEN & LOWE *in* LOWE 1835

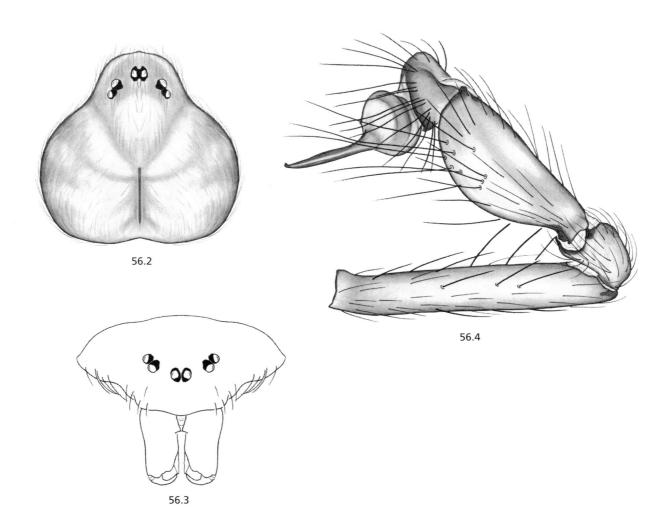

56.2

56.3

56.4

## Chapter 57
# SPARASSIDAE
3 genera, 9 species

Stephen E. Lew

**Common name —**
Giant crab spiders, huntsman spiders (*Heteropoda*).

**Similar families —**
*Olios* and *Heteropoda* may be mistaken for other large, dorsoventrally flattened spiders in families such as Selenopidae (p. 221) or Homalonychidae (p. 118). *Pseudosparianthis* is more likely to be mistaken for spiders in the Anyphaenidae (p. 66), Clubionidae (p. 77), Liocranidae (p. 162), or Miturgidae (p. 173).

**Diagnosis —**
Sparassids can be distinguished from other large, two-clawed spiders with laterigrade legs by the membranous trilobed dorsoapical extension of the metatarsus (Fig. 57.2).

**Characters —**
body size: 7-30 mm.
color: earthy browns and grays.
carapace: longer than wide, fovea longitudinal.
sternum: pentagonal, blunt to acuminate posteriorly.
eyes: 4 pairs in two rows, recurved in *Heteropoda*, variable in other genera.
chelicerae: both fang margins toothed.
legs: laterigrade; metatarsus with membranous trilobed dorsoapical extension (Fig. 57.2); tarsi may hyperextend dorsad of the metatarsal axis (Fig. 57.3); tarsi with 2 claws, claw tufts, and scopulae.
abdomen: longer than wide, somewhat dorsoventrally flattened.
spinnerets: 3 pairs, conical, contiguous; Colulus present only in *Pseudosparianthis*.
genitalia: entelegyne; **female** epigynum with an anteriorly bordered atrium (Figs. 57.6, 57.8, 57.10); **male** palpus with large RTA, bifid in heteropodinae genera (Figs. 57.7, 57.9) simple in *Pseudosparianthis* (Fig. 57.5).

**Distribution —**
*Olios*: western states, *Pseudosparianthis*: Florida. *Heteropoda*: established populations only in Florida, occasionally found syanthropically in other southern states and California.

**Natural history —**
Sparassids are cursorial and ambush predators. Their flattened morphology and laterigrade leg attachment allow these large spiders to fit into surprisingly small crevices. *Olios* and *Heteropoda* adapt readily to manmade structures and modified habitats, *Heteropoda venatoria* LINNAEUS 1767 being a common pantropical invader.

**Taxonomic history and notes —**
In addition to the valid name Sparassidae, this family has at least as often been called Heteropodidae (Thorell 1873). Jäger (1999a) established the priority of Sparassidae (Bertkau 1872). The family name Eusparassidae has also been used, largely by authors who considered *Sparassus* to be a junior synonym of *Eusparassus* (see discussion

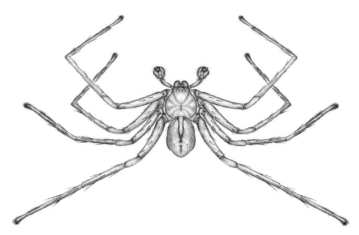

**Fig. 57.1** *Olios fasciculatus* SIMON 1880a

in Kaston 1974:49-50, also Bonnet 1958:4097). Bertkau's Sparassidae included only *Sparassus* (= *Micrommata*) and *Thanatus* (now in the Philodromidae, p. 192), whereas Thorell's Heteropodidae also included *Heteropoda*, and several genera now placed in the Selenopidae and the Gnaphosidae. Walckenaer (1837), describing *Olios*, did not place the new genus in a family, but allied it with the type genera of the Sparassidae, Thomisidae, Philodromidae, and Clubionidae. *Olios* was placed in the Sparassidae by Simon (1874b). Decades later, Simon (1897a) ranked the sparassids as a subfamily of the Clubionidae, based on their cheliceral armature. This placement was accepted by some but not by most (see bibliographic summary in Bonnet 1958: 4096-4097). The North American sparassids were reviewed by Fox (1937e). Fox described the genus *Tentabunda* for species of *Pseudosparianthis* with a single pair of metatarsal macrosetae, which includes the North American *Pseudosparianthis cubana* BANKS 1909b. Bryant (1940) synonymized *Tentabunda* back into *Pseudosparianthis*, presumably because *Pseudosparianthis cubana* is the type species and *Pseudosparianthis* has priority (Penney 2001). The worldwide fauna presently includes 989 species in 83 genera (Platnick 2005). It is very likely that there are many undescribed species of *Olios* in western North America.

**Genera —**
HETEROPODINAE
*Heteropoda* LATREILLE 1804b, *Olios* WALCKENAER 1837
SPARIANTHINAE
*Pseudosparianthis* SIMON 1887g

# Key to genera —
## North America North of Mexico

**Note:** *Eusparassus* is not included in this key. *Eusparassus hansii* SCHENKEL 1950b is the sole North American representative of the genus, and is known only from the type specimen. The author and editors have examined photomicrographs of this specimen and believe that it is misplaced in the Sparassidae, being more likely one of the two-clawed genera in the Tengellidae (p. 230).

1     Trochanters entire, colulus present, spinnerets raised upon sclerotized base (Fig. 57.4); genitalia as in Figures 57.5-57.6 .......................................... ***Pseudosparianthis***
           **Div.** 1 species: *Pseudosparianthis cubana* BANKS 1909b — **Dist.** FL — **Ref.** Fox 1937e

—     Trochanters notched, colulus absent, spinnerets set into abdomen in normal fashion ............................................ **2**

2(1)    Tibia I with 2 pairs of ventral macrosetae; genitalia variable, including as in Figures 57.7-57.8 ..................... ***Olios***
           **Div.** 7 species — **Dist.** CA, east to TX — **Refs.** O. Pickard-Cambridge 1896a, Chamberlin 1924b, Fox 1937e, Roth 1988a

—     Tibia I with more than 2 pairs of ventral macrosetae; genitalia as in Figures 57.9-57.10 ..................... ***Heteropoda***
           **Div.** 1 species: *Heteropoda venatoria* (LINNAEUS 1767) — **Dist.** Gulf states and CA, introduced — **Refs.** F.O. Pickard-Cambridge 1900, Fox 1937e, Edwards 1979b

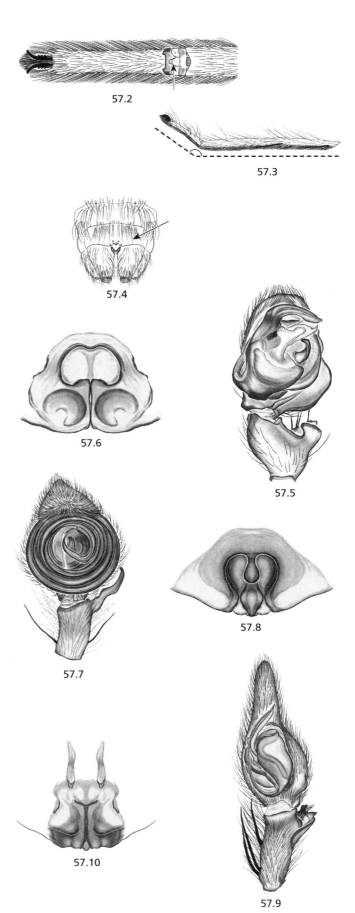

# Chapter 58
# SYMPHYTOGNATHIDAE
1 genus, 1 species

Jonathan A. Coddington

### Similar families —
Anapidae (p. 64), Mysmenidae (p. 175), Theridiosomatidae (p. 244).

### Diagnosis —
All instars of the Nearctic members of this orbicularian family can be distinguished by the possession of only four eyes (Fig. 58.1) and tiny size. Females lack pedipalps (Figs. 58.2-58.3). All similar families are larger, have pedipalps, and 6-8 eyes, although in anapids the pedipalps are reduced (see chapter 15). Males can be distinguished from similar groups by size and eye pattern. In the field, *Anapistula* spin small (2-4cm) flat orbs with central radii cut out, numerous accessory radii, and very numerous sticky spirals (Fig. 58.4).

### Characters —
**body size:** males ~ 0.4 mm; females ~ 0.5 mm.
**color:** carapace, legs and abdomen yellowish, with slightly darker extremities.
**carapace:** highest posterior to the fovea (Fig. 58.2).
**sternum:** nearly square, broadly truncate behind.
**eyes:** four, laterals contiguous, all median eyes lacking.
**chelicerae:** fused together from base to half their length.
**legs:** length order 1423, tarsi much longer than metatarsi.
**abdomen:** much higher than long.
**spinnerets:** six spinnerets with small, fleshy colulus.
**respiratory system:** book lungs absent.
**genitalia:** entelegyne.

### Distribution —
Symphytognathidae is apparently pan-tropical, but very poorly known. *Anapistula* has never been revised, so it is difficult to judge if the very large reported range is accurate (Australia, North America, South America, Africa, southeast Asia, and the Seychelles). The range of *Anapistula secreta* GERTSCH 1941a is the southern United States through the Greater Antilles to Colombia.

### Natural history —
At a fraction of a millimeter in adult size, Symphytognathidae includes the smallest known spiders (Baert & Jocqué 1993). As far as is known, web-building symphytognathids prefer to live in the interstices of leaf litter, hollow logs, and other small, extremely humid microhabitats, although under the right conditions they will spin webs at least a meter or two above the forest floor (pers. obs.). At least some *Curimagua* FORSTER & PLATNICK 1977 are commensals in the webs of much larger spiders (Vollrath 1978). To date, most new species have been collected by litter extraction methods, but the webs are not difficult to detect if clouds of cornstarch are gently wafted across likely patches of leaf litter, low-growing vegetation or moss. Animals seem to spend most of their time at the web hub, but if disturbed move slowly to the periphery. Like other symphytognathoid families (Theridiosomatidae, Anapidae, Mysmenidae), web construction (so far as is known) involves out-of-plane radii, but symphytognathids cut these radii during construction so that the final web is planar.

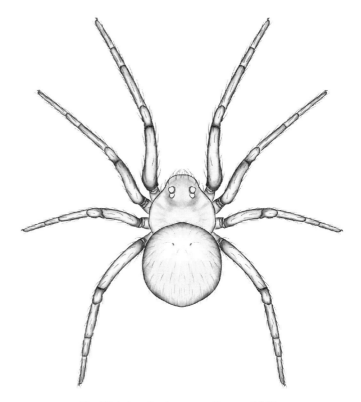

**Fig. 58.1** *Anapistula secreta* GERTSCH 1941a

Finished webs appear to have hundreds of radii (Fig. 58.4) but almost all of these are actually "accessory" radii added to the underside of the orb after sticky spiral construction. True and accessory radii can be distinguished because the paths of sticky spirals bend sharply at true radii, but continue straight across accessory radii. After accessory radius construction, the hub area is rebuilt to reduce the number of radial lines reaching the hub (Eberhard 1986, pers. obs.). Egg sacs are guarded at the margin of the web (Griswold & Yang 2003). Their small size, fine-meshed webs, and cheliceral fusion, has led to speculations (Vollrath pers. comm.) that they are vegetarian and rely on the web to sieve spores and pollen from the air. Although plausible, no direct evidence supports this speculation as yet.

### Taxonomic history and notes —
Hickman (1931) described the first specimen of Symphytognathidae from Tasmania and immediately recognized it as a new family. Ten years later Gertsch (1941a) described the second, *Anapistula secreta*, from specimens from Barro Colorado Island, Panama, which still later was collected from the United States. Thereafter Forster (1951, 1958, 1959) expanded the concept of the family to include a

number of genera of small spiders with reduced eyes, respiratory systems, and pedipalps. Forster & Platnick (1977) reviewed the family and transferred several genera back to families such as Anapidae, Mysmenidae, and Micropholcommatidae. At present Symphytognathidae includes six genera and about 50 species of animals with fused chelicerae and pedipalps lost (in females and juveniles). The phylogenetic placement of the family has been studied by Coddington (1990), Griswold *et al.* (1998), and Schütt (2003). These authors generally agree on the symphytognathoid clade (Theridiosomatidae, Anapidae, Mysmenidae, and Symphytognathidae), but the included genera are so poorly known that the delimitation of the latter three families and Micropholcommatidae – and therefore the number of clades within symphytognathoids – have been questioned (Schütt 2003).

**Genus —**

*Anapistula* Gertsch 1941a

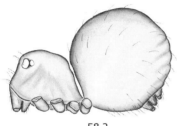
58.2

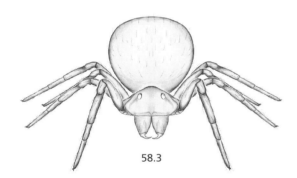
58.3

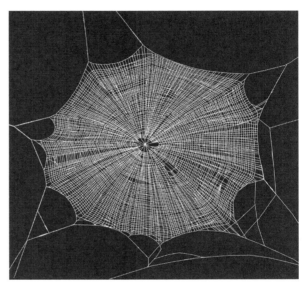
58.4

## Chapter 59
# TELEMIDAE
1 genus, 4 species

Robert G. Bennett
Joel M. Ledford

### Similar families —
Telemids may easily be misidentified as Ochyroceratidae (p. 179), Leptonetidae (p. 122) (especially *Archoleptoneta*), Oonopidae (p. 185), or small Pholcidae (p. 194). However, telemids alone possess the abdominal sclerotized ridge above the pedicel.

### Diagnosis —
Minute, three-clawed, ecribellate haplogyne spiders lacking book lungs; distinguished from other spiders by a transverse zigzag ridge anterodorsally on abdomen (R in Fig. 59.2) and, in females, a dorsal brush of distinctive short, stout and abruptly tapered macrosetae on the palpal tarsus (Fig. 59.3).

### Characters —
**body size**: minute, 1.0-1.7 mm.
**color**: uniformly pallid to orange in life, pallid to green in alcohol.
**cephalothorax**: oval, slightly longer than wide; cephalic region somewhat elevated; fovea inconspicuous.
**sternum**: subtriangular, narrowly extended between coxae IV.
**eyes**: six eyes (AME absent) in three pairs (Fig. 59.4) or blind.
**chelicerae**: unmodified-unfused, non-chelate, lamina lacking from inner margin; promargin of fang furrow with one tooth and a few denticles, retromargin with several denticles.
**mouthparts**: labium rebordered, not notched; endites longer than wide, convergent.
**legs**: long, thin and fragile, PT/C 1.8-3.5; Emerit's glands present dorsally on tibiae (Figs. 59.5, EG in Fig. 59.6), leg breakage usually occurring at coxa-trochanter joint.
**abdomen**: high and rounded with anterodorsal zigzag sclerotized ridge immediately above pedicel (R in Fig. 59.2).
**spinnerets**: ecribellate; colulus pentagonal and very large; ALS with short distal segment and widely separated by colulus (Fig. 59.7).
**respiratory system**: book lungs absent; transverse pair of tracheal spiracles located approximately equidistant from epigastric groove and spinnerets, a second transverse pair of spiracles located anterior to epigastric groove in at least some telemids (SP in Fig. 59.8), several pairs of tracheae leading from spiracles into cephalothorax.
**genitalia**: haplogyne; ***female*** (Fig. 59.10) with simple copulatory opening leading to single, poorly defined spermathecal chamber (SC); ***male*** (Fig. 59.9) with simple palpus, no patellar or tibial apophyses, bulb simple and attached to ventral base of cymbium.

### Distribution —
Telemids are widely distributed with most described species occurring in Africa and east and southeast Asia. *Usofila* is the only described Nearctic genus, and occurs from Alaska south to central California and east to montane Alberta and Colorado.

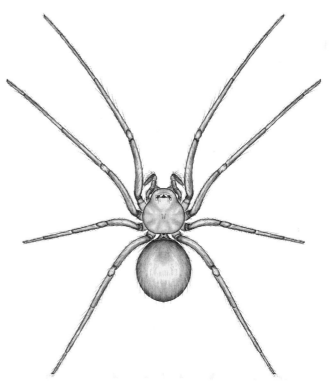

**Fig. 59.1** *Usofila oregona* CHAMBERLIN & IVIE 1942a

### Natural history —
Telemids are cryptozoic spiders found in moist conditions, in leaf litter, under rocks, and especially in caves where they can be very abundant. Telemids build small sheet webs (similar to those produced by leptonetids) underneath which they hang inverted. Mating has not been directly observed, however males and females may cohabit in webs for several weeks (Ledford, pers. obs.). Egg sacs are suboval in shape and deposited on the outer margins of the web. Troglobitic species produce limited quantities of large eggs, while their epigean relatives produce numerous smaller ones (D. Ubick, pers. comm.). Individuals may be reared to maturity on a diet of freshly killed fruit flies (*Drosophila*), as long as moisture levels are closely monitored (Ledford, pers. obs.).

### Taxonomic history and notes —
Telemidae is a small family with 7 genera and 22 species currently recognized (Platnick 2005). Telemids were originally placed in Leptonetidae (e.g., Comstock 1912, Simon 1893a) in a subfamily shared with ochyroceratids. Based on the anatomy of respiratory and circulatory systems, Petrunkevitch (1923, 1928b, 1933) classified telemids as a distinct (non-Nearctic) family in a separate suborder from leptonetids (which still included ochyroceratids). Brignoli (1973a) first recognized a Nearctic component of Telemi-

dae with the transfer of *Usofila* to that family from Ochyroceratidae. Recent analysis (Coddington & Levi 1991, Emerit 1985, Platnick 1986d, Platnick *et al.* 1991, Ramírez 2000) has supported the monophyly of Telemidae and, contrary to Petrunkevitch but similar to Simon (1893a) and others, a close relationship with both Leptonetidae and Ochyroceratidae.

Marx (1891) published a description of the single Nearctic genus, *Usofila*, for a single species (*Usofila gracilis*) described in unpublished papers of Keyserling. Subsequently Chamberlin & Ivie (1942a) described two more species (*Usofila flava*, *Usofila oregona*) and transferred (Chamberlin & Ivie 1947b) *Ochyrocera pacifica* BANKS 1894b to *Usofila*. Of these four species, *Usofila gracilis* and *Usofila oregona* are known from both sexes; the two others from females alone.

Emerit (1981, 1984, 1985) and Emerit & Juberthie (1983) first described and illustrated interesting glands, probably producing defensive "repugnatorial" secretions, on the tibia of telemids. Now known as Emerit's glands (Bennett 1989), these have since been described in leptonetids as a probable synapomorphy (Platnick 1986d) as well as in certain cybaeids and salticids (Bennett 1989 and included references).

Specimens representing a substantial number of undescribed species, including all the eyeless ones, await study in museum collections. This work was started by Gertsch, but left unfinished upon his death (D. Ubick, pers. comm.). *Usofila* is considered by some to be a junior synonym of the primarily Palaearctic genus *Telema* SIMON 1882c (e.g., Gertsch 1979, Roth 1994). In part because they are haplogyne and tiny, few illustrations exist that are of use for the separation of the Nearctic species. A revision of *Usofila* and *Telema* would be timely and welcomed. References for Nearctic Telemidae are Chamberlin & Ivie 1942a, Comstock 1912, and Marx 1891.

### Genus —
TELEMINAE
*Usofila* KEYSERLING *in* MARX 1891

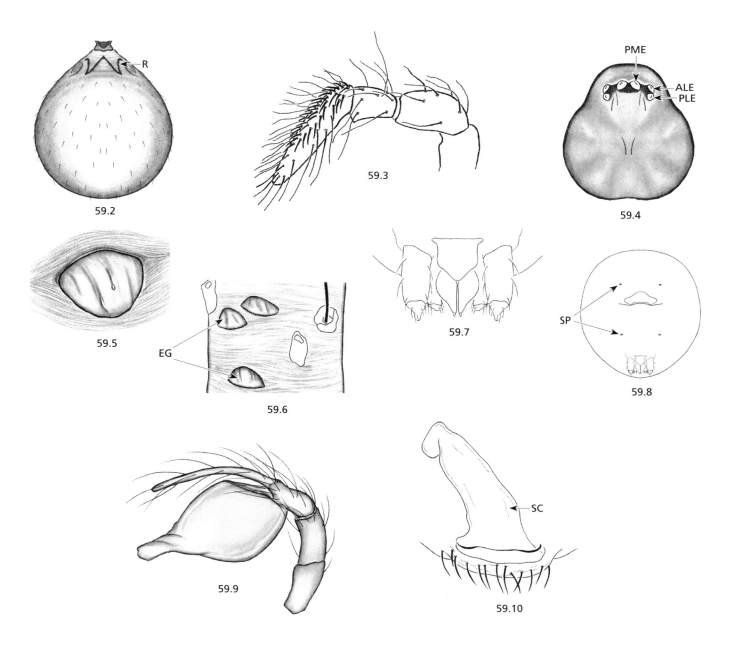

# Chapter 60
# TENGELLIDAE
5 genera, 26 species

Darrell Ubick
David B. Richman

## Similar families —
Ctenidae (p. 83), Zorocratidae (p. 258).

## Diagnosis —
Tengellids may be distinguished from other ecribellate, entelegyne spiders in having tarsi with claw tufts (or dense scopula, in *Lauricius*) and a vestigial third claw. These are large spiders with toothed cheliceral margins, eyes in two straight rows, notched trochanters (shallow in *Lauricius*), and spinnerets conical and contiguous. *Lauricius* has laterigrade legs with 3 pairs of ventral spines on the anterior tibiae; other genera have prograde legs with at least 5 pairs of ventral spines on the anterior tibiae.

## Characters —
**body size:** 7-20 mm.
**color:** usually concolorous brown to gray.
**carapace:** pyriform, with deep longitudinal fovea.
**sternum:** oval, posterioly pointed; precoxal triangles present (*Lauricius*) or absent (other genera).
**eyes:** eight in two rows, roughly subequal, small and nearly round.
**chelicerae:** stout with toothed margins; geniculate in *Lauricius*.
**mouthparts:** endites and labium longer than wide.
**legs:** generally stout; laterigrade in *Lauricius*, prograde in other genera; all trochanters with notches (shallow in Lauricius); anterior tibiae with 5-7 pairs of ventral spines (3 pairs in *Lauricius*), tarsi with dense claw tufts (with apical scopular extension in *Lauricius*); vestigial inferior claw present.
**abdomen:** usually concolorous; patterned in *Lauricius* and *Anachemmis*.
**spinnerets:** anterior spinnerets conical and contiguous.
**respiratory system:** one pair of book lungs; tracheal spiracle slightly anterior to spinnerets.
**genitalia:** entelegyne; *female* epigynum strongly sclerotized, with median lobe; *male* palpus with RTA having 1 to several prongs, median apophysis hook-like, tegulum and subtegulum with interlocking lobes (except in *Lauricius*).

## Distribution —
*Liocranoides* occurs in the Appalachian Mountains, *Lauricius* in Arizona to New Mexico and n into Colorado, and the remaining genera are Californian, with some extension to the north and south.

## Natural history —
Tengellids are ground dwelling nocturnal wandering hunters. They are largely forest dwellers and live amongst rock outcrops and in thick leaf litter. They are often found in caves and although species of most genera (*Titiotus*, *Liocranoides*, and *Anachemmis*) have been recorded from caves, none appears to be troglobitic. Specimens have been collected wandering on the ground at night, under rocks, and especially in pitfall traps. Egg sacs may be attached to the substrate and guarded by the female (*Lauricius*) or suspended from a thread (*Titiotus*) (Ubick, pers. obs.).

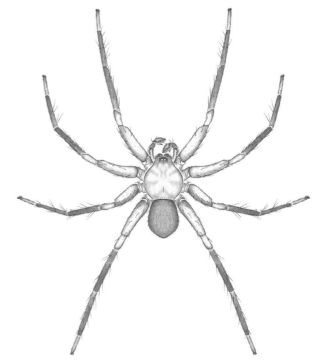

**Fig. 60.1** *Anachemmis sober* Chamberlin 1919b

## Taxonomic history and notes —
The genera currently placed in Tengellidae have been widely separated in the past. Simon (1897a) included his known genera in Clubionidae, although in different subfamilies: *Lauricius* in Clubioninae; *Liocranoides* in Liocraninae; and *Titiotus* in Cteninae. Chamberlin (1919b) described *Anachemmis*, which ended up in yet another subfamily, Micariinae. Lehtinen (1967) grouped the genera in Miturgidae: Tengellinae, subsequently elevated to family and where they now remain. *Lauricius* is quite distinct from the other genera and probably misplaced in the family, although its affinities are not known (Platnick 1999). The remaining genera are closely related to each other, but apparently not to *Tengella* F. Dahl 1901b. In fact, this "*Liocranoides*" complex was placed basal to a group including Zoropsidae, Zorocratidae, and Tengellidae (*sensu stricto*) by Silva Dávila (2003), suggesting that it may deserve family status.

## Genera —
LIOCRANOIDES COMPLEX
  *Anachemmis* Chamberlin 1919b, *Liocranoides* Keyserling 1882b, *Socalchemmis* Platnick & Ubick 2001, *Titiotus* Simon 1897a
INCERTAE SEDIS
  *Lauricius* Simon 1888b

# Tengellidae

## Key to genera —
### North America North of Mexico

**1** Trochanters with broad, shallow notches (Fig. 60.2); legs laterigrade; male palpus lacking tegulum-subtegulum locking process (Fig. 60.3); bulb huge, projecting posteriorly out of cymbium (Figs. 60.3-60.4); RTA bifid (Fig. 60.4); female epigynum with scape (Fig. 60.5) ............ ***Lauricius***
  **Div.** 2 species: *Lauricius hooki* GERTSCH 1941b and *Lauricius hemicloeinus* SIMON 1888b — **Dist.** AZ to NM and CO — **Refs.** Gertsch 1935b, Edwards 1958

— Trochanters with deep notches (Fig. 60.6); legs prograde; male palpus with tegulum-subtegulum locking process (Fig. 60.7); bulb smaller, contained within cymbium (Figs. 60.7-60.8, 60.10, 60.12, 60.14); RTA variable but not bifid; female epigynum lacking scape (Figs. 60.9, 60.11, 60.13, 60.15) .............................................................................. **2**

**2(1)** RTA of male palpus with 3 or more prongs (Figs. 60.7-60.8); epigynum with wide median lobe, > 2/3 epigynal width (Fig. 60.9) .................................... ***Titiotus***
  **Div.** 2 species described, *Titiotus californicus* SIMON 1897a and *Titiotus flavescens* (CHAMBERLIN & IVIE 1941b), and a few dozen undescribed (Roth 1994) — **Dist.** CA-s OR — **Refs.** Simon 1897a, Banks 1913, Chamberlin & Ivie 1941b

— RTA of male palpus with 1 or 2 prongs; epigynum with narrower median lobe (<2/3 epigynal width) ................. **3**

**3(2)** RTA of male palpus usually with only one prong, if with 2 prongs then dorsal prong shorter and narrower (Fig. 60.10); epigynum with median lobe triangular and with sharp edges (Fig. 60.11) ............................ ***Anachemmis***
  **Div.** 5 species — **Dist.** CA-AZ to nw MEX — **Refs.** Chamberlin 1919b, Platnick & Ubick 2005

— RTA of male palpus with 2 subequal prongs; epigynum with middle lobe neither triangular nor with sharp edges ....................................................................................... **4**

**4(3)** Male palpus with embolus having a large mesobasal process, appearing deeply bifurcate (Fig. 60.12); epigynum with median lobe parallel-sided or posteriorly pointed and with rounded edges, lacking anterior hood (Fig. 60.13) ...................................... ***Socalchemmis***
  **Div.** 16 species — **Dist.** CA-AZ to nw MEX — **Ref.** Platnick & Ubick 2001

— Male palpus with embolus lacking a large mesobasal process, but may have short, apical branches (Fig. 60.14); epigynum with median lobe reduced to a narrow septum and having a broad anterior hood (Fig. 60.15) .................. ................................................................................... ***Liocranoides***
  **Div.** 5 species — **Dist.** Appalachian Mountains — **Ref.** Platnick 1999

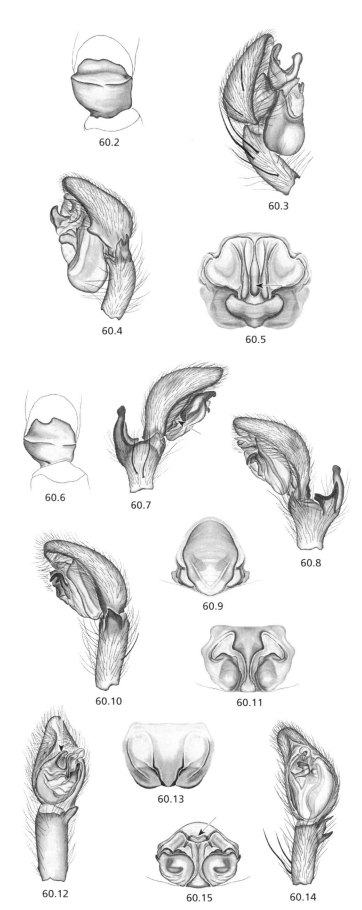

# Chapter 61
# TETRAGNATHIDAE
11 genera, 38 species

Herbert W. Levi

## Common name —
Longjawed orb weavers.

## Similar families —
Araneidae (p. 68), Pholcidae (p. 194), Pimoidae (p. 197), Theridiidae (p. 235).

## Diagnosis —
Three tarsal claws. No cribellum. Eight eyes in two transverse rows clustered into four ME forming a trapezoid and at some distance from LE (Figs. 61.3, 61.16, 61.20, 61.32). Tapetum, when present, differs from that of araneid orb weavers by being always canoe-shaped or lacking a tapetum altogether. Height of clypeus differs from the Linyphiidae by being usually less than three diameters of anterior median eye (Figs. 61.3, 61.20, 61.32). Labium longer than wide to wider than long with distal margin swollen. Endites longer and narrower (Figs. 61.4, 61.9) than the squarish ones of araneid spiders. First legs longest, third shortest. Fourth femur sometimes with trichobothria (Fig. 61.10-61.12). Abdomen oval to worm-shaped (Figs. 61.2, 61.5, 61.19, 61.22, 61.29). Colulus present and spinnerets contains aggregate glands producing viscid silk. Epigynum differs from that of most araneids by being flat, a depression, without ventral projections (Figs. 61.11, 61.18, 61.33) (except *Plesiometa* Fig. 61.12, *Meta* Figs. 61.29-61.30) or being lost (Figs. 61.21, 61.23). Male palpus differs from araneid palpus by lacking both radix and median apophysis and having the palpal tibia cone-shaped, longer than wide (Fig. 61.34), rarely as long as wide (Fig. 61.8).

Most spin orb-webs to catch prey.

## Characters —
**body size:** 1.2-25 mm.
**color:** alcohol soluble greenish or reddish, yellowish, or silver (*Leucauge*); yellowish in alcohol. *Meta* and *Metellina* are black, white and reddish.
**carapace:** longer than wide, cephalic region narrower in most. Height of clypeus usually less than three diameters of anterior median eye (Figs. 61.3, 61.16, 61.25).
**sternum:** longer than wide, shield-shaped, truncate behind in *Glenognatha* (Fig. 61.23).
**eyes:** ME in a trapezoid some distance from LE (Figs. 61.3, 61.16, 61.20, 61.32). PME and LE with canoe-shaped tapetum or tapetum lacking.
**chelicerae:** often enlarged with many strong teeth, especially in males (Figs. 61.3, 61.15, 61.20).
**mouthparts:** labium longer than wide (Fig. 61.4) to wider than long with distal margin swollen. Endites longer, narrower, distally widened (Figs. 61.4, 61.9) unlike squarish ones of araneid spiders.
**legs:** first legs longest, second slightly longer or much longer than fourth, third shortest (Figs. 61.5, 61.31). Without tarsal trichobothria but may have brushes of trichobothria on fourth femur (Fig. 61.11).
**abdomen:** abdomen oval to worm-shaped (Figs. 61.2, 61.10, 61.22, 61.29), rarely with one or two pairs of anterior humps (Figs. 61.13, 61.26).

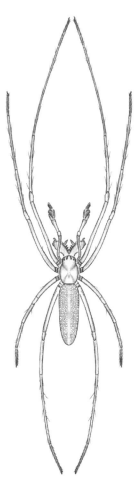

**Fig. 61.1** *Tetragnatha extensa* (LINNAEUS 1758)

**spinnerets:** colulus present and six spinnerets with aggregate glands producing viscid silk.
**respiratory system:** tetragnathids have two book lungs and a spiracle anterior of the colulus with four tracheae confined to the abdomen. This spiracle is moved anteriorly in *Glenognatha* (Fig. 61.23), with two trunk tracheae branching into numerous fine tracheae, some of which extend into the cephalothorax.
**genitalia:** entelegyne and haplogyne; ***female*** epigynum differs from that of most araneids by being flat, a depression, without ventral projections (except *Plesiometa* Fig. 61.12, *Meta* Figs. 61.29-61.30) or lost (*Tetragnatha, Pachygnatha, Glenognatha*, Figs. 61.21, 61.23); ***male*** palpus with paracymbium attached to cymbium (Fig. 61.34) and sometimes modified and enlarged. Embolus wrapped by conductor and without radix and median apophysis.

### Distribution —
Common, widespread.

### Natural history —
Tetragnathidae catch prey with orb webs which are frequently oriented horizontally and have an open hub. Many build near water or in vegetation above it. *Nephila* makes vertical, bright yellow webs. *Meta* builds in dark places, e.g., caves or wells. *Pachygnatha* construct webs only in very early instars, and subsequently become vagrant hunters.

### Taxonomic history and notes —
Tetragnathidae have been transferred in and out of the old families Epeiridae, Argiopidae and present Araneidae. Levi (1980a, 1981a) revised and illustrated the North American species. Marusik & Koponen (1992) found that the North American *Meta* "*menardi*" is distinct from the Eurasian species and gave it a new name. Dondale (1995) found an earlier name for the American *Meta*, *Meta ovalis* GERTSCH 1933c.

### Genera —
NEPHILINAE
  *Nephila* LEACH 1815
METINAE
  *Dolichognatha* O. PICKARD-CAMBRIDGE 1869b, *Meta* C.L. KOCH *in* HERRICH-SCHÄFFER 1835, *Metellina* CHAMBERLIN & IVIE 1941b, *Metleucauge* LEVI 1980a
LEUCAUGE GROUP (Alvarez pers. comm.)
  *Azilia* KEYSERLING 1882b, *Leucauge* DARWIN *in* WHITE 1841, *Plesiometa* F.O. PICKARD-CAMBRIDGE 1903a (**Note**: *Plesiometa* is used in the key, not in the World Catalog)
TETRAGNATHINAE
  *Glenognatha* SIMON 1887d, *Pachygnatha* SUNDEVALL 1823, *Tetragnatha* LATREILLE 1804b

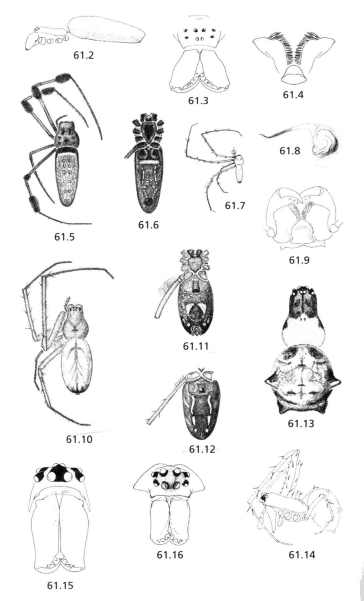

### Key to genera —
North America North of Mexico

| | | |
|---|---|---|
| **1** | Abdomen cylindrical, or tapering, 2-3 times as long as wide (Figs. 61.2, 61.5) | **2** |
| — | Abdomen globular or oval (Figs. 61.10, 61.12, 61.19, 61.24) | **3** |

**2(1)** Enlarged chelicerae with 5 to 9 teeth (Fig. 61.3); endites divergent (Fig. 61.4); epigynum absent ..... ***Tetragnatha***
  **Div.** 15 species — **Dist.** widespread — **Refs.** Levi 1981a

— Chelicerae not modified, with 3 teeth on margins (Fig. 61.9); endites convergent (Fig. 61.9) .................. ***Nephila***
  **Div.** 1 species: *Nephila clavipes* (LINNAEUS 1767) [Figs. 61.5-61.8] — **Dist.** subtropical, NC to Gulf Coast, FL, AL, LA, MS, TX — **Ref.** Levi 1980a

**3(1)** Fourth femur with a proximal cluster of trichobothria (Figs. 61.10-61.12) ............................................. **4**

— Fourth femur without trichobothria ............................ **5**

**4(3)** Female venter with pair of triangular areas (Fig. 61.11); epigynum flat (Fig. 61.11); male palpal tibia longer than palpal bulb ........................................................... ***Leucauge***
  **Div.** 1 species: *Leucauge venusta* (WALCKENAER 1841) [Fig. 61.10] — **Dist.** e NA to s CA — **Ref.** Levi 1980a

— Female venter with two longitudinal bands (Fig. 61.12); epigynum cone-shaped (Fig. 61.12); male palpal tibia shorter palpal bulb ........................................ ***Plesiometa***
  **Div.** 1 species: *Plesiometa argyra* (WALCKENAER 1841) — **Dist.** FL — **Ref.** Levi 1980a — **Note** *Plesiometa* was synonymized with *Leucauge* because there was only a single species, not very distinct from *Leucauge*. However since that time, other undescribed species have been found in South America, making *Plesiometa* a distinct genus in the Tetragnathidae

**5(3)** Lateral eyes separated (Figs. 61.15, 61.16) ................. **6**

— Lateral eyes adjacent (Figs. 61.20, 61.25) ..................... **7**

# Tetragnathidae

**6(5)** Posterior median eyes separated (Fig. 61.16); abdomen with single hump (Fig. 61.19) .................................. **Azilia**
  Div. 1 species: *Azilia affinis* O. Pickard-Cambridge 1893b [Figs. 61.16-61.19] — Dist. Gulf states FL, AL, LA, MS, TX — Ref. Levi 1980a

— Posterior median eyes adjacent (Figs. 61.13, 61.15); abdomen with two pairs of humps (Fig. 61.14) ........................ **Dolichognatha**
  Div. 1 species: *Dolichognatha pentagona* (Hentz 1850a) [Figs. 61.13-61.15] — Dist. se USA — Ref. Levi 1981a

**7(5)** Height of clypeus 2-3 times diameter of anterior median eyes (Fig. 61.20) ................................................. **8**

— Height of clypeus about equal to diameter of anterior median eyes (Figs. 61.25, 61.32) ...................... **9**

**8(7)** Posterior tracheal spiracle close to spinnerets (Fig. 61.21); female (Fig. 61.22) ...................... **Pachygnatha**
  Div. 8 species — Dist. widespread — Ref. Levi 1980a

— Posterior tracheal spiracle some distance from spinnerets (Fig. 61.23); female (Fig. 61.24) ............... **Glenognatha**
  Div. 4 species — Dist. e, sw USA — Refs. Levi 1980a, Hormiga & Döbel 1990

**9(7)** Abdomen widest anteriorly (Figs. 61.26-61.27); cheliceral retromargin with 3 teeth (Fig. 61.25) ............. **Metellina**
  Div. 3 species — Dist. Pacific provinces and states, to Rocky Mountain states — Ref. Levi 1980b

— Abdomen widest in middle (Figs. 61.28, 61.31); cheliceral retromargin with 4 teeth ............................... **10**

**10(9)** Abdomen almost vertical (Fig. 61.29), near oval (Figs. 61.28-61.30) ......................................... **Meta**
  Div. 2 species: *Meta ovalis* Gertsch 1933c and *Meta dolloff* Levi 1980a — Dist. ne provinces and states, Appalachian Mountains to AR, CA — Refs. Levi 1980b, Marusik & Koponen 1992, Dondale 1995

— Abdomen horizontal, dorsoventrally flattened (Figs. 61.31, 61.33) ............................................. **Metleucauge**
  Div. 1 species: *Metleucauge eldorado* Levi 1980b [Figs. 61.31-61.34] — Dist. CA — Ref. Levi 1980b

# Chapter 62
# THERIDIIDAE
32 genera, 234 species

Herbert W. Levi

### Common name —
Cobweb weavers.

### Similar families —
Nesticidae (p. 178).

### Diagnosis —
Three tarsal claws. No cribellum. Eight eyes in two transverse rows with four ME and two pairs of LE; LE are usually adjacent to each other (Fig. 62.3). Posterior median eyes and lateral eyes have a canoe-shaped tapetum. Height of clypeus usually more than two diameters of anterior median eyes (Fig. 62.3). Chelicerae usually weakly sclerotized (Fig. 62.3), but extending to a point underneath the clypeus. Endites usually converge distally. The abdomen is frequently spherical.

Theridiids differ from other araneoid families (except the tropical Synotaxidae) by not having the labium distally swollen (Fig. 62.2). As in Nesticidae and tropical Synotaxidae, there is a tarsal comb, a row of serrated setae on the fourth leg (arrow Fig. 62.4 and enlarged view), indistinct in some *Dipoena* and *Euryopis*. The paracymbium of the male palpus is near the tip of the cymbium (arrow in Fig. 62.42) or behind the bulb (arrow in Fig. 62.47 [palp with bulb removed]), whereas in related families the paracymbium is at the base of the cymbium. Unlike most araneoids, the colulus (arrow Fig. 62.7) is absent in many species (Fig. 62.5), or it may be reduced to two setae (Fig. 62.6). The theridiids are small to medium sized spiders and, compared to the linyphiids, are soft and slightly sclerotized, with great diversity.

### Characters —
**body size:** 0.8-12.0 mm, the largest *Latrodectus* sp.
**color:** drab brown to colorful greens and reds (the last two wash out in preserving fluid).
**carapace:** usually longer than wide (Fig. 62.37).
**sternum:** longer than wide, with pointed posterior tip (Fig. 62.2), truncate in *Crustulina*.
**eyes:** ME some distance from LE, which are adjacent (Fig. 62.3). Tapetum of PME and LE canoe-shaped.
**chelicerae:** weakly sclerotized, often without teeth (Figs. 62.3, 62.44) or with small teeth; teeth may be large in *Enoplognatha*, *Robertus* and *Pholcomma*; fang short (Figs. 62.3, 62.49) except in *Euryopis* and *Dipoena*, which have a drawn out long transparent fang (best seen by reflecting light on the fang) (Fig. 62.50).
**mouthparts:** labium lacks the distal swelling of other araneoids (Fig. 62.2); endites are longer than wide (arrow Fig. 62.2) and converge distally.
**legs:** 3-clawed; legs of variable length, some quite short; first usually longest, second or fourth next, third shortest (Fig. 62.20); legs usually have few or no thick setae and no femoral macrosetae; fourth legs of most with comb macrosetae (Fig. 62.4), indistinct in *Dipoena* and *Euryopis*.
**abdomen:** ball-or globe-shaped (Fig. 62.11), rarely elongate (Figs. 62.22, 62.71).
**spinnerets:** six spinnerets, the median ones hidden by

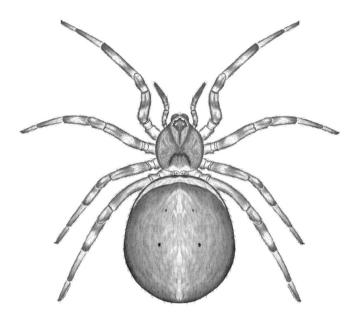

**Fig. 62.1** *Steatoda hespera* CHAMBERLIN & IVIE 1933a

other four (Figs. 62.5-62.7); colulus often absent (Fig. 62.5) or only 2 setae remaining (Fig. 62.6).
**respiratory system:** spiracle in front of spinnerets (Fig. 62.6) and two book lung covers are present.
**genitalia:** entelegyne; *female* epigynum has two seminal receptacles (Fig. 62.10), except four in *Latrodectus* and *Dipoena* and most *Euryopis* (Fig. 62.56); epigynum is usually a simple depression (Figs. 62.19, 62.48), (except in *Tidarren*, *Neottiura*, *Nesticodes* and some *Theridion* and *Thymoites* sp.); *male* palpal tibia conical, often short, with a row of large setae at the distal rim; palpus with paracymbium (arrow in Fig. 62.42) on the distal end of the cymbium or behind the bulb (Fig. 62.47); palpus usually with a conductor supporting the tip of embolus (E in Fig. 62.135), radix (R in Fig. 62.135) and median apophysis (M in Fig. 62.135), although the homology of the last two structures is uncertain; the sclerites shown in Fig. 62.91 from right to left, embolus (E), conductor (C), sclerotized median apophysis (M), radix (R) mostly hidden; a bulb cymbium lock mechanism is present in some genera, the median apophysis fitting under the paracymbium.

### Distribution —
Worldwide.

### Natural history —
Theridiids are sedentary and usually make an irregular web with sticky strands attached to the substrate. The

strands break when prey touches the line, pulling the prey toward the center of the web. The spider hangs, upside down, in the web. *Rhomphaea* and *Argyrodes* are found in the webs of other spiders, especially orb weavers, and may feed on these hosts, their eggs, or their prey. *Dipoena* and *Euryopis* are ant predators living on the ground.

*Latrodectus* species are venomous to people.

**Taxonomic history and notes —**

At a time when we could not even recognize most common American theridiid species, Archer (1946, 1950) created many new, inadequately illustrated genera. He also found that the type species of *Theridion* WALCKENAER 1805 did not belong to the genus, as usually understood, calling for more generic changes. The names were stabilized by the International Commission on Zoological Nomenclature by giving the genus *Theridion* a new type species (ICZN 1958, Opinion 517).

I was encouraged to start revisions with emphasis on illustrating genitalia to facilitate identification of species. I tried to avoid splitting. After all, the proliferation of genera in linyphiids has not helped in their study, identification, or study of phylogeny. Worse, it has resulted in constant name changes due to shifting genera. In retrospect it appears that to avoid splitting, I lumped. Here, several of the genera, I synonymized, *Arctachaea* LEVI 1957e, *Meotipa* SIMON 1894a, *Neottiura* MENGE 1868, *Nesticodes* ARCHER 1950, *Rhomphaea* L. KOCH 1872a, and *Wamba* O. PICKARD-CAMBRIDGE 1896a, are resurrected (in part following others), to avoid polyphyletic genera.

In the 1960's, B. Vogel (personal communication) made me aware that in labeling illustrations, I switched the names of palpal sclerites in some genera. This was subsequently corrected in Levi & Randolph (1975). Coddington (1990), studying the homology of the sclerites in theridiids and related families renamed the theridiid radix as the "theridiid tegular apophysis", but did not rename the theridiid median apophysis. Because homology with both radix and median apophysis of the araneid palpi is uncertain, I am keeping here the name radix. For purpose of discussion, they can be called theridiid radix and theridiid median apophysis.

Wunderlich (1987a, 1992a, 1995n) resurrected some of Archer's genera and made several new genera. Unfortunately, splitting *Theridion* has created genera of only a few species each, mostly based on males, instead of finding characters to reconfigure the large genus. Also, I find it difficult to key out Wunderlich's *Simitidion* which differs by having metatarsi III and IV without trichobothria, as I did not study the trichobothria of other genera. Another of Wunderlich's genera has a type species known only from a female. *Rugathodes* ARCHER 1950, previously synonymized with *Theridion* (Levi 1957a) but resurrected by Wunderlich, differs by modifications of the male chelicerae; but modified male chelicerae are also found in some other *Theridion* not belonging to the *Rugathodes* group of species (e.g., *Theridion leechi* GERTSCH & ARCHER 1942, *Theridion punctipes* EMERTON 1924b, *Theridion neomexicanum* BANKS 1901b, *Theridion pictipes* KEYSERLING 1884a).

Knoflach in Austria studied the behavior of some theridiid spiders and admirably illustrated the species studied (Knoflach & van Harten 2000b, Knoflach & Thaler 2000).

More recently Yoshida (2001b, c, d, 2002, 2003) revised the Japanese theridiids, adding generic names for some additional theridiid species groups, and also splitting *Dipoena* and *Euryopis*. Although Yoshida's splitting may better reflect phylogeny than Wunderlich's, the genera may not be needed and would make generic placement of females impossible without males. Roberts (1985a, b, 1987, 1995) and Zhu (1998), although citing Wunderlich in their treatment of European and Chinese theridiids, did not use Archer's and Wunderlich's genera.

Platnick's catalog (2005) lists all the species described including some of the genera here lumped (*Emertonella* BRYANT 1945b, *Laseola* SIMON 1881a, *Phycosoma* O. PICKARD-CAMBRIDGE 1879d, *Trigonobothrys* SIMON 1889b, *Yaginumena* YOSHIDA 2002, *Simitidion* WUNDERLICH 1992a, *Takayus* YOSHIDA 2001c, *Keijia* YOSHIDA 2001c, *Rugathodes* ARCHER 1950, but missing some other genera (*Meotipa* and *Arctachaea*). Thus the key does not follow Platnick's 2005 version of his catalog. Recently, Agnarsson (2004) proposed a phylogeny for the theridiid genera.

An excellent summary of theridiid external anatomy and natural history is in Knoflach & Pfaller (2004) and mating behavior in Knoflach (2004). Both have superb close-up colored photographs of theridiids.

I am thankful to I. Agnarsson, J. Coddington and C.D. Dondale for suggestions and improvements, I. Agnarsson also provided useful information on subfamilial arrangement.

**Genera —**

HADROTARSINAE
*Dipoena* THORELL 1869, *Emertonella* BRYANT 1945b, *Euryopis* MENGE 1868, *Lasaeola* SIMON 1881a, *Phycosoma* O. PICKARD-CAMBRIDGE 1879d, *Trigonobothrys* SIMON 1889b, *Yaginumena* YOSHIDA 2002

SPINTHARINAE
*Chrosiothes* SIMON 1894a, *Episinus* WALCKENAER in LATREILLE 1809, *Spintharus* HENTZ 1850b

LATRODECTINAE
*Crustulina* MENGE 1868, *Latrodectus* WALCKENAER 1805, *Steatoda* SUNDEVALL 1833b

PHOLCOMMATINAE
*Enoplognatha* PAVESI 1880, *Pholcomma* THORELL 1869, *Phoroncidia* WESTWOOD 1835, *Robertus* O. PICKARD-CAMBRIDGE 1879g, *Stemmops* O. PICKARD-CAMBRIDGE 1894a, *Styposis* SIMON 1894a, *Theonoe* SIMON 1881a

ARGYRODINAE (Note: revalidated synonyms)
*Argyrodes* SIMON 1864, *Faiditus* KEYSERLING 1884a, *Neospintharus* EXLINE 1950a, *Rhomphaea* L. KOCH 1872a

ANELOSIMINAE
*Anelosimus* SIMON 1891d

THERIDIINAE
*Achaearanea* STRAND 1929, *Arctachaea* LEVI 1957e, *Chrysso* O. PICKARD-CAMBRIDGE 1882c, *Coleosoma* O. PICKARD-CAMBRIDGE 1882c, *Keijia* YOSHIDA 2001c, *Neottiura* MENGE 1868, *Nesticodes* ARCHER 1950, *Paratheridula* LEVI 1957d, *Rugathodes* ARCHER 1950, *Simitidion* WUNDERLICH 1992a, *Takayus* YOSHIDA 2001c, *Tekellina* LEVI 1957d, *Theridion* WALCKENAER 1805, *Theridula* EMERTON 1882, *Thymoites* KEYSERLING 1884a, *Tidarren* CHAMBERLIN & IVIE 1934, *Wamba* O. PICKARD-CAMBRIDGE 1896a

## Key to genera —
### North America North of Mexico

Abbreviations: E = embolus, M = median apophysis, P = paracymbium, R = radix, Y = cymbium

1  Abdomen with a series of humps (Figs. 62.8-62.9), eye region projecting above clypeus (Fig. 62.8) ........................
............................................. ***Phoroncidia***
**Div.** 1 species: *Phoroncidia americana* (EMERTON 1882) [Figs. 62.8, 62.9], ♀ 1.7-2.7 mm, ♂ 1.3-1.6 mm — **Dist.** e USA — **Refs.** Levi 1955c, 1964d — **Note** previously placed in *Oronota* SIMON 1871b by Levi 1955c

—  Abdomen otherwise .......................................................... **2**

2(1)  Large colulus present (arrow Fig. 62.7) ............................ **3**

—  Colulus absent or not easily seen (Fig. 62.5), or only two setae remaining (Fig. 62.6) .............................................. **12**

3(2)  Fourth tarsi longer than metatarsi ..................... ***Theonoe***
**Div.** 1 species: *Theonoe stridula* CROSBY 1906, ♀♂ 0.8-1.2 mm [Figs. 62.10-62.12] — **Dist.** s CAN, e USA — **Ref.** Levi 1955a — **Note** previously placed in *Coressa* SIMON 1894a by Levi 1955a

—  Fourth tarsi equal to or shorter than metatarsi ................ **4**

4(3)  Carapace and sternum with large dumb-bell shaped granulations (Figs. 62.14-62.15) ....................... ***Crustulina***
**Div.** 2 species: *Crustulina altera* GERTSCH & ARCHER 1942 and *Crustulina sticta* (O. PICKARD-CAMBRIDGE 1861a) ♀♂ 1.5-2.7 mm [Figs. 62.13-62.16] — **Dist.** AK to FL, CA — **Ref.** Levi 1957b

—  Carapace and sternum smooth or sometimes finely granulated (Figs. 62.11, 62.21, 62.26) ........................................... **5**

5(4)  LE separated by their diameter; hourglass-shaped red mark on venter of abdomen (Fig. 62.20), red mark rarely separated into two parts; females with 4 seminal receptacles; no teeth on cheliceral margin .............. ***Latrodectus***
**Div.** 5 species, ♀ 6.0-12.0 mm, ♂ 2.0-8.0 mm [Figs. 62.17-62.20] — **Dist.** s CAN to MEX — **Refs.** Levi 1959c, 1967c, 1984b, McCrone & Levi 1964, Kaston 1970, Zhang *et al.* 2004

—  LE adjacent or only slightly separated; without hourglass-shaped red mark on venter; females with 2 seminal receptacles; usually teeth on cheliceral margin ......................... **6**

6(5)  Abdomen worm-shaped (Figs. 62.22-62.23), or higher than long (Fig. 62.29) with humps extending beyond spinnerets; fourth tarsi without comb; third claw longer than paired claws; male eye region or clypeal region swollen, projecting, or groove below eyes (Figs. 62.21, 62.26) ...... **7**

—  Abdomen oval to spherical; third tarsal claw shorter than paired claws; comb setae on fourth tarsus; male eye region and clypeus not modified ...................................... **8**

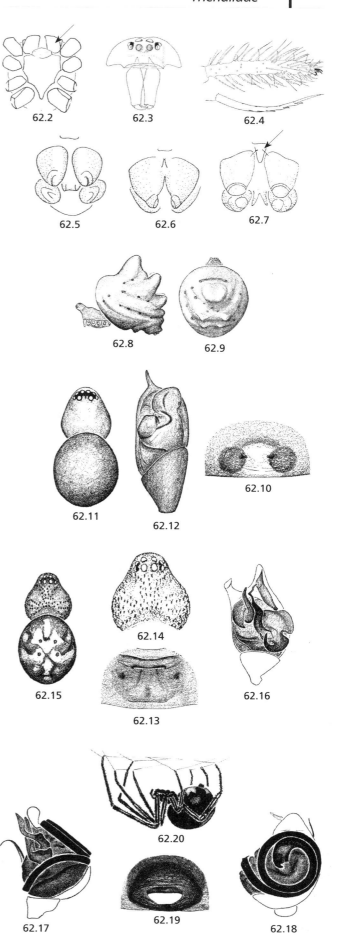

**7(6)** Female abdomen tapering to a single posterior tip, usually one or more spines on tip; abdomen 4-6 times as long behind spinnerets as anterior to it (Fig. 62.23); male carapace with a single cephalic projection or if lacking, clypeus slanting and projecting below (Fig. 62.21) .................... **Rhomphaea**
Div. 3 species, ♀ 4.0-12.0 mm, ♂ 3.0-6.6 mm [Figs. 62.21-62.25] — Dist. s CAN to FL — Ref. Exline & Levi 1962 — Note previously placed in Argyrodes by Exline & Levi 1962

— Abdomen otherwise (Fig. 62.29) and male cephalic region otherwise (Fig. 62.26) .................... **Argyrodes**
Div. 13 species, ♀ 2.0-5.2 mm, ♂ 1.0-4.3 mm [Figs. 62.26-62.29] — Dist. s CAN, e USA, s USA to CA — Ref. Exline & Levi 1962 — Note includes genera Faiditus and Neospintharus used by Agnarsson (2004)

**8(6)** AME smaller than other eyes (Figs 62.30, 62.34) and total length less than 1.5 mm .................... **9**

— AME about the same size as other eyes, if smaller, total length greater than 1.5 mm .................... **10**

**9(8)** AME vestigial, minute, less than 1/3 diameter of LE (Fig. 62.30) .................... **Styposis**
Div. 1 species: Styposis ajo Levi 1960, ♀♂ 1.6 mm [Figs. 62.30-62.31] — Dist. AZ — Ref. Levi 1960 — Note colulus in Styposis is difficult to see, and Styposis ajo keys out here and in couplet 21

— AME only slightly smaller than other eyes (Figs. 62.32, 62.34) .................... **Pholcomma**
Div. 3 species, ♀♂ 1.0-1.6 mm [Figs. 62.32-62.35] — Dist. e US — Refs. Levi 1957d, 1975a — Note colulus in Pholcomma spp. is difficult to see and keys out here and in couplet 21

**10(8)** Abdomen without pattern, uniformly colored (Fig. 62.37) .................... **Robertus**
Div. 15 species, ♀♂ 1.5-4.5 mm [Figs. 62.36-62.38] — Dist. AK to TN, IL, UT — Ref. Kaston 1946 — Note the generic name Robertus O. Pickard-Cambridge 1879g has priority over Ctenium Menge 1871 as ruled by the Opinion 1426 (ICZN 1987b). Kaston (1946) used Ctenium

— Abdomen with markings (Figs. 62.40, 62.45) .................... **11**

**11(10)** Female with tooth on cheliceral retromargin; male chelicerae (Fig. 62.39) larger than females, palpal paracymbium on margin of cymbium (arrow in Fig. 62.42); coloration various, to yellow-brown and black .................... **Enoplognatha**
Div. 9 species, ♀ 3.0-9.0 mm, ♂ 2.8-6.5 mm [Figs. 62.39-62.43] — Dist. AK to FL, CA — Refs. Levi 1957a, 1962a, Hippa & Oksala 1982

— Female without teeth on cheliceral retromargin; male chelicerae same as in female (Fig. 62.44); paracymbium dorsal in cymbium, usually hidden behind bulb (arrow in Fig. 62.47, which has bulb removed); coloration purple-brown often with white line around anterior of abdomen (Fig. 62.45) .................... **Steatoda**
Div. 20 species, ♀ 2.5-11.0 mm, ♂ 2.0-7.5 mm [Figs. 62.44-62.48] — Dist. AK to MEX — Refs. Levi 1957b, 1962a, Gertsch 1960b

**12(2)** Chelicerae with elongated flat fang and without teeth (Fig. 62.50); female usually with 2 pairs of seminal receptacles (except in some *Euryopis*) (Fig. 62.56); AME usually largest and farther apart than PME (Fig. 62.53) .................... **13**

— Chelicerae with normal fang (Fig. 62.49); female usually with only one pair of seminal receptacles .................... **14**

**13(12)** Abdomen subtriangular, pointed behind (Figs. 62.51, 62.53) ............................................................... ***Euryopis***
   **Div.** 20 species, ♀♂ 2.0-3.0 mm [Figs. 62.51-62.56] — **Dist.** widespread — **Refs.** Levi 1954b, 1963a — **Note** includes the genus *Emertonella* used by Yoshida (2003)

— Abdomen globose, rounded behind (Fig. 62.59); male carapace often high with grooves (Figs. 62.61-62.62) ....... ............................................................... ***Dipoena***
   **Div.** 17 species, ♀♂ 1.2-4.5 mm [Figs. 62.57-62.66] — **Dist.** se US — **Refs.** Levi 1953b, 1963a — **Note** includes the genera *Lasaeola, Trigonobothrys, Yanigumena* used by Yoshida (2003) and *Phycosoma* used by Platnick (2005)

**14(12)** Colulus represented by two setae (Fig. 62.6) ............... **15**

— Colulus completely absent, or not easily seen (Fig. 62.5) .. ............................................................................................ **22**

**15(14)** PME 2 1/2 to 3 diameters apart (Fig. 62.67); abdomen longer than wide, widest anteriorly (Figs. 62.67, 62.69) .... ............................................................................. ***Spintharus***
   **Div.** 1 species: *Spintharus flavidus* HENTZ 1850b, ♀ 3.0-5.4 mm, ♂ 2.0-3.1 mm [Figs. 62.67-62.70] — **Dist.** MA to se USA — **Refs.** Levi 1955b, 1963e

— PME at most 2 diameters apart (Figs. 62.71, 62.74, 62.79); abdomen shaped otherwise ........................................... **16**

**16(15)** Abdomen longer than wide, widest posteriorly with median posterior or lateral humps (Fig. 62.71) ................. .................................................................................. ***Episinus***
   **Div.** 2 species: *Episinus amoenus* BANKS 1911 and *Episinus cognatus* O. PICKARD-CAMBRIDGE 1893b ♀♂ 3.0-6.0 mm [Figs. 62.71-62.73] — **Dist.** MD to FL, TX — **Ref.** Levi 1955b

— Abdomen shaped otherwise ............................................. **17**

**17(16)** Venter of abdomen with anterior overhang black (Figs. 62.74-62.75); eyes often reddish; palpal conductor absent, cymbium or median apophysis supporting embolus (Figs. 62.76-62.77) ................................................ ***Chrosiothes***
   **Div.** 6 species, ♀♂ 1.3-3.0 mm [Figs. 62.74-62.78] — **Dist.** e, s USA — **Refs.** Levi 1954a, 1964f, Piel 1995 — **Note** *Theridiotis* LEVI 1954a is a junior synonym of *Chrosiothes* SIMON 1894a

— Venter of abdomen and anterior overhang otherwise; eyes not reddish; palpal cymbium never supporting embolus . ............................................................................................ **18**

**18(17)** Eyes large, closely grouped; eye region black except between PME (Fig. 62.79); abdomen often with white spot dorsal of spinnerets (Fig. 62.79); leg IV > leg I ................................................................................ ***Stemmops***

**Div.** 2 species: *Stemmops bicolor* O. Pickard-Cambridge 1894a and *Stemmops ornatus* (Bryant 1933a) ♀♂ 1.3-2.1 mm [Figs. 62.79-62.81] — **Dist.** NJ to se US — **Ref.** Levi 1955c, 1964f

— Eyes, eye region and abdomen otherwise; leg I > leg IV ............................................................................................ **19**

**19(18)** Total length greater than 2 mm; cheliceral margins with teeth; abdomen oval, longer than wide, often with median, dorsal, longitudinal band (Fig. 62.82) ........ ***Anelosimus***

**Div.** 2 species, *Anelosimus studiosus* (Hentz 1850b) and *Anelosimus analyticus* (Chamberlin 1924b) ♀♂ 2.0-5.0 mm [Figs. 62.82-62.84] — **Dist.** e USA, s CA — **Ref.** Levi 1956b

— Total length less than 1.8 mm ....................................... **20**

**20(19)** Abdomen wider than long (Fig. 62.85) ........... ***Tekellina***

**Div.** 1 species: *Tekellina archboldi* Levi 1957d, ♀♂ 0.9-1.3 mm [Figs. 62.85-62.87] — **Dist.** FL — **Ref.** Levi 1957d

— Abdomen spherical to longer than wide (Fig. 62.32) .. **21**

**21(20)** AME vestigial, minute, less than 1/3 diameter of LE (Fig. 62.30) ........................................................... ***Styposis***

**Div.** 1 species: *Styposis ajo* Levi 1960, ♀♂ 1.6 mm [Figs. 62.30-62.31] — **Dist.** AZ — **Ref.** Levi 1960 — **Note** colulus in *Styposis* is difficult to see, and *S. ajo* keys out here and in couplet 9

— AME only slightly smaller than other eyes (Fig. 62.32, 62.34) ........................................................... ***Pholcomma***

**Div.** 3 species, ♀♂ 1.0-1.6 mm [Figs. 62.32-62.35] — **Dist.** e US — **Refs.** Levi 1957d, 1975a — **Note** colulus in *Pholcomma* spp. is difficult to see and keys out here and in couplet 9

**22(14)** Females .......................................................................... **23**

— Males ............................................................................... **39**

**23(22)** Abdomen wider than long (Figs. 62.88, 62.92) ........... **24**

— Abdomen longer than wide, spherical or higher than long ................................................................................ **26**

**24(23)** Abdomen with tubercle on each side (Fig. 62.88) ............................................................................... ***Theridula***

**Div.** 3 species, ♀ 1.8-2.8 mm [Figs. 62.88-62.90] — **Dist.** s CAN to MEX — **Refs.** Levi 1954c,1966b

— Abdomen without distinct tubercle (Fig. 62.92) ......... **25**

**25(24)** Abdomen with dorsal, anterior, median black patch (Fig. 62.92) ................................................................ ***Wamba***

**Div.** 2 species: *Wamba crispulus* (Simon 1895g) and *Wamba congener* (O. Pickard-Cambridge 1896a) ♀ 1.2-3.6 mm [Figs. 62.91-62.93] — **Dist.** ME to TX, CA — **Ref.** Levi 1957a — **Note** species previously placed in *Theridion* by Levi 1957a. *Chindellum* Archer 1950 is a junior synonym

— Abdomen with dorsal, anterior, median white patch (Figs. 62.133, 62.137, 62.139) ....................... ***Theridion,*** *in part*

**Div.** 65 species, 2 key out here: *Theridion istokpoga* Levi 1957a and *Theridion positivum* Chamberlin 1924b, ♀♂ 1.2-3.6 mm [Figs. 62.133-62.134, 62.137-62.138, 62.140] — **Dist.** FL, s TX, CA — **Ref.** Levi 1957a

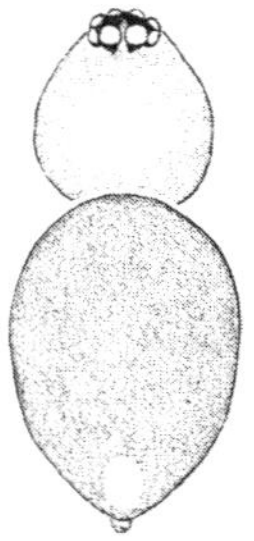
62.79

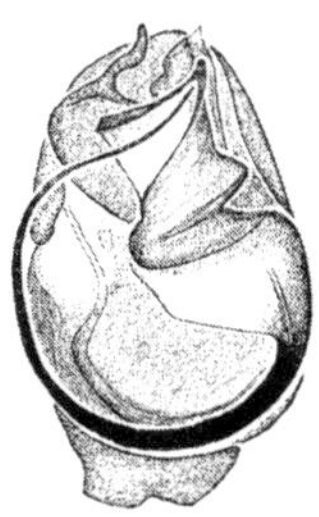
62.80

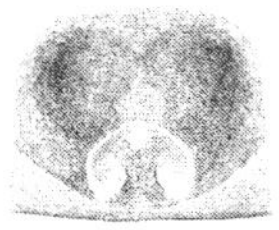
62.81

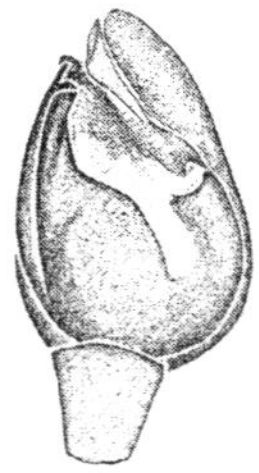
62.83

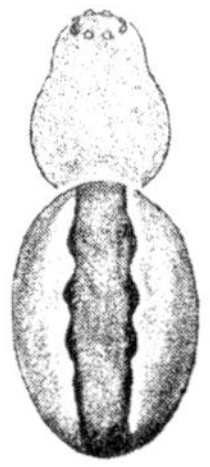
62.82

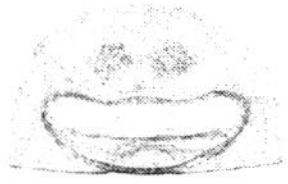
62.84

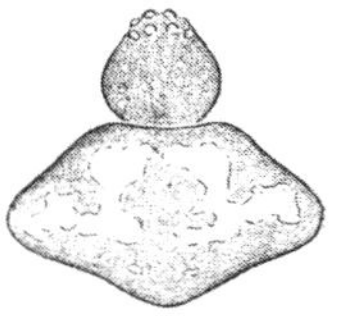
62.85

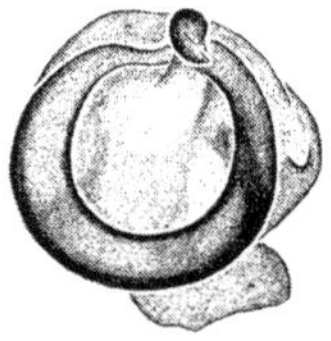
62.86

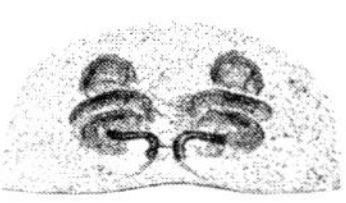
62.87

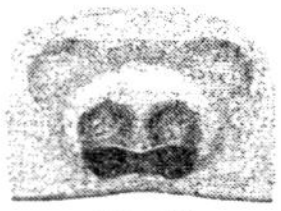
62.89

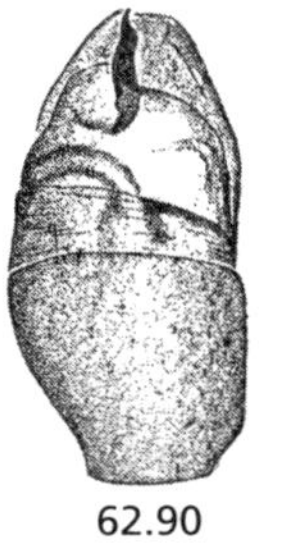
62.90

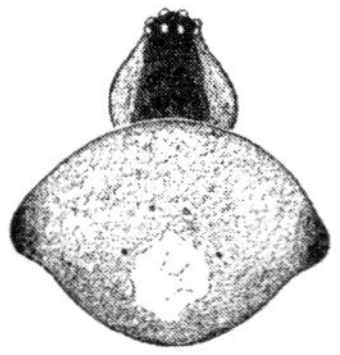
62.88

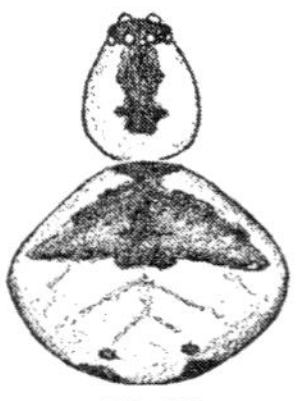
62.92

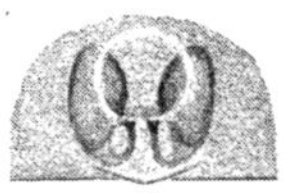
62.93

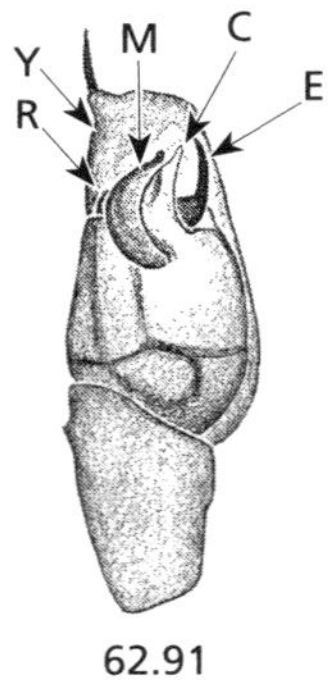

62.91

**26(23)** Abdomen higher than long (Figs. 62.95, 62.98, 62.101) .... .................................................................... **27**

— Abdomen spherical or longer than high ...................... **29**

**27(26)** Epigynum raised swelling (Fig. 62.97) .............. ***Tidarren***

**Div.** 2 species: *Tidarren haemorrhoidale* (Bertkau 1880a) and *Tidarren sisyphoides* (Walckenear 1841) ♀ 2.4-4.5 mm [Figs. 62.94-62.97] — **Dist.** KY, s US to CA — **Ref.** Levi 1957c

— Epigynum flat (Figs. 62.100, 62.102) ........................... **28**

**28(27)** Abdomen light, with large posterior, dorsal tubercle (Fig. 62.98) ............................................. ***Arctachaea,** in part*

**Div.** 2 species, one keys out here: *Arctachaea pelyx* Levi 1957e ♀ 2.7-3.5 mm [Figs. 62.98-62.100] — **Dist.** AK to CO — **Ref.** Levi 1957e — **Note** species previously placed in *Chrysso* O. Pickard-Cambridge 1882c by Levi 1962b

— Abdomen without tubercle, or if with tubercle, sides streaked with black (Figs. 62.101) ................... ***Achaearanea***

**Div.** 16 species, ♀ 1.7-8.0 mm [Figs. 62.101-62.103] — **Dist.** widespread — **Refs.** Levi 1955a, 1959b, 1963b, 1967c, 1980b

**29(26)** Abdomen longer than wide with posterior tubercle (Figs. 62.105, 62.108, 62.111) .................................................. **30**

— Abdomen oval to subspherical (Figs. 62.114, 62.129) ....... .................................................................... **33**

**30(29)** AK, CAN, Rocky Mountains ........... ***Arctachaea,** in part*

**Div.** 2 species, one keys out here: *Arctachaea nordica* (Chamberlin & Ivie 1947b) ♀ 2.7-3.5 mm — **Dist.** AK to CO — **Ref.** Levi 1957e — **Note** *Arctachaea nordica* was previously placed in *Chrysso* by Levi 1962b

— Southeastern USA ........................................................ **31**

**31(30)** Grooves on sides of abdomen (Fig. 62.105) ...................... .................................................................... ***Chrysso***

**Div.** 1 species: *Chrysso albomaculata* O. Pickard-Cambridge 1882c, ♀ 2.0-4.5 mm [Figs. 62.104-62.107] — **Dist.** NC to s USA — **Ref.** Levi 1957c, 1962b

— Without grooves (Figs. 62.108, 62.111) ....................... **32**

**32(31)** Abdomen often with feather-shaped setae (Fig. 62.108); epigynum with two separated dark circles (Fig. 62.109) ... .................................................................... ***Meotipa***

**Div.** 1 species: *Meotipa pulcherrima* (Mello-Leitão 1917a), ♀ 2.3 mm [Figs. 62.108-62.110] — **Dist.** FL — **Ref.** Levi 1962b — **Note** species previously placed in *Chrysso* O. Pickard-Cambridge 1882c by Levi 1962b

— Abdomen never with large setae (Fig. 62.111); epigynum with 2 longitudinal lines (Fig. 62.112) .............................. ........................................................ ***Coleosoma,** in part*

**Div.** 3 species , 1 keys out here: *Coleosoma acutiventer* (Keyserling 1884a), ♀ 2-3 mm [Figs. 62.111-62.112] — **Dist.** FL, Gulf coast — **Ref.** Levi 1959a

**33(29)** Epigynum raised and bulging (Figs. 62.115-62.116, 62.119-62.120) ............................................................ **34**

— Epigynum flat or with depression (Fig. 62.131) .......... **37**

**34(33)** S CAN, Pacific NW; epigynum tongue-shaped (Figs. 62.115-62.116) .................................................. ***Neottiura***

**Div.** 1 species: *Neottiura bimaculata* (Linneaus 1767) ♀ 2-3 mm [Figs. 62.113-62.117] — **Dist.** s CAN, WA, OR — **Refs.** Levi 1957a, Knoflach 1999 — **Note** species previously placed in *Theridion* by Levi 1957a

— Epigynum otherwise ..................................................... **35**

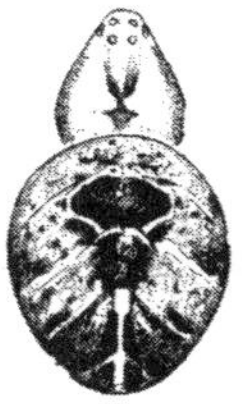
62.94

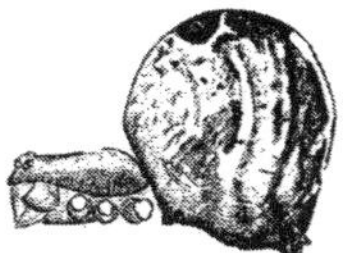
62.95

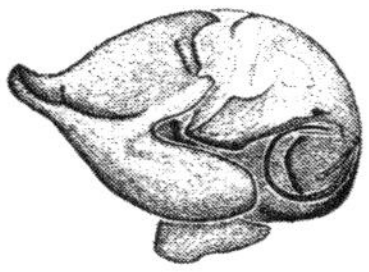
62.96

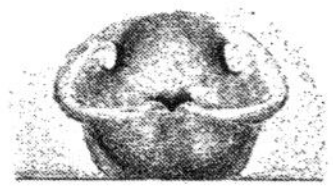
62.97

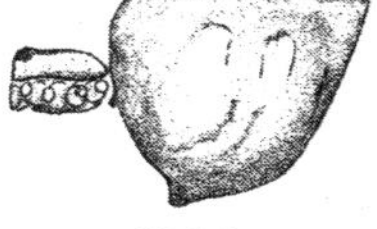
62.98

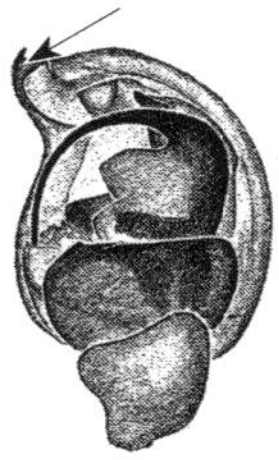
62.99

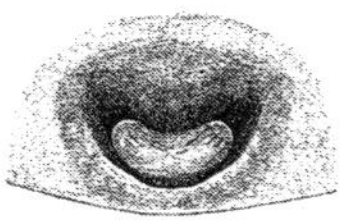
62.100

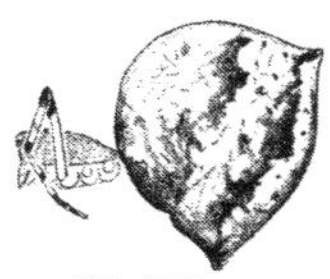
62.101

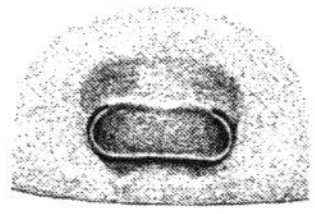
62.102

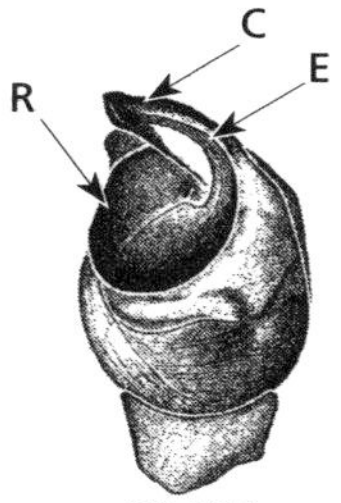

62.103

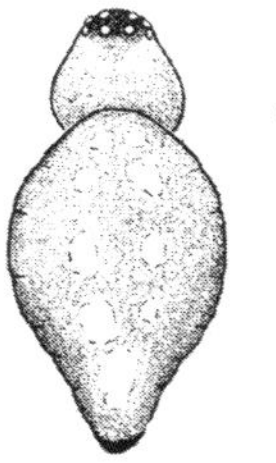
62.104

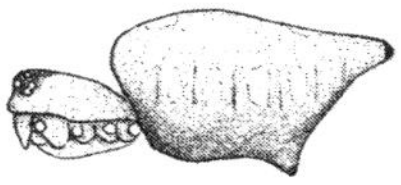
62.105

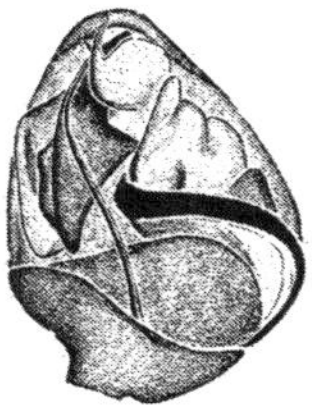
62.106

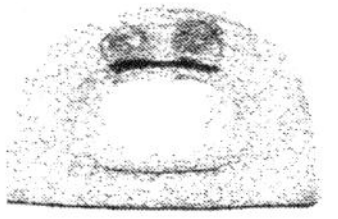
62.107

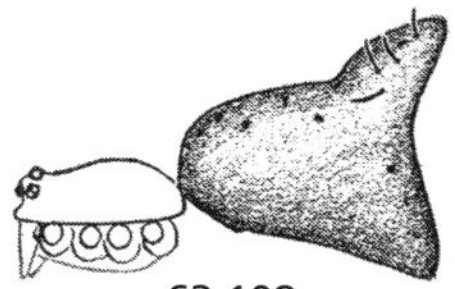
62.108

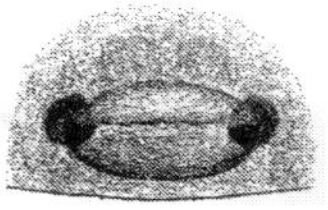
62.109

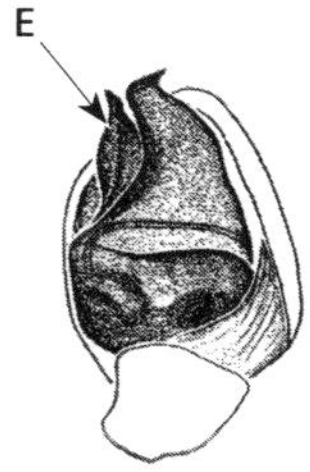

62.110

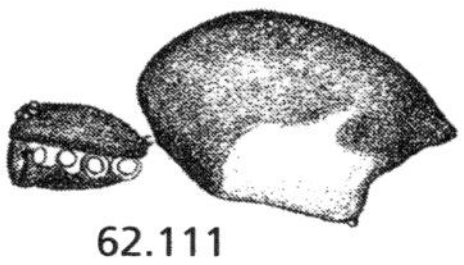
62.111

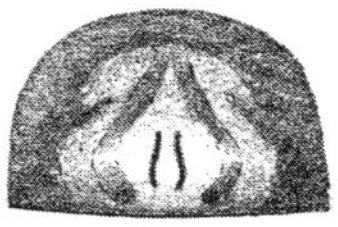
62.112

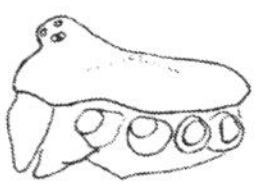
62.113

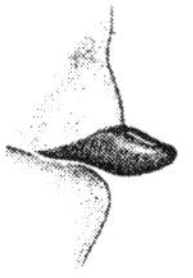
62.116

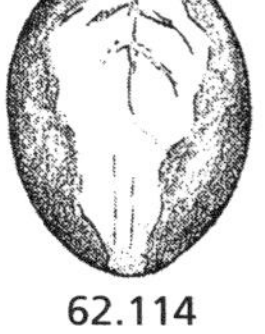
62.114

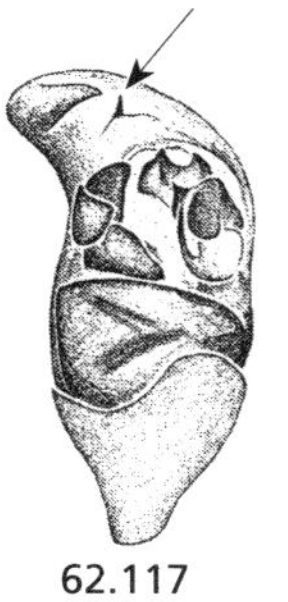
62.117

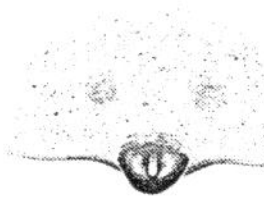
62.115

## Theridiidae

**35(34)** Epigynum a swelling with anterior opening (Figs. 62.119-62.120) .......................................................... ***Nesticodes***
   **Div.** 1 species: *Nesticodes rufipes* (Lucas 1846), ♀ 4.2-5.3 mm [Figs. 62.118-62.120] — **Dist.** CA, TX, FL — **Ref.** Levi 1957a — **Note** species previously placed in *Theridion* by Levi 1957a

— Epigynum otherwise ........................................................ **36**

**36(35)** More than 2.5 mm total length .......... ***Theridion,*** *in part*
   **Div.** 65 species, 2 key out here: *Theridion neomexicanum* Banks 1901b [Fig. 62.121] and *Theridion californicum* Banks 1904b [Fig. 62.122], ♀ 2.5-4.5 mm — **Dist.** w USA — **Ref.** Levi 1957a

— Less than 2.5 mm total length; usually carapace and legs orange ......................................................... ***Thymoites***
   **Div.** 13 species, ♀ 0.8-2.5 mm [Figs. 62.123-62.125] — **Dist.** CAN to MEX — **Refs.** Levi 1957a, 1959b, 1964b — **Note** species previously placed in *Paidisca* Bishop & Crosby 1926 (Levi 1957a) and *Sphyrotinus* Simon 1894c (Levi & Levi 1962), which are both synonyms

**37(33)** Abdomen with 6 black patches (Fig. 62.126); epigynum with two adjacent circles (Fig. 62.128) ... ***Paratheridula***
   **Div.** 1 species: *Paratheridula perniciosa* (Keyserling 1886b), ♀ 1.4-2.2 mm [Figs. 62.126-62.128] — **Dist.** Gulf coast — **Refs.** Levi 1957d, 1966b

— Abdomen and epigynum otherwise ............................. **38**

**38(37)** Abdomen with four black patches (Fig. 62.129), or two bands; epigynum a wide oval depression (Fig. 62.131) or with longitudinal lines (Fig. 62.112) ................................
........................................................ ***Coleosoma,*** *in part*
   **Div.** 3 species, 2 key out here: *Coleosoma floridanum* Banks 1900a, *Coleosoma normale* Bryant 1944, ♀ 1.0-3.0 mm [Figs. 62.112, 62.129-62.132] — **Dist.** greenhouses, se USA — **Refs.** Levi 1959a, Knoflach 1999

— Abdomen (Figs. 62.133, 62.137-62.139) and epigynum otherwise ............................................. ***Theridion,*** *in part*
   **Div.** 65 species, 61 key out here, ♀ 1.0-5.1 mm [Figs. 62.133-62.142] — **Dist.** AK to MEX — **Refs.** Levi 1957a, 1959b, 1963c, 1967c, 1969, 1980b — **Note** includes the genus *Simitidion* used by Wunderlich (1992a) and *Takayus, Keijia, Rugathodes* used by Yoshida (2002)

**39(22)** Abdomen with a sclerotized ring anteriorly (Fig. 62.130) ........................................................ ***Coleosoma,*** *in part*
   **Div.** 3 species, ♂ 1.0-3.0 mm [Figs. 62.111-62.112, 62.129-62.132] — **Dist.** greenhouses, se USA — **Refs.** Levi 1959a, Knoflach 1999

— Abdomen without such sclerotization ......................... **40**

**40(39)** Sclerites of palpus hidden behind large conductor and upright embolus as in Figs. 62.110 or 62.118 .............. **41**

— Palpus with other sclerites visible; conductor otherwise, usually filamentous or a short prong .......................... **42**

**41(40)** Palpus as in Fig. 62.110 ....................................... ***Meotipa***
   **Div.** 1 species: *Meotipa pulcherrima* (Mello-Leitão 1917a), ♂ 2.0 mm [Figs. 62.108-62.110] — **Dist.** FL — **Ref.** Levi 1962b — **Note** species previously placed in *Chrysso* by Levi 1962b

— Palpus as in Fig. 62.118 ................................. ***Nesticodes***
   **Div.** 1 species: *Nesticodes rufipes* (Lucas 1846), ♂ 2.8-3.7 mm [Figs. 62.118-62.120] — **Dist.** CA, TX, FL — **Ref.** Levi 1957a — **Note** species previously placed in *Theridion*

**42(40)** Palpus simple, only with tegulum and embolus (Figs. 62.90, 62.127) ..................................................... **43**

— Palpus with more structures (Figs. 62.99, 62.103) ....... **44**

**43(42)** Embolus straight (Fig. 62.127) ............... ***Paratheridula***
   **Div.** 1 species: *Paratheridula perniciosa* (Keyserling 1886b), ♂ 1.4-2.2 mm [Figs. 62.126-62.128] — **Dist.** Gulf coast — **Refs.** Levi 1957d, 1966b

— Embolus twisted or coiled (Fig. 62.90) ........... ***Theridula***
   **Div.** 3 species ♂ 1.2-2.8 mm [Figs. 62.88-62.90] — **Dist.** s CAN to MEX — **Refs.** Levi 1954c, 1966b

**44(42)** Palpus without median apophysis, with radix (R), conductor (C) and embolus (E), embolus thread-like (Fig. 62.103) .................................................... ***Achaearanea***
Div. 16 species, ♂ 1.5-5.0 mm [Figs. 62.101-62.103] — Dist. widespread — Refs. Levi 1955a, 1959b, 1963b, 1967c, 1980b

— Palpus with median apophysis (M in Fig. 62.135) ...... **45**

**45(44)** Cymbium projecting beyond bulb (Fig. 62.96, arrow in Figs. 62.91, 62.117) ............................................................ **46**

— Cymbium just covering bulb ........................................ **48**

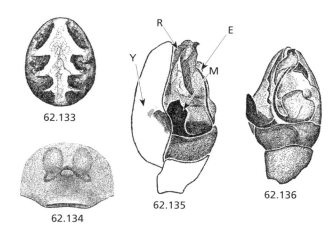

62.133
62.134
62.135
62.136

**46(45)** Male with only one palpus (Fig. 62.96); the other amputated ............................................................ ***Tidarren***
Div. 2 species: *Tidarren haemorrhoidale* (Bertkau 1880a) and *Tidarren sisyphoides* (Walckenear 1841) ♂ 0.9-1.4 mm [Figs. 62.94-62.97] — Dist. KY, s US to CA — Ref. Levi 1957c

— Male with two palpi ....................................................... **47**

**47(46)** Cymbium bare or with macroseta at tip (Fig. 62.91) or palpal tarsus twice as long as wide (Fig. 62.91) ................. ............................................................................................... ***Wamba***
Div. 2 species: *Wamba crispulus* (Simon 1895g) and *Wamba congener* (O. Pickard-Cambridge 1896a) ♂ 1.2-3.6 mm [Figs. 62.91-62.93] — Dist. ME to TX, CA — Ref. Levi 1957a — Note species previously placed in *Theridion. Chindellum* Archer 1950 is a junior synonym

— Cymbium with tooth on face of extension (arrow in Fig. 62.117) ............................................................ ***Neottiura***
Div. 1 species: *Neottiura bimaculata* (Linneaus 1767) ♂ 2-3 mm [Figs. 62.113-62.117] — Dist. s CAN, WA, OR — Refs. Levi 1957a, Knoflach 1999 — Note species previously placed in *Theridion*

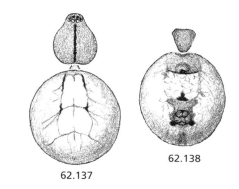

62.137
62.138

**48(45)** Cymbium with tooth on distal edge (Fig. 62.99) .............. ............................................................................................... ***Arctachaea***
Div. 2 species: *Arctachaea pelyx* Levi 1957e and *Arctachaea nordica* (Chamberlin & Ivie 1947b) ♂ 2.7-3.5 mm [Figs. 62.98-62.100] — Dist. AK to CO — Ref. Levi 1957e — Note species previously placed in *Chrysso* O. Pickard-Cambridge 1882c by Levi 1962b.

— Cymbium without such tooth ........................................ **49**

**49(48)** Palpus with an embolus loop, showing transverse and longitudinal sections of loop (Fig. 62.106) ........ ***Chrysso***
Div. 1 species: *Chrysso albomaculata* O. Pickard-Cambridge 1882c, ♂ 2.0-4.5 mm [Figs. 62.104-62.107] — Dist. NC to s USA — Refs. Levi 1957c, 1962b

— Palpus otherwise ............................................................ **50**

**50(49)** Legs short, first patella and tibia 0.9-1.8 times length of carapace; males often with clypeus very high (Fig. 62.123), bulging, projecting, or with groove in clypeus; less than 3.0 mm total length, often carapace and legs orange; male often with scuta on the abdomen .................. ***Thymoites***
Div. 13 species, ♂ 0.8-2.5 mm [Figs. 62.123-62.125] — Dist. CAN to MEX — Refs. Levi 1957a, 1959b, 1964b — Note species previously placed in *Paidisca* Bishop & Crosby 1926 (Levi 1957a) and *Sphyrotinus* Simon 1894c (Levi & Levi 1962), which are both synonyms

— Legs longer, first patella and tibia 1.5-3 times length of carapace; male clypeus not modified, males never with scutum on abdomen ........................................ ***Theridion***
Div. 65 species, ♂ 1.2-4.6 [Figs. 62.133-62.142] — Dist. AK to MEX — Refs. Levi 1957a, 1959b, 1963c, 1967c, 1969, 1980b — Note includes genus *Simitidion* used by Wunderlich (1992a) and the genera *Takayus, Keijia, Rugathodes* used by Yoshida (2002)

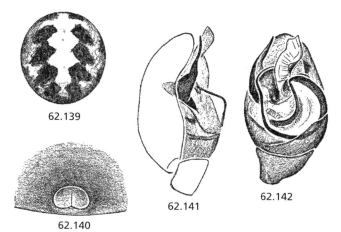

62.139
62.140
62.141
62.142

# Chapter 63
# THERIDIOSOMATIDAE
1 genus, 2 species — Jonathan A. Coddington

### Common name —
Ray spiders.

### Similar families —
Theridiidae (p. 235), Araneidae (p. 68), Linyphiidae (p. 124).

### Diagnosis —
This orbicularian family can be distinguished by the ventral prolateral sternal pits (Fig. 63.2), the conformation of the male palp [the topological arrangement of embolus (E), conductor (C), and median apophysis (MA) on the tegulum (Fig. 63.3)], the trajectory of the ejaculatory duct, the medially fused female spermathecae, web architecture (Fig. 63.4) and the extremely long third and fourth tibial dorsal trichobothria (Fig. 63.5) (Wunderlich 1980h, Coddington 1986a). With respect to the North American fauna, theridiosomatids are distinctive in being quite small spiders with extremely stout 1st femora, abundant macrosetae, chelicerae with robust teeth, and very large male genitalia. In the field, theridiosomatids spin modified but recognizable orbwebs with radii that join before reaching the hub, and a "tension" line extending from the hub to the substrate with which the animal distorts the web into a cone when ready to capture prey.

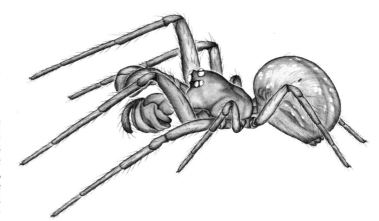

**Fig. 63.1** *Theridiosoma gemmosum* (L. KOCH 1877b)

### Characters —
**body size:** males 1.3-1.9 mm; females 1.9-2.4 mm.
**color:** abdomen dusky with light dorsal chevrons (*Theridiosoma gemmosum*) or a transverse white to silvery lunate band (*Theridiosoma savannum*). Legs uniform tan.
**carapace:** carapace nearly as high as long with the clypeus about 1.5 x an AME diameter.
**sternum:** sternum truncate behind, with glandular fossae or pores at junction of sternum and labium.
**eyes:** eight eyes in two rows, laterals juxtaposed, canoe tapeta in all except anterior medians.
**chelicerae:** robust with discrete, irregularly arranged teeth.
**legs:** tarsi 3-clawed; first and to some extent second femora unusually robust. 3rd and 4th tibial trichobothria 2-4 x tibia diameter, and distal 4th tibial trichobothrium especially long. Median tarsal claw elongate with a finely tapering tip. Female palpal tarsus clawless.
**abdomen:** smoothly globose; as high to higher than long.
**spinnerets:** six spinnerets with small, fleshy colulus.
**respiratory system:** paired book lungs and median tracheal spiracle, just anterior to spinnerets.
**genitalia:** entelegyne; *female* has a relatively large *bursa copulatrix* with ducts leading to medially fused spermathecae, a synapomorphy of the family, epigynum as in Fig. 63.7, posterior view; *male* has very large palpal bulbs with a mesal median apophysis (MA), and a complex embolic division hidden beneath the broad, blade-like conductor (C). The course of the ejaculatory duct is quite complicated, and several switchbacks are also synapomorphic for the family. (Coddington 1986a).

### Distribution —
Theridiosomatidae is a world-wide, but mainly tropical family. The two North American species (both *Theridiosoma*) occur, as far as known, only in eastern North America. *Theridiosoma* itself is also global. *Theridiosoma gemmosum* (L. KOCH 1877b) ranges from Michigan, southern Wisconsin, Kansas, and Arkansas in the west to all more eastern states, including southern Ontario in Canada (as well as a large range in northern Europe, where it may be introduced). *Theridiosoma savannum* (CHAMBERLIN & IVIE 1944) is a more southerly species, only common along the gulf coastal plain. The two species are broadly sympatric but prefer different microhabitats.

### Natural history —
The two known North American species occupy distinctly different habitats but are alike and unusual among orbweavers in being primarily diurnal predators. *Theridiosoma gemmosum* (Fig. 63.6) prefers extremely humid microhabitats that are almost always in closed canopy forest. Web sites include the undercut banks above, or the dense vegetation bordering, running streams, or interstices of wet rocks or stumps in swamps or seeps. Web orientation varies, but is most frequently vertical. The orbs are sparse with few (e.g., 6-12) radii and a like number of sticky spirals. Early in web construction, during the exploration of the website, one line is selected to serve as an "out-of-plane" radius, roughly perpendicular to the future plane of the orb. The non-sticky, or temporary, "spiral" is in the form of two concentric closed circles of silk, another behavioral synapomorphy of the family. Although the animal makes a relatively normal, complete orb initially, at the end of construction, it rebuilds the hub area, fusing radii so that only four lines reach the hub. These lines are held by the back four legs; the front four legs grasp the out-of-plane radius, or "tension" line, and by pulling on it distort the web into a cone (Fig. 63.8). The animals typically rest dorsal side up and with the front four legs fully flexed (unlike *Hyptiotes*, see chapter 67) maintain the tension in the web for hours

on end. It is unclear if the radial lines connect to each other. In some cases it seems that they do not, when the body and legs of the spider serve as the hub of the web (like *Hyptiotes*). The slack in the tension line produced by the distortion of the web is piled on top of the cephalothorax, obscuring the eye group. When prey contact the web, the spider releases the tension line and the web, with the spider, "pops," or springs backward into a planar configuration. By slamming the web against the prey, the velocity and force of impact between sticky lines and the prey is increased. Once the prey is entangled, the spider quickly attaches a new silk line to the unaffected radial lines and, paying out new silk behind itself, relaxes the sector of the web containing the prey. More of the lax sticky lines can thus contact the prey, and they are also harder for the prey to break. Attack behavior is always a bite followed by a wrap, whereupon the spider transports the prey bundle to the hub for feeding. Because the hub of the web tends to fall apart during prey capture, the web architecture degrades relatively quickly over the course of a day. Nematocerous flies (mosquitoes, midges, gnats) and collembola, which are apparently their principal prey, either do not adhere easily to viscid silk or fly slowly with so little momentum that impacting viscid silk does not result in entanglement.

*Theridiosoma savannum* spins horizontal orb webs on the surface of coarse (i.e. broadleaf) leaf litter in moist, closed-canopy southeastern forests, such as live oaks or gum trees. The web form is peculiar, with few, relatively long radii and a tightly spaced sticky spiral that occupies only about the central fourth of the area enclosed by the radii. The "tension" line extends from the hub more or less vertically down to the leaf litter. As in *Theridiosoma gemmosum*, the spider sits at the hub with the back four legs holding onto radii, and the front four legs grasping the tension line. Prey capture is similar, except that incoming prey are more likely hopping insects or insects attempting to land on the surface of the litter.

Both species make papery brown egg sacs suspended from a forked silk line (Fig. 63.9). The female abandons the egg sac, and, as it is often placed more prominently than the website, finding the characteristic egg sac is often the first clue that theridiosomatids are present.

### Taxonomic history and notes —

Theridiosomatidae was first described as the group Theridiosomatini or Theridiosomateae within the Argiopidae (Simon 1881a, 1895a). It was polyphyletic and included genera since dispersed to six other families. Wunderlich (1980h) was the first to point out the sternal pits, which to date all and only theridiosomatids possess. The current total is 12 genera and 72 species. For most of the 20th century it was most frequently treated as a subfamily of Argiopidae/Araneidae. Vellard (1924a) was the first to rank the group as a family, and various later authors agreed (e.g., Kaston 1948), but none of these opinions were justified phylogenetically. The family was revised worldwide at the generic level by Coddington (1986a), who reviewed its taxonomic history and recognized nine genera in four subfamilies, Platoninae, Epeirotypinae, Ogulninae, and Theridiosomatinae. His phylogenetic analysis placed Theridiosomatidae as sister to an informal group he called the "symphytognathoids" (Mysmenidae (Anapidae, Symphytognathidae)), a placement with which most subsequent analyses (e.g., Griswold *et al.* 1998) have agreed. Subsequently the term symphytognathoids was applied to all four families.

Symphytognathoid synapomorphies are mainly behavioral: all members of the group spin out-of-plane radii and radically remodel the hub region in similar ways after constructing the sticky spiral (Eberhard 1982, 1986, Coddington 1986b). They also "doubly attach" their egg sacs, which is to say the eggs are laid against a silk line attached at both ends to web and/or substrate. Some symphytognathoids subsequently cut one attachment so the egg sac hangs free. Symphytognathoids apart from theridiosomatids all spin at least some accessory radii after sticky spiral construction.

### Genus —

*Theridiosoma* O. Pickard-Cambridge 1879e

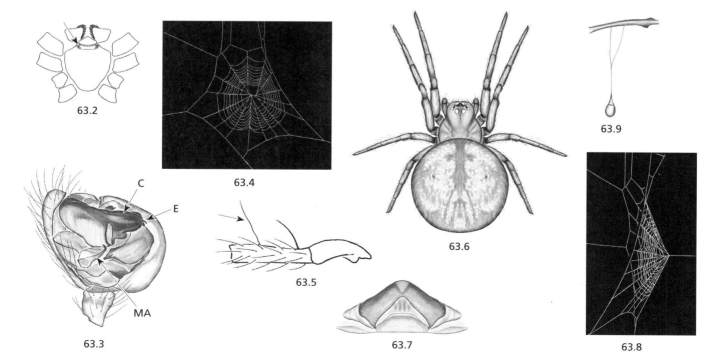

# Chapter 64
# THOMISIDAE
9 genera, 130 species

Charles D. Dondale

### Common name —
Crab spiders.

### Similar families —
Members of the family Thomisidae are the most crab-like of North American spiders, and are unlikely to be mistaken for anything else, except for Philodromidae (p. 192).

### Diagnosis —
Two-clawed, strongly crab-like spiders with legs I and II long and stout, III and IV short and slender (Figs. 64.1; 64.12-64.13). Lateral eyes situated on prominent pale tubercles, often larger than median eyes (Figs. 64.4-64.5; 64.8-64.9; 64.14).

### Characters —
**body size:** length 1.5-11.3 mm.
**color:** flower-inhabiting members brightly colored with green, yellow, red, and white; ground and litter representatives various tones of brown, gray, black, or dull red.
**carapace:** somewhat convex, approximately as wide as long, with scattered simple or clavate setae.
**eyes:** eight, in two transverse rows; lateral eyes situated on prominent pale tubercles and often larger than median eyes.
**chelicerae:** short, thick at base, tapered distally; retromargin lacking teeth or, rarely, with one weak tooth.
**legs:** strongly laterigrade, two-clawed; I and II longer and stouter than III and IV; lacking claw tufts and scopulae; femur I often with several strong erect macrosetae.
**abdomen:** round or oval in dorsal view, rather flat above, with scattered erect simple or clavate setae.
**genitalia:** entelegyne; *female* epigynum with deep atrium which is often subdivided longitudinally by a median septum, sometimes with hood; copulatory openings indistinct; *male* palpus with RTA, VTA, and sometimes ITA; tegulum flat, often round in ventral view, sometimes with hook-like apophyses; embolus short to long and hair-like

### Distribution —
Widespread, with a concentration in the southwest. At least one species is a true arctic spider, and a few others occupy subarctic/alpine tundra, a significant proportion of which extend into Siberia or even to Scandinavia (Dondale et al. 1997).

### Natural history —
The flower-inhabiting members (*Misumena, Misumenoides, Misumenops*) ambush pollinating insects. *Misumena vatia* (CLERCK 1757), which can change color in accordance with the substrate, is notorious for killing honey bees. Representatives of the genus *Coriarachne* live in crevices beneath loose bark on tree trunks, but can also be found on fence posts and wooden buildings. Representatives of the genera *Xysticus* and *Ozyptila*, with few exceptions, wander in or over litter or among grass roots, but are sometimes found under stones or logs, rarely on plants. Life histories of a few species are known [see, for example, Dondale (1977)]. Prey relationships are the subject of studies by Morse (1983, 1986, 1987) and by Morse & Fritz (1982). Schmalhofer (1999) investigated thermal responses.

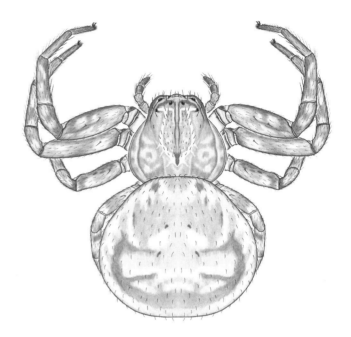

**Fig. 64.1** *Xysticus elegans* KEYSERLING 1880b

### Taxonomic history and notes —
Most thomisids are easily recognized as such, even when alive, though individuals of the genera *Xysticus, Ozyptila,* and *Coriarachne* tend to intergrade. Ono (1988) divided the four North American species of *Coriarachne* among two genera, one of which intergrades with one of the species groups of *Xysticus. Coriarachne floridana* BANKS 1896e, *Coriarachne utahensis* (GERTSCH 1932), and *Coriarachne versicolor* KEYSERLING 1880b were allocated to the genus *Bassaniana*, whereas *Coriarachne brunneipes* BANKS 1893a remained in *Coriarachne*, based on equivocal differences in microhabitat. Gertsch (1953), Schick (1965), Bowling & Sauer (1975), and Dondale & Redner (1978b) have all examined this question and are agreed that the four North American species are congeneric. Records of the genus *Isaloides* F.O. PICKARD-CAMBRIDGE 1900 and of *Majellula* STRAND 1932 from the southernmost USA (Gertsch 1953) have not been confirmed. Likewise, six species of the genus *Thomisus* WALCKENAER 1805, described from Georgia by Walckenaer (1837) and listed by Chamberlin & Ivie (1944) require confirmation.

### Genera —
*Coriarachne* THORELL 1870b, *Diaea* THORELL 1869, *Misumena* LATREILLE 1804b, *Misumenoides* F.O. PICKARD-CAMBRIDGE 1900, *Misumenops* F.O. PICKARD-CAMBRIDGE 1900, *Ozyptila* SIMON 1864, *Synema* SIMON 1864, *Tmarus* SIMON 1875a, *Xysticus* C.L. KOCH in HERRICH-SCHÄFFER 1835.

**Key to genera —**
North America North of Mexico

1 Carapace protruding in front, and abdomen with posterior tubercle (Fig. 63.2) .......................................... ***Tmarus***
   **Div.** 6 species — **Dist.** widespread — **Refs.** Schick 1965, Dondale & Redner 1978b

— Carapace not protruding (Fig. 64.3). Abdomen lacking tubercle ................................................................................. **2**

2(1) ALE approximately equal in diameter to AME (Fig. 64.4) ................................................................................. **3**

— ALE larger than AME (Fig. 64.5) ..................... **4**

3(2) Carapace with white transverse ridge on front (Fig. 64.4) ................................................ ***Misumenoides***
   **Div.** 1 species: *Misumenoides formosipes* (WALCKENAER 1837) — **Dist.** CA-FL, n to Great Lakes region — **Refs.** Schick 1965, Dondale & Redner 1978b

— Carapace lacking white transverse ridge ....... ***Misumena***
   **Div.** 2 species: *Misumena vatia* (CLERCK 1757) and *Misumena fidelis* BANKS 1898b — **Dist.** widespread — **Refs.** Gertsch 1939b, Schick 1965, Dondale & Redner 1978b

4(2) Carapace low (approximately as low at level of coxa III as at level of PER, Fig. 64.3) ........................... ***Coriarachne***
   **Div.** 4 species — **Dist.** widespread — **Refs.** Gertsch 1953, Schick 1965, Dondale & Redner 1978b — **Note** includes Nearctic species currently placed in *Bassaniana* STRAND 1928b

— Carapace higher (higher at level of coxa III than at level of PER, Fig. 64.6) ............................................................ **5**

5(4) Carapace shiny, almost devoid of setae (Fig. 64.7) ............ ................................................................................. ***Synema***
   **Div.** 3 species — **Dist.** s half of contiguous USA, except west coast — **Ref.** Gertsch 1939b

— Carapace only moderately convex and with numerous setae (Fig. 64.8) ................................................................. **6**

6(5) Lateral eyes on each side situated on completely conjoined tubercles (Fig. 64.9) ........................................... **7**

— Lateral eyes on each side situated on incompletely conjoined tubercles (Fig. 64.8) ............................................. **8**

7(6) Basitarsus I with 1 or more prolateral macrosetae (Fig. 64.10) ........................................................... ***Diaea***
   **Div.** 2 species: *Diaea livens* SIMON 1876e and *Diaea seminola* GERTSCH 1939b — **Dist.** CA, FL — **Refs.** Gertsch 1939b, Schick 1965

— Basitarsus I lacking prolateral macrosetae (Fig. 64.11) .... ................................................................................. ***Misumenops***
   **Div.** 25 species — **Dist.** widespread — **Refs.** Gertsch 1939b, Schick 1965, Dondale & Redner 1978b

8(6) Tibia I with more than 2 pairs of ventral macrosetae, and femur I rather slender (approximately 4 times as long as wide, Fig. 64.12). Body rarely with clavate setae ................ ................................................................................. ***Xysticus***
   **Div.** 67 species — **Dist.** widespread — **Refs.** Gertsch 1953, Schick 1965, Dondale & Redner 1978b, Reeves et al. 1984.

— Tibia I with only 2 pairs of ventral macrosetae, and femur I relatively short and stout (approximately 3 times as long as wide, Fig. 64.13). Body with many clavate setae (Fig. 64.14) ................................................................ ***Ozyptila***
   **Div.** 20 species — **Dist.** widespread — **Refs.** Gertsch 1953, Schick 1965, Dondale & Redner 1975a, 1978b

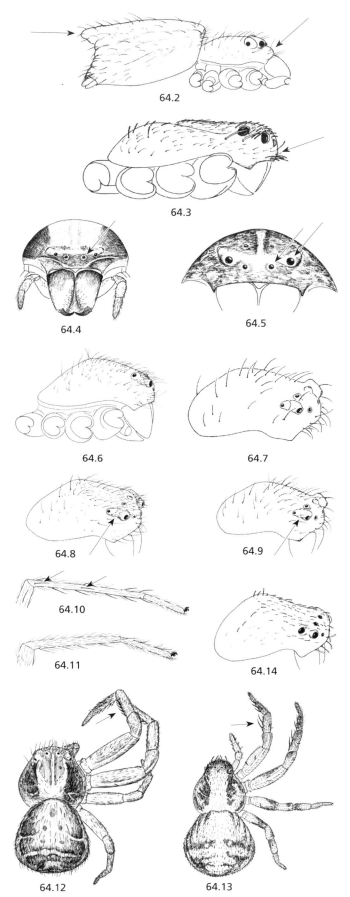

# Chapter 65
# TITANOECIDAE
1 genus, 4 species

Bruce Cutler

### Similar families —
Amaurobiidae (p. 60), Amphinectidae (p. 63), Dictynidae (p. 95), Tengellidae (p. 230), Zorocratidae (p. 258).

### Diagnosis —
In the field resemble small to medium sized amaurobiids. The following combination of characters will distinguish the females from those of other families: calamistrum >70% length of metatarsus IV (Fig. 65.2), cribellum divided (Fig. 65.3), legs with macrosetae, endites parallel or nearly parallel (Fig. 65.4). Males lack a calamistrum, but may be distinguished from other similar male cribellate spiders by the complex palpal tibial apophysis in which the prolateral process as seen dorsally resembles the head of a bird (Fig. 65.5).

### Characters —
**body size:** small to medium size, 4-8 mm body length.
**color:** carapace paler than abdomen, yellow to dark orange; abdomen gray to black, may have pale spots on anterior area; appendages color of carapace or slightly darker.
**carapace:** pyriform, longer than wide; fovea obscure.
**sternum:** truncate anteriorly, pointed posteriorly.
**eyes:** at anterior edge of carapace; eight in two rows; both eye rows nearly straight.
**chelicerae:** robust; cheliceral teeth-3 promarginal, 2 retromarginal; ectal face with row of short macrosetae in male.
**mouthparts:** endites parallel or nearly so.
**legs:** with a few macrosetae.
**abdomen:** oval.
**spinnerets:** cribellum divided, spinnerets moderate length.
**respiratory system:** two book lungs, tracheal spiracle near spinnerets.
**genitalia:** entelegyne; *female* epigynum anteriorly with tapering median septum, *bursa copulatrix* round, spermatheca elongate and medial (Fig. 65.7); *male* palpus with a complex tibial apophysis with several sclerotized processes, prolateral process as seen dorsally resembles a bird's head (Fig. 65.5), palpal tibia lacking protuberance or process, embolus, as seen in mesal view, thin and tapered (E Fig. 65.6).

### Distribution —
Unlike most small families, titanoecids are common and widely distributed in our region. They occur throughout the contiguous United States (except for peninsular Florida) north to Alaska and southeastern Canada.

### Natural history —
Our sole genus, *Titanoeca*, is usually found in more xeric microhabitats near the ground. Over much of our region, this detail is often a good clue as to whether one is observing this genus or an amaurobiid. Egg sacs are globular and suspended in the web.

### Taxonomic history and notes —
Titanoecidae consists of 4 genera and 44 species and is predominantly Palearctic with some tropical representatives. *Titanoeca* and related genera were removed from Amaurobiidae by Lehtinen (1967) to the new family Titanoecidae. A key to the North American species of the family is found in Leech (1972), who treated Titanoecidae as a subfamily (Titanoecinae) of the Amaurobiidae.

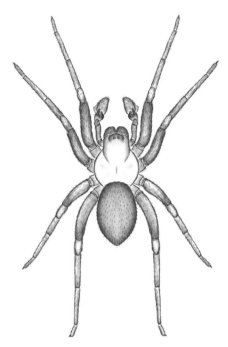

**Fig. 65.1** *Titanoeca americana* EMERTON 1888a

### Genus —
*Titanoeca* THORELL 1870b

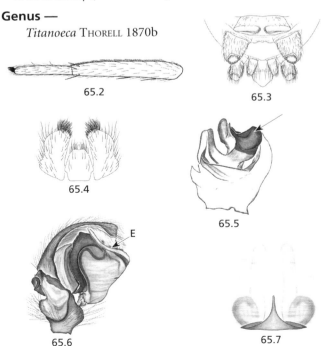

# Chapter 66
# TRECHALEIDAE
1 genus, 1 species

James E. Carico

**Common name —**
Long legged water spiders.

**Similar families —**
Lycosidae (p. 164), Pisauridae (p. 199).

**Diagnosis —**
Only a single genus in the family Trechaleidae, *Trechalea*, and one species, *Trechalea gertschi* CARICO & MINCH 1981, are in our region. This family can be distinguished from other North American lycosoid families by the combination of: the presence of a retrolateral tibial apophysis, somewhat laterigrade legs, flexible tarsi, recurved PE, flattened form of the egg sac, and manner of transporting the spiderlings on the egg sac while it is attached to the spinnerets.

**Characters —**
**body size:** 1.3-2 cm.
**color:** ground color varies from light to dark gray; carapace ocular area dark, serrated light submarginal bands and posterior light median band; varied darker markings and several dark setae on abdomen.
**carapace:** widest posteriorly, cephalic area distinct
**sternum:** truncate anteriorly and pointed posteriorly
**eyes:** 4 AE subequal eyes in slightly recurved row; larger 4 PE equal eyes in recurved row.
**chelicerae:** boss lacking; fang furrow margins toothed.
**legs:** somewhat laterigrade, formula is IV-II-I-III; pairs of macrosetae on venter of tarsi, metatarsi, and tibia; trichobothria on dorsum of all segments except femora.
**abdomen:** ovoid, moderately low.
**spinnerets:** 3 pairs, unmodified.
**respiratory system:** book lung spiracles at anterior third; tracheal spiracles near spinnerets.
**genitalia:** entelegyne; *female* broad median field of the epigynum overlapping the lateral lobes (Fig. 66.3), spermathecae (dorsal view) (Fig. 66.4); *male* retrolateral tibial apophysis prominent; median apophysis, tegulum, subtegulum, conspicuous on the ventral surface of a complex genital bulb (Fig. 66.2).

**Distribution —**
Distribution of *Trechalea gertschi* in the USA is limited to portions of the Gila River drainage in Arizona and New Mexico (Carico 1993) but the range extends southward into the Mexican states of Sonora and Chihuahua. The genus *Trechalea* is restricted to the western hemisphere from Arizona southward to Bolivia.

**Natural history —**
This species is found at the margins of permanent streams within the otherwise xeric Southwestern USA zone and is typically found perched head-down on the surfaces of rocks near the water margin. These spiders readily run across the water surface and occasionally crawl underwater by walking down the surface of a partly submerged rock, features shared with many species of North American *Dolomedes* (Pisauridae). Females carry the flattened,

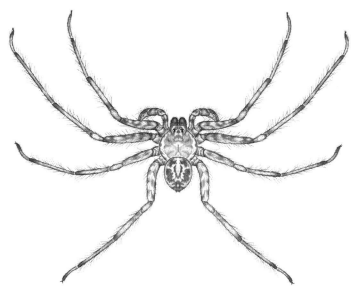

**Fig. 66.1** *Trechalea gertschi* CARICO & MINCH 1981

bivalved egg sac with their spinnerets, and spiderlings are carried upon the upper surface of the sac after emergence.

**Taxonomic history and notes —**
Trechaleids are widely diversified into several genera and are widespread from Arizona through Argentina with *Trechalea gertschi* as the northern-most representative of this family. The family was named by Simon (1890a) for the genus *Trechalea*, but he abandoned the family name immediately thereafter. Subsequently the family was first recognized as valid by Carico (1986) for a monophyletic group of genera and who subsequently (1993) redescribed it in a taxonomic revision of the type genus. The validity of the family has been confirmed by Sierwald (1990) (referred to as "*Trechalea* genus-group"), Coddington & Levi (1991), Carico (1993), and Griswold (1993).

**Genus —**
*Trechalea* THORELL 1869

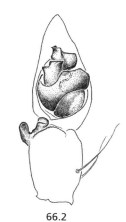
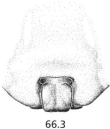

66.2
66.3
66.4

Chapter 67

# ULOBORIDAE
7 genera, 16 species

Brent D. Opell

**Common name —**
Hackled band orb-weavers.

**Similar families —**
Uloborids might be confused with small members of the family Araneidae (p. 68), but are easily identified by the presence of a cribellum and calamistrum.

**Diagnosis —**
Small spiders that lack cheliceral poison glands and their fang openings. Femurs II, III, and IV bear a row of long trichobothria. Females have an oval cribellum (undivided) and a calamistrum formed of a single row of setae. A row of short, ventral macrosetae extends from the distal one-half to one-third of metatarsus IV to the end of tarsus IV.

**Characters —**
**body size:** 4 – 8 mm.
**carapace:** pear or oval shaped in orb-weaving species, broad in *Hyptiotes* species, and rectangular in *Miagrammopes* species. Thoracic groove shallow, transverse, and situated in center or posterior third of carapace.
**sternum:** oval, widest between first and second coxae, becoming progressively narrower between successive coxae. Sternum entire in all genera but *Miagrammopes*, where it is divided into three regions by weakly sclerotized transverse bands (Opell 1979, 1984b).
**eyes:** orb-weaving genera have two rows of four eyes each. The ALE's of *Hyptiotes* species are very small and the posterior row is strongly recurved and features prominent PLE tubercles. *Miagrammopes* species lack an anterior eye row and have prominent PLE tubercles (Opell 1988, Opell & Cushing 1986, Opell & Ware 1987).
**chelicerae:** of average size with a few small teeth along the prolateral and retrolateral margins of the fang furrow. All species lack poison glands and their associated cheliceral fang openings.
**mouthparts:** length of endites about 1.5 times their width except in *Miagrammopes* species, where length is twice width. Labium free from sternum and not rebordered. Labium length and width similar except in *Miagrammopes* species, where length is at least twice width.
**legs:** first two pairs of legs directed anteriorly, last two pairs directed posteriorly. The legs of *Hyptiotes* species are short and robust; the legs of *Miagrammopes* species are elongated, particularly leg I. Femurs II, III, and IV bear a row of long trichobothria (Opell 1979). The calamistrum of females is formed of a single setal row (Opell 2001a). Like mature males of most other cribellate spiders, male uloborids lack a calamistrum. A row of short, ventral macrosetae extends from the distal one-half to one-third of metatarsus IV to the end of tarsus IV (Opell 1979).
**abdomen:** oval with greatest width and height reached at the anterior one-third, where there is often a prominent low tubercle or pair of tubercles. The abdomen of *Hyptiotes* species is nearly spherical, and that of *Miagrammopes* species nearly cylindrical (Opell 1989a).
**spinnerets:** conical, short, and tightly clustered; cribellum not divided.

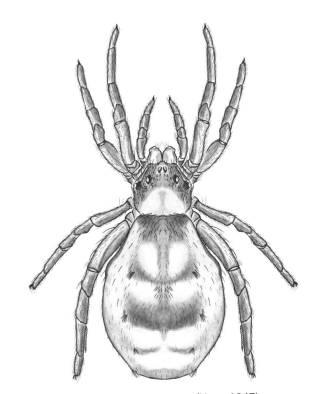

**Fig. 67.1** *Hyptiotes cavatus* (Hentz 1847)

**respiratory system:** a pair of stout tracheal trunks extend anteriorly from a common atrium that opens just anterior to the cribellum. Several smaller tracheae extend posteriorly from the atrium. In *Philoponella* species the two tracheal trunks divide into small branches that are confined to the abdomen. In members of the other genera the two tracheal trunks exhibit one of two patterns: 1. They extend into the cephaothorax before dividing into tracheoles that then enter the legs or 2. They first divide into smaller tracheoles before passing into the cephalothorax (Opell 1979, 1987, 1990, 1998).
**genitalia:** entelegyne; ***female*** of the North American representatives of this family possess an epigynum that features either a central, sclerotized crypt into which the sperm ducts open (*Philoponella*), a central raised area or hood with ducts opening on its posterior margin (*Hyptiotes*, *Miagrammopes*, and *Siratoba*), or a pair of median (*Zosis*) or posterior (*Uloborus* and *Octonoba*) lobes with duct openings on their posterior margins. The ***male*** palp of most North American genera features a large, concave cymbium and a large basal hematodocha, crowned by a disk-shaped tegulum. A dome-shaped median apophysis originates in the center of the tegulum and bears a claw-shaped or enrolled median apophysis spur. The embolus arises from beneath the median apophysis and extends

upward toward the median apophysis spur. In *Uloborus* and *Philoponella*, the embolus is guided by a conductor that originates from the base of the median apophysis. In *Zosis* and *Octonoba*, the conductor is absent and a tegular spur guides the embolus. The palpi of *Siratoba*, *Hyptiotes*, and *Miagrammopes* also feature a median apophysis with one or two apical spurs, a conductor, and, in the case of *Siratoba*, a radix (Opell 1979). However, in these three genera, these sclerites are often larger and more intertwined, making them more difficult to distinguish and identify.

## Distribution —

*Uloborus* and *Hyptiotes* are represented by several species throughout the United States and Canada, whereas *Miagrammopes mexicanus* O. Pickard-Cambridge 1893b is found only in southern Texas. *Siratoba referens* (Muma & Gertsch 1964) and *Philoponella* species are restricted to the Southwestern United States and Texas. *Zosis geniculata* (Olivier 1789) is found only in the Gulf Coast states. *Octonoba sinensis* (Simon 1880b) is an introduced Asian species with a wide, but patchy distribution east of the Rocky Mountains, where it appears to be confined to greenhouses and barns (Muma & Gertsch 1964, Opell 1979, 1983).

## Natural history —

The seven uloborid genera represented in North America construct capture webs of three types: orb webs (*Octonoba, Philoponella, Siratoba, Uloborus,* and *Zosis*), triangle webs (*Hyptiotes*) and simple webs that are formed of only a few capture lines and lack a stereotypic architecture (*Miagrammopes*) (Lubin et al. 1978, Opell 1979, 1982, 1994a, 1996). Each of these web forms contains cribellar capture threads (Opell 1994a, 1994b, 1999). Triangle webs are oriented vertically. In contrast to Araneoid orb-webs (except most Tetragnathidae), uloborid orbwebs are typically oriented horizontally. The web's hub remains intact and a linear or spiral stabilimentum is often present (Eberhard 1971, 1972, 1973). *Hyptiotes* females deposit their egg sacs on small twigs, *Octonoba* females place their egg sacs at the edge of the web, and *Zosis* females incorporate their egg sacs in the orb (Opell 1984c). *Uloborus* females attach a growing chain of egg sacs along a radius at the edge of the orb (Cushing 1989, Cushing & Opell 1990a, 1990b). *Philoponella* and *Miagrammopes* females hold their egg sacs with one first leg until spiderlings emerge (Opell 1989b, 2001b, Smith 1997).

## Taxonomic history and notes —

Walckenaer (1841) described the genus *Zosis*. O. Pickard-Cambridge (1870a) described the genus *Miagrammopes*, suggesting that it was related to *Zosis* and *Hyptiotes* and that these three genera were allied with the family Epeirides. However, Thorell (1869) was the first to formally recognize this by establishing the Uloborinae as a subfamily of Epeiridae. O. Pickard-Cambridge (1871c) subsequently established the family Uloborides for the genera *Hyptiotes* and *Uloborus* and the family Miagrammopides for the genus *Miagrammopes*. Simon (1874a) combined these two families to form the family Uloboridae. He later expanded the family to include members of the present day families Deinopidae and Dictynidae as well as the uloborid genus *Sybota* Simon 1892a, a name that he introduced to replace a preoccupied genus name introduced by Nicolet (1849). O. Pickard-Cambridge (1896a) added the genus *Ariston* to the Uloboridae. The family Uloboridae assumed its current identity when Pocock (1900a) removed the Dictynidae and Comstock (1912) removed the Deinopidae. Mello-Leitão (1917b) described the genus *Philoponella*. Strand (1934) introduced the genus name *Orinomana* to replace a preoccupied name introduced by Chamberlin (1916). Muma & Gertsch (1964) revised the uloborid species of the United States and Canada. Lehtinen (1967) examined the family's diversity and described the genera *Daramuliana, Polenecia, Purumitra,* and *Tangaroa*. Opell (1979) reexamined the family and described the genera *Octonoba, Ponella, Siratoba,* and *Waitkera*. The genera *Conifaber* and *Lubinella* were subsequently described (Lubin et al. 1982, Opell 1984a).

## Genera —

*Hyptiotes* Walckenaer 1837, *Miagrammopes* O. Pickard-Cambridge 1870a, *Octonoba* Opell 1979, *Philoponella* Mello-Leitão 1917b, *Siratoba* Opell 1979, *Uloborus* Latreille 1806, *Zosis* Walckenaer 1841.

## Key to genera —
## North America North of Mexico

1  Eight eyes of similar size distributed in two rows, PLE not on prominent tubercles, carapace pear-shaped or oval (Figs. 67.2-67.8), constructs orb-webs that are usually horizontally oriented .................................................. **2**

—  PLE on tubercles that extend to carapace margin, ALE very small or entire anterior eye row absent, carapace rectangular or with an abruptly narrowed cephalic region (Figs. 67.9-67.10), constructs triangular webs or irregular webs of only a few threads .................................................. **11**

2(1)  Female .................................................. **3**
—  Male .................................................. **7**

3(2)  Epigynum with a large, heavily sclerotized central crypt (Fig. 67.11), posterior eye row nearly straight (Fig. 67.2), spider hangs from hub of web with first legs flexed so that metatarsus and tarsus lie above sternum .................................................. ***Philoponella***
   **Div.** 3 species — **Dist.** TX and southwestern USA — **Refs.** Muma & Gertsch 1964, Opell 1979, Opell & Eberhard 1983

—  Epigynum with a single median hood (Fig. 67.12) or pair of posterior or median extensions that are weakly or strongly sclerotized (Figs. 67.13-67.15), posterior eye row recurved so that a line across the anterior margins of the lateral eyes passes along or posterior to the posterior margins of the median eyes .................................................. **4**

4(3)  Epigynum with a single broad, rounded median lobe (Fig. 67.12), carapace as in Fig. 67.4, spider hangs from hub of web with first legs extended and held together .................................................. ***Siratoba***
   **Div.** 1 species: *Siratoba referens* (Muma & Gertsch 1964) — **Dist.** southwestern U.S. — **Refs.** Muma & Gertsch 1964, Opell 1979, Opell & Eberhard 1983

—  Epigynum with a pair of posterior or median lobes (Figs. 67.13-67.15) .................................................. **5**

5(4)  Prominent setal brush on first tibia (Fig. 67.17), epigynum with two weakly sclerotized posterior lobes (Fig. 67.13), carapace as in Fig. 67.5, spider hangs from hub of web with first legs extended, held together, and flexed at tibia-metatarsus joint .................................................. ***Uloborus***
   **Div.** 5 species — **Dist.** throughout USA and southern CAN — **Refs.** Muma & Gertsch 1964, Opell 1979, Opell & Eberhard 1983

—  First tibia without conspicuous setal brush, epigynal lobes with distal sclerotization .................................................. **6**

6(5)  Epigynum with broad median lobes (Fig. 67.14), spider hangs from hub of web with first legs extended and widely spread .................................................. ***Zosis***
   **Div.** 1 species: *Zosis geniculata* (Olivier 1789) — **Dist.** Gulf Coast states — **Refs.** Muma & Gertsch 1964, Opell 1979, Opell & Eberhard 1983

—  Epigynum with narrow posterior lobes whose tips are well sclerotized (Fig. 67.15), a round crypt is found on the posterior surface of each lobe (Fig. 67.16), carapace as in Fig. 67.7, spider hangs from hub of web with first legs extended and slightly spread .................................................. ***Octonoba***
   **Div.** 1 species: *Octonoba sinensis* (Simon 1880b) — **Dist.** introduced Asian species, east of the Rocky Mountains in greenhouses and barns — **Refs.** Muma & Gertsch 1964, Opell 1979, Opell & Eberhard 1983

Spiders of North America — Uloboridae | 253

**7(2)** Carapace elongate or pear shaped (Fig. 67.6) .................. **8**

— Carapace nearly round (Figs. 67.3, 67.8) .......................... **9**

**8(7)** Simple palpus with a dome shaped median apophysis, a claw-shaped median apophysis spur, and a narrow conductor (Fig. 67.18) ........................ ***Uloborus***
   **Div.** 5 species — **Dist.** throughout USA and southern CAN — **Refs.** Muma & Gertsch 1964, Opell 1979

— Complex palpus with three large sclerites: a coiled radix that surrounds the embolus, a V shaped conductor, and a grooved median apophysis, Fig. 67.19), southwestern U.S.
   ................................................................ ***Siratoba***
   **Div.** 1 species: *Siratoba referens* (MUMA & GERTSCH 1964) — **Dist.** southwestern USA — **Refs.** Muma & Gertsch 1964, Opell 1979

**9(7)** Palpus with a broad, grooved median apophysis spur, no conductor present, and a very small tegular spur near the tip of the emboolus (Fig. 67.21), carapace as in Fig. 67.8, introduced Asian species, east of the Rocky Mountains in greenhouses and barns ............................. ***Octonoba***
   **Div.** 1 species: *Octonoba sinensis* (SIMON 1880b) — **Dist.** introduced Asian species, east of the Rocky Mountains in greenhouses and barns — **Refs.** Muma & Gertsch 1964, Opell 1979

— Palpus with a claw-shaped median apophysis and some form of well-developed embolus guide or conductor (Figs. 67.20, 67.22) ............................................................. **10**

**10(9)** Palpus with a large, somewhat transparent tegular spur that enfolds and supports the embolus (Fig. 67.20), lacks a conductor that extends from the base of the median apophysis, Gulf Coast States ............................... ***Zosis***
   **Div.** 1 species: *Zosis geniculata* (OLIVIER 1789) — **Dist.** gulf coast states — **Refs.** Muma & Gertsch 1964, Opell 1979

— Tegular spur absent, prominent conductor with conductor spur and basal lobe extend from the base of the median apophysis (Fig. 67.22), carapace as in Fig. 67.3
   ................................................................ ***Philoponella***
   **Div.** 3 species — **Dist.** TX and southwestern USA — **Refs.** Muma & Gertsch 1964, Opell 1979

**11(1)** Carapace trapezoidal with abruptly narrowed cephalic region (Fig. 67.9), anterior eye row present, but ALE very small, abdomen nearly round (Fig. 67.1), female calamistrum extends along proximal half of metatarsus IV, leg I short and stout, male tibia I without dorsal macrosetae, spider constructs triangular webs ..................... ***Hyptiotes***
   **Div.** 4 species — **Dist.** throughout USA and CAN — **Refs.** Muma & Gertsch 1964, Opell 1979

— Carapace rectangular (Fig. 67.10), anterior eye row absent, abdomen cylindrical, female calamistrum restricted to proximal one-third of metatarsus IV, leg I elongated, male tibia I with dorsal macrosetae, spider construct irregular webs of only a few strands .................... ***Miagrammopes***
   **Div.** 1 species: *Miagrammopes mexicanus* O. PICKARD-CAMBRIDGE 1894a — **Dist.** southern TX — **Refs.** Muma & Gertsch 1964, Opell 1979

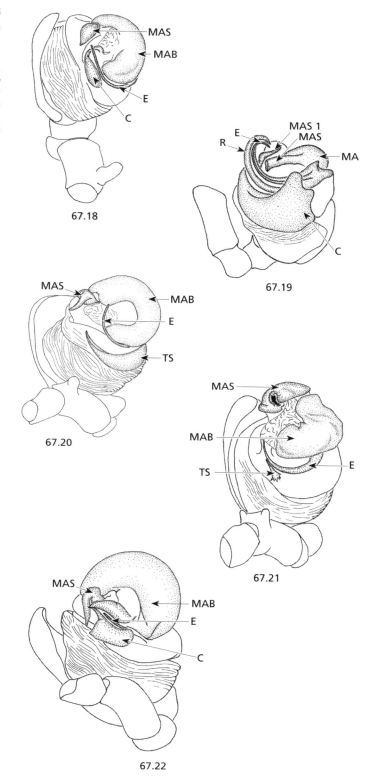

C   conductor
E   embolus
MA   median apophysis
MAB   median apophysis bulb
MAS   median apophysis spur
MAS 1   median apophysis spurs in the case of *Siratoba* where the MA has 2 extensions
R   radix
TS   tegular spur

## Chapter 68
# ZODARIIDAE
2 genera, 5 species

Darrell Ubick
Patrick R. Craig

### Similar families —
Amaurobiidae (p. 60), Corinnidae (p. 79).

### Diagnosis —
Small to medium-sized ecribellate, entelegyne spiders with three tarsal claws distinguished from other trionychans by the enlarged ALS and greatly reduced posterior spinnerets and the absence of a serrula.

### Characters —
**body size:** 2.0-14.0 mm.
**color:** yellowish to dark brown or with a purplish hue.
**carapace:** pyriform; cephalic area and clypeus high.
**sternum:** ovoid; precoxal triangles absent.
**eyes:** eight, in two straight rows (*Lutica*, Fig. 68.1) or with anterior row straight and posterior strongly procurved (*Zodarion*, Fig. 68.10).
**chelicerae:** margins lacking teeth (*Lutica*) or with distal toothed cusp (*Zodarion*).
**mouthparts:** endites convergent, lacking serrula; labium longer than wide.
**legs:** prograde; with (*Lutica*) or without (*Zodarion*) spines; tarsi with three claws, paired claws with teeth on mesal surface.
**abdomen:** oval to subspherical.
**spinnerets:** ALS greatly enlarged, usually cylindrical and nearly contiguous, other spinnerets reduced; colulus represented by setae (Figs. 68.4-68.6).
**respiratory system:** one pair of book lungs; tracheal spiracle near spinnerets.
**genitalia:** entelegyne; ***female*** epigynum strongly sclerotized with large median lobe (*Lutica*) or small and weakly sclerotized (*Zodarion*); ***male*** RTA present; cymbium distally produced, sometimes with modified apical setae (*Zodarion*); bulb with curved embolus and at least one tegular apophysis.

### Distribution —
Lutica is endemic to the southern coast and offshore islands of California south to Baja California; *Zodarion*, a native of the Palearctic region, has been collected in Pennsylvania, Colorado, British Columbia and Québec (*Zodarion rubidum* SIMON 1914a) and in California (*Zodarion* sp.).

### Natural history —
*Lutica* occurs along coastal sand dunes, where it lives in undergound silken burrows and captures prey by biting through the burrow walls (Ramírez 1995c). *Zodarion* is found on and beneath rocks, where it constructs retreats covered with sand grains, and feeds on ants (Jocqué 1991a); additional observations are given in Cushing & Santangelo (2002). Whereas *Zodarion rubidum* is nocturnal, the species in California, which was found in serpentine grassland habitats, was observed running along the ground and rocks even during the hottest part of the day (D. Ubick, pers. obs.).

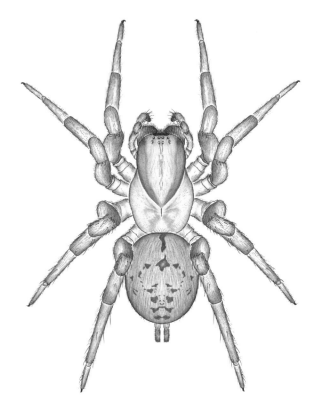

**Fig. 68.1** *Lutica maculata* MARX 1891

### Taxonomic history and notes —
*Lutica* was described by Marx (1891) from a specimen purportedly from Oregon. Gertsch (1961) corrected the type locality to Santa Rosa Island, off the southern coast of California, and described three additional species. A different species arrangement was suggested by a molecular analysis (Ramírez & Beckwitt 1995), but no modifications to the nomenclature were proposed. *Lutica* is placed in the relatively basal zodariid subfamily, Lachesaninae (Jocqué 1991a).

*Zodarion rubidum* was first recorded in Pennsylvania by Vogel (1968a), as *Zodarion fulvonigrum* (SIMON 1874a) (recognized as *Zodarion rubidum* by Bosmans 1994b), and then in Colorado by Cushing & Santangelo (2002).

### Genera —
LACHESANINAE
  *Lutica* MARX 1891
ZODARIINAE
  *Zodarion* WALCKENAER *in* AUDOUIN 1826

## Key to genera —
North America North of Mexico

1   Large spiders (10-14 mm); eyes subequal, PER straight to slightly recurved (Fig. 68.1); endites triangular; legs with spines; abdomen white with dark markings; genitalia Figs. 68.2-68.3 .................................................. **Lutica**
    **Div.** 4 species — **Dist.** s CA coast and islands — **Ref.** Gertsch 1961 — **Note** an undescribed species is mentioned in Roth (1994)

—   Small spiders (2 mm); AME twice diameter of others; PER strongly procurved (Figs. 68.9-68.10); endites elongate; legs without spines; abdomen dark; genitalia Figs. 68.7-68.8 .................................................. **Zodarion**
    **Div.** *Zodarion rubidum* Simon 1914a — **Dist.** PA, CO, QC, BC — **Refs.** Vogel 1968a, Cushing & Santangelo 2002, Bosmans 1994b, 1997 — **Note** an undetermined species of *Zodarion* is known from central CA and ID (Ubick, pers. obs.).

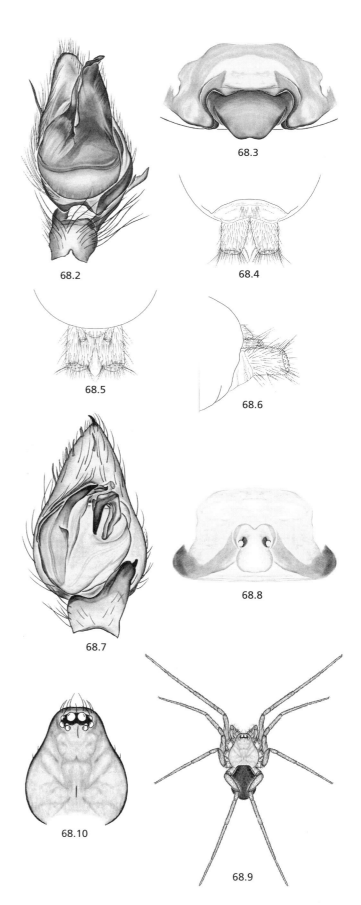

# Chapter 69
# ZORIDAE
1 genus, 2 species

Robert G. Bennett

**Diagnosis —**

Small, two-clawed, ecribellate, entelegyne spiders; eyes in three rows (4, 2, 2) in dorsal view, AER nearly straight, PER strongly recurved; tibiae I and II ventrally with 6-8 pairs of conspicuous elongate macrosetae.

**Similar families —**

Zorids will most likely be misidentified as Ctenidae (p. 83) or Lycosidae (p. 164), but members of those families lack the distinctive series of paired tibial macrosetae. From other spiders possessing numerous conspicuous pairs of tibial macrosetae (Phrurolithinae and other Corinnidae (p. 79), in particular, but also some Cybaeidae (p. 85), *Dictynidae* (p. 95), and Tengellidae (p. 230) zorids are distinguished by their eye arrangement.

**Characters —**

**body size**: small, 2.8-6.8 mm.
**color**: generally yellowish to orange-brown with heavily spotted legs and abdomen; cephalothorax with darker longitudinal marginal and paired medial bands (the latter extending posteriorly from PLE); sternum (Fig. 69.2) with single dark mark adjacent to each coxa.
**cephalothorax**: ovoid, slightly longer than wide, highest in region of fovea; fovea distinct, longitudinal; clypeus narrow, about one eye width wide.
**sternum**: (Fig. 69.2) pentagonal, nearly as wide as long, widest between coxae II and III and extended narrowly between coxae IV.
**eyes**: (Fig. 69.3) eight eyes appearing in three rows (4, 2, 2-AE, PME, PLE) in dorsal view, AE in relatively straight or slightly recurved row; PER about as wide as AER, strongly recurved; eyes sub-equal within rows with PE slightly larger than AE.
**chelicerae**: fang furrow usually with two retromarginal and three promarginal teeth.
**mouthparts**: labium quadrate, somewhat wider than long; endites parallel and about twice as long as labium.
**legs**: tarsi two clawed and with claw tufts and trichobothria, usually unspotted; macrosetae abundant; conspicuous series of paired, elongate, overlapping macrosetae present ventrally on metatarsi (2-3 pairs) and tibiae (6-8 pairs) of legs I and II (Fig. 69.4); legs in order of decreasing length-IV, I, II, III.
**abdomen**: oval, heavily spotted, posterior spinnerets usually visible in dorsal view.
**spinnerets**: colulus lacking; PLS appearing unisegmental (distal segment inconspicuous).
**respiratory system**: one pair of book lungs; tracheal spiracle close to spinnerets, tracheae not extending into cephalothorax.
**genitalia**: entelegyne; *female* epigynum (Figs. 69.6-69.7); atrium (AT) a shallow depression with anterolateral inconspicuous slit-like copulatory openings; connecting ducts (CD) short, broad, and poorly defined; spermathecae with short, anteriorly projecting heads (H) bearing simple vulval pores; spermathecal ducts simple, moderately convoluted, becoming bulbous basally (Fig. 69.7); complex vulval pores absent; *male* (Fig. 69.5) small RTA

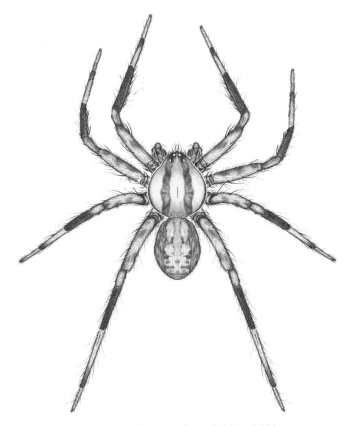

**Fig. 69.1** *Zora hespera* Corey & Mott 1991

present distally ; cymbium with basal retrolateral groove (G); genital bulb with two distally projecting apophyses (TA) membranously attached to tegulum: 1) a heavily sclerotized, conspicuous, anteriorly directed apical apophysis attached to midventral region of tegulum, 2) a membranous, inconspicuous "conductor" extending anteriorly from distal margin of tegulum; embolus (E) a simple, elongate structure curved clockwise (left palpus, ventral view) around outer margin of tegulum, thickened basally, rapidly narrowing, tip inconspicuously located distally in close proximity to tips of both apophyses.

**Distribution —**

Zoridae, as currently construed (Platnick 2005), has a worldwide distribution but is poorly represented in Africa and North America. Only two Nearctic species in one genus are known: *Zora hespera* Corey & Mott 1991 occurs in western North America from southern British Columbia to Idaho and south to California and Arizona, *Zora pumila* (Hentz 1850b) is widely distributed in the eastern USA from eastern Kansas to Massachusetts and south to eastern Texas and Florida.

## Natural history —

Bennett & Brumwell (1996), Bristowe (1958), Corey & Mott (1991), Guarisco (2001), and Kaston (1948) described various aspects of the natural history of zorids (primarily *Zora* spp.). *Zora* spiders are active, diurnal, ground and shrub dwelling hunters. Courtship and mating has been observed in late spring: courtship is brief and involves male leg I signaling; mating occurs in Kaston's (1948) "Type II" position. During the summer, individual loosely woven egg cases containing a variable number of eggs are constructed and attached under a flat sheet of silk to the undersides of objects; females usually remain with their egg cases. Adult females have been collected year-round; males from spring to early summer. *Zora* spiders are associated with a variety of wooded and disturbed habitats as well as being found indoors, but most often are encountered in open, sunny areas.

## Taxonomic history and notes —

Zoridae is a small family of debatable standing currently comprised of 13 genera and 73 species worldwide (Platnick 2005). Zorids have wandered about the "RTA clade" looking for a home since the mid-1800s and have yet to find a comfortable placement. Most recently, Silva Dávila (2003) provided support for a monophyletic Zoridae and a sister relationship between Zoridae + Miturgidae and Ctenidae. Previously zorids have been placed individually or in variously sized lumps within Gnaphosidae (Drassidae-F.O. Pickard-Cambridge 1893c), Clubionidae (e.g., Kaston 1948, Simon 1897a), Lycosidae (e.g., Dahl & Dahl 1927) and, most often, Ctenidae (e.g., Bonnet 1959, Roewer 1954a). Dahl (1912c) evidently first proposed elevation of Pickard-Cambridge's Zorinae to family status but Zoridae did not gain wide acceptance until Lehtinen's (1967) analysis placed it as a family (including zoropsids) separate from Ctenidae (but still within Lycosoidea). Silva Dávila's analysis (2003) places zoropsids with tengellids in a separate branch of the lycosoid lineage ("tengelloids"+ "true" lycosoids) with zorids placed as noted above in the sister branch with the "ctenoids." Among other recent works, Coddington & Levi (1991) suggested zorids are an unplaced group within Dionycha separate from Lycosoidea, and Griswold (1993) noted that zorids lack any of the proposed lycosoid synapomorphies and supported their eviction from Lycosoidea. However the former work was a review only and the latter did not specifically analyze zorid characters, as did Silva Dávila (2003).

C.L. Koch (1847a) described *Zora*. Of the two known Nearctic species, the eastern species, *Zora pumila*, was described early on [in *Katadysas* Hentz 1850b, subsequently transferred to *Zora* by Simon (1897a)] and is reasonably well known. Although about as widespread as its eastern relative, the western species, *Zora hespera*, was described only recently (Corey & Mott 1991). The Nearctic species were recently revised (Corey & Mott 1991) and subsequently further descriptive detail of *Zora hespera* was provided by Bennett & Brumwell (1996).

## Genus —

ZORINAE
   *Zora* C.L. Koch 1847a

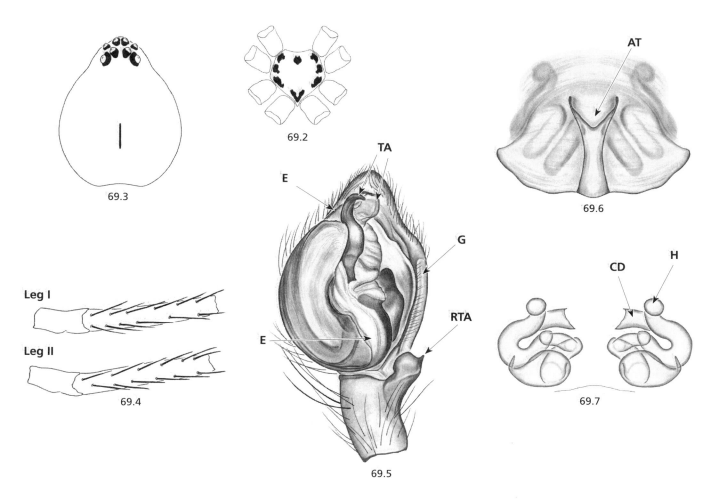

# Chapter 70
# ZOROCRATIDAE
1 genus, 6 species

David B. Richman
Darrell Ubick

## Similar families —
Amaurobiidae (p. 60), Amphinectidae (p. 63), Tengellidae (p. 230), Titanoecidae (p. 248).

## Diagnosis —
Medium to large sized (5-10 mm) cribellate, entelegyne spiders with eight eyes in two straight rows and posterior tarsi with only two claws.

## Characters —
**body size:** 5-10 mm.
**color:** cephalothorax and legs yellow to orange brown, eye region and chelicerae darker, abdomen gray with or without black markings dorsally.
**carapace:** pyriform, longer than wide.
**sternum:** oval, precoxal triangles absent.
**eyes:** eight in two rows, AER straight, PER straight to weakly procurved; subequal in size, small and nearly round.
**chelicerae:** stout; geniculate (more strongly so in female) with boss, margins with 3 teeth.
**mouthparts:** endites and labium longer than wide.
**legs:** prograde and moderately long, with leg IV longest; trochanters notched, posterior ones more deeply; anterior tibiae with 4-5 pairs of ventral spines; tarsi and anterior metatarsi densely scopulate; anterior tarsi with three claws, posterior with two; calamistrum an oval patch on mesobasal third of metatarsus IV.
**abdomen:** elongate oval.
**spinnerets:** cribellum present, divided; anterior spinnerets usually conical and nearly contiguous.
**respiratory system:** one pair of book lungs; tracheal spiracle close to spinnerets.
**genitalia:** entelegyne; *female* epigynum with median lobe usually bearing central atrium or hood and large lateral lobes usually with ectal pockets (Fig. 70.2); *male* palpus with single RTA; bulb with hook-shaped median apophysis and a slender embolus terminating between a hyaline conductor and a broad sclerotized tegular process (Fig. 70.3).

## Distribution —
*Zorocrates* is found from Arizona to Texas south to Panama.

## Natural history —
Little has been written on the natural history of *Zorocrates*. Gertsch & Riechert (1976) recorded a rare species from the lava beds of southern New Mexico but other species may be common in some riparian canyons in southern Arizona and thorn forests of Baja California. The spider constructs a small cribellate sheet beneath rocks and in crevices and burrows.

## Taxonomic history and notes —
*Zorocrates* was described by Simon (1888b) and placed in Zoropsidae, where it remained until Dahl (1913) elevated it to Zorocratidae. Lehtinen (1967) moved the genus to Miturgidae: Tengellinae and Griswold *et al.* (1999a) transferred it back to Zorocratidae. Silva Dávila (2003) considered it most closely related to *Tengella* (thus making Zorocratidae and Tengellidae monotypic as all other genera currently included in the families were transferred elsewhere) and placed it in the tengelloids, a transition group between the basal ctenoids and the higher lycosoids where family boundaries are particularly vague.

There are five species recorded from our region, and as many to the south. The Nearctic species were described by O. Pickard-Cambridge (1896a), Gertsch (1935a), Gertsch & Davis (1936), and Gertsch & Riechert (1976) as isolated descriptions and the family is badly in need of revision.

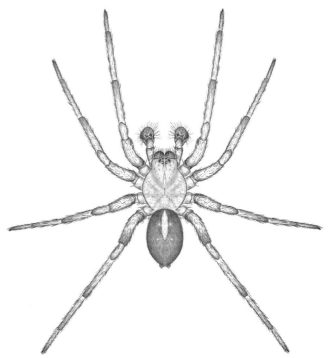

**Fig. 70.1** *Zorocrates unicolor* (Banks 1901a)

## Genus —
*Zorocrates* Simon 1888b.

70.2

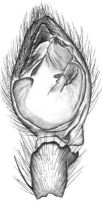

70.3

# Chapter 71
# ZOROPSIDAE
1 genus, 1 species

Darrell Ubick

### Similar families —
Ctenidae (p. 83), Lycosidae (p. 164), Zorocratidae (p. 258).

### Diagnosis —
Zoropsids are distinguished from other cribellate spiders by their large size (>10 mm), strongly recurved PER, straight AER, loss of the inferior claw, and body with dorsal markings.

### Characters —
**body size:** 10.0-17.0 mm.
**color:** carapace orange brown with dark paramedian markings; abdomen grayish brown with dark median stripe broken posteriorly into chevrons and lateral maculations; legs orange brown with dark bands and spots.
**carapace:** pyriform; fovea deep and longitudinal.
**sternum:** round, pointed posteriorly, truncate anteriorly.
**eyes:** eight in two rows, PER strongly recurved, AER straight in dorsal view.
**chelicerae:** geniculate, with boss; both margins with three large teeth.
**mouthparts:** endites longer than wide, broader apically; labium quadrate, with basal notch; serrula present.
**legs:** strongly spined; anterior tibiae with 6-7 pairs of ventral spines; metatarsus IV with oval calamistrum dorsally; tarsi with paired claws and dense claw tufts concealing vestigial unpaired claw.
**abdomen:** elongate oval, slightly flattened.
**spinnerets:** cribellum divided; ALS broad, conical, apically contiguous; others smaller, contiguous.
**respiratory system:** one pair of book lungs; tracheal system with single spiracle near spinerets.
**genitalia:** entelegyne; *female* epigynum strongly sclerotized with a pair of lateral lobes and median lobe represented by scape (Fig. 71.2); *male* palp with simple RTA (Fig. 71.3), bulb with hooklike median apophysis, hyaline conductor, membranous tegular process, and broad embolus (Fig. 71.4).

### Distribution —
Native to the Mediterranean region, *Zoropsis spinimana* has recently become established in the San Francisco region of California (Griswold & Ubick 2001).

### Natural history —
*Zoropsis spinimana* is a wandering hunter normally found on the ground and beneath rocks in forests and grasslands. In the Nearctic, it has been recorded only from human habitation. It makes little use of silk apart from constructing egg sacs, which are attached to the substrate and covered by a tangle of cribellate silk and guarded by the female (Griswold & Ubick 2001).

### Taxonomic history and notes —
Until recently, zoropsids contained only two Palearctic genera. Bosselaers (2002) described a third genus and Raven & Stumkat (2003) transferred two Australian genera to this family. The family belongs to the Lycosoidea, where it has been variously placed: within Zoridae (Lehtinen 1967), as

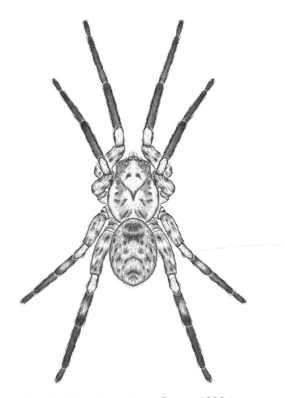

**Fig. 71.1** *Zoropsis spinimana* (Dufour 1820c)

most closely related to Acanthoctenidae (Griswold 1993), and as the sister group to Tengellidae plus Zorocratidae (Silva Dávila 2003).

### Genus —
*Zoropsis* Simon 1878a

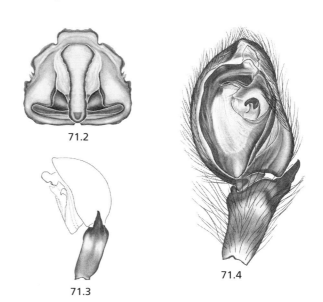

71.2

71.3

71.4

## Chapter 72
# Glossary and pronunciation guide

In the phonetic key to the scientific terms (derived from Beechhold 2000), the primary stress is indicated with capital letters. The sounds are as described below. See also page 331.

| | | | | | | | | |
|---|---|---|---|---|---|---|---|---|
| a | = | a of "bay" | ey | = | i of "bite" | s | = | s of "silly" |
| aa | = | a of "bat" | f | = | f or ph of "fast" or "phone" | th | = | th of "thug" |
| ah | = | a of "bar" | g | = | g of "bug" | u | = | u of "bull" |
| au | = | ou of "bout" | i | = | i of "bit" | uh | = | u of "but" |
| aw | = | ou of "bought" | j | = | j of "jug" | y + vowel | = | y of "you" |
| ch | = | ch of "chug" | k | = | c of "club" | z | = | s or z of "rings" or "zoo" |
| e | = | e of "bet" | o | = | o of "oh" | | | |
| ee | = | ee of "bee" | oo | = | oo of "shoo" | | | |

## A

**Abdomen** (= opisthosoma) — the posterior division of the spider body [Figs. 72.1, 72.4, 72.26]

**Accessory claws** — serrated setae on the tarsal tip of web building spiders, used to grip the silk line [Fig. 72.3]

**Aciniform** [aa-SIN-i-form] — form of a silk gland and its associated spigot type

**Acuminate** [aa-KYOO-min-at] — tapering to a point

**AER** — anterior eye row [Fig. 72.5]

**ALE** — anterior lateral eyes [Fig. 72.6]

**Agonistic** [AAG-o-NIS-tik] — combative or agressive behavior

**ALS** — anterior lateral spinnerets [Fig. 72.7]

**Alveolus** [al-vee-OL-uhs] — the concave ventral surface of the male palpal cymbium; the modified palpal tarsus of the male in which lies the genital bulb

**AME** — anterior median eyes [Fig. 72.6]

**AMS** — anterior median spinnerets

**Anal tubercle** — a small lobe posterior to the spinnerets that bears the anal opening [Figs. 72.4, 72.7]

**Angulate** — having an angular form

**Annuli** [AAN-yoo-lee] — transverse wrinkles on an epigynal scape, rings

**Anteriad** — see anterior

**Anterior** (= anteriad) — front (toward the front). Anterior view [Fig. 72.2]

**Anterior eye row (AER)** — the anteriormost row of eyes, usually consisting of the anterior median and anterior lateral eyes [Fig. 72.5]

**Anterior lateral eyes (ALE)** — eyes situated on the anterolateral portion of the eye group [Fig. 72.6]

**Anterior lateral spinnerets (ALS)** (= anterior spinnerets) — absent in most Mygalomorphae; large in Araneomorphae [Fig. 72.7]

**Anterior median eyes (AME)** (= primary eyes) — these are morphologically distinct from the other eyes and are often reduced or lost [Fig. 72.6]

**Anterior median spinnerets (AMS)** — present only in some Mesothelae; absent in Mygalomorphae; represented by cribellum or colulus in Araneomorphae

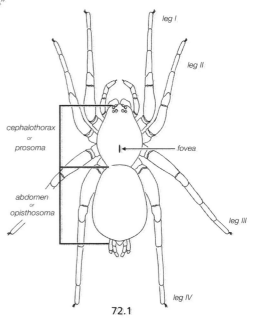

72.1

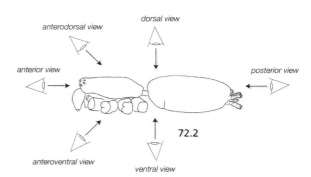

72.2

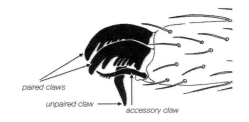

72.3

**Anterior spinnerets** — see anterior lateral spinnerets

**Anterodorsal** — towards the front of the dorsal (or top) surface of body or appendage. Anterodorsal view [Fig. 72.2]

**Anterolateral** — towards the front side portion of body or appendage.

**Anteromesal** — towards the midline front of a structure

**Anteroventral** — towards the front of the ventral (or underside) surface of body or appendage. Anteroventral view [Fig. 72.2]

**Apical** [AP-i-kuhl] (apicad) (= distal (distad)) — toward the terminal portion of an appendage (away from the body)

**Apodeme** [AAP-o-deem] — external skeletal process

**Apomorphic** — derived or advanced

**Apophysis** [ah-PAH-fah-sis] (pl: apophyses) — cuticular or sclerotized projection, common on palpal segments, including femur, patella or tibia of the palp [Fig. 72.8]. See also tibial apophysis (TA) and retrolateral tibial apophysis (RTA)

**Arboreal** — tree dwelling

**Arcuate** [AHR-kyoo-at] — curved like a bow or arc-shaped

**Attenuate** [ah-TEN-yoo-AT] — tapering into a long point

**Atrium** [A-tree-uhm] (pl: atria) — the enlarged opening of the gonopore in haplogyne females; a cavity in the epigynal plate of entelegyne females containing copulatory openings [Fig. 72.47]

**Autospasy** [aw-to-SPAA-see] — the loss of a leg or appendage at a locus of weakness; usually occurs at the coxa — trochanter joint but sometimes occurs instead at the patella — tibia joint

**Autotomize** [ah-TAW-to-MEYZ] — removal of appendage by spider

# B

**Basal** [BA-sil] (basad) = proximal (proximad) — toward the body or the base of an appendage

**Basitarsus** [BA-si-TAHR-suhs] (pl: basitarsi) see metatarsus

**Bidentate** [bey-DEN-tat] — with two teeth

**Bifid** — see bifurcate

**Bifurcate** [BEY-fuhr-KAT] (= bifid) — two-pronged

**Bipartite** [bey-PAHR-teyt] — in two parts; split into two sections

**Biseriate** [bey-SEER-ee-at] — arranged in two parallel rows

**Blumenthal's tarsal organ** — see tarsal organ

**Book lung** — a respiratory structure consisting of a rounded gas exchange portion with a posterior spiracle. Most Araneomorphae have a single pair located anterolaterally of the epigastric furrow [Fig. 72.26]; Mygalomorphae and Hypochilidae have a second pair posterior of the epigastric furrow

**Book lung cover** (= branchial operculum) — the plate covering the book lung [Fig. 72.4]

**Boss** (= condyle) — a rounded swelling located at the base of the chelicera [Fig. 72.9]

**Branchial operculum** [BRANK-ee-awl o-PUHR-kyoo-luhm] — see book lung cover

**Bristles** — small rigid hairs or small spines

**Bulb** (= bulbus = genital bulb = palpal organ) — the genital structure of the male spider containing the sperm reservoir and attached to the palpal tarsus [Fig. 72.12]; rarely fused to the tarsus, as in some Oonopidae, from which it can be differentiated by lacking setae

*Bursa copulatrix* — Latin term for female copulatory pouch

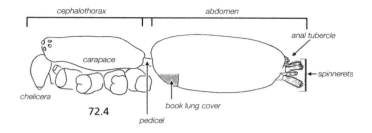

72.4

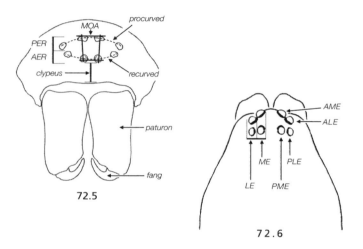

72.5

72.6

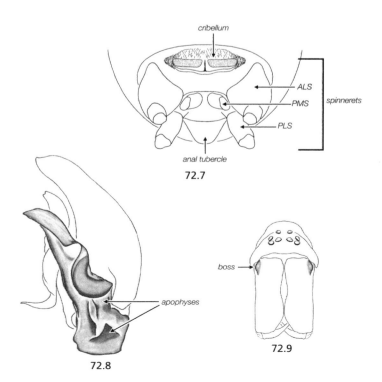

72.7

72.8

72.9

# C

**Calamistrum** [kaal-aa-MIS-truhm] — a row of curved thick setae (rarely arranged in an oval patch) located dorsally along metatarsus IV of cribellate spiders; used for combing out silk from the cribellum [Fig. 72.10]

**Capitate** [KAAP-i-TAT] — linear structure having a distal swelling

**Caput** [KA-puht] — see cephalic region

**Carapace** [KAR-ah-pas] — dorsal part of the cephalothorax [Fig. 72.4]

**Cavernicolous** [KAAV-uhr-NI-co-luhs] — cave dwelling

**Carina** [kahr-EE-nah] (pl: carinae) — ridge or keel

**Caudal** — towards the posterior end

**Cephalic region** (= caput = cephalon = *pars cephalica*) — the anterior (= head) portion of the carapace [Fig. 72.11]

**Cephalon** [sef-AAL-uhn] — see cephalic region

**Cephalothorax** [sef-ah-lo-THOR-aax] (= prosoma) — the anterior division of the spider body [Figs. 72.1, 72.4, 72.26]

**Cervical grooves** — shallow grooves separating the cephalic from the thoracic part of the carapace [Fig. 72.11]

**Chelate** [KEEL-at] — pincer-like; in spiders refers to the fused chelicerae of some Haplogynae where the fang and lamina form a pincer [Fig. 72.40]

**Chelicera** [kuh-LI-suh-rah] (pl: chelicerae [kuh-LI-suh-ree]) — the anterior appendages of a spider (the "jaws") consisting of a large basal segment (paturon) and an apical fang [Figs. 72.4, 72.26]

**Cheliceral extension** — a pointed basal projection behind the clypeus, conspicuous in some Theridiidae, Nesticidae, and Pholcidae [Fig. 72.18]

**Cheliceral furrow** (= fang furrow) — the groove of the chelicera into which the fang closes [Fig. 72.13]

**Cheliceral lamina** — see lamina

**Cheliceral teeth** — large and/or tiny toothlike projections on the cheliceral furrow margins [Fig. 72.13]

**Chemoreception** — sensing of chemical stimuli

**Chevron** — V-shaped pattern

**Chilum** [KEE-luhm] (pl: chila) — a small sclerite at the base of the chelicera (under the clypeus) [Fig. 72.17]

**Clade** — monophyletic group

**Cladogram** — diagram derived from phylogenetic analysis illustrating evolutionary relationships among groups

**Clavate** [KLAA-vat] — clubbed

**Clasping spine** — a type of mating spur consisting of an enlarged, curved spine that articulates against the leg segment, as in Mysmenidae [Fig. 72.16]

**Claw** — see tarsal claw

**Claw dentition** — the pectinate ventral surface of most claws, arranged in either a single (uniserial) or double (biserial) row of teeth

**Claw tuft** — a dense brush of hairs between the paired tarsal claws; almost always indicates the absence of the unpaired claw; tufts may be composed of simple hairs or thick broad ones (= tenent hairs) [Fig. 72.15]

**Clypeus** [KLEY-pee-uhs] — the space between the anterior edge of the carapace and the anterior eyes [Fig. 72.5]

**Cochlea** [KAHK-lee-ah] (= stretcher) — a pit at the tip of the epigynal scape found in some Linyphiidae

**Colulate** [KAWL-yoo-lat] — having a colulus

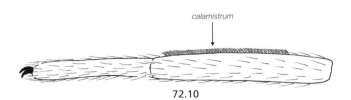

72.10

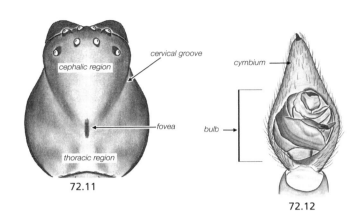

72.11

72.12

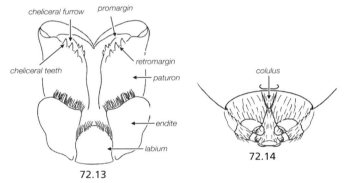

72.13

72.14

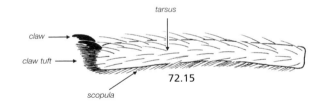

72.15

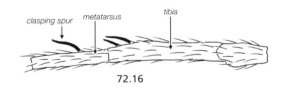

72.16

72.17

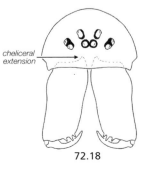

72.18

**Colulus** [KAWL-yoo-luhs] — a nonfunctional cribellum that may be as large as a cribellum but is more commonly reduced to a small fleshy lobe or even just a pair of setae [Fig. 72.14]

**Concolorous** [kawn-KO-lor-uhs] — of a uniform color

**Conductor** — the structure of the male palp associated with the embolus; may be sclerotized or hyaline [Figs. 72.29, 72.31, 72.51]

**Condyle** (KAWN-deyl) — see boss

**Congeneric** — of the same genus

**Conglomerate** — irregular aggregation or mass of objects

**Conical** — cone shaped

**Connecting ducts** — see copulatory ducts

**Conspecific** — of the same species

**Contiguous** — not separated

**Copulatory ducts** (= insemination ducts = connecting ducts = entrance ducts = spermathecal ducts) — ducts in the female genitalia which connect the copulatory pores to the spermathecae [Fig. 72.34]

**Copulatory openings** (= copulatory pores = epigynal openings) — external pores on the epigyna of entelegyne females [Figs. 72.33, 72.34]

**Copulatory pores** — see copulatory openings

**Cosmopolitan** — worldwide, occurring globally

**Coxa** [KAWX-uh] (pl: coxae) — the basal, or first, segment of a spider leg and palp [Figs. 72.19, 72.26, 72.75]

**Cribellate** [KRIB-e-lat] — having a cribellum

**Cribellum** [kri-BEL-uhm] — a broad flat spinning plate anterior to the remaining spinnerets of cribellate spiders; homologue of the AMS [Fig. 72.7]

**Ctenidia** [ten-ID-ee-ah] — structures on the male palpal tibia of some Dictynidae consisting of a short process bearing diminutive stout spines [Fig. 72.20]

**Cusps** — short thick spines on the legs of some spiders (as in Trachelinae) [Fig. 72.22]

**Cuspules** — small spines on the endites and labium of Mygalomorphae [Fig. 72.21]

**Cryptozoic** [KRIP-to-ZO-ik] — living in hidden or concealed habitats

**Cymbium** [SIM-bee-uhm] (pl: cymbia) — the adult male palpal tarsus, especially when modified as a spoon-shaped structure containing the palpal bulb [Fig. 72.12]

# D

**Declivity** [de-KLEV-i-tee] — sloping downward

**Dentate** — toothed

**Denticles** — small teeth

**Diad** — a pair of two contiguous eyes [Fig. 72.23]

**Diaxial** (= labidognathous) — downward projecting chelicerae with the fangs operating along the transverse axis (somewhat like salad tongs); present in Araneomorphae [Fig. 72.24]

**Dichotomous key** — an arrangement of diagnostic characters for the identification of organisms using alternative choices

**Digitiform** [dij-IT-i-form] — fingerlike

**Divergent** — structures whose distance apart increases distally

**Diverticula** [DEY-vur-TI-kyoo-law] — extensions of the digestive system

**Dionychous** [DEY-o-NEY-kuhs] — having two tarsal claws

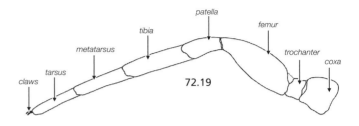

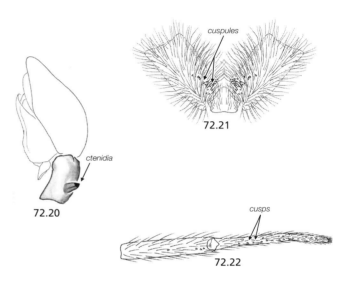

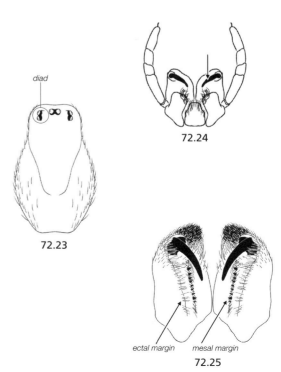

**Distal** (distad) (= apical = apicad) — toward the terminal portion of an appendage (away from the body)

**Distitarsus** [DIS-ti-TAHR-suhs] (pl: distitarsi) — see tarsus

**Dorsal** (dorsad) — upper surface (toward the top). Dorsal view [Fig. 72.2]

**Dorsal groove** — see fovea

**Dorsomesal** — toward the middle of the top of the body or appendage

# E

**Eclose** — to emerge from a egg

**Ecribellate** [ee-CRIB-ah-lat] — lacking a cribellum

**Ectal** (ectad) — away from the midline of the body

**Ectal margin** — the outer margin of the cheliceral furrow in Mygalomorphae [Fig. 72.25]

**Edentate** — untoothed

**Embolic division** — in the Linyphiidae, the terminal portion of the palpal bulb, consisting of the radix, embolus and various accessory sclerites. It is attached to the suprategulum by a narrow stalk

**Embolus** [EM-bo-luhs] — the intromittent organ of the male palp, connected to the sperm reservoir; often slender, but variable in shape and length [Figs. 72.29, 72.31, 72.51]

**Emerit's gland** — oval cuticular glands found on the appendages of some spiders (e.g., Cybaeidae, Leptonetidae, Telemidae)

**Endemic** — restricted to a particular region or habitat

**Endite** [EN-deyt] (= maxilla = gnathocoxa = gnathobase) — the expanded lobe of the palpal coxa situated laterally of the labium [Figs. 72.13, 72.36, 72.75]

**Endogean** [en-DAH-jee-in] — occurring beneath the surface (as opposed to epigean)

**Entelegyne** [en-TEL-i-jeyn] (= entelegynous condition) — the derived form of spider genitalia where the female has external copulatory openings, typically on a sclerotized epigynum, and the male has relatively complex palpi. Also refers to the clade, Entelegynae, which includes the vast majority of living spiders. See haplogyne

**Entrance ducts** — see copulatory ducts

**Ephemeral** — short lived

**Epigastric furrow** [EP-i-GAAS-trik] — a transverse groove across the anterior ventral part of the abdomen [Fig. 72.26, 72.34]

**Epigastric plates** — see book lung cover

**Epigean** [EP-i-JEE-in] — occurring on the surface (as opposed to endogean)

**Epigyne** [EP-i-GEYN] — see epigynum

**Epigynum** [i-PIJ-i-nuhm] (pl: epigyna) (= epigyne) — the sclerotized region of female spiders covering the internal genitalia and located between the book lungs and anterior of the epigastric furrow [Figs. 72.26, 72.32, 72.33, 72.47, 72.76]

**Eye formula** — represents the eye distribution starting with the anterior row (e.g., "4-2-2" means 4 eyes in the anterior row, 2 eyes in the middle row and 2 eyes in the posterior row)

# F

**Fang** — the apical segment of a chelicera; has the opening of the venom gland [Fig. 72.5]

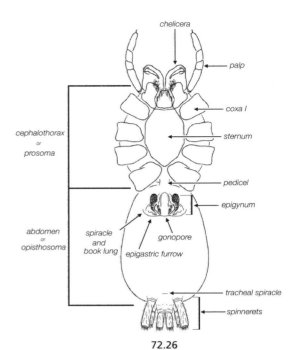

72.26

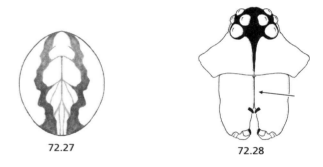

72.27   72.28

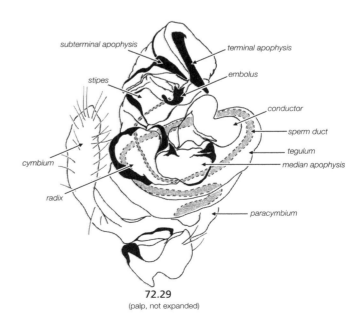

72.29
(palp, not expanded)

**Fang furrow** — see cheliceral furrow

**Femoral spot** — sclerotized spot of unknown function located ventrally and subapically on femur I and sometimes femur II, usually on females but can be present on males, as in Mysmenidae

**Femur** (pl: femora) — the third segment of the leg, outward from the body, located between the trochanter and patella [Figs. 72.19, 72.75]

**Fertilization ducts** — internal ducts through which females fertilize their eggs; in haplogyne spiders these are the same as copulatory ducts [Fig. 72.34]

**Fissidentate** [fis-i-DEN-tat] — teeth with multiple points

**Fickert's gland** — a swelling of the sperm duct within the embolus found in some Linyphiidae

**Filiform** [FIL-i-form] — thread-shaped

**Folium** — leaf-shaped markings on the abdominal dorsum of some Araneomorphae [Fig. 72.27]

**Fossa** (pl: fossae) — deep pit

**Fossorial** — digging or burrowing

**Fovea** [FOV-ee-uh] (= dorsal groove = thoracic groove = furrow = = pit) — the central depression in the carapace representing the invagination of the carapace where the stomach muscles attach [Fig. 72.1, 72.11]

**Fulcrum** [FUL-cruhm] — a structure that serves as a support for another structure

**Furrow** — see fovea

**Fused chelicerae** — chelicerae that are joined mesally, at least along the base (as in some Haplogynae) [Fig. 72.28]

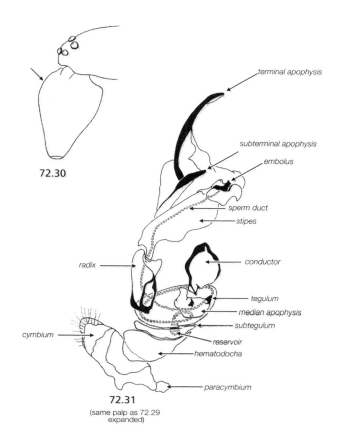

72.30

72.31 (same palp as 72.29 expanded)

# G

**Geniculate** [jen-IK-yoo-lat] — bent, knee-like; usually pertains to chelicerae in which the basal segment (paturon) extends beyond the clypeus giving the chelicerae a bent appearance [Fig. 72.30]

**Genital bulb** — see bulb [Fig. 72.12]

**Glabrous** [GLA-bruhs] — smooth and shiny; refers to cuticle without hairs or spines

**Globose** [GLO-bos] — rounded, spherical

**Gnathobase** [NAATH-o-bas] — see endite [Figs. 72.13, 72.75]

**Gnathocoxa** [naath-o-CAHX-ah] (pl: gnathocoxae) — see endite [Figs. 72.13, 72.75]

**Gonopore** [GAH-no-por] — the genital opening, located in the middle of the epigastric furrow [Fig. 72.26]

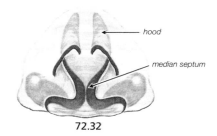

72.32

# H

**Habitus** [HAAB-i-tuhs] — general appearance

**Hackled** — combed out, in reference to cribellate silk

**Hair** — see seta

**Haplogyne** [HAAP-lo-jeyn] (= haplogynous condition) — the primitive form of spider genitalia where the female has the copulatory openings internally, within the gonopore, and typically lacks a sclerotized epigynum, and the male has relatively simple palpi. This form is found in the basal spiders, including Mygalomorphae, Hypochiloidea, and Haplogynae. See entelegyne.

**Hematodocha** [hee-MAA-to-DO-kah] (pl: hematodochae) — in male spiders, membranous, inflatable parts of the bulb which become distended during copulation [Fig. 72.31]

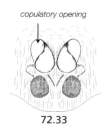

72.33

72.34

**Hemocyanin** [HEE-mo-SEY-uh-nin] — oxygen carrying molecule, pigment in the hemolymph

**Hemolymph** [HEE-mo-limf] — body fluid

**Hirsute** [her-SOOT] — hairy

**Hood** — a pocket-like structure at the anterior end of the epigynum (as in Lycosidae) *[Fig. 72.32]*

**Hyaline** [HEY-ah-leyn] — translucent or transparent

**Hydric** [HEY-drik] — characterized by a high amount of moisture (in contrast to mesic and xeric)

**Hygroreception** [HEY-gro-ree-SEP-shuhn] — sensing water or moisture gradients

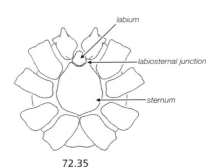

72.35

**Incrassate** [in-CRAA-sat] — thickened

**Inferior claw** — see unpaired claw

**Instar** — a developmental stage prior to adulthood

**Intercalary sclerite** — a sclerite located between the tegulum and the terminal apophysis and partially covered by the subtegulum *[Fig. 72.37]*

**Intercoxal sclerites** — narrow sclerites between the coxae *[Fig. 72.59]*

**Intromittent organ** — external male sexual organ

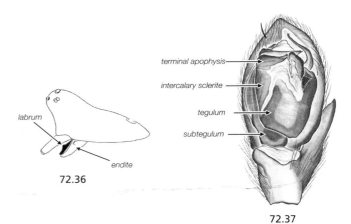

72.36

72.37

**Keel** — a serrated ridge, truncated outgrowth of the cheliceral margin *[Fig. 72.41]*

**L**

**Labidognathous** [lah-BI-dog-NA-thus] — see diaxial *[Fig. 72.24]*

**Labiosternal junction** [LEY-bee-o-STUR-nuhl] — the boundary between the labium and the sternum *[Fig. 72.35]*

**Labium** [LA-bee-uhm] (pl: labia) (= lip) — the anterior median extension of the sternum *[Figs. 72.13, 72.35]*

**Labral cone** (= labral spur) — a short projection from the labrum of certain spiders (Anapidae) *[Fig. 72.36]*

**Labrum** [LA-bruhm] — the upper lip (usually covered by the chelicerae and not easily visible) *[Fig. 72.36]*

**Lamella** [laam-EL-uh] — see lamina

**Lamelliform** [laam-EL-i-form] — leaf shaped

**Lamina** [LAM-in-ah] (pl: laminae) (= lamella) — sclerotized ridge on the cheliceral margin or mesal surface *[Fig. 72.40]*; also a sclerite in the male palp of many Linyphiidae

**Lanceolate** [LAAN-see-O-lat] — shaped like a lance

*Lapus calami* — Latin term used in nomenclature meaning literally "a slip of the pen" for an inadvertent textual error made by the author rather than a typographical error

**Lateral claws** — see paired claws

**Lateral eyes** (LE) *[Fig. 72.6]*

**Lateral ocular quadrangle** LOQ — see ocular quadrangle *[Fig. 72.39]*

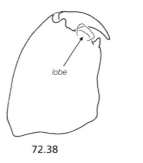

72.38

72.39

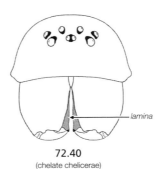

72.40
(chelate chelicerae)

72.41

**Laterigrade** [LAAT-er-uh-grad] — crab-like leg orientation with the legs directed outward from the body; these legs are twisted at the trochanter with the morphologically prolateral surface assuming a dorsal position (e.g., Thomisidae)

**LE** — lateral eyes *[Fig. 72.6]*

**Leg formula** — a series of four numbers (e.g., 4123) that give the relative leg lengths, from the longest to the shortest

**Leg numbering** — legs are numbered using Roman numerals, as I, II, III, IV, starting with the anterior leg; for example, tibia III refers to the tibia of leg III *[Fig. 72.1]*

**Lip** — see labium *[Figs. 72.13, 72.35]*

**Lobe** — a rounded outgrowth of the cheliceral margin *[Fig. 72.38]*

**LOQ** lateral ocular quadrangle — see ocular quadrangle

**Lorum** [LUHR-uhm] — a set of sclerites on the dorsal surface of the pedicel *[Fig. 72.42]*

**Lunate** [LOO-nat] — crescent-shaped

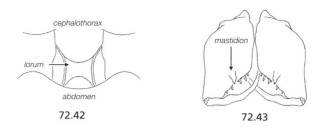

72.42

72.43

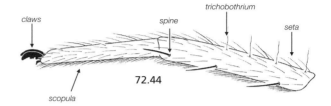

72.44

# M

**Macrosetae** [MAAK-ro-SEE-tee] — see spines *[Fig. 72.44]*

**Maculate** [MAAK-yoo-lat] — spotted

**Mastidion** [maas-TID-ee-awn] (pl: mastidia) — a tooth on the cheliceral face *[Fig. 72.43]*

**Mating spur** (= spur) — a rigid cuticular outgrowth, with or without apical spines, found on the anterior legs of some male spiders and used as a restraining structure during copulation *[Fig. 72.48]*; see also clasping spur and megaspine

**Maxilla** [maax-IL-uh] (pl: maxillae) — see endite *[Figs. 72.13, 72.75]*

**ME** — median eyes *[Fig. 72.6]*

**Mechanoreception** — sensing movement, tension or pressure

**Median apophysis** — on the male palp, a sclerotized structure in the lower center of the bulb *[Figs. 72.29, 72.31]*

**Median claw** — see unpaired claw *[Fig. 72.3]*

**Median lobe** — the lobe-like protuberance along the midline of some epigyna *[Fig. 72.76]*

**Median ocular area**, MOA — see median ocular quadrangle *[Fig. 72.5]*

**Median ocular quadrangle**, MOQ (= MOA = median ocular area = quadrangle) — area encompassed by the four median eyes *[Fig. 72.5]*

**Median septum** — a raised median longitudinal ridge on the floor of the epigynal atrium typically found in Lycosidae; a slender median lobe *[Fig. 72.32]*

**Median spinnerets** — see posterior median spinnerets

**Megaspine** — a rigid cuticular outgrowth with a large apical spine *[Fig. 72.45]*

**Merovoltine** [MER-o-VOL-teen] — >1 year/generation

**Mesal** [MEE-sahl] (mesad) — toward the midline of the body *[Fig. 72.49]*

**Mesal margin** — the inner margin of the cheliceral furrow in Mygalomorphae *[Fig. 72.25]*

**Mesic** [MEE-zik] — characterized by a moderate amount of moisture (in contrast to hydric and xeric)

**Metatarsus** (pl: metatarsi) (= basitarsus) — the 6th segment of the leg, located between the tibia and tarsus *[Fig. 72.19]*

**MOA**, median ocular area — see median ocular quadrangle

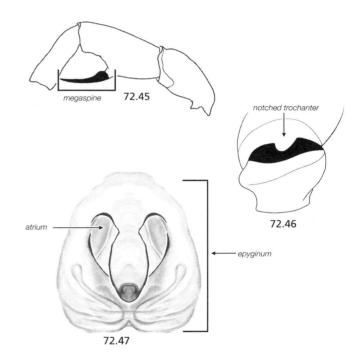

72.45

72.46

72.47

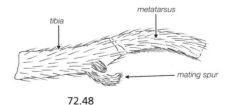

72.48

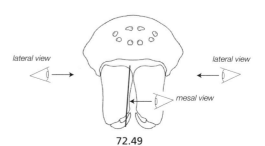

72.49

**Monotypic** — taxon with only one subordinate member, as a genus composed of only one species

**Monophyletic** — group of organisms sharing a common ancestor

**Morphology** — the form and structure or an organism

**MOQ** — see median ocular quadrangle

**Multivoltine** [MUHL-tey-VOL-teen] — >1 generation/year

**Mygalomorphae** [my-GAL-o-MOR-fee] — infraorder of spiders characterized by paraxial chelicerae and 2 pairs of book lungs

**Myrmecomorph** [meyr-MEE-ko-MOORF] — resembling ants

**Myrmecophile** [meyr-MEE-ko-FEYL] — associated with ants

# N

**Nearctic** — Temperate and arctic parts of North America, including Greenland

*Nomen dubium* (pl: nomina dubia) — Latin term for scientific names of unknown or doubtful application

*Nomen nudum* (pl: nomina nuda) — Latin term for scientific names that could not be associated with a recognizable biological entity

**Notched trochanter** — one with a shallow to deep ventroapical excavation [Fig. 72.46]

# O

**Ocular quadrangle**, OQ (= lateral ocular quadrangle LOQ) — the total area occupied by the eyes [Fig. 72.39]

**Onychium** [ahn-IK-ee-uhm] (pl: onychia) — apical extension of the tarsus that bears the tarsal claws, prominent in Oonopidae and Ochyroceratidae [Fig. 72.50]

**Opisthosoma** [o-PIS-tho-SO-muh] — see abdomen

**OQ** — see ocular quadrangle

**Orthognathous** [or-thog-NA-thuhs] — see paraxial

**Ostia** [AHS-tee-uh] — an aperture or opening

# P

**Paired claws** (= superior claws = lateral claws) — found at the tip of the leg tarsi of all spiders [Fig. 72.3]

**Palea** [PA-LEE-uh] (pl: paleae) — in the male palp, a pillow-like structure at the distal end of the bulb, may be membranous or sclerotized; in Lycosidae, the distal part of the tegulum where it gives rise to the terminal apophysis [Fig. 72.51]

**Palp** (= palpus = pedipalp) — the most anterior leg-like appendage in spiders [Figs. 72.26, 72.75]

**Palpal organ** — see bulb

**Palpus** (pl: palpi) — see palp [Figs. 72.26, 72.75]

**Paracymbium** [par-aw-SIM-bee-uhm] (pl: paracymbia) — an appendage of the cymbium; usually a rigid outgrowth but may be flexibly attached by a membrane [Figs. 72.29, 72.31, 72.52]

**Paramedian** — along longitudinal axis

**Paraphyletic** [PAAR-ah-fey-LET-ik] — a group of taxa that excludes one or more descendants of the common ancestor

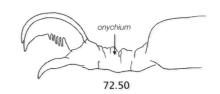

72.50

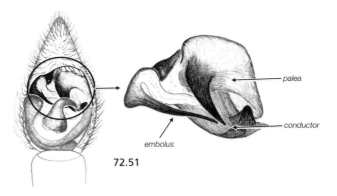

72.51

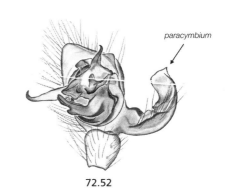

72.52

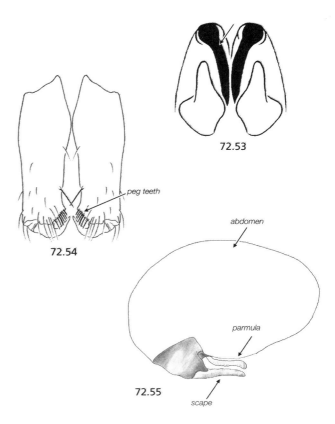

72.53

72.54

72.55

**Paraxial** [par-AAX-ee-uhl] (= orthognathous) — type of chelicerae that project forward with fangs articulating along the longitudinal (vertical) axis (like the teeth of a garden rake); present in Mesothelae and Mygalomorphae *[Fig. 72.53, 72.25]*

**Parmula** [par-MYOO-luh] — in Linyphiidae, the process arising from the dorsal wall of the epigynum; in some species dorsal to the scape *[Fig. 72.55: abdomen, lateral view]*. Also referred to as dorsal scape

*Pars cephalica* — cephalic region

*Pars thoracica* — thoracic region

**Parthenogenetic** — reproducing without males

**Patella** [pah-TEL-ah] (pl: patellae) — the 4th segment of the legs and palpi outward from the body; it is a short curved segment located between the femur and tibia *[Figs. 72.19, 72.75]*

**Paturon** [paa-TUR-awn] — the basal segment of a chelicera *[Figs. 72.5, 72.13]*

**Pectinate** [PEK-tin-at] — comblike

**Pedicel** [PED-i-sel] — the narrowed connection between the cephalothorax and abdomen; the "waist" of the spider *[Figs. 72.4, 72.26]*

**Pedipalp** [PED-ee-paulp] — see palp *[Figs. 72.26, 72.75]*

**Peg teeth** — spine-like teeth on the chelicerae situated in sockets *[Fig. 72.54]*

**PER** — posterior eye row *[Fig. 72.5]*

**Petiole** [PET-ee-ol] (= petiolus) — an enlarged, sclerotized pedicel

**Phylogeny** [fey-LAH-jo-nee] — the evolutionary history of a group of organisms

**Pit** — see fovea *[Fig.72.1, 72.11]*

**PLE** — posterior lateral eyes *[Fig. 72.6]*

**Plesiomorphic** [PLEE-zee-o-MOR-fik] — ancestral or primitive

**Pleural bars** [PLUR-awl BARS] — narrow sclerites between the carapace and coxae *[Fig. 72.57]*

**PLS** — posterior lateral spinnerets *[Fig. 72.7]*

**Plumose hairs** — recumbent hairs with long branches *[Fig. 72.56]*

**Pluridentate** [plur-i-DEN-tat] — with multiple teeth

**PME** — posterior median eyes *[Fig. 72.6]*

**PMS** — posterior median spinnerets *[Fig. 72.7]*

**Polyphyletic** — a group of taxa based on convergence that excludes the common ancestor

**Porrect** [por-EKT] — extending forward; in spiders refers to the projecting diaxial chelicerae of certain Araneomorphae, as Dysderidae, to differentiate them from the paraxial chelicerae of Mygalomorphae *[Fig. 72.60]*

**Posterior** (posteriad) — back; toward the back. Posterior view *[Fig. 72.2]*

**Posterior eye row (PER)** — *[Fig. 72.5]*

**Posterior lateral eyes (PLE)** *[Fig. 72.6]*

**Posterior lateral spinnerets (PLS)** (= posterior spinnerets) — typically well developed *[Fig. 72.7]*

**Posterior median spinnerets (PMS)** — (= median spinnerets) the innermost pair of spinnerets, typically small *[Fig. 72.7]*

**Posterior spinnerets** — see posterior lateral spinnerets

**Precoxal triangles** (= precoxal sclerites) — triangular sclerotized extensions from the sternum to the coxae *[Fig. 72.59]*

**Preening brush** — a cluster of setae at the ventral tip of posterior metatarsi *[Fig. 72.61]*

**Preening comb** — a transverse row of rigid setae located ventroapically on the ventral tip of the posterior metatarsi *[Fig. 72.62]*

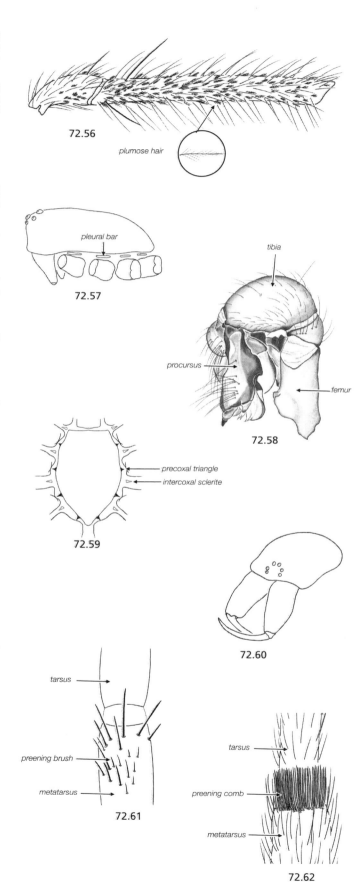

**Primary eyes** — see anterior median eyes [*Fig. 72.6*]

**Process** — outgrowth of surface, margin, or appendage

**Procursus** [pro-KUR-suhs] — retrolateral paracymbium found in pholcids [*entire palp, lateral view, Fig. 72.58*]

**Procurved eye row** — one in which the lateral eyes are positioned anterior to the median eyes [*Fig. 72.5*]

**Prograde** [PRO-grad] — typical leg orientation with the anterior pairs directed forward and the posterior pairs backwards (e.g., in Lycosidae)

**Prolateral** — on the anterior surface of an appendage when held at right angles to the body axis

**Promargin** — the anterior margin of the cheliceral furrow in Araneomorphae (mesal margin in Mygalomorphae) [*Fig. 72.13*]

**Promarginal** — on the anterior margin

**Prosoma** [pro-SO-muh] — see cephalothorax [*Figs. 72.1, 72.4, 72.26*]

**Proximal** (proximad) = basal (basad) — toward the body or the base

**Pubescent** [pyoo-BES-ent] — bearing fine setae

**Punctate** [PUHNK-tat] — with impressed points or punctures

**Pyriform** [PEER-i-form] — pear shaped

# Q

**Quadrangle** — see median ocular quadrangle [*Fig. 72.5*]

**Quadrate** — square-shaped

# R

**Radix** [RA-dix] — sclerotized apophysis of the male palp [*Figs. 72.29, 72.31*]

**Ramus** (pl: rami) — a branch of a structure (e.g., in Salticidae)

**Rastellum** [rah-STEL-uhm] (pl: rastella) — a rake-like structure near the cheliceral fang base of some Mygalomorphae [*Fig. 72.63*]

**Rebordered labium** — one which is thicker apically than basally [*Fig. 72.65*]

**Recumbent** [ree-KUHM-bent] — lying down against another structure, as in 'recumbent hair'

**Recurved eye row** — one in which the lateral eyes are positioned posterior to the median eyes [*Fig. 72.5*]

**Relict** [REL-ikt] — a persistant remnant of an otherwise extinct taxon

**Repugnatorial glands** [ree-PUHG-naa-TOR-ee-awl] — glands that secrete noxious compounds

**Reticulate** [re-TIK-yoo-lat] — covered with a network of lines or ridges; meshed

**Retrolateral** — on the posterior surface of an appendage when held at right angles to the body axis

**Retromargin** — the posterior margin of the cheliceral furrow in Araneomorphae (ectal margin in Mygalomorphae) [*Fig. 72.13*]

**Retrolateral tibial apophysis** (RTA) — a sclerotized process on the retrolateral face of the male palpal tibia [*Fig. 72.64*]

**RTA** — see retrolateral tibial apophysis [*Fig. 72.64*]

**Rugose** [ROO-gos] — wrinkled or corrugated in appearance

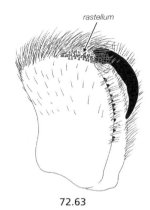

72.63

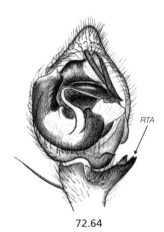

72.64

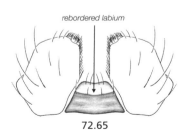

72.65

# S

**Scales** — flattened hairs of various shapes

**Scape** (= scapus) — an elongated process of some epigyna, usually posteriorly directed [Fig. 72.55: epigynum, lateral view]

**Sclerite** [SKLER-eyt] — a sclerotized plate

**Sclerotized** [SKLER-o-teyzd] — thickened and hardened

**Sclerotization** [SKLER-o-ti-ZA-shuhn] — the process of becoming sclerotized, or thickened and hardened

**Scopula** [SCAW-pyoo-lah] (pl: scopulae) — a brush of hairs, such as the dense pads along the ventroapical surface of the legs [Fig. 72.15, 72.44], where they may occur with or without claw tufts; also refers to the looser fringe of hairs on the anterior edge of the endite and labium [Fig. 72.66]

**Scutum** [SKOO-tuhm] (pl: scuta) (= scute) — sclerotized abdominal plate, usually dorsal, of some Araneomorphae [Fig. 72.67]

**Secondary eyes** — all eyes other than the anterior median eyes; only these may have the tapetum layer and appear shiny

**Seminal receptacle** (= spermatheca) — internal sclerotized sperm-storage sacs present in all female spiders [Fig. 72.34]

**Semidiaxial** — chelicerae intermediate between the diaxial condition of araneomorphs and paraxial of mygalomorphs, as in Hypochilidae

**Senescence** [sen-ES-enz] — the process of aging

***Sensu lato*** — in the broad sense

***Sensu stricto*** — in the strict sense

**Serrula** [SER-oo-luh] (pl: serrulae) — a row of tiny teeth along the anterior edge of the endite [Fig. 72.66]

**Seta** [SEE-tuh] (pl: setae) (= hair) — may be erect or recumbent; simple, serrate, feathery, to plumose, or clavate to broad scales [Fig. 72.44]

**Setose** — bearing setae

**Sigilla** [sig-IL-uh] (s = sigillum) — circular impressions or "dimples" on the abdomen and sternum of some spiders representing points of muscle attachment and endoskeletal attachments [Fig. 72.67]

**Sinuate** — S-shaped or wavy structure

**Slit sense organs** — stress or strain receptor in the exoskeleton

**Somatic** [so-MAAT-ik] — referring to non-reproductive structures

**Spatulate** [SPAA-choo-LAT] — apically enlarged, spoon-shaped

**Spermatheca** [spuhr-mah-THEE-kah] (pl: spermathecae) (= seminal receptacle) — internal sclerotized sperm-storage sacs present in all female spiders [Fig. 72.34]

**Spermathecal ducts** — see copulatory ducts [Fig. 72.34]

**Spigot** — tiny cylindrical extension of the spinnerets from which silk is extruded

**Spine** (= macroseta = spiniform) — a) pointed rigid structure situated in a socket and capable of articulating; as this is an enlarged hair, macrosetae is a more correct although less often used term in spider taxonomy[Fig. 72.44] — b) solid, pointed extension of the cuticle

**Spiniform** — see spine; spine-shaped

**Spinnerets** — finger-like or conical silk spinning organs located at the posterior end of the abdomen [Figs. 72.4, 72.7]

**Spinose** [SPIN-nos] — bearing spines

**Spinule** [SPIN-yool] — small spine

**Spiracle** [SPEER-ah-kul] — the opening into the book lung or trachea [Fig. 72.26]

**Spur** — see mating spurs [Fig. 72.48]

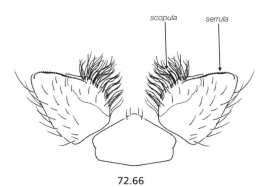

72.66

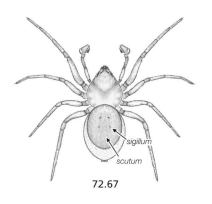

72.67

**Sternum** (pl: sterna) — the ventral sclerite of the cephalothorax [Figs. 72.26, 72.35]

**Stretcher** — see cochlea

**Stria** (pl: striae) — fine impressed lines

**Stridulatory file** [STRID-oo-luh-TOR-ee feyl] — a series of fine grooves used in conjunction with thorns to produce sound, typically located along the ectal margin of the chelicera [Fig. 72.68] or along the anteroventral margin of the abdomen [Fig. 72.69]. See also thorn

**Subtegulum** [suhb-TEG-yoo-luhm] — the basal sclerite of the bulb [Figs. 72.31, 72.37]

**Superior claws** — see paired claws [Fig. 72.3]

**Suprategulum** [SOOP-rah-TEG-yoo-luhm] — in male Linyphiidae, a sclerotized structure arising from the tegulum

**Synanthrope** [SIN-aan-THROP] — an organism associated with humans

**Synapomorphy** [sin-AAP-o-MOR-fee] — shared derived character

# T

**TA** — tibial apophysis

**Tailpiece** — in male Linyphiidae, a linear structure attached to the radix which generally points towards the proximal end of the unexpanded palp

**Tapetum** [taa-PEE-tuhm] — light reflecting crystalline structure behind the retinal cells of the secondary eyes

**Tarsal claw** — sharp curved structure at the tip of the tarsus, typically 1 on the palp and 2 or 3 on the legs [Figs. 72.3, 72.15, 72.75]. See also paired claws and unpaired claw

**Tarsal organ** — sense organs, usually pit-like, on the dorsal surface of spider tarsi

**Tarsus** (pl: tarsi) (= distitarsus) — the 7th or distal-most segment of the leg [Fig. 72.19], the distal segment of the palp [Fig. 72.75]

**Teeth** — conical, pointed, rigid outgrowths of the margins of the cheliceral furrow [Fig. 72.13]

**Tegulum** [TE-gyoo-luhm] — the middle sclerite of the male palp, that contains the sperm duct and bears the embolus [Figs. 72.29, 72.31, 72.37]

**Tegular apophysis** [TEG-yoo-lahr ah-PAH-fah-sis] (= tegular process) — sclerotized apophysis of the male palp of some spiders (= radix in Theridiidae) [Fig. 72.71]

**Tenent** [TEN-ent] — enlarged setae associated with tarsal claws

**Tergites** [TUR-geyts] — transverse sclerites on the abdominal dorsum in Mesothelae and some Mygalomorphae, often flanked by strong setae, and representing the original segmented condition [Fig. 72.70]

**Terminal apophysis** — in the male palp, a sclerotized or hardened structure in the apical part of the bulb [Figs. 72.29, 72.31]

**Thermoreception** — sensing temperature

**Thoracic groove** — see fovea [Fig. 72.1, 72.11]

**Thoracic region** — the posterior division of the carapace [Fig. 72.11]

**Thorn** — small thick pointed seta used with stridulatory file to produce sound [Fig. 72.72]

**Tibia** (pl: tibiae) — 5th segment of the leg and palp; located between patella and metatarsus of leg, patella and tarsus of palp [Figs. 72.19, 72.75]

**Tibial apophysis (TA)** — a process on the male palpal tibia, most commonly retrolateral, but may occur on other surfaces [Fig.

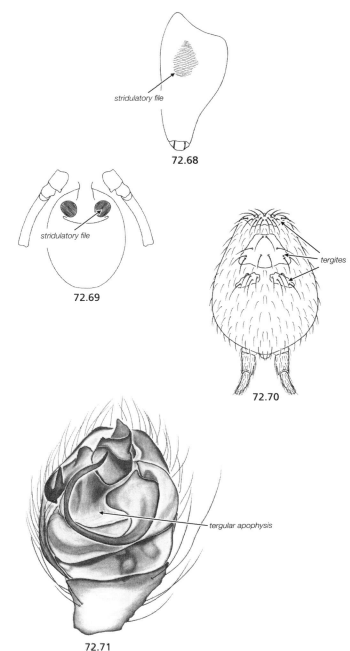

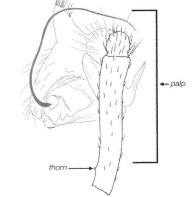

*72.64, 72.74]*. See also retrolateral tibial apophysis, ventral tibial apophysis

**Tracheal spiracle** — the opening to the tracheal respiratory system [Fig. 72.26]

**Triad** — a group of three contiguous eyes [Fig. 72.73]

**Trichobothrium** [trik-o-BAW-three-uhm] (pl: trichobothria) — a sensory hair (sensillum) arising from a round pit (bothrium); typically these are very slender and project vertically from the surface above the other hairs, rarely are they short and thick [Fig. 72.44]

**Trifid** [TREY-fid] — three-pronged

**Trionychous** [TREY-o-NEY-kuhs] — having three tarsal claws

**Trochanter** [TRO-kaan-ter] — second segment of the leg or palp between the coxa and femur, often a short ring-shaped segment [Figs. 72.19, 72.75]

**Troglobite** — an organism that only occurs in caves

**Troglophile** — a facultative cavernicole

**Truncate** [TRUHN-keyt] — squared off

**Tuberculate** [too-BUHR-kyoo-lat] — having tubercles or knob-like outgrowths on the body

**Type specimen** — the specimen that is used as the basis of a published species description

# U

**Unidentate** [yoo-ni-DEN-tat] — with a single tooth

**Univoltine** [YOO-ni-VOL-teen] — 1 generation/year

**Unpaired claw** (= inferior claw = median claw) — the third claw (not paired) located apically on the tarsus; absent from male palpi [Fig. 72.3]

**Urticating** [UHR-ti-ca-ting] — irritating or stinging; used to refer to barbed hairs of some mygalomorphs

# V

**Van der Waal's forces** — a force between non-polar molecules creating a temporary shift of orbital electrons between the molecules resulting in polarization and attraction

**Venter** — the underside or lower surface

**Ventral** (ventrad) — bottom (toward the bottom), ventral view [Fig. 72.2]

**Ventral tibial apophysis, (VTA)** — a sclerotized process on the ventral surface of the male palpal tibia, as in some Philodromidae [Fig. 72.74]

**Vestigial** [ves-TIJ-ee-uhl] — a structure that is a remnant

**Viscid** [VIS-kid] — sticky

**Vulva** [VUHL-vah] — a collective term for the internal components of the female genitalia associated with the epigynum including copulatory ducts, spermathecae, and fertilization ducts

# X

**Xeric** [ZEE-rik] — characterized by a low amount of moisture (in contrast to mesic and hydric)

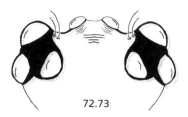

72.73

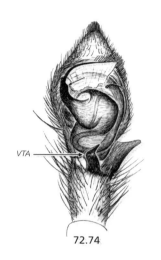

72.74

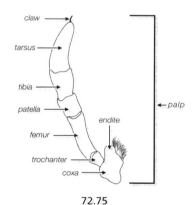

72.75

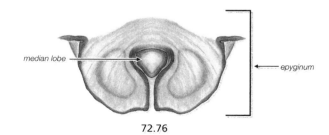

72.76

Chapter 73

# An etymological dictionary of North American spider genus names

by H.D. Cameron

Explaining the etymology of spider genus names takes more than looking up words in the Greek or Latin dictionaries. A professional knowledge of Greek and Latin is a *sine qua non*, to be sure, but it is only a beginning. It is essential to read carefully the original description of a genus, and often of the included species as well. The feature that inspired the name may be hidden, for example, in the description of the female genitalia of the type species (*Acantholycosa*).

Further, it is necessary to investigate all of the names originated by a given author to gain sufficient familiarity with the nomenclatural habits, preoccupations, quirks, and style of that author. It may be necessary to determine what reference books the author was using. For example it is essential to recognize that Eugene Simon was using an edition of the Natural History of Pliny the Elder, edited by J. Sillig (1851-55) (*Neozimiris*). We know that Crosby and Bishop had before them the Loeb Classical Library eight volume edition of the Geography of Strabo (*Soucron*). We must understand that George and Elizabeth Peckham read Kipling (*Messua*), and were interested in the history of Spain in the Napoleonic period (*Talavera*). Octavius Pickard-Cambridge knew Robert Southey's epic poem "Roderick, the Last of the Goths," and took a number of names from the dramatis personae of that poem (*Florinda*). He appears to have read deeply in the history of the Visigoths (*Teudis*).

Authors like Simon or Crosby and Bishop who were faced with the need to produce an abundance of names which they could be sure were not preoccupied, had recourse to surfing the lexicons for Greek or Latin words, or classical proper names, which they applied without any semantic connection with the spider. Among an abundance of examples see *Horcotes, Neon, Rhetenor, Sisis, Taranucnus, Teutana, Tigellinus, Phlattothrata*.

One must be aware of a host of incidental and out-of-the-way bits of information. One must take account of the way a French schoolboy would pronounce Greek (*Hyptiotes, Ozyptila*), or how a 19th century German or French schoolboy would spell his Latin (*Traematosis, Typhochraestus, Megamyrmaekion*). It helps to know the world political headlines in 1935 that influenced Crosby and Bishop (*Aduva*). We must take into account Baron Walckenaer's rough and ready grammar (*Attus, Sparassus, Thomisus, Drassus*). One must keep an eye out for names from the world's religions (*Anahita, Ambika, Aysha, Kukulcania*). Correct publication dates may be essential to deducing the correct etymology (*Pholcus*).

Occasionally some bold but disciplined guesswork is in order to deduce what was in the mind of an author (*Zygoballus, Wadotes, Ero, Larinia*).

R.V. Chamberlin may well have created more names than any other American arachnologist. His nomenclatural habits were unusually idiosyncratic, and present a considerable challenge to anyone trying to deduce their etymology. In order to protect against preoccupation Chamberlin used elements from a Native American language he knew, namely, Gosiute. Sometimes he even created hybrid names part Gosiute and part Greek or Latin (*Phrurotimpus, Tidarren, Astrosoga*).

Gosiute is a Uto-Aztecan language in the group called Numic, and is very closely related to Shoshoni and Paiute. The Gosiute Tribe was a small hunting and gathering tribe living southwest of the Great Salt Lake in the early 19th century. There are about 300 Gosiutes living in the Utah area today, and about 175 live on the Gosiute Reservation on the Utah-Nevada border.

Chamberlin became interested in Gosiute culture early in this century and published three papers on their language: "Animal and Anatomical Terms of the Gosiute Indians" in 1908b; "Ethnobotany of the Gosiute Indians" in 1911; and "Place Names and Personal Names of the Gosiute Indians" in 1913. Incidental items of Gosiute vocabulary are scattered through one or two of his earlier taxonomic papers, but he soon gave up the practice of explaining his obscure etymologies. He very kindly left that for me to do. See for example *Kibramoa, Pimoa, Piabuna*.

**Misdivision of morphemes and rhyming morphs.**

Latreille's 1804 genus name *Lycosa* is the genuine Greek feminine singular present participle of a verb λυκόω meaning 'tear like a wolf.' Grammatically the word is properly divided lyc-'wolf' and -osa, the feminine participle ending. C.L. Koch in 1846 created another name, *Trochosa*, from a genuine Greek feminine participle meaning 'running quickly.' In 1848 he created, on the analogy of *Lycosa*, the name *Pardosa*, which is not genuine Greek, but if it were it would mean something like 'acting like a leopard.' In the same year Koch created another analogical name on the rhyming principle, not real Greek, *Arctosa*, 'acting like a bear.' Simon got into the game in 1885 with *Alopecosa* 'acting like a fox.' Nathan Banks in 1900 now misdivided the words with *Allocosa* from Greek ἄλλος 'another' and a second element -cosa instead of -osa, meaning now by the rhyming principle 'another wolf spider,' where -cosa, which has no meaning itself, has become a conventional nomenclatural suffix meaning 'Lycosid.' R.V. Chamberlin in 1904 created *Schizocosa* from a Greek element meaning 'split' and the rhyming morph -cosa. The name, as is well known, was inspired by the excised transverse arms of the median septum of the epigynum. On the rhyming principal, then, there follows a cascade of names in -cosa: *Avicosa, Varacosa, Mustelicosa*; and in -osa: *Dingosa, Lynxosa, Crocodilosa*, etc.

Likewise, Sundevall (1833b) gave us *Dictyna* from a cult name of the Roman goddess Diana. As he explained himself, it is the epithet of the goddess of the hunting nets, from the Greek word for 'net.' It is properly divided as Dicty-na. Chamberlin on the rhyming principal misdivided the name as Dict-yna, and produced such names as *Emblyna, Phantyna, Tosyna*, where the conventional nomenclatural suffix -yna simply means 'Dictynid.' Similarly names with the rhyming morph -lena designate Agelenids (*Hololena, Calilena, Rualena, Tortolena*).

Although in some cases the old authors give us direct evidence of their intention in creating names, as, for instance, Menge and Sundevall, most of them do not. A certain detective ingenuity is required to second guess their intention based on all of the hints brought to bear. I am reasonably confident that most of the etymologies are soundly deduced and explained, but there are a handful, like *Epeira, Tekella* and *Ulesanis*, which rest on more adventurous guesswork.

When a quotation cites a particular reference (e.g., Bonnet) the blibliographic system used in this citation has not been adjusted to correspond with the bibliographic system used in this book.

**Note certain conventions:**

The asterisk * marks reconstructed or non-attested linguistic forms. The symbol (‡) marks invalid names. Single quotes set off translations of Greek or Latin words, and the meanings of the genus names. Double quotes set off verbatim quotations with translations in square brackets. Full names, works and editions of classical Greek and Latin authors can be found in the standard lexicons below, and are cited with the conventional reference numbers.

Liddell, H.G. and R. Scott, A Greek-English Lexicon revised and augmented by H.S. Jones and R. Mackenzie. Oxford 1968 [LSJ].

Lewis, C. and C. Short. A Latin Dictionary. Oxford 1879 [Lewis & Short].

Oxford Latin Dictionary ed. P.G.W. Glare. Oxford. 1982 [OLD].

*Acacesia* SIMON 1895a:795 (Araneidae)
From a mythological proper name. The Greek adjective ἀκακήσιος is an epithet of the god Hermes in Arcadia, used by the poet Callimachus (3rd century B.C.) and the guide-book writer Pausanias (2nd century A.D.). Simon gave it a feminine form. It is a common practice of Simon to scan the dictionary for unused words, which he applied without any semantic connection. The name means 'guileless, gracious'.

*Acanthepeira* MARX in HOWARD 1883:22 (Araneidae)
Formed on the genus name *Epeira* WALCKENAER, with the addition of a first member of the compound from the Greek feminine noun ἄκανθα 'thorn, spine' in reference to the cone-like protuberances on the abdomen of the type species *Acanthepeira stellata* (WALCKENAER). Marx created this name for *Plectana stellata* (WALCKENAER) in Howard's List of the Invertebrate Fauna of South Carolina, but abandoned it in his Catalog (Marx 1890:548), and returned the species to *Epeira*. Kaston (1938d:188; 1938f:259) revalidated the name *Acanthepeira*. See further discussion under *Marxia*.

*Acanthoctenus* KEYSERLING 1877a:693 (Ctenidae)
A combination of the Greek noun ἄκανθα 'thorn, prickle,' and the genus name *Ctenus* (which see). A reference to the nine pairs of spines under the first and second tibiae, which also inspired the name of the type species *spiniger*.

*Acantholycosa* DAHL 1908:197, 204, 367 (Lycosidae)
From the Greek feminine noun ἄκανθα 'thorn, spine' plus the genus name *Lycosa*. The type species *Acantholycosa norvegica sudetica* (L. KOCH) has a long thin thorn (Dorn) in the middle of the epigynum.

(‡) *Acartauchenius* SIMON 1884r:740 (Linyphiidae)
From the Greek two-ending adjective ἄκαρτος -ον 'unshaven' and the masculine noun αὐχήν (gen.) -ένος 'neck' plus an adjective forming suffix -ιος, yielding a masculine adjective meaning 'with unshaven neck'. The name refers to the fact that the frons of the male is elevated into a hairy cone whose posterior is covered with long spines [see Locket & Millidge (1953:256 fig. 155F) giving the impression of an unshaven neck].

*Achaearanea* STRAND 1929:11 (Theridiidae)
Replacement name for *Achaea* O. PICKARD-CAMBRIDGE 1882c:428, which was pre-occupied in the Lepidoptera, by combining Pickard-Cambridge's name with the Latin aranea 'spider'. Achaea is a part of Greece on the north coast of the Peloponnese on the Corinthian Gulf. Since Pickard-Cambridge's species came from tropical America, Japan and the Tonkin Gulf, there seems to be no semantic connection. He seems simply to have searched the dictionary for an available proper name. See under *Phycosoma*. Roewer (1942a) omits the name. Bonnet (1955:144) incorrectly rejects the substitution, arguing his principle (1945:106) that preoccupation in the Lepidoptera does not dictate a substitute. Brignoli re-established *Achaearanea* and Levi (1955a) placed the familiar *Theridion tepidariorum* in *Achaearanea*.

(‡) *Actinoxia* SIMON 1891g:318 (Cyrtaucheniidae)
Apparently from the Greek feminine noun ἀκτίς ἀκτῖνος 'ray, beam' but the second element is obscure, unless it is from the Greek adjective ὀξύς 'sharp' with a feminine suffix. There are no authentic Greek words of this pattern. And there is nothing in the description of the genus which suggests rays or beams. Roth (1994:25) synonymized this genus under *Promyrmekiaphila* SCHENKEL, but that synonymy was not recognized, and Bond & Opell (2002:518) consider it a junior synonym of *Aptostichus*.

*Aculepeira* CHAMBERLIN & IVIE 1942a:75 (Araneidae)
The type species is *Epeira aculeata* EMERTON (the species name means 'armed with spines') in which the distal end of tibia II in the male bears coarse spines. The genus name is confected, by the usual custom of Chamberlin and Ivie, from the first two syllables of the type species name *aculeata* and the genus name *Epeira* (which see).

*Admestina* PECKHAM & PECKHAM 1888:78 (Salticidae)
Perhaps derived with a suffix from the Greek one-ending adjective ἀδμής (gen.) -ῆτος 'untamed, unwedded'. This is not well-formed, and does not conform to strict Greek word-formation. There is no obvious semantic connection.

(‡) *Aduva* BISHOP & CROSBY 1936b:39 (Linyphiidae)
A political comment by the authors. Aduwa, a town in Tigre province, northern Ethiopia, was the site of a disastrous defeat for the Italians on March 1, 1898, by the Emperor of Abyssinia, Menelek II, thwarting for a generation Italy's colonial ambitions. In 1935 those ambitions were revived when Mussolini invaded Ethiopia, and Italian forces under Marshal DeBono took Aduwa on October 6. This was celebrated as a great victory in Italy, and Mussolini boasted that he had wiped out the "shame of Aduwa". The Emperor, Haile Selassie, aroused great sympathy for the plight of Ethiopia, but was forced to leave his country on a British battleship on May 2, 1936, the day after the publication of Crosby and Bishop's paper. Until the Munich settlement between Chamberlin and Hitler two years later, Aduwa served as the great symbol of appeasement. This genus name appears to represent an effort by Crosby and Bishop to keep the shame of Aduwa alive. It has been synonymized under *Islandiana* BRÆNDEGAARD by Ivie (1965:1).

*Agassa* SIMON 1901a:643 (Salticidae)
No discoverable etymology.

(‡) *Agathostichus* SIMON 1895a:885 (Araneidae)
Corrected spelling in Simon's index of *Agatostichus* SIMON. See under *Agatostichus*.

(‡) *Agatostichus* SIMON 1895a:885 (Araneidae)
The name is so spelled in the body of Simon's first volume, but the index of that volume corrects the name to Agathostichus (Simon 1895a:1070). This variation stems from the problem of the French schoolboy's pronunciation of Greek. See the discussion under *Rhecostica*. From the Greek adjective ἀγαθός -ή -όν 'good' and στίχος 'row, rank, line'. In Simon's personal nomenclatural convention -stichus comes to mean 'armed'. Hence Simon intends this to mean 'well-armed' with reference to the row of long acute, erect, tubercles on the cephalothorax which are 'better' than those of *Mastophora* (*Glyptocranium*) from which Simon wishes to distinguish this genus. Roewer (1942a:900) adopted the corrected spelling of Simon's index. Gertsch (1955:250), in accordance with the strict requirements of the Code, retained Simon's original spelling (Code Art.32c). But it is arguable that the two spellings in the original publication indicate that there is evidence of a *lapsus calami* and fulfill the criteria for emendation under 32c.ii. Brignoli (1983c:255) restored the correct spelling *Agathostichus*.

*Agelena* WALCKENAER 1805:51 (Agelenidae)
Agelena itself is not a Greek or Latin word, but appears to be derived from the Greek word ἀγέλη 'herd, company, shoal' by the addition of an adjectival suffix -ena. Walckenaer gave no indication of the etymology, and subsequent authors have tried to puzzle it out.
Simon in his youth, and when his command of Greek and Latin was particularly insecure (1864:211), was the first to attempt an etymology: "ἀ, privitif; γαλήνη, calme". That is, from the Greek negative prefix (alpha privative) plus the Greek noun γαλήνη 'calm' meaning 'without calm, active'. Simon was only sixteen at the time, and his etymologies are amateurish.
Thorell (1870b:132) also derived the name from an alpha-privative and the Greek feminine noun γαλήνη 'calm, tranquillity' arguing that it referred to "the animal's rapid and restless motions". To suit this derivation Thorell modified Walckenaer's original spelling to Agalena. He argues:

"To derive this name as some have done from ἀγέλη, herd, has no other foundation than the accidental similitude of the letters in the two words."

Thorell himself has committed the sin he excoriates, since it is only the accidental similarity to the Greek word for 'calm' that supports his etymology.
Bonnet (1955:178) says:

"Thorell (1870b p.132) thought he was correctly modifiying the word **Agelena** of Walckenaer to **Agalena** by establishing the etymology: a-[alpha privative] and γαλήνη ['calm, repose'] (without repose), referring to the rapid movements of that spider, when it is disturbed. But since Walckenaer did not himself give an etymology of the name, it is hardly probable that it was the character of extreme mobility which he wished to indicate in the name, especially since he gave as a behavioral characteristic of this genus that these spiders "held themselves immobile" in their tubes (Walck. 1805:51). Nothing prevents us from assuming that he took as the sense of his name **Agelena** the word ἀγέλη, meaning "flock, herd, drove" corresponding to the fact that these spiders often live gregariously in the same locality. Hence it is not necessary to pay any attention to Thorell's modification of the name of this genus. Menge had already in 1850 used the name **Agalena** for citing A. labyrinthica, but it is likely that this is merely a printer's error [une coquille d'imprimerie]."

See further Bonnet (1953a). Thorell was followed by Menge, Karsch, L. Koch, Lendl, Thorell, Pavesi, and Emerton.
Note that Walckenaer in the same paper (1805:59) listed another species *Epeïra agalena* WALCKENAER 1802:198 (*Aranea agalena*), and glossed it in French with agalène. These two words, agelena and agalena probably have nothing to do with one another. There is no evidence of misprint or mistake or confusion with Agélène Agelena (which is consistently so spelled three times on the same page, and passim e.g., 83, and in the index to the figures). The two words are carefully, and consistently accented differently in French.
All of these etymological efforts can be safely dismissed as well-meant but erroneous speculation. There is a better explanation.
A standard epithet for the goddess Athena in the Homeric poems is ἀγελείη traditionally interpreted to mean 'driver of spoil, forager' (e.g., Iliad 15.213). There are two possibilities. Walckenaer may have simply

copied the word out of the dictionary incorrectly, with nu instead of iota thus producing *Agelena* rather than *Ageleia. Or, since Athena is connected with weaving and spiders from the story of Arachne, Walckenaer may have created his name by conflating Athena's epithet ἀγελείη with the name Athena itself to produce the hybrid agel -ena. It is curious that no arachnologist employed the name Athena itself for a genus of spiders; it would seem to be a natural. It would have been available to Walckenaer in 1805, since *Athena* was first used in 1819 by Hübner in the Lepidoptera.

### *Agelenopsis* Giebel 1869a:250 (Agelenidae)

Derived from the genus name *Agelena* Walckenaer by the addition of the suffix formed from the Greek feminine noun ὄψις (gen.) -εως 'aspect, appearance'. Giebel originally printed the word as *Agenelopsis* an obvious printer's error, since he meant to derive the name from *Agelena*. Thorell (1877c:489) saw that it was a printer's error, and emended it to *Agalenopsis*, in accordance with his incorrect theory about the etymology of *Agelena*. See under *Agelena*. Simon (1898a:258) was the first to emend it to *Agelenopsis*. Strand incorrectly retained the typographical error of the original spelling (Strand 1929:11). The name is feminine in gender, and means 'looking like *Agelena*, having *Agelena*-aspect'. Chamberlin & Ivie (1935b:31) proposed that all American species of *Agelena* should be gathered into a distinct genus, and they resuscitated Giebel's old name. Giebel originally separated his new American species *albipilis* [now *Agelenopsis potteri* (Blackwall 1846a)] from *Agelena* on the basis of the eye arrangement, a criterion which has not held up. Now the two genera are distinguished on the basis of differences in the genitalia. See Kaston (1948: 286) for the details.

### (‡) *Agnyphantes* Hull 1932:106 (Linyphiidae)

The second element is the rhyming morph after such Linyphiid genera as *Lepthyphantes* and *Pityohyphantes*, from the Greek noun ὑφάντης 'weaver'. The first element is unexplained. It may be from the Greek two-ending adjective ἄγνως 'unknown'. Hence, 'the unknown Linyphiid'.

### *Agroeca* Westring 1861:311 (Liocranidae)

Simon's spelling (1877f:189) *Agraeca* is an error resulting from the French pronunciation of Greek. Bonnet says (1955:206):

> "Etymologically (ἀγροῖχος [sic; correctly ἄγροικος]), the name of this genus ought to be written **Agrœca**; it was an error on Simon's part to write **Agræca**, especially since Westring had spelled it correctly originally. Normally even, one should write **Agroeca**, separating the o and the e so as to pronounce the two letters separately, though it is not necessary to place a dieresis over the o as some authors have done. But following long custom of writing **Agrœca**, I retain that orthography."

The name is the latinized feminine of the Greek two-ending compound adjective ἄγροικος-ον 'dwelling in the field, farmer, bumpkin' a compound of ἄγρος 'field' and οἰκέω 'to dwell'. Bonnet is wrong about the dieresis, since in Greek (and in its latinization) the word has three syllables not four, as the Greek accentuation makes clear. The Greek diphthong οι is regularly latinized to oe.

### *Agyneta* Hull 1911a:583 (Linyphiidae)

Hull erected this genus for a group of *Micronemeta* Menge which he separated on the basis of their shorter legs. The second half of the name is taken from the second half of *Micronemeta* (which see), and the first part, Agy-is apparently taken from the first half of an obscure Greek adjective (used only once, in a veterinary treatise) ἄγυιος 'with weak limbs'. Hence, the name means 'close to *Micronemeta* with weak limbs'.

### (‡) *Aigola* Chamberlin 1921c:36 (Linyphiidae)

This is one of Chamberlin's hybrid names, part Gosiute Indian and part Latin. See the introduction of the chapter (p. 272) for Chamberlin's Gosiute studies, and see under *Astrosoga* and *Graphomoa* for Chamberlin's habit of creating hybrid names. The Gosiute Indian word aigo means 'tongue' (Chamberlin 1908b:101), or 'process or thorn' (Chamberlin 1911: 72). The last syllable -la is apparently the feminine of the Latin diminutive suffix -lus -la -lum. The type species is *Aigola crassimana* (Emerton 1882) which has travelled under several names, one of which was *Gongylidium tuberosum* Emerton 1915b. Chamberlin's type species was *Aigola tuberella* Chamberlin 1921c:37 [now *Oreonetides vaginatus* (Thorell 1872b)], by which he meant to pay homage to Emerton's name. Perhaps the genus name *Aigola* was influenced by this decision and Chamberlin formed it with the same ending -la. The Latin adjective tuberosus means 'provided with bumps or humps' and Chamberlin may have used the Gosiute word aigo 'tongue or process' to translate tuberosus. Chamberlin's description of the new genus concentrates on the lobes, spurs, and apophyses of the elements of the male palp, and the name may be intended to refer to these. If my analysis is correct, the name would mean 'little process'. Synonymized under *Oreonetides* Strand (Linyphiinae) by Holm (1945b:45).

### *Aliatypus* Smith 1908:211, 214, 231 (Antrodiaetidae)

Beginning from *Atypus* Latreille, O. Pickard-Cambridge created *Atypoides* (which see), meaning 'like *Atypus*', and Smith created *Aliatypus* meaning 'another *Atypus*' from Latin alius-a-um 'other, another' and Latreille's genus name *Atypus*. Smith can be excused from the charge of creating a hybrid name, half Latin and half Greek, since he treated *Atypus* as as Latin genus name, not a Greek word.

### (‡) *Allepeira* Banks in Banks et al. 1932:23 (Araneidae)

From the Greek adjective ἄλλος -η -ον 'another' and the genus name *Epeira*. A replacement name for *Hentzia* McCook, which Banks argued was preoccupied by *Hentzia* Marx in Howard 1883. Bonnet (1955:226), who retains *Hentzia*, argued:

> "Banks created the new name of the genus **Allepeira** for **Hentzia** *McCook*, 1894, because there existed **Hentzia** *Marx*, 1883, used by that author for **Hentzia** *palmarum*, which now proves to be the type of **Wala** *Keyserling*, 1884.-I find it useless to follow this modification, first because, if **Hentzia** was indeed used by Marx, he gave no diagnosis; moreover, by 1932 it had been 50 years since that **Hentzia** of Marx was used."

McCook described *Epeira basilica* in 1878 (1878a:124) but decided (1894:244) that it belonged in its own genus, for which he created *Hentzia* in ignorance of Marx's earlier use. The Basilica Spider travelled under Banks's name *Allepeira* until Levi (1968a:322) synonymized it under *Mecynogea* Simon.

### *Allocosa* Banks 1904d:113 (Lycosidae)

Erected for *Lycosa funerea* Hentz 1844. Banks created the name four years earlier (Banks 1900c:539) but it was a nomen nudum (cf. Dondale & Redner 1983b:933). Chamberlin (1908a:169, 284, 285) maintained the name, but Gertsch (1934a:3) subsumed *Allocosa* under *Arctosa* C.L. Koch, and was followed by Bonnet. But Kaston in the various editions of *How to Know the Spiders* distinguished between the two genera and was followed by Dondale and Redner. *Allocosa* is a rhyming name formed on the name *Lycosa* (which see) by misdivision of morphemes. The first part is from the Greek three-ending adjective ἄλλος -η -ον 'other'. Hence 'another *Lycosa*'.

### *Allocyclosa* Levi 1999:304 (Araneidae)

The Greek adjective ἄλλος 'another' to erect a genus close to *Cyclosa*. Cf. *Allocosa*.

### *Allomengea* Strand 1912b:346 (Linyphiidae)

From the Greek three-ending adjective ἄλλος -η -ον 'other' and the genus name *Mengea*, a feminine adjective formed on the name of the arachnologist Anton Menge (1808-1880), author of Die Preussische Spinnen published between 1866 and 1879. It is Strand's replacement name for three other preoccupied names. Bonnet neatly sorts this out (1955: 228 note 57):

> "**Pedina** Menge *1866:125*, being preoccupied by an Echinoderm, was replaced by F.O. Pickard-Cambridge in *1903d:35* with **Mengea**. Noticing the same thing, but ignorant of the change, Dahl in *1909c:6* created for **Pedina** Menge the new name **Pedinella**. Strand in 1912 realized that **Mengea** F.O. Pickard-Cambridge was preoccupied by a fossil Strepsipteran and created the genus **Allomengea**. Fortunately, **Pedinella** Dahl was itself preoccupied by a Protozoan, or else Strand's name **Allomengea** would have been invalid."

### *Alopecosa* Simon 1885g:10 (Lycosidae)

Formed by the convention of creating rhyming Lycosid names with the segment -osa by misdivision of morphemes and rhyming on the basis of *Lycosa* Latreille (which see). The first part is from the Greek feminine noun ἀλώπηξ -εκος 'fox'. For the tradition of naming Lycosids for predatory animals see under *Pardosa*. The second part is formed to rhyme with *Pardosa* and means it is a wolf spider; and the first part, Alopec- means it is named for a fox.

### *Alpaida* O. Pickard-Cambridge 1889d:52 (Araneidae)

This is a feminine medieval Frankish name (see the introduction of the chapter p. 272 for Pickard-Cambridge's Gothic names). It is one spelling variant of the name of a mistress of Pepin (II) of Heristal (died 714 A.D.), a Mayor of the Palace under the Merovingians. He was the father of Charles Martel, who was the hero of the Battle of Tours (732 A.D.), which halted Moslem expansion into Europe. Charles Martel was an illegitimate son of Pepin by Alpaida. Her name is also spelled Chalpaïda (Förstemann 1900:67). I have not succeeded in finding the immediate source from which Pickard-Cambridge was likely to have taken the name, but he was particularly interested in the history of the Goths and the Franks.

### *Amaurobius* C.L. Koch 1837g:15 (Amaurobiidae)

From Greek ἀμαυρός 'dark' and a combining form -βιος 'living' based upon the Greek noun βίος 'life' hence, 'living in the dark'. This is not a genuine Greek compound, but it is formed on the analogy of such authentic Greek compounds as ἀμφίβιος. For the relationship of this name with *Ciniflo* Blackwall see Bonnet (1955:272). Bristowe (1939:6), applying the law of priority in a Procrustean fashion, pointed out that in C.L. Koch's fascicule 141 he named the genus *Amaurobius* for the first time with two new species *roscidus* and *tigrinus*, which in fact belong in *Coelotes*. Hence Bristowe argued *Amaurobius* C.L. KOCH ought to replace *Coelotes* BLACKWALL 1841a as its senior synonym, and the name *Ciniflo* BLACKWALL 1841a ought to be resuscitated to replace *Amaurobius*. Bonnet argued that the date of the fascicule in question is wrong, and the new genus is properly established in the Uebersicht des Arachnidensystems 1837g, where it includes true *Amaurobius* along with some true *Coelotes* which have now been transferred out of the genus *Amaurobius*. See under *Ciniflo*.

(‡) *Ambika* LEHTINEN 1967:212 (Oecobiidae)
Synonymized under *Oecobius* LUCAS by Shear (1970:135). This is a Sanskrit word meaning 'mother, good woman'. As a proper name it has numerous applications: mother of Dhritarashtra in the Mahabharata; one of the female domestic deities of the Jain religion; a place in Bengal; the name of two rivers. Given Lehtinen's practice of using mythological names from the mythologies and religions of the world, he probably had the Jain deity in mind.

*Amphinecta* SIMON 1898a:228-30; 233-235 (Amphinectidae)
From the Greek prefix ἀμφι- 'around' and the Greek masculine noun νήκτης 'swimmer'. Hence, 'one who swims around'. This is properly latinized as a masculine a-stem noun, but Simon regarded it as feminine. He included it in his group Argyroneteae based upon the aquatic spider *Argyroneta aquatica* (CLERCK 1757). He says (p. 233):

"the behavior of **Amphinecta** of New Zealand must be analogous to that of **Argyroneta**, in view of the great organic resemblance of the two species, but it has not been observed."

While this explains Simon's name, sad to say, he was wrong about the habits of the spider, and its relationship to *Argyroneta*. *Amphinecta* is not aquatic. Forster says:

"These spiders live on the forest floor and are often collected in and under rotten logs."

It is the type genus of the family Amphinectidae erected by Forster & Wilton (1973:157). The North American genus *Metaltella* has been transferred here from the Amaurobiidae by Davies (1998a:242).

*Anachemmis* CHAMBERLIN 1919b:12 (Tengellidae)
Created from *Chemmis* SIMON (1898a:215) by the addition of the prefix ana-which is merely a differentiating prefix like para-, meta-, or epi-. It comes from the Greek prefix ἀνα- 'up, again'. Chemmis is a proper place name which Simon gleaned from the dictionary in accordance with his usual habit of searching for unused words that are available for names. Chemmis is an Egyptian town on the right bank of the Nile mentioned by Herodotus (2.91), or a floating island in the western delta of Egypt on which Isis bore the Egyptian Apollo (Horus) and Artemis (Bubastis) according to Herodotus (2.156). As a place name, Chemmis is properly feminine in Greek. Simon left the matter ambiguous by using a genitive patronymic species name (*frederici*), and Banks (1901a:583) used a two-ending adjective, *unicolor* which is also of ambiguous gender, but later (1914a:677) used a feminine species name *punctigera*. Chamberlin in 1919b used ambiguous species names for his species of *Anachemmis*, namely *sober* and *dolichopus* (which can be feminine), but later (1924b:667) was the first to make *Chemmis* unambiguously masculine with his species *monisticus*. But *Chemmis* is feminine in Greek and should be feminine as a genus name, although Chamberlin and Bonnet treated it as masculine. See *Socalchemmis*.

*Anacornia* CHAMBERLIN & IVIE 1933a:29 (Linyphiidae)
This genus agrees in general structure with *Cornicularia* MENGE, that is, the male cephalothorax has a short blunt process projecting over the clypeus. The name *Cornicularia* (which see) refers to a Roman helmet with a projecting horn, and the name *Anacornia* is meant to refer again to such a structure. The first part is from the Greek prefix ἀνα- meaning 'up, above, on top' and the second part is from the Latin neuter noun cornu 'horn' plus a feminine adjective-forming suffix -ia. Hence, 'having a horn on top'. It is a hybrid word, part Latin and part Greek.

*Anahita* KARSCH 1879g:103 (Ctenidae)
Anahita is an ancient Iranian goddess in the Avesta, the sacred book of Zoroastrianism. She was the female member of a divine triad with the male gods Ahura-Mazda and Mithra, and was the goddess of waters, fertility and procreation. Artaxerxes II, who reigned from 404 B.C. to either 363 or 357, fostered her cult throughout the Persian Empire, especially in Asia Minor. Karsch described the type species from Japan, and there is obviously no connection. He was fond of using proper names from ancient sources and, for example, used another Persian name, Mandane, mother of Cyrus of Persia, for a Clubionid genus.

*Anapis* SIMON 1895a:927 (Anapidae)
Simon's name *Anapis* (1895a:923) is a substitute for *Amazula* KEYSERLING, which was preoccupied in the Coleoptera. Anapis is the Latin proper name of a Sicilian hero, who, with his brother Amphinomus, saved their parents from an eruption of Mt. Aetna (Seneca de Beneficiis 3.37; Valerius Maximus 5.4). Anapis is also the name of a river flowing into the bay of Syracuse (Ovid Metamorphoses 5.417). There is no semantic connection. Simon was in the habit of searching the dictionary for usable names, regardless of meaning.

*Anapistula* GERTSCH 1941a:2 (Symphytognathidae)
Gertsch said he was using the family name Symphytognathidae in the broader

"sense used by Dr. Louis Fage to include, in addition to the typical genus **Symphytognatha** of Hickman, a small number of genera and species related to **Anapis** Simon."

Fage had combined the Anapidae and *Symphytognatha*. Forster & Platnick (1977) have separated them. Gertsch's name is clearly a feminine diminutive formed on the name *Anapis* SIMON, perhaps with a rhyming reminiscence of such names as *Tarentula*, *Sintula*, and *Tortula*.

*Anasaitis* BRYANT 1950:160 (Salticidae)
Derived from the genus name *Saitis* SIMON (1876a:168) by the addition of the Greek preverb ἀνα-. The author said:

"The genus **Anasaitis** is separated from the genus **Saitis** by the iridescent scales on the clypeus and mandibles of the male."

She also created *Parasaitis* in the same paper. Simon's name was gleaned, as was his habit, from the Latin dictionary, without regard to any semantic point. It means 'pertaining to the Saitic nome' in ancient Egypt.

*Anelosimus* SIMON 1891d:11 (Theridiidae)
These species were usually subsumed under *Theridion*. Simon himself in 1894 (1894a:550) synonymized it with *Theridion*, but F.O. Pickard-Cambridge (1902a:373, 394) maintained it as a separate genus, and it was revalidated by Levi (1956b:412). The name, so spelled, has no discoverable etymology, although it has the look of a Latin masculine superlative adjective. In the original article the name is spelled three times as *Anelosimus*, with an air of deliberate intention. However, it is important to note that Simon (1894a:548, 550 and in the index) later spelled it *Adelosimus*, again three times with equally deliberate intention. This later spelling should be regarded as Simon's deliberate correction of the name, not as a *lapsus calami*, as Bonnet would have it (1955:158, 1959:4437). *Adelosimus* at least has a possible etymology, namely from the Greek three-ending adjective ἄδηλος 'uncertain, not evident, invisible'. The Greek superlative of this adjective is ἀδηλότατος, but it seems that Simon irrationally put a Latin superlative ending on the simple adjective yielding adelosimus, analogous to altissimus 'highest'. I suggest Simon's intended meaning was 'very uncertain' and reflects the fact that he was uncertain whether this should be a separate genus or should be included in *Theridion*. As a legal matter, the meaningless original spelling has been maintained, despite Simon's attempt to emend his own name.

*Anibontes* CHAMBERLIN 1924a:6 (Linyphiidae)
The end of the name is formed with a rhyming morph on analogy with other Linyphiid names like *Bathyphantes*, *Pityohyphantes*, *Linyphantes*. The first part of the name is from the Gosiute Indian word ani 'ant' (Chamberlin 1908b:78). See the introduction of the chapter p. 272 for Chamberlin's Gosiute studies. Chamberlin says (1924a:6):

"This genus differs from **Microneta** particularly in its caudally narrowed cephalothorax, which gives the individual a typically ant-like appearance."

The new species name is *mimus* 'imitator' emphasizing again the ant-mimicry. The precise origin of the syllable -bo-is a mystery. In Gosiute it usually means 'position on the surface' (Chamberlin 1911:46) as in the compound toyan-bo 'on the mountain' (Chamberlin 1911:51). There is a Gosiute word anibo 'fly' (Chamberlin 1908b:86) but in view of the explicit comparison with an ant, this would not seem to apply.

(‡) *Anitsia* CHAMBERLIN 1921c:38 (Linyphiidae)
This is one of Chamberlin's Gosiute Indian names (see the introduction of the chapter p. 272). In Gosiute the suffix tsi is a common ending in the names of living things, especially in animal names (Chamberlin 1911:42). Compare the simple unu 'badger' (Chamberlin 1908b:79) and the compound yunutsi 'badger' (Chamberlin 1911:43). The first part ani means 'ant' (Chamberlin 1908b:78). So the name is a Gosiute compound meaning 'ant' with a Latin feminine ending. Synonymized under *Collinsia* O. PICKARD-CAMBRIDGE by Holm (1944:131), then *Collinsia* itself has been synonymized under *Halorates* HULL by Millidge (1977:11).

*Annapolis* MILLIDGE 1984a:147 (Linyphiidae)
Erected for *Sciastes mossi* MUMA 1945. The genus is named for the capital of Maryland where the type species was collected. The capital in turn was named in 1694 for Princess Anne of England, who chartered the town in 1708, after she became Queen.

*Antistea* SIMON 1898a:271-275 (Hahniidae)
Apparently from the Latin masculine noun antistes (gen.) antistitis 'one who stands in front of a temple, overseer, high priest'. The correct

feminine counterpart is antistita (gen.) antistitae, but Simon created a feminine in his own way. Simon gives no rationale for the name.

*Antrobia* TELLKAMPF 1844:321 (Linyphiidae)
Feminine Greek compound adjective formed from the neuter noun ἄντρον 'cave', and the combining form -βιος 'living' by analogy with such genuine Greek words as λιμνόβιος 'living in a lake'. Greek compound adjectives have no distinctive feminine form, but when they are latinized they may take a feminine form. Tellkampf orignaly spelled the name *Anthrobia* because of a German custom in his day of spelling some Latin words with th instead of simple t. Thorell (1869:41) correctly emended it to *Antrobia*. This spider, from Mammoth Cave, Kentucky, was originally named *Anthrobia monmouthia*, which was an obvious slip for *Anthrobia mammouthia* (Bonnet 1955:335, note 128). The interplay between th and t can also be seen in the variant German spelling in the title of Tellkampf's article, where he uses the term Gliederthiere 'arthropods' rather than the modern spelling Gliedertiere. Since there is internal evidence that *monmouthia* is an inadvertent error, *Anthrobia* should be so regarded in accordance with the Code Art. 32.5.1.

*Antrodiaetus* AUSSERER 1871a (Antrodiaetidae)
Notice this is misspelled *Anthrodiaetus* in Simon (1892a:81, 1895a:1070). Ausserer actually wrote *Antrodiaetus*. Waterhouse (1902:404) spelled it *Anthrodietus*, incorporating Simon's error, and using the reasonable spelling e for æ, permissible by some conventions, but the rules of zoological nomenclature privilege the original spelling, which happens also to be formally correct. A Greek compound adjective formed from Greek neuter noun ἄντρον 'cave' and feminine noun δίαιτα 'way of life' (cf. English diet). There are at least 20 regularly formed compound adjectives in Classical Greek with the second element -διαιτος, and indeed ἀντροδίαιτος itself occurs (Orph. Hymn. 32.3). Hence, 'dwelling in caves' as Ausserer made explicitly clear by glossing his new name "in höhlen wohnend".

*Anyphaena* SUNDEVALL 1833a:20 (Anyphaenidae)
From the Greek verb ἀνυφαίνω 'weave anew' Sundevall glossed the name: Ανυφαινω, retexo (The Greek in Sundevall's Conspectus is printed without accents or breathings), 'reweave'. Greek ὑφαίνω is the first person singular present form of the Greek verb meaning 'to weave'. and the preverb ἀνα-means 'again'. Latin ae is the conventional rendering of Greek ai and Latin y is the rendering for Greek u. There is no such authentic Greek word as *ἀνυφαίνη and Sundevall simply nominalized the first person present active of the verb by imposing a feminine suffix. In creating the genus he printed *Anyphœna*, which should be *Anyphæna*. Bonnet accurately explains the incorrect spelling (1955:336) as an artifact of the printing fonts where the italic æ had the form œ. Menge's incorrect notion (1873:333) that the name comes from the alpha privative 'not' and the word ὑφαίνω 'weave' ignores Sundevall's explicit explanation.

(‡) *Anyphaenella* BRYANT 1931:115 (Anyphaenidae)
A feminine Latin diminuative based on the name *Anyphaena* (which see). Bryant erected this genus for a group of *Anyphaena* with type species *Anyphaena saltabunda* HENTZ 1847. Synonymized under *Wulfila* O. PICKARD-CAMBRIDGE by Platnick (1974:242).

*Aphileta* HULL 1920:8 (Linyphiidae)
From the two-ending Greek compound adjective ἀφίλητος-ον 'unloved' formed from the negative prefix aj-the verb φιλέω 'love' and the verbal adjective forming suffix -τος. Greek compound adjectives have no distinctive feminine, but when latinized they may acquire a feminine form.

*Aphonopelma* POCOCK 1901c:553 (Theraphosidae)
The first element is from the Greek two-ending adjective ἄφωνος -ον 'silent, mute, dumb' and the second element from the Greek neuter noun πέλμα (gen.) -ατος 'sole of the foot'. Hence literally translated it would mean 'with silent sole of the foot'. But Pocock defines this new genus as lacking any plumose scopula on the posterior side of the trochanter of the palp. Another genus defined in this same paper, *Citharacanthus* (meaning 'having a harp of spines'), is characterized as having a "system of stout plumose stridulating-bristles upon the trochanter of the palp and first leg" as with genera *Acanthoscurria*, *Lasiodora*, *Phormictopus* (meaning 'harp-foot') etc. There is a rhyming convention of Theraphosid names with second element -pelma after *Eurypelma* C.L. KOCH. But the defining characteristic of *Aphonopelma* is the absence of such a stridulating organ. It seems clear that Pocock meant his name to signify 'the (silent) Theraphosid without a stridulating organ on the posterior side of the trochanter of the palp'. That is, we do not take the second element -pelma literally as meaning 'sole' but conventionally as meaning 'Theraphosid genus after *Eurypelma*'. Raven (1985a:152) placed *Aphonopelma* in synonymy with *Rhecostica* SIMON, but ICZN (1991) Opinion 1637 revalidates *Aphonopelma*.

*Apollophanes* O. PICKARD-CAMBRIDGE 1898a:252 (Philodromidae)
An invented masculine Greek proper name. A combination of the divine name Apollo and the suffix found in such proper names as Aristophanes. The simple name *Apollo* was preoccupied in the Mollusca (Schinz 1822).

*Apomastus* BOND & OPELL 2002:523 (Cyrtaucheniidae)
The authors say:
"Latinized form of the Greek apomastos meaning 'not provided with a lid'. This refers to the absence of a trap door on burrows constructed by this genus."
The Greek two-ending adjective ἀπώμαστος is a compound of the alpha privative prefix ἀ-'not' and the neuter noun πῶμα 'cover, lid' plus an adjective forming suffix -αστος.

*Appaleptoneta* PLATNICK 1986d:15 (Leptonetidae)
From the first two syllables of Appalachian plus the genus name *Leptoneta* (which see). Formed for a group of species from the area of the Appalachian Mountains.

*Apostenus* WESTRING 1851:46; 1861:322 (Liocranidae)
Westring says (1861:322) that the thorax is "antice angustatus" [narrowed anterad] and his name appears to be a Greek translation of angustatus, from the Greek verb ἀποστενόω 'contract, narrow' reformed into a masculine Latin adjective.

*Aptostichus* SIMON 1891g:317 (Cyrtaucheniidae)
*Aptostichus* has many rows of spines beneath the anterior metatarsi. The Apto-part seems to be from Greek ἁπτός, a verbal adjective based upon the verb ἅπτω 'touch' which would give haptos meaning 'touchable, palpable'. But Simon ignored the Greek aspirate, because as a French schoolboy he would never have pronounced it (see the discussion under *Rhecostica*). The second element -stichus is from Greek στίχος meaning 'row, line, rank' as in a rank of soldiers or a row of trees and is customarily used for a row of spines. Hence 'armed with a palpable row of spines'. Transferred to the Cyrtaucheniidae from the Ctenizidae by Raven (1985a:137).

*Arachosia* O. PICKARD-CAMBRIDGE 1882c:425 (Anyphaenidae)
The author says succinctly that this is a proper name. It is the name of a district, a town and a river in northeastern ancient Iran, mentioned in Pliny's Natural History (6.92). The city is now Kandahar in Afghanistan.There is no semantic connection.

*Araneus* CLERCK 1757:22 (Araneidae)
Latin has two words meaning 'spider': the masculine araneus and the feminine aranea. Originally araneus meant 'spider' and aranea meant 'spider web' but the first century B.C. poet Catullus (68.49) already used aranea to mean 'spider'. Before Linnaeus, authors simply used the words as an ordinary common designation for spiders without taxonomic formality, e.g., Aldrovandus and Ray (aranea), Lister, Moufet, Petiver (araneus), and Aldrovandus, for one, used both. The grandfather of all nomenclature codes, the Strickland Report of 1842 (Anonymous 1842) recommended (p. 109) that the starting point for binomial nomenclature be the 12th (1767) edition of the Systema Naturae. Subsequently this was moved back to the 10th edition of 1758, where all spiders were in the genus *Aranea*. But the Swede Carl Alexander Clerck published Aranei Suecici, dated 1757, in which he had already used the binomial system, and his one genus was *Araneus*. Prof. Pierre Bonnet circulated a petition among his contemporary arachnologists urging that the Clerck names be given priority over the Linnean names, and this was formally declared at the 1948 Paris meeting of the Commission on Zoological Nomenclature (ICZN 1948). It was finally incorporated into the Code as Article 3.1b. However, Platnick (2005) uses 1757 as the date of publication and we have chosen to follow it to avoid confusion among arachnologists. See comment in the beginning of the reference section p. 331.

*Araniella* CHAMBERLIN & IVIE 1942a:76 (Araneidae)
Feminine diminutive of *Aranea* erected for *Epeira displicata* HENTZ 1847. and those species of *Araneus* near *cucurbitinus* CLERCK. The type species is *Epeira displicata* HENTZ.

*Archerius* LEVI 1957d:114 (Anapidae)
Named in honor of Allan F. Archer, curator of Arachnida of the Alabama Museum of Natural History, University of Alabama, and a worker on orbweaving spiders. The genus, erected for *Archerius mendocino*, was synonymized under *Comaroma* BERTKAU by Oi (1960a:184), which in turn was placed in the Anapidae by Wunderlich (1986:93). Forster *et al.* (1990: 114) say:
"the California species, at least [viz. mendocino], is not an anapid. It resembles micropholcommatids in general appearance, but has the paracymbial form we take to be plesiomorphic within the Theridiidae."
Roth (1994:65) said *Comaroma* is not a North American genus. But it is now considered a senior synonym of *Archerius* in Platnick 2005.

*Archoleptoneta* Gertsch 1974:198 (Leptonetidae)
Formed on the genus name *Leptoneta* (which see) by prefixing a form derived from the Greek feminine noun ἀρχή (gen.) -ῆς 'beginning'. The usual combining form is archi-or arche-. Gertsch created this genus for a group of newly described species which "seem[s] to represent the kind of ancestral stock from which the stereotyped modern representatives were derived." Hence, 'archetypical *Leptoneta*'.

*Arctachaea* Levi 1957e:102 (Theridiidae)
Chamberlin & Ivie (1947b:25) described *Achaea nordica* from Alaska, which was the starting point for Levi's new genus name (although Levi made *pelyx* Levi 1957e the type species). *Achaea* O. Pickard-Cambridge (1882c:428) is a name taken from the geographical area of the Peloponnese, but it was preoccupied in the Lepidoptera, and Strand replaced the name with *Achaearanea* (Strand 1929:11). Levi's name then has the old *Achaea* as its second element, and the first element Arct-is taken from 'arctic' in homage to the *nordica* of Chamberlin and Ivie, and in view of its Alaskan provenance. It was then synonymized under *Chrysso* O. Pickard-Cambridge (Levi & Levi 1962:16).

*Arctella* Holm 1945b: 71 (Dictynidae)
Since the type species is *Arctella lapponica*, and Holm's paper is on the spiders of the Torneträsk region of northernmost Sweden, it is clear the name is a combination of "arctic" and the diminutive suffix -ella.

*Arcterigone* Eskov & Marusik 1994:42 (Linyphiidae)
From the prefix Arct-meaning 'northern, arctic' plus the genus name *Erigone*. The Greek noun ἄρκτος means 'bear' and refers to the constellation Ursa Major, the Great Bear or the Big Dipper.

(‡) *Arctilaira* Chamberlin 1921c:39 (Linyphiidae)
Bonnet (1955:640) correctly explains the name:
"*It seems that Chamberlin having created this genus for* **Hilaira glacialis** *of Kulczyński, wanted to form an arctic* **Hilaira***; consequently, the name* **Arctilaira** *with the syncope of the h is quite correctly formed, like* **Montilaira**."
That is, the first part of the name is from the term arctic (dropping the final c) and the second part is taken from the -laira part of *Hilaira* Simon. Synonymized under *Hilaira* by Holm (1960b:119).

*Arctobius* Lehtinen 1967 (Amaurobiidae)
The type species is *Amaurobius agelenoides* Emerton 1919a:106 which ranges from Finnish and Russian Lappland, through Alaska and Western Canada. From the Greek masculine noun ἄρκτος 'bear' specifically the constellation of the Great Bear, hence 'north' and the combining form -βιος 'living' by analogy with such genuine Greek words as λιμνόβιος 'living in a lake'. The name means 'living in the north'.

*Arctosa* C.L. Koch 1847a:123 (Lycosidae)
Formed to rhyme with *Lycosa* Latreille, but with the first element from the Greek masculine noun ἄρκτος 'bear'. This is the earliest of a series of Lycosid names formed with the name of a predatory animal as the first element and either -osa or -cosa as the rhyming second element. E.g. *Dingosa* Roewer, *Lynxosa* Roewer, *Pardosa* C.L. Koch (leopard), *Alopecosa* Simon (fox), *Crocodilosa* Caporiacco, *Hyaenosa* Caporiacco, *Mustelicosa* Roewer(weasel).

*Arcuphantes* Chamberlin & Ivie 1943:16 (Linyphiidae)
The second half is formed on analogy with other Linyphiid names such as *Bathyphantes* and *Lepthyphantes* (which see). Chamberlin and Ivie say that this new genus is related to *Lepthyphantes*, and in their paper it is placed between the discussions of *Lepthyphantes* and *Bathyphantes*. The new genus is based upon *Linyphia arcuata* Keyserling (1886b:74), whose specific name, 'bent like a bow' refers to the six or seven oblique bow-shaped ("bogenförmige") white bands on the black abdomen. From the first four letters of this specific name Chamberlin and Ivie formed the first half of their new name. This is a favorite strategy of Chamberlin. See under *Avicosa*, *Varacosa*. Hence the meaning is 'genus close to *Lepthyphantes* based upon *Linyphia arcuata* Keyserling'. Despite this Chamberlin and Ivie designated *Arcuphantes fragilis* Chamberlin & Ivie 1943 as the type species. All species included in the genus by Chamberlin and Ivie were new species except *A. arcuatulus* Roewer 1942a, and they apparently chose to honor the oldest species *arcuatulus* in the new genus name. *Arcuphantes arcuatulus* is a replacement name for *Linyphia arcuata* Keyserling, pre-occupied.

*Argenna* Thorell 1870b:123 (Dictynidae)
Thorell says this is derived from Ἄργεννος, a mythological proper name. Argennos was a fair young man beloved of King Agamemnon. He drowned in the river Cephissos in Boeotia. Thorell gleaned this from a mythological hand book, and latinized it with a feminine ending.

*Argennina* Gertsch & Mulaik 1936a:2 (Dictynidae)
Feminine diminutive of *Argenna* created for a new species *Argennina unica* Gertsch & Mulaik close to *Argenna*.

*Argiope* Audouin 1826: 121 (Araneidae)
Also known as Savigny *in* Savigny and Audouin 1825:121 (with the spelling [in French] Argyope); 1827:328 (with the spelling both [in French] Argyope and [in Latin] Argiope but with the species listed [in Latin] with the genus spelled Argyope). See introduction paragraph in the reference section p. 331.

Kaston (1948:221 note 2) says:
"*This name has often been given the orthography* **Argyope***, a form recently revived by Bristowe. It was Audouin's habit to prefix the Latin spelling of his names with a French form so that the heading appears* "*Genre* **Argyope**, **Argiope**". *The other French authors of the time had the same habit so that there is no more justification for substituting* **Argyope** *for* **Argiope** *than for substituting Walckenaer's* **Latrodecte** *and* **Agelene** *for* **Latrodectus** *and* **Agelena** *respectively.*"
This would be clearly right, except for the uncomfortable fact that after once spelling the name in Latin as Argiope, the author (Audouin, alone by this second edition, though he attributes the name to Savigny) proceeds to list the species in Latin with the genus spelled Argyope. In the face of this confusion we should invoke the clear fact that the proper Latin spelling of the name is Argiope, despite Audouin's vacillation, and it should be retained.
Ἀργιόπη is a proper name from Greek mythology, composed of ἀργι-, a combining form for ἀργής -ῆτος (Schwyzer 1.447) meaning 'white, gleaming, bright' and -οπη feminine adjective formation (nominalized) built on the combining form -οψ/-ωψ (Schwyzer 1.426) meaning 'face or eye' (not to be confused with an unrelated homonymous form -οy meaning 'voice'). Thus the name means 'with bright face or with bright eye'.
There are several Argiopes to choose from:
1) A nymph who, rejected by her lover Philammon, wandered about until she gave birth to Thamyris (Apollod. 1.16; Paus. 4.33.3.)
2) Wife of Orpheus (instead of the usual Eurydice) (Athenaeus 13.597).
3) Daughter of King Teuthras of Mysia, given in marriage to Telephus, the son of Heracles, once it is revealed by Auge, wife of Teuthras, that Telephus is the son of Heracles by Auge herself (Diodorus Siculus 4.33).
4) Mother of Cadmus, founder of Thebes (Hyg. Fab. 6.178).
Since all but a few of Savigny's names (there is good internal evidence to believe it was Savigny who found the names) are taken from the Fabulae of Hyginus, it is probably the fourth possibility he had in mind. Linnaeus had initiated the practice of taking names from Hyginus. See Heller (1945).
Berland's etymology (1929c:50), though he ignores the mythological reference, is fundamentally correct despite Bonnet's criticism (Bonnet 1955:667, note 475).

*Argyrodes* Simon 1864:253 (Theridiidae)
Walckenaer (1841:282) described a spider from Georgia on the basis of Abbot's manuscript drawings (Abbot 1792:12, figs. 99, 100) and named it *Linyphia argyrodes*. Abbot had labelled it "Silver-spider" and Walckenaer described the abdomen as "de couleur métallique d'argent le plus brilliant" that is, of the most brilliant metallic silver color. It is clear that Walckenaer meant the species name to mean "silvery." There is in fact a Greek two-ending adjective ἀργυρώδης -ες 'rich in silver (referring to a place)' used by Xenophon. It is an adjective compounded of the masculine noun ἄργυρος 'silver' and the adjective forming suffix -ωδης -ες which denotes any kind of similarity as well as possession. Note that Greek adjectives of this formation have the masculine and feminine identical, ending in long ês, and the neuter distinguished only by ending in short es. When these adjectives are latinized they lose even that exiguous distinction of genders, and the masculine, feminine, and neuter forms of the adjective will be spelled the same, since in ordinary Latin spelling there is no distinction between long ê and short e. Hence, Walckenaer's grammar is flawless, and the feminine noun *Linyphia* is modified by the correct feminine form of the adjective *argyrodes*.
Simon (1864:253) then created a new genus for the spiders which Walckenaer had included in his group (famille) Linyphies épéirides (Walckenaer 1841:281), and which had recently been treated by Vinson (1863:269, 274) as Les Épéirides comprising two "races" – one named Les Parasites (Parasitae). In this group Vinson included *Linyphia parasita* Vinson, *L. zonata* Walckenaer, and *Linyphia argyrodes* Walckenaer. The other "race" Vinson named Les Indépendentes, comprising *Linyphia viridis* Vinson, and *Linyphia inaurata* Vinson. Simon concluded these species should be separated into a new genus, so he took Walckenaer's species name *argyrodes* and elevated it to a genus name. When Greek s-stem adjectives in -ης/-ες become transliterated into Latin the length of the Greek vowels becomes hidden in Latin -es, so the Latinized adjective may be any of the three genders, when it is used as a noun. The choice is up to the original author. Simon had to choose one of the three available genders for his new name, and he unambiguously chose feminine. He listed, after Vinson, the species included in his new genus: *Argyrodes parasita* Vinson, *Argyrodes zonata* Walckenaer, *Argyrodes epeirides* Walckenaer, *Argyrodes viridis* Vinson, and *Argyrodes inaurata* Vinson. Now Simon's *Argyrodes epeirides* is in fact Walckenaer's *Linyphia argyrodes*. In order to avoid the

jingle of *Argyrodes argyrodes*, Simon changed Walckenaer's species name to epeirides and attributed it to Walckenaer, because it was taken from Walckenaer's "family" name, Les Épéirides -a reasonable step by Simon's lights, but forbidden by modern practice.

The suffix of Greek s-stem adjectives in -ωδης -ες comes from an Indo-European root meaning 'smell' found, for example in the first syllable of Latin odor, and in Homeric adjectives ευώδης 'sweet-smelling' and θυώδης 'smelling of incense'. As time went on, the meaning of the Greek suffix was extended from smell to general sensible similarity as in φλογώδης 'fiery' and then to possession as in λιθώδης 'possessing stones, stony'.

From quite a different, and unrelated word, εἶδος 'form' there exist s-stem adjectives in -οειδής -ές, such as δρεπανοειδής 'sickle-shaped, sickle-like'. Because of the similarity in meaning these two forms begin to trespass on each other's territory. So, for instance, there are two distinct words ἀραχνοειδής and ἀραχνώδης both meaning 'like a cobweb'. When Greek adjectives in -οειδής come into Latin as loan words they are transliterated as -oides, because the Greek diphthong ei comes into Latin as an i. So from Greek σησαμοειδής 'like a sesame seed' comes Latin sesamoides. So with zoological names directly from a Greek adjective in -ωδης, as with Walckenaer's argyrodes, there will be no i in them and they will not be related to the Greek word εἶδος 'form', but if they come from a Latin word borrowed from a Greek word in -οειδής it will take the form -oides and will be related to Greek εἶδος 'form'. See Buck & Peterson (1948:698) and Chantraine (1933:429-432).

All of this is not generally understood, and has led to some nomenclatural turbulence. In particular, erroneous etymologies have led to mistakes in gender. The etymologies found in the literature are as follows.

1) Simon, in the first edition of the Histoire naturelle des araignées (1864:253), thoroughly misunderstood what Walckenaer meant by the name and imagined the second part came from (as he prints it) "οἶδος gonflé." There is a rare Greek word οἶδος meaning 'swelling' but it has nothing to do with Walckenaer's name, and does not form adjectives in either -ειδης or -ωδης. Simon's knowledge of Greek and Latin, never very good, was at this point very shaky indeed. We must remember that he was only sixteen!

2) Thorell (1869:80) incorrectly derived the second element of the name from εἶδος 'form, appearance'.

3) Bonnet (1955.704) says: "The name of this genus has been treated as masculine (most often) or feminine. According to the etymology of the name [ἄργυρος "silver" and οἶδος "swelling, (according to Simon), or εἶ δος 'form' (according to Thorell)] it ought to be neuter! In the face of this anomaly, and to standardize with the least possible disturbance, I think it preferable to declare *Argyrodes* always masculine."

All of these etymologies and Bonnet's discussion are in error. Bonnet is wrong (1955:36) when he says that forms in -odes is incorrect. He lumps -ides together with the other suffixes, although it has nothing to do with Greek εἶδος, belongs to a different declension (first declension rather than third), and has a grammar and etymology totally different from the other suffixes. Further, Bonnet is wrong in his principle that the gender of a compound should be taken from the lexical root of the second member. He fails to distinguish between nouns and adjectives. Despite the fact that the noun εἶδος is neuter, the adjective δρεπανοειδής 'sickle-shaped' cannot be neuter because of its form, which makes it either masculine or feminine.

It is clear that Simon in 1864 meant the gender to be feminine. Simon himself changed it to masculine (1873a:129), and he was followed by Thorell, Keyserling, Rainbow, Banks and others. Octavius Pickard-Cambridge vacillated between masculine and feminine, sometimes in the same paper, but settled on feminine in his 1889 Biologia Centrali-Americana volume (1889d). His nephew Frederick in his later Biologia volume (1902a:401-406) made it masculine. Banks (1910:23) made it neuter occasionally, as did Bösenberg & Strand (1906:134) in one case (*crucinotum*). There has been a great deal of fluctuation, and Bonnet is correct in saying that masculine predominates. Priority dictates it should be feminine. Since 1961 the ICZN has arbitrarily decreed – albeit upon a false grammatical notion – that all genus names compounded with the suffixes -oides or -odes shall be masculine (Art. 30b).

The problems of synonymy are explained by Exline & Levi (1962:75):

> "*Strand (1928)* noted that **Argyrodes** *was preoccupied and proposed the new name* **Argyrodina**. **Argyrodes**, *however, continued to be used for spiders until the 1930's. At that time* **Conopistha** *Karsch, 1881, with the type* C. bonadea *Karsch was recognized as a synonym of* **Argyrodes**, *and* **Conopistha** *has generally been used for this genus for the last 20 years. We now consider* **Ariamnes** *(Thorell 1869) and* **Rhomphaea** *(L. Koch 1872) both proposed before* **Conopistha**, *to be a synonym of* **Argyrodes**. *If we follow strictly the laws of priority, the genus should be called* **Ariamnes** *(Doleschall 1857). However, those who disagree with us may still consider* **Conopistha** *the correct generic name. Of the two recent catalogers, Roewer uses* **Argyrodina** *(in his last volume he points out that* **Conopistha** *should have been used);*
>
> *Bonnet employs* **Argyrodes**, *strongly favored by usage. The senior homonym* **Argyrodes** *Guenée 1845, a moth of the family Perlididae, is a junior objective synonym of* **Eucarphia** *Heubner 1825 [and is not available for a Lepidopteran genus]."*

**Argyrodes** was validated by Levi & Levi (1962:16) and they made it masculine in accordance with the Code.

### *Argyroneta* LATREILLE 1804b:134 (Cybaeidae)

Martyn (1793:67) had observed: "When this Spider is in the water, the abdomen appears like quicksilver, or polished steel." This is no more than a translation of Clerck (1757:143): "Quoad araneus erat in aquis, speciem argenti vivi vel etiam chalybis expoliti alvus referebat." It is clear that these passages give us the source for Latreille's thinking. He says (1806: 94) "abdomine, bulla aerea, et splendore argenteo, incluso" [abdomen enclosed in an air bubble with a silver shine].

So this name is formed from the Greek masculine noun ἄργυρος 'silver' but the second element presents a puzzle. It appears to be the combining form -νήτης -ου, which does not occur independently in Greek, but in compounds is the masculine agentive noun-forming suffix related to the verb νέω 'to spin'. Hence, 'spinner of silver'. There is an authentic, but rare Greek masculine noun χρυσονήτης -ου 'spinner of gold thread'. Many spider names have been formed on analogy with *Argyroneta*, so that the suffix -neta with the araneological meaning 'spinner' has developed a life of its own. E.g. *Leptoneta, Microneta, Malloneta, Oreoneta, Poeciloneta*. It forms Latin first declension masculines, but despite that, Latreille and everybody else has treated it as a feminine. This peculiarity of gender, and the fact that Greek has two homonymous verbs νέω 'spin' and νέω 'swim' has led Bonnet, following Thorell, to pursue another line of argument. I quote Bonnet (1955:723):

> "*Thorell (1870b, p. 236) in the table of contents gives the name* **Argyronecta** *vide* **Argyroneta** *without explanation; he seems simply to have reproduced the spelling used by Heer in 1865, for the fossil species* longipes. *Perhaps he also wanted to indicate that the correct spelling for* **Argyroneta** *was* **Argyroneta**. – *In fact one generally gives as the etymology of* **Argyroneta**: ἄργυρος *'silver';* νέω *'spin or swim'. It is certain, after all, that it is not the verb* νέω *(=spin=weave) which is involved, for it does not refer to a spider which spins silver. –* Latreille, *in creating the name* **Argyroneta** *wanted to embody in the name the characteristic fact about that spider, which swims in the water, and to refer to the silvery appearance as a result of the bubble of air which surrounds its body. But the verb* νέω *(=swim) would not yield the form -neta necessary to form correctly the word* **Argyroneta**; *however there exists the Greek word* νηχτός *(=swimmer) [correctly* νηκτός *or* νήκτης*] which accords very well, but which gives the spelling* **Argyronecta**, *which was invented by Heer, in the same fashion as one writes* **Notonecta**. -*But it is not my purpose 150 years after its creation and use under the form* **Argyroneta** *to try to impose this correction; there is a rule against it:* **Argyroneta**, *a name incorrectly formed, should be conserved."*

Strict Greek grammar would suggest the name should mean 'spinner of silver thread' but when all is said and done, I think the most likely intended meaning, given the quotes from Martyn, Clerck, and Latreille is 'silver swimmer'. The Greek words for 'swimmer' are a masculine a-stem νήκτης which would give Latin necta, and the verbal adjective νηκτός which in the feminine would also give Latin necta. The fact that Latreille treated the word as feminine argues that he meant it to be the feminine of the verbal adjective νηκτός -η -ον. Walckenaer (1805:84) used this name without any reference to Latreille, one of the many indications that Latreille and Walckenaer exchanged information in Paris, and that the priority of ideas is not the same as the priority of publication. It is conceivable that it was Walckenaer who actually invented this name. Transferred to the Cybaeidae by Grothendieck & Kraus (1994:272).

### *Ariadna* AUDOUIN 1826:109 (Segestriidae). See comment p. 331

Proper name from mythology. The daughter of Minos of Crete, who helped Theseus negotiate the Labyrinth and kill the Minotaur. She escaped with him to go to Athens, where he abandoned her on the island of Naxos, where the god Dionysus came to her rescue. Her name, Ariadne, is the Cretan form for Ariagne, which means 'preëminently holy' from the bound prefix ἀρι-, which intensifies the meaning of the compound, and ἀγνος -η -ον 'holy'. The story is told in many places, principally: Apollodorus Bibliotheca Epit. 1.7 ff.; Plutarch Life of Theseus 17-20; Hyginus Fab. 41,43; Ovid Metamorphosis 8.174 ff. *Ariadna* is the correct Latin form of her name, and that spelling is the original form in Savigny and Audouin. Latreille subsequently (1829b 5:548) used the Greek spelling, transliterated, *Ariadne*, and was followed by Sundevall, C.L. Koch and Thorell.

*Artema* Walckenaer 1833:438 [*nomen nudum*] 1837:656 (Pholcidae)
This should be from the Greek neuter noun ἄρτημα (gen.) ἀρτήματος 'hanging ornament, ear-ring'. The problem is that Walckenaer has made a mistake in gender. In 1833 he listed no species under this genus, apparently forgot what the gender should be, and in 1837:656, 657 he added species giving them feminine specific names *mauriciana* and *atlanta*.

*Ariston* O. Pickard-Cambridge 1896a:216 (Uloboridae)
Pickard-Cambridge specifies that this is a proper name (1896a:216 footnote), hence it must be the masculine Greek proper name Ἀρίστων -ovoς (rather than the neuter of the adjective ἄριστος -η -ον 'best'). There are several people of that name to choose from: an illegitimate son of the tragedian Sophocles, a peripatetic philosopher from Alexandria, or a Stoic philosopher from Chios. The North American species *Ariston referens* Muma & Gertsch (1964:17) has been placed in the new genus *Siratoba* Opell 1979:486-490.

(‡) *Asagena* Sundevall 1833b:19 (Theridiidae)
From Greek σαγήνη 'large fishing net or hunting net' with an alpha-privative, hence, 'lacking a net'. This is the etymology Sundevall himself gives: "Σαγήνη rete, praefixo a privato: Carens reti [Sagene net, with an alpha privative prefixed: Lacking a net]." He considered it a member of the Tribe of Drassi rather than a web builder. Synonymized under *Steatoda* Sundevall (Levi 1957b:375).

*Ashtabula* Peckham & Peckham 1894:139 (Salticidae)
The name is taken from the town of Ashtabula, Ohio, but the reason is not given.

(‡) *Astrosoga* Chamberlin 1940b:5 (Cyrtaucheniidae)
This genus was erected for a trap-door spider sent to Chamberlin for identification by Prof. J.C. Cross, who found it near Kingsville, Texas. The new species name Chamberlin supplied was *rex*, a Latin masculine noun in apposition meaning 'king' for the type locality of Kingsville, on the King Ranch. The genus name is a typically complicated and playful Chamberlin name for Texas, constructed as follows. The first part is from the Greek noun ἀστήρ (gen.) ἀστέρος (with combining form ἀστρο-) 'star'. The second half is from part of a Gosiute Indian word sokup 'earth, ground' (Chamberlin 1913:4). The combining form is found in the word pasoga from pa 'water' + soga 'ground' meaning 'swampy ground' (Chamberlin 1913:8). The morpheme soga is also found in his genus *Calisoga* (Chamberlin 1937b:4) which means 'California country'. Hence, *Astrosoga* means 'star country' and is Chamberlin's invention for the 'Lone Star State'. So, the combination of the genus name and the species name *commemorates* the type locality, Kingsville, Texas. Maintained temporarily as a genus and placed in the Cyrtaucheniidae by Raven (1985a:137), but considered a junior synonym of *Eucteniza* Ausserer by Roth (1985:9, 1994: 25) and synonymised by Bond & Opell (2002:509).

(‡) *Atopogyna* Millidge 1984b:264 (Linyphiidae)
Erected for *Centromerus cornupalpis* (O. Pickard-Cambridge 1875c). From the Greek two-ending compound adjective ἄτοπος -ον 'out of place, extraordinary, paradoxical, strange' and the Greek feminine noun γυνή 'woman'. Hence, 'strange epigynum'. Millidge says:

> "The genus is characterised by the form of the epigyne. The ventral plate is drawn out posteriorly into a broad scape, which holds the genital openings and the ducts. The scape is split posteriorly into two short arms, and arising from the median notch there is a narrow subsidiary scape which carries a well-defined socket."

*Attidops* Banks 1905a:275 (Salticidae)
Derived from the family name Attidae (see *Attus* Walckenaer) by the addition of the suffix -ops, from Greek ὤψ (gen.) ὠπός 'face, eye, look'. Removed from the synonymy of *Ballus* C.L. Koch by Edwards (1999a:8).

(‡) *Attus* Walckenaer 1805:23 (Salticidae)
Thorell (1870b:218) derived the name from the Greek verb ἄττω which is an Attic form equivalent to the more common Ionic form ἀίσσω 'leap about, dart'. The genus name is not well-formed, since the double tau is peculiar to the present stem of the verb, and would not appear in authentic nominal forms. Nouns would have to have a velar, e.g., πολυάιξ 'impetuous'. But Walckenaer's classical learning was rickety, and he very likely would not have cared about this. If Thorell is right (1870b:147, 174) Walckenaer formed *Drassus, Sparassus*, and *Sphasus* on this same principle, that is on the present stems of verbs. The accumulation of such examples leads me to believe Thorell was indeed right. Walckenaer (1837: 489) defended his name against the challenge of *Atta* Jurine [actually, Fabricius], for a genus of fungus-growing ants. He points out that his name is two years older than Jurine's, and it is the Hymenopteran genus which must change to avoid confusion, not his own (see under *Salticus*) Of course, the modern Code protects both of them (Art. 56b). Jurine's term is from a Latin proper name: Caius Quintius Atta was a Roman actor mentioned by Horace (Epistles 2.1.79). The name is supposed to refer to people who walk on the tips of their shoes. The parallelism of the Walckenaer names *Drassus, Sparassus*, and *Sphasus* convinces me that the derivation from ἄττω is the correct one.
Bonnet (1955:781) says:

> "The genus **Attus** created by Walckenaer in 1805, comprised the 46 Salticids recognized up to then without its being possible to assign a particular species as type, the first named, **morsitans** being a nomen nudum. Walckenaer himself considered his genus **Attus** identical to the genus **Salticus** Latreille (1804); the later having priority it is then **Salticus** which should be maintained and **Attus** is its synonym."

But species continued to be described under *Attus* until as late as 1903 (Bösenberg), all of which have subsequenty been transferred to other genera.

*Atypoides* O. Pickard-Cambridge 1883:354 (Antrodiaetidae)
From the genus name *Atypus* (which see) and the Latin suffix -oides 'like, similar' from the Greek adjectival suffix -οείδης -ες, which forms two-ending adjectives. See the discussion of this suffix under *Argyrodes*. The gender in such Greek adjectives when transliterated into Latin is always ambiguous; it can be masculine, feminine or neuter. Since the type species name is a genitive (*riversi*), Pickard-Cambridge does not solve the problem for us. Banks, whose Latin and Greek were very unsure, treated it as feminine. Bonnet arbitrarily regarded it as masculine, and the Code (Art. 30b) now arbitrarily decrees that such words are to be treated as masculine.

*Atypus* Latreille 1804b:135 (Atypidae)
Thorell (1870b:165) said the name was derived from an alpha privative plus the verb τυπόω 'form' and explained the name as 'unshapely' quoting Latreille "laid de figure" [ugly in form]. That is not technically correct, since the name is taken directly from the adjective ἄτυπος, not from the related verb. Latreille (1816:71) glossed the genus name with the French word difforme. Walckenaer's corresponding name *Oletera* (1805:7) is from the Greek feminine agent noun ὀλέτειρα 'destroyer, murderer'. Note that Walckenaer, who gives first French names and then Latin names, lists under his genus Olétère (*Oletera*) the species Olétère difforme (*Oletera atypus*). It is clear that Walckenaer took the word atypus to mean 'misshapen'. It is also clear that Walckenaer was not careful about gender, using a masculine adjective with a feminine noun. He did later correct it to *atypa* (1826). The relationship between Walckenaer's (1805) Tableau des Araignées and Latreille's (1804b) Tableau méthodique des Insectes is fairly incestuous. They were both in Paris, and there are a lot of indications that they knew what each other was doing, traded information, and influenced each other, so that the priority of ideas between them is not the same as the priority of publication. Latreille has been a little inaccurate with the meaning of atypus. The Greek two-ending adjective ἄτυπος means 'stammering', and occurs as a loan-word adjective atypus in Latin with that meaning. This clearly has nothing to do with Latreille's name. However, in the Greek medical writers ἄτυπος means 'conforming to no distinctive type (of a disease)'.

(‡) *Aulacocyba* Simon 1926:518 (Linyphiidae)
Based upon *Tapinocyba parisiensis* Simon 1884r. From the Greek feminine noun αὖλαξ (gen.) αὔλακος 'furrow' and the Greek feminine noun κύβη 'head' (an obscure noun found only in an ancient dictionary, the Etymologicum Magnum), with a reference to the relationship with *Tapinocyba*. Hence, 'with furrowed head'. Bonnet (1957a:2882) synonymized it under *Microctenonyx* Dahl, which had priority. See under *Tapinocyba*.

(‡) *Avicosa* Chamberlin & Ivie 1942a:38 (Lycosidae)
Erected as a subgenus of *Schizocosa* Chamberlin for those species grouped around *avida* (Walckenaer 1837). The last element of the name is the rhyming morph-cosa which is customary for names in the Lycosidae, formed by misdivision of morphemes from the name *Lycosa* (which see). The first element of the name is formed by the rhyming principle and misdivision of morphemes from the first two syllables of the species name *avida*. Synonymized under *Schizocosa* Chamberlin by Dondale & Redner (1978a:146).

*Aysha* Keyserling 1891:83, 129 (Anyphaenidae)
Keyserling said this is a proper name. The famous Aysha (variously spelled Aisha, Ayesha or Ayeshah) was the favorite wife of the prophet Mohammed. The North American species have been transferred to *Hibana* by Brescovit (1991b).

*Azilia* Keyserling 1882b:270 (Tetragnathidae)
This genus is defined as being very close to *Zilla* C.L. Koch, and appears to be constructed by prefixing a negative (alpha privative) and giving it an adjectival form. Hence, 'not *Zilla*'.

(‡) *Bactroceps* Chamberlin & Ivie 1945c:223 (Linyphiidae)
From the Greek neuter noun βακτρόν 'staff, rod' and the Latin combining form -ceps (from caput capitis 'head') forming one-ending compound adjectives as in biceps 'two-headed', pisciceps 'fish-headed' etc. Thus, it is a hybrid term, half Latin and half Greek, of the kind we are warned against.

Hence, 'rod-head'. Synonymized under *Spirembolus* CHAMBERLIN by Millidge (1980b:110).

*Badumna* THORELL 1890a:322 (Desidae)
Thorell (1890a) says in his note (322 note 3) that this is a "Nom. propr. mythol." that is, a proper mythological name. He treated it as feminine. No such mythological name can be found. From its construction the name looks to be Germanic, for badu-in Germanic means 'war, strife' and Tacitus reported (Annals 4.73) that among the Frisians there was a sacred grove of the war goddess Baduhenna. The second part of that name means 'rage, passion'. It seems likely that Thorell's handwriting may have led a typesetter to produce the variant Badumna. Or Thorell may have simply misremembered the name from his schoolboy reading of Tacitus. Considered a senior synonym of *Phryganoporus* SIMON and *Ixeuticus* DALMAS by Gray (1983a:249). Placed in the family Desidae by Forster (1970b:21).

*Bagheera* PECKHAM & PECKHAM 1896:88 (Salticidae)
The type species is *kiplingi* which confirms that the Peckhams named this genus for the fostering black panther in The Jungle Books. They also took the Salticid genus names *Messua* and *Nagaina* from The Jungle Books.

*Ballus* C.L. KOCH 1850:68 (Salticidae)
Etymology not obvious. There is no such Greek or Latin word. But there is a Greek noun βαλλισμός 'a jumping about, dancing' (Athenaeus 13. 362b), and a verb βαλλίζω 'dance, jump about' (Epicharmus 79; Sophron 11.12; Athenaeus 8.362b). Koch may have confected a rough and ready name from them. Thorell (1870b:212) derived the name from the Greek verb βάλλω 'throw', but that is unconvincing both morphologically and semantically. C.L. Koch first created it as a subgenus of *Attus*, without any explanation of the derivation, along with the other parallel subgenera names *Ciris* and *Rhanis*. The Ciris is a Latin poem in hexameters, doubtfully attributed to Vergil, and found in the *Appendix Vergiliana*. It tells the story of Scylla, daughter of King Nisus, who caused her father's death, and was changed into a sea-bird, the ciris. Rhanis is a sea nymph in the train of Diana (Ovid Metamorphoses 3.171). Koch was fond of creating subgenera with names of a given category (see under *Corythalia*). This suggests that *Ballus* too might be a proper name from mythology or literature, but I am unable to find such a name. But the name Ballus does exist in Renaissance Latin literature (e.g., Antonio Beccadelli's imitations of Martial). Simon (1864:311) made it feminine *Balla*, in accordance with a youthful notion he had to make everything feminine in order to agree with aranea.

*Barronopsis* CHAMBERLIN & IVIE 1941a:601 (Agelenidae)
The type species of the subgenus *Barronopsis* is *Agelenopsis barrowsi* (GERTSCH 1934d:23). The species name honors the collector W.M. Barrows, an archaeologist from Ohio State University, and Barron is meant to be a latinization of the name Barrows, avoiding a non-Latin w and using a euphonic link phoneme n. The -opsis part is a rhyming morpheme to indicate its relationship to *Agelenopsis*.

*Baryphyma* SIMON 1884r:694 (Linyphiidae)
From the Greek three-ending adjective βαρύς -εῖα -ύν 'heavy' and the Greek neuter noun φῦμα (gen.) φύματος 'growth, tumor, tubercle'. There is a misprint in Bonnet's note (1955:851, note 4), namely a psi for a phi. The name, meaning 'heavy tubercle' refers to the large and obtuse cephalothorax of the male with a "very low lobe."

*Bassaniana* STRAND 1928b:42 (Thomisidae)
Replacement name for *Bassania* O. PICKARD-CAMBRIDGE 1898a:249, which was preoccupied in the Lepidoptera. Pickard-Cambridge said succinctly that Bassania is a proper name without further explanation. Bassania is the Latin name of a town in Illyria (Livy 44.30.7), now Elbasan in east central Albania.

*Bathyphantes* MENGE 1866:111 (Linyphiidae)
From the Greek three-ending adjective βαθύς -εῖα -ύ 'deep,' and the Greek masculine first declension agent noun ὑφάντης -ου 'weaver' from the verb ὑφαίνω 'weave'. Menge translated the name "Erdweber" that is, 'earth weaver'. He says:

> "Because they make their webs close to the ground and stand near **Bolyphantes**, I have tried to indicate their relationship by a similar sounding name."

Menge has been free with the Greek adjective βαθύς which means 'deep in the sea or deep in the woods, or deep in the soul' not 'close to the ground' for which the adverb χαμαί would have provided a better prefix.

(‡) *Bathyphantoides* KASTON 1948:133 (Linyphiidae)
Formed from the genus name *Bathyphantes* (which see) by the addition of the Latin suffix -oides (see the discussion under *Argyrodes*) 'similar to'. Hence, 'similar to *Bathyphantes*'. Synonymized under *Bathyphantes* by Hackman (1954:13).

*Beata* PECKHAM & PECKHAM 1895:167 (Salticidae)
From the feminine of the Latin adjective beatus -a -um 'blessed, happy'.

*Bellota* PECKHAM & PECKHAM 1892:67 (Salticidae)
This is not a genuine Greek or Latin word, but apears to be based on the Latin adjective bellus -a -um 'beautiful, with an invented feminine suffix.

*Bianor* PECKHAM & PECKHAM 1885c:284 (Salticidae)
A masculine mythological proper name. *Bianor* is the legendary founder of Mantua according to Vergil Eclogues 9.60.

*Blabomma* CHAMBERLIN & IVIE 1937:219 (Dictynidae)
Bonnet (1955:888) is correct when he says that because the name is formed with the second element of the Greek noun ὄμμα 'eye', which is neuter, the genus name ought to be neuter. Chamberlin and Ivie treated it as feminine, presumably because it ended in -a. The first element is from the Greek feminine noun βλάβη 'harm, damage' with reference to the fact that in this genus "the anterior median eyes [are] extremely small or missing." Transferred to the Dictynidae by Lehtinen (1967:222).

*Blestia* MILLIDGE 1993a:126 (Linyphiidae)
In honor of A.D. Blest in recognition of his work on mynoglenines.

*Bolyphantes* C.L. KOCH 1837g:9 (Linyphiidae)
The first part of the name is from the Greek masculine noun βόλος 'net' and the second part of the name is from the Greek masculine agent noun ὑφάντης 'weaver'. Hence, 'net weaver'.

*Bothriocyrtum* SIMON 1891g:315 (Ctenizidae)
By convention the Greek neuter noun βόθριον diminutive of the masculine βόθρος 'pit or trench' is used to indicate the cephalothoracic fovea. The Greek adjective κυρτός -ή -όν means 'swollen, bulging, convex'. But what can it mean to say the fovea is swollen? The fovea in this genus is a deep procurved semilunar depression (Gertsch & Wallace 1936:23). The word is also strangely formed; the parts are reversed from the expected (grammatically correct) order which would have yielded *Cyrtobothrium. This name cannot be etymologized literally and straightforwardly. Simon described this genus, by differentiating it from the closely allied *Cyrtocarenum* AUSSERER (that name is correctly formed from κυρτός -ή -όν 'swollen, bulging, convex' and the Greek neuter noun κάρηνον 'head' hence 'with convex head'.) The type species of Simon's *Bothriocyrtum* is *Cteniza californicum* O. PICKARD-CAMBRIDGE *in* MOGGRIDGE 1874, which Lucas (1884:142) had included in *Cyrtocarenum*. It is clear to me that Simon, in creating a new name for his new genus, wished to indicate its closeness to *Cyrtocarenum* by using part of that name, irrespective of its literal meaning, to form his new one.

(‡) *Brachybothrium* SIMON 1884m:314 (Ctenizidae)
From the Greek u-stem adjective βραχύς -εῖα -ύ 'short' and the neuter diminutive noun βόθριον 'little pit' or in arachnological terms 'fovea'. Nothing in Simon's description of *Brachybothrium* suggests reference to the fovea, but Raven 1985a:124 in his diagnosis of the Antrodiaetidae (= Simon's Bracybothriinae) says the fovea is a small closed pit or longitudinal groove. It is clear from Coyle (1971:332, 381) that the fovea is a short longitudinal groove. Now synonymized under *Antrodiaetus* AUSSERER 1871a.

*Brachypelma* SIMON 1891h:338 (Theraphosidae)
From the Greek three-ending adjective βραχύς -εῖα -ύς 'short' and the neuter noun πέλμα 'sole of the foot' which conventionally refers to the scopula. Simon says, "The metatarsus of the fourth pair of legs armed with a thick scopula extending about to the middle." Hence, 'with a short scopula'.

*Brachythele* AUSSERER 1871a:173 (Nemesiidae)
Ausserer gives the etymology: βραχύς kurz und θηλή Warze. That is, from the Greek three-ending adjective βραχύς -εῖα -ύ 'short' and the feminine noun θηλή 'teat, nipple' which conventionally in arachnology came to designate a spinneret. Indeed it appears to be Ausserer who initiated this convention. In Ausserer's diagnosis of the genus he says, "Die Spinnwarzen sehr kurz, The spinnerets very short." Raven (1985a:160) transferred this genus to the Nemesiidae from the Ctenizidae. The American species *theveneti* SIMON 1891f, originally described in *Brachythele*, was transferred to *Calisoga* CHAMBERLIN.

*Bredana* GERTSCH 1936a:20 (Salticidae)
There is no indication in the original description that this genus is close to *Breda* PECKHAM & PECKHAM 1894, although both *Breda* and *Bredana* belong to the subfamily Marpissinae. Gertsch says that the genus is near *Salticus*, also a Marpissine. It seems reasonable to assume that this name is derived from Peckhams' genus *Breda*, which itself is apparently named for the city in the Netherlands (there is another one in Iowa). The point of the Peckhams' name is obscure, although they had a habit of using place names for genera. Cf. *Ashtabula* and *Fuentes*.

*Brommella* Tullgren 1948 (Dictynidae)
Considered a synonym of *Pagomys* Chamberlin 1948b:16, which is preoccupied. *Brommella* was revalidated by Braun (1964:152). A feminine diminutive formed on the proper name Bromme, perhaps the T. Bromme of Gistel & Bromme (1850).

*Calilena* Chamberlin & Ivie 1941a:604 (Agelenidae)
See the general remarks on Agelenid names in -lena under *Hololena*. Since the first described species of this group was *Agelena californica* Banks 1896b:89, and all the species treated by Chamberlin and Ivie in the original paper are from California, it is clear that the morpheme Cali-is a shortened form of California. Hence, 'the Agelenid from California'.

*Calileptoneta* Platnick 1986d:15 (Leptonetidae)
Platnick explains, "The generic name is a contraction of Californian Leptoneta and is feminine."

*Calisoga* Chamberlin 1937b:4 (Nemesiidae)
The first part of the name is taken from the first letters of California, the type locality. The second half is the combining form of a Gosiute Indian word sokup 'earth, ground' (Chamberlin 1913:4). The combining form is found in the word pasoga from pa 'water' + soga 'ground' meaning 'swampy ground' (Chamberlin 1913:8). Hence, the name means something like 'California country'. The morpheme -soga is also found in in his genus *Astrosoga* (Chamberlin 1940b:5)

*Callilepis* Westring 1874:43 (Gnaphosidae)
From the Greek neuter noun κάλλος 'beauty' which yields the combining form kalli-, and the Greek feminine noun λεπίς -ίδος 'scale (of fish) or flake (of metal)'. Hence, 'with beautiful scales'. Presumably because of the color of the type species *Aranea nocturna* Linnaeus 1758. Platnick (1975b) mentions six white spots on the abdomen of some males, but gives no color information on *nocturna*.

(‡) *Callioplus* Bishop & Crosby 1935a:45 (Amaurobiidae)
The type species is *Callioplus hoplites* (Bishop & Crosby 1935a), where the species name is from the Greek masculine noun ὁπλίτης -ου, 'heavily armed warrior'. Other species in the genus are armipotens Bishop & Crosby 1935a, euoplus Bishop & Crosby 1935a, hoplomachus Bishop & Crosby 1935a, pantoplus Bishop & Crosby 1935a, ('strong in arms, well-armed, battler in arms', and 'completely armed', respectively). These species names (except for the Latin armipotens) are derived from the Greek word ὅπλον 'tool, ornament, armor, weapon' and in the conventions of zoological nomenclature its derivatives are used to mean 'armed'. So the second element is so derived and means 'armed'. The first element is from the Greek word κάλλος 'beauty', which appears in the combination καλλιοπλία 'possession of fine armor'. Bishop and Crosby appear to have made a masculine noun based upon this Greek word. My guess is that they meant it to refer to the complex process on the male carpoblem (the tibial apophysis), hence 'beautifully armed' would be a reasonable translation. Synonymized under *Cybaeopsis* Strand by Yaginuma (1987a:461).

*Callobius* Chamberlin 1947:6 (Amaurobiidae)
Erected for the species of the *Amaurobius bennetti* Blackwall 1846a group, which Chamberlin and Ivie had treated in an earlier paper (Chamberlin & Ivie 1947a). The second element is a rhyming form to connect the genus with *Amaurobius* and the first element is a rhyming form to connect the genus with *Callioplus*. The author points out that the females of both *Callobius* and *Callioplus* have a "median area or lobe embraced by 2 large lobes which meet behind."

*Calodipoena* Gertsch & Davis 1936:8 (Mysmenidae)
Originally in Theridiidae, transferred to the Symphytognathidae as a synonym of *Mysmena* Simon by Levi & Levi (1962:17). Brignoli (1980j: 729) revalidated the name. The first part is from the Greek adjective καλος -ή -όν 'beautiful' and the second part from the Theridiid genus name *Dipoena* (which see). Hence, 'the beautiful *Dipoena*'

*Calponia* Platnick 1993a:2 (Caponiidae)
"The generic name is a contraction of Californian *Caponia*, and is feminine in gender." It differs from most Caponiids, and "resembles only the African type genus *Caponia* Simon in retaining eight eyes." The type species *Calponia harrisonfordi* is known only from the coastal mountain ranges of central California. Platnick has honored the screen actor Harrison Ford because of his efforts on behalf of the American Museum of Natural History.

*Calymmaria* Chamberlin & Ivie 1937:211 (Hahniidae)
The first part is from the Greek neuter noun κάλυμμα 'hood, veil covering' related to the Greek verb καλύπτω 'cover, hide'. Chamberlin and Ivie erected this genus to embrace most of the Nearctic species which had been included in *Tegenaria* Latreille, and this is their attempt to render *Tegenaria* into a name with a corresponding Greek root. They reasoned that *Tegenaria* contains the Latin root teg-'cover' and that could be translated by the Greek noun κάλυμμα. The second element -aria is taken over to rhyme with *Tegenaria*. See *Tegenaria*. Transferred to the Hahniidae by Lehtinen 1967:221.

*Camillina* Berland 1919b:458 (Gnaphosidae)
A replacement name by adding a feminine diminutive Latin suffix for *Camilla* Tullgren (1910:105), which was pre-occupied in the Diptera and Coleoptera. Camilla is the name of the Volscian warrior maiden in the 11th book of the Aeneid of Vergil.

*Caponina* Simon 1891c:573 (Caponiidae)
This genus differs from *Caponia* in having only two eyes, but is similar to *Caponia* in having a cordiform sternum, by which the two can be separated from *Nops* MacLeay 1839. *Caponia* (Simon 1887d:cxciv) is a replacement name for *Colophon* O. Pickard-Cambridge (1874c:170) which was preoccupied in the Coleoptera (Gray 1832). Colophon is a geographical name, an ancient Greek city in Asia Minor. There is no semantic rationale for the name; Pickard-Cambridge often had recourse to irrelevant proper names for his genera. Simon was equally arbitrary, and *Caponia* has no obvious meaning. We may reasonably conjecture, since the only species known to Simon was *Caponia natalensis*, and he gave its distribution as Southern Africa, Caponia may be derived from the Cape of Good Hope.

*Carorita* Duffey & Merrett 1964:573 (Linyphiidae)
Named "after Miss Rita Carol Jones in recognition of her services to the work of the senior author."

*Castianeira* Keyserling 1880a:334 (Corinnidae)
Bonnet (1956a:959) says:
"Keyserling created the genus in 1880 with the spelling **Castianeira** saying it was a proper name. Etymologically it is not possible to support that spelling. But Keyserling in 1887:442 for the species bivittata wrote **Castaneira**. In 1891 Keyserling again used the name **Castianeira**."

Bonnet concludes that Keyserling's 1887b spelling was a *lapsus calami*. But Bonnet is incorrect in saying Keyserling's spelling cannot be supported etymologically. Its origin is perfectly clear. Καστιάνειρα is one of the wives of Priam in Homer Iliad 8.305. It is a name like other Greek names for women such as Dianeira (wife of Hercules) with the second element -aneira, which forms feminine nouns derived from the noun ἀνήρ 'man, husband'. The first element is that found in several Greek women's names such as Epicaste, or Jocaste (wife and mother of Oedipus). Cf. Schwyzer 442. Pokorny (Indogermanisches etymologisches Wörterbuch, Bern, 1959, vol. 1, p. 517). The first element is related to a root meaning 'shine' found in the Homeric perfect participle κεκασμένος and in the name Kastor (of Castor and Pollux, the Twins), and interprets the name to mean 'endowed with a pre-eminent husband'. It is clear that *Castianeira* is the etymologically correct form. The family Corinnidae was proposed by Lehtinen (1967) and finally accepted by Wunderlich (1986:28).

(‡) *Catabrithorax* Chamberlin 1920a:198 (Linyphiidae)
This name is in imitation of Crosby's (1905a) genus *Oedothorax* 'with swollen thorax'. From the verb καταβρίθω 'to be heavily laden' hence, 'with heavy thorax'. We might have expected something like *Catabrithothorax but Crosby, perhaps wisely from a euphonic standpoint, has syncopated the name. Synonymized under *Collinsia* O. Pickard-Cambridge by Hackman (1954:22).

*Cavernocymbium* Ubick 2005 (Amaurobiidae)
The first part of the name is from the Latin feminine noun caverna 'cave, hollow'. The author explains, "The generic name refers to the hollow (caverna L.) cymbium of the male palp."

*Caviphantes* Oi 1960a:178 (Linyphiidae)
The second half of the name -phantes is a rhyming morph after such Linyphiid genera as *Bathyphantes*, *Lepthyphantes* and *Linyphantes*. The first half is from the Latin adjective cavus-a-um 'hollow' or as a masculine noun 'cave'. Oi does not give an explicit etymology for *Caviphantes*, but the type species *samensis* "occurs deep in recesses of the cave." Hence the name means 'Linyphiid found in a cave'.

*Centromerita* F. Dahl 1912b:616 (Linyphiidae)
Note that there is a misprint in Bonnet (1956a:982 line 18) where he seems to attribute the name *Centromerita* to Strand instead of the correct *Centromeria*. This matter is correct in Bonnet's note (1956a:981 note 42) which reads:
"Strand in 1901[b:32] wishing to separate **Centromerus bicolor** from the other species of that genus created for it the new genus **Centromeria**. Dahl in 1912 established that the name was preoccupied by a Rhyncbote [i.e. Hemiptera] (Stal, 1870) and created in turn the name **Centromerita**. In 1928 Strand made the same discovery [that his earlier name was preoccupied], and ignorant of Dahl's modification created the name **Centromerides**. The following year Simon (whom we must scold for having ignored Dahl and Strand) fabricated yet another name **Centromerinus**. There is no doubt that it is **Centromerita** of Dahl which ought to endure, and it is the one which Neave has not recorded!"

So, *Centromerita* is a replacement name for the preoccupied *Centromeria*, which was a new related genus derived from *Centromerus* DAHL (which see).

## *Centromerus* F. DAHL 1886:69, 73 (Linyphiidae)
From Greek neuter noun κέντρον 'goad, spur' and μηρός 'thigh, limb'. The name seems to refer to the fact that in this genus all tibiae have one or two spines, and in some species the femora have one or two spines. (Locket & Millidge 1953:349, Simon 1894a:702-703).

## (‡) *Centromeroides* SCHENKEL 1950b:59 (Linyphiidae)
Formed from the genus name *Centromerus* (which see) by the addition of the Latin suffix -oides (see the discussion under *Argyrodes*) 'similar to'. Hence, 'similar to *Centromerus*. Synonymized under *Linyphantes* CHAMBERLIN & IVIE by van Helsdingen (1973c:10).

## (‡) *Cephalethus* CHAMBERLIN & IVIE 1947b:30 (Linyphiidae)
The head of the male in this genus is "extended forward over the chelicerae in two peculiar lobes" and the type species is *birostrum* (sic) meaning 'with double snout'. The genus name probably is meant to refer to this feature, and the first part comes from the Greek feminine noun κεφαλή 'head'. The second part is not at all clear. First, the species name *birostrum* should indicate that Chamberlin and Ivie meant the genus to be neuter, but they are not always careful about such things, and the word rostrum may have been too compelling for them to change the name to *birostrus* to agree with the ending of *Cephalethus* (cf. genuine Latin repandirostrus 'with recurved beak' used by Quintilian to describe dolphins). If they were serious about neuter gender, the second part may come from the Greek neuter noun ἦθος 'habit, manner' which would form neuter compounds. But it is hard to see what such a compound 'Head-manner?' would mean, and Chamberlin and Ivie are not given to such grammatical punctilio. The only genus name I can find upon which it might be modelled is *Boethus* THORELL (Salticidae), but that does not seem convincing. Boethus was a philosopher (not the better known Boethius). The morpheme etho-may occur in Chamberlin's genus *Ethobuella*, an equally puzzling name. Given the penchant of Chamberlin for making up arbitrary morphemes, I would conclude that the second part of Cephalethus is just made up. I do not find any likely rationale to derive it from Gosiute, and it does not conform to Gosiute phonology. The problem remains what the gender should properly be. Since the end of the word must be regarded as an arbitrary combination of letters, even though it looks masculine, it probably must be considered neuter with the original authors. The problem has been eliminated by Tanasevitch (1985a:56) who makes *Cephalethus* a junior synonym of *Savignia* BLACKWALL 1833a.

## *Ceraticelus* SIMON 1884r:595 (Linyphiidae)
(See Bonnet 1956a:1008). The story begins with the genus *Ceratina* MENGE 1868:170, which is pre-occupied in the Hymenoptera. *Ceratina* is from the feminine of the Greek adjective κεράτινος -η -ον 'made of horn' because this genus is characterized by an abdominal scutum, as Menge explains. Emerton (1882:32) replaced Menge's pre-occupied name with the derived *Ceratinella*, adding some American species to the genus, which had previously contained only European species. Simon separated the European and American species from one another, reserving the name *Ceratinella* for the European species and creating the new genus *Ceraticelus* for the American species. For the second element of *Ceraticelus*, Simon used the Greek adjective ἴκελος -η -ον a poetic word meaning 'like, resembling'. Hence Simon intended the word to mean 'like *Ceratina*'. Subsequently some American species were also included in *Ceratinella*.

## *Ceratinella* EMERTON 1882:32 (Linyphiidae)
A feminine Latin diminutive of *Ceratina* MENGE 1868:170 (See *Ceraticelus*).

## *Ceratinops* BANKS 1905a:309 (Linyphiidae)
Banks (1892a) described the type species *annulipes* in *Ceratinella* EMERTON and later erected this derived genus for it. The second element -ops comes from Greek adjectives meaning 'looking like' such as μῆλοψ 'looking like an apple' οἶνοψ 'looking like wine'. This suffix then becomes conventional in zoological nomenclature for forming the names of derived genera. Hence, 'looking like *Ceratina*' (see *Ceraticelus*.) Forms in -ops are one ending adjectives (the masculine, feminine and neuter have the same form). Banks used another one-ending adjective *annulipes* as species name, therefore leaving the gender obscure. Crosby and Bishop were the first to disambiguate in it in 1933b, choosing feminine with *carolina*, *crenata*, *inflata*, *lata*, and *rugosa*. Bonnet on mistaken principles (1955:35) changed them all to masculine. The first species *annulipes* is ambiguous as to gender. Crosby and Bishop treated the genus as feminine, which is certainly permissable grammatically. The Code (30.a.ii) now arbitrarily specifies that names in -ops are masculine irrespective of etymology or treatment by the author.

## *Ceratinopsidis* BISHOP & CROSBY 1930:28 (Linyphiidae)
A name derived from *Ceratinopsis* EMERTON. Bishop and Crosby seem to have assumed that the stem of the Greek feminine noun ὄψις (gen.) ὄψεως 'appearance' would be *ὀψιδος. This is incorrect grammatically. Hence, the last morpheme of the name, that is -opsidis, is not an authentic Greco-Latin morpheme. See under *Ceratinopsis*. The authors treated the name as feminine.

## *Ceratinopsis* EMERTON 1882:36 (Linyphiidae)
(See Bonnet 1956a:1017 note). Derived by the addition of the suffix -opsis (from the Greek feminine noun ὄψις (gen.) ὄψεως (stem) *ὀψι -'appearance'. The suffix -opsis is added to the preoccupied genus name *Ceratina*, see under *Ceraticelus*. Hence, 'looking like *Ceratina*'.

## *Cercidia* THORELL 1869 (Araneidae)
This name was invented by Thorell to substitute for Menge's name *Cerceis* pre-occupied in the Crustacea. Cerceis is a mythological proper name (Greek Κερκηΐς) a sea nymph, daughter of Ocean and Tethys (Hesiod Theogony 355) a name related to the Greek noun κερκίς (gen.) -ίδος 'weaver's shuttle, or any tapering rod'. Thorell's name would be the actually occurring diminutive neuter noun κερκίδιον 'small weaver's shuttle' made into a feminine. It was Thorell's habit to provide substitute names with a close resemblance to the preoccupied name.

## *Cesonia* SIMON 1893a:375 (Gnaphosidae)
This is the feminine of the Roman proper name Caesonius, spelled with the medieval substitution of long e for ae. Simon probably found this by searching the dictionary, although he might have known it from from the sixth satire of Juvenal (6.616), where Caesonia, the wife of the Emperor Caligula, is supposed to have poisoned him with an aphrodisiac which drove him mad. Cf. Suetonius's Life of Caligula 50. This is just an arbitrary use of a convenient proper name and there is no necessary semantic rationale, unless the husband-poisoner Caesonia seemed to Simon an appropriate person for whom to name a spider.

## *Chalcoscirtus* BERTKAU 1880c:284 (Salticidae)
From the Greek noun χαλκός (gen.) -οῦ 'copper, bronze' and the Greek proper name Σκίρτος 'Leaper' related to the verb σκιρτάω 'leap, spring, bound'. Scirtus is the name of a satyr in the Greek Anthology, and the attendants of the god Dionysus are called the Scirtoi. Hence, 'copper leaper'. Presumably this is a reference to the bronze color of the female of *Chalcoscirtus infimus* (SIMON 1868b), the type species.

## *Chasmocephalon* O. PICKARD-CAMBRIDGE 1889a:45 (Anapidae)
From the Greek neuter noun χάσμα (gen.) χάσματος 'yawning, gulf, chasm' and the Greek feminine noun κεφαλή 'head'. This forms an invented compound adjective *chasmocephalos meaning 'possessing an indented head'. Pickard-Cambridge then chose to use the neuter of that fabricated adjective. Pickard-Cambridge says:

> "The ordinary oblique indentations dividing [the caput] from the thorax are greatly exaggerated, forming a deep chasm or cleft on each side."

Hence, 'with head having a deep cervical groove'. Gertsch (1960a:5) had described the only known Nearctic Anapid, *Chasmocephalon shantzi*, from California. Platnick & Forster (1990:108) show that it is not congeneric with the Australian type species of *Chasmocephalon*, and erect the new genus *Gertschanapis* (which see) for it in honor of Willis Gertsch.

## *Chaunopelma* CHAMBERLIN 1940a:30 (Theraphosidae)
A subgenus of *Aphonopelma* POCOCK. The second element is from Greek πέλμα (gen.) -ατος 'sole of the foot' referring to the tarsal scopulae in the Theraphosids, but here used conventionally as a rhyming morph after *Eurypelma* C.L. KOCH, to indicate a Theraphosid (see discussion under *Aphonopelma*). The first element is from the Greek adjective χαῦνος -η -ον 'soft, spongy, loose-grained' referring to the defining feature of the genus, fine, soft, prone hairs clothing the face of the first coxa. Hence 'the Theraphosid distinguished by fine soft hairs'.

## *Cheiracanthium* C.L. KOCH 1839a:9 (Miturgidae)
From the Greek feminine noun χείρ (gen.) χειρός 'hand' and the neuter diminutive noun ἀκάνθιον 'little spine'. (There are misprints in Bonnet's note [1956a:1047, note 97]) Hence 'hand spinelet' referring to the characteristic backwardly directed process on the cymbium of the male. Koch's original spelling was *Cheiracanthium* but Thorell (1869) objected that this was an incorrect rendering of Greek into Latin, and "improved" it to *Chiracanthium* [anticipated by Agassiz]. It is a question of whether to transliterate Greek into Latin, which by the strictest convention would give Cheirakanthion, or to "domesticate" the Greek word into Latin, as the Romans often did. "Domestication" reflected the Roman pronunciation of Greek, rather than strictly the Greek letters. By the first century B.C. the Greek diphthong ei was pronounced by Greeks like a Latin long i (that is like the English vowel in meet), hence Greek χειρουργικός 'surgical' becomes the "domesticated" Latin word chirurgicus. Koch mixed the conventions, keeping the Greek diphthong ei but rendering the Greek kappa as a Latin c and using the Latin termination -ium rather than the Greek termination (transliterated) -ion. The question is confused by the fact that Koch often used a German school convention to aid in remembering whether a Latin vowel was long or short. For instance Latin long e was written æ (hence Koch's incorrect writing Væjovis [Scorpionida] for correct Latin Vêjovis), and occasionally Latin long i was written ei for the same reason. Koch's spelling may reflect this convention.

Thorell purified the name, and is followed by Bonnet. Strand (1929:16) tried to re-establish the original spelling, but Bonnet's practical position to retain Thorell's improvement has prevailed with many authors, despite the fact that the Code unambiguously privileges the original spelling even if demonstrably ill-formed (Art. 32). Transferred to the Miturgidae by Ramírez et al. (1997:45).

*Cheliferoides* F.O. PICKARD-CAMBRIDGE 1901a:254 (Salticidae)
Formed by addition of the suffix -oides 'shaped like' to the genus name *Chelifer* (Pseudoscorpionida). See the comments on -oides under *Argyrodes*. The name *Chelifer* GEOFFROY is from the Greek feminine noun χήλη 'claw' and the Latin adjective-forming suffix -fer -fera -ferum 'bearing'. The common pseudoscorpion *Chelifer cancroides* (LINNAEUS) is the familiar "book-scorpion." The reference here, however, is not to the pseudoscorpion. Pickard-Cambridge wanted a genus name which meant "claw-bearing" and the choicest one *Chelifer* was already taken. So he derived a secondary name. The first leg in this genus is "very incrassate, the tibia being almost as broad as long, and with the protarsus forming a distinct chela." A name of this form may be either masculine, feminine or neuter (see the discussion under *Argyrodes*), but Pickard-Cambridge clearly treated it as masculine.

(‡) *Chemmis* SIMON 1898a:215 (Corinnidae)
Chemmis is a place name which Simon gleaned from the dictionary in accordance with his usual habit of searching for unused words that are available. Chemmis is an Egyptian town on the right bank of the Nile mentioned by Herodotus (2.91), or a floating island in the western delta of Egypt on which Isis bore the Egyptian Apollo (Horus) and Artemis (Bubastis) according to Herodotus (2.156). As a place name, Chemmis is properly feminine in Greek. Simon left the matter ambiguous by using a genitive patronymic species name (*frederici*), and Banks (1901a:583) used a two-ending adjective, unicolor which is also ambiguous in gender. Chamberlin (1924b:667) was the first to make it masculine (*monisticus*). Since Diodorus Siculus (1.63) gave Chemmis as the name of the king who built the Great Pyramid (Cheops), one could argue that the genus name can be masculine. Transferred to the Liocranidae by Reiskind (1969:165) and considered a junior synonym of *Megalostrata* by Bonaldo (2000).

*Cheniseo* BISHOP & CROSBY 1935c:260 (Linyphiidae)
This name appears to reflect Crosby and Bishop's customary habitat in central New York. The name is probably an alternative spelling of the Seneca Indian place name Tyo'nesíyo', an important settlement near the modern Geneseo, New York. The closest attested spelling I can find is Chenissios. See Hodge (1906) and Beauchamp (1907:106). The name means "There it has a fine valley" in Seneca, an Iroquoian language.

*Chinattus* LOGUNOV 1999a:145 (Salticidae)
Since the type species is *Habrocestoides szechwanensis* PRÓSZYŃSKI 1992a, from Szechwan province, it is clear that the name is a combination of "China" and Walckenaer's old abandoned salticid name *Attus* (which see).

*Chiracanthium* (Clubionidae)
See *Cheiracanthium*.

*Chisosa* HUBER 2000:124 (Pholcidae)
Latinized from the name of the Chisos Mountains in Texas.

(‡) *Chocorua* CROSBY & BISHOP 1933b:164 (Linyphiidae)
Geographical name. A town in New Hampshire, Carroll County, where Elizabeth Bryant collected one of the specimens of *Chocorua cuneata* (EMERTON), the type species. Included in *Diplocephalus* by Hackman (1954:28).

*Chorizomma* SIMON 1872a:221 (Dictynidae)
From the Greek verb χωρίζω 'separate, divide' and the Greek neuter noun ὄμμα -ατος 'eye'. The name refers to the fact that the posterior median eyes are separated farther away from each other than each is from the laterals. Chamberlin & Ivie (1942a:20-21) took three species from *Chorizomma* and placed them in their new genus *Yorima*. At the same time they placed *Chorizomma* as a subgenus of *Cicurina* (Roth 1956a:2) with the type species *Chorizomma subterraneum* SIMON. Revalidated by Brignoli (1983c:518). Transferred to Dictynidae by Lehtinen (1967:222).

(‡) *Chorizommoides* CHAMBERLIN & IVIE 1937:218 (Dictynidae)
Derived from *Chorizomma* by the addition of the suffix -oides 'like, in the form of'. Bonnet (1956a:1077) wrongly and arbitrarily decided that the gender should be neuter, even though Chamberlin and Ivie treated it as masculine. Grammatically it can be masculine, and the Code (Art. 30.b) now, just as arbitrarily, and without real grammatical justification, decrees such names must be masculine. Lehtinen (1967:222) considered it a junior synonym of *Blabomma* CHAMBERLIN & IVIE.

(‡) *Chorizops* AUSSERER 1871a:144 (Ctenizidae)
Ausserer explained that his name is derived from the Greek verb χωρίζω 'remove to a distance (entferne)' and the feminine noun ὤψ 'eye, face, countenance' (note that there is a misprint in Ausserer's Greek namely ὤψ [meaningless] for ὤψ). Note that Bonnet is wrong in believing that a compound always takes its gender from the second element (1955:26). If it is a compound noun it will do so, but compound adjectives do not. Bonnet is also wrong in believing (1955:34-35) that when -ops means 'eye'; it is masculine, but when it means 'face' it is feminine. Chorizops (Greek χωρίζωψ) is a one-ending compound adjective, and can be masculine, feminine or neuter. Ausserer chose masculine with the species *loricatus* (C.L. KOCH 1842c). Now the Code (Art. 30.a.ii) arbitrarily decrees that gender names endng in -ops, whatever the etymology and regardless of the treatment by the original author, shall be masculine. The point of the name is that Ausserer regarded the eye arrangement to be similar to that of *Actinopus* (C.L. KOCH), but the anterior eye row is "removed (entfernt)" from the anterior margin of the cephalothorax by at least the length of the posterior eye row. Hence, 'distanced eye-row'. Gertsch & Platnick (1975:5-6) sunk *Chorizops* as a junior synonym of *Cyclocosmia* AUSSERER.

*Chrosiothes* SIMON 1894a:521 (Theridiidae)
This is not a well-formed Greek word, but it might be explained if we assume Simon made a simple mistake. A clue to the meaning may be garnered from the description of the type species *Chrosiothes silvaticus* Simon, described on the same page as the genus name. The cephalothorax is "fulvo-rufescens" or reddish yellow. There is a good Greek two-ending adjective ἰώδης -ες 'rust-colored, ferrugineous'. Since Simon not infrequently made mistakes in spelling based on the French pronunciation of Greek, it is reasonable to suppose, I think, that he would have confused in his mind a delta and a theta. This would explain the last part of the name. The first part is straightforward, and comes from the Greek masculine noun χρώς (gen.) χρωτός 'color of the skin' a word Simon was fond of using when he created names. Cf. *Chrosioderma, Sterrhochrotus*. These examples show that Simon was not punctilious in choosing the correct combining form. Greek has no compounds with χρώς as the first element, but a host of one-ending adjectives with χρώς as the second element, such as μελάγχρως (gen.) -ωτος 'with black skin'. Hence, if my reasoning is correct, this name means 'with integument rust-colored'

*Chrysso* O. PICKARD-CAMBRIDGE 1882c:429 (Theridiidae)
Bonnet (1956a:1084) said:

> "O. Pickard-Cambridge in creating the name **Chrysso** did not indicate its etymology. (Perhaps he took it from Greek χρυσός to refer to the yellow color of the abdomen; in that case, he should have written **Chryso** and made the species masculine.) But in our ignorance of what the author thought we will maintain the spelling **Chrysso** and regard it as feminine, since that is what he decided."

Bonnet's claim that the word should be spelled with one s and be masculine is not true. Many familiar feminine Greek mythological names have this form: Calypso, Leto, Melantho, Clotho, Erato, Clio. Such names are often formed with geminate consonants, as Ἀγαθθώ, Θεοκκώ, Σαπφώ. Pickard-Cambridge says of his first species *albomaculata*: "This pretty little Spider is of an orange-yellow color." Hence, he created a Greek feminine name of the familiar pattern based on the word for gold, as with the woman's name "Goldie."

*Cicurella* CHAMBERLIN & IVIE 1940:15 (Dictynidae)
Formed by the addition of a Latin feminine diminutive suffix -ella. A subgenus of *Cicurina* (see Gertsch 1992)

*Cicurina* MENGE 1871:271 (Dictynidae)
Fabricius (1793:410) had named a species *Aranea cicurea*. The species name is an incorrect feminine form of the Latin one-ending adjective cicur (gen.) -uris 'tame, mild'. Fabricius' reason for the name was that the spider lived in German houses (Habitat in Germaniae domibus). The species was transferred by C.L. Koch to *Tegenaria*, and finally Menge erected the new genus *Cicurina* for it by adding the Latin adjective forming suffix -ina to the Latin adjective cicur (Bonnet 1949). Menge translated the genus name as "Zahmspinne" that is, 'tame spider' and said that the name was formed from cicur mansuetus 'That is, from the Latin adjective cicur 'tame' which Menge glossed with another Latin adjective mansuetus 'tame'. The Cicurines were transferred to the Dictynidae by Lehtinen 1967.

*Ciniflo* BLACKWALL 1841a:607 (Amaurobiidae)
From the Latin masculine noun ciniflo -onis 'hair-curler, one who curls hair'. The word occurs in the Satires of Horace, the Roman poet of the first century B.C. (Horace Satires 1.2.98). Porphyrio, the ancient commentator on Horace, says:

> *ciniflones et cinerarii. ab officio calamistrorum in cinere calefaciendorum quibus matronae capillos crispabant.* "Hair-curlers and curler-heaters [are so called] from their duty of heating up in the ashes [cinere] the curling-irons [calamistra] with which women used to curl their hair."

The link in Porphyrio's note between ciniflones and calamistra indicates the source of the name. Blackwall must have been very learned to have known the word, and Porphyrio's note on the passage. He obviously meant *Ciniflo* to mean 'one who wields the calamistrum'. Blackwall was the first to describe the cribellum and the calamistrum (Blackwall 1833e: 471-485, 767-770 [Read Jan.18, and Feb. 15, 1831]). But he named the calamistrum first in a later article (Blackwall 1841b:223-224 [read June 5th, 1838]). Blackwall did not give a name to the "additional mammulae" he found in *Clubiona atrox*. The term cribellum (the Latin word meaning 'sieve') is due later to Ludwig Koch.

Bristowe (1939:6, 13) argued that species now included in the European genus *Coelotes* should be transferred to *Amaurobius*, and species now in *Amaurobius* should be transferred to *Ciniflo* BLACKWALL 1841a. Kaston (1948:516 note 1) explained:

> "Apparently C.L. Koch used the name **Amaurobius** for spiders of different genera in two of his approximately contemporary publications. The usual citation [for **Amaurobius** is Koch 1837 [Übersicht der Arachnidensystem I:15] in which the genotype is **Aranea atrox** DeGeer. This is a true cribellate spider of Europe. But Koch also used the name **Amaurobius** in Heft [141] of Herrich-Schaeffer's continuation of Panzer's "Deutsche Insekten" [Koch, C.L. 1937b. Arachniden in Panzer, Faunae Insectorum Germaniae initia. Heft 141. Regensburg] in association with only roscidus and tigrinus [Agelenidae]. According to Bristowe, 'though congeneric with tigrinus, the identity of roscidus is uncertain, so tigrinus must be selected as the type. Tigrinus is a synonym of **atropos** Walck., so **atropos** Walck. [an Agelenid spider of Europe, now included in **Coelotes**] becomes the type of **Amaurobius**'. Bristowe believes that Heft [141] appeared on October 1, 1836, and thus antedates the Uebersicht Arachn. Syst., in which a cribellate spider was placed with the name **Amaurobius**."

So if *Amaurobius* was first used for an Agelenid and should replace *Coelotes* BLACKWALL 1841a, then *Ciniflo* BLACKWALL is the first available name for the cribellate spider. The issue apparently turns on which of Koch's publications has precedence.

Bonnet, however (1955:272 note 92), argued that the date of Heft 141 is indeed 1837, not 1836; further, that in Heft 141 the genus *Amaurobius* is merely named, without description or species, and it is only in the Übersicht der Arachnidensystem 1837g where Koch characterized the genus *Amaurobius*, and included both the cribellate and agelenid species, so Bristowe's argument from the dates of the two publications is irrelevant. And, finally, Bonnet argued that Bristowe's proposal would just make unnecessary trouble, and

> "it would be inconceivable today, for a question of priority of three or four months, to go so far as to introduce a modification as profound as that proposed by Bristowe. The strict application of the Law of Priority could only lead us into greater confusion, because it would be necessary now to call a hundred **Amaurobius** by the new name **Ciniflo**, and call fifty **Coelotes** by the name **Amaurobius**."

(‡) *Citharoceps* CHAMBERLIN 1924a:607 (Segestriidae)
From the Greek feminine noun κιθάρα -α 'lyre' and the Latin combining form -ceps 'head.' "The genus *Citharoceps* was erected on the basis of the remarkable large stridulating patches on the carapace" (Beatty 1970: 478). Junior synonym of *Ariadna* AUDOUIN 1826 (Beatty 1970: 454).

*Clavopelma* CHAMBERLIN 1940a:30 (Theraphosidae)
A subgenus of *Aphonopelma* POCOCK. The second element is from Greek πέλμα (gen.) -ατος 'sole of the foot' referring to the tarsal scopulae in the Theraphosids, but here used conventionally as a rhyming morph after *Eurypelma* C.L. KOCH, to indicate a Theraphosid (see discussion under *Aphonopelma*). The first element is from the Latin feminine noun clavis (gen.) clavis 'club'. The genus is differentiated by having the "dorsal surface of trochanter of palpus of male densely clothed with hairs which in the genotype are flattened, straight, and moderately clavate."

(‡) *Clitolyna* SIMON 1894a:673 (Linyphiidae)
The etymology is obscure. Simon created an argiopid genus *Clitaetra* (1889b:226) and *Clitistes* (1902h:20; 1903a:996) which he says is close to *Clitolyna*. None of these have a discoverable etymology.

*Clubiona* LATREILLE 1804b:134 (Clubionidae)
The Greek lexicon will give words κλουβός 'kiln', κλουβίον 'small cage' κλουίον 'crate' none of which can directly be the source for Latreille's name, since they come from papyrological texts, which he could not have known. But in ecclesiastical Greek texts that could conceivably have been known to him κλουβός means 'a coop; a monk's cell' (Lampe 1961) This genus name may simply be the neuter diminutive *κλούβιον 'little cell' with a feminine ending tacked on. The point is that these spiders spin a retreat. Walckenaer (1805:41) created the group Les Caméricoles (Camerariae), that is 'those who live in rooms'. In this group he included the genus *Clubiona* and described their habits saying: "Spiders which spin threads around the chamber of fine white silk, where they remain enveloped." Walckenaer colorfully divided the genus into the Dryads, Hamadryads, Furies, Nymphs, Satyrs, and Fates. In French and Latin:Les Dryades (Dryadae); Les Hamadryades (Hamdryadae); Les Furies (Furiae); Les Nymphes (Nymphae); Les Satyres (Satyri); Les Parques (Parcae) (Walckenaer 1805:41-44; 1826:111-154). There is abundant evidence that Latreille and Walckenaer, good friends, both in Paris, exchanged information in the years 1804-5, and the publication dates do not always reflect the real priority of ideas. So it is easily possible that Latreille could have anticipated Walckenaer's classification in his name.

Erichson's notion (Erichson 1845:4) that it comes from κλέος 'fame' and βιόω 'live' is to be rejected. Menge's notion (1873:350) that the name comes from Greek γλύφω 'engrave' is plumb crazy. Thorell (1870b:144) says the derivation is unknown, dismisses Erichson's suggestion, and cautiously suggests, "Perhaps the name is formed from κλωβίον, a bird-trap (with reference to the sack-like tube which these spiders inhabit)." This is essentially correct. The word κλωβίον is just another spelling of κλουβίον, and these words seem to be borrowed from Hebrew kelub 'cage, basket'.

(‡) *Clubionoides* EDWARDS 1958:375 (Clubionidae)
Derived from the genus name *Clubiona* by the addition of the suffix -oides 'like, shaped like', see the discussion under *Mythoplastoides* and *Argyrodes*. Erected for *Clubiona excepta* L. KOCH 1866f. Junior synonym of *Elaver* O. PICKARD-CAMBRIDGE, which was revalidated by Brescovit et al. (1994:36).

*Cnephalocotes* SIMON 1884r:699 (Linyphiidae)
This appears to be from the Greek neuter noun κνέφαλλον 'flock of wool, cushion, pillow.' But there is nothing in the description of the genus nor of the type species (*Cnephalocotes obscurus*) to suggest such a name. Blackwall (1864:298) says that *obscurus* has an abdomen "thinly coated with hairs," but that seems hardly sufficient to generate the name. The second part -cotes seems to have been arbitrarily invented by Simon.

(‡) *Cochlembolus* CROSBY 1929:79 (Linyphiidae)
From the Greek masculine noun κόχλος 'spiral sea shell, murex' and the anatomical term embolus. Hence, 'with twisted embolus'. The type species of this genus is *alpinus* BANKS 1896a, which had been included in *Tortembolus* CHAMBERLIN (which means 'with twisted embolus'). Crosby in differentiating his genus from Chamberlin's sought a name that meant the same, using a Greek word instead of Chamberlin's Latin. See *Tortembolus*. Junior synonym of *Scotinotylus* SIMON (Millidge 1977:30).

*Coelotes* BLACKWALL 1841a:618 (Amaurobiidae)
The Greek feminine noun κοιλότης -τητος means 'a hollow or concavity'. Blackwall transferred his 1833d:436 *Clubiona saxatilis* ('*Clubiona* of the rock') to this genus, which suggests he meant his name to refer to the fact that the spider lived in the hollow of a rock. Oddly in Blackwall's summary of his article (1841a:619) he called the species *Cavator saxatilis*, which would mean something like 'excavator of the rock'. It looks as though Blackwall thought that κοιλότης was a masculine agent noun 'hollower out' rather than a feminine abstract noun 'a hollow'. The two Greek suffixes, the masculine agent suffix -της (gen.) -του (second declension) and the feminine abstract suffix -της (gen.) -τητος (third declension) are quite different in origin. There are Greek masculine nouns such as δεσμώτης 'prisoner, one bound' or μισθώτης 'one who pays rent' from a verb μισθόω 'to pay rent'. So if Blackwall made up a supposed agent noun from the verb κοιλόω 'hollow out' he might well have produced something like *κοιλώτης. The substitute name Cavator, a correct Latin agent noun, which would be a direct translation of his invented *κοιλώτης, may represent either second thoughts, or a prior manuscript name. In any case it is a junior synonym of *Coelotes*. Blackwall (p. 618) described *Coelotes*, but on the next page (619), he slipped and called his new genus *Cavator*, which reveals something of his thought processes.

Note that Blackwall's species name *saxatilis* is ambiguous in gender, either masculine or feminine. Subsequent authors have all assumed that Blackwall meant his name to be masculine.

No North American species remain in *Coelotes*. Transferred to the Amaurobiidae by Wunderlich (1986:24).

*Coleosoma* O. PICKARD-CAMBRIDGE 1882c:426 (Theridiidae)
From the Greek neuter noun κολεόν 'scabbard' and the neuter noun σῶμα (gen.) σώματος 'body'. The male abdomen is described as "of a cylindrical form, drawn out into a kind of sheath, into which the distinct pedicle uniting the cephalothorax and abdomen is inserted." Hence, 'with scabbard shaped body'.

*Collinsia* O. PICKARD-CAMBRIDGE 1913:135 (Linyphiidae)
Derived from the proper name Collins. Pickard-Cambridge gave no hint who this is. The only likely candidate I can find is W.P. Collins, a Fellow of the Royal Microscopical Society and an expert on diatoms (Cassiro 1890:4).

*Coloncus* CHAMBERLIN 1949:522 (Linyphiidae)
Chamberlin & Ivie (1944:58) had described *americanus* in Simon's genus *Araeoncus* SIMON 1884r:631. Simon's name is from the Greek three-

ending adjective ἀραιός -ά -όν 'slender, narrow' and the masculine noun ὄγκος 'bulk, mass, body', apparently a reference to the fact that the male carapace is elevated smoothly without a definite lobe. Chamberlin and Ivie erected this new genus saying it was "apparently related to *Araeoncus*" and signalled that relationship by constructing a name using -oncus. They give no clue about the origin of the first part. One of their new species, *cascadeus*, is described from Colorado, and it is probable that the first part of their new name, Col-is from Colorado. This would be in line with their practice. See *Calisoga*.

*Colphepeira* ARCHER 1941b:12 (Araneidae)
    Erected for *Epeira catawba* BANKS 1911, listed formerly in the Theridiosomatidae, but placed in the Araneidae by Levi (1978a:420). A name built on the genus name *Epeira* by the addition of a prefix taken from Greek κόλφος a spelling variant of the more usual κόλπος 'gulf'. Archer meant this to mean 'the *Epeira* from the Gulf States'.

*Comaroma* BERTKAU 1889:74 (Anapidae)
    H.W. Levi's species *Archerius mendocino* (Levi 1957d:115) described originally in the Theridiidae, was transferred to the Anapid *Comaroma* by Oi (1960a:184). Originally *Comaroma* was placed by Bertkau in the Linyphiidae:Erigoninae, and was first transferred to the Theridiidae by Oi, and subsequently to the Anapidae by Wunderlich (1986:93). Forster *et al.* (1990:114) say that Levi's species, at least, is not an Anapid, but shows features which link it with the Theridiidae. Roth (1994:65) said that *Comaroma* is not a North American genus. Bertkau's original description is not available to me, and I have relied on the description of *Comaroma simoni* in Chyzer & Kulczyński (1894:50), where the integument of the abdomen is described as being "grosse impresso-punctata" that is, 'thickly punctate with impressions' and 'with many horny little plates or scales, with corresponding impressions. These are similar to the "sclerotized spots" of *Archerius* Levi (1957d:114). A ventral view of the abdomen of *Comaroma* is given in Chyzer & Kulczyński's plate II, 15a, which is very similar to Levi's fig. 43 (1957a), and they both confirm the impression that the abdomen has the texture and look of a strawberry. The Greek word for strawberry is the feminine noun κόμαρος (actually the Arbutus unedo), and it is borrowed into Latin as a neuter noun comaron 'the fruit of the arbutus tree' (Pliny. Natural History 15.24.99) related to the strawberry (fragum). The two are considered identical in the 4$^{th}$ century A.D. Herbarium traditionally attributed to Apuleius. It seems clear that Bertkau meant the name to refer to the similarity of the texture of the abdomen to a strawberry. He seems to have taken a partially latinized spelling comarom and added a feminine ending, although with his patronymic species name *simoni* Bertkau left the gender ambiguous. Bonnet (1956a:1195) assumed it was neuter because the ending is similar to the familiar Greek neuter ending in -μα (gen.) -ματος, but that clearly has nothing to do with Bertkau's formation. I think the name should be grammatically feminine, and Oi treated it so.

*Conigerella* HOLM 1967:19 (Linyphiidae)
    A derived feminine diminutive from the Latin adjective coniger 'cone-bearer' which itself was unavailable because it was preoccupied in the Coleoptera. A junior synonym of *Mecynargus* KULCZYŃSKI *in* CHYZER & KULCZYŃSKI 1894.

(‡) *Coniphantes* IVIE 1969:57 (Linyphiidae)
    As a new subgenus of *Bathyphantes* MENGE. Formed by the rhyming principle from the last two syllables of *Bathyphantes* for the second half and from the first two syllables of the species name *conicus* from *Tmeticus conicus* EMERTON 1914a = *Bathyphantes* (*Coniphantes*) *pullatus* (O. PICKARD-CAMBRIDGE 1863), which has been placed in *Kaestneria* WIEHLE by Merrett (1963b), followed by Millidge (1984b).

(‡) *Conopistha* KARSCH 1881c:39 (Theridiidae)
    From the Greek masculine noun κῶνος 'pine-cone; geometrical cone' and the combining form ὀπισθο-'behind, after'. The name refers to the conical extension of the abdomen. Kaston(1948:87) says:

> "The name **Argyrodes** *Simon 1864 was shown by Strand to be pre-occupied [Guenée 1845, an unused name] and the name* **Argyrodina** *proposed to replace it. But according to Miss Bryant,* **Conopistha** *has priority [over* **Argyrodina***], as it had been considered a synonym of* **Argyrodes** *by Simon."*

Levi & Levi (1962:16) argued that "usage strongly favors *Argyrodes* Simon since the senior homynym (*Argyrodes* GUENÉE) has not been used" and they revalidated *Argyrodes* SIMON instead of *Argyrodina* STRAND or *Conopistha* KARSCH.

*Coras* SIMON 1898a:258 (Amaurobiidae)
    A Latin proper name from Vergil's Aeneid (7.672), one of a pair of Argive twins, called acer Coras that is 'eager Coras'. In the need to find more and more useable names Simon often searched Roman poetry for the lesser characters, e.g., Cupiennius from Horace, Satires 1.2.36. The word is an a-stem masculine with a Greek nominative ending.

*Coreorgonal* BISHOP & CROSBY 1935c:217 (Linyphiidae)
    Bishop and Crosby give no hint about the etymology, and this name does not have the form of a Greek or Latin word. With the species name *bicornis* (simply taken over from Emerton's *Delorrhipis bicornis* EMERTON 1923) they indicate that the gender must be masculine or feminine. But no Latin word will end in l unless it is neuter. The other species name *monoceros* (simply taken over from Keyserling's *Erigone monoceros* KEYSERLING *in* SIMON 1884r) would also have to be masculine or feminine (the neuter would be monoceron from Greek μονόκερων). So this name cannot be construed as a Greek or Latin construction, and should be regarded as an arbitrary combination of letters. The decision whether to treat it as masculine or feminine must necessarily be arbitrary, and Bonnet's decision to treat it as masculine is reasonable. Crosby and Bishop list a specimen of *monoceros* taken in Tellamook Co., Oregon, Aug. 20, 1931. Because these species are distinguished with a clypeal horn in the male, I wonder if the name is a combination of the Latin corn-'horn' and an anagram of "Oregon" plus an ending. That is Cor-eorgon-al meaning something like "the horned spider from Oregon." This is admittedly a counsel of desperation, but at the very least I think we can conclude the name is an anagram of something. Millidge (1981a:168) synonymized it under *Scotinotylus* SIMON. Crawford (1988:17) revalidated it.

(‡) *Coressa* SIMON 1894a:647 (Theridiidae)
    This appears to be a made up word, with a feminine suffix -essa applied to the Greek word κόρη 'girl'. There is no semantic explanation. Synonymized under *Theonoe* SIMON (Levi & Levi 1962:29).

*Coriarachne* THORELL 1870b:186 (Thomisidae)
    From the Greek feminine noun κόρις -ιος, (Attic) -εως 'bug, bedbug, Cimex lectularius plus the Greek feminine noun ἀράχνη 'spider' hence 'a spider that looks like a bedbug' referring to the fact that this genus is differentiated from *Xysticus* by its flattened cephalothorax. Its type species was *Xysticus depressus* (C.L. KOCH 1837d). The species name *depressus* obviously means 'flattened'. Simon (1895a:1035) placed Keyserling's American species *C. versicolor* in *Xysticus* and Gertsch (1932:1) gathered the American species in a new genus and argued for a new name *Platyxysticus* ('flat *Xysticus*'). He reversed himself (1934b:1) and concluded:

> "A study of the European **Coriarachne depressa** shows that the American species are fully congeneric and are not referable to a distinct generic category."

*Corinna* C.L. KOCH 1841e:17 (Corinnidae)
    A Greek feminine proper name. Corinna was a lyric poetess of the 6$^{th}$ century B.C., who wrote in Boeotian dialect, a contemporary of Pindar. The Roman poet Ovid adopted her name for his fictional love in his elegaic poems the Amores.

(‡) *Cornicularia* MENGE 1869:226 (Linyphiidae)
    From the Latin neuter u-stem noun cornu 'horn' with several suffixes: first a diminutive suffix giving corniculum 'little horn;' then an adjective-forming suffix giving cornicularius 'endowed with a little horn;' then made feminine cornicularia (probably to reflect the fact that spiders are feminine, that is aranea is feminine). But there is more to it. Corniculum is a technical Latin military term for a horn-shaped ornament upon the helmet, as a reward for bravery. Then a cornicularius is a soldier who has been so rewarded, and the word comes finally to designate a rank, the adjutant of a centurion. Pliny (Natural History 10.60.124) tells a story of one Crates, nick-named Monoceros, who had trained ravens, used for hunting, which would sit on the corniculum of his helmet. Monoceros is Greek (μονόκερως) for 'having a single horn' that is 'unicorn'. Menge translated the genus name as "Hornspinnchen" that is, 'little horn spider'. Menge's type species was *Cornicularia monoceros*, which might suggest Menge knew Pliny's story, but actually he took the species name over from Wider *in* Reuss (1834:236 *Theridion monocerus*). Menge's species was synonomized with O. Pickard-Cambridge's 1861 species *Walckenaeria unicornis* becoming eventually *Cornicularia unicornis* O. PICKARD-CAMBRIDGE 1861a. The genus is characterized by a bump or horn on the cephalothorax of the males nicely illustrated by Kaston 1948: plates XXI-XXII). My sentimental favorite is *Cornicularia pinocchio* KASTON 1945b. Synonymized under the lacklustre *Walckenaeria* BLACKWALL by Millidge (1983:106).

(‡) *Coryphaeolanus* STRAND 1914a:163 (Linyphiidae)
    *Coryphaeolanus* is an unjustified emendation by Simon (1926:469, 474, 531) for Strand's original *Coryphaeolana*. This is a substitute name for the preoccupied *Coryphaeus* F.O. PICKARD-CAMBRIDGE (1894b:87), which is from the Greek masculine noun κορυφαῖος 'headman, leader'. Simon, and Bonnet following him (1956a:1229), believed that Strand should have provided a name with masculine gender to correspond with the original name. There is, of course, no such rule. The suffix -lanus is a combination of the Latin diminutive suffix -olus -a -um (as in filiolus) and the Latin adjective forming suffix -anus -a -um (as in the name Coriolanus, the man from Corioli). Millidge (1977:11) synonymized it under *Halorates* HULL.

*Corythalia* C.L. KOCH 1850:67 (Salticidae)
    The Greek proper name Κορυθαλία is an epithet of the goddes Artemis at Sparta derived from the feminine noun κόρυς (gen.) κόρυθος 'helmet'. Perhaps Koch chose the name to refer to the high cephalothorax of this genus ("der Kopf hoch"), but with such proper names there is often no

semantic connection, and it is unnecessary to assume one here, since it is obviously one of a series of proper names. Koch originally established *Corythalia* as a subgenus of *Euophrys* C.L. KOCH, and his other names for the subgenera are: Phoebe, Ino, Pandora, Dia, Pales, Maturna, Parthenie, Thore, Freya, Frigga, Aphirape, and Trivia, that is, proper names from mythology, either Greek or Teutonic (Freya, Frigga, and perhaps Thore, a feminized form of Thor). Maturna is probably a mistake for the Roman goddess Matura; I cannot find an origin for Aphirape.

### *Creugas* THORELL 1878b:175 (Corinnidae)
Thorell says this is a proper name from mythology. From the Greek masculine proper name Κρεύγας. Actually Creugas of Epidamnus was a victor in the boxing competition at the Nemean games. He was killed unfairly by his opponent Damoxenos of Syracuse who violated the rules. Consequently Creugas was awarded the victory. Pausanias, Description of Greece 8.40.3.

### *Crossopriza* SIMON 1893a:460 (Pholcidae)
Apparently the first element is from the Greek masculine noun (occurring only in the plural) κρόσσοι 'fringe, lappet'. Since the males of this genus are characterized as having the femur of the first leg serrate beneath, and since this genus was erected for *Artema pristina* SIMON 1890a, where the species name is not the Latin word pristina but rather a invented adjective from the Greek masculine noun πριστήρ 'saw' it is reasonable to conclude that the second element has something to do with sawing. One Greek verb meaning 'to saw' is πρίζω, and this would be the source for Simon's invention. Hence, 'fringe saw'. The species *Crossopriza lyoni* (BLACKWALL 1867a) has been introduced into Arizona, Texas, Florida and Louisiana.

### *Crustulina* MENGE 1868:168 (Theridiidae)
From the Latin neuter noun crustum 'pastry' with first a diminutive suffix added giving crustulum 'a small pastry, cookie' and finally the derivative suffix -ina. Hence, 'like a little cookie'. This is a reference to the crescentic elevations on the cephalothorax, which give it a somewhat crusty appearance. Menge translated the name as "Krustenspinne" that is 'crust spider' and explains it is taken from the crustlike skin of the cephalothorax.

### *Cryphoeca* THORELL 1870b:131 (Hahniidae)
From the Greek adverb κρύφα 'secretly' and a combining form of the Greek οἰκέω 'dwell' forming a Greek compound two-ending adjective κρύφοικος -ον, after the model of such genuine Greek adjectives as ἔνοικος 'inhabiting'. Then, when the name was Latinized it was given a feminine termination. Hence, 'dwelling in secret'. Transferred to the Hahniidae by Lehtinen (1967:225).

### (‡) *Ctenium* MENGE 1871:292 (Theridiidae)
From the Greek neuter noun κτένιον 'little comb' a diminutive of the masculine noun κτείς (gen.) κτενος 'comb'. Menge translated the name as "Tasterkämmchen" that is, 'little palp comb' and explained that it refers to the comb-shaped claw on the female palp. Preoccupied in the Lepidoptera. Simon (1884a:195) recognized that Menge's *Ctenium pingue* was synonymous with *Neriene livida* BLACKWALL (1836:486), which eventually was included in *Robertus* O. PICKARD-CAMBRIDGE 1879g:103 (cf. Levi & Levi 1962:27). Bonnet (1956a:1265) argued that the butterfly genus was *Ctenia*, and therefore not identical with *Ctenium*, which, he claimed, should have been preserved.

### *Cteniza* LATREILLE 1829:230; 1831:506 (Ctenizidae)
Ultimately from the Greek feminine noun κτείς (gen.) κτενος 'comb' but more immediately from the derived verb κτενίζω 'to comb'. Latreille took the verb and added a feminine nominative ending to make it into a noun. Walckenaer also created nouns directly from the present stem of verbs, as, for instance, *Drassus*. A purist would have liked Latreille to use the feminine active participle κτενίζουσα 'she who combs'. The name refers to the rastellum, that rake-like process on the chelicerae which characterizes this genus.

### *Ctenus* WALCKENAER 1805:18 (Ctenidae)
This should be from the Greek neuter noun κτῆνος -εος 'domestic animal (like an ox or sheep)', and Thorell (1870b:195) thought this derivation probable. But Walckenaer clearly regarded it as masculine instead of neuter. Still, his classical languages were wobbly, and he may well have made such an error. Walckenaer described this genus from a specimen sent to him from Cayenne, which he said was an "araignée assez grosse" [a fairly large spider]. It may be reasonable for Walckenaer to call a very large spider an 'ox'. It is clear that the state of the specimens explains the name. Walckenaer says:

> "The characters of this genus are taken from a very large spider sent from Cayenne to the Société d'histoire naturelle at Paris; but the abdomen and the fourth pair of legs are missing. But I possess a drawing of the late Oudinot, after a spider found in the vicinity of Paris, where the eyes are positioned in the same way, and which should be of this genus. There is another in Albin, p.34, fig. 167; but I have never seen either one."

Walckenaer then gave the animal the species name dubius 'doubtful' which certainly meets the case. Then the whole name *Ctenus dubius* would mean 'doubtful beast' The fact that Walckenaer got the genus wrong has confused us about his intention ever since. There is also a lesson here for anyone trying to etymologize zoological names: the genus name alone may not provide the answer.

Oudinot was for many years the Chief Illustrator (peintre en chef) of the Muséum d'histoire naturelle in Paris (Walckenaer 1837:369). Albin is Eleazar Albin, A Natural History of Spiders and Other Curious Insects. London, 1736.

### *Cybaeina* CHAMBERLIN & IVIE 1932:28 (Cybaeidae)
Derived from *Cybaeus* L. KOCH. by the addition of a Latin feminine diminutive suffix.

### *Cybaeopsis* STRAND 1907b:563 (Amaurobiidae)
Derived from the genus name *Cybaeus* L. KOCH by the addition of the suffix -opsis (from the Greek feminine noun ὄψις 'aspect, appearance'.) The genus was originally described as an Agelenid. Strand originally used a masculine species name (*typicus*), but the gender was corrected by Bonnet (1956a:1298) to feminine. Yaginuma (1987a:461) demonstrated that this is a senior synonym of *Callioplus* BISHOP & CROSBY. Transferred to the Amaurobiidae by Lehtinen (1967:226).

### *Cybaeota* CHAMBERLIN & IVIE 1933b:3 (Cybaeidae)
Derived from *Cybaeus*. The -ota of Chamberlin is meaningless, and is apparently just an arbitrary suffix to serve the splitting of the cybaeines into *Cybaeozyga*, *Cybaeina*, etc. Maybe Chamberlin had in mind such names as *Grammonota*, *Bellota*, *Cyrtonota*, etc.

### *Cybaeozyga* CHAMBERLIN & IVIE 1937:226 (Cybaeidae)
Derived from *Cybaeus* by addition of the adjective suffix -zygus-a-um from the Greek neuter noun ζυγόν 'yoke'. Hence, 'joined to *Cybaeus*' "The one male, on which this genus is based, resembles a small *Cybaeus* in general apearance."

### *Cybaeus* L. KOCH 1868:46 (Agelenidae)
From the Latin adjective cybaeus -a -um 'pertaining to a kind of merchant ship, namely a cybaea'. The words are found in Cicero's orations against Verres. These Latin words are borrowed from the Greek κυβαία 'merchant ship', but the Greek word has only been known to modern scholarship recently from papyri. L. Koch could only have know the passage from Cicero. Thorell (1870b:127) explained that the spider is named for this ship of burden because it is "thick and bellied (as such a ship)." Menge's notion (1871:286) that the name comes from Greek *κίβη 'head' is to be disregarded. First, there is no such Greek word, and it must be a misprint. There is an obscure word κύβη found only in an ancient lexicon (Etymologicum Magnum) which does mean 'head' and that is what Menge must have meant. He translated the name as "Krummkopfspinne" that is, 'crooked-head spider'. This is mere desperate guessing.

Note that Koch's original spelling appears to have been *Cybœus*, but the œ is merely a German typographical convention for æ.

The family Cybaeidae was proposed by Forster (1970b:13).

### *Cyclocosmia* AUSSERER 1871a:144 (Ctenizidae)
Ausserer explained that this name is from the Greek masculine noun κύκλος 'circle' and the Greek verb κοσμέω 'adorn'. Finally a feminine adjective-forming ending has been added. Hence, 'characterized by a circular ornament'. The name refers to the cork-like nearly cylindrical, truncated abdomen which characterizes this genus.

### *Cyclosa* MENGE 1866:73 (Araneidae)
From the Greek omicron-contract denominative verb κυκλόω 'to move in a circle' from which is formed a feminine active participle κυκλῶσα 'circling (spider)', referring, as Menge made clear, to its motions while spinning an aerial orb-web. Menge translated the name "Kreisspinne" that is, 'circle spider'.

### *Cyllodania* SIMON 1902d:363 (Salticidae)
The first part is from the Greek three-ending adjective κυλλός -ή -όν 'deformed, contracted'. The second part is a mystery. But it is clear that the name refers to the oddly misshapen male chelicerae illustrated in Simon 1903a:1049, fig. 1118.

### *Cyrtauchenius* THORELL 1869:37 (Cyrtaucheniidae)
A replacement name for *Cyrtocephalus* LUCAS (1843:318) which was preoccupied in the Coleoptera (Audouin 1834). Lucas's name is from the Greek three-ending adjective κυρτός -ά -όν 'arched, bent, swollen' and the combining form -κεφαλος from κεφάλη 'head'. Thorell substituted for the second element the Greek word αὐχήν -ένος 'neck' adding a latin adjectival suffix. Hence, 'with arched neck'. The cephalothorax is highly elevated in front.

### *Cyrtophora* SIMON 1864:262 (Araneidae)
From the Greek three-ending adjective κυρτός -ά -όν 'arched, bent, swollen' and the combining form -φορος 'bearing'. The young Simon explained his own name with some inaccuracy as from "κύρτος [sic],

bosse; φέρω, porter. Simon used the name for his sixth group of Araneids defined as having "parties supérieure et postérieure de l'abdomen couvertes de tubercules irréguliers [superior and posterior parts of the abdomen covered with irregular tubercles]". Hence, he meant the name to mean 'tubercle bearing'.

*Dactylopisthes* SIMON 1884r:592 (Linyphiidae)
Simon had described *Erigone digiticeps* (1881g:249), whose male has a striking fingerlike forward directed projection from the top of the cephalothorax, which bears the posterior median eyes ("*une longue pointe reculée, coudée en avant et portant les yeux médians de la seconde ligne*" [a long thin process removed from the margin, bent forward horizontally and bearing the median eyes of the second row]. The species name *digiticeps* is from the latin word digitus 'finger" and the combining form -ceps from caput -itis 'head'. Hence, the name means 'finger-head'. When Simon later erected a new genus for this species, he carried this idea forward into Greek. The genus name is from the Greek masculine noun δάκτυλος 'finger, toe' and the Greek adverb ὄπισθε 'behind' which in real Greek never is found as the second member of compounds. It is the term reculée 'behind' that prompted the second element based on ὄπισθε. Simon then gave it one of his favorite arbitrary endings analogous to other genera like *Tomopisthes*. While Simon's pseudo-Greek form obscures the gender, it is clear from the second species Simon included in this genus, *D. pauper* that he regarded the name as masculine.

*Deinopis* MACLEAY 1839:8 (Deinopidae)
Original spelling *Deinopis*. From the Greek three-ending adjective δεινός -ή -όν 'terrible' and the Greek combining form -ωπις 'having such and such eyes'. Bonnet (1956a:1468) says:

> "*Authors have sometimes given masculine species names to this genus and sometimes feminine. Their indecision seems to come from the fact that the term* ὤψ (*ops*), *on which* **Dinopis** (sic) *is formed, is feminine when it means view, face, aspect and masculine when it means eye. MacLeay in creating this genus did not give the etymology of the name, but he says that of the eight eyes there are two enormous ones which occupy half the frons. It is therefore logical to think that* **Deinopis** (sic) *in the spirit of the author means 'hideous face;' one should then admit that it is feminine.*"

Bonnet is clearly wrong here from a grammatical standpoint, since compound adjectives have all three genders. His idea was unfortunately perpetuated in the 1961 Code (Art. 30a), but the 1985 Code (Art. 3030a.ii) now decrees that names in -ops will always be masculine, irrespective of the etymology or earlier treatment. Note that this provision does not apply to *Deinopis*.

MacLeay's first species was *lamia*, which he spelled with a capital letter, according to the convention of the time, to indicate it was a proper name in apposition, and consequently gives no evidence for solving the problem of gender. Lamia was the daughter of Belus, king of Libya, who bore Zeus several children, all of whom were killed by Hera, except Scylla, the squid-like monster of the straits, whom Odysseus encountered in Book 12 of the Odyssey. Zeus is said to have marked his appreciation for Lamia's favors by granting her the power of plucking out and replacing her eyes. Because her own children had been destroyed, she turned mean and set about on a career of destroying the children of others. She became so cruel that her face turned into a hideous mask. There is a genuine Greek adjective applied in Sophocles' Oedipus at Colonus (84) to the Erinyes, δεινώψ (gen.) -ῶπος (a one-ending adjective) meaning 'fierce-eyed'. There are compound feminine adjectives in -ῶπις such as χρυσῶπις (gen.) -ιδος meaning 'with golden eyes'. It is difficult to find genuine Greek or Latin words which are strictly parallel with MacLeay's creation, and it is more likely that he was simply following the convention of zoological nomenclature which had begun to proliferate names in -opis. The next author to describe a species in this genus was C.L. Koch in 1846a, who regarded it as masculine with the species name *cylindraceus*. The first to make it feminine was Gerstäcker in 1873 with *cornigera*. The genus should be regarded as feminine, on the basis of the morphology of the word. Bonnet is so far right, but not for the reasons he gives. There is no reason to say that ὤψ is feminine when it means 'view, face, aspect' and masculine when it means 'eye'. The mistake comes from the fact that compound adjectives in -ωψ are one-ending adjectives, and can have any of the three genders when nominalized. On the discussion whether the name should be spelled *Deinopis* or *Dinopis*, see under *Cheiracanthium*. The question of gender can be settled by appeal to the parallel of genuine Greek feminine compound adjectives in -ῶπις, such as χρυσῶπις (gen.) -ῶπιδος, and by appeal to MacLeay's original species, a feminine noun in apposition, *Lamia*, which indicates he thought of the genus as feminine.

*Delopelma* PETRUNKEVITCH 1939b:567 (Theraphosidae)
The Greek neuter noun πέλμα -ατος means 'the sole of the foot' and is traditionally used in the Theraphosidae to refer to the scopula on the tarsi, which is the most important synapomorphy for the family (Raven 1985a: 37). The first element is from the Greek adjective δηλος -ή -όν meaning 'clear, obvious, visible, conspicuous, plain'. Hence 'with a conspicuous scopula'. But since -pelma is conventionally used as a rhyming morph to indicate a Theraphosid (see discussion under *Aphonopelma*) it probably should not be taken literally here. Since Petrunkevitch differentiated this genus from those close to it by "the complete absence of plumose hair" he may mean the first element to mean 'plain' and the name to indicate 'the Theraphosid with plain, that is, non-plumose, hair'. Chamberlin (1940a: 26) made *Delopelma* a subgenus of *Aphonopelma* POCOCK, but Raven (1985a) synonymized it under *Rhecosticha*

*Dendryphantes* C.L. KOCH 1837b:31 (Salticidae)
From the Greek neuter noun δένδρον 'tree' and the masculine noun ὑφάντης 'weaver', an agentive formation from the verb ὑφαίνω 'weave'. Hence 'tree weaver'. Many species once described in this genus have been distributed into various genera, and the only North American species remaining in it is *nigromaculatus* KEYSERLING 1885.

*Desis* WALCKENAER 1837:611 (Desidae)
No obvious etymology, and there is nothing in the description to give a hint. There is a Greek feminine noun δέσις 'a binding' but there seems to be no rationale for such a meaning. It should be noticed that Walckenaer gave the name in French with an acute accent, Désis, and that suggests the vowel e is long, not short, and then the name might possibly be taken from the feminine noun δῆξις 'a biting', which happens to be the word Aristotle uses for the bite of a spider in his chapter on spiders in the Historia Animalium (623a). Walckenaer's type species *dysderioides* is of ambiguous gender, but Thorell synonymized it under Fabricius's (1793) *Aranea maxillosa*, retaining the feminine species name and thus disambiguating the gender.

*Diaea* THORELL 1869:37 (Thomisidae)
A new name for *Diana* SIMON (1864:432), the Latin proper name of the Roman goddess, which was pre-occupied in the Fishes and the Coleoptera. Thorell habitually formed substitute names so as to be recognizably close in spelling and sound to the original (cf. Thorell 1869:37). He explained (1869:37) that Διαίος (sic) is a Greek masculine proper name. Δίαιος (so accented) was a Greek politician from Megalopolis in the mid second century B.C., who was opposed to Rome, mentioned by Polybius (38.10.8). Diaea is a Latin feminine formed upon it.

*Dictyna* SUNDEVALL 1833b:16 (Dictynidae)
This is a proper name, an epithet of Artemis the goddess of the hunt, usually spelled with two n's. Dictynna (Greek Δίκτυννα) is a feminine name derived from the neuter noun δίκτυον 'net (either fishing or hunting)'. Artemis was the patroness of hunting with nets. Sundevall said: "cognomen Dianae, retibus venatoriis utentis [epithet of Diana, who employed hunting nets]".

*Dicymbium* MENGE 1868:193 (Linyphiidae)
From the Greek prefix δι-'two, double' and the anatomical term cymbium from the Greek neuter diminutive noun κύμβιον 'little cup'. In this genus the male palpal tibia has a long, broad cup-shaped apophysis, which appears to envelop the cymbium itself. The illustration in Locket and Millidge 1953:210 text-fig.129 A, B, D makes this clear. Cf. Kaston1948:171, and plate XXIII, figs. 458-459. Hence, 'having a double cymbium'. Menge translated the name "Doppelschiff" that is 'double ship'. This is odd since he seems to be contradicting himself. There is no question that he understood that κύμβιον means 'little cup' since he glossed the Greek word as Latin acetabulum 'little vinegar cup'. Menge called the arachnological term cymbium (strictly speaking 'a little cup') a "Schiffchen" by confusion with the Latin word cymba (Greek κύμβη) meaning 'little boat' (Menge 1866:24).

*Dietrichia* CROSBY & BISHOP 1933b:160 (Linyphiidae)
Named for Henry Dietrich, the collector of the paratypes, an entomologist, who was a colleague of Crosby and Bishop's at Cornell (Cattell 1944: 450).

*Diguetia* SIMON 1895f:106 (Diguetidae)
Derived from a proper name. Simon established this genus in an article entitled "Sur les Arachnides recueillis en Basse-Californie par M. Diguet." Léon-Jacques-Gustave Diguet (1859-1926) went to the copper mines of Boléo, Baja California Sur, Mexico, as a chemist in 1889. The mines of Boléo are on the Gulf coast near San Ignazio. He undertook over a dozen scientific expeditions under the auspices of the Muséum d'Histoire Naturelle de Paris. His first was a 16 month expedition in Baja California in 1893-1894, and his collections, which ranged over the animal and plant kingdoms, were reported on at the meetings of the Museum naturalists by the various experts. Simon reported on the arachnids on January 29, 1895, and Diguet gave a general report to the Réunion des naturalistes on 26 February 1895.

Henry McCook had described *Segestria canities* (1890:135), and Simon, from the material collected by Diguet, concluded it was not a Segestriid, and erected the genus *Diguetia* in honor of the collector. It had been placed in the Sicariidae until Gertsch reestablished the family Diguetidae (1949:234, 266), which had first been established as a subfamily by F.O. Pickard-Cambridge in 1899a.

*Diplocentria* Hull 1911a:585 (Linyphiidae)
From the Greek three-ending adjective διπλόος -η -ον 'double' and the Greek neuter noun κέντρον 'goad, spur' plus a Latin feminine adjectival suffix. Emerton (1882:56) described the species *Tmeticus bidentatus*, and Bishop & Crosby (1938:87) placed this species in their new genus *Scotoussa*. Holm (1945b:19) showed that the European species *Tmeticus rivalis* O. Pickard-Cambridge (1905:61) is a synonym of Emerton's *bidentatus*. Since Hull selected *rivalis* as the type species of *Diplocentria* in 1911a, it has priority over *Scotoussa*. Kaston (1948:212) said:

> "The palpal tibia has a broad square tipped process on the dorsal side and smaller pointed one on the side nearest the paracymbium. From the side (fig. 658) both processes look sharp and tooth-like."

This feature was the point of Emerton's species name *bidentatus* and the point of Hull's *Diplocentria*. Hence, 'with double spur'.

*Diplocephalus* Bertkau in Förster & Bertkau 1883:228 (Linyphiidae)
From the Greek three-ending adjective διπλόος -η -ον 'double' and the feminine noun κεφαλή 'head' forming a two-ending compound adjective *διπλοκέφαλος 'two-headed' (not a genuine Greek word.) In several species of this genus, such as *cristatus*, the cephalic lobe appears to provide the spider with a twofold head. Kaston (1948:170) said of *cristatus*:

> "In the male the front is produced into two small lobes, of which the anterior bears the Anterior Median Eyes and the posterior the Posterior Median Eyes."

(‡) *Diplostyla* Emerton 1882:65 (Linyphiidae)
Emerton created this genus to replace *Stylophora* Menge 1866:125, which was pre-occupied in the Coelenterates and the Diptera. From Greek adjective διπλόος -a-on 'double' and the masculine noun στῦλος 'column, pillar, pole'. By convention the style of spiders is the embolus, hence, 'with a double embolus'. This appears to refer to the close parallel structures of the embolus and the conductor, best seen in Ivie 1969: 55, figs. 107, 108. To accept this derivation we must assume Emerton ignored Menge's explanation of the original name *Stylophora* as referring to the fine, pen-shaped scape of the epigynum from the other meaning of Greek στῦλος 'stilus, pen'. Menge's etymology has no element meaning "double." Emerton made it feminine to avoid the masculine name *Diplostylus* Salter in the Crustacea.

*Diplura* C.L. Koch 1850:75 (Dipluridae)
Koch had described a species *Mygale macrura* (C.L. Koch 1841f:38) 'the long-tailed mygale' with reference to the conspicuously long posterior spinnerets. When he erected this new genus for it, he again emphasized the two long posterior spinnerets. From Greek adjective διπλόος -α -ον (Attic διπλοῦς -ῆ -οῦν) 'double' and the feminine noun οὐρά 'tail'. Hence 'with a double tail'.

*Dipoena* Thorell 1869:91 (Theridiidae)
Thorell says "Deriv.: Δίποινος, proper name." The brothers Dipoinos and Skyllis of Crete were sculptors. Pliny (Nat. Hist. 36.4.9) dates them to about 580 B.C. Thorell turned the name Dipoinos into a feminine (because spider, that is aranea, is feminine), and latinized it. There seems to be no particular semantic connection with the name; Thorell probably scanned the index to an edition of Pliny to find a usable name. The name Dipoenos would mean something like 'twice-emancipated' or 'twice-avenged'.

*Dirksia* Chamberlin & Ivie 1942a:24 (Hahniidae)
A subgenus of *Ethobuella* erected for a new species *Ethobuella* (*Dirksia*) *anyphenoides*, named for the collector Jane C. Dirks. Transferred to the Hahniidae by Lehtinen (1967:230).

*Disembolus* Chamberlin & Ivie 1933a:20 (Linyphiidae)
The authors differentiate this genus from *Spirembolus* as follows:

> "The embolus is not in a single distally enlarging spiral as in **Spirembolus**, but is tortuously coiled, and is prolonged caudally into a stout tail piece somewhat similar to that of **Ceratinella**, **Grammonota**, etc."

This tail piece, which makes the embolus double and not single, as the description emphasizes, is surely the stimulus for the name. The second element, embolus, is obvious; the first element is the Latin prefix *dis* which etymologically is derived from the Indo-european word for 'two', but as an inseparable Latin prefix means 'asunder, in two parts, in different directions'. In its various avatars we find it in such Latin words as *distraho* 'pull apart', *divido* 'divide' etc. The problem for the purist is that this prefix is never used to form noun compounds, but is always attached to verbs. So *Disembolus* is not properly formed. On the other hand, the intention of Chamberlin and Ivie to designate an embolus which has two parts directed in different directions is prefectly clear.

*Dismodicus* Simon 1884r:563 (Linyphiidae)
The type species is *Dismodicus bifrons* Blackwall 1841a ('with a double forehead'), and a junior synonym is *Dicyphus* Menge ('with double hump') both referring to the fact that the head of the male is "elevated into a large lobe, bifid longitudinally" (Locket & Millidge 1953:221). From the Greek adjective σμωδικός -ή -όν 'pertaining to a weal or bruise' an adjective derived from the feminine noun σμῶδιξ -ιγγος 'weal or bruise' and the Greek prefix δι-'two'. Hence, 'having two weals or bruises' referring to the two large swollen lobes of the male head.

*Dolichognatha* O. Pickard-Cambridge 1869b:388 (Tetragnathidae)
From Greek adjective δολιχός -ή -όν 'long' and noun γνάθος (feminine second declension) 'jaw'. Hence, 'having long jaws' a reference to the long chelicerae of the males. This is more than just a rhyming form, because has a straightforward semantic interpretation. It does use the feminine form in -a invented by Latreille (see *Tetragnatha*.)

*Dolomedes* Latreille 1804b:135 (Pisauridae)
From the Greek masculine noun δόλος 'trick, stratagem' and -μηδης (gen.) -μηδεος a combining form from the neuter s-stem noun μῆδος 'thought, intention, plans, counsels' forming the adjective δολομήδης (gen.) -εος 'wily, contriving'. Actually δολομήδης is a false reading in the poems of Simonides of Ceos [(Simonides 43 (Bergk), 70 (Page)] for the textually correct δολομήτης (gen.) -ου. It is hard to determine whether Latreille got this from a text or made it up himself on the analogy of such names as Diomedes, Archimedes and words like θρασυμήδης 'rash, impulsive'.

*Drapetisca* Menge 1866:140 (Linyphiidae)
The Greek masculine noun δραπετίσκος is the diminutive of the noun δραπέτης (gen.) -ου 'runaway, fugitive'. It is found only in Lucian's satire The Fugitives (Lucian Fug. 33). Lucian of Samosata (ca. 115 A.D. -ca. 200 A.D.) wrote a large number of humorous sketches, philosophical dialogues of an ironic and humorous bent, fantastic tales (he is the father of science fiction), miraculous legends, debunking essays on religious conmen, parodic send-ups of contemporary philosophers, and a scurrilous narrative of the sexual adventures of a man turned into a donkey. Menge translated the genus name as "Fluchtspinne" that is, 'fugitive spider'.

*Drassinella* Banks 1904b:334 (Corinnidae)
Feminine diminutive derived from *Drassus* Walckenaer 1805.

*Drassus* Walckenaer 1805:45 (Gnaphosidae)
Junior synonym of *Gnaphosa* Latreille. Bonnet (1956a:1594) said:

> "For nearly 100 years the genus **Drassus**, created by Walckenaer in 1805 for his **Cellulicoles** (1802:220), has been used as a proper genus, without considering that one year earlier Latreille (1804a:134) had created for these same **Cellulicoles** of Walckenaer, the genus **Gnaphosa**. [That is, Latreille specifically stated that his genus **Gnaphosa** corresponded to Walckenaer's **Cellulicoles**] But it is not until Simon (1893a:340 note) that this synonymy is established; he, in order to replace the now unusable name **Drassus**, made use of the genus **Drassodes** created by Westring in 1851 for the **Drassus** of the type **lapidosus**. It would take another twenty years for the name **Drassodes** to replace it. And since the genus is one of the most numerous, the lengthy use of the first name explains the great number of **Drassus** which should today be named **Drassodes**."

Cf. Bonnet 1956a:1557.

Thorell (1870b:147) referred the name to the Greek verb δράσσομαι 'sieze, catch' but such a formation would be impossible grammatically. The double sigma is peculiar to the present stem of the verb, and derived nouns would not have them. (A correct noun form, for instance, would be δραγμος 'act of seizing') But Walckenaer obviously did not care about this nicety. He was in the habit of constructing such grammatical anomalies from the present tense of verbs as *Sparassus*, *Attus*, *Thomisus* and *Sphasus*. (which see). So Thorell is correct about the source of the name, although it must be noted that Walckenaer created it in violation of the rules of Greek noun derivation. He presumably meant it to mean something like 'catcher'.

*Drassodes* Westring 1851:48 (Gnaphosidae)
From the genus name *Drassus* by the addition of the Greek adjective forming suffix (two-ending) -ωδης -ωδες 'like, similar'. See the discussion under *Drassus*, and the discussion of names in -odes under *Argyrodes*. Westring created the name *Drassodes* for the species of *Drassus* of the type *lapidosus* (Walckenaer 1802) calling it, however, *lapidicola* (after Latreille 1806) which he separated from the remainder of the genus of the type *lucifugus* (now the type species of *Gnaphosa*). The confusing array of species names (*lapidosa*, *lapidicolens*, *lapidicola*, *lapidaria*) is sorted out by Bonnet (1956a:1570 notes 120 and 121).

*Drassyllus* Chamberlin 1922:148 (Gnaphosidae)
Diminutive formed on the name *Drassus* Walckenaer on the analogy of *Herpyllus* Hentz.

(‡) *Drepanotylus* Holm 1945b:25 (Linyphiidae)
From the Greek neuter noun δρέπανον 'sickle' and the masculine noun τύλος 'knob, callus'. Hence, 'with sickle-shaped knob'. The type species is *Erigone uncata* O. Pickard-Cambridge (1873b:546), which has a large sickle-shaped apophysis on the male palpal tibia. Synonymized under *Notiomaso* Banks by Wunderlich (1978c:253), but removed from synonymy by Merrett *et al.* 1985:397.

(‡) *Drexelia* McCook 1892:127 (Araneidae)
Named, said McCook:

"*in recognition of the noble contribution to scientific and industrial education made by our fellow townsman, Mr. Anthony J. Drexel.*"

The wealthy Philadelphia family Drexel owned the brokerage firm Drexel and Company. Anthony Joseph Drexel (1826-93) was co-owner of the Philadelphia Public Ledger, as well as the founder and benefactor of the Drexel Institute of Technology, founded in 1892. Synonymized under *Larinia* Simon by Levi (1975b:102).

(‡) *Dugesiella* Pocock 1901c:551 (Theraphosidae)
A feminine diminutive noun based upon the proper name Dugès. Alfredo Dugès was the collector of the type species *Dugesiella crinita* Pocock, collected at Guanajuato, Mexico. He is surely the A. Dugès whom we repeatedly encounter in the pages of the arachnid volumes of the Biologia Centrali-Americana as a collector of spiders, scorpions and whip scorpions from Mexico, especially Guanajuato and Vera Cruz. Bonnet's bibliography lists him as having published a few articles in the 1890's. Pocock presumably did not have in mind the rather more famous Antoine Dugès, the renowned physician from Montpellier, who published several important articles on spiders in the 1830's, and collaborated with H. Milne Edwards on the Arachnid section in Cuvier's Le Règne Animal 3ᵉ edition in 1836. Raven (1985a:152) placed *Dugesiella* in synonymy with *Rhecostica* Simon, which itself has been invalidated in favor of *Aphonopelma* by Opinion 1637 (ICZN 1991).

*Dysdera* Latreille 1804b:134 (Dysderidae)
Latreille took this name over from Aldrovandus (1602:603, 618, Dysderi), who took it directly from the 2nd century B.C. Hellenistic poet Nicander, (Theriaka 738). The form in Aldrovandus is a latinization of the neuter singular of a Greek i-stem adjective δύσδηρις (m.f.) -ι (n.) (gen.) -ιος meaning 'hard to fight with'. Aldrovandus seems to have regarded it as the name of the spider. He says: Quod autem Dysderi, sive Sphecion appelatur, rutulum satis, Vespae crudis vescenti simile [Which however is called the Dysderi, or the Sphecion, is very red, and similar to the wasp that eats raw meat.]. Aldrovandus's Latin is a literal, and rather clumsy translation of the line from Nicander, Therika 738: " Ἄλλο γε μὴν δύσδηρι, τὸ δὴ σφήκειον ἔπουσι /πυρσὸν ἅλις, σφηκὶ προσαλίγκιον ὠμοβορῇ [But another kind is an aggressive foe, the one men call the wasp-spider, reddish and like the ravenous wasp. (tr. Gow and Scholfield)]."

Here δύσδηρι is not a name, but only a descriptive adjective. Nicander of Colophon (2nd cent. B.C.) wrote two didactic poems: the Theriaka, on poisonous animals, and the Alexipharmaka, on antidotes to poisons. Latreille simply turned Aldrovandus's name Dysderi into a manageable Latin noun with a feminine ending. The Greek adjective is formed from the prefix δυσ- meaning un-or mis-, with the implication of 'hard, unlucky' and the feminine i-stem noun δῆρις 'battle' hence 'hard to fight against'. Pliny N.H. 29.86, obviously following this passage of Nicander, mentioned a spider which is like a wasp, except without wings. Attempts to identify it in modern terms are unconvincing.

*Ebo* Keyserling 1884b:678 (Philodromidae)
Keyserling gives us no hint about the etymology. He produced one other creation of this shape, Odo, which is a Teutonic proper name, and he created over a dozen genera taken from Germanic antiquity, such as Agobardus, Heribertus, Wendilgarda etc. There was a Bishop of Limoges named Ebo (752-786 A.D.) and the historian Saxo Gammaticus (1150-120 A.D.) has several figures called Ebbo. (Saxo, writing in Latin often geminated consonants in names, writing Oddo for Odo, for instance). The name Ebo is widespread in medieval Germany (Förstemann 1900:435). It is a shortened form of Ebur, meaning 'prince, lord'. very frequent in Old Icelandic poetry in the form jöfurr. Keyserling left the gender obscure by using a species name (latithorax 'with wide thorax'), which is a Latin one-ending adjective, and therefore can be masculine, feminine or neuter. But if the derivation I have suggested is right, the gender must be masculine. The situation has become confused as Bonnet says (1956a:1641):

"*Simon after having used* **oblongus** *in creating that species, called it* **oblonga** *the same year. Is* **Ebo** *masculine or feminine? Keyserling in creating this name gave no explanation and his species* **latithorax** *does not yield a solution to the question Furthermore the name* **Ebo** *is not a Latin word. Consequently, the only course is to maintain the gender imposed by Simon when he created the species* **oblongus** *and conclude that the genus* **Ebo** *is masculine.*"

Bonnet argued on the grounds of priority, and I argue on the grounds of etymology, because it is a masculine Germanic name, but both arguments conclude the gender ought to be masculine.

(‡) *Eboria* Falconer 1910:83 (Linyphiidae)
William Falconer was a member of the Yorkshire Naturalist's Union, became its president in 1927, and was particularly occupied with the spiders of Yorkshire (Bonnet 1945:53). This genus name is derived from the Roman name for York, Eboracum, usually abbreviated Ebor. Falconer seems to have made a feminine name by adding -ia to this abbreviation. The type species *caliginosa* is a very rare spider found only in two localities in Yorkshire (Marsden and Scammonden) and on Scawfell Pike (Locket and Millidge 1953:317). Holm (1973:182) synonymized it under *Semljicola* Strand 1906a. It was revalidated by Crawford (1988:14), and sunk again by Eskov & Marusik (1994:67).

*Eidmannella* Roewer 1935:195 (Nesticidae)
Feminine diminutive formed on the proper name Eidmann. Hermann August Eidmann (1897-1949) was a specialist in forest entomology and social insects, especially ants. He was Professor and Director of the Zoologische Insitut at Münden.

*Eilica* Keyserling 1891:29 (Gnaphosidae)
Keyserling says this is a proper name, without further explanation. It is one of his Teutonic names, a feminine, frequently found in 11ᵗʰ century texts (Förstemann 1900:29). On Keysering's penchant for medieval German names see under *Ebo*.

*Elaver* O. Pickard-Cambridge 1898a:238 (Clubionidae)
Elaver is the Latin name of the river Allier, a tributary of the Loire, in France. Julius Caesar mentions it in his Gallic Wars. There is a neuter noun, but O. Pickard-Cambridge considered the name to be feminine. There is no semantic connection; his species are all from Mexico. Note that O. Pickard-Cambridge also created the genus *Liger*, which is the Latin name for the Loire. *Elaver* was formally revalidated by Brescovit *et al.* (1994:36).

*Emblyna* Chamberlin 1948b:9 (Dictynidae)
Erected for "about 75 known North American species" with *Dictyna completa* Chamberlin & Gertsch 1929 as type. The second part of the name is a rhyming morph from the last two syllables of *Dictyna* (which see). The first part of the name Embl-is taken from the term embolus, since the males in this genus possess a "characteristically thickened embolus." Compare *Tosnyna, Phantyna, Varyna*. Synonymized under *Dictyna* as a "section" by Chamberlin & Gertsch (1958:48), but it was revalidated by Lehtinen (1967:232).

*Emertonella* Bryant 1945b:182 (Theridiidae)
Miss Bryant had described *Euryopis emertoni* in 1933 in honor of the American arachnologist James Henry Emerton (1847-1931) with whom she had studied. This became the type species of the new genus, which honors Emerton.

(‡) *Enidia* Smith 1904:115 (Linyphiidae)
A new name for *Dicyphus* Menge (preoccupied). From the female proper name Enid. Synonymized under *Hypomma* Dahl (Bonnet 1956a:1656).

*Enoplognatha* Pavesi 1880:325 (Theridiidae)
From Greek ἔνοπλος -η -ον 'armed' and the feminine second declension noun γνάθος (gen.) γνάθου 'jaw'. Compound adjectives formed on the noun γνάθος, become two-ending second declension adjectives such as πλατύγναθος -ον 'with broad jaws', without a distinct feminine form. But on the model of *Tetragnatha* (which see) such zoological names acquire a distinctive feminine form. It means 'with armed mandible' and refers to the long teeth of the male chelicerae. The type species is *Theridion mandibulare* Lucas (1846), which initiates the convention of stressing the chelicerae in the naming.

*Entelecara* Simon 1884r:617 (Linyphiidae)
Perhaps from the Greek two-ending adjective ἐντελής -ές 'perfect, complete, full' and the neuter noun κάρα (gen.) κάρητος 'head'. Hence, 'with perfect or entire head'. In this genus the male head is "elevated and excavated at sides into well-defined lobe" (Locket & Millidge 1953: 210). Hence, 'with well-defined head'. Since this name appears to end in a neuter Greek noun by Art.30a of the Code it would be neuter. But it has been universally regarded as feminine, and Simon in the original publication clearly intended it to be feminine. The type species is *Theridium acuminatum* Wider in Reuss 1834, and Simon deliberately gave it the name *Entelecara acuminata*, so declaring his new genus to be feminine. The species name *acuminata* refers to the fact that the frons of the male has a cone-shaped lobe.

*Entychides* Simon 1888b:213 (Cyrtaucheniidae)
*Lapsus calami* for *Eutychides*, probably just a case of a reversed piece of type. Bonnet said (1956a:1846 note 118):

"*It is certainly the consequence of a typographical error that this name is written* **Entychides** *in the articles of 1888 and 1891. -Simon subsequently (1892, 1902)* wrote Eutychides, *but without mentioning the original error."*

*Eutychides* is a Greek proper name meaning roughly 'son of good luck'. The famous Eutychides was a sculptor, a pupil of Lysippos, mentioned by Pliny the Elder (34.51). Simon was fond of gleaning names from Pliny. Despite the evidence of an original error, the rules privilege the original spelling, which is meaningless.

### *Epeira* WALCKENAER 1805:53 (Araneidae)

Nobody has ever figured out satisfactorily what Walckenaer meant by this name. It has frustrated all attempts at analysis, and the several suggestions made in the past (catalogued below) are certainly to be rejected. The etymologies which have been suggested are all seriously flawed.

1. Erichson's derivation (1845:5) from the Greek verb ἐπείρομαι 'examine' is fanciful and without merit.
2. The young Simon (1864:259) derived it from a supposed Greek verb ἐπείρω "*faire un tissu* [make a web]" but there is no such verb.
3. Menge (1866:41) derived the name from πείραω [try, attempt], or said it stands for ἔμπειρα [experienced]. He said the derivation from ἤπειρος 'mainland' "*scheint mir unzulässig, weil sie keinen Sinn hat* [seems to me unacceptable because it has no semantic connection]". Neither of his suggestions is convincing.
4. Thorell 1869:53 said "Derivation unknown." In his footnote 1 he continued:

> *"In Agassiz' Nomenclator Zoologicus [i.e. Erichson] it is derived from* 'ἐπείρομαι *examinor" a derivation which appears to me to be destitute of all grounds. According to Simon,* **Epeira** *comes from* 'ἐπείρω *faire un tissue [make a web]" which verb I have not been able to find in any Greek Lexicon to which I have access.--May not the name perhaps be formed of* ἐπί, *on, and* εἶρος *wool (with reference to the circumstance of the female's being usually found, after laying her eggs, sitting beside or upon the wool-like cocoon?"*

It has to be said, sadly, that even Thorell, who had far more talent and knowledge about these questions than the others, still made a desperate and unconvincing suggestion.

At the risk of being included myself in some subsequent list of desperate and unconvincing solutions, I offer my best guess.

Any solution must take into account the peculiar way Walckenaer spelled the name both in French and in Latin. He is consistent for decades in spelling the name Épeïre in French and *Epeïra* in Latin, showing that he intended the word to be pronounced with three syllables in French and with four syllables in Latin. Notice that the French version has an accent aigu on the first and second e.

Walckenaer named his division Orbiculariae because aerial orb webs are characteristic of this genus, and he may have derived his name from the Greek adjective ἐπηέριος 'aerial, borne through the air'. This would explain the syllabification, but only one of the accents aigus. It has the advantage of giving the most satisfactory semantic connection.

There is perhaps a further hint in Clerck's Aranei Suecici (Clerck 1757: 16). He divided the spiders into two "agmines" [bands, or troops]: the Aranei Aërei or Luft-Spindlar and the Aranei Aquatici or Vattu-Spindlar, that is the aerial spiders and the aquatic spiders (by the latter he meant *Argyroneta aquatica*). The Aranei Aëri he divided into two "classes" the Retiarii or web-builders, and the Saltatores, or the wandering spiders. It is conceivable that Walckenaer was influenced by Clerck's term in forming Epeïra.

Kaston (1948:224) had originally followed Petrunkevitch who separated *Aranea* and *Epeira* on the basis of the shoulder humps. Kaston changed his mind, finally (1977:27) persuaded by two articles by Bonnet (1950, 1953b).

### *Eperigone* CROSBY & BISHOP 1928a:46 (Linyphiidae)

For reasons explained below Crosby and Bishop abandoned *Parerigone*, and this is a substitute for it, derived from *Erigone* by addition of the Greek prefix ἐπί 'on top of, over, in addition to'. Bonnet (1958:3433) explained:

> *"Crosby and Bishop created the genus* **Parerigone** *in 1926 with* **Erigone probata** *O.P.C. as the type. Subsequently they realized that they had confused two different species in that form:* **Erigone probata** *proper (=***Catabrithorax probatus***) and* **Tmeticus maculatus** *(=***Eperigone maculata***); in view of that fact, they placed* **Parerigone** *in synonymy with* **Catabrithorax** *and created the genus* **Eperigone** *for the remainder of their old genus* **Parerigone***. There is no cause to revise their decision, but I think that, in the case of a genus created with a type which contains within it two different species, it was possible to retain that genus for the second form without having to create a new term. In any case I have discovered that* **Parerigone** *is also preoccupied by a Dipteran (Brauer, 1898)"*

Berland (1932a:76) had created a genus *Parerigone*, but during the printing of his paper realized it was preoccupied by the Crosby and Bishop genus, and himself substituted the name *Anerigone* (1932d:119).

### *Epiceraticelus* CROSBY & BISHOP 1931:380 (Linyphiidae)

Derived from *Ceraticelus* by addition of the Greek prefix ἐπί 'on top of, over, in addition to'.

### *Episinus* WALCKENAER *in* LATREILLE 1809:371 (Theridiidae)

Apparently from the Greek s-stem adjective ἐπισινής -ές 'injurious, liable to be injured;' (which would give episines) or possibly from the rare Greek second declension two-ending adjective ἐπίσινος -ον 'plotting mischief' (which would give episinius). In any case some morphological adjustment is necessary to get Walckenaer's name, and the semantic connection is slender. Thorell (1869:79) said: "Probably ἐπισινή, hurtful (σίνομαι, plunder, injure.)"

### *Eridantes* CROSBY & BISHOP 1933b:146 (Linyphiidae)

From the Greek masculine agent noun ἐριδάντης (gen.) -ου 'wrangler'. Crosby and Bishop liked to browse through the Greek lexicon for usable words (see under *Soucron*), and my guess is, since the type species is *Lophocarenum erigonoides* EMERTON 1882, they wanted a name suggestive of *Erigone* and hit upon this.

### *Erigone* AUDOUIN 1826: 115 (Linyphiidae)

See also Savigny 1817:pl. 1 (in French Erigones). See comment p. 331
Mythological proper name. Erigone was the daughter of an Athenian named Icarius, who entertained the god Dionysus as a guest, and in return received the gift of wine. Icarius shared the wine with his neighbors, who got drunk and killed him. His daughter, Erigone, with her faithful dog Maera, searched for his body. She found it beneath a tree and, in a transport of grief, hanged herself on it. As a reward for her piety she was placed in the heavens as the constellation Virgo. Apollod. 3.191 ff.; Hyg. Fab. 130; Ov. Met. 10.451. Another Erigone was the daughter of Aegisthus and Clytaemnestra, Hyg. Fab. 122. The name means "born of the dawn, early born." Of the nine genus names attributable to Savigny, six of them come from Hyginus, which also was a favorite source of names for Linnaeus. See also Heller (1945).

### *Erigonella* F. DAHL 1901b:263 (Linyphiidae)

Feminine diminutive of *Erigone*. Considered a synonym of *Troxochrus* SIMON 1884r; revalidated by Denis (1948a).

### *Eriophora* SIMON 1864:261 (Araneidae)

From the Greek neuter noun ἔριον 'wool' and the combining form -φορος 'bearing'. These form a compound two-ending adjective *ἐριόφορος -ον 'wool-bearing'. Greek compound adjectives do not have a distinctive feminine form, but Simon latinized this name with a feminine ending. Simon says: "The body is of solid color, and covered with abundant, dull, white wooly hairs" The name was originally used by Simon for a "groupe" of the genus *Epeira*, parallel with his other "groupes" *Nuctenea* and *Neoscona* and he does not even mention the name in the second edition of the Histoire naturelle des araignées. It is F.O. Pickard-Cambridge (1903a:461) who elevated it to a genus.

### *Eris* C.L. KOCH 1846f:189 (Salticidae)

Eris is the Greek goddess who personifies discord and strife and excites war. Greek Ἔρις (gen.) -ιδος, 'strife, discord'. It was Eris who threw the Apple of Discord into the crowd at the wedding of Achilles's parents, Peleus and Thetis, inscribed "To the Fairest", which initiated the beauty contest among Hera, Athene, and Aphrodite. Paris, the son of Priam king of Troy, chosen as judge by the three goddesses, gave the prize to Aphrodite in return for her promise to give him the most beautiful woman in the world for his wife, namely Helen, currently married to Menelaus, king of Sparta. Paris abducted Helen, which started the Trojan War.

### *Ero* C.L. KOCH *in* HERRICH-SCHÄFFER 1836b: Heft 138, pl. 3, 4 (Mimetidae)

Etymology obscure. There is a masculine Latin noun aero (gen.) -onis which is sometimes spelled ero meaning 'braided or wicker basket, hamper'. Could this refer to the armature of the interior spines on the forelegs, which could be conceived as a capture basket? But the genus has been consistently treated as feminine. Still, Koch may simply have taken over De Geer's feminine species name from *Aranea tuberculata* (De Geer 1778b:226) without thinking. But on the other hand Koch did not mention the characteristic feature of the interior spines on the forelegs in his descriptions. Thorell (1869:89) suggested it is probably a proper name. The youthful Simon (1864:182) said it was from "εἴρω nouer" [tie with a knot]. But that verb means 'string together in rows' and, further, a first person verb will not become a nominative noun. As with most of Simon's juvenile etymologies this must be rejected. Menge (1866:146) could not decide whether the name is the mythological name Hero, or from the verb ἐράω 'love' or the feminine noun ἔρα 'earth'. None of these suggestions is satisfactory. There is a Germanic proper name Ero (Förstemann 1900: 453), but it appears to be masculine.

*Eskovia* MARUSIK & SAARISTO 1999:125 (Linyphiidae)
Named for K.Y. Eskov, who described the type species *Minicia exarmata* ESKOV 1989a.

*Estrandia* BLAUVELT 1936:164 (Linyphiidae)
Bonnet (1956a:1803) says:

> "This genus being dedicated to Strand, I [Bonnet] had the intention of modifying the name from the incorrect **Estrandia** to **Strandia**, then I discovered that the latter had already been taken by another arachnid (Roewer 1910) -Under those conditions I believed it preferable to let **Estrandia** stand, rather than invent another name."

Blauvelt published this genus in the Festschrift for Embrik Strand obviously meaning to honor Strand in the name. I am sure the initial E is meant to be the initial of Embrik, and is not an error. The name honors E. Strand. After all, Strand himself (1929:18) created a genus *Idastrandia*, presumably for his wife.

*Ethobuella* CHAMBERLIN & IVIE: 1937:224 (Hahniidae)
The authors give no hint about the etymology of this name. This genus is characterized as "a genus of small Agelenids" and the name is a feminine diminutive in -ella. But that leaves us with the puzzling remnant Ethobu-. The name remains an enigma. Transferred to the Hahniidae by Lehtinen (1967:234)

*Euagrus* AUSSERER 1875:135, 160 (Dipluridae)
As a subgenus of *Diplura*. It is Simon (1892a:185) who raised it to full generic status. Ausserer explained that the name is from the Greek two-ending compound adjective εὔαγρος -ον 'lucky in the chase, affording good sport', from the prefix εὐ-'good, well' and the feminine noun ἄγρα 'hunt'. In form the name could be either masculine or feminine, and Ausserer treated it as masculine with the species *mexicanus*. Whether it should be spelled with a u or with a v has been a much vexed question. Latin makes no distinction between u and v, so the two spellings are really identical. Simon (1892a:185) was the first to spell it *Evagrus*. Bonnet (1945: 133) in a desire to achieve some consistency, recommended using u before a consonant and v before a vowel. He offered the word evangelium 'gospel', as a justifying example. But this makes some Greek words look funny, and disguises the prefix eu-. Ausserer spelled this genus name with a u and the Code privileges the original spelling (Art. 32.b), and in any case the Code regards the two spellings as homonymous (Art. 58.4). Raven (1979:634) and Coyle (1988:213) properly argue for *Euagrus*.

Note: The single Latin phoneme /u/ has two allophones, one vocalic and one consonantal. The ancient Latin grammarians (Charisius; Priscian) distinguished phonetically between the two allophones, although they were not distinguished in the writing system. This is only natural, since writing systems as a rule only distinguish between phonemes, not between allophones. It is in the middle ages when the vernacular languages had a distinction between phonemes /u/ and /v/, and Latin was pronounced like the local vernacular, that a scribal tradition developed of writing consonantal /u/ with the grapheme v. This medieval tradition descended to modern philological and editorial practice, specifically to the editors of the Latin dictionary that Bonnet would have used.

*Eucteniza* AUSSERER 1875:148 (Cyrtaucheniidae)
As a subgenus of *Cteniza* (which see). Ausserer said it was necessary to divide the genus *Cteniza* into two subgenera *Cteniza sensu stricto*, and *Eucteniza*, and gave four differentiating features. The name is created with a familiar nomenclatural practice by prefixing the morpheme eu-from the Greek adverb, adjective and prefix εὐ-'well, good'. It is apparently Simon (1892a:110) who raised it to full generic status. Transferred to the Cyrtaucheniidae by Raven (1985a:137).

*Eulaira* CHAMBERLIN & IVIE 1933a:14 (Linyphiidae)
The authors explained that their new genus is "close to *Montilaira* and *Catabrithorax*." The name is derived from *Montilaira* by adding the prefix εὐ-'genuine, true, typical' to the second half of the name. *Montilaira* CHAMBERLIN 1921c:41 itself was constructed for the species *Hilaira uta* CHAMBERLIN 1919a *Hilaira* is a genus of Simon's (1884a:374) from the Greek feminine adjective (it occurs only in the feminine) ἱλάειρα 'mildly shining'. So Chamberlin created *Montilaira* from the Latin masculine noun mons, montis 'mountain' and the last five letters of *Hilaira* SIMON. Then, from *Montilaira* Chamberlin created the new name *Eulaira*. It is clear from the internal evidence that the spelling *Eularia* on the same page is a misprint. Simon's genus *Hilaira* is misprinted *Hilaria* on p. 532 of the same paper.

*Euophrys* C.L. KOCH in HERRICH-SCHÄFFER 1834: heft 123, pl. 7, 8 (Salticidae)
From the Greek two-ending u-stem adjective εὔοφρυς -υ 'with fine eyebrows' formed from the prefix εὐ-'good, well' and the Greek feminine u-stem noun ὀφρῦς -ύος 'eyebrow'. As with all two-ending adjectives, the gender is formally ambiguous, either masculine or feminine. This is a case where the rule that the compound takes its gender from the gender of the final element (Code Art. 30) does not work, since the gender of ὀφρῦς is overridden by the morphology of adjective formation. Koch clearly treated it as feminine (Bonnet 1956a:1867 note 131) with such species as *ancilla, tigrina, hamata, incompta* etc. Those authors who treated it as masculine (e.g., Emerton, O. Pickard-Cambridge) are wrong. For the question whether it should be spelled with a u or with a v see the discussion under *Euagrus*. Koch's original spelling with a u should be preserved. The name refers to the type species *Euophrys frontalis* WALCKENAER (1802: 246) which has the anterior eyes in the male fringed with vivid orange hairs.

*Euryopis* MENGE 1868:174 (Theridiidae)
From the Greek three-ending, u-stem adjective εὐρύς -εῖα -ύ 'broad, wide' and the Greek combining form -ωπίς -ωπίδος which forms compound feminine nouns meaning 'having such and such a face'. Hence 'having a broad face'. Menge explained that the name is from εὐρύ 'broad' and ὄπτομαι 'I see' and that it refers to the size and separation of the anterior median eyes. He translated the name as "Weitgesicht" that is, 'wide face'. Menge's first species was *flavomaculata*, hence he treated it as feminine.

*Eurypelma* C.L. KOCH 1850:73 (Theraphosidae)
From the Greek three-ending u-stem adjective εὐρύς -εῖα -ύ 'broad, wide' and the neuter noun πέλμα (gen.) -ατος 'sole of the foot'. Hence 'with a wide sole of the foot' referring to the extensive tarsal scopula which characterizes the Theraphosidae. In Koch's summary of characters he says, "Die Sammetbürste der Fussohlen sehr breit (the velvet brushes of the foot sole very wide)." Raven (1985a:146, 153) demonstrated that *Eurypelma* is an objective synonym of *Avicularia* LAMARCK. North American species decribed in *Eurypelma* are now placed in *Aphonopelma* or *Dugesiella* (Chamberlin 1940a:3). Jung argued there is only one genus in the United States (University of Florida unpublished thesis). For a detailed review of the history of the name see Petrunkevitch (1939b). Subsequently, just as lycosid names are formed on a rhyming principle with the suffix -cosa, theraphosid names came to be generated with the segment -pelma which comes to mean merely "theraphosid." E.g. *Aphonopelma* (which see), *Aphantopelma, Calopelma, Dryptopelma, Lasiopelma*.

*Eustala* SIMON 1895a:795 (Araneidae)
This is originally a species name, which was raised to a genus name by Simon. Walckenaer (1841:33, 37) named several species of *Epeira* from Abbot's drawings among which were *Epeira anastera* and *Epeira eustala* [Abbot 1792 fig. 381 (*anastera*) and figs. 119, 120 and 361 (*eustala*)]. Both are now included in *Eustala anastera*. The word anastera is from the poetic form of the Greek adjective ἀνάστερος -α -ον used by the astronomical poet Aratus (3rd century B.C.) meaning 'unmarked by stars'. Walckenaer apparently meant to contrast this species with those which did have "stars." I suggest that the species with stars would be his *Epeira spatulata* (1841:44) or his *Epeira eustala* (1841:37), now both included in *Eustala anastera*. The species *spatulata* was described from Abbot's figure 366, which shows the spider with a narrow spatulate dark band on the abdomen, with the sides of the abdomen sprinkled with dark spots (tigrés de taches noires). A photograph of Abbot's figure 366 can be seen in Chamberlin & Ivie (1944:11). I suggest these might be the "stars" with which the pattern of Abbot's fig. 381 was contrasted. Or his *Epeira eustala* might serve equally well for the contrast, since he described it as having the abdomen "moucheté sur les côtés de taches brunes comme celles du Léopard" that is, "speckled on the sides with brown spots like those of the leopard." So by the species name *anastera* Walckenaer meant to designate the "unspeckled" form, contrasting with the speckled forms *spatulata* and *eustala*. *Eustala* is an incorrect feminine latinization from the Greek two-ending adjective εὐστάλης -ες 'well-equipped, neat, compact, well-behaved'. McCook (1888b:199, 1894:173) sorted out the synonymy of this most variable species, which Walckenaer had described under eleven different names. McCook said (1894:173):

> "The more thorough studies which I gave Abbot's MSS. in the summer of 1892 showed me that the first description in order [of Walckenaer's] is given under the name of E. **anastera** on page 33, and is No. 4 of Walckenaer's descriptions of the Orbweavers. This corresponds with Abbot's No. 381, which is sufficiently accurate to be recognized as the species under consideration."

Simon, taking into account McCook's synonymies, and being faced with the plethora of names for this species in Walckenaer, separated it from *Epeira*, put it into a new genus, simply raised Walckenaer's species name *eustala* to a genus name for it, and retained the other name *anastera* for the species name, since it was the first in Walckenaer's order.

*Evarcha* SIMON 1902a:397 (Salticidae)
The Greek two-ending adjective εὔαρχος -ον 'easily governed or beginning well' from the prefix εὐ-'good, well' and the feminine noun ἀρχή 'beginning, government'. Simon has had to invent a latinized feminine form, because the original Greek two-ending adjective has no distinctive feminine. C.L. Koch (1850:65) established a new subgenus *Maturna*, which is not a genuine Latin word, but appears to be a mistake for the Roman goddess Matura, patroness of ripening fruits (see under *Corythalia*).

My guess is that Simon in including part of Koch's *Maturna* in his new *Evarcha* thought that the meaning 'with a good beginning' is a reasonable equivalent for Koch's 'early in the morning' and was paying quiet homage with this name. On the question whether it should be spelled with a u or a v see the discussion under *Euagrus*. In this case Simon originally spelled it with a v which should be preserved by Art. 30.b of the Code.

*Faiditus* KEYSERLING 1884a:160 (Theridiidae)
A medieval Latin noun meaning 'one who is involved in a feud, an exile.' There is no discoverable semantic rationale. Revalidated by Agnarsson 2004:478.

*Falconina* BRIGNOLI 1985a:380 (Corinnidae)
Replacement name for *Falconia* SCHENKEL which was preoccupied. Schenkel's name is in honor of the British arachnologist William Falconer.

*Filistata* LATREILLE 1810:121 (Filistatidae)
Probably from Latin filum 'thread' and an agentive combining form meaning 'standing'. The combining form -stata is not native Latin, but is borrowed from Greek as in such Latin words as orthostata (ὀρθοστάτης) 'standing straight (an architectural term for a pillar)', apostata (ἀποστάτης) 'standing apart, deserter, apostate'. There are at least 42 authentic Greek compounds in -στάτης. The name would then mean 'standing on a thread'. It may be objected that Latreille has created a hybrid name, half Latin and half Greek. There is another problem. If the etymology I have proposed is correct, then *Filistata* ought to be a masculine noun of the Latin first declension, yet it has universally been taken to be feminine. Latreille's first species (1810:121) was *testacea* (now *insidiatrix* FORSKÅL 1775), which is unambiguously feminine. The other possibility, that the second element is the feminine perfect passive participle of the verb sisto 'put, place, cause to stand', is unsatisfactory because the form status-a-um has the specialized meaning 'appointed, determined' and cannot mean 'standing'. Despite the fact that the grammatical niceties are a little wobbly, I think we may conclude that Latreille meant his name to mean 'standing on a thread'. The American species were gathered into a new genus *Kukulcania* by Lehtinen (1967:242, 300).

*Filistatinella* GERTSCH & IVIE 1936:1 (Filistatidae)
Feminine diminutive from *Filistata* by the addition of two diminutive suffixes, the Latin -ina followed by the Latin -ella.

*Filistatoides* F.O. PICKARD-CAMBRIDGE 1899a:47 (Filistatidae)
Derived from *Filistata* by the additioin of the suffix -oides (see the discussion under *Argyrodes*).

*Floricomus* CROSBY & BISHOP 1925b:33 (Linyphiidae)
The authors described a new species *Floricomus floricomus* (found in the stomach of the toad *Bufo quercicus*), which was later found to be identical with *Pholcomma rostratum* EMERTON 1882. So the type species of their new genus became *Floricomus rostratus* (EMERTON 1882). They gathered together in *Floricomus* species that have a hardened sclerite on the abdomen and in which the males lack cephalic pits. From the Latin adjective floricomus -a -um 'crowned with flowers' from flori-combining form of the feminine noun flos floris 'flower' and the feminine noun coma 'hair' with a three-ending adjectival termination. "The species placed here agree in having the clypeus protuberant and clothed with hair" (Bishop & Crosby 1935b:33). See the figures in Kaston (1948) figs. 412, 418, 419. The name refers to this crown of hairs.

*Florinda* O. PICKARD-CAMBRIDGE 1896a:164 (Linyphiidae)
A literary reference. Florinda is a character in Robert Southey's poem "Roderick, the Last of the Goths" published in 1814. Pickard-Cambridge also took the genus names *Favila*, *Ervig*, *Witica*, *Wamba*, *Pelayo*, and *Egilona* from Southey's poem. Pickard-Cambridge was in the habit of using Gothic and other Teutonic names.
Roderick was the last Visigothic King of Spain before the Arab invasion. He had usurped the throne in 710 A.D. His reign is enveloped in legends, specifically the legend of Florinda, or La Cava, daughter of Count Julian (or Urban). The sad story is that King Roderick seduced Florinda, and in revenge for his daughter's dishonor Count Julian aided the Arabs in conquering Spain and deposing King Roderick. The story is told by Voltaire and by Gibbon, and is a frequent theme of Spanish folk songs, some of which had been translated into English by Lockhart and published in 1823. Cervantes said that no Spaniard would name his daughter Florinda, reserving the dishonored name for dogs.

*Frontinella* F.O. PICKARD-CAMBRIDGE 1902a:420 (Linyphiidae)
Diminutive formed on *Frontina* SIMON (1884a:206) with type species *bucculenta*. Bonnet (1956a:1922):

> "The name of the genus **Frontina** being pre-occupied in Diptera (Meigen 1838) Simon himself replaced it in 1887 by **Floronia**." (Simon 1887e:clviii)

Keyserling described *Frontina calcarifera* (1886b:107), referring it to *Frontina* SIMON, and F.O. Pickard-Cambridge (1902a:420) creating the genus *Frontinella* with the type *laeta* said:

> "The species here referred to this genus may be recognized by the characters given in the Table [sternum longer than broad, and eye groups widely separated], and **Frontina calcarifera** Keys. probably belongs to it."

So Pickard-Cambridge derived this name from *Frontina* on the basis of Keyserling's *Frontina calcarifera*. Simon's name *Frontina* is a feminine form taken from the Roman surname *Frontinus*, and there seems to be no particular semantic connection. Simon repeatedly used proper names from antiquity as a convenient supply of new names, without any concern for semantic appropriateness. The famous Frontinus wrote a treatise on the water supply of the city of Rome in the first century A.D. (De Aquis). Bonnet (1956a:1922-1923) subsumed both *Frontinella* and *Floronia* under *Linyphia*, as did Roewer (1954b:1511), but *Frontinella* was revalidated "with no comments" by Gertsch & Davis (1946:1).

*Fuentes* PECKHAM & PECKHAM 1894:113 (Salticidae)
This is the plural of the Spanish feminine noun fuente 'fountain, spring' treated as a Latin singular. The Peckhams left the gender ambiguous with the type species *pertinax*, a one-ending Latin adjective. Simon (1903a: 846) with *taeniola*, Mello-Leitão (1945b:282) with *sexpunctata* and Kaston (1938d:195; 1948:444) with *lineata* treated it as feminine, but Banks (1929: 63) treated it as masculine with *punctatus*. Since Simon was the first to disambiguate the gender as feminine, and since the original Spanish word is feminine (albeit plural) it is reasonable to regard the genus name as a feminine singular. The type species was collected in Belize, British Honduras. The Peckhams left several indications that they were interested in Spanish history, and this seems to be a case in point. Fuentes is a Spanish place name. Of the towns in Spain called Fuentes, the most famous would be Fuentes de Onoro, where there was an important battle on May 5, 1811, in which the Duke of Wellington defeated the French general Masséna. This explanation is reinforced by other evidence that the Peckhams were reading Spanish history. They named another genus *Talavera* for the indecisive Battle of Talavera which took place on January 28, 1809, in which Sir Arthur Wellesley (later the Duke of Wellington) prevailed against Marshal Soult, the French commander. See under *Talavera*. The other possibility is that Fuentes is a Spanish surname. The species from the United States have been transferred to other genera (see Platnick 1993c:761).

*Gamasomorpha* KARSCH 1881c:40 (Oonopidae)
From the genus name *Gamasus* LATREILLE 1802b (Acari) of parasitic mites plus the Greek feminine noun μορφή -ῆς 'form, shape' combined into a feminine Latin adjective. Hence, 'having the form of a Gamasid mite'. Latreille's name *Gamasus* has no obvious etymology.

(‡) *Garritus* CHAMBERLIN & IVIE 1933a:9 (Theridiidae)
A genus from the Raft River Mountains, Utah, "close to *Robertus*", so the last part of the name is formed on the analogy of *Robertus*. Chamberlin more than once used garrina as a species name. For example, *Hilaira garrina* from Pike's Peak (Chamberlin 1949:532), and *Cicurina garrina* (1919a:256) also from the Raft River Mountains. He also created a genus *Garricola* (1916:231) from Peru, meaning 'mountain dwelling', which he explained is derived from the Gosiute Indian noun garri or gari 'mountain' (cf. Chamberlin 1908b:2) and the Latin suffix -cola 'dweller'. Hence, *Garritus* means 'montane form close to *Robertus*'. Kaston (1946:1) synonymized this genus under *Ctenium* MENGE, and Levi and Levi (1962: 27) restored *Robertus*, saying, "Usage strongly favors *Robertus*, although *Ctenium* is an older available synonym." *Robertus* has prevailed.

*Gasteracantha* LATREILLE in SUNDEVALL 1833b:14 (Araneidae)
Levi (1978a:434), who gives Sundevall 1833b as author, says:

> "Sundevall cited Latreille,1831 as author of the name **Gasteracantha**. However, **Gasteracanthe** Latreille (Cours d'Entomologie, p.530) is a nomen nudum since no species are included; it is thus an invalid name since it lacks an indication (ICZN, Art. 16, V). Bonnet (1957) also erroneously cites Latreille (1831) as author."

Bonnet (1956a:1934) said:

> "Certain authors attribute the name to Latreille, others to Sundevall. In reality, it is indeed Latreille who created it in 1831 for Walckenaer's ninth family of **Epeira**, but with the French word Gastéracanthe; Sundevall in 1833 did nothing but give the word its Latin form **Gasteracantha**, and recognized in so doing that the name was Latreille's. It is then logical to attribute the name to Latreille in spite of the slight original fault of which he stands accused; moreover, if one wanted to be both strict and correct one ought to say **Gasteracantha** Latreille in Sundevall 1833."

From Greek γαστήρ 'belly' and ἄκανθα 'thorn', referring to the characteristic spines on the abdomen. Bonnet (1956a:1934) criticized the formation of the name:

> "Gastéracanthe and **Gasteracantha** are also both badly formed; according to the rule for forming names from Latin or Greek, it is always the root derived from the genitive which ought to be used, the root from the

*nominative being often irregular. That is the case with* gaster, gastros; *and just as one says* Gastropod *(and not* Gasteropod*) one ought to have said* **Gastrocantha** *(and not* **Gasteracantha** *). This is just the correction which Agassiz [Erichson] made, but he has not been followed by the araneologists who have always said* **Gasteracantha***; and that unanimity of the authors who have used for 100 years a malformed word, obliges me, according to the rule which I have established for myself, not to make useless changes, however natural and logical they may be. We will continue then to write* **Gasteracantha** *since all the authors are of that opinion."*

I think it would comfort M. Bonnet to learn that the formation *Gasteracantha* is formally without blemish. The Greek word γαστήρ has two attested genitives γαστρός and γαστέρος, and furthermore, there are attested Greek compounds built on the second alternative, namely, γαστεροπλήξ 'glutton' and γαστερόχειρ 'living by the work of one's hands' literally 'belly hand'.

## *Gaucelmus* KEYSERLING 1884a:99 (Nesticidae)

This appears to be a macaronic confection from the German word Gaukler 'juggler, trickster, mountebank'. Usually treated as a junior synonym of *Nesticus* THORELL, but revalidated by Gertsch 1971a:96.

## (‡) *Gayennina* GERTSCH 1935b:21 (Anyphaenidae)

Diminutive formed on *Gayenna* NICOLET *in* GAY 1849:450, from which it differs "in having the tracheal spiracle situated midway between the genital furrow and the spinnerets." Now included in *Oxysoma* NICOLET *in* GAY 1849 (Platnick 1974:259). Nicolet's genus was named in honor of Claudio Gay whose Historia fisica y politica de Chile includes Nicolet's work.

## *Gea* C.L. KOCH 1843b:101 (Araneidae)

This is a late spelling for the Latin *Gaea* (e for ae), which is a Latin spelling for the Greek Γαῖα 'the goddess Earth'. Hentz's species name *heptagon* 'a figure with seven angles' is a noun in apposition, since the genus name is feminine.

## *Geolycosa* MONTGOMERY 1904:293 (Lycosidae)

Montgomery erected this genus for *Lycosa arenicola* SCUDDER 1877 [preoccupied; now *Geolycosa pikei* (MARX 1881)]. "In this group the burrowing habit is so far developed that, excepting adult males during the mating season, their whole life is passed underground." From the Greek γεω-, a combining form of Attic γῆ, γῆς 'Earth', and the genus name *Lycosa*. Hence, 'the underground *Lycosa*'. Montgomery, in fact, probably just appropriated the English prefix geo-as in geology, geometry, geodesy etc. without recourse to the particulars of Greek word-formation.

## *Gertschanapis* PLATNICK & FORSTER 1990:109 (Anapidae)

Gertsch (1960a:5) had described the Nearctic Anapid, *Chasmocephalon shantzi*, from California. Platnick and Forster showed that it is not congeneric with the Australian type species of *Chasmocephalon* and erected this new genus for it in honor of Willis Gertsch.

## (‡) *Gertschia* KASTON 1945a:17 (Salticidae)

Kaston erected this genus for *Synemosyna noxiosa* Hentz 1850b

*"in the belief that Hentz's* S. **noxiosa** *differs sufficiently from* **picata** *and* **americana** *to warrant its removal from the genus* **Peckhamia** *in which it has been placed by Peckham & Peckham [1909:370] and later American authors who have erroneously considered it a synonym of* **scorpionia***. The name honored Willis J. Gertsch. It is now placed in* **Synageles** *SIMON".*

[Cutler *in* Kaston [1977:72]).

## *Gertschosa* PLATNICK & SHADAB 1981b:176 (Gnaphosidae)

Named for Willis J. Gertsch on the rhyming analogy of *Gnaphosa*.

## *Ghelna* MADDISON 1996:239 (Salticidae)

Erected with the type *Metaphidippus castaneus* (HENTZ 1846). Maddison (pers. commun.) has a private code associating letters of the alphabet with colors. *Ghelna* is a rather somber-colored Dendryphantine unlike most others which have striped males and spotted females. The combination of the letters g and h in this code indicate a sombre earthy tone. The rest is a euphonic combination of letters. The name is treated as a feminine.

## *Gladicosa* BRADY 1987:285 (Lycosidae)

Erected for species close to the type *Lycosa gulosa* WALCKENAER 1837. "The generic name is a combination of gladius (Latin for sword) referring to the unique sword-shaped embolus of the male palpus, and cosa derived from the generic name *Lycosa*."

## *Glenognatha* SIMON 1887d:194 (Tetragnathidae)

The second element is a rhyming morph to associate this genus with *Tetragnatha*. The first element is from Greek γλήνη 'eyeball, socket of a joint'. Simon described the type species of the genus from a male holotype from Arizona (Levi 1980a:66), which has a noticeably globular palpal bulb, illustrated in Simon's fig. 791. There the bulb certainly looks like an eyeball. Levi mentions the spherical tegulum of the male palp. Hence, in the convention of the rhyming morph the name means 'close to *Tetragnatha* with a spherical tegulum in the male palp'.

## *Glyphesis* SIMON 1926:348 (Linyphiidae)

Not itself an authentic Greek word, but cobbled together from the Greek feminine noun γλυφή -ῆς 'carving' plus the ending found in such words as poiesis, paresis. Hence, 'carving, engraving' with reference to the sulcus behind each posterior lateral eye.

## *Gnaphosa* LATREILLE 1804b:134 (Gnaphosidae)

In the article Drasse in the second edition of the Nouveau Dictionnaire d'histoire naturelle, Latreille (1817: Tome 7:575) explained his view of the status and meaning of this name:

*"This genus* [**Drassus**] *is mentioned in the 24th volume of the first edition of this work, under the name* **Gnaphosa***, which means obscure in Greek; but I have subsequently adopted the name which M. Walckenaer has given it [namely,* **Drassus***]."*

He further explained under the article *Gnaphosa* (Latreille 1817: Tome 13:266) that he preferred Walckenaer's name, despite the admitted priority of his own "because it is easier to pronounce" The explanation that it means 'obscure' rules out any connection with the Greek masculine noun γναφεύς 'cloth-dresser, fuller, cloth-carder'. It is clear that Latreille constructed his name on the Greek masculine noun δνόφος 'darkness' which has a collateral form γνόφος, occurring in later prose. The Septuagint (a Greek translation of the Old Testament circa 200 B.C.) has a verb γνοφόω 'darken' (Lamentations 2:1: "How the Lord in his anger has darkened the daughter of Zion") which would give a feminine present participle \*γνοφῶσα 'darkening'. The name seems constructed on analogy with *Lycosa* (which see). The fact that it its spelled *Gnaphosa* rather than the expected \**Gnophosa* is probably explained by confusion with the words for 'card' or perhaps merely by similarity of pronunciation in Latreille's mind. Since the word means 'cause to be dark' he is not using it with precision. He seems to mean by obscur either 'dark in color' or 'living in dark places' and does not make clear which. But since the genus is based on *Aranea lucifuga* WALCKENAER 1802:221 = *Drassus lucifugus* WALCKENAER 1805:45, which means 'the spider which flees the light' it is clear that Latreille intended the name to mean 'living in the dark'.

## (‡) *Gnathantes* CHAMBERLIN & IVIE 1943:5 (Linyphiidae)

A combination of the Greek feminine noun γνάθος -ου 'jaw' used in arachnology customarily to designate the chelicerae, and the last two syllables of the Linyphiid genera *Bathyphantes, Lepthyphantes, Hyphantes* etc. Chamberlin and Ivie described this genus as distinctive in having the "chelicerae very long and widely divergent distally extending for more than half their lengths beyond the tips of the endites." Hence, by the rhyming principle the name means 'Linyphiid genus with distinctive chelicerae'. Crawford (1988:18) synonymized it under *Meioneta* HULL, which itself is now synonymized under *Agyneta* Hull (Buckle *et al.* 2001:99).

## *Gnathonargus* BISHOP & CROSBY 1935c:220 (Linyphiidae)

Otherwise spelled incorrectly *Gnathonagrus*. The first part is from a comic proper name Γνάθων (gen,) -ονος 'with full mouth' (derived from γνάθος -ου 'jaw'), used for the parasite or moocher, a stereotypical comic character in ancient drama. The second part would be from the masculine noun ἄγρος -ου 'field, country'. The resulting compound, a masculine noun, would mean something like 'gluttonous, mooching bumpkin'. However, the name is spelled *Gnathonagrus* on page 220, and *Gnathonagrus* in the legend to plate XVII. Crosby and Bishop, tended to group these genera in series, and this genus goes with *Gnathonarium* KARSCH (which see), the type species of which, in turn, was once in Walckenaer's genus *Argus* (named for the mythological creature with innumerable eyes). So it is clear that the last part of *Gnathonagrus* is a slip for -argus, and Crosby and Bishop meant to refer to Walckenaer's genus. Bonnet (1957a:2023) spelled it *Gnathonagrus*; but Roewer (1942a:623) correctly spells it *Gnathonargus* as do Brignoli and Platnick.

## *Gnathonarium* KARSCH 1881a:10 (Linyphiidae)

The latinized form of a comic proper name, Γναθωνάριον, occurring in the Greek novelist Longus (4.16) in his tale of Daphnis and Chloe. It is a diminutive of the name Γνάθων -ονος 'Muncher' used by Plutarch and Roman comic dramatists for the parasite or moocher, a stock figure in ancient comic drama. It is a comic name based on the Greek feminine noun γνάθος 'jaw' (see under *Gnathonargus*).

## *Gnathonaroides* BISHOP & CROSBY 1938:84 (Linyphiidae)

Derived from the genus name *Gnathonarium* by the addition of the suffix -oides (see discussion under *Argyrodes*). Such compounds may be masculine, feminine or neuter, and Crosby and Bishop simply took over the species name from *Araeoncus pedalis* EMERTON when they set up this new genus for it. With the species name in this form, namely a Latin third declension two-ending adjective, they left it unclear whether the name was supposed to be masculine or feminine. Bonnet unfortunately meddled with this name in two ways (Bonnet 1957a:2024 note 88). First, he emended it to *Gnathonarioides*, on the grounds that it is built on the stem of *Gnathonarium* and the i ought to be there. While Bonnet is technically correct, this is an unjustified emendation, from the standpoint of the Code. Second, Bonnet held erroneously (1955:36-37) that any derived compound ought to have the same genus as the base word, and since *Gnathonarium* was neuter, the derived name should also be neuter. This is an erroneous principle, at least from the standpoint of Greek and Latin grammar. Currently the Code arbitrarily decrees (again without any justification from Greek or Latin grammar) that all compounds in -oides or -odes shall be masculine. (Code Art. 30.1.4.4) unless the original author explicitly made another choice of gender clear. Since they did not, the name is masculine.

### *Gonatium* MENGE 1868:180 (Linyphiidae)
From the Greek neuter diminutive noun γονάτιον 'little knee, hip-joint, groin' the diminutive of γόνυ (gen.) γόνατος 'knee'. Menge translated the name as "Kniespinne" that is "Knee-spider" and explained that it refers to the inward bowed knees and shins of the palps and first two pairs of legs (that is the patellae and tibiae).

### *Goneatara* BISHOP & CROSBY 1935c:259. (Linyphiidae)
Etymology unknown. The authors give no hint. "This genus is characterized by the peculiar development of the head in the male." They considered the name masculine (with *Goneatara platyrhinus*). Since the species had been included in their genus *Oedothorax* 'swollen thorax' we might assume that the new name somehow reflected that. The first element may be from Greek γωνία 'angle' reflecting the shape of the male cephalothorax. The remainder is a mystery.

### *Gongylidiellum* SIMON 1884r:600 (Linyphiidae)
Diminutive derived from *Gongylidium* MENGE (1868:183) (which see). Hence, a diminutive of a diminutive meaning 'very tiny *Gongylidium*'.

### *Gongylidium* MENGE 1868:183 (Linyphiidae)
From the Greek three-ending adjective γογγύλος -η -ον 'round' from which Menge created a neuter diminutive with the addition of the suffix -ίδιον, hence, 'little round thing'. He apparently did not take the name from the existing Greek neuter noun γογγυλίδιον 'little pill'. Menge translated the name as "Rundknie" that is, 'round-knee' and explained that it refers to the rounded patella of the male palp. There is, however, no reference to the patella in the Greek word. Banks consistently misspelled this name as *Gonglydium*, and was followed by Emerton. Bonnet pointed out (1957a:2045) that in the genus as now defined, none of the species that Banks included remain. Roth (1994:113) reports that Millidge *in litt.* told him that the two species usually included are in the wrong genus.

### (‡) *Gosipelma* CHAMBERLIN 1940a:4 (Theraphosidae)
A subgenus of *Aphonopelma* POCOCK. The second element is from the Greek neuter noun πέλμα (gen.) -ατος 'sole of the foot' but used here as a rhyming morph after *Eurypelma* C.L. KOCH, to indicate a theraphosid (see *Aphonopelma*). The first element is from the name of the Gosiute Indians, a special study of Chamberlin's. The name Gosiute, or the etymologically more correct Kutsipiutsi, is made up of the noun kutsip 'ashes, parched earth, desert' and iutsi 'people' (Chamberlin 1913:2). Chamberlin meant his name to mean 'desert theraphosid' using a combining form of the Gosiute noun kutsi /gosi in its basic meaning 'desert'.

### (‡) *Gosiphrurus* CHAMBERLIN & IVIE 1935b:38 (Corinnidae)
The first element, Gosi-, is from the name of the Gosiute Indians, a special study of Chamberlin's. The name Gosiute, or the etymologically more correct Kutsipiutsi, is made up of the noun kutsip 'ashes, parched earth, desert' and iutsi 'people' (Chamberlin 1913:2). Chamberlin and Ivie say: "This genus is in the vicinity of *Phrurolithus*, though probably not belonging to the immediate group." Hence the last part of the name is dissected from the first part of *Phrurolithus*, to indicate the close relationship to that genus. Rather than making explicit reference to the Gosiute Indians, it may be that Chamberlin meant his name to mean 'desert Phrurolithine' using a combining form of the Gosiute noun kutsi /gosi in its basic meaning, 'desert'. It was originally placed in the Clubionidae, and Brignoli (1983c:545) had classified it in the Gnaphosidae: Micariinae. Roewer (1954a:570) placed it in the Liocranidae: Phrurolithinae. Roth (1985, 1994) synonymized *Gosiphrurus* under *Drassinella* BANKS. Transferred to the Corinnidae from the Liocranidae by Bosselaers & Jocqué (2002:265).

### *Grammonota* EMERTON 1882:38 (Linyphiidae)
From the Greek combining form γραμμο-(as in γραμμοποίκιλος 'striped') of the neuter noun γράμμα (gen.) -ατος 'letter of the alphabet, drawing' and the Greek neuter or masculine noun νῶτον or νῶτος 'back'. The compound is made into a Latin feminine adjective. In Greek such compounds, being two-ending adjectives, never have a distinct feminine form. Hence, 'with pictured or inscribed dorsum'. The type species, originally described in *Erigone*, is *pictilis* O. PICKARD-CAMBRIDGE 1875c, which means 'embroidered'. So Emerton apparently was trying to maintain a reference to Cambridge's species in his new genus name. "The abdomen is gray, usually with conspicuous lighter areas in paired spots forming a distinct pattern" (Kaston 1948:198).

### *Graphomoa* CHAMBERLIN 1924a:7 (Linyphiidae)
The first part of the name is from the Greek combining form γραφο- 'write, mark' from the feminine noun γραφή 'writing, drawing'. The second part is from the Gositute Indian noun moa 'leg' (Chamberlin 1908b:91). The single new species described under this new genus *Theridioides* has the "legs pale, deeply ringed with black." Hence, 'with marked legs'.

### (‡) *Habrocestum* SIMON 1876a:131 (Salticidae)
The Greek kestos is the embroidered Girdle of Aphrodite, with which she could induce the subtle power of love. Hera borrowed it for an afternoon to aid her in seducing Zeus in order to divert his mind from the battle in the 14th book of the Iliad. The Greek adjective habros means 'graceful, delicate'. So the neuter compound noun Simon invented means 'having a delicate girdle'. In the original description of the genus Simon (1876a) says "plastron petit allongé plus étroit que les hanches intermédiaires." Simon (1874a:3) defined 'plastron' as "pièce principale du sternum." So it seems clear that Simon intended *Habrocestum* to mean 'having a delicate elongated narrow sternum'.

### *Habronattus* F.O. PICKARD-CAMBRIDGE 1901a:241 (Salticidae)
From the Greek adjective ἁβρός -ή -όν 'graceful, delicate, pretty' and the old genus name *Attus* WALCKENAER. The n is strange, and appears to be merely a euphonic device to avoid either the hiatus of *Habroattus or the ill-favored mouthful *Habrattus. Hence 'the pretty Salticid'.

### *Hackmania* LEHTINEN 1967:236, 356 (Dictynidae)
Proposed for some *Argenna*. In honor of Walter Hackman of the Zoological Museum, Helsingfors, author of "The Spiders of Newfoundland" (Hackman 1954).

### *Hahnia* C.L. KOCH 1841a:61 (Hahniidae)
Named in honor of Carl Wilhelm Hahn (1786-1836). His premature death interrupted his great work Die Arachniden, after he had finished only two volumes. It was continued by Carl Ludwig Koch. Bonnet is moved to remark (1957a:2079) that Koch's type species *Hahnia pusilla* is one of the rare instances of a spider which has retained its original name unchanged for 115 years (in 1957). Chalk up 165 years by now, and counting. "Le fait est assez rare pour qu'il soit remarqué" says Bonnet. I cannot refrain from remarking on it myself.

### (‡) *Hahnistea* CHAMBERLIN & IVIE 1942a:27 (Hahniidae)
Erected for a new species *Hahnistea longipes*. The genus is close to *Hahnia* C.L. KOCH, and the name is confected from the first syllable of *Hahnia* and from the last three syllables of the Hahniid genus *Antistea* (which see). Synonymized under *Hahnia* by Opell & Beatty (1976:420).

### *Halorates* HULL 1911a:581 (Linyphiidae)
There is an obscure Greek masculine noun from the ancient encyclopedia called the Suda, ἁλωρήτης (gen.) -ου 'watcher of salt'. It is a compound of ἅλς (gen.) ἁλος 'salt' and an agent noun in -της from the verb ὁράω 'see'. The point is that this genus occurs in salt flats, salt marshes and the seashore (Locket & Millidge 1953:325). Hull, having decided to reflect the salt flats in his name, could easily have found this word in the lexicon browsing among the "salt" words in the vicinity of ἅλς.

### *Hamataliwa* KEYSERLING 1887b:457 (Oxyopidae)
Keyserling seems fond of confounding us with exotic names for which he gives no explanation, or only favoring us with a laconic nom. prop. [proper name]. This name occasions one of Bonnet's most peculiar notes. Bonnet, our universally acknowledged hero, our chalcenteric tower of strength, the inspiration and lodestar of all of us who care about intelligent precision in nomenclature, was driven to his wits end by Keyserling's vagaries here. Bonnet says (1957a:2082):

> "Keyserling in describing this genus and its species twice wrote **Hamataliwa** without giving any explanation of the etymology. Marx (1890b:592) and Simon (1898a:380) give no reason for their writing **Hamataliva**; for their part it is probably a lapsus. Still Comstock (1912:660) tries to maintain that spelling saying that **Hamataliwa** is "obviously" a typographical error. I do not see what he bases his evidence on, for the fact that the name is written **Hamataliva** in [Keyserling's] table of contents (p. 489) proves nothing. It is just as easy to regard that as the error. After having written that note I thought that perhaps Comstock's idea was that **Hamataliva** meant 'living at Hamata' liva being the latinization of the English word "living." But Keyserling did not give the name of the locality; he simply says "Nordamerika", and my most detailed atlas gives no geographical name similar to that in North America. (There is only a Djebel Hamata in Upper Egypt). Moreover, it would appear extraordinary if Keyserling, who wrote in German, had used an English word to form the name **Hamataliwa**. Once again it is better to maintain the original spelling

*without seeking to understand what it means, when the author has said nothing about it."*

I quote this at length as an illustration of the imaginative gymnastics sometimes necessary to try to solve these problems. I am not sure I am any more successful than the revered Bonnet. I have the advantage of having figured out that Keyserling was in the habit of using names from the Teutonic middle ages, such as *Odo, Agobardus, Ebo, Wala, Wendilgarda* etc. and *Liwa* is such an old Germanic name, meaning 'dear, beloved' (Schönfeld 1911:155; Förstemann 1900:1018-1021). Cf. the Ostrogothic queen Gudeliva (Schönfeld 1911:114) from gude 'good' and -liva meaning 'gracious'. cf. the Visgothic king Liwigildus (Schönfeld 1911:156). Furthermore there are many Teutonic names beginning with the element Hama-meaning 'armed soldier' (Schönfeld 1911:125; Förstemann 1900: 743). There is also a name Amata (Schönfeld 1911:17-18). Unfortunately, I have not been able to find a Teutonic compound name *Hamataliwa*. I am convinced, however, we are on the right track, and someday scanning the Middle High German romances we will run across this name. Liwa is the name of a Visigothic king, and is masculine. Such a compound, as I expect we will find, would be masculine. But Keyserling clearly treated the name as feminine with the species *grisea*. In the index Keyserling (1887b: 489) spelled the name *Hamataliva*.

## *Haplodrassus* CHAMBERLIN 1922:148 (Gnaphosidae)

Derived from *Drassus* by the addition of a prefix from the Greek adjective ἁπλόος -η -ον 'simple, single, plain'. Chamberlin erected this genus for *Drassus hiemalis* EMERTON and allied species, characterized by the fact that the males have a flattened and dorsally shifted retrolateral tibial apophysis, which is contrasted with the bifid or laterally expanded structures of closely related genera (Platnick & Shadab 1975b:1, 5). Hence, 'the *Drassus* with simple rather than complex palpal tibial apophysis'.

## *Hasarius* SIMON 1871a:329 (Salticidae)

Simon tells us in a footnote without further explanation that this is a proper name. I have been unable to find such a name.

## *Hebestatis* SIMON 1903d:21 (Ctenizidae)

This name cannot be analyzed straightforwardly as a Greek or Latin word. Simon has perpetrated some kind of hanky-panky in creating it. *Hebestatis theveneti* was originally described as a *Cyclocosmia* (Simon 1891g:313) because it had "Abdomen magnum postice ampliatum et truncatum" [abdomen large broadened and truncate behind]. So the idea of bluntness, would seem to lie behind the name. The Latin one-ending adjective hebes (gen.) hebetis 'blunt, dull' probably figures in the first part of the name, but it is impossible to rationalize the second part grammatically. Maybe the name has something to do with the Latin verb hebeto 'make blunt or dull'. There is a line of Ovid in the Metamorphoses (7.210): vos mihi taurorum flammas hebestatis 'you have blunted for me the flames of the bulls'. (from Medea's prayer to the Night, the Earth and the Moon reminding them of their past favors to her, when she helped Jason subdue some fire-breathing bulls). We know Simon used Ovid, and he was capable of such anagrams, intentionally or not. Hebestastis is the second person plural perfect active (syncopated) of the verb hebeto, which Simon might conceivably have metathesized, intentionally or not, into *Hebestatis*.

Bonnet says the name is masculine, despite the fact that the only species is *theveneti*, a genitive (named for the collector Thevenet), which does not help us determine gender. Bonnet has decided arbitrarily, and in accordance with ICZN Art. 30d the name should be treated as masculine by default.

## (‡) *Hedypsilus* SIMON 1893a:486 (Pholcidae)

This name would appear to be a combination of the Greek three-ending adjective ἡδύς -εῖα -ύ 'sweet' and the three-ending adjective ψιλός -ή -όν 'bare'. There is no obvious semantic rationale. *Hedypsilus culicinus* SIMON 1893d was introduced into Texas. *Hedypsilus* is now included under *Modisimus* SIMON 1893d (Huber 1997a:238.).

## *Helophora* MENGE 1866:126 (Linyphiidae)

From the Greek masculine noun ἧλος (gen.) -ου 'nail' and the adjective forming combining form -φορος -α -ον 'bearing' from the verb φέρω 'bear'. Hence, 'nail-bearing'. In Menge's terminology (1866:27) "Nagel" is the scape of the epigynum. Menge translates the name as "Nagelspinne" that is, 'nail spider' explaining that it refers to the rod-shaped extension (stabförmige Verlängerung) on the epigynum, which reaches beyond the middle of the abdomen. Kaston (1948:126) says:

> "Hull (1920) proposed Scaptophorus to replace **Helophora**, on the supposition that the latter was preoccupied. However **Helophoria** Agassiz 1846 and **Helophorus** Illiger 1801 and Gistl 1848 do not invalidate **Helophora**, according to Article 36 [now 56b, the "one letter difference" rule] of the Rules of Zoological Nomenclature."

Hull's proposed substitution *Scaptophorus* would mean 'rod-bearing' from Greek σκῆπτρον 'scepter, rod' (some Greek spelling variations are involved; there is a real Greek word σκαπτόφορος). Hull, too, meant his name to refer to the long scape of the epigynum, as did Menge, and was trying to provide a new name which would be reminiscent of the old.

## *Hentzia* MARX *in* HOWARD 1883:26 (Salticidae)
## *Hentzia* MCCOOK 1894:244 (Araneidae)

Both Marx and McCook named genera in honor of Nicholas Marcellus Hentz (1797-1856) "the father of American arachnology." *Hentzia* MCCOOK is now *Mecynogea* SIMON with the type *basilica* MCCOOK 1878a (Levi 1968a:322). See under *Allepeira*. Hentz described what was to become the type species of *Hentzia* as *Epiblemum palmarum* (HENTZ 1846: 366; his 1835 description is a *nomen nudum*), and *Epiblemum* HENTZ is in part a junior synonym of *Salticus* LATREILLE. So Marx followed the tradition of naming the new genus for Hentz, the describer of its type species. The species of *Hentzia* MARX are sometimes listed under *Wala* KEYSERLING (1885:516). Bonnet said (1959:4804) that although *Hentzia* MARX has priority over *Wala* KEYSERLING, and *Hentzia* MCCOOK, it has been forgotten for 50 years. He objected (Bonnet 1957a:2157) to Banks's resurrecting the name in 1932, because *Hentzia* MARX is a *nomen nudum*, which Marx himself abandoned in his 1890 catalogue (Marx 1890), not even citing it in synonymy. Bonnet, as usual, argued for sinking names unused for 50 years, but the strict rule of priority has ultimately prevailed and *Wala* KEYSERLING is suppressed.

## *Herpyllus* HENTZ 1832:102 (Gnaphosidae)

The Greek masculine noun ἕρπυλλος means 'tufted thyme' (Thymus sibthorpii)' related to the verb ἕρπω 'creep, crawl' presumably because it is a creeping plant. The word occurs in a 4th century A.D. Latin botanical work the Herbarium of Lucius Apuleius, who also wrote the famous novel The Golden Ass. Hentz did not give us any rationale for the choice.

## *Hersilia* Audouin 1826:114 (Hersiliidae)

Proper name. Wife of Romulus, founder of Rome, and its first legendary king. Ovid. Metamorposes 14.830 ff; Livy 1.11.2. There is no semantic rationale for the name.

## *Hesperocosa* GERTSCH & WALLACE 1937:4 (Lycosidae)

From Greek ἕσπερος 'evening, west' and the rhyming morph -cosa from *Lycosa* to indicate that it is a lycosid. Erected for *Schizocosa unica* GERTSCH & WALLACE (1935:9) from Texas and New Mexico. Hence, 'western lycosid'.

## *Hesperocranum* UBICK & PLATNICK 1991:10 (Liocranidae)

"The generic name is a contraction of hesperos (Greek for western) and *Liocranum*, and is neuter in gender." The genus is closely related to *Liocranum* and was erected for the new species *Hesperocranum rothi* from California and Oregon.

## *Heterodictyna* F. DAHL 1904:118 (Dictynidae)

From the Greek three-ending adjective ἕτερος -α -ον 'other' and the genus name *Dictyna*. It means 'the other *Dictyna* of the pair'. Bonnet said (1956a:1459 note 50):

> *"Dahl in 1904 created the genus* **Heterodictyna** *for the species of* **Dictyna** *with trichobothria, but he did not name the species which he included in this genus. Elsewhere in 1906 [1906d:60], he himself placed the new genus in synonymy with* **Dictynina** *Banks [1904a:342], which was also a genus of 1904, but which is employed with a species [pallida]. Also, in this case, as it is difficult to know which name has priority in reality, there is reason to grant it to* **Dictynina***."*

Further (1957a:2183):

> *"But in 1924 Dahl [Dahl 1924:154] employed* **Heterodictyna** *with the species* **Dictyna flavescens** *Walckenaer, which can be considered the type species [of* **Heterodictyna** *Dahl 1924] which on that account would place it in synonymy with* **Dictyna***."*

However, Gertsch (1946a:7) included *Dictynina pallida* BANKS 1904b in the genus *Mallos*, and Chamberlin & Gertsch (1958:47) reactivated the name *Heterodictyna* with the type species *flavescens* WALCKENAER 1830. Lehtinen (1967:252) created the new name *Nigma* with the type species *flavescens* WALCKENAER. Brignoli (1983c) and Platnick (2005) accepted *Nigma*, as a replacement name for *Heterodictyna*.

## *Heteroonops* DALMAS 1916:203 (Oonopidae)

From the Greek three-ending adjective ἕτερος -α -ον 'other, different' plus the genus name *Oonops* (which see).

## *Heteropoda* LATREILLE 1804b:135 (Sparassidae)

From the Greek adjective ἕτερος -α -ον 'other, different, unequal, uneven' and the combining form of the masculine noun πούς ποδός 'foot' plus an adjective-forming ending, here feminine. Hence, 'with unequal legs'. There is an authentic Greek one-ending adjective ἑτερόπους (gen.) -ποδος 'with uneven feet, halting, lame'. This may refer to the marked difference between the male and the female palp, i.e. the sexes have differing legs, or to the fact that the legs are of different lengths (II, I, IV, III).

For a full discussion of the complicated relation between *Sparassus* and *Heteropoda* see Fox (1937e:461).

*Hexura* SIMON 1884m:315 (Mecicobothriidae)
At first glance this seems to be simply from the Greek numeral ἕξ 'six' and the feminine noun οὐρά 'tail'. Hence it would mean, 'having six spinnerets (tails)'. But this feature is not particularly distinctive, since lots of Mygalomorphs have six spinnerets. Simon in discussing his group Hexurinae which he included among the Atypidae (1892a:195) said they are like the other Atypidae in their endites, but in other respects are like the group *Macrothele* of the Dipluridae, because their spinnerets are arranged like those of Dipluridae and elongated like *Macrothele*. It seems that Simon wanted to draw the comparison with the Dipluridae in the name. *Hexathele* ('with six spinnerets') had already been used by Ausserer in 1871a, and Simon said (1892a:187) that *Hexathele* differs from *Macrothele* only in having six spinnerets. All of these elements, it seems, enter into his choice of a name. It appears that he wanted to link this genus with *Diplura* by the ending of his new name -ura, and with *Hexathele* by the beginning, Hex-. So *Hexura* does not mean simply 'with six spinnerets' as we would suppose from Greek alone, but rather 'sharing features with both *Diplura* and *Hexathele*'.

*Hexurella* GERTSCH & PLATNICK 1979:27 (Mecicobothriidae)
Latin feminine diminutive from *Hexura* SIMON.

*Hibana* BRESCOVIT 1991b:730 (Anyphaenidae)
An arbitrary combination of letters treated as feminine.

*Hilaira* SIMON 1884a:374 (Linyphiidae)
From the Greek feminine adjective (it occurs only in the feminine) ἱλάειρα 'mildly shining'. The type species *excisa* is described (p. 375) as having "plastron brun, finement chagriné, brilliant" [sternum brown, finely chagrined, shining].

*Hillhousia* F.O. PICKARD-CAMBRIDGE 1894b:89 (Linyphiidae)
Derived from an English surname. Pickard-Cambridge does not tell us who this Hillhouse is, but the only likely candidate is William Hillhouse (1850-1910), who published several works on the botany of Bedfordshire, was a Fellow of the Linnaean Society, and was Professor of Botany at the Mason Science College, Birmingham, and Professor in the University of Birmingham. He was one of the editors of the English journal The Midland Naturalist, and translated E. Strasburger's Handbook of Practical Botany. See Cassiro (1885:220) and Freeman (1980:167).

*Hogna* SIMON 1885g:9 (Lycosidae)
(as a new subgenus). Simon used this subgeneric name for the species close to *Lycosa radiata* LATREILLE (1817:347). Roewer (1954a:247) elevated it to full generic status. Dondale & Redner (1990) have placed the North American *Lycosa* species in *Hogna*. Simon gave no rationale for the name. It could possibly be be a Latin form of the place name Hogne in Belgium. It is not clear why Simon chose it. Another possibility is that it is somehow derived from the Teutonic mythological name Hogni, father of one of the Valkyries. Neither of these guesses is very convincing.

*Holocnemus* SIMON 1873a:48 (Pholcidae)
From the Greek two-ending compound adjective ὁλόκνημος -ον 'with the whole shin (of a ham containing the whole leg, ham and shank)' which is in turn from the three-ending adjective ὅλος -η -ον 'whole, complete, entire' and the feminine noun κνήμη -ης 'limb' Hence, 'with whole leg'. The point of the name is not clear. But, since the genus is distinguished by the presence of a row of macrosetae on the femur and tibia of the male, it is possible that Simon really meant the name to be *Hoplocnemus, which would mean 'with armed tibia'. The European species *Holocnemus pluchei* (SCOPOLI 1763) has been introduced into Arizona and California.

*Hololena* CHAMBERLIN & GERTSCH 1929:105 (Agelenidae)
Elevated to genus by Chamberlin & Ivie 1942b:203. The agelenid genus names in -lena are formed to rhyme with *Agelena* (Hololena, Rualena, Tortolena, Calilena, Novalena) by misdivision of morphemes (cf. the Lycosid names in -cosa). The morpheme -lena has no meaning (except the taxonomic conventional meaning 'derived from *Agelena*'). The first element is from the Greek adjective ὅλος -η -ον 'entire, whole'.

*Homalonychus* MARX 1891:29 (Homalonychidae)
From the Greek adjective ὁμαλός -ή -όν 'even, level, uniform (of a surface)' and the masculine noun ὄνυξ (gen.) ὄνυχος 'claw, talon'. This refers to the fact that the tarsal claws are without teeth.

*Horcotes* CROSBY & BISHOP 1933b:15 (Linyphiidae)
From the Greek masculine a-stem noun ὁρκώτης -ου 'the officer who administers the oath' an agent noun related to the verb ὁρκόω 'to swear an oath'. There is no particular semantic point to the name. Crosby and Bishop were in the habit of searching the Greek Lexicon for Greek words as yet unused in zoological nomenclature. See *Hormathion*.

(‡) *Hormathion* CROSBY & BISHOP 1933b:161 (Linyphiidae)
From the Greek neuter diminutive noun ὁρμάθιον 'little string or chain of beads' from ὅρμαθος 'string of beads'. Crosby and Bishop were in the habit of searching the dictionary for available Greek words, and for this paper they were in the omicron -rho section and found both *Hormathion* and *Horcotes*. There is no particular semantic rationale. Bonnet's emendation to *Hormathium* (Bonnet 1957a:2236) is unjustified, according to the current Code. Bonnet was trying to regularize the nomenclature by Latinizing all original Greek spellings. See Bonnet 1945:134. Synonymized under *Thyreosthenius* SIMON by Hackman (1952b:73).

*Hybauchenidium* HOLM 1973:85 (Linyphiidae)
Erected for *Erigone aquilonaris* L. KOCH 1879c. From the Greek masculine noun ὕβος (gen.) -ου 'camel's hump' and the Greek masculine noun αὐχήν (gen.) -ένος 'neck' plus the diminutive ending -ίδιον. Hence, 'little creature with a humped neck'.

*Hybocoptus* SIMON 1884r:708 (Linyphiidae)
From the Greek masculine noun ὕβος (gen.) -ου 'camel's hump' and the verbal adjective κοπτος -ή -ον 'chopped off'. Refers to the fact that in the male the large cephalic hump is divided in front from the surface of the clypeus by a deep incisure. The type species is *decollatus* SIMON, which means 'decapitated'. See Locket & Millidge 1953:228, text-fig. 139C, and Roberts 1987: text-fig. 17d. Hence, 'with the hump chopped off'.

(‡) *Hyctia* SIMON 1876a:18 (Salticidae)
This is neither a Greek nor Latin word, and has no discoverable etymology. Simon gives us no clue. Synonymized under *Marpissa* C.L. KOCH by Barnes (1958:15).

*Hypochilus* MARX 1888a:161 (Hypochilidae)
From the Greek prefix ὑπό 'under' and the neuter noun χεῖλος (gen.) χείλους 'lip'. There are authentic Greek compound two-ending adjectives in -χειλος like ἀγκυλόχειλος 'with curved lip' λαγώχειλος 'with harelip' πρόχειλος 'with prominent lips'. Regularly in Greek grammar neuter nouns like χεῖλος form compound adjectives which may be masculine. Marx explained the derivation as: "Hupo below, cheilos lip: from the position of the labium." In the description of the genus he says: "Labium broad, short and straight, situate below the maxillae which stand upon it." Marx intended *Hypochilus* to mean something like 'with labium beneath the maxillae'. Marx says that he gave the spider this name at the suggestion of Thorell, whom he then honored with the species name *thorelli*.

*Hypomma* F. DAHL 1886:87 (Linyphiidae)
From the Greek preposition ὑπό 'under' and the neuter noun ὄμμα (gen.) ὄμματος 'eye'. In the male the head is elevated into a large, longitudinally divided lobe, but all of the eyes are below the lobe. See Locket & Millidge 1953:224,Text-fig. 137E, F, G, and Roberts 1987:50-51, Text Fig. 16j, 17a, b. Hence, 'with eyes beneath the cephalic lobe'. Dahl originally incorrectly treated the name as feminine. It was corrected to neuter by Kulczyński (1894:325).

*Hypselistes* SIMON 1894a:671 (Linyphiidae)
This is not a real Greek or Latin word, but Simon favored names in -istes (Argistes, Clitistes, Typhistes) even though none of them is morphologically correct from the strict Greek standpoint. The Greek suffix -ιστης is an agent suffix, which forms nouns related to verbs ending in -ιζω, such as ἀγωνίστης from ἀγωνίζω, and in Greek it would not be attached to adjectives. The first element is from the Greek three-ending adjective ὑψηλος -ή -όν 'high, lofty, proud'. Emerton (1882:46) put the type species now *Hypselistes florens* into *Lophocarenum* MENGE (1868) which means 'with a crest on the head' (Greek λόφος 'crest'; κάρηνον 'head'). And in creating this new genus for it Simon apparently intended to refer to the very high cephalic lobe of the male, which was also the point of Menge's name. The simple name *Hypselos* was unavailable to Simon, because it was preoccupied in the Coleoptera (Schönherr 1843), so Simon seems to have created this with his favorite suffix -istes as the next best thing.

*Hypsosinga* AUSSERER 1871b:827 (Araneidae)
(as subgenus) Levi (1972a:240) said:

> "Both Wiehle (1931) and Roewer's Katalog der Araneae spell the generic name **Hyposinga**. Roewer changed the spelling of many generic names from long accustomed usage to the spelling of that of the original author. But here Roewer changed the original spelling. Ausserer consistently spelled the name with "s" and also indicated that a main character of the subgenus is the high clypeus (hypso, Greek for high)."

In the diagnosis Levi went on to say:

> "The clypeus height in **Hypsosinga** is 1.5 to 3 diameters of the anterior median eyes, but only about one diameter of the anterior median eyes in **Singa**."

The forms ὑψι-ὑψο- 'high' form adverbs, nouns adjectives and verbs in Greek. *Hypsosinga* would be analogous to such a genuine Greek word as ὑψόφωνος 'with high-pitched voice'. Hence, because it has a higher clypeus than *Singa*, the meaning is 'high *Singa*'.

*Hyptiotes* WALCKENAER 1837:279 (Uloboridae)
Walckenaer originally spelled this name *Uptiotes* since the French, in pronouncing Greek, habitually omit aspiration, and the h of the latinization is represented in the Greek only by a diacritical mark. Furthermore, the Greek letter upsilon can be transliterated with a u as well as with a y depending upon the convention. Bonnet (1957a:2269) said it was corrected to *Hyptiotes* first by Erichson in Agassiz (Erichson 1845), but that is not true. It is first corrected in Agassiz (1848:553), a fact which Thorell (1869:67) recognized. Strict interpretation of the Code (Art. 32.c.ii) would require using Walckenaer's spelling, since "incorrect transliteration or latinization" is not grounds for emendation, and one cannot even argue that Walckenaer's spelling is incorrect. It merely follows an alternative convention. On the other hand, because *Hyptiotes* has been consistently used since Thorell 1869, it would be folly to meddle.
The formation of *Hyptiotes* is not entirely clear. Walckenaer's species (1837:277) was *anceps* meaning 'ambiguous' and it is a one-ending adjective so we cannot tell whether it is masculine or feminine. There is a real Greek feminine abstract noun ὑπτιότης -τητος meaning 'supine position' used to describe leaves by the botanical writer Theophrastus (ca. 300 B.C.), so *Hyptiotes anceps* might then mean 'ambiguous upsidedownness'. But Walckenaer might well have made up a word as if from the Greek verb ὑπτιόω 'to be on one's back' by the addition of a masculine agent suffix -της (gen.) -του yielding *ὑπτιώτης, which would mean something like 'the one lying on his back'. This obviously refers to the characteristic position of the spider, dorsum down, holding the single thread between its triangular web and the attachment point. It is apparently Thorell (1869: 67) who first made it unambiguously masculine by including C.L. Koch's species *Mithras paradoxus* in the genus as *Hyptiotes paradoxus*.
Further, when Walckenaer first created the name (1833:438) it was a *nomen nudum* and therefore C.L. Koch's name *Mithras* (C.L. KOCH *in* HERRICH-SCHÄFFER 1834: heft 123, fig. 9) would seem to have priority over Walckenaer's 1837 substantiation of the name with his species *anceps* (now *paradoxus*). But Bonnet (1957a:2269), following Thorell (1869), pointed out that Koch's *Mithras* was preoccupied in the Lepidoptera, whereby Walckenaer's name was revalidated.

*Icius* SIMON 1876a:54 (Salticidae)
C.L. Koch had created the genus *Icelus* (C.L. Koch 1846e:174 *nomen nudum*; 1850:55) from the Greek adjective ἴκελος -η -ον 'similar, like'. Koch meant to indicate that this genus is like *Phidippus* ("eine in allen Formen mit *Phidippus* übereinstimmende Gattung [a genus in all respects agreeing with *Phidippus*]"). A purist would have preferred him to spell it *Hicelus. Since the name *Icelus* is preoccupied in Pisces, Simon replaced it with a similar name, which is not a Greek nor Latin word. *Icius* is a confection formed by removing the -elus from Icelus (as if it were a suffix) and adding the Latin adjective suffix -ius. The North American species have been transferred to other genera (see Platnick 1993c:770).

*Idionella* BANKS 1893a:130 (Linyphiidae)
Based upon *Ceratinella* EMERTON (which see). Erected by Banks for his *Ceratinella formosa*, which he had described in 1892 (Banks 1892a:33). The first part of the name is from the Greek three-ending adjective ἴδιος -α -ον 'one's own'. The second part is a rhyming morph from *Ceratinella*. Banks says:"My *Ceratinella formosa* is not a true *Ceratinella*, I propose for it *Idionella* distinguished by the position of the horny shield." Hence, Banks is calling this genus 'my own *Ceratinella*'.

*Improphantes* SAARISTO & TANASEVITCH 1996b:177 (Linyphiidae)
The type species is *Lepthyphantes improbulus* SIMON 1929. The genus name is a combination of the first part of the species name plus the rhyming morph -phantes taken from *Lepthyphantes*. Simon's species name *improbulus* is a rare Latin adjective used by Juvenal meaning 'somewhat wicked'.

*Incestophantes* TANASEVITCH 1992:45 (Linyphiidae)
The type species is *Lepthyphantes incestus* L. Koch 1879c, and the genus name is a combination of the first part of Koch's species name plus the rhyming morph -phantes taken from *Lepthyphantes*. Koch's species name is a Latin adjective meaning 'impure, defiled, sinful'.

*Isaloides* F.O. PICKARD-CAMBRIDGE 1900:163 (Thomisidae)
Derived from the genus name *Isala* L. KOCH 1876a:796 by the addition of the suffix -oides (see under *Argyrodes*). F.O. Pickard-Cambridge somewhat doubtfully believed this genus to be close to Koch's Australian genus *Isala* (etymology unknown).

*Ischnothyreus* SIMON 1893a:296-298 (Oonopidae)
Simon created the name *Ischnaspis* (Simon 1891c:562) from the Greek adjective ἰσχνός -ή -όν 'dry, withered, lean, spare, reduced' and the feminine noun ἀσπίς (gen.) -ίδος 'shield'. Hence, 'with reduced shield'. This genus belongs to Simon's group Oonopidae loricatae or "Armored Oonopids" in which the abdomen is "armored" with plates or scuta (shields) both dorsally and ventrally in the typical forms. But in this genus the ventral scutum is reduced, and does not extend beyond the epigastric furrow (Simon 1893a:296). However, the name *Ischnaspis* was preoccupied in the Hemiptera, so Simon substituted *Ischnothyreus* meaning roughly the same thing, from the Greek masculine noun θυρεός 'a shield shaped like a door, equivalent to the Roman scutum'. Hence, 'with reduced scutum'.

*Islandiana* BRAENDEGAARD 1932:22 (Linyphiidae)
From the fact that Braendegaard described his type species *Islandiana princeps* from Iceland.

(‡) *Isohogna* ROEWER 1954a:261; 1960d:571 (Lycosidae)
From the Greek three-ending adjective ἴσος -η -ον 'equal' and the genus name *Hogna* SIMON.

(‡) *Ivesia* PETRUNKEVITCH 1925b:320 (Nesticidae)
In honor of J.D. Ives, who wrote on the cave fauna of Tennessee. A series of the type species *tennesseensis* was taken by J.D. Ives from a mound of bat guano in the zone of total darkness one-quarter of a mile inside Indian Cave, near New Market, Tennessee. Synonymized under *Nesticus* THORELL by Gertsch (1984:23).

*Iviella* LEHTINEN 1967:241 (Dictynidae)
In honor of Wilton Ivie, talented American arachnologist who collaborated with R.V. Chamberlin and W. J. Gertsch. Erected for *Tricholathys ohioensis* CHAMBERLIN & IVIE 1935b, and *Tricholathys reclusa* GERTSCH & IVIE 1936.

*Ivielum* ESKOV 1988b:682 (Linyphiidae)
Erected for the type species *Ivielum sibiricum* ESKOV 1988b. Presumably named in honor of Wilton Ivie. Note the spelling with one l, *Ivielum*, not to be confused with *Iviella* LEHTINEN 1967 (Dictynidae).

(‡) *Ixeuticus* DALMAS 1917a:329 (Desidae)
From Greek ἰξευτικός an adjective meaning 'pertaining to one who catches birds with birdlime'. Ixeutica (neuter plural) was the title of a lost poem on birdcatching by the ancient Greek didactic poet Oppian (3rd cent. A.D.), who also wrote a poem on the fishes of the Black Sea, the Halieutica, which is extant. The word is derived from Greek ἰξός 'glue', which is cognate with Latin viscus 'sticky'. Synonymized under *Badumna* THORELL by Gray (1983a:249). The genus *Badumna* was placed in the family Desidae by Forster (1970b:21).

*Jacksonella* MILLIDGE 1951:561 (Linyphiidae)
Formed from the English proper name Jackson with the addition of a Latin feminine diminutive suffix -ella. Erected for *Maro falconeri* which was described by A.R. Jackson (1908:61).

*Jalapyphantes* GERTSCH & DAVIS 1946:7 (Linyphiidae)
The first part of the name is geographical, for the city of Jalapa, Vera Cruz, Mexico (although none of the species comes from there), and the second part is a rhyming morph to associate the new genus with other Linyphiid genera such as *Lepthyphantes*, *Bathyphantes*, and *Pityohyphantes* (which see). Since the second half is actually hyphantes purists would have preferred Gertsch and Davis had written *Jalaphyphantes, but the original spelling, of course, prevails. See the discussion under *Lepthyphantes*

*Kaestneria* WIEHLE 1956:272 (Linyphiidae)
Named for the zoologist A. Kaestner, famous for his work on pseudoscorpions, Pedipalpi and Palpigradi, and his textbook, Lehrbuch der speziellen Zoologie.

*Kaira* O. PICKARD-CAMBRIDGE 1889d:56 (Araneidae)
This could be from Greek noun καιρός 'right proportion, right time, right place' or from the Greek noun καῖρος 'the webbing or thrums to which the threads of the warp are fastened'. Semantically the latter seems preferable for an orb-weaver, but in neither case would the feminine form be justified grammatically. Pickard-Cambridge gave us no clue. Simon (1895a:894) emended the name to *Caira* thinking thus to latinize it properly. Such an emendation is not permitted by the Code, although Bonnet argued for it (Bonnet 1955:924; see also 1945:134).

*Keijia* YOSHIDA 2001c:169 (Theridiidae)
The author says, "dedicated to my father the late Keiji Yoshida."

*Kibramoa* CHAMBERLIN 1924b:589 (Plectreuridae)
Chamberlin (1919b:4) described a new spider from California as *Plectreurys suprenans* which became the type species for this genus. The name is one of Chamberlin's Gosiute Indian names. The Gosiute adjective kiberant (with combining forms kibera, kibr) means 'high' (Chamberlin 1913:7) and the noun moa means 'leg' (Chamberlin 1908b:91). Hence, 'with high legs'. The genus is distinguished from the closely related *Plectreurys* partly by "its much longer and more slender legs".

*Kukulcania* LEHTINEN 1967:300 (Filistatidae)
Erected for *Filistata hibernalis* HENTZ 1842b. Kukulcan was one of the three great culture heroes of the Maya Indians. The name means 'plumed serpent' and Kukulcan is equivalent to the Aztec god Quetzalcoatl. Both were reputed to come from afar, brought cultural and technical benefits

including writing, disappeared, and were expected to return in time. Kukulcan was represented with protruding teeth, a pendulous nose, and lolling tongue [MacCulloch & Gray 1916-1932(11): 134-136]. Lehtinen was fond of creating names from the mythologies of the world. Cf. Devendra, Pandava.

(‡) *Labuella* CHAMBERLIN & IVIE 1943:6 (Linyphiidae)
This is not a genuine Greek or Latin word, although it ends in the common Latin feminine diminutive suffix -ella. Chamberlin and Ivie established the new genus *Pimoa* (which see) in the same paper as *Labuella*. The type species of *Pimoa* was *Labulla hespera* GERTSCH & IVIE (1936:16). It seems clear that Chamberlin and Ivie built the name *Labuella* on the linyphiid genus name *Labulla* SIMON (1884a:262). Synonymized under *Oreonetides* STRAND by Saaristo (1972:69).

*Labulla* SIMON 1884a:262 (Linyphiidae)
This is a Latin feminine proper name invented by the Roman poet Martial (1st century A.D.) for an adulteress in two of his entertaining salacious epigrams. There is no semantic connection. Simon was in the habit of gleaning random convenient proper names from antiquity, and there is every reason to expect Simon to have known Martial. The reference is Martial Epigrammata 4.9 and 12.93. The Nearctic species of *Labulla* have been transferred to *Pimoa* CHAMBERLIN & IVIE (1943:9).

*Larinia* SIMON 1874a:115 (Araneidae)
Simon gives no explanation of the name. It might appear to be connected with the Greek adjective λαρῑνος -ή -όν 'fatted, fat' but it escapes me why a conspicuously slender spider should be named 'fat'. Thorell, for his part, saw nothing fat in it, for he created the genus *Lipocrea* for some of these spiders (Thorell 1878b:6), specifically *phthisica* L. KOCH 1871. Thorell's name is from the Greek two-ending adjective λιπόκρεως -ων 'wasted, thin' literally, 'losing flesh' and Koch's species name means 'wasted, phthisic'. It is more likely that Simon, as was his frequent habit, searched the dictionary for usable names, and came up with the ancient Italian town Larinum, well known from Cicero's speech Pro Cluentio, a famous murder case. Simon seems to have stuck a feminine suffix on the base of this geographical name. Simon (1887d:clxxxvii) synonymized Thorell's *Lipocrea* under his own *Larinia*. He later (1929:692) treated *Larinia* as a subgroup of *Araneus*, but Bonnet (1957a:2346) argued for keeping it distinct. Many species have been transferred to other genera (see Platnick 1993c:440).

*Larinioides* DI CAPORIACCO 1934b:14 (Araneidae)
From the genus name *Larinia* plus the suffix -oides 'like, similar' (see the discussion under *Argyrodes*). Since forms in -oides can be masculine, feminine, or neuter, according to strict grammar, and Caporiacco's species name, a genitive, obscures the gender of the genus name, we have no clear notion of his intent. But now the Code arbitrarily decrees that such names shall be masculine, irrespective of the original publication (Code Art. 30b). Grasshoff (1983) argued that this is the oldest unambiguous available name for the so-called *Araneus cornutus* group. Levi had resurrected Simon's old subgeneric name *Nuctenea* SIMON 1864 (Levi 1974b:300), but included *umbraticus* CLERCK 1757. Now *Nuctenea* applies to Old World spiders related to *umbraticus*, and *Larinioides*, with type species *Larinioides folium* (SCHRANK 1803), contains only spiders of the *cornutus* group.

*Lasaeola* SIMON 1881a:136 (Theridiidae)
Note that Simon's original spelling was *Lasaeola*. It was spelled *Laseola* by F.O. Pickard-Cambridge (1894c:168), which Bonnet (1956a:1501) considered a *lapsus calami*, but it was picked up by several authors. The spelling which looks like it might be *Lasoeola* on Simon's page 137, is nothing more than a typographical convention where the ligature for ae looks as if it might be oe. The etymology is not obvious. The name might be constructed from the Greek three-ending adjective λάσιος -α -ον 'shaggy, bushy, wooly'. Some of the included species might be so described: *Laseola auberti* SIMON 1881a has legs with long hairs; *Laseola braccata* C.L. KOCH 1841b has an abdomen with scattered long yellow hairs; *Laseola sericata* (SIMON 1879a) has an abdomen "garnished with fine hairs closely spaced" *erythropus* has an abdomen with scattered fine yellow hairs; also *testaceomarginata, prona, inornata, auberti, convexa,* and *pyramidalis*. Since the majority of the species included by Simon have some noticeable shagginess of the abdomen, it is likely that the name comes from the Greek word λάσιος. Simon then added a Latin feminine diminutive suffix -ola. The diphthong ae cannot be formally explained; it probably was generated by the way Simon pronounced his Greek. Simon himself synonymized *Lasaeola* under *Dipoena* THORELL in the Histoire Naturelle des Araignées (1894a:563). It was removed from synonymy with *Dipoena* THORELL by Wunderlich (1988:148).

*Lathys* SIMON 1884o:321 (Dictynidae)
This name is Simon's substitute for *Lethia* MENGE (1869:249), which was preoccupied in the Lepidoptera. Menge's name is from the obscure Greek two-ending adjective λήθιος -ον 'causing forgetfulness' which the ancient lexicographer Hesychius says is equivalent to the two-ending adjective λαθραῖος -ον 'secret'. These words are all related to the name of the river Lethe (Greek Λήθη), the River of Forgetfulness in the Underworld, from which the souls drank before being resurrected. The adjective Menge chose, λήθιος -ον, has no distinctive feminine form, and the word he cites *λήθια, does not actually exist. It is true that such Greek words sometimes do develop a distinctive feminine when latinized. Menge translated the name as "Hüllspine" that is, 'covered spider' and explained that it refers to the spider's concealment under moss and lichen. Simon obviously tried to find a replacement which would reflect Menge's name, and based it upon another form of the same Greek root λαθ-as in the Doric feminine noun λαθοσύνα (used by Euripides) meaning 'forgetfulness'. There is no Greek word which would give the Latin form *Lathys*, but the closest would be a neuter noun λάθος (gen.) λάθεος 'escape from detection'. Perhaps the name is constructed on analogy with such names as *Plectreurys* SIMON, *Poecilarcys* SIMON, *Trigonobothrys* SIMON, and *Arcys* WALCKENAER. Morphologically the gender of *Lathys* is indeterminate, but Simon chose to make it feminine.

(‡) *Latithorax* HOLM 1943:22 (Linyphiidae)
From the Latin adjective latus-a-um 'wide, broad' and the Greek masculine noun θώραξ (gen.) -ακος 'trunk, chest, corselet'. Since thorax can be regarded as a Latin anatomical term, the name avoids the opprobrium of being a hybrid. The genus was erected for *Sintula faustus* O. PICKARD-CAMBRIDGE 1900a, hence, 'with broad thorax'. Synonymized under *Semljicola* by Saaristo & Eskov 1996

*Latrodectus* WALCKENAER 1805:81 (Theridiidae)
This name has suffered many variant spellings and misunderstandings. Bonnet (1957a:2364) said:

> "In accordance with his custom Walckenaer in creating the genus **Latrodectus** (1805:81) did not give the etymology of the name. Subsequently Agassiz (1848 [actually Erichson 1845:7]) by accepting the etymology λάτρον 'merces' [profit] δέκτος 'acceptus' [accepted] [N.B. Bonnet's transcription is incorrect] and Simon (1864:177) by accepting the etymology λάτρον '[of the] worker' and δήκτης 'biting' justified the spelling **Latrodectus**. In 1868 Canestrini and Pavesi [1868:782] gave the spelling **Latrodectes**, but without explanation, and in 1869, Thorell, rejecting the interpretations of Agassiz [actually of Erichson] and Simon, established that one ought to write **Lathrodectus** which has the etymology λάθρα 'secretly' and δήκτης 'biting'. Finally Kobert (1888:442) invented the spelling **Lathrodectes** claiming that it was the correct orthography."

First, there is no doubt that Walckenaer's original spelling *Latrodectus* is to be preserved, irrespective of any etymological argument. The Code (Art. 32.c.ii) is clear on that point. Of the suggested etymologies, Erichson's (in Agassiz) is to be dismissed without a second thought. As usual Erichson had no idea what he was doing, and his suggestion is ignorant and completely without merit. Simon's etymology is slightly better, since he was right about δήκτης, but there is no Greek word λάτρος meaning 'worker'. The word is λατρεύς, and it would not combine this way. Thorell's etymology is fundamentally correct, but needs some adjustment Thorell (1869:95) said:

> "The name "**Latrodectus**" is evidently formed of λάθρα [secretly (adv.)] and δήκτης [biter] in the same manner as e.g., **Lathrobium** of λάθρα and βιόω (to live), and ought therefore to be written **Lathrodectus**."

He further said that the derivations in Agassiz and Simon are "very improbable, as yielding no rational meaning for the name." There is a genuine Greek word, cited by the grammatical writer Phrynicus, λαθροδήκτης 'biting secretly' from the combining from λαθρο-'secret' and the first declension masculine agent noun δήκτης -ου 'biter' related to the verb δάκνω (δήκω) 'bite'. Walckenaer, as is frequently the case with the French, did not pronounce the Greek aspirated theta, and rendered it with a t. See the discussion under *Rhecostica*. As to the termination, Thorell has appealed to the existence of a real Greek word, λαθροδήκτης, of which, in fact, Walckenaer may well have been unaware, since the word occurs in an obscure text. But there are at least 18 Greek compounds of the passive verbal adjective -δηκτος 'bitten' such as κροκοδιλόδηκτος 'bitten by a crocodile' or καρδιόδηκτος 'gnawing the heart'. This last is from the Agamemnon of Aeschylus, and therefore not obscure, and in addition, the usually passive -δηκτος has an active meaning in the Aeschylus passage, namely, 'biting'. So Walckenaer's invention *λαθρόδηκτος 'biting in secret' can, in fact, be justified morphologically.

*Lauricius* SIMON 1888b:208 (Tengellidae)
Although this name is not from an authentic Greek or Latin word, it appears to be an adjective derived from Latin laurus 'laurel'. The reason is obscure. Transferred to the Miturgidae by Lehtinen (1967:243) and to the Tengellidae by Wolff (1978:139).

*Lepthyphantes* MENGE 1866:131 (Linyphiidae)
From the combining form λεπτ-of the Greek three-ending adjective λεπτος -ή -όν 'fine, thin, delicate' and the Greek masculine first declen-

sion agent noun ὑφάντης -ου 'weaver' from the verb ὑφαίνω 'weave'. Hence, 'weaving a delicate web'. as Menge, in fact, indicated. Menge clearly erected *Bathyphantes* and *Lepthyphantes* (thick weaver and slender weaver') to contrast with one another in the same paper. The rules of Greek internal sandhi (rules of phonetic combination within words) require this to result in λεφθυφάντης which is properly transliterated *Lephthyphantes* 'fine weaver'. And that should be that. Unfortunately Menge first wrote *Lepthyphantes* on page 131 and *Leptyphantes* on page 133, neither of which is strictly correct (Bonnet 1957a:2404). Thorell (1869:82) corrected it to *Lephthyphantes*. Purists will want *Lephthyphantes*; champions of original spelling will want *Lepthyphantes* on the plea of page priority. Bonnet (1945:135) in his "petite leçon de grec" understood this perfectly, and complimented Thorell's punctilio, but he preferred the least justifiable form *Leptyphantes* on a principle of "simplification" based upon the fact that French sometimes simplifies the word rhythme to rythme. Unfortunately when Bonnet says, "Rien ne s'oppose alors à ce que *Leptyphantes* soit aussi correct" he is wrong. Just as Bonnet, even though disagreeing, complimented Thorell for being careful about such things. I, even though disagreeing, pay homage to Bonnet, our heroic model for caring about such things.

The gender of the Greek word ὑφάντης is masculine, and Menge treated its compounds as masculine, even though for this particular genus the evidence may be somewhat misleading. Menge's species were *muscicola* ('dweller in the moss') and *crypticola* ('dweller in the secret place'). Compounds of the Latin suffix -cola 'inhabitant' are first declension masculine nouns, despite the fact that they look feminine. See the Code (Art. 30.a.i example). So Menge's specific names are masculine nouns in apposition.

*Leptoctenus* L. KOCH 1878a:994 (Ctenidae)
From the Greek three-ending adjective λεπτός -ή -όν 'slight, thin' and the genus name *Ctenus* (which see). Hence, 'thin *Ctenus*'. The type species *Leptoctenus agalenoides* L. KOCH 1878a has long thin legs. The body length is 70mm, and the length of the first leg is 135mm.

*Leptoneta* SIMON 1872b:477 (Leptonetidae)
The first element is from the Greek three-ending adjective λεπτός -ή -όν 'fine, thin, delicate' and the second element is taken by analogy from the second element of *Argyroneta* (which see). The morpheme -neta is a latinization of the Greek first-declension masculine agentive suffix -νήτης from the verb νέω 'spin' as in the authentic, but rare, Greek word χρυσονήτης 'spinner of gold thread'.

One should not be misled by the fact that the Greek lexicon will show a word λεπτόνητος, as if from λεπτός and the combining form -νητος, which would be the passive verbal adjective 'spun' from the verb νέω 'spin'. Unfortunately that is a ghost word resulting from a scribe's copying error in a single manuscript where a fragment of the Athenian comic writer Eubulus is quoted in the 3rd century A.D. encyclopedic writer Athenaeus. I cannot believe Simon would have known this. Furthermore there are no authentic Greek words of this formation.

Simon treated this name as feminine, not the correct masculine, with the type species *infuscata* SIMON 1872b, as did Latreille with *Argyroneta*. Hence, the intended meaning is 'slender spinner'.

*Leptorhoptrum* KULCZYŃSKI in CHYZER & KULCZYŃSKI 1894: 5, 79 (Linyphiidae)
As Kulczyński explicilty states (1994:76, note 1) from the Greek three ending adjective λεπτός -ά -όν 'slender' and the neuter noun ῥόπτρον 'the wood in a mouse trap which falls and catches the mouse' This latter perhaps refers to the process of the male palp, illustrated in Table III, 20a, which looks somewhat like the bar of an old fashioned mousetrap. It is decribed as "processus duo longi attenuati basi inter se connati" [two long slender processes joined at the base].

*Lessertia* F.P. SMITH 1908:329 (Linyphiidae)
In honor of the Swiss arachnologist Roger de Lessert, who worked on the spiders of Switzerland, and became one of the great specialists on the spiders of Africa.

*Leucauge* DARWIN in WHITE 1841:473 (Araneidae)
White (p. 474) specified that Darwin proposed this name as a subgenus of *Linyphia* in his manuscript. From the Greek adjective λευκός -ή -όν 'light, bright, clear, white' and the noun αὐγή 'light (of the sun), gleam, sheen, (of a bright object)'. Hence, 'with a bright gleam'. The Greek ending in -e is retained, instead of latinizing the ending to -a. Since the type species is *argyrobapta* meaning 'dipped in silver' from Greek ἄργυρος 'silver' and βαπτος -ή -όν 'dipped' the subgenus name was apparently suggested by this species name. The name *Leucauge* was forgotten for 60 years. Emerton (1884:331) erected the genus *Argyroepeira* ('silver *Epeira*') for *Epeira hortorum* HENTZ (1835:551), now *Leucauge venusta* (WALCKENAER 1841), and *Argyroepeira* was used by Thorell, Keyserling, Marx, McCook and several others. The first to resurrect *Leucauge* as a genus seems to be Waterhouse (1902:198). While Bonnet admitted *Leucauge* had priority, he grumbled about its replacing *Argyroepeira* (Bonnet 1957a:2460, note 91):

> "This term [*Leucauge*] having been forgotten for 60 years, it was needless to resuscitate it to replace **Argyroepeira** which had been generally employed; the rule ought to have worked in favor of the latter."

Yet some will surely rejoice at preserving the only spider name which can be attributed to Charles Darwin.

*Linyphantes* CHAMBERLIN & IVIE 1942a:45 (Linyphiidae)
Erected for *Linyphia ephedrus* CHAMBERLIN & IVIE 1942a and 17 related new species. The genus is close to *Bathyphantes* MENGE, and the name is confected from the first part of *Linyphia* and the second part of *Bathyphantes*. As it turns out, the name is well-formed (or nearly) because there are authentic Greek words λινυφαντεῖον 'workshop for weaving flax' and a doubtful word λινοφάντη"ς 'weaver of flax'.

*Linyphia* LATREILLE 1804b:134 (Linyphiidae)
From the Greek neuter noun λίνον 'anything made of flax, line, thread, net' and the root ὑφ-'weave'. There is no real Greek word *λινύφια and Latreille confected this name by adding a feminine adjective ending -ia. Hence, 'thread-weaver'.

(‡) *Linyphiella* BANKS 1905a:311 (Linyphiidae)
A Latin feminine diminutive built on the genus name *Linyphia* LATREILLE. Erected for *Linyphia coccinea* HENTZ 1850a. Synonymized under *Florinda* O. PICKARD-CAMBRIDGE by Gertsch & Davis (1946:5).

*Liocranoeca* WUNDERLICH 1999a:67 (Liocranidae)
Separated from *Agroeca* with the type species *Agroeca striata* KULCZYŃSKI 1882b. Hence the name is a combination of *Liocranum* and *Agroeca*.

*Liocranoides* KEYSERLING 1882b:290 (Tengellidae)
Derived from *Liocranum* L. KOCH by the addition of the zoological suffix -oides 'like, similar in form'. See the discussion under *Argyrodes* for the origin of this suffix. Hence '*Liocranum*-like'. Keyserling said *Liocranoides* is very much like *Liocranum*. The name of the genus could be either masculine, feminine or neuter, and the name of the type species, *unicolor*, is a one-ending adjective, hence ambiguous as to gender. The Code now arbitrarily decrees that names in -oides are masculine (Art. 30b). Transferred to the Miturgidae Tengellinae by Lehtinen (1967:244). Tengellidae raised to family rank by Wolff (1978:139).

*Liocranum* L. KOCH 1866a:2 (Liocranidae)
From the Greek three-ending adjective λεῖος -α -ον 'smooth, plain, level (as a piece of ground)' and the neuter of the combining form -κρανος -η -ον 'head' as in the authentic Greek adjective ψιλόκρανος 'bald-headed'. Hence, 'level-headed' from the fact that in this genus the cephalothorax is very flat.

(‡) *Liodrassus* CHAMBERLIN 1936a:4 (Gnaphosidae)
From the Greek three-ending adjective λεῖος -α -ον 'smooth, plain, level (as a piece of ground)' and the genus name *Drassus*. Chamberlin says:

> "In general structure close to **Herpyllus** from which most readily distinguished in lacking teeth on the margins of the furrow of the chelicerae, the lower margin being wholly smooth."

It is apparent that it is the smoothness of the lower margin of the chelicerae which is emphasized in the name. The genus has been synonymized under *Nodocion* CHAMBERLIN (Ubick & Roth 1973a:6, Platnick & Shadab 1980a:1-2).

*Lithyphantes* THORELL 1869:94 (Theridiidae)
According to Thorell (1869:94-95) the name is derived from λίθος 'stone' and ὑφάντης 'weaver'. But it does not mean 'stone-weaver'. Thorell said:

> "If from C. Koch's heterogenous genus **Phrurolithus** we detach some not allied forms, as for instance **Ph. trifasciatus**, which is a **Singa**, **Ph. ornatus**, which seems to be the young of **Steatoda bipunctata**, as also **Ph. festivus** and **minimus**, which belong to the Drassoidae, the remaining Theridioidae form a perfectly natural group, which has accordingly been acknowledged by Ohlert as a separate genus, and by him characterized in a satisfactory manner. Westring has however as early as 1851 reserved the name **Phrurolithus** to the above named Drassoidae which Koch had referred to this genus, so that Ohlerts **Phrurolithus** requires a new name. We have chosen the name **Lithyphantes**, as indicating the habits of the various species belonging to this genus."

Note that for Thorell both *Phrurolithus* as newly conceived, and his new *Lithyphantes* were placed in the Theridiidae (or Theridioidae in Thorell's terms). So Thorell's 'stone-weaver' is meant to indicate a spider once included in the old genus *Phrurolithus* ('stone-warder, stone guarder'.) When Thorell referred in his explanation to "the habits of the various species belonging to this genus" he was referring to the fact that *Lithyphantes* builds "webs near the ground under boards and stones" (Kaston 1948:78). Synonymized under *Steatoda* SUNDEVALL by Levi (1957b:375).

*Litopyllus* CHAMBERLIN 1922:147 (Gnaphosidae)
This is not an authentic Greek word, but a rhyming name by analogy with the neighboring genus *Herpyllus*. The second element, then, is from *Herpyllus* by misdivision of morphemes. The first element could come either from the Greek noun (accusative case -no nominative exists) λῖτα 'linen cloth', or from the Greek three-ending adjective λιτός -ή -όν 'simple, unadorned, paltry, small'. Since *Litopyllus* is distinguished from *Herpyllus* "by the cheliceral promargin bearing a carina not divided into distinct teeth" (Platnick & Shadab 1980a:17), I think the second possibility is likely. Hence, 'close to *Herpyllus* but with untoothed (unadorned) carina.'

(‡) *Lophocarenum* MENGE 1868:198 (Linyphiidae)
From the Greek masculine noun λόφος 'crest (of the head or of a hill)' and the Greek neuter noun κάρηνον 'head' referring to the large cephalic lobe of the males. Menge translated the name as "Hügelkopf" that is, 'hill-head'. This genus is a junior synonym of *Pelecopsis* SIMON (Locket et al. 1974:85). Bonnet (1957a:2563) explained that Menge created the genus *Lophocarenum* in 1868 for 11 species, and Simon in 1884 recognized the genus with 16 species, of which only three were those which Menge had included. Simon placed the remainder of Menge's species into several neighboring genera. Simon maintained the genus *Lophocarenum* with *parallelum* WIDER *in* REUSS 1834 as type. In 1864:196 Simon had created the subgenus *Pelecopsis* for *Micryphantes inaequalis* C.L. KOCH 1841c, but that name turned out to be a synonym of *Lophocarenum elongatum* WIDER *in* REUSS 1834. Hence *Pelecopsis* SIMON 1864 has priority over *Lophocarenum* MENGE 1868. F. Dahl (1901c:262) resurrected the genus *Pelecopsis* and was followed by many workers. So both by the principles of priority and usage, *Pelecopsis* should replace *Lophocarenum*. Authors still use the subfamily name Lophocareninae SIMON 1884a, and have not replaced it with Pelecopsinae Bonnet 1957a. The Code (Art. 40) specifies that after 1960 a family group name based on an invalid type-genus name is not to be replaced. But before 1961, if a family-group name has been replaced, and the replacement has won general acceptance, then the replacement is to be maintained. Bonnet just gets in under the wire. The question is whether Pelecopsinae has won general acceptance. If it has, then it is to be maintained; if it has not, then Lophocareninae is to be maintained.

*Lophomma* MENGE 1868:209 (Linyphiidae)
From the Greek masculine noun λόφος 'crest' and the neuter noun ὄμμα (gen.) ὄμματος 'eye'. From the fact that in the male the posterior median eyes are at the summit of the cephalic lobe (Kaston 1948:182). Menge translated the name as "Hügelauge" that is 'hill-eye'. The gender is neuter.

*Loxosceles* HEINEKEN & LOWE *in* LOWE 1835:321 (Sicariidae)
From the Greek adjective λοξός -ή -όν 'slanting, crosswise, oblique' and the combining form -σκελης which is the adjective forming combining form related to the neuter noun σκέλος (gen.) -ους 'leg'. These combine to form a two-ending Greek adjective *λοξοσκελής -ές 'with slanting legs' like ἰσοσκελής -ές 'with equal legs'. When transliterated into Latin *Loxosceles* could be any of the three genders, but Lowe's species was *citigrada* (now *rufescens*) so we may conclude he meant the name to be feminine. Strand and Neave (as reported by Bonnet 1957a:2573) argued for the spelling *Loxoscelis*, but they are wrong. Bonnet demonstrated that Lowe did not spell it that way, and furthermore forms in -scelis (Greek -σκελίς -ίδος) represent quite a different construction (feminine noun instead of s-stem adjective), and not merely a variation in spelling. The name presumably refers to the fact that the legs of *Loxosceles* are "un peu latérales" (Simon 1893a:271), that is, slightly laterigrade.

*Lupettiana* BRESCOVIT 1997a:65 (Anyphaenidae)
An arbitrary combination of letters treated as feminine.

*Lutica* MARX 1891:32 (Zodariidae)
Marx stated that this is an Indian name for 'spider' but he did not tell us to which Amerindian language he was referring. I have not succeeded in tracking it down. Marx described *Lutica maculata* MARX 1891 from a specimen sent to him from Lake Klamath, Oregon, but Gertsch (1961) corrected the type locality to Santa Rosa Island, California. See Ramírez & Beckwitt (1995).

*Lycosa* LATREILLE 1804b:135 (Lycosidae)
From the Greek masculine noun λύκος -ου 'wolf'. The comparison to a wolf goes back at least to Aristotle in the 4th century B.C. (Aristotle Historia Animalium 623a2), who described τὸ τῶν καλουμένων λύκων γένος 'the kind of spiders called wolves' which he divided into three kinds, "the small, the larger, and the brightly colored." Attempts to identify these with modern species are unconvincing (see Keller 1913: 462-464, Gil 1959:49). Nicander (Theriaka 734) mentioned "another spider, the huntsman, like a wolf in form (or like the wolf-spider), destroyer of flies, and he ambushes bees, fig-wasps, horse-flies, and whatever comes to his toils." Philumenus (De Ven. An. 15, 1-2) said, "the wolf is the second kind of spider, which lives in the toils of the web, catches flies and eats them." There is some confusion among the ancients whether these wolf spiders build webs or not. Pliny (Natural History 29.27.85) said, "among the poisonous spiders is one the Greeks call 'wolf' and again (11.28.80) "of the wolf spiders the smallest do not weave webs" but then went on to describe the "third kind of wolf spider" in terms that make it clear it is an orb-weaver. Aldrovandus (1602:602, 604.) also listed among the spiders a Lupus (i.e. 'wolf').
Thorell (1870b:190) was correct in deriving the name from the denominative verb λυκόω 'tear like a wolf'. Then the name is the correct feminine active present participle λυκῶσα 'one (feminine) who tears like a wolf'. Latreille formed at least one other name on this pattern, viz. *Gnaphosa*.
Subsequently this name has spawned a host of related names by the principle of the rhyming morph and by misdivision of morphemes, to produce Lycosid names in -osa and -cosa, such as *Arctosa* C.L. KOCH or *Schizocosa* CHAMBERLIN.

*Lyssomanes* HENTZ 1845:197 (Salticidae)
From the Greek two-ending adjective λυσσομάνης -ες 'raving mad' from λύσσα 'madness, rage' and the combining form -μάνης related to μανία 'madness'. While the transliterated Latin *Lyssomanes* could be either masculine, feminine or neuter, and Hentz's species was *viridis* which does not resolve the ambiguity of gender, Peckham and Peckham (1888), the first revisers, unambiguously used masculine species names (Taczanowski in 1873 used *longipes* which is also ambiguous.) Presumably it refers to the frenzied activity of the spider.

*Macrargus* F. DAHL 1886:76 (Linyphiidae)
From the Greek three-ending adjective μακρος -ά -όν 'long' and the genus name *Argus* WALCKENAER. Hence, 'long *Argus*'. Walckenaer's name *Argus* (1841:344) was preoccupied at least six times. He employed this genus (Bonnet 1955:702) for 67 species belonging now to 40 different modern genera. Wider *in* Reuss (1834:223) had described *Theridion rufum*, which Walckenaer (1841:348) called *Argus rufus*, which Dahl then designated as the type species of his new genus *Macrargus*. Argus is the proper name of the mythological creature with innumerable eyes, which was sent by the jealous Hera to stalk Zeus's unfortunate paramour Io, who had been turned into a cow.

*Maevia* C.L. KOCH 1846h:69 (Salticidae)
A Roman proper name. One Maevius was secretary to Verres, the peculant governor of Sicily pilloried by Cicero, and another was a wretched poetaster mentioned by Vergil and Horace. Cicero, Against Verres 2.3.75(175); Vergil, Eclogues 3.90; Horace, Epodes 10.2. Koch simply used the feminine form of the name. I doubt that Koch knew of the woman Maevia mentioned in the Digest, a compendium of Roman legal opinions, although he may well have dug her out of a dictionary in a search for useable names. There is no particular semantic point.

*Majellula* STRAND 1932:140 (Thomisidae)
This is a substitute name formed by the addition of the diminutive suffix -ula to the O. Pickard-Cambridge name *Majella* (O. Pickard-Cambridge 1896a:191), which was preoccupied in the Crustacea. Cambridge said *Majella* is a proper name without further explanation.

*Mallos* O. PICKARD-CAMBRIDGE 1902a:308 (Dictynidae)
From the Greek masculine noun μάλλος (gen.) -ου 'flock of wool'. The type species is *niveus* which means 'snowy' in Latin. The abdomen is white dorsally clothed with gray-white hairs. All of this would be convincing, except that Pickard-Cambridge told us in a footnote that this is a proper name without specifying more exactly. It could be the name of the city Mallos, mentioned in the Apocrypha (2 Macc. 4:30), the geographer Strabo (14.675), Ptolemy (5.7.4) and Pliny the Elder (5.9.1). The Rev. Mr. Pickard-Cambridge may have run across this proper name in his reading of the Apocrypha, and may have appreciated the pun on the Greek for 'flock of wool'.

*Mangora* O. Pickard-Cambridge 1889d:13 (Araneidae)
Etymology unknown. Pickard-Cambridge gave us no hint about this name, and does not even mention the characteristic tuft of cilia on Tibia III. His nephew Frederick first mentioned that distinctive feature.

*Marchena* PECKHAM & PECKHAM 1909:512 (Salticidae)
Not a Greek or Latin word, and the Peckhams gave no explanation of the name. Possibly it is built upon the proper name March on the analogy of such names as *Agelena* and *Storena*.

(‡) *Marmatha* CHAMBERLIN & IVIE 1942a:40 (Theridiidae)
Erected as a subgenus of *Enoplognatha* PAVESI for those forms in which the males have a stout, sharp pointed, relatively short fang, and the posterior margin of the chelicerae with a heavy, double-toothed process, as in the type species *marmorata* (HENTZ 1850b). The first element of the name, marm-, is formed from the first syllable of *marmorata*, and the second element, -atha, is taken from the last two syllables of *Enoplognatha* by misdivision of morphemes and the rhyming principle. Hence, *Marmatha* the genus close to *Enoplognatha* based upon the species *marmorata*. Now synonymized under *Enoplognatha* (Levi 1957a:6).

*Maro* O. PICKARD-CAMBRIDGE 1906a:77, 86 (Linyphiidae)
The cognomen of the great Roman poet P. Vergilius Maro, author of the Aeneid.

*Marpissa* C.L. Koch 1846d:60 (Salticidae)
This name has had what you might call a checkered history, and Bonnet's note on its vagaries occupies a whole page (Bonnet 1957a:2718). Koch created the name, and used it consistently, spelled with an i, for 14 different species, and we may reasonably conclude he knew what he was doing. But Thorell, ever the purist (and this is generally to his credit) asserted (Thorell 1870b:213) that Koch "undoubtedly" derived the name from the Greek mythological figure Μάρπησσα, a woman whom Apollo tried unsuccessfully to abduct (Apollodorus 1.7.8; Plutarch Agis 9). Marpessa's father Evenus (who provided Simon with a spider name in 1877) tried to keep her a virgin, by beating all her suitors in a chariot race. If they lost they forfeited their lives. Apollo disapproved and challenged Evenus to a race himself. But Marpessa's true human lover Idas snatched her away from everybody. To Thorell's chagrin the name *Marpessa* was preoccupied in the Mollusca, so he proposed as a substitute the new name *Marptusa* (1877b:561). The point of this was that *Marpessa* is transparently related to the Greek verb μάρπτω 'seize' and the feminine active present participle is μάρπτουσα '(she) seizing'. Thorell was trying to keep as close as he could to the previous name, and maintain a sort of philological elegance. Karsch (1878d:28) realizing that *Marpessa* was preoccupied, but unaware of Thorell's substitute, supplied his own substitute *Marfisa*, a proper name taken, Bonnet informed us, from the Orlando Furioso of Ariosto, one of the books that drove Don Quixote over the brig. The youthful Simon (1868b:17), animated by a curious boyish notion (he was 19) that he would regularize the nomenclature of the salticides by making all the names masculine, produced *Marpissus*. This tangle of meddlesomeness is cause for applauding the rigor of the Code in sanctifying original spellings against the reformational zeal of purists. This is all explained in one of Bonnet's most spectacular and ingenious notes, and the reader would find both joy and self-improvement in consulting the original.

(‡) *Marxia* McCook 1894:192 (Araneidae)
In honor of George Marx (1838-1895). McCook said:

> "I have thought it necessary to make a new genus to receive the species originally described by Walckenaer as **Plectana stellata**. Subsequent writers have relegated this species to **Epeira**."

The subsequent writers were Hentz, Keyserling, Emerton, McCook himself in his Vol. I, and Marx in his 1890 Catalogue (Marx 1890:548). McCook listed *Marxia stellata* (Walckenaer 1805), and *Marxia nobilis* (Walckenaer 1841) and added a new species *Marxia grisea* McCook 1894 from Biscayne Bay, Florida, from the Marx collection. Even if McCook had been aware that Marx had already in 1883 created the genus *Acanthepeira* for the species *stellata* (see under *Acanthepeira*), he also knew that Marx had abandoned *Acanthepeira* in his 1890 Catalogue. Kaston (1938d:188; 1938f:259) resuscitated the name *Acanthepeira*, and sank *Marxia*.

*Masikia* Millidge 1984a:152 (Linyphiidae)
From the locality of the type species (*atra*), the Masik River, Banks Island, Northwest Territories, Canada.

*Maso* Simon 1884r:861 (Linyphiidae)
A Roman cognomen (genitive Masonis). Several members of a family mentioned by Pliny and Cicero are named C. Papirius Maso. There is no particular semantic connection. Simon was in the habit of searching the dictionary for usable names.

*Masoncus* Chamberlin 1949:536 (Linyphiidae)
Chamberlin meant this name to be derived from *Maso* because it is close to that genus, which he treated in the same paper on the previous page. He also appears to be making a reference to the genus *Araeoncus* Simon. Hence the name is a combination of the genus *Maso* Simon and the genus *Araeoncus* Simon.

*Masonetta* Chamberlin & Ivie 1939a:64 (Linyphiidae)
Feminine diminutive derived from *Maso* Simon.

*Mastophora* Holmberg 1876:143 (Araneidae)
From Greek μαστός 'breast' and the feminine combining form -φορα 'bearing', hence 'bearing breasts'. The name refers to the two prominent humps on the abdomen which characterize several species of the genus (e.g., *cornigera* Hentz 1850a). Simon (1895a:883) proposed a new genus *Glyptocranium* for Hentz's *Epeira cornigera*. *Glyptocranium* is from the Greek three-ending verbal adjective γλυπτός -ή -όν 'carved' and the Greek neuter noun κρανίον 'upper part of the head, skull'. Latinized to Glyptocranium it means 'having a carved head' and refers to the pair of large protuberances on the cephalothorax. Holmberg's obscure publication had not come to the attention of European workers. But South American workers, Mello-Leitao, Brèthes, and Canals, had always used Holmberg's name, and Mello-Leitao's revision in 1931b settled the matter. See Gertsch 1955:231. Bonnet (1957a:1995) said:

> "Simon appears to have been unaware of the genus **Mastophora** Holmberg (1876), and in 1895 he created the genus **Glyptocranium** which is a synonym. The priority would revert to **Mastophora** but the name, which South American authors have adopted, can be considered preoccupied, according to the old conception, by the Nematode name **Mastophorus** (Diesing, 1853). -I would let **Mastophora** stand if **Glyptocranium** did not exist, but since it does exist, it is better to use it and continue to consider **Mastophora** to be preoccupied.--There is no question of replacing **Glyptocranium** with another name of Holmberg's, **Heterocephala** (Holmberg 1876: 112) [used by Mello-Leitão] which would have priority [over **Glyptocranium**]; in the first place, there is a rule against it, and in the second place, the name is also preoccupied by two names **Heterocephalus**, once in the Mammals (Rueppell, 1842) and once in the Nematodes (Marion, 1870)."

Bonnet let his better judgement lapse here. By the "one letter difference" rule (Code:Art. 56.b; but see Appendix D.I.3). *Mastophora* is not preoccupied.

*Maymena* Gertsch 1960a:30 (Mysmenidae)
Gertsch erected this genus for *Nesticus mayanus* Chamberlin & Ivie (1938:134), described from southern Mexico, and named for the Maya. Gertsch considered this genus close to *Mysmena* Simon, and constructed the name partly from the first syllable of the type species name, *mayanus*, and partly from the last two syllables of the genus name *Mysmena*.

*Mazax* O. Pickard-Cambridge 1898a:275 (Corinnidae)
A Latin word gleaned from the dictionary designating a nomadic people living around the Sea of Azov. The word is used in Lucan Pharsalia (4.681)(1st cent. A.D.) an epic poem about the war between Caesar and Pompey. There is no semantic connection with the spider. The word is masculine, but O. Pickard-Cambridge treated it as feminine by naming the type species *spinosa*.

*Mecynargus* Kulczyński *in* Chyzer & Kulczyński 1894:121 (Linyphiidae)
From the Greek verb μηκύνω 'lengthen' and the genus name *Argus* Walckenaer. The type species is longus Hence, 'lengthened *Argus*'. See the discussion of Walckenaer's name under *Macrargus*.

*Mecynogea* Simon 1903d:25 (Araneidae)
Based upon the genus *Gea* with the addition of a prefix formed from the Greek verb μηκύνω 'lengthen'. The name refers to the long parallel-sided abdomen. Hence, the name means roughly 'lengthened *Gea*'. The word is not well-formed; no authentic Greek compound would have this shape. Simon (1895a:759) included *Mecynogea* in the Argiopinae along with *Gea* and *Argiope*. Levi (1968a:322, 1980a:11, 14) removed it ultimately to the Araneinae. The North American species has travelled under the names *Linyphia lemniscata* Walckenaer 1841, *Hentzia basilica* McCook 1878a, *Allepeira basilica* by Banks *et al.* 1932, and finally was placed in *Mecynogea* by Levi (1968a:322) who followed Exline (1948:309) in retaining Walckenaer's species name *lemniscata*.

*Megahexura* Kaston 1972a:60 (Mecicobothriidae)
From the Greek irregular three-ending adjective μέγας μεγάλη μέγα 'large' plus the genus name *Hexura* (which see). Created for *Hexura fulva* Chamberlin 1919b, which is about one and a half times larger than the species of *Hexura*, as well as differing in the spination and the spinnerets. Hence, 'the big *Hexura*'.

*Megalepthyphantes* Wunderlich 1994b:168 (Linyphiidae)
The prefix Mega-is from the Greek neuter adjective μέγα 'large'. Erected for the *nebulosus* group of *Lepthyphantes*. Comstock observed long ago (Comstock 1912:393) that "most of the species [of *Lepthyphantes*] are of small size, but [*nebulosus*] is larger measuring one sixth of an inch in length."

*Megalostrata* Karsch 1880c:374, 377 (Corinnidae)
A feminine noun from the Greek combining form μεγαλο- 'large, great' and στρατος 'army, host' hence 'great army'.

*Megamyrmaekion* Wider *in* Reuss 1834:217 (Gnaphosidae)
This name has a complicated and surprising etymology. The first part is from the Greek irregular three-ending adjective μέγας μέγα 'large'. The second part is a derived form from the Greek masculine noun for ant μύρμηξ (gen.) -ηκος. There are two meanings for the derived Greek neuter noun μυρμήκειον. The first, which does not figure in Wider's name, designates a species of spider (a φαλάγγιον) in Nicander Theriaca 747. Nicander said:

> "The antlet -now mark-which in truth resembles the ant, has a fiery neck, though its body is dust-colored; its broad and spangled back is all speckled, and its dusky head is raised but little on its neck, yet it inflicts as much pain as the spiders aforementioned."
>
> (translation Gow & Scholfield 1953:79).

It has been identified with *Salticus formicarius*, with the Solpugid *Galeodes araneoides*, and with more likelihood with the muttilid *Mutilla europea* (Gow & Scholfield 1953:185). Pliny (Nat. Hist. 29.87) summarized Nicander, and latinized the name as myrmecion. The reason Wider spelled it with the diphthong ae is due to a German school convention of distinguishing a Latin long e by writing it as the diphthong ae. The same thing happens in the scorpion name *Vaejovis* for *Vêjovis*. Bonnet sorted out the history of the spelling as follows (Bonnet 1957a:2750, note 39):

> "The name was written **Megamyrmaekion** by Wider at the time of its creation, correctly latinized to **Megamyrmecium** by Agassiz, modified to **Megamyrmecion** and **Megamyrmaecion** by Simon. The original spelling, as Strand maintains, cannot be retained, and one properly writes **Megamyrmecium**, as one writes **Chiracanthium**, **Theridium**, or **Zodarium** This is in accord with the Rules of Nomenclature (Appendix, § F.) which correspond with the Latin spelling myrmex, -ecis; the diphthong ae in the word is not conceivable."

Pliny, of course, violated the Rules of Nomenclature by latinizing it myrmecion. Bonnet's heroic campaign to regularize the nomenclature has triumphed in spirit, though not in detail. The new rule of regularity is to preserve the original spelling ruat caelum (legal jargon for "even if the heavens fall"). So the "correct" spelling is *Megamyrmaekion*

The meaning, in Nicander is 'the antlike spider' but Wider *in* Reuss (1834:217), as it turns out, intended the second, quite different meaning of μυρμήκιον namely, 'wart, tubercle (Warze)' used in the Greek medical writers, so called because the irritation caused by a wart was likened to the creeping of ants (formication). And specifically Reuss meant the Greek word μυρμήκιον to mean 'Spinnwarze' that is 'spinneret'. So what Wider really intended his name to mean was 'possessing large spinnerets'. He says:

> "Abdomen with 4 blunt spinnerets, of which the inferior are more than twice as thick, and a quarter longer than the superior."

His figure (Wider *in* Reuss 1834: Tab.XVIII, fig. 12) makes this very clear.

Note further that Latreille had named a genus of spiders *Myrmecium* in 1824. This *myrmecion* should also not be confused with the *myrmekia* in *Myrmekiaphila* ATKINSON, where it means 'ants' nest'. It is worth quoting Bonnet's further note (Bonnet 1957a:2750, note 40);

> "Walckenaer changed the name **Megamyrmaekion** [to **Dyction**] because he found it too long! What would he have done had he lived until the epoch of **Parapallaseakotylodermogammarus**.? [a crustacean genus named by Dybrowski in 1926]"

*Megamyrmaekion californicum* SIMON 1893a was declared a *nomen dubium* by Ubick & Roth (1973b:1) and by Platnick & Shadab (1976b:17), and the rest of the North American species have been transferred to *Scopoides* PLATNICK 1989b (see Platnick 1993c:664).

### *Meioneta* Hull 1920:9 (Linyphiidae)

We must consider a succession of spider names in order to understand how this one comes about. The type species *rurestris* C.L. KOCH 1836e:84 was first placed in Koch's genus *Micryphantes* meaning 'small weaver'. Simon (1884a:404) included it in Menge's 1869 genus *Microneta* 'small spinner'. Hull then included it in his new genus, the name of which is formed from the Greek two-ending comparative adjective μείων μείον 'lesser, smaller' and the rhyming morph -neta by analogy with such names as *Argyroneta* (which see), meaning 'spinner'. Hence, 'smaller spinner'. See the discussion under *Leptoneta* and *Argyroneta*.

### *Melocosa* GERTSCH 1937:6 (Lycosidae)

A rhyming name analogous to *Lycosa* with the first element rhyming with the irregular three-ending Greek adjective μέλας μέλαινα μέλαν 'black'. The word is not formed with strict morphological rules. A genus created for *Lycosa fumosa* EMERTON 1894, a particularly dark wolf spider. Hence, 'black wolf spider'. Bonnet and Roewer list this genus under Pisauridae; transferred to the Lycosidae by Leech (1969b:890).

### *Melpomene* O. PICKARD-CAMBRIDGE 1898a:285 (Agelenidae)

Greek proper name for the muse of tragedy. Considered a synonym of *Agelena* WALCKENAER until revalidated by Chamberlin & Ivie (1942b:237).

### *Menemerus* SIMON 1868b:662 (Salticidae)

The second half of the name is from the Greek masculine noun μηρός-οῦ 'thigh' and this is in accord with Simon's usual practice. The first part is puzzling. Since Simon's Greek was not altogether secure at this age (he was 20), he tried to construct a compound using the Greek word μήνη 'moon'. Strictly speaking it should form compounds in μηνο-it is clear from Simon's description that he means the name to refer to the shape of the male palpal femur. He said:

> "The femur is swollen, almost as broad as long, but is a little narrowed at its extremities; it is in the form of a crescent, that is to say being strongly convex on the external side and a little concave interiorly."

This "crescent" thigh is clearly the feature that Simon was trying to describe in his name, which means something like 'moon thigh'.

### *Meriola* BANKS 1895d:81 (Corinnidae)

Etymology unknown. Perhaps a personal surname. Removed from the synonymy of *Trachelas* by Platnick & Ewing (1995:8).

### *Messua* PECKHAM & PECKHAM 1896:93 (Salticidae)

A name taken from The Jungle Books of Rudyard Kipling. Messua is the mother of Mowgli in the chapter titled "Tiger-Tiger." The Peckhams also took the Salticid genus names *Bagheera* and *Nagaina* from The Jungle Books.

### *Meta* C.L. KOCH *in* HERRICH-SCHÄFFER 1835:Heft. 134, pl. 12 (Tetragnathidae)

Thorell (1869:35(1)) correctly said this is a mythological proper name Μῆτα. She is the daughter of Hoples and the first wife of Aegeus, legendary King of Athens (Apollodorus 3.15.6).

### *Metacyrba* F.O. PICKARD-CAMBRIDGE 1901a:252 (Salticidae)

The Peckhams (1888:75) had included *Attus taeniola* HENTZ 1846:353 in the genus *Cyrba* SIMON (1876a:165). Cyrba, although it looks like a Greek word, is not one actually. Two possibilities exist. The Greek noun κύρβας (gen.) -αντος is a shortened form of Κορύβας 'Corybant, celebrant in the mysteries of the goddess Cybele' and Simon conceivably might have latinized this as Cyrba. Since Corybants leap and carry on, there would be a conceivable semantic connection. The second, less likely, possibility is that Simon derived it from the Greek feminine plural noun κυρβείς 'a three-sided pyramid for posting announcements'. It is hard to see in this case what the semantic connection would be. Pickard-Cambridge said ad loc "I cannot find that this species is congeneric with Simon's type of Cyrba, and therefore a new genus is required for it." He created a new name by the usual device of prefixing meta-.

### *Metagonia* SIMON 1893d:318 (Pholcidae)

This is not a case of the prefix meta-used to form a related genus name, for there is no genus *Gonia. Simon says *Metagonia* is distinguished by its "long abdomen markedly widened behind, with a truncate apex obtusely bifid." So Simon means meta, from the Greek preposition μετά, to mean 'behind, after'. The second element is formed from the Greek feminine noun γωνια 'angle corner'. Simon (1893a:467) says that the abdomen has "deux angles un peu saillants" [two angles somewhat projecting]. So the name means 'with an angled behind'.

### *Metleucauge* LEVI 1980a:44 (Araneidae)

Formed from the genus name *Leucauge* (which see) by the addition of the prefix meta-.

### *Metaltella* MELLO-LEITÃO 1931c:94 (Amphinectidae)

Built with the Meta-morpheme upon the Dictynid genus *Altella* SIMON (1884o:321), a feminine diminutive of the Latin adjective altus -a -um 'high, tall, deep'. The name *Altella* may refer to the convex cephalothorax of the female.

### *Metaphidippus* F.O. PICKARD-CAMBRIDGE 1901a:258 (Salticidae)

Derived from *Phidippus* by the addition of the prefix meta-. Pickard-Cambridge wanted to break up the large genus *Phidippus* into more manageable sections, although he said that it is difficult on structural grounds to separate *Phidippus*, *Paraphidippus*, and *Metaphidippus*. He included in *Metaphidippus* those species which the Peckhams had included in *Dendryphantes*, and said "It is very difficult to remove them from *Paraphidippus*, and if they happened to be of the same size there would be still less inclination to do so."

### *Metazygia* F.O. PICKARD-CAMBRIDGE 1904:501 (Araneidae)

Formed from the genus name *Zygia* by the addition of a prefix meta- for the genus created to accomodate *Epeira wittfeldae* McCOOK 1894:168. Incidentally, the species name *wittfeldae*, a feminine genitive proper name, is for Mrs. Anna Wittfeld of Merrit Island, Florida. F.O. Pickard-Cambridge (1902d:15) replaced C.L. Koch's *Zygia*, which was preoccupied, with *Zygiella* (see under *Zygiella*). Both *Zygiella* species and *Metazygia* species share a darkened cephalic area of the cephalothorax, and this may be sufficient for Cambridge to have associated them, although he does not say as much. See *Zygia*.

### *Metellina* CHAMBERLIN & IVIE 1941b:14 (Araneidae)

A double diminutive derived from the genus name *Meta*. C.L. Koch first by adding the Latin diminutive suffix -ella to which is then added the diminutive suffix -ina. The double diminutive seemed necessary because *Metella* had already been used by Fage (1931:195) for a linyphiid. Fage was using the Roman feminine proper name, *Metella*, and not a diminutive of *Meta*, and also *Metella* was preoccupied in the Fishes.

*Metepeira* F.O. Pickard-Cambridge 1903a:457 (Araneidae)
"The species referred to this genus form a small but distinct group, which has more affinities with the Metinae than with any of the present subfamily [Araneinae]." Presumably the similarity is due to the Metinae and Metepeira having markedly long metatarsi. So the *Metepeira* is a combination of the genus name *Meta* C.L. Koch and the genus name *Epeira* Walckenaer, and not a compound with the usual prefix meta-.

*Metopobactrus* Simon 1884r:748 (Linyphiidae)
From the Greek neuter noun μέτωπον 'forehead, space between the eyes' and the neuter noun βάκτρον 'staff, stick, cudgel' with a masculine adjectival ending. There is a genuine Greek two-ending adjective of this formation παράβακτρος -ον 'like or near a staff'. The name refers to the rod-like process of the male cephalothorax. "Frons maris haud lobata sed acuminata et saepe in processum baciliformem supra oculos producta" [Front of the male not lobate but acuminate and often produced into a rod-shaped process above the eyes] (Simon 1894a:657). Hence, 'with a rod-shaped process on the forehead'.

*Mexigonus* Edwards 2003a:69 (Salticidae)
The author says:
"A combination of the country which seems to be the center of origin of the genus (Mexico), and the Greek compounding form -gonos (in its latinized form) meaning "born" i.e. born in Mexico; gender is masculine."
Real Greek abounds in such compounds such as, παλαιόγονος 'born long ago', θαλασσίγονος 'born of the sea' etc.

*Mexitlia* Lehtinen 1967:248, 359 (Dictynidae)
Lehtinen proposed this genus for the *trivittatus* group of *Mallos* (Chamberlin & Gertsch 1958:45) which occurs in Mexico and SW USA. Mexitli is a name given to the Aztec war god Huitzilopochtli, from which, according to some authors, the name of Mexico is derived, see Simeon (1885). Lehtinen simply added an *a* to give it a feminine Latin ending.

*Miagrammopes* O. Pickard-Cambridge 1870a:400 (Uloboridae)
This genus is distinguished from all other Uloboridae by the lack of the anterior eye row. The anterior eyes are either absent or not perceivable. The name is from the feminine form of the Greek numeral μία 'one', plus the Greek feminine noun γραμμή 'line, row' plus the combining form -ώπης 'face, eye' as in the authentic Greek two-ending adjective κυανώπης -ες 'dark-eyed'. Hence, 'having one row of eyes'. The form is ambiguous as to gender, and the type species is a genitive (*Miagrammopes thwaitesi* O. Pickard-Cambridge 1870a) which does not resolve the question. Pickard-Cambridge added species *Miagrammopes lineatus* (1894a:137) and *Miagrammopes mexicanus* (1893b:116) making clear he considered the name to be masculine. But he had added a species *Miagrammopes brevicauda* (1882c) as if he considered it feminine. Bonnet (1957a:2830) decided for masculine even though there has been some variation among the authors. It is reasonable to extend to names in -opes the arbitrary rule of the Code (Art. 30.a.ii) that a name in -ops "is to be regarded as masculine regardless of its derivation or of its treatment by its author."

*Micaria* Westring 1851:47 (Gnaphosidae)
C.L. Koch in 1835 created the genus *Macaria*, which is the feminine of the Greek three-ending adjective μακάριος -α -ον 'blessed, happy' but that was preoccupied in the Lepidoptera. Westring replaced it by changing a single letter. The resulting name happens to be the feminine of a Latin adjective micarius-a-um 'gathering crumbs, frugal, economical'. I bet Westring was tickled by this little trick. Thorell (1870b:146) suggested it is derived from the Latin verb micare 'shine'. There is some merit in this, since the type species is *Micaria fulgens* (Walckenaer 1802), and *fulgens* means 'shining'. Thorell then is surely right in supposing Westring thought the Latin adjective micarius-a-um meant 'shining'. From the Latin feminine noun mica 'tiny fragment, crumb, mote, grain' comes the verb micare 'to be quick in motion, vibrate, have a tremulous motion' and then the verb comes to mean 'glisten'. Although the adjective micarius means 'picking up small fragments' it was natural for Westring to think it was connected with shining. It may be that he knew very well what the true Latin meaning was, but chose to make a pretty pun. Westring (1861:331) said explicitly that he changed the name to *Micaria* taken from the verb micare. His description of the genus begins (p. 330): Quoad pubescentiam sericeam vel metallo-micantem abdominis [As far as the silky pubescence and the shining metallic pubescence of the abdomen]. So Westring actually used the verb micare in his description.

*Micrargus* F. Dahl 1886:79 (Linyphiidae)
From the Greek three-ending adjective μικρός -ή -όν 'small' and the genus name *Argus* Walckenaer. See the discussion of Walckenaer's name under *Macrargus*.

*Micrathena* Sundevall 1833b:14 (Araneidae)
From the Greek adjective μικρός -ή -όν 'small' and the name of the goddess Athena, patroness of domestic arts and handicrafts, especially spinning and weaving. Compare the story of Arachne and Athena told in Ovid's Metamorphoses 6.5 ff. By chance three authors discovered this same genus in the same year, 1833, and gave it three different names. Sundevall named it *Micrathena*, Perty (1833) named it *Acrosoma* and Walckenaer named it *Plectana*. Sundevall's first species was *Epeira clypeata* Walckenaer 1805, and Perty's first species is his own *swainsonii* (designated as the type by F.O. Pickard-Cambridge 1904:525), but Walckenaer's 1833 genus is a *nomen nudum*. He first included species in 1841. The situation is very complicated, and is sorted out in Bonnet's long, involved note (Bonnet 1957a:2858-2859). Bonnet pointed out that Sundevall has the merit of making clear the distinction between *Gasteracantha* and *Micrathena*, whereas *Acrosoma* and *Plectana* include both Gasteracanths and Micrathenes. It is for this reason that Bonnet gave the prize to Sundevall. When Walckenaer (1841:172) finally gave a diagnosis of *Plectana*, he did so in such a way as to make his genus a junior synonym of *Gasteracantha*. The question of *Acrosoma* is, as Bonnet said, perhaps with intended wit, "a little more prickly" but Bonnet argued that Perty made the same mistake as Walckenaer and his name is also a junior synonym of *Gasteracantha*.
Sundevall (1833b:14) explained his name as follows:
"Μικρα, parva et Αθηνα, nomen graecum Minervae, et armatae et texentis. (The Greek in Sundevall's Conspectus is printed without accents) [the Greek name for Minerva, who was both armor-clad and was a weaver]"
Sundevall's description of *Micrathena* said: "abdomen magnum, coriaceum, armatum. [abdomen large, leathery, and armed]" He was elegantly appealing to Athena's two attributes as an armed weaver. Nice!

(‡) *Microcentria* Schenkel 1925d:297 (Linyphiidae)
From the Greek three-ending adjective μικρός -ά -όν 'small' and the second half of the genus name *Diplocentria* (which see) from which it was differentiated. Schenkel said:
"In general habitus the new genus is so closely related to **Diplocentria** (Hull) that the two genera could be combined."
The type species is *pusilla* 'tiny' because it is smaller than the following species in Schenkel's paper with which it is compared namely, *Diplocentria rivalis* (O. Pickard-Cambridge 1905). Hence, 'small spider close to *Diplocentria*'. Synonymized under *Diplocentria* Hull by Wunderlich (1970:407) followed by Millidge (1984a:154).

*Microctenonyx* F. Dahl 1886:80 (Linyphiidae)
From the Greek three-ending adjective μικρός -ά -όν 'small' plus the stem κτεν- of the Greek masculine noun κτείς (gen.) κτενός 'comb' plus the Greek masculine noun ὄνυξ (gen.) ὄνυχος 'talon, claw'. Hence, 'small spider with pectinate claws'.

*Microdipoena* Banks 1895d:84 (Mysmenidae)
Originally included in the Theridiidae, and considered near *Dipoena*. From the Greek three-ending adjective μικρος -ά -όν 'small' and the genus name *Dipoena* (which see). Levi (1956a:8) included it in the genus *Mysmena* still in the Theridiidae. Mysmeninae was given sub-family status in the Theridiidae, by Petrunkevitch (1928b:115), and Mysmenidae was given family status by Forster & Platnick (1977:2). The genus *Microdipoena* was revalidated by Saaristo (1978:125).

*Microhexura* Crosby & Bishop 1925a:145 (Dipluridae)
From the genus name *Hexura* (which see) with a prefix from the Greek three-ending adjective μικρός -ά -όν 'small'. Crosby and Bishop in error named their first species montivagus, a masculine adjective meaning 'wandering the mountain' whereas the genus name is feminine, and they should have written montivaga. Bonnet (1957a:2884) corrected it in accordance with the Code (Art. 31.b).

*Microlinyphia* Gerhardt 1928:629 (Linyphiidae)
Derived from *Linyphia* by the addition of the prefixed Greek adjective μικρός -ά -όν 'small' for the species *Linyphia pusilla*, for which the species name is Latin for 'very small'. Usually considered a synonym of *Linyphia*, but accepted by Wiehle (1956:297).

*Microneta* Menge 1869:227 (Linyphiidae)
From the Greek three-ending adjective μικρός -ά -όν 'small' and the combining form -neta the correct latinization of a Greek masculine agent noun -νήτης 'spinner'. See the discussion under *Argyroneta*. By rights this would be a Latin first declension masculine, but from Latreille's day such names have been treated as feminine. Menge clearly indicated the etymology above, so it is initially puzzling that he translated the name of this genus as "Ringspinne" that is, apparently, 'Ring spider'. The trick is that ring here has nothing to do with "rings" but is a dialect adjective meaning 'slight' related to the standard German word gering, and the meaning of the genus name is 'slight spinner'.

*Micropholcus* DEELMAN-REINHOLD & PRINSEN 1987:73 (Pholcidae)
From the Greek three-ending adjective μικρός -ά -όν 'small' plus the genus name *Pholcus* (which see). Erected for *Pholcus fauroti* SIMON 1887e = *Pholcus unicolor* PETRUNKEVITCH 1929b.

(‡) *Micryphantes* C.L. KOCH *in* HERRICH-SCHÄFFER 1834: Heft. 121:19 (Linyphiidae)
From the Greek three-ending adjective μικρός -ά -όν 'small' plus the Greek masculine noun ὑφάντης 'weaver' an agent noun built on the verb ὑφαίνω 'weave'. It is the earliest of the genus names in -hyphantes on which the others were modelled. Brignoli (1983c:322) considered it a *nomen dubium*. Bertkau (1878:384) created the family Micryphantidae, but it has generally been considered a part of Linyphiidae BLACKWALL (1859b:261). Bonnet (1957a:2913) said:

> "The genus **Micryphantes** *being in the family of the Linyphiidae, the family Micryphantidae naturally becomes a synonym [of Linyphiidae]. From that fact Micryphantidae cannot be used for the Erigones, although it has priority over Erigonidae."*

Kaston said in the Spiders of Connecticut (1948:142 note):

> "*Petrunkevitch (1939) for reasons which are not clear, has substituted the name Erigonidae for this family. This name had been first used in 1923 by Gerhardt, who, however, in the same paper also used* **Micryphantidae** *interchangeably with it. The genus* **Micryphantes** *is considered unrecognizable by Simon (1929) and species which had been previously placed in it were by him placed in his newly erected* **Aprologus** *and* **Ischnyphantes***. But the latter genus is apparently a synonym of* **Meioneta** *Hull 1920, overlooked by Simon. Yet Hull and other English and Continental araneologists apparently still consider* **Micryphantes** *as a valid name, though there may be some difference of opinion as to what is its type species. It seems to me preferable to retain the name Micryphantidae until further light is thrown on this question."*

Locket & Millidge (1953:173) in British Spiders said:

> "*Two good reasons seem to us to exclude the use of the name Micryphantidae; first, the genus* **Micryphantes** *is of doubtful validity and no longer recognized by us; and second, if the genus* **Micryphantes** *were used in its most recently adopted form it would clearly fall within the subfamily Linyphiinae."*

*Mimetus* HENTZ 1832:104 (Mimetidae)
Strictly this should be the Greek three-ending adjective μιμητος -ή -όν 'to be imitated or copied' but it is hard to see the point of that. Apparently Hentz played fast and loose with the Greek masculine noun μιμητής -οῦ 'imitator, actor, impersonator' by roughly giving it a Latin masculine ending. He says that *Mimetus* spins a web resembling that of *Theridium tepidariorum*, and seems to take over the webs of other spiders. Presumably, this is what Hentz meant the name to refer to. Hentz (1847:472 = 1875:112) in his discussion of *Epeira labyrinthea* said:

> "*The first time I found this spider, I also found the first* **Mimetus***, which had invaded the web of one of these and taken its place, so that for a period I thought this species a transition to that sub-genus [i.e.* **Mimetus***]."*

And in his discussion of *Mimetus* itself (Hentz 1832:106 = 1875:8) said:

> "*It is evident that in a perfectly natural arrangement,* **Theridium** *should be placed near* **Epeira***, and this genus [i.e.* **Mimetus***] between the two. There is a true* **Epeira***, E. labyrinthea, which is found in the same locality; and I first suspected that this was a* **Theridium** *which had taken possession of the web of that* **Epeira***. The resemblance of habits in these two species shows, however, the close affinity between the two genera and this."*

And further (Hentz 1850a:32 = 1875:138):

> "*The* **Mimetus** *can make a web like that of* **Theridion***, but prefers prowling in the dark, and taking possession of the industrious* **Epeira***'s threads and home, or the patient* **Theridion***'s web, after murdering the unsuspecting proprietor."*

It seems clear from these remarks that Hentz meant the name to mean 'imitator'.

(‡) *Mimognatha* BANKS 1929:90 (Tetragnathidae)
Erected for *Theridion foxi* MCCOOK 1894. From the Greek noun μῖμος -ου 'imitator, mime, actor' and the second element of the genus name *Tetragnatha* by the rhyming principle. Banks placed *foxi* into its own genus because of its small size, its shorter chelicerae with only a few small teeth, and because of the large spherical palpal tegulum (Levi 1980a:64). Levi placed it in *Glenognatha* SIMON.

(‡) *Minyrioloides* SCHENKEL 1930:10 (Linyphiidae)
Derived from the genus name *Minyriolus* SIMON (which see) by the addition of the suffix -oides. See the discussion under *Argyrodes*. Synonymized under *Baryphyma* SIMON by Millidge (1977:19)

*Minyriolus* SIMON 1884r:787 (Linyphiidae)
This appears to be a diminutive constructed by Simon on the Greek three-ending adjective μινυρός -ή -όν 'whimpering, whining, tweeting' or simply 'small'. The idea of 'tweeting' seems to have no particular semantic point. Simon appears merely to be digging for more Greek words that mean 'small' especially since the type species is *pusillus*, which is Latin for 'very small'. This species was first described by Wider *in* Reuss (1834: 243) as *Theridion pusillum* 'the itty-bitty *Theridium*'. As is often the case, the new genus name was semantically derived from the species name of the type. Simon added the diminutive suffix -iolus because the simple name *Minyrus* was preoccupied in the Coleoptera (Schönherr 1836). It is a palaearctic genus.

*Misumena* (Thomisidae) LATREILLE 1804b:135
From Greek μισουμένη the feminine present middle participle (with passive meaning) of the verb μισέω 'to hate' hence '(she) being hated'. Clerck originally (1757:128) named the type species *Araneus vatius* which is Latin for 'bow-legged spider'. Linnaeus's name (Linnaeus 1758:620) was *Aranea calycina* which means 'spider in the calyx of a flower' with a Latinized adjective from the Greek noun κάλυξ (gen.) κάλυκος 'cup, or calyx'.

*Misumenoides* F.O. PICKARD-CAMBRIDGE 1900:136 (Thomisidae)
Derived from *Misumena* by the addition of the combining form -oides the latinization of the the Greek two-ending adjectival suffix -οειδής -ες 'having the form of' derived from the Greek s-stem neuter noun εἶδος 'form'. The Latinized form is ambiguous as to gender and can be masculine, feminine or neuter but Pickard-Cambridge treated it as masculine. Bonnet (1957a:2948) is quite wrong in treating it as feminine. His reason (1955:37) is one he simply invents, namely, that forms in -oides must take the same gender as the base word on which they are formed. This is unsupportable grammatically, and contrary to actual usage in zoology. The Code (Art. 30.b) quite arbitrarily and without any grammatical justification decrees that all words in -oides must be masculine.

*Misumenops* F.O. PICKARD-CAMBRIDGE 1900:134 (Thomisidae)
Derived from *Misumena* by the additon of the suffix -ops, the latinization of the Greek adjective forming suffix -οψ (gen.) -οπος which forms Greek one-ending adjectives meaning 'looking like' such as μῆλοψ 'looking like an apple' and οἶνοψ 'looking like wine'. The forms are ambiguous as to gender and can be either masculine feminine or neuter; but Pickard-Cambridge treated this name as masculine. The Code (Art. 30.a.ii) arbitrarily and without grammatical justification decrees all genus names ending in -ops are "to be treated as masculine, regardless of its derivation or of its treatment by its author." Bonnet's note (1957a:2961) is unfortunately in error. It is based upon the misconception that the compound adjective in Greek has only one gender, and that is determined by the gender of the uncombined second member of the compound. Of course all adjectives have all three genders, and the gender of the underlying second member of the compound is irrelevant.

*Miturga* THORELL 1870c:375 (Miturgidae)
From the Greek masculine noun μίτος 'thread' and the combining form -ουργος 'worker' as in κάκουργος 'evil worker, thief'. There is a genuine Greek word μιτόεργος 'working the thread' of which μίτουργος is the regular contraction. Thorell gave it a feminine ending.

*Modisimus* SIMON 1893d:485 (Pholcidae)
A puzzling name. The only Greek or Latin word that remotely might connect with it is the very obscure Greek masculine noun μοδισμός -οῦ 'measuring by bushels (modii)'. First, there seems to be no semantic connection. Second, it requires postulating that Simon misremembered and misspelled the word. All of this is very unconvincing. This is probably merely a made up word.

(‡) *Montilaira* CHAMBERLIN 1921c:41 (Linyphiidae)
This genus was erected for the species *Hilaira uta* (CHAMBERLIN 1919a: 253) *Hilaira* is a genus of Simon's (1884a:374) from the Greek feminine adjective (it occurs only in the feminine) ἱλάειρα 'mildly shining'. So Chamberlin created *Montilaira* from the Latin masculine noun mons, montis 'mountain' and the last five letters of *Hilaira* SIMON. For the vicissitudes of its validity see Buckle *et al.* 2001:122, who synonymized it under *Halorates* HULL.

*Mumaia* LEHTINEN 1967:250, 395 (Uloboridae)
A new genus erected for *Miagrammopes corticeus* SIMON 1893b in honor of the American arachnologist Martin H. Muma. Synonymized under *Miagrammopes* by Opell (1984b:238).

*Mygale* LATREILLE 1802b:345; 1802a:49 (Theraphosidae)

From the Greek feminine noun μυγαλή (gen.) -ῆς 'shrew, field-mouse' which itself is a compound of μῦς 'mouse' and γαλέη 'weasel'. From the mouse-like size and shape of the large Theraphosids. This has been a designation for a spider since antiquity. Nicander (Theriaka 816) said the bite of the mygale is poisonous, and Pliny (N.H. 8.227) said, "In Italia muribus araneis venenatus morsus est" [In Italy the spider-mice have a venomous bite]. He is clearly referring to the passage of Nicander (Gow & Scholfield 1953:817).

For the history of the name I can do no better than quote Bonnet (1957a:2989):

> "There is no doubt that it was Walckenaer who actually created the name; Latreille himself recognizes this (Latreille 1804b:149), and even criticizes Walckenaer for it because Aristotle and Cuvier had already used the name in the Quadrupeds. But for Latreille "the mistake was without remedy" (!) and he continued to use the name to designate spiders.
>
> "The first time Walckenaer mentioned these spiders in writing (Walckenaer 1802:249) he named them "les Mygales" (in French) without giving them their scientific name, which he had certainly created earlier in a memoire read to the Société philomatique de Paris, but which has disappeared without a trace, at least I have not found it.
>
> "In short, it is Latreille who, the same year (1802a:345 and 1802b:49) produced the term *Mygale*, and it is to him that the name must be credited, even though he did not want it so! But let the soul of Latreille rest in peace, for the nomenclators of our day have found the remedy! By the law against homonymy and by applying the law of priority *Mygale* (Araneae) cannot be maintained. By that fact it falls into synonymy with *Avicularia* Lamarck, the first *Mygale* cited by Latreille, being itself the species *avicularia*. Likewise, if one attributes the name to its true father Walckenaer, the first species of that author was *fasciata*, then *Mygale* would fall in synonymy with *Poecilotheria* Petrunkevitch.
>
> "But what is funny is that *Mygale* (Araneae) has disappeared without any advantage to the genus *Mygale* which Cuvier created in 1800 to designate the musk-rat (Leçons d'Anatomie comparée, Paris, an vii t.1, table 1).The name, which Brandt had modified to *Myogale* (1836), was still-born, by the fact that since 1777 there existed already the name *Desman* created by Guldenstadt for this animal."

*Myrmarachne* MACLEAY 1839:10 (Salticidae)

The name of this ant-mimic genus is from the Greek masculine noun μύρμηξ (gen.) μύρμηκος 'ant' and the Greek feminine noun ἀράχνη 'spider'. The earliest described species included (not the type) was *Aranea formicaria* DE GEER 1778b. That species name is a Neolatin adjective based upon the Latin noun formica 'ant'. Since the name *Myrmarachne* is not properly formed grammatically (it should be *Myrmecarachne* based upon the stem) several purists (Agassiz, Scudder, Banks, Neave, Sherborn) have emended it, but Bonnet defended Simon's preserving the original spelling "which is surely more agreable to the ear than the correct form (1957a: 2997 note 211)."

*Myrmecotypus* O. PICKARD-CAMBRIDGE 1894:123 (Corinnidae)

From the stem of the Greek masculine noun μύρμηξ (gen.) μύρμηκος 'ant' and the masculine noun τύπος 'impression, die, form, type'. Hence 'ant-type' because these spiders are ant mimics.

*Myrmekiaphila* ATKINSON 1886:131 (Cyrtaucheniidae)

The Greek word for 'ant' is the masculine noun μύρμηξ (gen.) -ηκος and the Greek word for 'ant-hill, ants nest' is the derived feminine noun μυρμηκία (gen.) -ίας. There is a problem with the connecting vowel in this genus name. Compound two-ending adjectives formed with the second element -φιλος -ον 'loving, beloved' will be connected to their first elements in several ways determined morphologically, usually with -o-, but never in authentic Greek with -a-. So, strict grammar would require *Myrmekiophilos. Such compounds may have a passive meaning as πάμφιλος 'beloved by all'; or an active meaning as πονηρόφιλος 'fond of bad men' (from πονηρός 'evil'); or may be adjective-noun compounds like κακόφιλος 'a bad friend' a word which could also mean 'possessing bad friends'.

Atkinson meant his name to mean 'fond of ants nests' and explicitly gave this etymology (1886:131). He collected his specimens while hunting for ants, and they were always found in close proximity to ants nests. He said (116-117):

> "From the location of their nests it is quite evident that members of the species are extremely fond of ants and seek to build their nests either directly in an ants nest or in close proximity to it."

Hentz (1850b:286 = 1875:159) had named the type species *Mygale fluviatilis* 'the mygale from the river' because it was found in the water during an inundation of the Tennessee River. Atkinson established his genus with the description of *Myrmekiaphila foliata* (now *fluviatilis* HENTZ 1850b). This adjective in classical Latin means 'leafy'. It comes to have a specialized meaning in araneology, namely, 'having a folium'. A folium ('leaf') is the area on the abdominal dorsum of a spider, marked by scalloped margins, which is thought to look leaf-like in outline. Atkinson (1886:133) described the folium of his new species and refered to it as the "foliation."

Bonnet (1957a:3015) discussed the variations in spelling of the genus name:

> "Atkinson in creating this genus in 1886 named it **Myrmekiaphila** (μυρμηκία ant-hill; φιλέω 'love'), a spelling correct in itself, but which Simon emended in 1891 to **Myrmeciophila**, but without giving the least explanation. It is easy to see that Simon in doing so transliterated the Greek kappa with a c (which the Rules of Nomenclature require; see volume 1, p. 134 [of the Bibliographia aranearum], and used the connecting vowel o which serves to unite the two terms of a word composed of Greek roots, like Hydrophila. -**Myrmeciophila** is then a correct emendment and Strand (1934) is wrong to retain the original spelling **Myrmekiaphila**, as is Petrunkevitch in changing it to **Myrmekiophila**.The name of the [spider] genus **Myrmecium** gives the sense of the normal spelling of words constructed on that Greek root."

Bonnet is right that *Myrmeciophila* is a more elegant spelling, and less open to philological criticism, but the Code privileges original spellings even if the names are badly formed (Code Art. 32.c.ii). Further, Greek two-ending compound adjectives do not have a distinctive feminine form, so the ending in -a, though not philologically correct, is in accordance with zoological convention. Transferred to the Cyrtaucheniidae by Raven (1985a:137).

*Mysmena* SIMON 1894a:586-588 (Mysmenidae)

This is not a Greek or Latin word, although it has a vague similarity to such names as *Misumena* and has an ending like a Greek feminine middle participle. The problem is that Greek would never have connected the two parts of such a compound without a vowel between mys and mena. Simon created this genus for *Dipoena leucoplagiata* SIMON 1881a, where the species name means 'with white sides' referring to the two white spots on the abdomen, as if there were two full moons. If one looks up the word for moon in the Greek lexicon, one will find μείς μηνός (nominative and genitive). I would not put it past Simon to have made a name out of this combination, with a change in spelling (y instead of ei) based upon pronunciation, to indicate the two "moons" implied by *leucoplagiata* ( i.e. 'moon-moon'). Then he made it feminine because spiders are feminine. This would be in accordance with the custom of making a new genus name based on the type species name.

*Mysmenopsis* SIMON 1897d:865 (Mysmenidae)

Derived from *Mysmena* by the addition of the suffix -opsis from the Greek feminine noun ὄψις (gen.) ὄψεως 'aspect, appearance'. Hence, 'with the appearance of *Mysmena*'.

*Mythoplastoides* CROSBY & BISHOP 1933b:142 (Linyphiidae)

From the Greek masculine noun μυθοπλάστης (gen.) -ου, 'coiner of legends' with the addition of the Latin adjective forming suffix -oides corresponding to the neuter noun εἶδος 'form'. They chose the derived name in -oides, because mythoplastes would seem to be preoccupied by *Mythoplastis* in the Lepidoptera (Meyrick 1919). Crosby and Bishop gave no indication of a rationale for the name, but it was their habit to search the Greek lexicon for usable names without concern for any semantic rationale. The Latin form of the word is ambiguous as to gender, and could be masculine, feminine or neuter, and Crosby and Bishop treated it as masculine. The Code arbitrarily, and without grammatical justification decrees that all names in oides are masculine (Code. Art. 30.b). This genus is was synonymized under *Entelecara* SIMON by Hackman (1954), but was revalidated by Crawford (1988:14).

(‡) *Nanavia* CHAMBERLIN & IVIE 1933a:26 (Linyphiidae)

From the language of the Gosiute Indians, nanavi 'tall woman' (Chamberlin 1913:13) with -a added to give it a Latin feminine ending. This word is used as a proper name in Gosiute, and there may be no explicit semantic point. On the other hand, the authors specify that in the type species *monticola*, the only species described in the paper, the female is similar to the male except larger. Synonymized under *Leptorhoptrum* KULCZYŃSKI by Eskov & Marusik (1994).

*Naphrys* EDWARDS 2003a:68 (Salticidae)
The author states that the name is "a contraction of North American *Euophrys*." It is feminine as is *Euophrys* C.L. KOCH (which see).

(‡) *Nemesoides* CHAMBERLIN 1919b:1 (Cyrtaucheniidae)
Formed by the addition of the the Latin adjective forming suffix -oides' having the form of', which forms Latin one-ending adjectives, to the genus name *Nemesia* AUDOUIN 1826. The Latin suffix is a transliteration of the Greek form -οείδης -ες 'having the shape of' (not to be confused with a quite different Greek suffix -ώδης -ες). There is a pitfall regarding the name *Nemesia*. If we were to take it at face value, it would appear to be from the Greek neuter plural noun Νεμέσια, a festival of the dead at Athens mentioned by the 4th century orator Demosthenes (41:11), and also found in the Suda, a Byzantine encyclopedia. But this is not to be expected from what we know of Savigny's nomenclatural habits, and more generally the habits of his generation. Incidentally, there is good internal evidence that Savigny was the member of the team of Savigny and Audouin who made up the names. Of the nine genus names due to him, six are taken from the Fabulae of Hyginus. Linnaeus had initiated the practice of gleaning mythological names from Hyginus, which is a veritable treasure house for the busy systematist. Since the divinity Nemesis is mentioned in the preface of Hyginus, it is likely that from this beginning Savigny gave it a clearly feminine form *Nemesia*. See Heller (1945).

The name *Nemesoides* is ambiguous as to gender; it can be masculine, feminine or neuter. Chamberlin regarded it as feminine. The Code (Art. 30.b) arbitrarily and without grammatical justification decrees that all names in -oides are masculine.

Bonnet (1958:3040) makes one of his very rare mistakes in reporting that *Nemesia cellicola* was called *Mygale cellicola* in Audouin 1826, and was first called *Nemesia cellicola* by Ausserer in 1871.

*Neoanagraphis* GERTSCH & MULAIK 1936a:11 (Liocranidae)
Formed from the Greek three-ending adjective νεός -ά -όν 'new' and the genus name *Anagraphis* SIMON 1893a:352, an invented word apparently from Greek feminine noun γραφίς γραφίδος meaning a 'stilus for writing, an engraving tool, a needle'. Since the legs in this genus are strongly aculeate, it is just conceivable that Simon might have meant something like 'needled up'. The ἀνα-part means 'up'.

*Neoantistea* GERTSCH 1934c:18 (Hahniidae)
From the genus name *Antistea* SIMON (which see) by adding the prefix neo-'new' from the Greek adjective νεός -ά -όν 'new'. Gertsch explained the rationale (1934c:2):

> "Although all the Nearctic species [of Hahniids] were originally described in **Hahnia**, a survey shows that three genera are represented. A new generic name **Neoantistea**, is proposed for a group of four species, including **Hahnia riparia** Keyserling, which was placed in **Antistea** by Simon."

*Neoapachella* BOND & OPELL 2002:514 (Cyrtaucheniidae)
The authors say the name is "in honor of the Apache Indian Nation that has a reservation near the type locality."

*Neocryphoeca* ROTH 1970:115 (Hahniidae)
From the the Greek three-ending adjective νεος -ά -όν 'new' and the genus name *Cryphoeca* (which see). Erected for two new species from Arizona, closely related to *Cryphoeca* with type *Neocryphoeca gertschi* ROTH 1970. Transferred to the Hahniidae by Brignoli 1983c:509.

*Neoleptoneta* BRIGNOLI 1972i:134 (Leptonetidae)
From the Greek three-ending adjective νεός -ά -όν 'new' and the genus name *Leptoneta* (which see).

*Neon* SIMON 1876a:208 (Salticidae)
When Simon established this genus he named the type species *Neon reticulatus* (BLACKWALL 1853), that is, he regarded the name as masculine. So we must conclude that the name is not, as we might have suspected, the neuter of the Greek three-ending adjective νεος -ά -όν 'new'. Simon gave no further indication of the source of the term, but Bonnet correctly deduced (1958:3050) that the word is the proper name of a Boeotian, that is Νέων (gen.) -ωνος. This would be the Theban, Neon, the son of Askondas, who was leader of the pro-Macedonian party in Boeotia about 230-220 B.C., and is mentioned in Polybius's history (Polybius 20.5.5). Simon commonly used proper names from antiquity without any semantic point. The North American species *nelli* Peckham & Peckham 1888 is named in honor of Mr. Philip Nell (Peckham & Peckham 1888:88).

*Neonella* GERTSCH 1936a:23 (Salticidae)
Feminine diminutive formed on the genus name *Neon* SIMON (which see).

*Neoscona* SIMON 1864:261 (Araneidae)
Simon used this name to designate a subgenus (actually "group") of *Epeira* WALCKENAER. It was apparently F.O. Pickard-Cambridge (1904:466) who first raised it to genus level. This is one of Simon's juvenile names, created when he was only sixteen, and his Greek, never too sure at any stage, was very shaky then. Simon said it is derived from νέω filer [spin]; σχοῖνος roseau [reed]. His intention is clear, and we must take his word for it, but his morphology is faulty. There is no grammatical way that Simon's name could be formed from the Greek words he cited. νέω is the first person singular of a verb, and could not form the first element of a compound in that form. In the second element he has ignored the aspirate c which should be transliterated with ch (cf. under *Rhecostica*). Secondly, he incorrectly transliterated the diphthong oi which should become oe. The proper formation would be *σχοινονήτης (with an agentive nominal combining form), a masculine compound noun, which should be latinized as either *schoenonetes or *schoenoneta (masculine a-stem). Such a compound would mean 'spinning a reed' and that is not what Simon meant to indicate. He said, "spiders living at the edge of water" and he surely meant the name to mean 'spinning among the reeds'.

Thorell (1869:34) amended the name to *Neoschoena*, but no one followed him. Hence Bonnet (1958:3055), although agreeing with Thorell on the etymology, decided Simon's spelling is to be retained "even though incorrectly written." Thorell understood what the young Simon was trying to do, and amended the name to make clear it was derived from the Greek masculine (or feminine) noun σχοῖνος 'reed, rush' on the analogy of such two-ending compound adjectives as βαθύσχοινος -ον 'deep grown with rushes'.

*Neospintharus* EXLINE 1950a:112 (Theridiidae)
From Greek νεο- 'new' and the genus name *Spintharus* (which see).

*Neotama* BAEHR & BAEHR 1993:68 (Hersiliidae)
From Greek νεο- 'new' and the genus name of the type species *Tama variata* Pocock 1899d. *Tama* SIMON 1882b:256 is a Latin feminine noun meaning 'a certain swelling of the feet.' Simon says the legs are strongly unequal (valde inequales), and this may have been the rationale for the name.

*Neottiura* MENGE 1868:163 (Theridiidae)
Menge explained that this name was from Greek νεοττιά nidus [nest] and οὐρέω custodio [I guard], "from the habit of the spider of carrying around and guarding its egg-sack." His German translation of the name is "Nesthüterin" i.e. she who guards the nest. The type species is *Neottiura bimaculata* (LINNAEUS 1767). There is a very rare Greek verb οὐρέω 'guard, watch' a denominative from the masculine noun οὖρος 'watcher, guardian."

*Neozimiris* SIMON 1903a:984 (Prodidomidae)
From the Greek three-ending adjective νεός -ά -όν 'new' and the genus name *Zimiris* SIMON (1882b:239). Simon said that *Zimiris* is a geographical name. Simon was fond of gleaning usable names from the Natural History of Pliny the Elder (1st century A.D.) and Pliny (36.129) gives the name of a city "in the sandy regions of Ethiopia" as Zimiris. The type species of *Zimiris* was collected at Aden, Yemen, and clearly Pliny is the source of Simon's name. He used an edition of Pliny published prior to 1851, since later editors have emended the name to Zmiris. Although in the original publication Simon left the gender a mystery with the patronymic species name *doriai* (named for the Italian naturalist C. Doria), subsequently (1884b:cxli) he treated it as feminine (*indica*). Banks (1898b:214) had described *Zimiris pubescens* from Mexico, and Simon (1903a:984) erected *Neozimiris* for it. Hence, the meaning is not only 'new *Zimiris*' but 'New World *Zimiris*'. Family Prodidomidae revalidated by Platnick 1990a:36.

*Nephila* LEACH 1815:133 (Tetragnathidae)
An invented feminine adjective as if from a Greek two-ending adjective *νήφιλος -ον 'fond of spinning'. The combining form νη-is derived from the neuter noun νῆμα (gen.) νήματος 'thread' a noun related to the verb νέω 'spin'. The second element is the combining form -φιλος 'loving'. Hence, 'thread-loving or fond of spinning'. Greek does not have a distinctive feminine form of compound adjectives, so this acquires a feminine form when latinized. Erichson's derivation (1845:9) from νεφέλη 'cloud' is fanciful, ignoring morphology, phonology and semantics. The name of the American species *clavipes* (LINNAEUS 1767) is an invented Latin compound meaning 'club-footed' and refers to the club-shaped tufts on the legs. Levi (in Roth 1985) put the Nephilines under Tetragnathidae.

*Neriene* BLACKWALL 1833c:133 (Linyphiidae)
From the Latin mythological name of the wife of Mars. This word in classical Latin has a Greek form with a nominative in-e and a Greek form of genitive Nerienes. It is said to be a Sabine word meaning 'bravery' (Aulus Gellius 13.23). Blackwall probably got it from Plautus Truculentus 515: "Mars peregre adveniens salutat Nerienem uxorem suam" [Mars, arriving from afar greets his wife Neriene]. The type species is *Neriene clathrata* SUNDEVALL 1830. Bonnet (1958.3087) gave a long list of synonyms, spiders described in *Neriene* now in other genera. He argued that *Neriene* was not usable as a genus name, and criticized Chamberlin for using it. It has been long considered a synonym of *Linyphia* LATREILLE, but was revalidated by van Helsdingen (1969:73). Locket et al. (1974:119) considered it a subgenus of *Linyphia*.

*Nesticodes* ARCHER 1950:22 (Theridiidae)
   From the genus name *Nesticus* (which see) and the suffix -odes 'like, similar' (see the discussion under *Argyrodes*). Erected for *Theridion rufipes* LUCAS, synonymized under *Theridion* by Levi (1957a:19), revalidated by Wunderlich (1987a:214).

*Nesticus* THORELL 1869:88 (Nesticidae)
   From the Greek three-ending adjective νηστικός -ή -όν 'having to do with spinning' an adjective derived form the verb νέω 'spin'.

(‡) *Nidivalvata* ATKINSON 1886:129 (Antrodiaetidae)
   From Latin masculine noun nidus 'nest' and the feminine adjective valvata meaning 'provided with double doors' from the noun (used in the plural with singular meaning like 'trousers' and 'scissors') valvae 'door with two leaves, folding door'. Atkinson himself gave this etymology explicitly. He incorrectly described the trap-door as follows:
   
   "Each door is a surface of a half circle, is hung by a semicircular hinge, and the two meet when closed in a straight line over the middle of the hole."
   
   Actually the burrow has a flexible collar which extends above the soil surface, and the spider closes it by pulling it on opposing sides collapsing the two sides inward (Coyle 1971:276) A junior synonym of *Brachybothrium* SIMON (1884m:314), which itself is a junior synonym of *Antrodiaetus* AUSSERER 1871a.

*Nigma* LEHTINEN 1967:252 (Dictynidae)
   Replacement name for *Heterodictyna* DAHL (which see). Etymology unknown. Treated as feminine.

*Nodocion* CHAMBERLIN 1922:147, 154 (Gnaphosidae)
   Bonnet was right (1958:3105) when he supposed that the end of the word is constructed to rhyme with the neighboring genus *Megamyrmecion* WIDER in REUSS 1834. Chamberlin treated *Megamyrmecion californicum* SIMON 1893a just one page previously. So the -cion is a rhyming morph. *Nodocion* is characterized by the lack of teeth on the retromargin of the chelicerae. The first part of the name is from the Greek three-ending adjective νωδός -ή -όν 'toothless'. Chamberlin had used this Greek word earlier in his species *Wala noda* CHAMBERLIN 1916: 296 from Peru where he explained that the species name is from νωδος which he glossed as 'edentate'. Hence, *Nodocion* means 'no teeth, close to *Megamyrmecion*'. Of course, Chamberlin used Simon's spelling of *Megamyrmecion*. Bonnet was concerned that Chamberlin gave his new genus masculine gender, even though the second element of the name is taken from a neuter genus name. Bonnet concluded that, since Chamberlin did this intentionally, and not in ignorance, Chamberlin's gender should be maintained. One could argue from the Code (Art. 30a iii) that it should be neuter. But since it has universally been treated as masculine, and that was clearly Chamberlin's intention, it is best to leave well enough alone.

*Nops* MACLEAY 1839:2 (Caponiidae)
   This is the Greek one-ending adjective νώψ (gen.) νώπος 'purblind', which is constructed from an apocopated negative prefix ν(η) and the feminine noun ὤψ (gen.) ὠπός 'eye'. There are other Greek one-ending adjectives of this formation like μονώψ (gen.) μονώπος 'one-eyed'. Νώψ is found only once in all of extant Greek literature-in the ancient lexicographer Hesychius. MacLeay prided himself on his Cambridge classical education, and the fact that he knew this prodigiously obscure word suggests his pride was not undeserved. Like all one-ending adjectives it is ambiguous as to gender, and can be masculine, feminine or neuter. MacLeay's type species name is the genitive of a place name in Cuba, guanabacoae, which gives no clue about the gender of the genus name. Bonnet (1958:3114) was unjustified in his principle (Bonnet 1955: 35) that names in -ops, where it means 'eye', should be masculine. He misunderstood the nature of Greek adjective formation. It is not true that a Greek compound adjective takes its gender from the gender of the second element. Keyserling in 1878c cagily added the species *variabilis*, a name which is also ambiguous as to gender. It was Hasselt in 1887a who was first bold enough to resolve the question with the feminine species name *glauca*. If priority in these matters has any force, the genus should be treated as feminine. But the Code (Art. 30a ii) now arbitrarily decrees that "a genus-group name ending in -ops is to be treated as masculine, regardless of its derivation or of its treatment by its author." Derived genus names are: *Orthonops* CHAMBERLIN, *Tarsonops* CHAMBERLIN, *Nopsides* CHAMBERLIN, *Indonops* TIKADER, *Bruchnops* MELLO-LEITÃO.

(‡) *Notiomaso* BANKS 1914b:78 (Linyphiidae)
   From the Greek three-ending adjective νότιος -α -ον 'moist, southern' and the genus name *Maso* SIMON (which see). The type species is *australis* BANKS 1914b and the locality is South Georgia, so it is clear that Banks intended this name to mean 'the southern *Maso*'. Transferred from the Erigoninae to the Linyphiinae by Wunderlich 1978c:253. This is not a North American genus, but for a long time Banks's species *Notiomaso australis* was listed as a North American species in the mistaken belief that "South Georgia" meant the southern part of the American state of Georgia. It means South Georgia Island in the South Atlantic.

(‡) *Notionella* BANKS 1905a:312 (Linyphiidae)
   From the Greek three-ending adjective νότιος -α -ον 'moist, southern' plus a diminutive ending from Latin on the analogy of *Ceratinella* EMERTON. In the latter name the n is etymological (diminutive from *Ceratina* MENGE, which is from the feminine of the Greek adjective κεράτινος -η -ον 'made of horn'), but in Banks's name the n results from misdivision of morphemes. Banks's specimen of *Notionella interpres* (1905a:312) came from North Carolina, and presumably he meant the name to mean something like 'southern spider close to *Ceratinella*'. Synonymized under *Ceratinopsis* EMERTON by Bishop & Crosby (1930:15).

*Novalena* CHAMBERLIN & IVIE 1942b:224 (Agelenidae)
   Previously in Chamberlin & Ivie 1941b:9, *nomen nudum*. See the general remarks on Agelenid names in -lena under *Hololena*. The Nova-morpheme means 'new' from the Latin adjective novus-a-um. Hence, 'new Agelenid genus in the series *Hololena*, *Ruelena*, etc'.

*Nuctenea* SIMON 1864:261 (Araneidae)
   Simon used this name for a "groupe" of his subgenus *Epeira* WALCKENAER. From the Greek feminine noun νύξ (gen.) νυκτός (stem) νυκτ- 'night' and the verb νέω 'spin'. This is one of Simon's juvenile names (he was only 16) and is not perfectly formed. A French schoolboy would have pronounced the high front rounded Greek vowel upsilon with the French high front rounded vowel u. Hence the spelling with u rather than with the more usual y. (Compare Simon's 1864 genus *Nuctobia* 'night living'). Furthermore the connecting vowel would more strictly be an i as in genuine Greek compounds like νυκτινόμος 'feeding at night'. For the second half of the compound, in the absence of an agent noun built on this verb, Simon appears simply to have removed the personal ending from the verb νέω and added a feminine ending. At any rate, he clearly meant this to mean 'night spinner' and it referred to the fact that spiders of this group, as he said, "construct their webs at night and destroy them for the daytime." Grasshoff (1983) showed that *Larinioides* DI CAPORIACCO 1934b is the earliest unambiguous available name for the *Araneus cornutus* group. *Nuctenea* had been applied to a heterogeneous group. Levi had resurrected Simon's old subgeneric name *Nuctenea* (Levi 1974b:300) for the *cornutus* group, but had also included *umbraticus*. Now *Nuctenea* applies to the Old World spiders related to *Nuctenea umbraticus* (CLERCK 1757) and *Larinioides* with type species *folium* (SCHRANK 1803) includes only the *cornutus* group.

*Oaphantes* CHAMBERLIN & IVIE 1943:7 (Linyphiidae)
   A genus separated from *Bathyphantes* for *Bathyphantes pallidulus* BANKS 1904b, so the last part of the name is a rhyming morph -phantes from *Bathyphantes* (which see). The gender ought to be masculine, since the Greek noun ὑφάντης 'weaver' is masculine. But Banks and Emerton got this wrong by making *Bathyphantes* feminine. Chamberlin and Ivie correctly made their genus masculine. The first part of the name is from the Gosiute Indian adjective oabit 'yellow, brown' with combining form oa-as in oawuda 'brown bear' (Chamberlin 1908b:79). Compare Chamberlin's species name *oatimpa* (*Coryphaeolana*) from oabit 'yellow' and timpi 'stone' for a species from Yellowstone Park (Chamberlin 1949:527; 1908b: 82). The form oa, then, is the Gosiute translation of the type species name *pallidula* 'light yellow'. Hence, the name means 'yellow spider close to *Bathyphantes*'.

*Ochyrocera* SIMON 1891c:565 (Ochyroceratidae)
   The first element is from the Greek three-ending adjective ὀχυρός -ά -όν 'stout'. The second element is a neo-Latin suffix -cerus-a-um, common in zoological names, which is formed by analogy from Greek two-ending adjectival combining form -κερως -ων (gen.) -ωτος, which is derived from the neuter noun κέρας (gen.) κέρατος 'horn'. Authentic Greek two-ending adjectives like ὑψίκερως -ων 'with lofty antlers' or δίκερως -ων 'with two horns' get transliterated into Latin with -kerôs as in monocerôs-ôtis 'unicorn' and this gets transferred into the second declension in neo-Latin as *monocerus*. Hence, the name means 'with stout horns'. This does not refer, as one might suppose, to horns on the head, but to the very long, arcuate spine of the palpal bulb. The type species is *arietina* SIMON 1893a, which means 'like a ram's head' and indeed the palpal bulb is shaped like a ram's horn (Simon 1893a:281, fig. 243.) Hence, the name means 'with a remarkable long, spine on the palpal bulb which looks like the stout horn of a ram'.

*Ocrepeira* MARX in HOWARD 1883:22 (Araneidae)
   Marx introduced this name in a list of the spiders of South Carolina for *Wixia ectypa* (WALCKKENAER 1842), without any description. It was validated by Levi 1993b:64. The second half of the name is the old genus *Epeira*, but there is no sure explanation for the first half Ocr-, unless Marx came across the Greek feminine noun ὄκρις 'a jagged point or prominence' with reference to the abdomen, which projects markedly forward over the cephalothorax.

*Octonoba* OPELL 1979:512 (Uloboridae)
   The type species for this genus is *Uloborus octonarius* MUMA 1945. Opell states that the genus name is an arbitrary combination of letters, but the first part Octon- is obviously meant to reflect the type species name

*octonarius.* The Latin adjective *octonarius* means 'consisting of eight'. Muma (1945:93) says: "This species differs from the other American forms in that the abdomen of the female bears four pairs of tubercles." These eight tubercles are the inspiration for the name.

*Oecobius* Lucas 1846:101 (Oecobiidae)
From the Greek two-ending adjective οἰκόβιος -ον 'domestic' from the masculine noun οἶκος 'house' and the adjective forming suffix -βιος -ον 'living' from the masculine noun βίος 'life'. Hence, 'living in the house'.

*Oedothorax* Bertkau *in* Förster & Bertkau 1883:228 (Linyphiidae)
From the Greek combining form οἰδο-(Cf. οἰδέω 'swell' οἶδμα 'swelling' οἶδος 'swelling'), and the Greek masculine noun θώραξ (gen.) θώρακος 'breastplate, corselet, trunk'. Hence, 'with swollen thorax'. There are authentic one-ending Greek adjectives such as χαλκεοθώραξ 'with bronze breastplate'. The name is formally ambiguous as to gender, but Bertkau treated it as masculine. The name refers to the fact that the male has a rounded hump behind the eyes with a groove in front of the lobe, see Kaston 1948 fig. 490. The type species is *gibbosus* Blackwall 1841a: 653, which means 'hump-backed'.

*Olios* (Sparassidae) Walckenaer 1837:202
From the cult name of Apollo, Ἀπόλλων Ὅλιος = Οὔλιος = ὀλοός 'baleful, deadly'. Used of the Dog-star Hom. Il. 11.72. Epithet of Artemis and Apollo. I wonder whether Hogg knew this when he named a species *Olios artemis* Hogg 1915a. It is unclear to me why Walckenaer used this inscriptional form rather than the usual literary forms. The type species *Sparassus argelasius* Walckenaer (1805:40) was sent from Bordeaux "par l'habile entomologiste Argelas" [by the skilled entomologist Argelas] about whom nothing more is known.

(‡) *Onondaga* Peckham & Peckham 1909:491 (Salticidae)
The Peckhams were fond of using place names for their genera. This genus is named either for the lake in central New York or for the Iroquoian tribe, which gave the lake its name. The Onondaga were an important tribe of the Iroquois Federation, and their name means 'on top of the mountain'. Synonymized under *Fuentes* Peckham & Peckham 1894 by Kaston (1938d:195), and subsequently by Barnes (1958:3) under *Marpissa*.

*Oonops* Templeton 1835:404 (Oonopidae)
From the Greek neuter noun ᾠόν 'egg' and the feminine noun ὤψ (gen.) ὠπός 'eye'. There are authentic Greek one-ending adjectives of this formation like μονώψ (gen.) μονώπος 'one-eyed'. Hence, 'egg-eyed'. This refers to the appearance of the six large oval eyes. Strictly speaking, the n, the nominative neuter ending, does not belong in the name, but the correct formation would have been *Oops, a name that serious science could hardly maintain (cf. Thorell 1869:12 fn.2.). Templeton is to be praised for allowing euphonic taste to outweigh grammatical punctilio. The name, like all one-ending adjectives, is ambiguous as to gender; it could be masculine, feminine or neuter. Templeton treated it as masculine.

*Opopaea* Simon 1891c:560 (Oonopidae)
Probably from the Greek feminine noun ὀπωπή 'sight, eye, eyeball' or the derived word ὀπωπία 'bones of the eyes'. This explanation does not account for the diphthong ae, but Simon was not always accurate about such things. Since the close genus *Oonops* means 'egg-eyed' it is reasonable for Simon to have chosen another 'eye' word for his new genus.

*Orchestina* Simon 1882c:237 (Oonopidae)
Feminine diminutive formed upon the Greek masculine noun ὀρχήστης (gen.) -ου 'dancer'. Simon had separated this group from the genus *Schoenobates* Blackwall, which means 'rope-dancer' and the species names *saltabunda* Simon 1893b 'frequently jumping' and *saltitans* Banks 1894b 'jumping' suggest that Simon named this genus for its active behavior, and to reflect the old genus of Blackwall in his new name.

*Oreonetides* Strand 1901b:29 (Linyphiidae)
Formed with the Greek patronymic suffix -ιδης 'son of' upon the genus name *Oreoneta* Kulczyński *in* Chyzer & Kulczyński 1894, which is from the Greek s-stem neuter noun ὄρος (gen.) ὄρεος, 'mountain' which yields a combining form ὀρεο-, plus the arachnological suffix -neta 'spinner'. For the formation of -neta see the discussion under *Argyroneta*. Hence, 'mountain spinner' and 'son of mountain spinner'. Bonnet incorrectly claimed that *Oreonetides* is feminine (Bonnet 1958:3204) on the grounds of his unjustifiable principle that the derived genus ought to be given the gender of the original genus. Strand clearly treated it a masculine. Furthermore, from a morphological standpoint, the Greek suffix -ιδης 'son of' can never be feminine. The feminine patronymic suffix in Greek is -ις (gen.) -ιδος, as in Chryseis, daughter of Chryses, in the Iliad. In the accusative case she is Chryseida, and becomes Chaucer's and Shakespeare's Cressida.

*Oreophantes* Eskov 1984a:667 (Linyphiidae)
Erected for *Bathyphantes recurvatus* (Emerton 1913a), which was transferred from *Oreonetides* Strand. The second part -phantes is the customary rhyming morph to designated Linyphiids, after *Bathyphantes* etc. The first half is taken from the first half of the genus name *Oreonetides* (which see) from which Eskov transferred the type species.

*Origanates* Crosby & Bishop 1933b:154 (Linyphiidae)
This name appears to be derived from the Greek word for oregano, a feminine second declension noun ὀρίγανος. Further it appears to be a mistake for the associated adjective ὀριγανίτης 'flavored with oregano'. Crosby and Bishop give no shadow of a hint why this should be, but it was their habit to scan the dictionaries for usable names without regard to semantics. They may have found also it in the Latin dictionary under origanites. They treated it as masculine.

*Orodrassus* Chamberlin 1922:163 (Gnaphosidae)
Derived from *Drassus* by the addition of a prefix from the Greek neuter noun ὄρος (gen.) ὄρεος 'mountain'. Chamberlin erected this genus for the type species *Drassus coloradensis* Emerton 1877, collected from Gray's Peak, Colorado. Hence, 'the mountain *Drassus*'.

(‡) *Oronota* Simon 1871b:vii (Theridiidae)
From the Greek neuter noun ὄρος -εος 'mountain' plus the Greek neuter noun νῶτον 'back' (or sometimes masculine νῶτος) combined into an invented three-ending adjective *ὀρόνωτος -η -ον, on the analogy of such genuine Greek compound adjectives as ἀκανθόνωτος 'spine-backed' ὀστρακόνωτος 'having the back covered with a hard shell' and ἀστερόνωτος 'having a starry back (epithet of heaven)'. Hence, 'mountain-backed' with reference to the peculiar hard, lumpy, abdomen with many humps. Simon later (1873a) created another name for this genus, *Oroodes* 'shaped like a mountain'. Bonnet (1958:3210) said:

> "Simon, after having created in 1871 the genus name *Oronota*, for *Epeira paradoxa* of Lucas, -doubtless not remembering afterward that he had done so -again in 1873 created, for the same species, a new term *Oroodes*! Then, he discovered that the genus had a synonym *Ulesanis* L. Koch, which, having been created in 1872, had priority [over *Oroodes*], and on those grounds has, in effect, been used consistently up to the present, without anyone thinking about *Oronota*, which itself had priority over *Ulesanis*. Today we will invoke the law of prescription to leave *Oronota* in oblivion."

The "law of prescription" is the provision that long unused names (the period of limitation varies from 100, to 80, to 50 years) are exceptions to the Law of Priority and are *nomina oblita*. Bonnet argued for this exception (1945:103-106; 1955:12-14) and it was formalized in the 1961 Code (Art. 23a), but is no longer permitted by the ICZN Code (ICZN 1999) without application to the Commission for a ruling (Art. 23b).

Levi & Levi (1962:31) sank *Ulesanis* (as well as *Oronota*) under *Phoroncidia* Westwood 1835.

*Orthonops* Chamberlin 1924b:597 (Caponiidae)
It has become a custom to form new Caponiid genera on the basis of *Nops* MacLeay 1839 (which see). Here the first element is from the Greek three-ending adjective ὀρθός -ή -όν 'straight'. Chamberlin treated this genus as masculine, but see under *Nops*. Chamberlin created two contrasting genera in this paper, *Orthonops*. and *Tarsonops*. *Orthonops* is described as "agreeing essentially with *Nops* excepting that the anterior tarsi have an inferior claw strongly developed and wholly lack the paired membranous laminae of the latter genus." Further *Orthonops* has "a distinct false suture dividing the anterior tarsus into two principle segments." *Tarsonops* by contrast has the "tarsi divided by several pale lines of false sutures." It is clear that *Tarsonops* is named to reflect this feature of the tarsi. *Orthonops* is apparently named to contrast with it by the possession of a single straight false suture.

*Ostearius* Hull 1911a:583 (Linyphiidae)
Hull gave no indication of the meaning of the name. It appears to be misspelled from the Latin masculine noun *ostiarius* 'doorkeeper, porter' perhaps because the type species (*melanopygius* O. Pickard-Cambridge 1879d) is often found indoors in England.

*Oxyopes* Latreille 1804b:135 (Oxyopidae)
From the Greek three ending u-stem adjective ὀξύς -εῖα -ύ 'sharp' and the combining form -ωπης which forms two-ending adjectives meaning 'having such and such eyes' as in the adjectives κυανώπης -ες 'having dark (or blue) eyes' and κυνώπης -ες 'dog-eyed, i.e. shameless'. In form it could be either masculine or neuter when latinized. The feminine forms of these adjectives are distinctive namely, κυανῶπις -ιδος and κυνῶπις -ιδος. Latreille clearly treated it as masculine (*heterophthalmus*). Hence it means 'having sharp eyes'. Simon's unjustified emendation to *Oxyopa* (1864:386) is the result of his imperfect schoolboy Greek. He mistook Latreille's Greek s-stem adjective for an a-stem masculine noun, which would be latinized into a first declension masculine (like πειράτης = Latin *pirata* 'pirate').

(‡) *Oxyptila* Simon 1864:527 (Thomisidae)
Original spelling *Ozyptila* corrected by Thorell 1869:36 to *Oxyptila*.

Bonnet (1945:139) said:

> "Simon created the genus **Ozyptila** in 1864; he himself realized in 1875 that this word was incorrect, that it was a juvenile orthographic mistake (he was 16 years old when he published the first edition of the Histoire naturelle des Araignées), and that he should have written **Oxyptila**. Following that all authors wrote **Oxyptila** and the word has been thus used many hundreds of times. It is therefore inadmissable that Banks in 1907, appealing no doubt to the false interpretation of the rule that 'no name, once published may be rejected' should come to re-establish the name **Ozyptila**, and was followed by some American authors and Professor Strand."

Bonnet's well-known championship of long-used names has been overruled in this case by the Code (Art. 33), and Dondale & Redner (1975a: 132-133) have adopted the original spelling (albeit with apologies for its philological defects). The name is from the Greek three-ending u-stem adjective ὀξύς -εῖα -ύ 'sharp' and the neuter noun πτίλον 'down, underfeathers'; these then are formed into a three-ending Latin adjective, oxyptilus, of which this name is the feminine. But semantically this would seem to make little sense: 'possessing sharp down'. since the critical character of the genus is its clavate setae, and presumably the young Simon (he was 16 remember) may have meant to refer to this in the name and got confused with his Greek -something he did frequently in 1864. Thorell (1870b:185), never gentle with the whippersnapper Simon, complained of this semantic fault saying:

> "The name **Oxyptila** can moreover hardly be retained, on account of its signification.which is absolutely the reverse of the characteristic feature (the club-like thickening of the bristles towards the apex)."

There are authentic Greek compound adjectives like χλωρόπτιλος 'with green feathers' and τετράπτιλος 'with four wings'. Simon (1864:439) gave the name as Ozyptile in French and Ozyptila in Latin, but explicitly derived the name from ὀξύς piquant [sharp] and πτίλος duvet [down]. He referred to the only species *Ozyptila claveata* (WALCKENAER 1837) as "la thomise à clous [the Thomisus with nails]" and emphasized the "épines fines, dures et un peu élargies à leur extrémité [fine spines, hard and somewhat enlarged at their extremity]." Regarding the variation in spelling, Simon would be neither the first nor the last schoolboy to confuse the two curly Greek letters zeta (ζ) and xi (ξ). French schoolboy pronunciation of Greek may have contributed to his confusion.

*Oxysoma* NICOLET in GAY 1849:513 (Anyphaenidae)
From Greek ὀξύς "sharp" and σῶμα "body", hence "having a sharp or pointed body." Neuter.

*Ozyptila* SIMON 1864:527 (Thomisidae)
See *Oxyptila*.

*Pachygnatha* SUNDEVALL 1823:16 (Tetragnathidae)
To rhyme with *Tetragnatha*. From the Greek three-ending adjective παχύς -εῖα -ύ 'thick'. Hence "with thick jaws" in reference to the thick powerful divergent chelicerae which distinguish this genus. Sundevall (1830:208) specified that the chelicerae are "multo crassiores [much thicker]" than those of the neighboring genera *Tetragnatha* and *Linyphia*.

(‡) *Pachylomerides* STRAND 1934 (Ctenizidae)
See *Pachylomerus* AUSSERER.

*Pachylomerus* AUSSERER 1871a:145 (Ctenizidae)
Ausserer explained that the name is from the Greek three-ending adjective παχυλός 'thick' and the Greek masculine noun μηρός 'thigh, femur'. The name refers to the swelling of the femurs of the posterior pair of legs ("ihre Schenkel in der unteren Hälfte bauchig aufgetrieben."). Bonnet (1958:3288) said:

> "Since the name of the genus **Pachylomerus** was preoccupied by a Coleopteran (Kirby, 1832), Strand, recognizing this in 1934, replaced it with **Pachylomerides**. But in accord with my proposition (volume I, p.110), I retain **Pachylomerus** as a genus of spider, seeing that it has been used for a very long time without the least confusion with the other. Furthermore, **Pachylomerides** was unneccesary, because **Pachylomerus** already had a junior synonym in **Ummidia** Thorell, which ought to have been used."

Roewer (1954b:1613) properly sank it in favor of *Ummidia* THORELL. Raven (1985a:142) retained Pachylomerinae as a subfamily name, after Simon's group, Pachylomereae (1889m:178), as permitted by Art. 40a of the Code.

*Pacifiphantes* ESKOV & MARUSIK 1994:49 (Linyphiidae)
The second element is the rhyming morph after such Linyphiid genera as *Lepthyphantes* and *Pityohyphantes*, from the Greek noun ὑφάντης 'weaver'. The first element is from the Pacific Ocean reflecting the distribution of the genus from Russia and Canada.

(‡) *Pagomys* CHAMBERLIN 1948b:16 (Dictynidae)
From the Greek masculine noun πάγος -ου 'crag or rock' and the Greek masculine noun μῦς (gen.) μυός 'mouse'. Hence, 'rock mouse' presumably from the terrain of its type locality in the Uintah Mountains, Utah. Cf. Chamberlin's name *Zanomys*. Preoccupied by Gray 1864 in the Pinnipedia. A junior synonym of *Brommella* TULLGREN 1948 (Braun 1964:152).

(‡) *Paidisca* BISHOP & CROSBY 1926:178 (Theridiidae)
From the Greek feminine noun παιδίσκη 'young girl, maidservant, prostitute'. There is no particular semantic point. Crosby and Bishop often merely searched the dictionaries for usable names. Now synonymized under *Thymoites* KEYSERLING (Levi & Levi 1962:25).

*Paracornicularia* CROSBY & BISHOP 1931:375 (Linyphiidae)
Derived from *Cornicularia* CROSBY & BISHOP (which see).

*Paradamoetas* PECKHAM & PECKHAM 1885b:78 (Salticidae)
Differentiated, by the prefix Para-, from *Damoetas* Peckham & Peckham (1885a:277), which was created for an Australian species *nitidus* (L. Koch 1880). Damoetas is a proper name from Vergil (Eclogues 2.37, 39; 3.1, 58; 5.72). It is a first declension masculine Bonnet observed (1958:3325 note 29)

> "Peckham, creating in 1885 the genus name **Damoetas**, with the species **nitidus**, considered it masculine; it should be the same with **Paradamoetas**"

This is correct and the Peckhams were wrong to make this genus feminine with the species *formacina*. The prefix para-is from the Greek preverb and preposition παρά 'beside'.

*Paramaevia* Barnes 1955:7 (Salticidae)
As subgenus of *Maevia* C.L. Koch. Upgraded to a genus by Barnes (1958:49). Sunk again under *Maevia* by Roth (1982, 1985). Derived from *Maevia* by the conventional prefix Para-.

*Paramarpissa* F.O. PICKARD-CAMBRIDGE 1901a:252 (Salticidae)
Separated from *Marpissa* to which it is similar, except it has only one or two, or no spines beneath the first tibia. Synonymized under *Pseudicius* SIMON. Maddison (1987) suggested that the genus *Paramarpissa* should be revalidated for *Icius piraticus* PECKHAM & PECKHAM 1888, because tibia I has no ventral spines.

*Paraphidippus* F.O. PICKARD-CAMBRIDGE 1901a:273 (Salticidae)
Erected for those species close to *Phidippus* in which "the eyes of the posterior row are not or scarcely wider than those of the anterior row." Kaston (1973:11) revalidated *Eris* C.L. KOCH for those species often listed under *Paraphidippus*. Bonnet had said (1956a:1786 note 76):

> "**Eris aurigera** (type of the genus) having been placed in synonymy with **Paraphidippus marginatus** by the American authors, there is no doubt that the whole genus **Eris** ought to be shifted to the genus **Paraphidippus**; but since **Eris** is available and has priority, it is the inverse which ought to occur, or at least both ought to be placed under **Dendryphantes** according to the opinion of Simon. But I follow the contemporary American authors who maintain this generic distinction, but applying the name **Eris** by priority, as I have just explained."

*Parasyrisca* SCHENKEL 1963:261 (Gnaphosidae)
Formed from *Syrisca* by the distinguishing prefix Para-. Ovtsharenko and Marusik (1988:214) consider *Orodrassus orites* CHAMBERLIN & GERTSCH 1940 to be a *Parasyrisca*.

*Paratheridula* LEVI 1957d:105 (Theridiidae)
Levi said: "The oval abdomen of *Paratheridula* females distinguishes them from the closely related *Theridula*. Differentiated from *Theridula* (which see) by the prefix Para-from the Greek preverb and preposition παρά 'beside'.

*Paratheuma* BRYANT 1940:387 (Desidae)
Differentiated from *Theuma* SIMON by the prefix Para-which is from the Greek preverb and preposition παρά 'beside'. Bryant does not indicate any close connection with *Theuma* in the original description. *Theuma* SIMON (1893a:348-350) is a substitute name for the preoccupied *Scylax* SIMON (1889l:384). Scylax is the Greek form of a Carian proper name. He was a famous explorer of the 5th century B.C., sent by Darius from the mouth of the Indus around the Arabian peninsula. Theuma is a place name, a village in Thessaly mentioned by Livy (32.13), who apparently treated the name as neuter, although the evidence is not decisive. Modern editions of Livy print the name of the village as Teuma, but the dictionaries of Simon's day would have spelled it *Theuma*. There is no semantic rationale for the name, and Simon merely gleaned it by scanning the Latin

dictionary for unused words available for genus names. He unambiguously treated the name as feminine in 1893a, and Bryant followed suit by making *Paratheuma* feminine. *Paratheuma* was originally described in the Gnaphosidae, and was transferred to the Desidae by Platnick (1977c: 199).

*Paratrochosina* ROEWER 1960d:931 (Lycosidae)
  Formed from *Trochosina* SIMON by the prefix para-. One of the included species was *Trochosina arctosaeformis* DI CAPORIACCO 1940c. *Trochosina* SIMON (1885g:10) was a subgenus erected for *Trochosa terricola* by the addition of the diminuitive suffix -ina.

*Parazanomys* UBICK 2005 (Amaurobiidae)
  The author explains, "The generic name refers to the presumed relationship with the genus *Zanomys*." Which see.

*Pardosa* C.L. KOCH 1847a:100 (Lycosidae)
  Formed to rhyme with *Lycosa* LATREILLE, but with the first element from the Greek and Latin words for 'leopard'. The usual Greek word for 'leopard' is the feminine πάρδαλις, but there is a later masculine form, πάρδος, used by Aelian, which is borrowed into Latin as pardus. This is the second in a series of Lycosid names formed with the name of a predatory animal as the first element and either-osa or -cosa as the rhyming second element. E.g. *Dingosa* ROEWER (dingo), *Lynxosa* ROEWER (lynx), *Arctosa* C.L. KOCH (bear), *Alopecosa* SIMON (fox), *Crocodilosa* CAPORIACCO (crocodile), *Hyaenosa* CAPORIACCO (hyaena), *Mustelicosa* ROEWER (weasel).

(‡) *Parerigone* CROSBY & BISHOP in CROSBY 1926:4 (Linyphiidae)
  Bonnet (1958:3433) explained:
  
  "Crosby and Bishop created the genus **Parerigone** in 1926 with **Erigone** probata O. Pickard-Cambridge as the type. Subsequently they noticed that they had confused two different forms under that name:*Erigone probata* proper (=*Catabrithorax probatus*) and *Tmeticus maculatus* (=*Eperigone maculata*); in view of that fact they placed **Parerigone** in synonymy with **Catabrithorax** and created the genus **Eperigone** for the remaining species of their **Parerigone**."
  
  See also *Eperigone*. Synonymized under *Halorates* HULL by Millidge (1977:11).

*Parogulnius* ARCHER 1953:20 (Theridiosomatidae)
  Archer treated his new genus, with type species *Parogulnius hypsigaster* ARCHER 1953 from Alabama, just following his treatment of *Ogulnius* O. PICKARD-CAMBRIDGE 1882c:432 (Theridiosomatidae), and clearly meant them to be similar. Ogulnius is a Roman family name (Livy 10.6; 27.3). The most famous member of the family was Quintus Ogulnius Gallus, who, as tribune of the plebs, brought about the passage of the Lex Ogulnia in 300 B.C., which admitted plebeians to the major priesthoods.

*Peckhamia* SIMON 1901a:494-496 (Salticidae)
  In honor of George Williams Peckham (1845-1914) and his wife Elizabeth Gifford Peckham, the great American workers in the Salticidae.

*Pelecopsidis* CROSBY & BISHOP 1935b:31 (Linyphiidae)
  The 1933b:108 reference is a *nomen nudum*. Derived from *Pelecopsis* SIMON by the addition of a suffix. The suffix is not formed with strict correctness, since the Greek word ὄψις (gen.) ὄψεως is not a dental stem, and the -ιδισ suffix is made out of thin air, on the analogy of such real dental stems as ἀσπις -ιδος. The authors leave the gender ambiguous with the Latin two-ending adjective frontalis as the species name. It could be either masculine or feminine. Roewer (1942a:617, 1954b:1740) incorrectly printed it as *Pelecopsides* (arguably slightly better morphologically but still an unjustified emendation), and was followed by Brignoli (1983c: 320, 744) and Platnick (1989b:215, 662), but later Platnick (1993c:238) restored the original spelling. Bishop and Crosby said:
  
  "This genus is related to **Pelecopsis** in having hardened sclerites on the abdomen and in having the embolic division of the spiral type with a long tail-piece. It differs in having all the eyes borne on the cephalic lobe."

*Pelecopsis* SIMON 1864:196, 477 (Linyphiidae)
  Simon created this as a subgenus of *Microphantus* [sic] C.L. KOCH containing *Micryphantes inaequalis* C.L. KOCH 1841c (now *Pelecopsis elongata* WIDER in REUSS 1834), and gave as the etymology of the name πήληξ 'casque' ὄψις 'front [i.e. forehead]' explaining that "the two superior eyes [i.e. the posterior medians] are placed in the male on a tubercle which has the shape of a little helmet." Bonnet (1958:3446) rightly questioned whether the combination can mean what Simon intended. The first element, the Greek feminine noun πήληξ (gen.) πήληκος 'helmet, serpent's crest' is all right, but the second element is from the Greek feminine noun ὄψις (gen.) ὄψεως 'appearance, countenance, vision, eyes' and does not really mean 'forehead'. Crosby & Bishop (1931:381) said:
  
  "Simon established **Pelecopsis** as a sub-genus of **Micryphantus** [sic] Koch, placing only one species under it. [*Micryphantes inaequalis* C.L. Koch 1841] In his later work Simon ignored **Pelecopsis** and placed its type and other related species in **Lophocarenum** Menge (1868). Smith, F.P. (Jour. Quekett Microscopical Club Ser.2 Vol. 9 No. 58, Apr. 1906) seems to have been the first to revive Simon's name for the group."

*Pelegrina* FRANGANILLO BALBOA 1930:44 (Salticidae)
  The full name of this Spanish arachnologist is Pelegrin Franganillo Balboa. Bonnet (1958:3455 note 123) rather sniffily observed, "It is very rare to see an author dedicate a genus to himself." Franganillo Balboa began his career in Spain, and moved to Cuba in 1918. Pelegrina is a medieval Latin spelling of classical peregrinus 'foreigner, pilgrim'. Franganillo Balboa gave his forename a feminine ending.

*Pelicinus* SIMON 1891c:559 (Oonopidae)
  Bonnet (1958:3455) explained:
  
  "Simon, after having created in 1891 the genus **Pelicinus** must have thought in 1893 [1893a:303] (perhaps with good reason) that this term was a lapsus for **Pelecinus**; therefore, since the latter term was preoccupied by a Hymenopteran (Latreille 1800) he replaced it with **Philesius** [1893a:303]. Today, we would allow **Pelicinus** to stand, but since the change has been made, and it was done no doubt for good reason, there is no cause for its reinstatement."
  
  Simon indeed spelled the name *Pelecinus* in 1893a, indicating that he regarded the earlier spelling as a *lapsus*. The Greek masculine noun πελεκῖνος means 'pelican' and is derived from the noun πελεκάν -ᾶνος 'pelican' which in turn derives its name from its axe-shaped beak, from the noun πέλεκυς -εως 'axe'. There is nothing in the description which seems to explain the choice of the name 'pelican'. Simon's substitute name Philesius is from the Greek masculine noun φιλήσιος an epithet for the god Apollo, meaning roughly 'friendly' which Simon gleaned out of the lexicon. Brignoli (1983c:190) restored the original spelling and said:
  
  "This name [Pelicinus] is not prae-occ. as stated by Bonnet (BA:3455); the nom. nov. **Philesius** Simon 1893 (accepted by Bonnet but not by Roewer) is therefore unjustified."
  
  Roth (1985:10; 1994:141) on the basis of an unpublished manuscript by Muma considered *Pelicinus* a junior synonym of *Opopaea* SIMON 1891c.

*Pellenes* SIMON 1876a:90 (Salticidae)
  A substitute name for *Pales* (C.L. Koch 1850:64), which he created as a subgenus of *Europhrys* (see the discussion under *Corythalia*). Pales, the mythological name of a Roman god, protector of shepherds and cattle, was preoccupied in the Diptera. Simon's name *Pellenes* appears to be derived from the name of the Greek city Pellene (Πελλήνη) in Achaea. The strictly correct form would be Πελληνεύς rather than *Πελληνής. There is no particular semantic point to the name, but it reflects the fact that Simon was trying to approximate Koch's old name *Pales* with this substitute.

(‡) *Pelopatis* BISHOP 1924b:20 (Pisauridae)
  This elegant name is back-formed from existing Greek words. From the Greek masculine noun πηλός 'mud' and the verb πατέω the Greeks formed a compound verb πηλοπατέω 'walk on mud'. There also exists, only in the plural, a feminine noun πηλοπατίδες 'galoshes, mud-treaders, a kind of boot with thick soles' and Bishop has correctly formed the feminine singular of this noun *πηλοπατίς (gen.) -ίδος meaning 'mud-treader'. Bishop says "specimens from the Okefinokee swamp in Georgia were taken in the low ground bordering the islands." It is a sad loss to nomenclatural elegance that in the interest of phylogenetic accuracy this genus has been sunk under *Pisaurina* SIMON (Carico 1972a:297).

(‡) *Peregrinus* TANASEVITCH 1982:1501 (Linyphiidae)
  From the latin word meaning 'stranger, one from foreign parts'. Preoccupied, see *Perregrinus*.

(‡) *Perimones* JACKSON 1932b:11 (Linyphiidae)
  Although this name has a very convincing look to it, there is no Greek or Latin word or name to which it corresponds. Synonymized under *Satilatlas* KEYSERLING by Millidge (1981c:243).

*Perregrinus* TANASEVITCH 1992:50 (Linyphiidae)
  A replacement name for *Peregrinus* TANASEVITCH 1982.

*Perro* TANASEVITCH 1992:49 (Linyphiidae)
  A replacement name for *Pero* TANASEVITCH (1985a:55) which was preoccupied in the Lepidoptera (Herrich-Schäffer 1885). Since *Pero* is regarded as feminine, it is most probably a mythological name. Pero was the daughter of King Neleus of Pylos, who promised to give her in marriage to the man who brought her the cattle of Iphiclus. Spelling the replacement name with two r's is a convenience, and has nothing to do with the Spanish word for dog.

*Peucetia* THORELL 1869:37 (Oxyopidae)
A substitute name for *Pasithea* BLACKWALL (1858a:427) which was preoccupied in several groups (Worms, Molluscs, Crustaceans). Pasithea is the name of one of the three Graces (Charites) of Greek mythology. Peucetios (Greek Πευκέτιος) is a masculine mythological proper name taken from a list in Apollodorus (3.8.1) of the fifty sons of Lycaon, the son of Pelasgus. It comes in the list just after *Nyctimus*, a name Thorell used in 1877. In accordance with his usual habit Thorell found a replacement name for Blackwall's *Pasithea* which would preserve as many of its phonological features as possible: beginning with p, having the same number of syllables, with the stress on the antepenult, having a feminine ending, and sounding rather the same. So he turned the name of one of Lycaon's sons into a feminine noun.

*Phanetta* KEYSERLING 1886b:124 (Linyphiidae)
Keyserling gives no hint about the etymology of this name, although it appears to be a feminine diminutive formed on the Greek word φανή or φανός 'torch'. Hence 'little torch'. While the only species, *subterranea* EMERTON 1875, found in several caves in Kentucky, is yellow and white in color, that hardly seems enough brightness to warrant the name.

*Phanias* F.O. PICKARD-CAMBRIDGE 1901a:251 (Salticidae)
A common Greek masculine proper name Φανίας. The one Pickard-Cambridge was probably familiar with was a poet whose epigrams we have in the Greek Anthology, also called the Palatine Anthology (Anth. Pal. 6:294, 295, 297, 299).

*Phantyna* CHAMBERLIN 1948b:15 (Dictynidae)
Erected for *Dictyna micro* CHAMBERLIN & IVIE 1944. The second part of the name is a rhyming morph from the last two syllables of *Dictyna* (which see). The first part of the name Phan-may be from the Greek three-ending adjective φανός -ή -όν 'bright' but except for the orange-yellow legs of the type species there is nothing particularly bright about the spider. Chamberlin in another paper (1949:553) uses a species name *phana* (*Tapinocyba*) for a species which is light yellowish brown, which tends to confirm this interpretation. Compare the names *Tosnyna*, *Tivyna*, *Varyna*. Synonymized under *Dictyna* by Chamberlin & Gertsch (1958:48). Revalidated by Lehtinen (1967:271).

*Phidippus* C.L. KOCH 1846d:125; 1850:53 (Salticidae)
Phidippus is a Greek aristocratic proper name Φείδιππος. There are several possibilities. A grandson of Heracles is mentioned in the Iliad; an Athenian citizen of that name is mocked in the comedies, and is mentioned in the Greek orators; a slave who was physician to King Deiotaros is mentioned in Cicero's oration De Rege Deiotaro (17 ff; 30 ff.). This last is the most likely source for Koch to have known the name, and would have given it to him in the Latin spelling Phidippus. The name is composed of the root of the Greek verb φείδομαι 'spare' and the masculine noun ἵππος 'horse'. Hence, 'one who spares the horses'. Note that the Latin accent is correct and the word is pronounced *Phidíppus*, with the stress on the penult. Bonnet clarified the history of the name (1958:3510):

> "Koch and Berendt created the genus **Phidippus** in 1845 for some fossil Salticids, but in that year it was a nomen nudum; the following year, in 1846, Koch used the name for existing, that is, non-fossil, forms, which he believed were in the same genus, and he gave a diagnosis in1851; in view of that, it is **Phidippus** the genus of the existing forms which has priority; this was the understanding of Menge in 1854 when he created the genus **Gorgopis** for the fossil **Phidippus** of Koch and Berendt."

*Philesius* SIMON 1893a:303 (Oonopidae)
See *Pelicinus* SIMON.

*Philodromus* WALCKENAER 1826:86 (Philodromidae)
The name is the Greek two-ending adjective φιλόδρομος 'loving the course' from the combining form φιλο- 'loving' and the masculine noun δρόμος 'race'. Bonnet (1958:3541) clarified the dating of the name:

> "The genus described by Walckenaer does indeed date from 1826; the dates 1824, 1825 given by certain authors are in error; now, if Latreille was able to cite Philodromus in 1825, it is because our two savants communicated their discoveries, and, in view of that fact, Latreille had known of the creation of the genus before its official birth in fascicules 11-12 of the Faune française."

*Philoponella* MELLO-LEITÃO 1917b:6-8 (Uloboridae)
Derived from the genus name *Philoponus* THORELL (1887:127) by the addition of the Latin feminine diminutive suffix -ella. *Philoponus* THORELL is from the Greek two-ending compound adjective φιλόπονος -ov 'laborious, industrious' from the combining form φιλο- 'loving' and the masculine noun πόνος -ov 'labor'. Bonnet (1958:3589) was wrong in saying that the name ought to be *Philoponellus* in the belief that a compound ought to have the same genus as the second member. This is an idiosyncratic notion of Bonnet's, and is grammatically incorrect. The genus *Philoponella* was considered a synonym of *Uloborus* LATREILLE until revalidated by Lehtinen (1967:258) and it corresponds to the *Uloborus republicanus* SIMON 1891d group of Muma & Gertsch (1964:29) cf. Opell (1979:518).

*Phlattothrata* CROSBY & BISHOP 1933b:165 (Linyphiidae)
Crosby and Bishop were having a little quiet fun here. Φλαττόθρατ is actually a genuine bit from Aristophanes' Frogs (1285-1295), in the contest between two tragic poets, where the character Euripides is parodying the high-flown, obscure poetic style of the character Aeschylus, which he thinks is vacuous, exaggerated, and meaningless, hence, 'blah, blah, blah'. Euripides uses certain well known lines of Aeschylus's Agamemnon capping them mockingly with this tag of transparently onomatopoetic nonsense:

> "How the twin-throned power of Achaians, youth of Hellas
> To phlattothrattophlattothrat
> did send the Sphinx unlucky presiding bitch
> To phlattothrattophlattothrat etc."

Crosby and Bishop made a feminine noun out of it. Their genus was downgraded to a subgenus of *Tapinocyba* SIMON by Chamberlin & Ivie (1947b:50), but Crawford (1988:14) restored it to full generic status.

*Phlegra* SIMON 1876a:120 (Salticidae)
A place name, in Greek Φλέγρα, an ancient name reported by Herodotus and Strabo for Pallene in Thrace. Because the name is clearly related to the verb φλέγω 'burn' it presumably refers to the volcanic nature of the area. There seems to be no particular semantic reason for the spider name.

*Pholcomma* THORELL 1869:98 (Theridiidae)
From the genus name *Pholcus* (which see) and the Greek neuter noun ὄμμα (gen.) ὄμματος 'eye. Hence, "*Pholcus* eyed". The genus name is neuter. Thorell confessed he had not seen the species on which he founded this new genus, but Pickard-Cambridge's excellent description sufficed. The type species is *Theridion projectum* O. PICKARD-CAMBRIDGE (1862a: 7962). Thorell said:

> "According to Cambridge, the two centre eyes of the front row are very minute, almost contiguous; on each side of these is a group of three almost contiguous eyes, in the form of an equilateral triangle. The eyes of these two groups are disproportionately large compared with the size of the spider. The male has a projecting ridge around the abdomen. 'By the position of the eyes this species seems to be allied to the genus **Pholcus**, though in general form and appearance it is much more like the true **Theridia**' (CAMBR.)."

*Pholcophora* BANKS 1896e:57 (Pholcidae)
From the genus name *Pholcus* (which see) and the combining form -phorus-a-um ultimately from Greek -φορος -ov 'bearing'. This is an illogicaly formed compound, which would have to mean 'carrying a *Pholcus*'. I think Banks, whose Latin and Greek was rather rough-and-ready, simply picked up a common suffix to create a name which appeared to be derived from *Pholcus*.

*Pholcus* WALCKENAER 1805:80 (Pholcidae)
From Greek adjective φολκός which the Greek lexicon (Liddell-Scott-Jones) tells us probably means 'bandy legged'. It is found only once in Greek literature, and is applied to the despised character Thersites in the Iliad (2.217). The true meaning is uncertain, and the ancient commentators on Homer said it meant 'squinting, squint-eyed'. It is this interpretation which Thorell (1869:98, 101) used. The meaning 'bandy-legged' was first suggested and argued for by Philipp Karl Buttmann (Lexilogus zu Homer und Hesiod, Berlin 1818 ff.) It is clear that Walckenaer could only have thought the word meant 'squint-eyed' since the 'bandy-legged' meaning postdates his 1805 publication. So we may be confident in assuming Walckenaer meant to refer to the eye arrangement of *Pholcus*, namely six eyes arranged in triangular groups of three on each side of the cephalothorax. plus two anterior median eyes. While Walckenaer's description ("eight eyes, almost equal, occupying the front of the cephalothorax") would not lead us to see the connection, his figure (pl. 8, fig. 80) shows the triangular groups of eyes each on a stumpy stalk directed away from one another, and well might suggest 'squint-eyed'.

*Phormictopus* POCOCK 1901c:542, 545 (Theraphosidae)
From the Greek masculine noun φορμικτής 'lyre-player' and the Greek masculine noun πούς 'foot'. The name refers to the "stridulating-organ between the palpus and the first leg."

*Phoroncidia* WESTWOOD 1835:452 (Theridiidae)
From the Greek combining form -φορος 'bearing' and the Greek masculine noun ὄγκος 'barbed points of an arrow' plus the diminutive suffix -ίδιον which is then given a feminine form. Hence, something like 'little barb bearer'. The word is not well formed, since the combining form -

φορος never comes first in real Greek compounds. The name refers to the fact that in the type species (*aculeata* Westwood 1835) the abdomen is armed with long spines.

### *Phrurolithus* C.L. Koch 1839e:100 (Corinnidae)
From the Greek masculine noun φρουρός 'watcher, guard' and the masculine noun λίθος 'stone'. The resulting two-ending compound adjective *φρουρολίθος 'guarding the stone' has a parallel in authentic Greek in the two-ending adjective φρουροδόμος -ον 'guarding the house' an epithet applied to dogs. Hence, the meaning is 'guarder of stones' and refers to the habits of the spider. Koch said (1839e:101):

> "Her abode, as Hahn already observed, is always under stones. Underneath them the female lays her eggs. in a spherical lump and spins over them a light, transparent web, on which the mother awaits the hatching of her young."

The name is obviously meant also to suggest the mother's guarding of her eggs under the stone. Transferred to the Corinnidae from the Liocranidae by Bosselaers & Jocqué (2002:265).

### *Phruronellus* Chamberlin 1921a:69 (Corinnidae)
Originally created for a number of distinct species grouped around *Phrurolithus formica* Banks 1895d, and included by Chamberlin in the Clubionidae. Chamberlin formed the name on the rhyming principle by taking the first two syllables of *Phrurolithus* and adding a diminutive suffix -nellus probably in imitation of *Scotinella*. Later it was pointed out by Gertsch that *Phrurolithus formica* Banks is congeneric with *Phrurolithus festivus* C.L. Koch in Harrich-Schäffer 1835, the genotype of *Phrurolithus*, and thus *Phruronellus* became a synonym of *Phrurolithus*. Roth (1985: 10; 1994:83) considered it a junior synonym of *Scotinella* Banks, which had been itself considered a junior synonym of *Phrurolithus* until Kaston (1972a:230, 1977:48) established that the American species are distinct from the European *Phrurolithus*, but do belong in *Scotinella* and he has been followed by recent authors. Transferred to the Corinnidae from the Liocranidae by Bosselaers & Jocqué (2002:265).

### *Phrurotimpus* Chamberlin & Ivie 1935b:34 (Corinnidae)
Created for a group of species grouped around *Phrurolithus alarius* (Hentz 1847). Derived from *Phrurolithus* C.L. Koch (which see) by retaining the first element φρουρο-'guarding' and by translating the second elemement -λιθος 'rock, stone' into the Gosiute Indian word for 'stone' timpi (Chamberlin 1908b:82), and giving it a Latin ending. Hence, 'guarding stone'. Transferred to the Corinnidae from the Liocranidae by Bosselaers & Jocqué (2002:265).

### *Physocyclus* Simon 1893a:470 (Pholcidae)
From the Greek verb φυσιόω 'puff up' related to the noun φῦσα 'belows' and the masculine noun κύκλος 'circle'. An authentic Greek parallel would be the name Φυσίγναθος 'Puff-cheek' a frog warrior in the mock epic the Batrachomyomachia, "The Battle of the Frogs and Mice." Hence, 'with puffed circle' referring to the globose abdomen, a feature which is also marked in the type species name *globosus* (Taczanowski 1874).

### *Phycosoma* O. Pickard-Cambridge 1879d:692 (Theridiidae)
A combination of the theridiid genus name *Phycus* O. Pickard-Cambridge 1870d plus the Greek neuter noun σῶμα 'body'. The author says "The whole spider bears considerable resemblance to the genus *Phycus* Camb." Hence, 'with a body like the genus *Phycus*.' Pickard-Cambridge (1870d:742) explained that *Phycus* was a proper name, and he would have easily found it perusing the Latin dictionary. There is no semantic rationale. *Phycus* is the Latin name of the most northern point of the Libyan coast. It was a habit of O. Pickard-Cambridge to use geographic names irrespective of any semantic connection. Examples of such geographical Pickard-Cambridge names are: *Achaearanea, Amphissa, Bedriacum, Caledonia, Elaver, Liger*.

### *Piabuna* Chamberlin & Ivie 1933a:40 (Corinnidae)
This is another Chamberlin name which is derived from the Gosiute Indian language. This genus is close to *Phrurolithus* and *Phruronellus* differing in the structure of the eyes, where the anterior median eyes are large. The name is aparently derived from the Gosiute adjective piup, with combining form pia 'big' (Chamberlin 1908b:81) and the noun bui 'eye' (Chamberlin 1908b:85) and a suffix -na which occurs in many animal names. Hence, 'with large eyes'. Originally described in the Clubionidae, incorrectly placed in the Gnaphosidae by Brignoli, included in Clubionidae: Phrurolithinae by Roth (1985), in Liocranidae, by Platnick (1989b: 433), and transferred to the Corinnidae from the Liocranidae by Bosselaers & Jocqué (2002:265).

### *Pimoa* Chamberlin & Ivie 1943:9 (Pimoidae)
One of Chamberlin's names derived from the Gosiute Indian language. From the Gosiute adjective piup, with combining form pi- 'big' (Chamberlin 1908b:79), and the noun moa 'leg' (Chamberlin 1908b:91). Probably a reference to the fact that the "legs of the male in this genus are longer and more slender that those of the female." The family Pimoidae was established by Hormiga (1993:534).

### *Pimus* Chamberlin 1947:19 (Amaurobiidae)
This is derived from the Gosiute Indian adjective piup with combining form pi- 'big' (Chamberlin 1908b:79). Chamberlin often created hybrid names from Gosiute and Latin or Greek (see *Astrosoga*). For instance *Ceratinella tosior* Chamberlin 1949:506 has a species name made up of the Gosiute word tosabit 'white, gray' with combining form tosa as in tosawuda 'gray bear' (Chamberlin 1908b:79), and the Latin masculine comparative ending -ior yielding 'rather gray'. He described a Dictynid species *Tosyna pior* Chamberlin 1948b:16 where the species name like tosior is made up of the Gosiute adjective pi-'big' and the Latin masculine comparative ending -ior yielding 'bigger' described by Chamberlin as "one of the largest forms." From these considerations, we may confidently deduce that *Pimus* and the type species name *pitus* are both Latinized hybrids built on the Gosiute adjective 'big' plus the Latin adjectival suffixes -mus and -tus. Although these spiders are not particularly big (5-7 mm long), the type species is noticably bigger than its congeners, and this may be enough to generate the name.

### *Pippuhana* Brescovit 1997a:110 (Anyphaenidae)
An arbitrary combination of letters treated as feminine.

### *Pirata* Sundevall 1833b:24 (Lycosidae)
Walckenaer (1805:14) had already named a section of the genus *Lycosa* the Piraticae (also with the colorful French designation Les Corsaires) after the species *Araneus piraticus* Clerck (1757):102. Clerck said he caught his specimen running into the water between reeds and the shore, and that it was busy catching prey. Presumably then, from the aquatic predatory habits, the name 'pirate' seemed appropriate. From the Greek a-stem masculine noun πειράτης (gen.) -ου 'pirate' which comes into Latin as a first declension masculine pirata-ae. The genus is masculine, but because it ends in -a much confusion has resulted and many authors have given it feminine species names incorrectly. Bonnet (1958:3654) correctly treated it as masculine. Sundevall (1833a:192) created a section Piratae containing Clerck's species *Araneus piraticus* which he named *Lycosa piratica*. It was in 1833b that Sundevall created the genus *Pirata*.

### *Pisaura* Simon 1885d:354 (Pisauridae)
Simon erected this new genus for *Ocyale mirabilis* (Clerck 1757). Audouin's genus *Ocyale* (1826:149). The feminine of the Greek adjective ὠκύαλος -η -ον 'swift on the sea' a Homeric epithet of ships, which is a compound of the adjective ὠκύς -εῖα -ύ 'swift' and the noun ἅλς (gen.) ἁλός 'salt, sea'. The name refers to the capacity of this spider to run over the surface of the water. Simon's name is from the Latin name of the Pesaro River in Italy, Pisaurum. Strictly the adjectival form ought to be *Pisauria*. Simon in 1877 listed this spider (*mirabilis* the type species) in a collection from the island of Ischia. While the spider occurs in Italy, Simon did not indicate why he singled out Pesaro for the name. Probably, as was his habit, he was simply searching through the Latin dictionary for an unused word, without any regard to a semantic connection.

### *Pisaurina* Simon 1898a:295 (Pisauridae)
Derived from *Pisaura* (which see) by the addition of the Latin feminine diminutive suffix -ina.

### *Pityohyphantes* Simon 1929:620, 623, 741 (Linyphiidae)
From the combining form of the Greek feminine noun πίτυς (gen.) πίτυος 'pine tree' and the Greek masculine first declension agent noun ὑφάντης -ου 'weaver' from the verb ὑφαίνω 'weave'. Hence, 'pine weaver'. Simon (1929:620) called them "Araignées arboricoles" and again (1929: 623) said of *phrygiana*, "Assez commun sur les conifères, dans les régions montagneuses et subalpines [very common on conifers in mountainous and subalpine regions]." Koch collected the spider in Regensburg, but named it phrygiana which ought to mean 'from Phrygia (in Asia Minor)'. Walckenaer (1841:260) may have solved this mystery, when he translated the species name as "embroidered (brodée)". Strictly speaking that would be in Latin phrygionius -a -um, but Koch may simply have been playing with the fact that Phrygians were famous for their embroidery.

### *Platoecobius* Chamberlin & Ivie 1935c:270 (Oecobiidae)
From the Greek adjective πλατύς -εια -ύν 'flat' and the genus name *Oecobius* (which see). Erected for *Thalamia floridana* Banks 1896e, which had been placed in *Oecobius* by Petrunkevitch, and which has a markedly flattened carapace. From a strict grammatical standpoint the name is not well-formed, and ought to have been *platyoecobius, as the parallel authentic Greek compound πλατύουρος 'flat-tailed' makes clear.

### *Platycryptus* Hill 1979:215 (Salticidae)
From the Greek adjective πλατύς -εῖα -ύ 'flat' and the verbal adjective κρυπτός -ή -όν 'hidden, secret'. The author said:

> "*Platycryptus*' is a combination of Greek words for 'broad, flat' and 'hidden'. The combined term refers to the cryptic appearance and behaviour of these flattened spiders, which frequently wedge themselves into crevices."

Erected for *Metacyrba undata* (De Geer 1778b) and *Marptusa californica* (Peckham & Peckham 1888). Hence, 'both flat and cryptic'.

(‡) *Platyxysticus* Gertsch (1932:1) (Thomisidae)
See *Coriarachne*.

*Plectreurys* Simon 1893d:266 (Plectreuridae)
From the Greek neuter noun πλῆκτρον 'plectrum, implement for striking a lyre, spear point, cock's spur' and the Greek u-stem three-ending adjective εὐρύς -εῖα -ύ 'broad'. The name is not well-formed. It violates a rule of Greek compound formation by placing the adjectival element after the nominal element. The correct formation would be *εὐρύπληκτρος. There are no authentic Greek compounds ending in -ευρυς. Despite the fact that the name ends in the masculine form of an adjective, Simon regarded it as feminine. This tongue-twister is intended to mean something like 'having a broad spur' with reference to the coupling spur at the extremity of the first tibia of the male. Simon said (1893d:266):

> "The male, hardly smaller than the female, has a remarkable sexual character, in which he resembles those of the Theraphosids: the anterior tibia are provided externally near the extremity with a short and blunt tubercle, prolonged into a strong spine resembling the spur of **Diplura**."

It is clear that Simon meant his name to mean 'having a broad cock's spur'.

*Plesiometa* F.O. Pickard-Cambridge 1903a:438 (Tetragnathidae)
From the Greek three-ending adjective πλήσιος -α -ον 'near, close to' and the genus name *Meta* (which see). Pickard-Cambridge created the genus *Pseudometa* a few pages later (444) and Simon had created *Mecynometa* (1895a:737) as they separated the Metine genera close to *Leucauge* and *Meta*. Hence, 'close to *Meta*'. Synonymized under *Leucauge* by Levi 1980a:23.

*Plexippus* C.L. Koch 1846d:107 (Salticidae)
From the Greek two-ending adjective πλήξιππος -ον, an epithet of epic heroes which means 'striking or driving horses' from the combining form of the verb πλήσσω 'strike' and the noun ἵππος 'horse'. Meant to be parallel with *Phidippus* (which see). Koch (1850:52) said:

> "The spiders belonging here have great similarity in their forms with the genus **Phidippus**."

Koch obviously wanted another "horse" compound to signal this relationship.

*Pocadicnemis* Simon 1884r:713 (Linyphiidae)
From the stem of the Greek feminine noun ποκάς (gen.) ποκάδος 'wool, hair' and the feminine noun κνημίς (gen.) κνημῖδος 'greave, shin-guard, legging'. The connecting vowel -i- is Simon's invention; Greek would most likely have used -o-. Hence, 'having wooly shanks'. Although Simon described the genus as having the body and limbs very setose, his description does not particularly focus on the tibiae, as the name would lead us to expect. He says:

> "The species of this genus are remarkable for the length of the hairs and spines which clothe the legs and often also the abdomen."

*Poecilochroa* Westring 1874:45 (Gnaphosidae)
From the Greek three-ending adjective ποικίλος -η -ον 'many-colored, spotted, pied' and the feminine noun (Attic dialect) χρόα 'skin'. Hence, 'having varicolored skin'. There are no authentic Greek compounds of this form. The proper formation would have been a one-ending adjective ποικιλόχρως which actually exists, used by Aristotle. The type species *Pythonissa variana* C.L. Koch 1839d:65 has a bright yellowish-red cephalothorax, and a brilliant, satiny, somewhat metallic black abdomen with white bands and spots. Koch's species name *variana* 'variegated' reflects this bright color scheme. It seems clear that Westring meant to translate Koch's *variana* with the Greek adjective ποικίλος which also means 'variegated'. The North American species are now in *Sergiolus* (Platnick & Shadab 1981e).

*Poeciloneta* Kulczyński in Chyzer & Kulczyński 1894:71 (Linyphiidae)
From the Greek three-ending adjective ποικίλος -η -ον 'many-colored, spotted, pied' and the rhyming morph -neta by analogy with such names as *Microneta* Menge. The rhyming morph -neta is a Latin first declension masculine, as if from a Greek agent noun -νήτης 'spinner'. See the discussion under *Argyroneta*. Despite the etymology, the genus has been traditionally regarded as feminine. The type species is *Neriene variegata* Blackwall (1841a:650), which, as its name indicates, is variegated in color. Kulczyński correctly translated Blackwall's species name into Greek as ποικίλος and added the rhyming morph -neta. Hence, 'variegated spinner'.

*Porrhomma* Simon 1884a:202, 353, 874. (Linyphiidae)
From the Greek adverb πόρρω 'forward, further' and the Greek neuter noun ὄμμα (gen.) ὄμματος 'eye'. No authentic Greek word would be formed this way. The genus is neuter, and the species names which seem to violate this are either proper names in apposition (e.g., *proserpina*), or nouns in apposition (e.g., *cavernicola*). The spelling porrh-with an h requires some comment (see Bonnet 1958:3763, note 337). It was a convention of the Byzantine grammarians to write a double rho with a smooth breathing mark over the first, and a rough breathing mark over the second, and this was reflected in the Latin spellings with an h. For example the adjective πύρρος 'red-headed' becomes the latinized name Pyrrhus. It is this convention which Simon's name reflects. Modern lexica may not adopt this convention. L. Koch in creating *Porropis* (1876a:807), for instance, did not employ it. Koch's use of the Greek adverb πόρρω may aid in determining Simon's intention. Koch's name *Porropis* refers to the fact that the eyes are far apart (distantes, longe inter se sejunctos dispositi). Simon (1892a:185) created the Diplurid genus *Porrhothele* separating it from *Macrothele* Ausserer. The -thele part of Simon's name is a rhyming morph to associate it with the genus from which it was separated. The Porrho-part refers to a difference in the spacing of the eyes from *Macrothele*. In his key to the Linyphiinae (1884a:202) the couplet that leads to *Porrhomma* says "Yeux petits et largement séparées [eyes small and widely separated]." It seems clear that Simon, taking his lead from L. Koch, took the Greek adverb πόρρω to mean 'separated' (in Greek it may mean 'at a distance, but not quite 'separated'), and he intended his name to mean 'with widely separated eyes'.

*Poultonella* Peckham & Peckham 1909:576 (Salticidae)
Diminutive formed on a proper name. Probably for Edward B. Poulton, a British entomologist.

*Procerocymbium* Eskov 1989b:75 (Linyphiidae)
From the Latin adjective procerus 'long, extended' and the anatomical term cymbium ('little boat'). Hence, 'with a long extended cymbium'.

*Prodidomus* Hentz 1847:466 [=1875:105] (Prodidomidae)
Hentz proposed this as a subgenus of *Cyllopodia* (now *Hyptiotes*). Hentz said (1847:467): "taken in the recess of a large box in a dark cellar, hiding itself in holes." The name is not well-formed, but I presume it is a combination of the first part of such Latin words as proditio 'betrayal, discovery' proditor 'betrayer' and the Latin feminine noun domus 'home'. Hence, something like 'betraying its home'. Hentz clearly regarded the name as masculine (despite the feminine gender of domus). There are no authentic Latin compounds of domus. The family Prodidomidae was downgraded to a subfamily of Gnaphosidae by Platnick & Shadab (1976c), but subsequently Platnick (1990a:36) revalidated the family.

*Promyrmekiaphila* Schenkel 1950b:28 (Cyrtaucheniidae)
Derived from the genus name *Myrmekiaphila* Atkinson (which see) by the addition of the Latin prefix pro-. Transferred from the Ctenizidae to the Cyrtaucheniidae by Raven (1985a:137).

(‡) *Prosopotheca* Simon 1884r:829 (Linyphiidae)
From the Greek neuter noun πρόσωπον 'face, countenance' and the feminine noun θήκη 'case, chest, tomb'. Hence, something like 'face-box' apparently referring to the male cephalic lobe. Downgraded to a subgenus of *Walckenaeria* Blackwall by Wunderlich (1972a:379). Bonnet's note said (1958:3781):

> "If **Spiropalpus spiralis** Emerton is a **Prosopotheca** it is evident that the name **Spiropalpus** has priority over the other which has the date 1884. I will not use it because the American authors do not appear to have fixed its identity. Thus Crosby after having established the identity in 1905 reversed himself in 1926, accepted it again in 1928 and reversed himself in 1931. Now **Spiropalpus spiralis** is considered a **Walckenaeria**; but if someday it is established that it is a **Prosopotheca** it will be necessary to retain that name over **Spiropalpus**."

Millidge (1983:106) included *Spiropalpus spiralis* Emerton 1882, and all of *Prosopotheca* under *Walckenaeria*.

(‡) *Pselothorax* Chamberlin 1949:541 (Theridiidae)
Since this genus was described as having "the cephalothorax greatly elevated and shaped like a truncated cone, conspicuously depressed and pitted above with a sharply defined curved furrow in the depression setting off the head" we would expect this name somehow to refer to these features. It follows the genus *Oedothorax* Bertkau in Förster & Bertkau ('with swollen thorax') in Chamberlin's paper, and he would seem to be following suit with the name. But there is no appropriate Greek word from which it is likely to have been formed. The closest candidates would have double l and would mean something like 'stuttering' (Greek ψελλός). There is nothing in the description of the new genus to explain the name. Chamberlin's intention remains a mystery, unless he has made a mistake in Greek. The description of the sole species, atopus, begins, "cephalothorax dusky." This leads me to believe Chamberlin meant to use the Greek word ψόλος 'soot, smoke'. If I am right his intention was to create a name meaning 'with sooty cephalothorax' which would have yielded Psolothorax, but simply misspelled the Greek word. Synonymized under *Dipoena* Thorell by Ivie (1967:126).

*Pseudeuophrys* F. Dahl 1912b:589 (Salticidae)
A compound of the combining form of Greek ψευδ-'false' and the genus name *Europhrys* (which see).

*Pseudicius* Simon 1885e:28 (Salticidae)
From the Greek adjectival combining form ψευδο-'false' and the genus name *Icius* Simon (which see). Hence, 'the false *Icius*'.

*Pseudosparianthis* Simon 1887g:472 (Sparassidae)
From the Greek adjectival combining form ψευδο-'false' and the genus name *Sparianthis* Simon (1880a:339), which itself is a confection made up of the first syllable of the genus name *Sparassus* and a Greek adjective ιάνθινος -ον 'violet colored' from ἴον 'violet' and ἄνθος 'flower'. Simon put a feminine ending on it. Hence, 'the violet Sparassid'. Unfortunately the type species (*granadensis* Keyserling 1880b) is rather brownish than violet. It is possible, given the fact that Simon in the same paper created *Spariolenus*, that he simply made up combining forms Σπαριο-and Σπαρια-and there are no violets involved. In that case *Sparianthis* would mean something like '*Sparassus* flower'

*Psilochorus* Simon 1893a:482 (Pholcidae)
From the Greek three-ending adjective ψιλός -ή -όν 'bare, smooth, unarmed, simple' and the masculine noun χόρος 'dance'. This genus belongs to Simon's group Blechrosceleae (from Greek βληχρός 'faint, gentle, slight' and σκέλος 'leg'), in which the legs are unarmed in both sexes (1893a:479), and this seems to be the point of Simon's name. It should mean something like 'bare dance' as if the second part referred to the unarmed legs.

(‡) *Pusillia* Chamberlin & Ivie 1943:23 (Linyphiidae)
Fashioned from the Latin three-ending adjective pusillus-a-um 'tiny, slight' by the additon of a feminine suffix -ia. Erected for *Linyphia mandibulata* Emerton (1882:64), but the authors were aware that *Linyphia pusilla* Sundevall (1830:214), which they considered distinct, but limited to Europe, was the oldest species in their genus. Therefore, they honored Sundevall's species in the genus name, while designating Emerton's *mandibulata* as the type species. Gerhardt (1928:629) had placed *pusilla* in his new genus *Microlinyphia*. *Pusillia* has been synonymized under *Microlinyphia* by Wiehle (1956:297).

*Rabidosa* Roewer 1960d:581 (Lycosidae)
Roewer in his Katalog created many new genera of Lycosidae, but Brignoli said (1983c:432) "it is apparent that most students of this group give little value to most of the genera described by Roewer in 1954a, b and 1960d". This genus was erected for the single species *Lycosa rabida* Walckenaer, and the name is formed from the rhyming morph -osa, which is customary for Lycosid genera on analogy with *Pardosa* (which see), and the first part of the species name rabida. Roewer in 1954 erected the genus without any explanation of its characters, but later (1960d) remedied this omission. Dondale & Redner (1990:41) subsumed the species under *Hogna* Simon. Roth (1994:127), on advance notice from Brady, treated *Rabidosa* as a valid genus with 5 species, and Brady & McKinley (1994: 138) formally revalidated the genus.

(‡) *Rachodrassus* Chamberlin 1922:160 (Gnaphosidae)
Derived from *Drassus* by the addition of a prefix from the Greek feminine second declension noun ῥαχός (gen.) -οῦς 'thorn hedge'. It is improperly transcribed, and a strict spelling should be *Rhachodrassus. But the Code privileges the original spelling even if improperly latinized (Code Art. 32.c.ii). The type species is *Rachodrassus echinus* Chamberlin 1922:160, and echinus is a noun in apposition meaning 'hedgehog' (Greek ἐχῖνος), hence the spininess, prickliness of this animal is being emphasized. But nothing in the original description of the genus or of the type species explicitly indicates this. Platnick & Ovtsharenko (1991:115) synonomized it under *Talanites* Simon.

*Reo* Brignoli 1979k:923 (Mimetidae)
An anagram of the Mimetid genus *Ero* C.L. Koch erected by Brignoli for a new species *Reo latro* Brignoli 1979k from Kenya. Latro is a Latin masculine noun in apposition meaning 'robber, thief'. Roth (1994:132) said: "Brignoli (1979k) suggests that some North American species of *Mimetus*, at least *Mimetus eutypus* Chamberlin & Ivie 1935b, belong in *Reo*, but this seems very unlikely."

(‡) *Rhaebothorax* Simon 1926:454 (Linyphiidae)
From the Greek three-ending adjective ῥαιβός -ή -όν 'crooked, bent' and the Greek masculine noun θώραξ (gen.) θώρακος 'breastplate, corselet, trunk'. Hence, 'with crooked thorax'. Simon's usage of this element can be paralleled in the name of his genus *Rhaeboctesis* (1897a:143) which differs from *Liocranum* by its more convex cephalothorax. The second half of the name is from the Greek feminine noun κτῆσις 'possession' so the name would mean something like 'bend possession'. We may conclude that Simon used the form rhaebo-to refer to a convex cephalothorax. The male of the type species, *paetulus* O. Pickard-Cambridge 1975f:333 (Latin 'slightly blink-eyed') has the cephalic part very convex. Synonymized under *Mecynargus* Kulczyński *in* Chyzer & Kulczyński by Millidge (1977: 17).

*Rhecostica* Simon 1892a:162 (Theraphosidae)
Ionic dialect form for Attic ῥαχός, a feminine word meaning 'thorn hedge'. Why the Ionic form? Because that is the way it occurs in Herodotus, where it would most likely be encountered by the reader. The Attic form is much less likely to be encountered. Simon described his group Homoeommateae (Simon 1892a:161) as having the anterior legs armed with numerous spines, and I assume this is what the rhecho-is supposed to refer to. The -stica part is trickier. There is no such element in either Latin or Greek. The French often make no distinction in pronunciation between the Greek unaspirated stop row (p, t, k) and the Greek aspirated stop row (ph, th, ch). This fact of the French pronunciation of ancient Greek generates a mistake in spelling, which Simon made often. For instance, the Araneid genus, now "correctly" spelled in the catalogues *Agathostichus* meaning 'having a good row' [of cephalothoracic tubercles]. It was spelled by Simon (1895a:882, 883) as *Agatostichus*, but corrected in the Index to *Agathostichus* (Simon 1895a.1070). Furthermore, *Rhecostica* itself is "corrected" (by some hand, maybe not Simon himself) in the Index to volume I to *Rhechosticha* (Simon 1895a:1081). Therefore I conclude that -stica is Simon's spelling error for the feminine of the combining form -stichus. Stichus is from Greek στίχος meaning 'row, line, rank' as in a rank of soldiers or a row of trees. In the first instance, Simon seems to use the word senso stricto, as in *Aptostichus*, which has many rows of spines beneath the anterior metatarsi. The Apto-part seems to be from Greek ἅπτος, which would give haptos meaning 'touchable, palpable', but here also Simon ignored the Greek aspirate, because as a French schoolboy he would never have pronounced it. So stichus is used for a row of spines. But in the Araneid genus *Dicrostichus* (meaning 'with a bifid, forked or cloven tubercle') or in *Agathostichus* above, it no longer means 'row of spines' in the strict sense, but anything spinelike or even blunt, like the bifid cephalic tubercle of *Dicrostichus*. I think what has happened is that, irrespective of the strict Greek meaning, Simon has come to use the combining form -stichus to mean 'armed' in the general sense. So, *Rhechostica* appears to mean (in Simon's nomenclatural conventions) 'armed with a hedge of spines'. Raven (1985a:149, 152) placed both *Aphonopelma* Pocock and *Dugesiella* Pocock in synonymy with *Rhecostica*, but the priority was reversed by ICZN Opinion 1637, and *Aphonopelma* was revalidated (ICZN 1991).

*Rhetenor* Simon 1902d:406 (Salticidae)
A literary reference. Rhetenor was a companion of Diomedes in Ovid Metamorphoses 14.504. Since the name is not a proper Greek compound, and it has no discoverable etymology, modern texts of the Metamorphoses amend to Rhexenor (meaning 'man breaker') instead. There is no semantic connection; Simon got it from searching the dictionary.

*Rhomphaea* L. Koch 1872a:289 (Theridiidae)
From the Greek feminine noun ῥομφαία 'a large broad sword used by the Thracians'. It is the word used for the sword of Goliath in the Septuagint (I Kings 17:51). The name apparently refers to the spider's long vermiform abdomen, but Koch does not even mention this distinctive feature in the diagnosis of the genus. Synonymized under *Argyrodes* Simon (Levi & Levi 1962:27).

(‡) *Ritalena* Chamberlin & Ivie 1941a:613 (Agelenidae)
Constructed with the rhyming morph -lena from *Agelena* and the name of the type species rita. Chamberlin and Ivie created several genus names with the rhyming morph -lena to indicate their relationship to other Agelenids, such as *Calilena*, *Tortolena*, *Rualena* etc. (see under *Hololena*). The authors stated that this genus is "close to *Agelenopsis* in color and structure." The type locality is Madera Canyon in the Santa Rita mountains in Arizona, whence the species name. Synonymized under *Melpomene* O. Pickard-Cambridge by Roth & Brame (1972:45).

*Robertus* O. Pickard-Cambridge 1879g:103 (Theridiidae)
While this is obviously a latinization of the proper name Robert, Pickard-Cambridge gave no indication of who it is he intended to honor.

*Rualena* Chamberlin & Ivie 1942b:233 (Agelenidae)
The second part of the compound -lena is taken by misdivision of morphemes from the genus name *Agelena*. See the general remarks on Agelenid names in -lena under *Hololena*. The first part of the compound, Rua-is adopted from the species name of the type species *Agelena rua* Chamberlin (1919b:9). The species name rua is the Gosiute Indian word meaning 'son' (Chamberlin 1911:50) used as if it were a Latin feminine adjective.

(‡) *Rugatha* Chamberlin & Ivie 1942a:42 (Theridiidae)
Erected as a subgenus of *Enoplognatha* Pavesi for forms with long chelicerae in the male with a large spur on the mesal side. The type species is *Enoplognatha pikes* Chamberlin & Ivie 1942a (from Pike's Peak), and it is similar in structure to *Enoplognatha rugosa* Chamberlin & Ivie 1947b. The name of the subgenus is formed from the first syllable of *rugosa* ('wrinkled'), and the last two syllables of *Enoplognatha*. Included in *Enoplognatha* Pavesi by Levi (1957a:6).

*Rugathodes* ARCHER 1950:24 (Theridiidae)
Erected as a subgenus of *Theridion* WALCKENAER with type species *sexpunctatum* EMERTON 1882. Formed from the genus name *Rugatha* CHAMBERLIN & IVIE with the addition of the Greek adjective forming suffix -odes ( as in the genuine Greek adjective ἐυώδης -ες see under *Argyrodes*). Archer in a discussion of the genus *Theridion* said that *Enoplognatha*, and *Rugatha* (among others) were synonyms of *Theridion*, and that *Rugatha* was "hardly an outstanding subgenus but merely a species group." Presumably Archer meant a nod to this sunk genus in the construction of his new name. It had been treated as a junior synonym of *Theridion* by Levi (1957a:19), but was elevated to genus rank by Wunderlich (1987a:213).

*Saaristoa* MILLIDGE 1978b:123 (Linyphiidae)
Erected with type species *Neriene abnormis* BLACKWALL 1841a:649 and named in honor of the Finnish arachnologist M.I. Saaristo. Crawford (1988:20) transferred the American species *Lepthyphantes sammamish* LEVI & LEVI 1955 to this genus.

*Salticus* LATREILLE 1804b:135 (Salticidae)
From the Latin adjective salticus-a-um 'dancing' related to the verb salto -are 'to dance'. Walckenaer and Latreille both working in Paris in 1804-5 were friends, and often exhanged information, but apparently did not always check with each other about their names. Although Latreille gracefully preferred Walcknaer's name *Drassus* to his own *Gnaphosa* (see under *Gnaphosa* ), here he stuck to his guns and insisted on his own *Salticus* over Walckenaer's *Attus*. He pointed out that his own name was in fact first, and that the similar name *Atta* had been used by Fabricius for a genus of ants, and would cause confusion (Latreille 1819).

*Saltonia* CHAMBERLIN & IVIE 1942a:23 (Dictynidae)
Erected for a new species, *Saltonia imperialis* CHAMBERLIN & IVIE 1942a collected at Fish Springs, Salton Sea, California. The genus name is a latinization of Salton. Transferred to the Dictynidae by Lehtinen (1967:263).

*Sarinda* PECKHAM & PECKHAM 1892:40 (Salticidae)
A sarinda is a folk form of the fiddle in North India. "Its body is a deep, waisted wood shell with a skin belly. The upper half is left open, its sides forming pointed down curving barbs." (see Encyclopedia Brittanica vol. 8 [Micropaedia] p. 896). There is a line drawing of a sarinda in Apel, Willi, 1944. The Harvard Dictionary of Music, p.800. The scant biographical information about the Peckhams (Mallis 1971:348-351) makes no mention of any musical interests on their part, but we may deduce from some of their spider names that they were interested in opera and in India. Their genus name *Nilakantha* (Peckham & Peckham 1901b:8) is the name of Lakmé's father in Delibes's opera Lakmé. The narrow waisted form of the body of the musical instrument is surely the basis of comparison with the narrow-waisted form of this ant-mimicking spider. The Peckhams treated the name as feminine.

*Sassacus* PECKHAM & PECKHAM 1895:176 (Salticidae)
Proper name. Sassacus was the last chief of the Pequot Indians, an Algonkian tribe of the Connecticut Valley, who under his leadership fought a losing war with the English settlers in 1637. Sassacus fled and was killed by the Mohawks, who sent his scalp to the governor of Massachussetts. His name, Sassakusus in the Massachussetts language, is supposed to mean 'he is wild'. See Hodge (1906:468-469).

*Satilatlas* KEYSERLING 1886b:127 (Linyphiidae)
This appears to be built upon the mythological name Atlas, Atlantis from the Greek Ἄτλας Ἄτλαντος, the Titan who held up the heavens. Although Keyserling, with his patronymic species name *marxii* gave no hint about the gender, it is apparently masculine. Millidge (1981c:243) assumed it is masculine under Article 30d of the Code There is no obvious source for the first part of the name, and Keyserling gave us no hint.

*Savignia* BLACKWALL 1833a:104 (Linyphiidae)
Named for Jules César Savigny, who with Victor Audouin published the natural history sections of the Description de l'Égypte. Bonnet spelled this *Savignya* appealing to the Rules of Nomenclature of his day (Art. 8, §h, note b), but the current code privileges the original spelling (Artt.31a.iii; 32).

*Scaphiella* SIMON 1891c:561 (Oonopidae)
A feminine diminutive formed on the Greek feminine noun σκάφη 'trough, tub, basin, skiff'. The name refers to the little trough on the abdominal dorsum resulting from the fact that in the female the ventral abdominal scutum, greatly dilated, curves up around the sides of the abdomen. See Simon 1893a:288, fig. 258.

(‡) *Scaptophorus* HULL 1920:8 (Linyphiidae)
From the Greek two-ending compound adjective σκαπτόφορος, which is made up of the Greek neuter noun σκῆπτρον 'scepter, rod' (some Greek spelling variations are involved) and the combining form -φορος 'bearing'. Hence, 'rod-bearing. This is a substitute name for *Helophora* MENGE (which see) which means 'nail-bearing' and is meant to refer to the long scape of the epigynum. Kaston (1948:126, note 1) explained:

> "Hull (1920) proposed **Scaptophorus** to replace **Helophora** on the supposition that the latter was preoccupied. However **Helophoria** Agassiz 1846 and **Helophorus** Illiger 1801 and Gistl 1848 do not invalidate **Helophora**, according to Article 36 [now 56b] of the Rules of Zoological Nomenclature."

Hence *Scaptophorus* HULL is a junior synonym of *Helophora* MENGE. There is a mischievous misprint in Bonnet (1957a:2484), *Scatophorus*, which was unfortunately followed by Roth (1988b:47).

*Schizocosa* CHAMBERLIN 1904a:178 (Lycosidae)
From the Greek combining form σχιζο-'split' related to the verb σχίζω 'to split', as in authentic Greek compounds like σχιζόπους 'with parted toes (not webbed)' and the rhyming morph -cosa from *Lycosa* (which see). The name refers to the fact that in this genus the epigynum has the transverse arms of the median septum with the tips excised. Hence, 'the Lycosid with the split T formation'. Chamberlin had before him other derived genera from *Lycosa* such as *Pardosa* and *Arctosa* C.L. KOCH, *Allocosa* BANKS, and *Alopecosa* SIMON, and he wished to follow suit with a rhyming name to indicate the relationship.

There is a problem with Chamberlin's earlier name *Schizogyna*, and Bonnet (1958:3944) explained:

> "Chamberlin indicated p. 178, but without explanation that **Schizogyna** should be replaced by **Schizocosa**. Since the first term is correct and is not preoccupied, the author did not have the right to modify it, and that is doubtless why Banks [1910:58] maintained it. But today because **Schizocosa** is adopted by all the American authors, there is no reason to return to the prior name."

*Schizogyna* (Chamberlin 1904a:147) obviously also referred to the split transverse arms of the median septum of the epigynum. Dondale & Redner (1978a:144) solved the problem definitively:

> "Our view is that **Schizocosa** is available under the International Code of Zoological Nomenclature, whereas **Schizogyna** is not. The name **Schizogyna** failed to satisfy Article 10 (a) dealing with interrupted publications [now 10b], and Article 12 [now still Article 12] requiring a description, definition, or indication, and is therefore a **nomen nudum**. Chamberlin (1904) casually denoted it by the ambiguous term "group" in a discussion of lycosid genitalia, and, as first reviser of the genus **Schizocosa**, did not apply **Schizogyna** to any taxon in the family Lycosidae (Chamberlin 1908)."

*Sciastes* BISHOP & CROSBY 1938:75 (Linyphiidae)
From the Greek masculine noun σκιάστης an epithet of Apollo used by the obscure poet Lycophron (3rd cent. B.C.). The meaning is uncertain, but it seems to refer to some holy arbor of the god. Bishop and Crosby had a habit of scouring the Greek dictionary for obscure and unused words to serve their insatiable appetite for names. It is unnecessary to look for any semantic connection. See under *Sisis*.

*Scirites* BISHOP & CROSBY 1938:70 (Linyphiidae)
Another of the fruits of Crosby and Bishop's scouring of the dictionary. This word is found just across the page in the Greek lexicon from Sciastes. The Greek masculine noun σκιρίτης means 'a worker in stucco'; or also a special, light-armed division of the Spartan army were called the Σκιρῖται, the singular of which is Σκιρίτης. It is not necessary to seek a semantic connection. See under *Sisis*.

*Scironis* BISHOP & CROSBY 1938:72 (Linyphiidae)
This was a good day in Ithaca for the sigma-kappa page of their lexicon (see under *Sisis*). Σκίρων is the name of the wind that blew in from the direction of Corinth and the Scironian rocks, and also the name of the highwayman who haunted those rocks preying on travellers. He was killed by Theseus as a public service. The Scironian rocks are in Greek Σκιρωνίδες πέτραι where Σκιρωνίδες is a feminine adjective meaning 'Scironian'. The road past those rocks from Athens to Megara, the road infested by the robber Sciron, is called Σκιρωνὶς ὁδός where Σκιρωνίς is the feminine singular of that adjective (ὁδός 'road' despite its look, is in fact feminine). This feminine adjective, Σκιρωνίς, then, is the source of the name. The type species is *tarsalis* which is ambiguous as to gender. Bonnet (1958:3952) decreed the name is masculine, but without justification. It is feminine.

*Scoloderus* SIMON 1887d:clxxxvii (Araneidae)
A substitute name for *Hypophthalma* TACZANOWSKI 1873:284 preoccupied. Simon emphasized the strong gibbosity of the pars cephalica (pars cephalica maxima et valde gibbosa) giving the spider a marked humpbacked profile. If the genus name were *Scoliodeirus or *Scoliodirus we could derive it from the Greek adjective σκολιός -ά -όν 'curved, bent' and the adjectival combining form -δειρος 'neck'. There are at least ten classical Greek compounds ending in -δειρος, such as δολιχόδειρος 'long-necked'. The adjective σκολιόδειρος 'with crooked neck' actually appears in

an inscription of the 2nd century B.C., but it is very unlikely Simon could have known it. I think Simon meant the name to mean 'with crooked neck' although the philological details slipped away from him.

*Scolopembolus* BISHOP & CROSBY 1938:63 (Linyphiidae)
From the Greek masculine noun σκόλοψ (gen.) σκόλοπος 'anything pointed, a stake, thorn' and the arachnological term embolus, which itself is from the Greek masculine noun ἔμβολος 'anything pointed so as to be thrust in, a peg, a stopper'. Hence, 'having a pointed embolus'. Bishop and Crosby said: "In this genus, the tail-piece of the embolic division is long, slender, undulating and gives rise directly to pointed embolus."

(‡) *Scopodes* CHAMBERLIN 1922:156 (Gnaphosidae)
The second element is the Greek two-ending adjective forming suffix -ώδης -ώδες 'like, similar to'. The first element is the Greek masculine noun σκόπος 'one that watches, look-out, scout'. Chamberlin described *Tetragnatha scopus* CHAMBERLIN 1916:239 from Peru and he explained that the species name came from the Greek word meaning 'a watcher'. This would give the meaning 'like a look-out'. The genus name *Scopus* was preoccupied in Aves, Mollusca and Ephemeroptera, so it was not available. The original spelling -odes is correct, and Chamberlin probably was imitating *Drassodes*, but *Scopodes* turns out to be preoccupied in the Coleoptera (Erichson 1845), so Platnick (1989b:482) substituted *Scopoides* using the Latin form of another Greek suffix -οειδης with a similar meaning On the differences between the latinized Greek forms -odes and -oides. See the discussion under *Argyrodes*.

*Scopoides* PLATNICK 1989b:482 (Gnaphosidae)
A replacement name for *Scopodes* CHAMBERLIN (1922:156) which is preoccupied in the Coleoptera. The North American species once placed in *Megamyrmaekion* WIDER in REUSS 1834 were placed in *Scopodes* CHAMBERLIN by Ubick & Roth (1973a, c).

*Scotinella* BANKS 1911:442 (Corinnidae)
Feminine Latin diminutive in -ella formed on the palaearctic genus *Scotina* MENGE, which is from the feminine of the Greek three-ending adjective σκοτεινός -ή -όν 'dark'. It refers, as Menge explained (1873:337), both to the habits of the spider and to the body color. Menge translated the name as "Düsterspinne" that is, 'dark spider'.

*Scotinotylus* SIMON 1884r:501 (Linyphiidae)
From the Greek three-ending adjective σκοτεινός -ή -όν 'dark, obscure' and the Greek masculine noun τῦλος -ου 'knob, knot, swelling'. The male frons (of the type species *antennatus*) is not lobate, but is elevated. Presumably this elevation is dark colored, hence, 'dusky swelling'. See Millidge (1981a). Later Simon (1894a:659) misspelled his genus name *Scotynotylus*, an unjustified emendation.

(‡) *Scotolathys* SIMON 1884o:321 (Dictynidae)
The second element is the genus name *Lathys* (which see) and the first element is from the Greek masculine noun σκότος 'darkness'. Hence, 'the Lathys of the darkness'. This may refer to the loss of the anterior median eyes in this genus. Synonymized under *Lathys* SIMON by Gertsch (1946a: 1-3).

*Scotophaeus* SIMON 1893a:364-71 (Gnaphosidae)
From the Greek masculine noun σκότος 'darkness' and the three-ending adjective φαιός -ή -όν 'gray'. Hence, 'dark gray'. The fact that the young Simon (1864:117) had called one of the species eventually included in this new genus, namely, *Scotophaeus scutulatus* (L. KOCH 1866b:93) by the name *Melanophora fusca*, that is 'black-bearing dark colored' suggests the reason he chose a name meaning 'dark gray' for his new genus. The North American species have been transferred to other genera (Platnick 1993c:664).

(‡) *Scotoussa* BISHOP & CROSBY 1938:87 (Linyphiidae)
This is another product of the lexicon searching of Bishop and Crosby that has yielded so many names in sc-from contiguous pages. This is a feminine place name Σκοτοῦσσα in Thessaly. It is a contracted form of the feminine adjective σκοτόεσσα 'dark, gloomy'. Since the type species (*bidentata* EMERTON 1882) is orange-yellow, no semantic connection was intended. As Bonnet (1958:3976) correctly pointed out, this name is etymologically identical to *Scotussa* SIMON 1898a (Agelenidae). *Scotoussa* is transliterated into Latin letter for letter; *Scotussa* is latinized according to pronunciation. They are exactly the same word. But the Code states that names differing by one letter are not homonyms (Code Art. 57.f.) and the alternation of u with ou is not among the listed variant spellings deemed to be identical (Code Art. 58). Still, Holm (1945b:19) synonymized *Scotoussa* under *Diplocentria* HULL and was followed by Millidge (1984a:154). *Scotussa* SIMON is now transferred to the Hahniidae and synonymized under the new name *Simonida* by Schiapelli & Gerschman (1958a:217) because it was preoccupied in the Diptera.

*Scylaceus* BISHOP & CROSBY 1938:91 (Linyphiidae)
From the Greek masculine noun σκυλακεύς (gen.) -έως 'puppy'. The more usual word for puppy, σκύλαξ was preoccupied in the Hemiptera. There is no particular semantic point. Bishop and Crosby habitually searched the dictionary for unused words, and they cut a wide swath in the letter sigma. See under *Sisis*.

*Scyletria* BISHOP & CROSBY 1938:89 (Linyphiidae)
From the very obscure feminine Greek noun σκυλητρία 'she who strips a slain enemy' used only once in all Greek literature by the euphuistic Hellenistic poet Lycophron. It is related to the verb σκυλάω 'despoil'. It may not even be a legitimate Greek word, since the text is dubious at that point, and it may only be a false reading. Another case of Bishop and Crosby's habitual dictionary searching. See above on *Scylaceus*, and below under *Sisis*.

*Scytodes* LATREILLE 1804b:134 (Scytodidae)
From the Greek s-stem two-ending adjective σκυτώδης -ες 'like leather' from the masculine s-stem noun σκῦτος (gen.) -εος 'leather, skin, hide' and the suffix -ώδης -ες 'like, similar to'. The young Simon's incorrect form *Scytoda* (1864:451) was the result of his confusing the s-stem form with an a-stem form in -ης which would come into Latin as a first declension masculine. The gender of the adjective *Scytodes* is morphologically ambiguous. It could be masculine, feminine or neuter. Latreille erected the genus to accommodate *Aranea thoracica* LATREILLE 1802a:56 and he retained the feminine species name in *Scytodes thoracica*. Thus, by Latreille's original choice, the genus is properly feminine. Yet the Code (Art. 30b) now arbitrarily decrees that names in -odes shall be masculine.

*Segestria* LATREILLE 1804b:134 (Segestriidae)
From the Latin feminine noun segestria-ae 'a covering wrapper of straw or hides for shielding goods or persons from the weather' derived from the feminine noun seges-etis 'grainfield, standing grain'. Apparently a reference to the characteristic tubular retreat of this genus. Menge's notion (1872:299) that it comes from the Greek noun σαγήνη 'net' is to be disregarded, since that is phonologically and morphologically impossible.

*Selenops* DUFOUR *in* LATREILLE 1819:579 (Selenopidae)
From the Greek feminine noun σελήνη 'moon' and the suffix -ωψ (gen.) -ωπος 'face, eye' (or its alternate form -οψ [gen.] -οπος), which forms one-ending compound adjectives. The name is morphologically ambiguous; it can be masculine, feminine or neuter. Dufour and Latreille made it masculine. Latreille (1819:579) explained that he intended the name to mean 'eyes in a crescent'. Bonnet (1958:4016 note 1) explains: "It is certainly Latreille who published the diagnosis of this genus, but in all his writings from 1817 to 1831 he honorably attributed it to Dufour who was its author."

*Semljicola* STRAND 1906a:448 (Linyphiidae)
The type locality is Novaya Zemlya, the islands in the Arctic Sea north of Russia, so the name is a combination of a latinization of Zemlya to Semlja, plus the Latin masculine combining form -cola 'inhabitant'. The gender is properly masculine (it is a first declension masculine noun) but Strand considered it feminine. Bonnet (1958:4025) corrected the gender. Hence, 'inhabitant of Novaya Semlya'.

*Septentrinna* BONALDO 2000:83 (Corinnidae)
The author explains:

> "The generic name, feminine, comes from the combination of the words septentrionis [sic], north in Latin, and *Corinna*, in allusion to the geographical distibution of the species of the group of the type species."

The type species is *Septentrinna bicalcarata* from the southwest United States and northern Mexico, the northernmost area treated in the paper. The Latin word for 'north' septentriones, is the name of the constellation of the Great Bear, literally 'the seven plow oxen'. See *Corinna*.

*Sergiolus* SIMON 1891c:573 (Gnaphosidae)
This name is ultimately derived from the Roman family name Sergius, whose best known representative is Cicero's enemy, Lucius Sergius Catilina, because *Sergius* itself was thought to be preoccupied by *Sergia* STIMPSON 1860 (Crustacea) and *Sergia* STÅL 1864 (Hemiptera), Simon created a derived name by adding the Latin derivative suffix -olus which forms diminutive nouns like gladiolus 'little sword'.

*Sibianor* LOGUNOV 2001a (Salticidae)
A combination of the genus name *Bianor* (which see), and the geographical name Siberia.

*Sidusa* PECKHAM & PECKHAM 1895:175 (Salticidae)
Origin unknown. It does not seem likely that the name would have anything to do with the Greek geographical name Sidus (Σιδοῦς) a place near Corinth where pomegranates grew. As a long shot we might guess that the Peckhams might have tried to make a feminine noun out of the Latin neuter noun sidus sideris 'star' perhaps with reference to its iridescent coloring. The origin remains unclear. The only US species has been transferred to *Tylogonus* SIMON (see Platnick 1993c:809, 820).

*Silometopoides* Eskov 1990a:52 (Linyphiidae)
Erected on the basis of the type species *Minyriolus pampia* Chamberlin 1949 by adding the Greek adjective-forming suffix -oides (see the discussion under *Argyrodes*) to the erigonine genus name *Silometopus* Simon (1926:353). The second member of the name *Silometopus* is from the Greek neuter noun μέτωπον 'brow, forehead'. The first element is the Latin adjective silus-a-um 'having a broad turned-up nose, pug-nosed, snub-nosed'. All but one of the species included in this genus by Simon have a large low cephalic lobe rounded in the rear and narrowed in front. The name *Silometopoides* is ambiguous as to gender, and could be masculine, feminine or neuter. Chambelin's species name *pampia* must be regarded as a noun in apposition, and does not resolve the ambiguity. Now the Code (Art. 30b) arbitrarily decrees that all names in -oides shall be masculine.

*Simitidion* Wunderlich 1992a:417 (Theridiidae)
The type species is *Theridion simile* C.L. Koch 1836d:62, and Wunderlich's genus is created from the simi-of simile, the -idion of *Theridion* plus a euphonic connector -t-.

*Singa* C.L. Koch 1836a:42 (Araneidae)
Probably a feminized form of the geographical name Singos, a city on the west coast of the Chalcidean peninsula at the north end of the Aegean Sea, found in Herodotus, Thucydides, Pliny and, Strabo. Thorell (1869: 58) said that it is a proper geographical name.

*Siratoba* Opell 1979:486 (Uloboridae)
An arbitrary combination of letters, according to the author. But Opell named a new species *Siratoba sira* saying the species name is derived from the Greek term for cellar. That would be the Greek masculine noun σίρος -ου 'pit for keeping grain'. The author has treated it as if it were an adjective and given it a feminine ending. It is clear that the first part of *Siratoba* is based upon the species name. The second part rhymes with *Octonoba*. The type species, originally *Ariston referens* Muma & Gertsch 1964:17, became *Siratoba referena*. But the species name is properly *referens*, since in the Latin third declension adjectives the masculine and feminine forms are identical.

*Sisicottus* Bishop & Crosby 1938:57 (Linyphiidae)
See *Sisis*.

*Sisicus* Bishop & Crosby 1938:61 (Linyphiidae)
See *Sisis*.

*Sisis* Bishop & Crosby 1938:66 (Linyphiidae)
Sisis is the father of Antipater, a king of Lesser Armenia. His name is recorded in the Geography of Strabo (12.3.28). Crosby and Bishop found it by scanning the index volume of the Loeb Classical Library edition of Strabo. See under *Soucron*. They were then able to derive successively the genera *Sisicus* (non-existent in Greek) from Sisis and *Sisicottus* (non-existent in Greek) from *Sisicus*

*Sisyrbe* Bishop & Crosby 1938:86 (Linyphiidae)
The name of an Amazon found in the Geography of Strabo (14.1.4). See under *Soucron* and *Sisis*.

*Sitalcas* Bishop & Crosby 1938:74 (Linyphiidae)
Another result of Bishop and Crosby's gleanings from the sigma pages of the lexicon (see under *Sisis*). Σιτάλκας (an a-stem masculine) is an epithet of Apollo at Delphi. The form of the name is in the Phocian dialect of Delphi, rather than Attic. When latinized it is a first declension masculine noun.

*Sitticus* Simon 1901a:581 (Salticidae)
This appears to be an adjective derived from the Roman name Sittius. Publius Sittius was a friend of Cicero's and one of his correspondents. It is a habit of Simon's to derive names from Cicero's friends, perhaps influenced by *Salticus*.

*Smeringopus* Simon 1890a:94 (Pholcidae)
From the Greek feminine noun σμήριγξ (gen.) -ιγγος 'hair on the thighs and necks of dogs' and the masculine noun πούς (gen.) ποδός 'foot'. Hence, 'hairyfoot'. There is nothing in Simon's description of the genus which seems to warrant this name, nor is there anything in Vinson's original description of the type species *Pholcus elongatus* Vinson 1863:135. It is a mystery why Simon chose this name.

*Smodigoides* Crosby & Bishop 1936b:52 (Linyphiidae)
Derived from the genus name *Smodix* by the addition of the Latin suffix -oides 'similar, like, having the shape of' (see the discussion of the suffix under *Argyrodes*). Strictly speaking it should be *Smodingoides* for the stem of σμώδιξ is σμωδιγγ-. In defining the genus Crosby and Bishop said: "This genus is closely related to *Smodix*." The discrepancy in dates is easily explained. Although *Smodix* was not published until 1938, they had obviously used it in manuscript, so it was before their eyes when they created this derived genus name. The name is morphologically ambiguous as to genus, and can be masculine, feminine or neuter. Crosby and Bishop established it as masculine with *Smodigoides retionax* Crosby & Bishop 1936b. Millidge (1984a:154) synonymized it under *Diplocentria* Hull.

*Smodix* Bishop & Crosby 1938:94 (Linyphiidae)
From the Greek feminine noun σμῶδιξ (gen.) -ιγγος 'welt, bruise'. Erected for *Tmeticus reticulatus* Emerton 1915b. Here is one of Bonnet's few real errors. Bishop and Crosby correctly treated the genus as feminine in accordance with Greek grammar, and listed the type species as *Smodix reticulata*. Bonnet reported this incorrectly (Bonnet 1958:4090) as *Smodix reticulatus*. There is probably no semantic connection, and this is another case where Crosby and Bishop were gleaning names from the lexicon (wee under *Sisis* and *Soucron*). Their 1938 paper has a dozen such sigma names. But, on the other hand, the name may refer to the "slight hump back of the cervical groove" of the type species *reticulata* (Bishop & Crosby 1938: 95).

*Socalchemmis* Platnick & Ubick 2001:2 (Tengellidae)
The authors explain that "the generic name is a contraction of southern Californian *Chemmis*." See under *Chemmis*.

(‡) *Sosilaus* Simon 1898a:350 (Lycosidae)
Greek proper name Σωσίλαος meaning 'savior of the people'. Formed on the analogy of his earlier name *Sosippus* Simon 1888b. While this looks like a Greek proper name, and is perfectly formed, I am unable to find an actual Greek so named. Maybe Simon made it up. Synonymized under *Pirata* Sundevall by Wallace & Exline (1978:9).

*Sosippus* Simon 1888b:205 (Lycosidae)
From the Greek masculine proper name Σώσιππος meaning 'saving the horses'. There is an obscure writer of Greek comedies, now lost, named Sosippus. It is a compound of the combining form σωσι-as in the adjective σωσίπολις 'saving the city', (also a proper name, for the sacred snake kept at a shrine in Olympia in northwest Greece) and the noun ἵππος 'horse'.

*Sosticus* Chamberlin 1922:146 (Gnaphosidae)
From the Greek three-ending adjective σώστικος -η -ον 'able to save, maintain or preserve'. Chamberlin gave no hint why he chose this name.

(‡) *Sostogeus* Chamberlin & Gertsch 1940:1 (Gnaphosidae)
The authors stated that "this genus is related to *Sosticus* in general characteristics." So the first part of the name is derived from the genus name *Sosticus*. The second part of the name may be inspired by another Chamberlin Gnaphosid genus *Geodrassus*. In Chamberlin's paper (1922: 159-160) the new genus *Geodrassus* immediately precedes the new genus *Sosticus*. Platnick & Shadab (1976b:9) synonymized *Sostogeus* under *Sosticus*.

*Soucron* Crosby & Bishop 1936b:60 (Linyphiidae)
Faced with the need to generate many new names which were not preoccupied, Crosby and Bishop were in the habit of surfing through lists of Greek and Latin proper names which they used without regard for any semantic connection with the spider. In the 1932 Loeb Classical Library eight volume edition of Strabo's Geography they turned to the index volume (vol. 8, p. 481) and found a nest of names in Sou-, from *Soucron* to *Soulgas* below. They just ran down the index, which is why these names have an alphabetical relationship. The index spells them in the Latin manner with a u, but they looked up the Greek spelling ou in the body of the text. Soucron (Strabo 3.4.6) is a river in Spain, now the Jucar. Although it is masculine in the Greek text of Strabo, the authors made it neuter.

(‡) *Soudinus* Crosby & Bishop 1936b:59 (Linyphiidae)
A famous Chaldean philosopher from Strabo (16.1.6). See *Soucron* above.

*Souessa* Crosby & Bishop 1936b:57 (Linyphiidae)
An ancient city in Latium from Strabo (5.3.10). See *Soucron* above.

*Souessoula* Crosby & Bishop 1936b:62 (Linyphiidae)
A city of Campania from Strabo (5.4.11). See *Soucron* above.

*Sougambus* Crosby & Bishop 1936b:61 (Linyphiidae)
A misspelling of Sougambrus, from the name of a Germanic tribe, the Sugambri, from Strabo (4.3.4). See *Soucron* above.

*Souidas* Crosby & Bishop 1936b:57 (Linyphiidae)
An obscure historian who wrote a history of Thessaly, from Strabo (7.7.12). This has nothing to do with the name Σουίδας, the traditional designation for a Byzantine encyclopedia dated in the third quarter of the 10th century A.D. The name is an a-stem masculine noun. See *Soucron* above.

*Soulgas* Crosby & Bishop 1936b:55 (Linyphiidae)
A river in Gaul, now the Sorgue, from Strabo (4.1.11). See *Soucron* above. Crosby and Bishop regarded it as masculine with the type species *corticarius*.

(‡) *Sparassus* Walckenaer 1805:39 (Sparassidae)
Thorell (1870b:176) derived this name from the Greek verb σπαράσσω 'tear apart'. This would be an ungrammatical formation, since the double sigma is peculiar to the present stem of the verb, and would not appear in any proper noun derivatives. But Walckenaer clearly cared nothing for this grammatical nicety. Indeed, he more than once committed the sin of creating nouns directly from present verb stems in the names *Attus, Sphasus, Thomisus,* and *Drassus*. For a full discussion of the complicated relation between *Sparassus* and *Heteropoda* see Fox (1937e:461).

*Spermophora* Hentz 1841a:117 (Pholcidae)
From the Greek two-ending compound adjective σπερμοφόρος -ον 'seed bearing', from the combining form from σπέρμα (gen.) -ατος 'seed' and the adjectival form related to the verb φέρω 'bear'. Greek compound adjectives do not have distinctive feminine forms, so Hentz Latinized this into a feminine. He meant this to refer to the fact that the female carries her eggs in her mouth: "The female carries in its mandibles its eggs glued together without any silk." Hentz later realized that 'seed' is not accurate in meaning, and changed the name to *Oophora* with the first element from the Greek neuter noun ᾠόν 'egg' (Hentz 1850b:285). Bonnet was wrong (1958:4113) when he suggested that *Spermophora* is incorrectly formed, since σπερμοφόρος is an authentic Greek adjective. Bonnet was right in pointing out that the Code does not permit the author to change his own name (Code Art. 33).

*Sphecozone* O. Pickard-Cambridge 1870d:733 (Linyphiidae)
From the masculine Greek noun σφήξ σφηκός 'wasp' and the feminine noun ζωνή 'girdle, belt, waist' hence 'wasp-waisted'. The genus is close to *Ceratinopsis*, but differs by the possession of a long, thin pedicel, whence the name. Millidge (1993d:168) placed the American species *magnipalpis* in this genus

*Sphodros* Walckenaer [1833a:439 *nomen nudum*]; 1835:638 (Atypidae)
From the masculine of the Greek three-ending adjective σφοδρός -ά -όν 'violent, vehement, excessive'. Walckenaer did not latinize the name to *Sphodrus* as would be expected, but retained its Greek ending. Gertsch & Platnick (1980:15) have properly maintained the original spelling.
Ausserer (1871a:143) substituted the name *Madognatha* meaning 'with bald chelicerae' (from μαδός 'bald' and γνάθος 'jaw') in the belief that Walckenaer's name was preoccupied by *Sphodrus* Clairville 1806 in the Coleoptera. Before the "one letter difference" rule (Code Art. 56.b) came into effect they certainly would be homonymous, since the one is merely the latinization of the other. If *Theridion* and *Theridium* are homonyms, then so are *Sphodros* and *Sphodrus*. But the "one letter difference" rule makes Walckenaer's name available.
Long considered a synonym of *Atypus* Latreille, the name *Sphodros* was revalidated by Gertsch & Platnick (1980:15), who point out that, although Walckenaer first used the name in 1833, it did not become available until particular species were associated with it in 1835.

(‡) *Sphyrotinus* Simon 1894c:524 (Theridiidae)
From the Greek neuter noun σφυρόν -οῦ 'ankle' and the verb τείνω 'stretch'. Perhaps referring to the fact that the anterior legs of the male are very much longer than the others. Hence, 'stretch-ankle'. If this etymology is correct, then the word is not well formed, since the present tense of the verb τείνω would not form compounds in this way. Synonymized under *Thymoites* Keyserling (Levi & Levi 1962:28).

*Spintharus* Hentz 1850b:283 (= 1875:156) (Theridiidae)
There are three related Greek words meaning 'spark' σπινθήρ (gen.) -ῆρος (masc.). σπινθαρίς (gen.) -θαρίδος (fem.) σπινθάρυξ (gen.) -υγος(fem.). But Hentz did not use any of these. He found in Cicero's letters the Romanized name of his Greek slave and amanuensis Spintharus (Cicero Letters to Atticus 13.25.3), which is obviously derived from the words meaning 'spark'. Although Hentz's type species *flavidus* is brightly colored, orange, yellow and red, Hentz give no hint that he meant to evoke the idea of a spark in this name. He was merely using a proper name from the Latin dictionary.

*Spirembolus* Chamberlin 1920a:197 (Linyphiidae)
From the first part of the Latin two-ending adjective spiralis -e and the arachnological term embolus, which itself is from the Greek masculine noun ἔμβολος 'anything pointed so as to be thrust in, a peg, a stopper'. Hence, 'having a spiral embolus'. The embolic division consists of a spiral embolus of several turns with a short coiled tailpiece (Millidge 1980b: 110-111).

(‡) *Spiropalpus* Emerton 1882:39 (Linyphiidae)
From the Latin two-ending adjective spiralis -e 'spiral' and the arachnological anatomical term palpus, hence 'with spiral palpus'. Emerton's type species was *spiralis*, and the names refer to the large embolic coil of the male palp. Emerton said:

> "This spider resembles closely those of the last genus [**Grammonota**], but has entirely different male palpi,
> with the tubes long and stiff and coiled in a flat spiral."

Synonymized under *Walckenaeria* Blackwall by Millidge (1983:106).

*Steatoda* Sundevall 1833b:16 (Theridiidae)
The Greek neuter noun στέαρ (gen.) στέατος means 'fat', but in the sense of 'tallow or suet, which congeals as opposed to mere fat (πιμελή), which does not'. The Greek two-ending s-stem adjective στεατώδης -ες means 'having tallow or suet' (literally 'like suet'), and the word is used by Aristotle (Hist. Anim. 520a14) to designate animals like sheep whose mutton is tallowy, as opposed to pork which is fatty. Sundevall's description said (1833b:17):

> *"Abdomen magnum, plerumque depressum, suborbiculare, speciem pinguetudinis vel sevi praebens.* [Abdomen large, mostly depressed, suborbicular, giving an appearance of fatness or tallow]."

Sundevall had before him Walckenaer's 2nd family of *Theridion*, the Rotundatae, and this perhaps inspired the name. My guess is that Sundevall meant this to mean 'fat' in the sense of 'rotund' for which the proper Greek word would have been πίων or παχύς as well as suggesting the texture or firmness of the abdomen. Sundevall made a grammatical error by treating this s-stem adjective (in -ης -ες), which would be latinized steatodes, as if it were an a-stem masculine noun (in -ης, genitive -ου), which would be latinized with an a-ending, steatoda. Then, because it ended in a, he treated it as a Latin feminine noun. It is a wonder that the purist Thorell never called him on this kind of hanky-panky. Thorell (1869:93) quoting Sundevall said: "στεατώδης saevum referens [sebum referens referring to tallow]". Sundevall spelled the Latin word usually spelled sebum 'tallow' with the alternative spelling sevum, and since the e is long, Thorell used the schoolboy spelling ae (see under *Megamyrmaecium*). Hence, the name means 'tallowy'.

*Stemmops* O. Pickard-Cambridge 1894a:125 (Theridiidae)
From the Greek neuter noun στέμμα (gen.) -ατος 'wreath, garland, chaplet' and the combining form -ωψ (gen) -ωπο" 'eye'. This forms a one-ending compound adjective, which can have any of the three genders as a genus name. Pickard-Cambridge left it ambiguous by using a one-ending species name (*bicolor*) and Simon followed suit with *concolor*. The code (Art. 30.A.ii) now arbitrarily specifies that all names in -ops shall be masculine regardless of their derivation or of their treatment by its author. A purist would have preferred that Pickard-Cambridge had formed Stemmatops, which the strict rules of compounding require. In describing the genus Pickard-Cambridge said: "Eyes large, subequal, those of the hind-central pair the largest, seated on a large subconical prominence and forming a kind of irregular ring round it." It is this irregular ring of eyes which has given birth to the name 'crown of eyes'. See Pickard-Cambridge's illustration Tab. 17 figs. 5b, 5c.

*Stemonyphantes* Menge 1866:138 (Linyphiidae)
From the Greek masculine noun στήμων (gen.) -ονος 'thread' and the masculine a-stem noun ὑφάντης -ου 'weaver'. Hence, 'thread-weaver'. The form -hyphantes is also a rhyming morph to associate this with other Linyphiid genera such as *Bathyphantes* etc. Menge translated the name as "Fadenspinne" that is 'thread spider' and explained that the name refers to the thread-like embolus "Eindringer" of the male palp.

*Stenoonops* Simon 1891c:564 (Oonopidae)
Formed on the genus *Oonops* (which see) with the first element from the Greek three-ending adjective στενός -ή -όν 'narrow'. The name refers to the fact that the ocular area is small occupying only half the front. The gender is masculine after *Oonops*.

*Storena* Walckenaer 1805:83 (Zodariidae)
This appears to be derived from the Latin feminine noun storea 'a mat made of straw, rushes or rope' with the addition of Walckenaer's favorite suffix -ena (Cf. *Agelena, Missulena, Delena*). Since these spiders are sand burrowers, and do not spin mat webs, the point of Walckenaer's name is puzzling, until we read his observation (1805:83) that "it is impossible for me make any conjecture about the habits of the araneids of this new genus." Since he knew nothing about them, the fact that they do not make mats does not disturb the etymology. Of the only species listed for the United States Gertsch (1979:213) said:

> "Another Zodariid, **Storena americana**, attributed to our fauna by George Marx on the basis of a specimen from Cohuta Springs, Georgia, more than likely was imported from the Bahama Islands."

He further remarked that the headquarters of the genus is Australia.

*Strotarchus* Simon 1888b:210 (Miturgidae)
This is the Lesbian-Boeotian dialect form of the Greek masculine noun στράταρχος (Lesbian-Boeotian στρόταρχος) meaning 'general, commander of an army'. Some editions of the Boeotian poet Pindar (5th cent. B.C.) may have spelled it this way although modern editions print the Attic form. I cannot explain why Simon chose the dialect form, unless

he was just scanning the dictionary. Transferred to the Miturgidae by Lehtinen (1967:321).

*Styloctetor* SIMON 1884r:733 (Linyphiidae)
From the Greek masculine noun στῦλος 'pillar, pole, writing instrument' and the masculne noun κτήτωρ 'possessor, owner'. Hence 'possessing a pen-like process'. The male palp of the type species, *inuncans* SIMON 1884r:735, has a "bulbe complexe, pouvu à l'extrémité d'un fort stylus détaché [a complex bulb with a strong stylus at the extremity]".

(‡) *Stylophora* MENGE 1866:128 (Linyphiidae)
From the Greek masculine noun στῦλος 'pillar, post, stilus' and the combining form -φορος 'bearing'. Menge translated this as "Stilspinne" that is "Stylus spider" and explained that it refers to the fine stylus-shaped scape of the epigynum. Bonnet (1958:4192) explained that the name *Stylophora* was preoccupied in the Coelenterata (Schweigger 1819) and in the Diptera (Robineau-Desvoidy 1830), and Emerton (1882:65) replaced it with *Diplostyla* (which see).

*Styposis* SIMON 1894a:590-592 (Theridiidae)
There is no obvious etymology for this name, and there is no such genuine Greek word. It does not seem to derive from the Greek noun στύπος 'stem, stump' nor does it seem to come from the Greek adjective στυφός -ή -ον 'astringent'. There is nothing in the description to give a hint what Simon meant. With the type species *flavescens* SIMON 1894a, a Latin two-ending adjective, Simon left the gender ambiguous. Bonnet treated it, reasonably, on the basis of the ending, as feminine.

*Subbekasha* MILLIDGE 1984a:138 (Linyphiidae)
"From Subbekashe, the spider in 'The Song of Hiawatha' by Longfellow [XIII.61]." The Eastern Ojibwa word for spider is e:ssepikke~ related to a verb assapikke 'make webs'. The suffix -ikke:'make, gather' is added to noun stems, so from assapy-'net' is formed e:ssepikke~ 'spider'. See Bloomfield (1956: 64, 77)

*Symmigma* CROSBY & BISHOP 1933b:159 (Linyphiidae)
From the Greek neuter noun σύμμιγμα (gen.) -ατος 'commixture'. Another case of dictionary searching. See under *Sisis*. Crosby and Bishop may be being very wicked here, since this word, and its close relatives also have something to do with sexual intercourse, and promiscuity. They correctly treated it as neuter.

*Symphytognatha* HICKMAN 1931:1322 (Symphytognathidae)
From the Greek three-ending adjective σύμφυτος -η -ον 'grown together' from the preverb σύν 'with' and an adjectival form of the verb φύω 'grow' and the feminine arachnological element -gnatha as in *Tetragnatha* (which see) designating the chelicerae, from the Greek feminine o-stem noun γνάθος 'jaw'. Hence, 'with the chelicerae grown together'. The name refers to the fact that the chelicerae are fused medially for most of their length (Forster & Platnick 1977:5).

*Synema* SIMON 1864:433 (Thomisidae)
Simon treated this as a subgenus of *Thomisa* (sic) WALCKENAER. He gave as its etymology "σύν indique ressemblance, couleur etc; αἷμα sang" that is, he intended the name to mean "blood colored". In his description he says, "une belle teinte rouge sanguin sur les parties latérales" [a beautiful blood red tint on the sides]. Simon originally regarded this name as a feminine adjective, and gave as the type species *Synema rotundata* [(Walckenaer's *Aranea rotundata*, now *Synaema globosum* (FABRICIUS 1775)]. Simon's original spelling was *Synema* in which he used the acceptable medieval spelling of the Greek diphthong (consider the medical term anemia). Thorell (1869:36) "corrected" the spelling to the "purer" *Synaema*, and furthermore treated it, not as a compound adjective as Simon had, but as a compound neuter noun in -ma (gen.)-matos. Bonnet's note (1958:4202) said:

> "Since the name is incorrectly spelled, according to the very etymology the author [Simon] gave, Thorell emended it to **Synaema**, which ought to be used; Simon himself accepted that emendation and one does not comprehend the insistence of a number of authors upon retaining the faulty name. Furthermore it is not Article 32 [on original spelling] of the Rules which is in question here, but Article 19 [incorrect spelling has no status] which ought to be applied to rectify a name which presents a fault in transcription."

First, Bonnet is wrong that this is misspelled; it is merely an alternative latinization. Besides, incorrect latinization does not justify emendation (Code Art. 32.c.ii). Second, Simon (1875a:202) listed the type species as *Synema globosa*, but then (1875e:cxcvii) changed it to *Synaema globosum*, and thus accepted Thorell's change, even though his original spelling was not in fact wrong. The authors have vacillated from one to the other over the years, both in the spelling and in the gender.

The original spelling was undoubtedly *Synema* and it was feminine, and objections to its correctness do not stand up to scrutiny. Platnick (1989b: 526) rightly concludeed that Thorell's *Synaema* is an unjustified emendation, but he retained the neuter gender, which is in accord with Greek grammar, αἷμα is a neuter noun.

*Synageles* SIMON 1876a:14 (Salticidae)
This name is not from any genuine Greek word, but seems to be cobbled together from some identifiable fragments. The Greek verb συναγελάζομαι means 'herd together' and is a denominative verb built on the feminine noun ἀγέλη 'herd' with the preverb σύν 'with, together'. There is no noun *συναγέλη, and no noun or adjective in -ης is derived from ἀγέλη. Simon's confection, which he treated as masculine, would mean something like 'belonging to a herd'. I do not see the point of the name, for these ant-mimicking spiders seem fairly solitary.

*Synaphosus* PLATNICK & SHADAB 1980a:21 (Gnaphosidae)
The authors explained this as an arbitrary combination of letters. But the second part-aphosus is clearly a rhyming morph analogous to *Gnaphosa* and the first part is from the Greek preposition and preverb σύν 'with' taken from the name of the type species *Nodocion syntheticus* CHAMBERLIN 1924b. Hence, 'the Gnaphosid genus based upon *syntheticus*.

*Synemosyna* HENTZ [1835:552 *nomen nudum*]; 1846:367 (=1875:72) (Salticidae)
This is a straightforward latinization of the Greek feminine noun συνημοσύνη 'ties of friendship'. Hentz gave no reason for the choice of the name.

*Syrisca* SIMON 1885d:375 (Miturgidae)
This is Latin for 'Syrian woman' frequently the name of female slaves in Roman comedy, e.g., Terence Andria 5.1.1.

*Syspira* SIMON 1895c:135 (Miturgidae)
This could be an invented feminine noun from the Greek verb συσπειράω 'contract, shrink' but there is nothing in the description of the genus to justify this guess. It may be Simon's spelling of an invented noun formed from the Latin verb suspiro 'sigh, breathe' but this is no more likely a guess than the other. Transferred to the Miturgidae by Lehtinen (1967:266).

*Tachygyna* CHAMBERLIN & IVIE 1939a:61 (Linyphiidae)
Formed on analogy with their earlier name *Tunagyna*. From the Greek three-ending adjective ταχύς -εῖα -ύ 'fast, quick' and the latinized feminine noun γύνη 'woman'. The reference could be to some feature of the epigynum, or the name may simply mean 'quick female'. The earliest species was described as *Tunagyna tuoba* three years earlier (1933a:23) from the female alone. Tuoba is Gosiute for 'woman' (Chamberlin 1913:19).

*Takayus* YOSHIDA 2001c:165 (Theridiidae)
Erected for the type species *Theridion takayense* SAITO 1939. Formed by giving a Latin ending to the locality of the type species. The author says, "The generic name is derived from Takayu Spa, and is masculine in gender." The type locality of Saito's species is Takayu, Yamagata Prefecture, Japan.

*Talanites* SIMON 1893a:363 (Gnaphosidae)
This name has the apparent form of a Greek word, but nothing corresponds exactly. Greek masculine a-stem nouns in -ites tend to form ethnica like Aberites 'man from Abdera' or Sybarites 'man from Sybaris'. But no place name seems to correspond to Talanites. The male syntype of *Talanites fervidus* SIMON 1893a, the type species, was collected in the area of the Dead Sea (Platnick & Ovtsharenko 1991:116). Perhaps the name is meant to relate to the Greek verb ταλανίζω 'be miserable, woeful' and Simon made up a passable noun intending it to mean 'wretch' or the like. The North American genus *Rachodrassus* CHAMBERLIN has been synonymized with the Eurasian *Talanites* by Platnick & Ovtsharenko (1991).

*Talavera* PECKHAM & PECKHAM 1909:575 (Salticidae)
The Peckhams give no explanation of this term, but it is a place name in Spain, taken it seems from the name of a large urticaceous tree, the tala, i.e. 'true tala'. The spider was first described from Long Island, and there are no associations with Spain or Spanish America. But the Peckhams have done this before, that is, name a spider for a place name in Spain, i.e. *Fuentes*. There is evidence that they were reading Spanish history. On January 28, 1809, the indecisive Battle of Talavera took place in which Sir Arthur Wellesley (later the Duke of Wellington) prevailed against Marshal Soult, the French commander. This derivation is reinforced by the fact that the Peckhams named the genus Fuentes from the Battle of Fuentes de Onoro, May 5, 1811, in which the French general Masséna was defeated by Wellington.

*Tama* SIMON 1882b:256 (Hersiliidae)
Simon said this is a geographical name without further explanation. There is a place in Algeria, whence the type species *edwardsi* LUCAS 1846 comes, called Tamada, and this may be the source. The Latin word tamae 'a kind of swelling of the feet and legs', is not the source of the genus name.

*Tapinauchenius* Ausserer 1871a:200(Theraphosidae)
From the Greek three-ending adjective ταπεινός -ή -όν 'low, humble, downcast' and the Greek masculine noun αὐχήν (gen.) -ένος 'neck' plus an adjective forming ending -ιος added to the stem αὐχέν-. There are authentic Greek compounds ending in -αὐχένιος. Hence, 'with a low neck' as Ausserer himself explains: "Cephalothorax länglich oval, sehr niedrig mit deutlicher Rückengrube". Note that the Greek diphthong ει is regularly transliterated with Latin long i. It refers to the fact that the cephalothorax is lower than in *Avicularia*. Simon (1891g:325-326) described two species from the United States *caerulescens* Simon 1891g and *texensis* Simon 1891g from immature specimens, and their status is doubtful.

*Tapinesthis* Simon 1914b:88 (Oonopidae)
This word appears to be compounded of the Greek adjective ταπεινός -ή -όν 'low, humble' and a suffix Simon had used once before in the genus name *Xenesthis* (1891h:333). It is not clear what he meant by it. In the description of *Xenesthis*, Simon remarked on the beautiful color of the pubescence, and it is conceivable that Simon combined the Greek adjective ξένος -η -ον 'unusual, strange' with his peculiar spelling of the Greek feminine noun ἐσθής -ῆτος 'raiment, clothing' with the resulting meaning 'unusual vestiture'. Then Simon might have meant *Tapinesthis* to mean something like 'humble vestiture'. This yellowish gray spider might warrant such a name. Kaston (1948:61) said that the only specimen [of *Tapinesthis inermis* (Simon 1882c)] ever found in the United States was that from the cellar of the Boston Society of Natural History building, and presumably came from Europe hidden in a package shipped to the Society.

*Tapinocyba* Simon 1884r:778 (Linyphiidae)
From the Greek three-ending adjective ταπεινός -ή -όν 'low, humble, downcast' and the Greek feminine noun κύβη 'head' (an obscure noun found only in an ancient dictionary, the Etymologicum Magnum). Note that the Greek diphthong ει is regularly transliterated with Latin long i. The name, meaning 'low head' refers to the fact that the males almost always lack a cephalic lobe. "Frons maris impressionibus linearibus impressis, sed gibbo fere semper carens [the front of the male with linear impressions but almost always without a hump]" (Simon 1894a:657).

*Tapinopa* Westring 1851:38 (Linyphiidae)
From the Greek three-ending adjective ταπεινός -ή -όν 'low, humble, downcast' and the Greek suffix -όπη as in *Argiope* (which see) meaning 'face'. Note that the Greek diphthong ei is regularly transliterated with Latin long i. Hence, 'with low face'. Westring said (1861:142): "The clypeus, because the anterior median eyes are close to the mouth, is very low (valde humilis)."

*Taranucnus* Simon 1884a:248(Linyphiidae)
The Latin name of a Celtic deity occurring in a Latin inscription found near Heilbronn. Simon came upon it by dictionary searching. The deity is now considered equivalent to Taranis, a Celtic thunder god, whose name is cognate with the Old Irish word torann 'thunder'. The inscription was published in the Corpus Inscriptionum Latinarum 13.6478.

*Tarsonops* Chamberlin 1924b:598 (Caponiidae)
MacLeay (1839:2) had created the genus *Nops* from the Greek one-ending adjective νώψ (gen.) νώπος (all three genders have the same form) meaning 'without eyes, purblind'. Then, following on this example, Chamberlin created the genera *Nopsides*, *Orthonops*, and *Tarsonops*. The genus *Tarsonops* is distinguished by having a bulging membranous pouch at the base of the flexed anterior tarsi, and it contrasts with *Orthonops* which has stiff and straight tarsi. Hence, 'straight *Nops*' and 'Tarsi *Nops*'.

*Tegenaria* Latreille 1804b:24.134 (Agelenidae)
From a medieval Latin masculine noun tegenarius meaning 'mat-maker'. The word occurs in a medieval glossary of Greek words as the Latin gloss for the Greek noun ψιαθοποιός meaning 'mat-maker'. It will be found in the standard Latin dictionaries under its alternative form tegetarius, and is derived from the Latin feminine noun teges -etis 'mat' which in turn is related to the Latin verb tego 'to cover'. Latreille gave it a feminine form, presumably to suggest it was parallel to the feminine noun aranea 'spider'. Walckenaer (1805:49) created the group Les Tapitèles (Vestiariae) that is, 'the carpet weavers' (for the French) and 'the clothes dealers' (for the Latin). In this group he placed Latreille's genus Tegenaria, and described the habits of the genus as follows:

> "Sedentary spiders, forming in the interior of buildings, caves, and subterranean cavities a horizontal web, very large and closely woven, on the superior part of which there is a cylindrical tube where they remain immobile."

The "closely woven web" is clearly what suggested the name 'carpet weaver or mat-maker'.
The suggestions made by Thorell (1870b:129) from τέγος 'roof' or τήγανον 'pan' perhaps with reference to the form of the web, or by Simon (1864:201) 'roof-raiser' a completely crazy suggestion, or Erichson (1845:12) from Τέγεα, a geographical name are all to be disregarded.

*Tekellina* Levi 1957d:107 (Theridiidae)
Derived from the genus name *Tekella* Urquhart 1894:211 by the addition of a Latin feminine diminutive suffix -ina. The origin of *Tekella* is obscure. Could it be from the handwriting on the wall at Belshazzar's feast (Daniel 5:27)? Tekel is the part the prophet interpreted as meaning, "Thou art weighed in the balance and found wanting." Could this be a jocular reference to the small size of the spider?

*Telema* Simon 1882c:204 (Telemidae)
A literary reference. The feminine form of Telemus, the name of a soothsayer in Ovid Metamorphoses 13.770.

*Teminius* Keyserling 1887b:421 (Miturgidae)
Origin obscure. The author gave no hint about the origin of the name, and it does not correspond exactly to any Greek or Latin word or proper name. *Teminius* was removed from synonymy with *Syriscus* Simon by Platnick & Shadab (1989:1).

*Tengella* F. Dahl 1901b:247 (Tengellidae)
Etymology unknown. Perhaps formed on the Scandinavian surname Tenger.

*Tennesseellum* Petrunkevitch 1925c:173 (Linyphiidae)
Formed with a neuter diminutive suffix on the name of the state Tennessee, which is the type locality.

(‡) *Tentabunda* Fox 1937e:463 (Sparassidae)
The feminine of the Latin adjective tentabundus-a-um meaning 'trying here and there' which would be appropriate to the hunting behavior of these spiders. Roth's spelling *Tentabuna* (1985) is a *lapsus calami*. Roewer synonymized the genus under *Pseudosparianthis* Simon (Roewer 1954a: 683; 1954b:1674).

*Tenuiphantes* Saaristo & Tanasevitch 1996b:180 (Linyphiidae)
Erected for many species formerly in *Lepthyphantes* with the type species *tenuis* Blackwall 1852a:18. The first element of the name is taken from the species name tenuis, a Latin adjective meaning 'slender'. The first element of *Lepthyphantes* is from the Greek adjective λεπτός 'slender'. The second part of the genus name is a rhyming morph taken from *Lepthyphantes*.

*Terralonus* Maddison 1996:239 (Salticidae)
Erected with the type *Metaphidippus mylothrus* (Chamberlin 1925c: 134). While most of the Dendryphinae are foliage dwellers, *Terralonus* is a ground dweller, and the name is designed to reflect that. From the Latin feminine noun terra 'land, ground' plus a euphonic second element which is an arbitrary combination of letters (Maddison, pers. comm.).

*Tetragnatha* Latreille 1804b:135 (Tetragnathidae)
This word is formed from Greek τέτρα 'four' and the feminine second declension noun γνάθος (gen.) -ου 'jaw'. They combine to form a genuine ancient Greek two-ending adjective τετράγναθος -ον meaning 'four jawed'. Aristotle (Hist. Anim. 622b and 555b) mentioned a spider called the φαλάγγιον, which was considered poisonous. The plural of this word is φαλάγγια The 1st century B.C. geographical author Strabo (16.4.12) and the 2nd cent. A.D. author Aelian, in On Animals (17.40) used this word in the plural to designate Indian spiders "with four jaws" that is φαλάγγια τετράγναθα, where τετράγναθα is the neuter plural of that adjective to agree with φαλάγγια. The evidence from the 2nd century B.C. geographer Agatharchides (fragment 59) indicates he treated the word as feminine (ἡ τετράγναθος). In the first century A.D. Pliny (Natural History 29.87) latinized the word into the neuter diminutive tetragnathium, or perhaps a masculine tetragnathius (the syntax obscures which) In the Renaissance Aldrovandus picked this word up from the ancient authors, and listed among the spiders (1601:603) two kinds of tetragnathium (using Pliny's form). He further cited Aelian and Strabo (using a different form) as follows (1601:604):

> "In another passage this same Aelian (17.40) has written that some Phalangia are called **Tetragnatha**, which Strabo mentions (Book 16) as follows: 'Beyond the Libyans and the Zephyri there is a desert region, which has large open meadows, which are uninhabited on account of the multitude of scorpions and spiders, which are called **Tetragnatha**, that is 'four jaws'.'"

The Latin word Tetragnatha in this passage of Aldrovandus (he, has of course, translated from the Greek of Aelian and Strabo) is a neuter plural. It is likely that Latreille got the word from Aldrovandus in this form and thought it was a feminine singular. As was his habit, Latreille created a French word Tétragnathe to go with the Latin Tetragnatha from Aldrovandus.
Attempts to identify the Tetragnathos of antiquity are not particularly convincing. Some argue from the four jaws that it has to be a Solpugid, others argue from its poisonous reputation that it must be a *Latrodectus*. Latreille obviously had no concern for this question. See Scholefield

(1972:374), Müller (1841-1870:190), Peck (1970:200), Gil Fernandez (1959:93), Keller (1913:468), Jones (1963:238), and Jones (1930:326).

*Teudis* O. Pickard-Cambridge 1896a:198 (Anyphaenidae)
Pickard-Cambridge does not favor us with a note about this peculiar name, but it clearly comes from his reading in Visigothic history. Theudis was a 6th century Visgothic king (Förstemann 1900:1409; Schönfeld 1911:234), whose reign in Spain lasted from 531-548 A.D. He is mentioned in Jordanes Historia Gothorum, and by Procopius in several works. The spelling Teudis actually occurs in some manuscripts of Jordanes. Theudis was a military commander under Theodoric the Ostrogoth, and engineered the death of Theodoric's predecessor King Amalaric, who was hated and despised by all. Theudis himself was murdered in 548. The etymology of the name makes it clear it is masculine, and it has been consistently so treated by subsequent authors.

(‡) *Teutana* Simon 1881a:161 (Theridiidae)
Proper name of an Illyrian queen who reigned from 230-228 B.C. after the death of her husband Agron. She was a colorful pirate queen and gave the Romans considerable trouble. She is more often known by the short form of her name, Teuta. Simon probably picked her up by cruising through the Latin dictionary, and meant no semantic connection with the spider. Levi (1957b:375) synonymized the genus under *Steatoda* Sundevall.

(‡) *Thalamia* Hentz 1850a:34 [= 1875:140] (Oecobiidae)
From the Greek three-ending adjective θαλαμιός -ή -όν 'belonging to the bedroom'. Hentz said, "small, forming a tubular dwelling of silk in the crevices of walls." Presumably it was first found in a bedroom. Simon (1892a:247) included Hentz's species *parietalis* under *Oecobius* Lucas, but Lehtinen (1967:269) revalidated the genus. It was again synonymized under *Oecobius* Lucas by Shear (1970:135).

*Thaleria* Tanasevitch 1984a:382 (Linyphiidae)
Named in honor of the arachnologist K. Thaler of Innsbruck, a specialist on the Linyphiidae. A species *Thaleria leechi* was described from Alaska by Eskov & Marusik (1992a:411).

*Thallumetus* Simon 1893b:241 (Dictynidae)
A proper name. Thallumetus was a slave of Cicero's friend, Titus Pomponius Atticus, who is mentioned in Cicero's letters to Atticus 5.12.2. Simon found this by surfing the dictionary, and there is no semantic connection.

*Thanatus* C.L. Koch 1837g:28 (Philodromidae)
This is the latinization of the Greek masculine noun θάνατος 'death'. There is no obvious semantic point to the name.

*Theonoe* Simon 1881a:130 (Theridiidae)
Greek feminine proper name. Theonoe was a sister of the prophet Calchas, official seer of the Greeks in the Trojan War. Her story, a complicated melodrama of mistaken identity, is told by Hyginus, Fabulae 190. Another Theonoe is the sister of the Egyptian king Theoclymenus in Euripides's Helen. The name means something like 'mind of god'. Hyginus, writing in Latin, retained the Greek form of this name. Bonnet (1959:4422) pointed out that the name was preoccupied in the Coleoptera (actually for an Hemipteran nymph), but nobody noticed, so there is no point in changing now. Bonnet argued that Article 34 [now 54] of the Code should be amended to read, "No genus name is to be rejected as a homonym unless it is used for another genus in the same major group of animals or in the same range." Insects and Arachnids would be different major groups. Levi & Levi (1962:29) pointed out that the putative antecedent *Theonoe* Phillipi 1865 has never been used, so the question is settled in favor of the spider.

*Theotima* Simon 1893a:439 (Ochyroceratidae)
A new name for *Theoclia* Simon, which was preoccupied. It comes from a feminized form of the Greek two-ending adjective θεότιμος -ον 'honored by the gods'.

*Theraphosa* Walckenaer 1805:2 (Theraphosidae)
Walckenaer proposed this name not as a genus, but as a suborder (although he did not call it a suborder, of course, but a "groupe" Les Théraphoses). It was Eichwald (1830:73) who first used it as a genus name (misspelled *Teraphosa*), although it is customarily credited to Thorell (1870b:161-163). Thorell argued that *Theraphosa* ought to replace the preoccupied *Mygale* (Cuvier in 1800 had used it for a mammal), and pointed out that it was wrong to assume Walckenaer's name was a neuter plural (i.e. as the name of a suborder), but claimed it was a true feminine singular. He further recognized Eichwald's use of it as a genus name. There is no straightforward way to analyze *Theraphosa* into a Greek or Latin word. Erichson's notion (Erichson 1845:13) that it is from Greek θήρ 'beast' and the verb ἀφοσιόω 'consecrate' is worthless. The name looks like it might be the feminine active participle *θηραφῶσα of a putative verb *θηραφάω, which, if it existed, might mean something like 'hunt' but this requires too much conjecture. Furthermore, I do not think that the Greek word θηράφιον, which occurs only once in all of Greek literature, in a text so obscure (Damocrates, a medical poet quoted in Galen) that it is inconceivable that Walckenaer would have known it, has anything to do with this genus name. I suggest it is a rhyming name, built on the analogy with *Gnaphosa* Latreille, which had been published the year before. Then we could connect it with the Greek feminine noun θήρα 'the hunt, the chase'. That is, the first element is a real Greek word meaning 'the hunt' and the second element is a rhyming morph formed by a misdivision of the morphemes of *Gnaphosa* (gna + phosa rather than the correct gnaph + osa). See *Gnaphosa*

*Theridion* Walckenaer 1805:72 (Theridiidae)
From the Greek neuter diminutive θηρίδιον 'tiny beast, animalculus' from the neuter noun θήριον 'wild beast'. Walckenaer used the Greek spelling of this word (1805) while Leach (1816:438) used the Latin form *Theridium*. In that great age when, as Dr. Johnson remarked, every gentleman got as much Greek as he had lace on his sleeves, this would have puzzled no one. They are the same word, and the difference in spelling would have excited no more comment than the difference between oe and œ. In the 20th century, when gentlemen with lace on their sleeves did not go into araneology, a stupendous donnybrook arose over which spelling was correct. Bonnet (1945:134; 1959:4431) championed *Theridium* by a more or less arbitrary rule which would leave nobody in doubt, while others championed the original spelling ruat coelum. The matter was settled legally in 1958, over Bonnet's staunch objections, when the original spelling *Theridion* was placed on the Official List of Generic Names in Zoology, name no. 1272. The later invalid emendations: *Theridium* Leach 1824, and *Theridio* Simon 1864, are listed in the Official List of Rejected and Invalid Generic Names in Zoology, names no. 1159, 1160. See Opinion 517 (ICZN 1958). See Levi & Levi 1962:30.

*Theridiosoma* O. Pickard-Cambridge 1879e:193 (Theridiosomatidae)
From the genus name *Theridion* Walckenaer (which see) and the Greek neuter noun σῶμα (gen.) σώματος 'body'. Hence, 'having the body of a *Theridion*'.

*Theridula* Emerton 1882:25 (Theridiidae)
Feminine diminutive formed on *Theridion* Walckenaer with the Latin diminutive suffix -ulus-a-um as in such words as adulescentulus 'little youth'. Purists might bristle at this word, first because it is part Greek and part Latin, and second because *Theridion* is already a diminutive, and the diminutive of a diminutive seems to be guilding the lily.

*Thiodina* Simon 1900b:392 (Salticidae)
An invented derivative with the diminutive suffix -ina from the Greek two-ending adjective θειώδης -ες 'sulphureous, yellow' from the neuter noun θεῖον 'brimstone, sulphur'. "Les femelles sont d'un jaune très pâle" [the females are very pale yellow] (Simon 1901a:457). Why did Simon not use the simpler form *Thiodes* which would have been available?

*Thomisus* Walckenaer 1805:28 (Thomisidae)
This name is not well formed, but follows the pattern of many of Walckenaer's creations. He did not always confine himself to authentic Greek morphology. He made several genus names by adding a noun ending to a present verb stem, ignoring the rules of strict noun formation, as *Sparassus*, *Attus*, *Sphasus* and *Drassus* (which see). There is a dialectical form of a Greek verb, given only by the ancient lexicographers Hesychius, and the Suda, θωμίσσω meaning "whip, scourge, tie, bind" (the Attic form is θωμίζω), and apparently this gave Walckenaer the basis for a noun. In describing the behavior of his new genus *Thomisus*, Walckenaer says they "spy out their prey stretching out a single thread of silk to catch it." It is from this single thread or "whip" that the name comes. Walckenaer created the noun *Thomisus* just as he created *Sphasus* from the verb σφάζω 'slaughter'. Thorell (1870b:183) essentially gave this explanation also, citing the dialectical form of the verb θωμίσσω found in Hesychius and the Suda. Perhaps we do not have to rely on Walckenaer's knowing the dialectical form, since the Attic form θωμίζω may have served just as well, given the example of *Sphasus* from σφάζω.

*Thymoites* Keyserling 1884a:161 (Theridiidae)
A Greek proper name Θυμοίτης (gen.) -ου, from the Iliad (3.146), the brother of King Priam. In Vergil Aeneid (2.32 ff.) he advised the Trojans to bring the Wooden Horse inside the walls.

*Thyreosthenius* Simon 1884r:744 (Linyphiidae)
From the Greek masculine noun θύρεος 'an oblong shield shaped like a door, the Roman scutum' plus the neuter noun σθένος (gen.) 'strength' plus an adjective ending -ιος. There actually exists a rare Greek adjective σθένιος -α -ον 'strong'. Hence, 'with a strong scutum'. But this genus does not have an abdominal scutum. The type species *biovatus* (O. Pickard-Cambridge 1875b:215) is remarkable for the strongly elevated cephalic lobe markedly divided into two egg-shaped lobes. The spider from head on might be imagined to be presenting a shield. See the illustration in Simon 1894a:622 fig. 693.

*Tibellus* SIMON 1875a:307 (Philodromidae)
This corresponds to no Greek or Latin word or name that I can discover. Simon regularly used proper names which he gleaned out of the dictionaries, and the closest proper name he might have gleaned is that of the Augustan Latin poet Tibullus (ca. 60-19 B.C.). While the type species is *oblongus* WALCKENAER 1802, Simon lists the species in the following order: *macellus* n.s. p.308; *propinquus* n.s. p.309; *oblongus* Walck. p.311; *oblongiusculus* n.s. p.312. It appears that the dictionary name Tibullus, became corrupted by association with the first species *macellus* and thus became *Tibellus*. Simon's spellings are often influenced by the way he pronounced Latin and Greek.

*Tibioplus* CHAMBERLIN & IVIE 1947b:51 (Linyphiidae)
From the Latin feminine noun tibia -ae 'shin' used here as the morphological term tibia plus an adjective-forming combining form of the Greek neuter noun ὅπλον 'tool, armament'. There are genuine Greek compounds in -όπλος such as πάνοπλος 'fully armed'. "Armed" is meant here in its technical arachnological sense as 'provided with spines or apophyses'. Hence, 'with tibia armed'. The genus has the male palpus "with a large dorsally directed spur on tibia." This is of course a hybrid word, half Latin and half Greek in origin, the kind we are advised against.

*Tidarren* CHAMBERLIN & IVIE 1934:4 (Theridiidae)
This is one of Chamberlin's hybrid Gosiute-Greek names, like *Astrosoga* (which see). The first part of the name is the Gosiute word tidutsi (combining form tida-) 'short, small' (Chamberlin 1911:46), and the second part is the Greek two-ending adjective ἄρρην -εν (gen.) ἄρρενος 'male, masculine'. Hence, the Gosiute-Greek compound name means 'small male'. A remarkable feature of this genus is "the extremely small size of the male as compared to the female."

Chamberlin and Ivie sometimes treated this name as masculine and sometimes as neuter. They listed species *fordum*, *mixtum*, *oblivium*, *passivum*, indicating neuter, but *minor*, indicating masculine (it should have been *minus* for the neuter). It seems clear that the authors meant this genus to be neuter, and simply made a grammatical mistake with *minor* instead of *minus*. While it may seem strange to give neuter gender to a word ending with the Greek adjective for 'male' this is perfectly permissable in ancient Greek. It is possible to apply the adjective 'male' to a noun which is grammatically feminine, for instance, ἄρρην σπορά 'male seed' (Euripides, Troades 503) or ἄρρην νηδύς 'male womb' (Euripides Bacchae 527, of Zeus giving birth to Dionysus from his thigh). The neuter expression τὸ ἄρρεν 'the male (thing)' is perfectly possible in Greek (Aeschylus Eumenides 737). Since *Tidarren* is differentiated by Chamberlin and Ivie from *Theridion*, and they have taken over the original neuter species names for eight previously described species which they include in their new genus, including the type species *Theridion fordum* KEYSERLING 1884a:23, they appear to have kept the neuter gender of their new name to imply 'small-male *Theridion*'. Bonnet (1959:4619) states that the genus is masculine, but without any reason. Perhaps he is appealing to his erroneus principle (1945:134) that all names ending in -n ought to be masculine. Levi (1957c:69-75) rightly chose neuter. The trouble-maker *minor* is now synonymized under *fordum* (Levi 1957c:73).

(‡) *Tigellinus* SIMON 1884r:838 (Linyphiidae)
A Latin proper name. Tofonius (or Ofonius) Tigellinus was the notorious prefect of the praetorian guard under the Roman emperor Nero. He took his own life in 69 A.D. after Nero's fall and suicide. Moviegoers may remember the affecting scene in "Quo Vadis" where Nero, played by Peter Ustinov, pretending to mourn the death of a friend, and affecting the pious gesture of preserving the imperial tears, speaks the unforgettable line, "Tigellinus, the weeping vial." This is regarded by some as the origin of the use of the glass vial for preservation of zoological specimens. Synonymized under *Walckenaeria* BLACKWALL by Millidge (1983:106).

*Tiso* SIMON 1884r:506 (Linyphiidae)
Etymology obscure. Simon treated it as masculine. Probably a proper name.

*Tinus* F.O. PICKARD-CAMBRIDGE 1901a:310 (Pisauridae)
The Latin masculine noun tinus -i is the name of a plant, the viburnum. There does not seem to be any particular point to that. Although it is not F.O. Pickard-Cambridge's usual habit to do so, he may have made up this word. The real etymology remains obscure.

*Titanoeca* THORELL 1870b:124 (Titanoecidae)
From the Greek three-ending adjective τίτανος -η -ον 'white' used as a masculine noun meaning 'white earth, gypsum, chalk, lime' plus the Greek masculine noun οἶκος 'house' The resulting compound is then made into a Latin feminine adjective. Thorell (1870b:124) explained that it is derived from τίτανος 'limestone' and οἰκέω 'dwell'. He further says p. 125):

> "I have found several examples of this species at Kissingen in Bavaria, but only females and young males, under stones in dry chalky declivities."

Hence, 'living in limestone'. The family Titanoecidae was established by Lehtinen (1967:380-382).

*Titiotus* SIMON 1897a:113 (Clubionidae)
This appears to be a made-up word. Simon (1884q:329) had created *Titurius* (Pisauridae) from a Roman proper name (an officer of Caesar's in the Gallic Wars), and *Titidius* (Thomisidae) (Simon 1895a:991-995) which appears to be a made-up word, perhaps inspired by the former. The concoction of *Titiotus* was perhaps inspired by the others.

*Tivyna* CHAMBERLIN 1948b:16 (Dictynidae)
Erected for seven species with type *floridana* BANKS 1904c:125, apparently equivalent to the *spathula* GERTSCH & DAVIS 1937 group of Chamberlin & Gertsch (1958:51). The second part of the name is a rhyming morph from the last two syllables of *Dictyna* (which see). The first part of the name Tiv-is a form of Gosiute ti-'small' (Chamberlin 1913:12; cf. Shoshoni ty-Crapo 1976:178). Chamberlin & Ivie (1932:19) gave a small variety of *Cybaeus reticulatus* SIMON 1886d:56 the name *tius* CHAMBERLIN & IVIE 1932, combining the Gosiute adjective with a Latin ending (cf. *Pimus pitus* CHAMBERLIN 1947:19), and described *Lycosa tivior* CHAMBERLIN & IVIE 1936a:17 combining the Gosiute adjective with the Latin comparative ending -ior (cf. *Tosyna pior* CHAMBERLIN & IVIE 1949:16). Compare the similarly formed genus names *Tosnyna, Phantyna, Varyna*. Hence, "small Dictynid". *Tivyna* was synonymized under *Dictyna* by Chamberlin & Gertsch (1958:48) and it was revalidated by Lehtinen (1967:271).

*Tmarus* SIMON 1875a:259 (Thomisidae)
The name of a mountain in Epirus (now Mt. Tomaros in northwestern Greece) mentioned in Vergil Eclogues 8.44. There is no semantic connection. This is a case of dictionary gleaning by which the pages of the lexicon are scanned for words available for new zoological names irrespective of any semantic connection. This is a frequent strategy of Simon's.

*Tmeticus* MENGE 1868:184 (Linyphiidae)
From the Greek three-ending adjective τμητικός -ή -όν 'cutting, piercing'. Menge translated this as "Scharfzahn" 'with sharp tooth' and explained that the fang has a sharp groove beneath ("die Klauen unten mit einer rinnenartigen Vertiefung versehen").

(‡) *Tortembolus* CROSBY in CHAMBERLIN 1925c:115 (Linyphiidae)
From the Latin adjective tortus -a -um 'twisted' and the araneological term embolus. Hence 'with twisted embolus' (see *Spirembolus*). Millidge (1980b:110) synonymized this genus under *Spirembolus* CHAMBERLIN, but said (114) that *Tortembolus approximatus* CHAMBERLIN 1949 "is an easily recognized species that does not belong in *Spirembolus*."

*Tortolena* CHAMBERLIN & IVIE 1941a:614 (Agelenidae)
See the general remarks on Agelenid names in -lena under *Hololena*. The morpheme Torto-is the combining form of the Latin adjective tortus 'twisted'. This appears to refer both to the double coiled embolus, and the spiral ridge or flange of the epigynum in this genus.

(‡) *Tosyna* CHAMBERLIN 1948b:16 (Dictynidae)
Established for ten North American species, "mostly small pale forms" with *apachea* CHAMBERLIN & IVIE 1935b:28 as type. The second part of the name is a rhyming morph from the last two syllables of *Dictyna* (which see). The first part of the name Tos- is from the Gosiute Indian word tosabit 'white, gray' with a combining form tosa as in tosawuda 'gray bear' (Chamberlin 1908b:79). Hence, 'pale Dictynid'. Compare Chamberlin's names *Tivyna, Phantyna, Varyna*. Synonymized under *Dictyna* by Chamberlin & Gertsch (1958:48).

*Trabea* SIMON 1876a:357 (Lycosidae)
Simon later changed the spelling to the incorrect *Trabaea* (1889a:338), which is not permitted by the Code. From the Latin feminine noun trabea 'a robe of state appropriate to augurs, kings, knights etc'. Simon clearly treated the name as feminine, so it cannot be from the masculine first declension proper name Trabea, an obscure Roman comic poet, unless, of course, Simon made a grammatical error. There seems to be no particular semantic connection. The spelling *Trabaea* is incorrect. Bonnet (1959:4663) discussed Simon's spelling change, dismissed the idea that it is a *lapsus calami* yet maintained the emendation, because Simon intended it (an invalid reason), and because many authors have adopted it. Brignoli repeated the emendation; Platnick correctly returned to the original spelling. The only American species, *aurantiaca* EMERTON 1885, is now included in *Trabeops* ROEWER (1959b:169).

*Trabeops* ROEWER (1959b:169) (Lycosidae)
Formed on the name *Trabea* by the addition of the nomenclatural suffix -ops.

*Trachelas* L. KOCH 1866a:2 (Clubionidae)
This is the Greek Τραχηλᾶς, an epithet of the emperor Constantine the Great (274-337 A.D.) meaning 'bull-necked' used by the 11th century Byzantine historian Georgius Cedrenus (1.472). The nickname is related to τράχηλος 'neck'. These masculine a-stems with nominative in -ᾶς form hypocoristics, or pet names, designating the possessor, maker or seller of something. So the meaning is roughly like 'Necky' (see Chantraine 1933: 31). The word was latinized by the 4th century A.D. Roman historian Aure-

lius Victor (Epit. 41) as a masculine first declension noun Trachala-ae, and will be found in this form in Lewis and Short's Latin dictionary. But L. Koch used the correct Byzantine Greek spelling (latinized, of course). How in the world did L. Koch know this? Was there a Byzantinist down the hall in Nürnberg, whom he could consult? The name refers to the fact that in this genus the cephalothorax is very convex.

The unfamiliar form of this name has occasioned some confusion. Bonnet's note (1959:4665) said:

> "Certain authors have considered **Trachelas** to be masculine, others feminine, and some have employed both Although L. Koch did not specify the gender of this word and the name of the first species (**minor**) does not indicate, it, there is no doubt that **Trachelas** formed on the Greek word τράχηλος is masculine."

The word is masculine, but not for the reason Bonnet gave. That is, it is not masculine because the related word τράχηλος is masculine, but because the name *Trachelas* itself has a masculine form.

There is a dispute about the authorship. Simon (1903a:1038) attributed the name to O. Pickard-Cambridge (1872a:256) and said:

> "The indication of the genus **Trachelas** by L. Koch in his table of genera of the Drassidae is not valid, because it is not accompanied by the citation of a species. The characters of the genus were given for the first time by O. Pickard-Cambridge in the specific description of **Trachelas minor**."

Bonnet (1959:4665) countered by saying:

> "The genus **Trachelas** L. Koch is perfectly valid, because there is a diagnosis in the table of determination and because the species minor taken by our author to designate this genus, is named by Canestrini and Pavesi since 1868, and they attribute it to Koch. Cambridge himself, in describing that species for the first time, did not fail to recognize it, and reproduced the figure of the male palp after the drawing by L. Koch. The attribution of this genus to O. Pickard-Cambridge (Simon 1903a, p. 1038) is not valid; moreover in 1932 our great arachnologist [i.e. Simon] re-established L. Koch as the author of the genus **Trachelas**."

### *Trachyzelotes* LOHMANDER 1944:13 (Gnaphosidae)

As a subgenus of *Zelotes* until elevated to a genus by Platnick & Murphy (1984:3). From the Greek three-ending adjective τραχύς -εῖα -ύ 'rough, rugged, jagged' plus the genus name *Zelotes* (which see). This genus can be distinguished by the presence of "a cluster of stiff setae on the anteromedian surface of the chelicerae" (Platnick & Murphy 1984:3). It is this "roughness" to which the name refers. Hence, 'the rough *Zelotes*'.

### *Traematosisis* BISHOP & CROSBY 1938:65 (Linyphiidae)

From the stem of the Greek neuter *noun* τρῆμα (gen.) -ατος 'hole, perforation' plus the genus name *Sisis* BISHOP & CROSBY. The authors said:

> "This genus is very close to **Sisis** but the male is provided with cephalic pits."

Hence, '*Sisis* with perforations'. Here is one of the very rare instances where Bonnet got something wrong (1959:4677). He assumed the word was formed as if it were from a (non-existent) Greek feminine noun *τρημάτωσις 'being pierced by holes'. by analogy with authentic Greek words like ἐκτυμπάνωσις 'swelling out like a drum'. So Bonnet either "corrected" it, thinking there was a misprint, or unconsciously transcribed it incorrectly as *Traematosis*. Strictly speaking, the name should be latinized as *Trematosisis*. The ae is a bit of hyperelegance, following the old German convention of writing long e as ae. See under *Megamyrmaekion*.

### *Trebacosa* DONDALE & REDNER 1981b:107 (Lycosidae)

The authors explain that the name is from the Latin one-ending adjective trebax-acis 'cunning, sly, crafty' plus the rhyming morph -osa characteristic of Lycosid genera. See under *Lycosa* and *Pardosa*. They further intend that the morph -osa means 'full of or prone to' as in Latin adjectives bellicosus-a-um 'warlike' or furiousus-a-um 'full of fury'. But in the antecedent names upon which such Lycosid genus names are modelled, namely, *Lycosa* and *Trochosa* the morph -osa represents the Greek feminine participle ending.

### *Trechalea* THORELL 1869:37 (Trechaleidae)

A new name for *Triclaria* C.L. KOCH (1848a:101) which was preoccupied. Thorell created the new name to sound as much like the old one as possible. Triklaria (τρικλαρία) is a personal epithet of the goddess Artemis used in Achaea (Pausanias 9.17.1; 7.22.1). Trechalea is from the feminine of the Greek three-ending adjective τρηχάλεος -α -ον 'rough, rugged, prickly'. Koch (1847i:101-102) described the genus as having long, fine "Stachelborsten" [spine-bristles](1847h:65) and this may be the inspiration for Thorell's choice of his substitute name. For the family status of Trechaleidae see Platnick (1993c:523).

### *Triaeris* SIMON 1891c:561 (Oonopidae)

This is a peculiar spelling of the Latin feminine noun trieris 'trireme, ship with three banks of oars'. The e in this word is long, and it is a 19th century European school tradition to spell long e with ae. The semantic connection, if any, is unclear.

### *Tricholathys* CHAMBERLIN & IVIE 1935b:26 (Dictynidae)

From the stem of the Greek feminine noun θρίξ (gen.) τριχός 'hair' and the genus name *Lathys* SIMON (which see). This genus has

> "a row of trichobothria along dorsal side of tarsus, metatarsus, and tibia. With a growth of long seta (sic), which stand at right angles to the surface; the setae underlaid with a growth of shorter and more appressed hairs."

Hence 'hair-*Lathys*'.

### *Trichopterna* KULCZYŃSKI *in* CHYZER & KULCZYŃSKI 1894:117 (Linyphiidae)

From the stem of the Greek feminine noun θρίξ (gen.) τριχός 'hair' and the feminine noun πτέρνη 'heel, lower part of the foot'. These are combined into a compound adjective analogous to real Greek adjectives such as ταχύπτερνος -ον 'with swift heels'. Hence, 'with hairy heels'. Probably a reference to the fact that in this genus metatarsus IV has a trichobothrium. The Nearctic species transferred to *Pelecopsis* SIMON by Crawford 1988

### *Trochosa* C.L. KOCH 1847a:95 (Lycosidae)

Formed on analogy with *Lycosa* LATREILLE as a rhyming companion genus. This name is authentic Greek, the feminine active present participle τροχῶσα of the Greek verb τροχάω (=τροχάζω) meaning 'run quickly'.

### *Trogloneta* SIMON 1922:200, 348 (Mysmenidae)

In the body of the article itself there is a *lapsus calami Troglonata*, which is corrected in the Table of Contents, p. 348. This happens also in the Histoire Naturelle des Araignées 2nd ed., where more than once the name is mispelled in the body of the work, but corrected in the indices. From the Greek masculine noun τρώγλη 'hole, cave' and the combining form -νήτης 'spinner' the compounded form of an agent noun from the verb νέω 'spin'. See under *Argyroneta*. Hence, 'cave-spinner'.

### *Troxochrus* SIMON 1884r:645 (Linyphiidae)

From the Greek masculine noun τρώξ (gen.) τρωγός 'gnawer, weevil' and the combining form -χρους 'color, hence 'weevil colored'. Westring (1851:42) described the species *Erigone scabricula* as having the "superficies thoracis concinne coriacea" [surface of the thorax thickly leathery]. The species name is a derivative of the Latin adjective scaber 'scaly, scurfy, rough'. Simon (1884r:647) characterized this species, now *Troxochrus scabriculus* as having "Céphalothorax brun-rouge foncé fortement chagriné [cephalothorax dark red-brown, strongly shagreened i.e. leathery with small round granulations]. The resemblance to the integument of a weevil inspired Simon's name. Strict grammatical formation, in which Simon was not always strong, would have required *Trogochrus. Although Crosby (1905a:303) reported an undescribed new species from Florida, there are apparently no Nearctic species.

### (‡) *Tuganobia* CHAMBERLIN 1933b:112 (Nesticidae)

The second part of the compound is a feminine form of the Greek adjectival suffix -βιος -α -ον 'living' as in such Greek compounds as ἀμαυρόβιος 'living in the dark'. The first element of the compound is the Gosiute Indian word meaning 'darkness' (Chamberlin 1913:19). The type locality is Potter Creek Cave (hence the type species *potteria*), Shasta Co. Calif, and the genus name, a Greek-Gosiute hybrid meaning 'living in darkness' is semantically appropriate for a cavernicole species. Chamberlin described the erigonid species *Ceraticelus tuganus* CHAMBERLIN 1949: 494 for its dark, "dusky" color and also (1948b:17) created the genus *Tugana* for *Scotolathys cavatica* BRYANT from Cuba, also a cave species (Bryant 1940:300). The genus name *Scotolathys* SIMON means roughly 'hiding in darkness'. So Chamberlin was creating a new genus name by using the Gosiute word for 'darkness' instead of the Greek word for 'darkness'. Transferred to the Nesticidae by Lehtinen & Saaristo (1980:51), synonymized under *Nesticus* by Gertsch (1984:16).

### *Tunagyna* CHAMBERLIN & IVIE 1933a:23 (Linyphiidae)

The second part of the compound refers to the epigynum, which is "conspicuously prolonged caudad in a median finger-like process". It is this feature which differentiates the females of this genus from the closely related *Oedothorax*. The first part of the name is from the Gosiute Indian language. Chamberlin had created the Pisaurid genus *Tunabo*, also from a Gosiute word (Chamberlin 1916:278), and there explained that the name is compounded from Gosiute tuna 'straight' and nabo 'mark', referring to the "sharply defined median longitudinal stripe over the entire length of the carapace and abdomen." From this explicit evidence.we may safely conclude that *Tunagyna* means 'with straight epigynum'. Chamberlin &

Ivie used *tunagyna* as a species name in *Helophora* (1943:6), for a species which has a very long epigynum.

## *Tusukuru* ESKOV 1993:50 (Linyphiidae)

In the language of the Ainu, the aboriginal people of Japan, Sakhalin and the Kurule Islands, the word means 'shaman' and is treated as a masculine.

## *Tutaibo* CHAMBERLIN 1916:237 (Linyphiidae)

Chamberlin's footnote on this page explicitly explained that tutaibo is the Gosiute word meaning 'a negro'. The type species from Peru *Tutaibo debilipes* has the carapace, sternum, legs, and abdomen entirely black. Millidge (1991:165) transferred the North American species *Ceratinopsis anglicana* (HENTZ 1850b:275) to the genus *Tutaibo*.

## *Tutelina* SIMON 1901a:554 (Salticidae)

This is a diminutive of the Latin feminine noun tutela 'guardianship, tutelage, keeper'. *Tutelina* (sometimes in ancient authors spelled *Tutilina*) when used as a proper name is a tutelary deity of Rome, or in some authors the tutelary deity of the grain harvest. There is no intended semantic connection; Simon, as usual, was gleaning useable names from the Latin dictionary.

## *Tylogonus* SIMON 1902d:384 (Salticidae)

The second member of the compound, -gonus, comes from the adjectival combining form of the feminine noun γωνία 'angle' which forms compounds in -γωνος -ον like ἑξάγωνος 'hexagon'. The first part of the name is from the Greek masculine noun τύλος -ου 'callus, knob, lump'. This name, then, means something like 'having a knob at the angle'. This interpretation is confirmed by the parallel Simon name, *Glyptogona* (1884j:326), a genus which is described as having the abdomen "découpé-lobé" that is, 'with incised lobes or angles' (1895a:866). By analogy then, *Tylogonus* means 'with swollen angles'. The name was apparently formed to reflect the shape of the chelicerae, which in the male are "long, vertical, narrow, flat, incurved and keeled on the interior side where they are armed with a pointed apophysis, directed directly inside or downward; below they appear strongly incurved and their internal apical angle is projecting and obtuse" (Simon 1903a:783). It is this "swollen" internal apical angle of the chelicerae which inspired the name.

## *Typhochraestus* SIMON 1884r:583 (Linyphiidae)

The second half is from the Greek three-ending adjective χρῆστος -η -ον 'good, useful' but the source of the first half is not altogether clear. It may be from the Greek masculine noun τύφος -ου 'fever, delusion, nonsense, humbug' or from the proper name of the giant Τύφως 'Typhon', the gigantic enemy of Zeus. In either case the meaning is enigmatic. The closest real Greek word is Πυθόχρηστος 'delivered by the Pythian god Apollo' where the second element has the technical meaning 'rendered in oracle'. Could it be that Simon misremembered this word from Greek tragedy, and changed it from *Pythochrestus* to *Typhochrestus* by metathesis? It is just the kind of mistake Simon might well have made. There would be no semantic rationale for such a word, but often Simon simply searched the dictionary for useable words without concern for semantic applicability. Simon originally spelled the name *Typhochrestus*, but on later pages (629, 663, 664) and consistently afterwards Simon spelled the name *Typhochraestus*. Bonnet's note (1959:4744 note 405) said:

> "Simon clearly wrote **Typhocrestus** in creating the name of the genus, but consequently, and consistently, spelled it **Typhocraestus**. Although he did not give any reason for this new spelling, it seems that it was intentional; moreover, the etymology of the term confirms the second spelling, for χρῆστος written with an eta should be transliterated with ae."

Bonnet is incorrect on this last point. The Greek eta should be transliterated with an e as Simon did in such names as *Menemerus*. However there was a schoolboy convention to distinguish long e from short e by writing long e with an ae. For instance Simon spelled the Latin word lêvis 'smooth' (with a long e) as laevis (e.g., Simon 1902d:384) to distinguish it from levis 'light in weight' (with a short e) Bonnet, Roewer, Brignoli and Platnick all accept *Typhochraestus*.

## (‡) *Ulesanis* L. KOCH 1872a:242 (Theridiidae)

Bonnet said (1959:4751):

> "I have not found any meaning for the term **Ulesanis**; with the species personata, L. Koch specified that it is feminine; there is no reason to alter that decision and Simon was wrong to treat it as masculine."

I have had no more luck than Bonnet. The name is not a Greek or Latin word, nor is it constructed like one. I find no place name, nor personal name that might be its source. Nothing in Koch's original description explains the name. The type species, *personata*, was collected from Upolu, one of the islands of West Samoa, but a large scale map of reveals no place name from which *Ulesanis* might be derived. Yet the Samoan language has an abundance of words beginning with ulu-or olu-, and o le is the article when the word begins the phrase. There is a Samoan word ulusina which means 'old man' and 'having gray hair'. Koch never travelled himself to Samoa. He found the spider in the Museum Godeffroy in Hamburg, but the collectors, namely, Herr Dr. Ed. Graeffe and Mr. Andrew Garret (L. Koch 1872a:viii) who sent the specimen may have reported a local name for the spider.

Simon's elegant name *Oronota* 'mountain back' is nicely descriptive of the bumpy abdomen of this genus (Simon 1871b:vii), but as Bonnet pointed out (1958:3210):

> "Simon after creating in 1871 the name **Oronota** for **Epeira paradoxa** Lucas -without remembering that he had done so certainly -recreated in 1873, for the same species a new name **Oroodes** ['mountain-shaped']! Then he discovered that his [1873] genus [Oroodes] had a [senior] synonym in **Ulesanis** L. Koch, which, created in 1872, had priority [i.e. over **Oroodes**] and therefore has been used constantly since then, without anyone thinking of **Oronota**, which actually had priority over **Ulesanis**. Today we will bring to bear the law of prescription to sink **Oronota**."

Cf. Code Art. 23.b. Levi (1955c) resurrected *Oronota*. The question has now become inconsequential by the fact that *Ulesanis* has been synonymized under *Phoroncidia* WESTWOOD 1835 (Levi & Levi 1962:31, Levi 1964d).

## *Uloborus* LATREILLE 1806:126 (Uloboridae)

From the Greek two-ending adjective οὐλοβόρος -ον 'with deadly bite' from the three-ending adjective οὖλος -η -ον 'dire, destructive, baneful, cruel' and the combining form -βόρος 'biting, eating'. The word οὐλοβόρος is from the Theriaka of Nicander (826) the 3rd century B.C. poem on noxious beasts and serpents, where it is an adjective referring to deadly snakes. The early authors frequently took over names from Nicander, Pliny, or Aristotle, often through the medium of Aldrovandus, without regard to the authentic ancient meaning. Cf. under *Dysdera*.

## *Ummidia* THORELL 1875a:102 (Ctenizidae)

This word is the feminine of a Roman family name Ummidius. Caius Ummidius Quadratus was a friend and pupil of the younger Pliny, and his son and grandsons are also known to us. Ummidia Quadratilla, the grandmother of Pliny's friend, died in 107 A.D. at the age of 80, and is memorialized in Pliny's letters (Epistulae 7.24). The stuffy Pliny was somewhat scandalized that she loved the performances of pantomime more than a Roman matron should have. It is doubtless from this source that Thorell would have known the name.

The original spelling was *Ummidia*. Roewer's *Umidia* (1942a:148) is an error corrected in Roewer (1954b:1702). Ausserer (1871a:145) established the genus *Pachylomerus* ('thick thighs') for *Aranea nidulans* FABRICIUS 1787 and *Pachyloscelis audouinii* LUCAS 1835b, but it was preoccupied in the Coleoptera. Strand realized this in 1934 and replaced it with *Pachylomerides*. Bonnet argued (1958:3288) that *Pachylomerus* ought to be retained by his principle (1945:110) that Article 34 (on homonymy; now Articles 54 and following) should allow homonymous names in different Classes. He further argued that if *Pachylomerus* is not maintained then the next available name was *Ummidia* THORELL. Roewer (1954b:1612-1613) synonymized *Pachylomerus* and *Pachylomerides* under *Ummidia*.

## *Urozelotes* MELLO-LEITÃO 1938a:111 (Gnaphosidae)

The second half is the genus name *Zelotes* (which see). The etymology of the first part is unknown. There is nothing in the original description which would tie the name to the Greek masculine noun οὖρος 'guardian, watcher' nor to the feminine noun οὐρά 'tail'. Mello-Leitão's intention remains a mystery.

## *Usofila* KEYSERLING in MARX 1891:35 (Telemidae)

Marx said:

> "Count Keyserling, who examined this interesting spider, named it **Usofila**, and intended to publish the description in the eighth number of his "Neue Spinnen aus Amerika" but was prevented by his untimely death. I received from his publishers, amongst his other manuscripts, also the description of **Usofila**, and present it herewith in translation with that of the only species U. **gracilis**."

This appears to be another one of Keyserling's Teutonic names, although I have not been able to find it explicitly. There are names beginning with Us-and names ending in -fila, but the exact combination eludes me.

Gertsch 1973a:154. considered *Usofila* a junior synonym of *Telema* SIMON (Roth 1994:167). See Platnick (1993c:124).

## *Varacosa* CHAMBERLIN & IVIE 1942a:36 (Lycosidae)

Erected as a subgenus of *Trochosa* C.L. KOCH, for those species wherein the crosspiece of the epigynum has the ends curved far forward, as in the type species of this subgenus *avara* KEYSERLING 1877a. The last element of the name is the rhyming morph -cosa which is customary for names in the Lycosidae, formed by misdivision of morphemes from the name *Lycosa*, and the first element of the name is formed by the rhyming principle and misdivision of morphemes from the last two syllables of the type species name *avara* KEYSERLING 1877a.

(‡) *Varyna* CHAMBERLIN 1948b:17 (Dictynidae)
Erected for three species with *mulegensis* CHAMBERLIN 1924b as type. The second part of the name is a rhyming morph from the last two syllables of *Dictyna* (which see). The first part of the name Var-is apparently taken from Chamberlin's own middle name, Vary. Compare *Tosnyna, Phantyna, Tivyna*. Synonymized under *Dictyna* by Chamberlin & Gertsch (1958: 48).

*Vermontia* MILLIDGE 1984a:166 (Linyphiidae)
"Derived from the name of the state (Vermont, U.S.A.) in which the type species was discovered." The type species is *Tmeticus thoracicus* EMERTON 1913a, collected from Mt. Mansfield, Vermont.

*Verrucosa* MCCOOK 1888a:79; 1894:199 (Araneidae)
McCook created this genus for *Epeira verrucosa* HENTZ (1835:551) [*nomen nudum*]; 1850a:19 = 1875:121), which he recognized was the same as *Epeira arenata* WALCKENAER 1842. The Latin adjective verrucosus-a-um means 'warty' and Hentz clearly was referring to the "six or eight tubercles more or less prominent at the apex" of the abdomen. So McCook created a new genus name out of Hentz's specific name, and used Walckenaer's specific name, which antedated Hentz's (1850a, since the 1835 listing was a *nomen nudum*). McCook inadvertently misspelled the name as *Verucosa* in 1888a, and corrected it in 1894.

*Wabasso* MILLIDGE 1984a:149 (Linyphiidae)
Millidge explained, "*Wabasso*, the North in 'Song of Hiawatha' by Longfellow. The name is masculine." It refers to the northern distribution of the genus. One is immensely grateful for this kind of etymological openness. If only the Peckhams or Octavius Pickard-Cambridge had been this forthcoming. But then it takes some of the fun out of the detective work. The lines to which Millidge makes reference (Song of Hiawatha II. 5-7) read:

> *From the regions of the Northwind,*
> *From the kingdom of Wabasso,*
> *From the land of the White Rabbit."*

It is understandable that Millidge would conclude that *Wabasso* meant 'north' but in fact it means 'rabbit'. This is made clear in another passage (Song of Hiawatha X. 248):

> *And the rabbit, the Wabasso,*
> *Scampered from the path before them.*

Bloomfield (Eastern Ojibwa, Ann Arbor, 1956, p. 267) glossed the Ojibwa word wa.po.so. as 'rabbit'.

*Wadotes* CHAMBERLIN 1925b:38, 122 (Agelenidae)
Erected for those forms originally placed in *Coelotes* BLACKWALL, which have two teeth on the retromargin of the fang furrow. The second element -otes is a rhyming morph to associate it with *Coelotes*. My guess about the first element, judging from Chamberlin's nomenclatural habits, is that it is an abbreviation of a locality name. While the designated genotype is dixiensis Chamberlin, the oldest species is calcaratus (KEYSERLING 1887b: 470), and Keyserling's species was collected in Washington, D.C. Thus the type locality of the oldest species of *Wadotes* is Washington, D.C., and I conjecture that Chamberlin formed the first part of this new name from Wa (for Washington) and d (for D.C.) to distinguish it from the state of Washington). Add to this the fact that this new genus is first published in the Proceedings of the Biological Society of Washington (D.C.) and we can also credit Chamberlin with a pretty compliment to that society.

*Wagneriana* F.O. PICKARD-CAMBRIDGE 1904:497 (Araneidae)
A replacement name for *Wagneria* MCCOOK (1894:203) which, as Pickard-Cambridge points out, is preoccupied five times in zoology. McCook erected the genus to receive *Epeira tauricornis* O. PICKARD-CAMBRIDGE 1889d, and named the genus for Professor Waldemar Wagner of Moscow (Vladimir Alexandrovich Wagner, 1849-1934). See Bonnet 1945:49 for a biographical sketch.

(‡) *Wala* KEYSERLING 1885:516 (Salticidae)
Keyserling was in the habit of forming genus names from old Teutonic personal names. There are at least 19 Keyserling names, such as *Ebo, Odo, Eilica* etc. that fall into this category. *Wala* is a medieval feminine Germanic name frequent in the 9th century (Förstemann 1900:1515). This may be a reference to Erde in Wagner's Siegfried. Wotan at the beginning of the third act addresses her as *Wala*. For the synonymy see the discussion under *Hentzia*, which has priority.

*Walckenaeria* BLACKWALL 1833a:105 (Linyphiidae)
Named for Baron Charles Athanasie Walckenaer (1771-1852), the great, polymath and arachnologist "grand érudit, philosophe, naturaliste, géographe, romancier, artiste", said Bonnet (1945:31). Blackwall later (1841b: 221 note) said explicitly "This generic name, which, through my own inadvertency, has hitherto been printed *Walckenaria* is now corrected [to *Walckenaera*]. By Blackwall's own evidence there was a *lapsus calami*, which should mean that Balckwall's correction should prevail. Bonnet (1956:4807), in the interest of stability, chose the emendation *Walckenaera* over the original spelling, and was followed by Brignoli (1983c: 363), but Roewer(1942a:668), Millidge (1983:105), and Platnick (2005) returned to the original spelling, on the grounds that Blackwall's emendation was unjustified.

(‡) *Walmus* CHAMBERLIN 1947:10 (Amaurobiidae)
Etymology unknown. Perhaps from a proper name. Synonymized under *Amaurobius* C.L. KOCH by Leech (1972:70).

*Wamba* O. PICKARD-CAMBRIDGE 1896a:190 (Theridiidae)
A masculine proper name, as Pickard-Cambridge informs us in his note. Wamba (died 683) was a king of the Visigoths in Spain. Pickard-Cambridge took this and other Visgothic names from the dramatis personae of Robert Southey's long poem "Roderick, the Last of the Goths" published in 1814. See under *Florinda*.

(‡) *Wideria* SIMON 1864:196 (Linyphiidae)
This name is formed in honor of the mysterious figure Wider. His career and even his first name was unknown to Bonnet (1945:53). Wider described 59 species in the Zoologischen Miscellen of August Emanuel Reuss. In Reuss's Vorwort (1834:26) he refered to Wider as "Herr Oberpfarrer F. Wider, in Beerfelden im Odenwalde" that is Rector of the parish of Beerfelden, a village about 40 km east of Mannheim in Hesse in the Oden Forest.
Simon (1864) erected *Wideria* as a subgenus of *Microphantus* C.L. KOCH (sic, Simon's *lapsus calami* for *Micryphantes*), and spelled it originally Viderius but on a later page (477) changed it to *Widerius*. Still later (1884r: 799) he changed it again, without comment, to *Wideria*. Bonnet (1959: 4796 note 11) argued for maintaining the latest form *Wideria*, despite the fact that F.O. Pickard-Cambridge (1901c:180, 1902d:10) advocated a return to the original *Viderius*, and F.P. Smith (1905a:245) championed *Widerius*. Roewer and Brignoli printed *Wideria*. Synonymized under *Walckenaeria* BLACKWALL by Wunderlich (1972a:391) followed by Millidge (1983:106).

(‡) *Willibaldia* KEYSERLING 1886b:277 (Linyphiidae)
From the German proper name Willibald. Keyserling gives no hint whether this refers to any particular person. It may well be another of Keyserling's medieval Teutonic names (Förstemann 1900:1594). Synonymized under *Porrhomma* SIMON by Thaler (1968:385).

*Willisus* ROTH 1981:102 (Hahniidae)
Named in honor of Willis J. Gertsch with the type species *Willisus gertschi* ROTH 1981 appropriately published in the Festschrift for Willis Gertsch on the occasion of his 75th birthday.

*Wixia* O. PICKARD-CAMBRIDGE 1882c:483 (Araneidae)
A proper name, as Pickard-Cambridge tells us, without specifying who it is. Simon (1895a:1.819) emended to *Vixia*, on the grounds that Latin has no w. Such an emendation is not justified by the Code. See Bonnet 1959: 4787 on Simon's policy of changing all W's to V's. The name is presumably Wix. Bonnet (1959:4827) worried whether this should be a masculine like the names of Gothic chieftains, *Witica*, and *Wulfila* and so forth, that Pickard-Cambridge was enamored of. But Latin masculine a-stems would never end in -ia so we are safe in concluding this is a feminine of the derived adjective wixius-a-um 'belonging to or referring to Mr. Wix'. The Dictionary of National Biography lists a Rev. Samuel Wix 1771-1861, a controversialist of high church views, and Fellow of the Royal Society, but there is no hint that Pickard-Cambridge knew him.

*Wubana* CHAMBERLIN 1919a:252 (Linyphiidae)
This genus is described as having the "pars cephalica rising behind the eye area: piligerous in a median longitudinal stripe with a brush of longer setae just behind the eyes." This is one of Chamberlin's Gosiute Indian names, made up of the noun wu 'hair' (Chamberlin 1908b:88) and a form pana/bana meaning 'on top, high, front surface'. Chamberlin himself does not give us the second element, but it can be deduced from the comparative evidence of other closely related Uto-Aztecan languages (Miller 1967: 41; Crapo 1976:67). Chamberlin (1911:46) did give us the affix -bo meaning 'on top'. If I have correctly analyzed the second element the name means 'having hair on top'.

*Wulfila* O. PICKARD-CAMBRIDGE 1895a:159 (Anyphaenidae)
Pickard-Cambridge's note said, "Nom. propr.: a Gothic Chief." The famous Wulfila or Ulfilas (311?-381 A.D.) was missionary bishop to the Visigoths, and translated the Bible into Gothic. Its text, preserved in the famous and unique Codex Argenteus in Uppsala, is the earliest specimen of a Germanic language. Pickard-Cambridge seems to have been on a roll, since he used the names Ervig, a Gothic noble (1895a:151), Favila (1895a: 156), a Gothic duke, Witica (1895a:160), a Gothic prelate, and Enrico (1895a:157), who he claims was a Gothic ruler. The gender of the name is properly masculine (an a-stem Latin masculine), and Pickard-Cambridge so treated it by using masculine species names. Simon (1897a:95) spelled

it *Wulfila* but treated it as feminine. Later (1897a:102, 103) he changed the spelling to *Vulfila* also treated as feminine. F.O. Pickard-Cambridge (1900: 105) has the correct masculine species names and the correct spelling *Wulfila*. Simon's error has unfortunately been widely followed.

### *Xeropigo* O. Pickard-Cambridge 1882c:423 (Corinnidae)
The author succinctly says that this is a proper name. The closest I can find are two villages in Greece, Xeropyrgo in Boeotia and Xeropigadho in Phocis.

### *Xysticus* C.L. Koch in Herrich-Schäffer 1835:Heft 129, pl. 16, 17 (Thomisidae)
The word occurs in Latin as the adjective xysticus-a-um, a loan word from Greek. The Greek word is a three-ending adjective ξυστικός -ή -όν meaning 'athlete' or 'pertaining to a ξυστός 'a walking place or gymnasium'. These are related to the verb ξύω 'to scrape'. The point is that athletes in antiquity, before the invention of the shower, would clean by rubbing themselves with olive oil and scraping it off along with the attendant grime by means of an implement called a strigil in Latin and a ψήκτρα or στλεγγίς in Greek. In the Vatican Museum there is a famous Roman copy of Lysippos's statue of an athlete cleaning himself, called the the Apoxyomenos ('Ἀποξυόμενος 'The Scraper'), which illustrates the process. Simon's spelling (1864:427) *Xystica*, the corresponding feminine, is an unjustified emendation in accord with the young Simon's theory that spider names ought to be feminine to agree with *Aranea*.

### *Yabisi* Rheims & Brescovit 2004a: 233 (Hersiliidae)
Erected for the type *Tama habanensis* Franganillo Balboa 1936b. The authors say, "The generic name is a noun in apposition taken from the Taino Indian language, which means 'tree' and refers to the most common habitat of these species." The Taino are the extinct peoples of the Greater Antilles and Bahamas.

### *Yaginumena* Yoshida 2002a:11(Theridiidae)
Named in honor of the great Japanese arachnologist Takeo Yaginuma. Erected for the type *Dipoena castrata* Bösenberg & Strand 1906a.

### *Yorima* Chamberlin & Ivie 1942a:19 (Dictynidae)
This may be a place name, but the type species *Chorizomma sequoiae* Chamberlin & Ivie 1937 was collected at Felton, California. Since this genus was erected for species once placed in *Chorizomma* Simon 1872a and is close to *Blabomma* it is conceivable that the -ma is a rhyming morph. But I find no clues to the source of Yori-. Transferred to the Dictynidae by Lehtinen (1967:275).

### (‡) *Yukon* Chamberlin & Ivie 1947b:52 (Linyphiidae)
Erected for *Yukon majesticum* Chamberlin & Ivie 1947b:52 from Alaska. The name is simply the unmodified name of the Alaskan river regarded as a neuter noun. Synonymized under *Scotinotylus* Simon by Millidge (1981a:168).

### *Zanomys* Chamberlin 1948b:17 (Amaurobiidae)
The second element -mys is the Greek noun (nominative) μῦς "mouse." Compare Chamberlin's name *Pagomys*. The first element may be the combining form of the Doric and Ionic dialect variant for the name Zeus, that is, Ζάν (genitive) Ζανός. I can discover no rationale for this choice.

### *Zelotes* Gistel 1848:xi (Gnaphosidae)
A replacement name for *Melanophora* C.L. Koch in Herrich-Schäffer 1833 (Heft. 120, pl.20, 21) which was preoccupied. L. Koch, unaware of this, replaced it in 1872 (1872a:139) with *Prosthesima*, which was used widely until Waterhouse resuscitated *Zelotes* (1902:399), and was followed by Simon and Banks. *Melanophora* would mean 'black-bearing'; *Prosthesima*, a defectively formed word, would mean 'addition' perhaps a reference to the substitution of the name. *Zelotes* is from the Greek masculine a-stem agent noun ζηλωτής -ου 'emulator, zealous admirer'. It is the nickname of the apostle Simon the Zealot (Luke 6:15; Acts 1:13).

### *Zilla* C.L. Koch in Herrich-Schäffer 1834: heft 124, pl.11 (Araneidae)
Some North American *Zygiella* were described under *Zilla*. This is an Old Testament proper name. Zillah, as it is spelled in the English Bible, was the wife of the patriarch Lamech (Genesis 4:19,22). In the Vulgate her name is spelled Sella, and in Luther's German Bible it is spelled Zilla, the obvious source for Koch's name. No North American species any longer remain in *Zilla*.

### *Zodarion* Walckenear in Audouin 1826:135 (Zodariidae)
This is Walckenaer's replacement name for *Enyo* Audouin (1826:135) which is preoccupied. Enyo is the name of one of the Graiae, and also a name for a Greek goddess of war. Bonnet says (1959:4970 note 60):

> *"Savigny and Audouin in creating the genus* **Enyo** *say that it is that 'which M. Walckenaer recently established under the name* **Zodarion***; logically they ought to have named their genus* **Zodarion***, instead of imagining another term (***Enyo***) which moreover was preoccupied;*
> *this suppresses all discussion of the priority of* **Zodarion** *which, in my opinion, would have been substantial. In any case, it has not been possible for me to find in the works of Walckenaer any indication of the name before 1837."*

The name is the Greek diminutive neuter noun ζῳδάριον 'animalcule, grub'. Walckenaer used the plural *Zodariones* (1837:635) as if the name were an n-stem masculine, but that is a grammatical error. The name is neuter. It is possible that Walckenaer's *Zodariones* is French not Greek, in which case he has not committed a grammatical error. Thorell (1869:107) "corrected" the spelling to the latinized form *Zodarium*, an unjustified emendation. He was followed by Bonnet (1959:4970).

### *Zora* C.L. Koch 1847a:102 (Zoridae)
A substitute name for *Lycaena* Sundevall (1833a:265) which was proccupied in the Lepidoptera. Sundevall's name means 'female wolf' as he himself explained "on account of the similarity, at least with respect to the eye arrangement to wolf spiders (*Lycosa*, *Dolomedes*)." *Zora* appears to be the feminine of the Greek three-ending adjective ζωρός -ά -όν 'pure, sheer, unmixed'. There is no obvious semantic point to the name, and neither Sundevall nor Koch gave any helpful hint. Thorell (1870b:140) interpreted the Greek adjective to mean 'strong, fiery' but that is a mistake. The adjective is used for undiluted wine, hence strong drink, or stiff drink, but not for strong or fiery behavior. The spider name may be taken directly from the feminine proper forename Zora, and have nothing to do semantically with the Greek adjective meaning 'pure' (the proper name itself, of course, may derive from the Greek adjective.)

Bonnet (1957a:2580) pointed out that Sundevall the next year (1833b) printed the name of the genus as *Lycodia* instead of *Lycaena*, "doubtless a typographical error" since *Lycaena* is preoccupied, the next available name would be *Lycodia* irrespective of the fact that it resulted from a typographical error, and not *Zora*. Bonnet argued that *Lycodia* has not been used for a hundred years, and *Zora* universally employed in that time.

Sundevall (1833b:22) did not explicitly give the synonymy, nor did he list the species (*spinimana*). We must deduce that Sundevall meant to substitute *Lycodia* for *Lycaena* from the order of the genera in the Conspectus Arachnidum, where *Lycodia* goes between *Argyroneta* and *Drassus*, which is where *Lycaena* was inserted in the appendix of Svenska Spindlarmes (1833a:265). There is no real evidence that *Lycodia* resulted from a typographical error, as Bonnet argued. He apparently assumed Sundevall meant to print *Lycodes*, or *Lycoides*, but that guess is insufficient to demonstrate that there was really a typographical error.

### *Zornella* Jackson 1932a:106 (Linyphiidae)
Etymology unknown. The author gives no hint about the origin. It is a feminine diminutive on the analogy of such names as *Dugesiella*, *Batesiella*, *Spencerella* (all Pocock names) and *Simonella*, *Keyserlinigiella* (Peckham names). On the basis of this pattern I suggest that *Zornella* is built upon the Scandinavian surname Zorn, although I have not been able to find a likely honorand. The fact that this genus is described from Lapland, and that Jackson expresses gratitude toward several Scandinavian workers, lends credence to the suggestion.

### *Zorocrates* Simon 1888b:211 (Zorocratidae)
The name is invented as a combination of the genus *Zora* with Greek names in -crates like Socrates, and Hippocrates. It is an s-stem masculine. The family Zorocratidae originally proposed by F. Dahl (1913) was revived by Griswold et al. (1999:59) to include several genera formerly misplaced in the Miturgidae and Tengellidae.

### *Zoropsis* Simon 1878a:325 (Zoropsidae)
Simon created the genus *Zoropsis* to accommodate part of *Zora* C.L. Koch, and Bertkau (1882:337) created the family Zoropsidae (he spelled it Zoropsididae and Simon emended it in 1890).

### *Zosis* Walckenaer 1841:231(Uloboridae)
From the Greek feminine noun ζῶσις (gen.) ζώσεως 'girding, cincture'. Considered a junior synonym of *Uloborus* Latreille, but revalidated by Lehtinen (1967:277). If the etymology is correct, it is clearly feminine, and was so treated by Simon, Bonnet, and Taczanowski. Opell (1979:508) and Lehtinen (1967:277) consider it masculine. Walckenaer's type species was caraïbe and his specimen was from Martinique, so the species name is a French word, still in French form, meaning 'Carib'. Simon (1864:247) first latinized it to *caraiba*. Later Walckenaer put his species in *Uloborus* (1841:231) with the name *Uloborus zosis*, where the species name is clearly the old genus name in apposition. If the etymology above is correct, the gender is feminine; if it is an arbitrary combination of letters it is arguable that it may be masculine. Unfortunately, there is nothing in the descriptions which cries out to be interpreted as a "cincture" to confirm the etymology.

### *Zygiella* F.O. Pickard-Cambridge 1902b:15 (Araneidae)
A new name for *Zygia* C.L. Koch in Herrich-Schäffer 1834: Heft 123, p. 17, 19) preoccupied in the Coleoptera. *Zygia* is from the feminine of the Greek three-ending adjective ζύγιος -α -ον 'belonging to the yoke,

draught animal'. As a proper name Ζύγια is an epithet of the goddess Hera, as patroness of marriage (conceived of as a "yoking"). This has nothing to do with another Greek feminine noun ζυγία 'maple tree'. Menge (1866:77) desperately suggested it referred to the long rein-like (zügelähnlich) palp of the male, but it is a long leap from yoke to reins. It is unclear which of these possibilities Koch intended. My guess is that he is using the proper name, the epithet of Hera, since it is a custom of his to employ them. Consider the genera *Alcmena, Cocalus, Corinna, Janus*. Pickard-Cambridge constructed a feminine diminutive in -ella to substitute for Koch's preoccupied name.

### *Zygoballus* PECKHAM & PECKHAM 1885b:81 (Salticidae)

This would appear to be from the Greek neuter noun ζυγόν 'yoke' and the genus name *Ballus* C.L. KOCH 1850 (which see). But the point of such a combination is not instantly obvious. I think that we can make something of the the point that the chelicerae of the males in this genus have two apophyses, one strong and vertical, coming down from the inner proximal end of each chelicera, and a heavier, hammer-headed one on the outer side (Peckham & Peckham 1909:578). The fact that the Greek name for the hammer-headed shark (Aristotle Hist. Anim. 506b10) is ζύγαινα suggests that this may be the source of the name. I can imagine the Peckhams wanting a Greek word meaning 'hammer-headed' and finding it by looking up what Aristotle called the hammer-headed shark. They then cobbled together a name. So, I suggest this genus name means 'the Salticid, close to *Ballus*, whose males have a hammer-headed apophysis on the chelicerae'.

### *Zygottus* CHAMBERLIN 1949:560 (Linyphiidae)

The second part of the word, -ottus is a rhyming morph to link this genus with *Sisicottus* BISHOP & CROSBY (which see), as Chamberlin makes clear, "A genus in some respects suggesting *Sisicottus*." The first part should be from the Greek neuter noun ζυγόν 'yoke' but there seems to be no semantic point to that, unless Chamberlin meant to indicate that this genus was "linked or yoked" to *Sisicottus*. Compare *Cybaeozyga* CHAMBERLIN.

# Appendix
# Pronunciation guide for taxa

In the phonetic key to the taxonomic names (derived from Beechhold 2000), the primary stress is indicated with capital letters. The sounds are as described below. Pronunciation is provided for all family names and names of common genera. Common genera are defined either as those containing five or more North American species as listed in "Spiders of North America" at (Bradley 2002) or those containing widespread and commonly encountered species such as *Pholcus phalangioides*. Common names from Breene (2003).

| | | |
|---|---|---|
| a | = | a of "bay" |
| aa | = | a of "bat" |
| ah | = | a of "bar" |
| au | = | ou of "bout" |
| aw | = | ou of "bought" |
| ch | = | ch of "chug" |
| e | = | e of "bet" |
| ee | = | ee of "bee" |
| ey | = | i of "bite" |
| f | = | f or ph of "fast" or "phone" |
| g | = | g of "bug" |
| i | = | i of "bit" |
| j | = | j of "jug" |
| k | = | c of "club" |
| o | = | o of "oh" |
| oo | = | oo of "shoo" |
| s | = | s of "silly" |
| th | = | th of "thug" |
| u | = | u of "bull" |
| uh | = | u of "but" |
| y + vowel | = | y of "you" |
| z | = | s or z of "rings" or "zoo" |

Pronunciations vary among arachnologists. The strictly correct pronunciations can be found in John Henry Comstock's The Spider Book (1940). However, these strictly correct pronunciations have, in some cases, been displaced by common usage. For instance, the correct pronunciation of the genus *Phidippus* is fid-DIP-uhs, but many workers have come to use FID-i-puhs. Particularly troublesome are the names based on *Tetragnatha*. The word comes ultimately from Aristotle, meaning "having four jaws," and he probably meant it to apply to a solifugid (another type of arachnid). The correct pronunciation is te-TRAAG-nuh-thuh, and this is properly indicated in Comstock. But that pronunciation has generally fallen out of use in favor of the common pronunciation te-truhg-NA-thuh. Likewise, *Pachygnatha* and *Enoplognatha* have been changed from paa-KIG-nuh-thu to paak-ig-NA-thuh, and from en-o-PLOG-nuh-thu to en-o-plog-NA-thuh. For *Phrurolithus* Comstock recommends the virtually unpronounceable fru-RO-lith-uhs, which has yielded to the common fru-ro-LITH-uhs. Therefore, in some instances, we have given two pronunciations: the strictly correct one followed by the one in common use.

## A

*Achaearanea* .................... aa-kee-ah-RA-nee-uh
AGELENIDAE .................... a-je-LEN-i-dee
*Agelenopsis* .................... a-je-len-AWP-sis
*Agroeca* .................... aag-ro-EE-kuh
*Allocosa* .................... aa-lo-KO-suh
*Alopecosa* .................... aa-lo-pe-KO-suh
AMAUROBIIDAE .................... a-mor-o-BEE-i-dee
*Amaurobius* .................... a-mor-O-bee-uhs
AMPHINECTIDAE .................... am-fin-EK-ti-dee
ANAPIDAE .................... aanAAP-i-dee
ANTRODIAETIDAE .................... aan-tro-dey-EET-i-dee
*Anyphaena* .................... aan-i-FEE-nuh
ANYPHAENIDAE .................... aan-i-FEE-ni-dee
*Appaleptoneta* .................... aap-aal-ep-to-NE-tuh
*Araneae* .................... a-RA-nee-ee
ARANEIDAE .................... aar-a-NEE-i-dee
*Araneus* .................... uh-RA-nee-uhs
*Arctosa* .................... ahrk-TO-sah
*Arcuphantes* .................... ahr-kyoo-FAHN-teez
*Argiope* .................... ahr-JEY-o-pee
*Argyrodes* .................... ahr-ji-RO-deez

## B

*Bathyphantes* .................... baath-i-FAAN-tees
*Blabomma* .................... blaa-BO-muh

## C

*Calilena* .................... kaa-li-LEE-nuh
*Calileptoneta* .................... kaa-li-lep-to-NET-uh
*Callilepis* .................... kaa-li-LEP-is
*Callobius* .................... kaa-LO-bee-uhs
*Calymmaria* .................... kaa-li-MAH-ree-uh
CAPONIIDAE .................... kaa-po-NEE-i-dee
*Castianeira* .................... caas-tee-uh-NEE-ruh
*Centromerus* .................... sen-tro-MER-uhs
*Ceraticelus* .................... ser-aat-I-sel-uhs
*Ceratinella* .................... ser-aat-in-EL-uh
*Ceratinops* .................... ser-AAT-in-ahps
*Ceratinopsis* .................... ser-aat-in-AHP-sis
*Cesonia* .................... ses-O-nee-uh
*Chrosiothes* .................... kro-see-O-theez
*Cicurina* .................... sik-uhr-EY-nuh
*Clubiona* .................... kluh-bee-O-nuh
CLUBIONIDAE .................... kluh-bee-AWN-i-dee
*Coleosoma* .................... ko-lee-o-SO-muh
*Coras* .................... KOR-uhs
CORINNIDAE .................... kor-IN-i-dee
CTENIDAE .................... TEN-i-dee
CTENIZIDAE .................... ten-IZ-i-dee
CYBAEIDAE .................... sey-BEE-i-dee
*Cybaeopsis* .................... sey-bee-AHP-suhs
*Cybaeus* .................... sey-BEE-uhs
*Cyclosa* .................... sey-KLO-suh
CYRTAUCHENIIDAE .................... sir-tawk-en-EE-i-dee

## D

DEINOPIDAE .................... dey-NO-pi-dee
DESIDAE .................... DES-i-dee
*Dictyna* .................... dik-TEY-nuh
DICTYNIDAE .................... dik-TIN-i-dee
*Diguetia* .................... di-GWET-ee-uh
DIGUETIDAE .................... di-GWET-i-dee
*Diplocephalus* .................... dip-lo-SEF-uh-luhs
DIPLURIDAE .................... dey-PLUR-i-dee
*Dipoena* .................... dip-EE-nuh
*Disembolus* .................... dis-EM-bo-luhs
*Dolomedes* .................... do-lo-MEE-dees
*Drassinella* .................... draas-in-EL-uh
*Drassodes* .................... draas-O-dees
*Drassyllus* .................... draas-IL-uhs
*Dysdera* .................... diz-DER-uh
DYSDERIDAE .................... diz-DER-i-dee

## E

*Ebo* .................... EE-bo
*Eidmannella* .................... eyd-maan-EL-uh
*Elaver* .................... el-AAV-uhr
*Emblyna* .................... em-BLEY-nuh
*Enoplognatha* .................... en-o-PLOG-nuh-thuh *or* en-o-plo-GNA-thuh
*Eperigone* .................... ep-er-IG-o-nee
*Erigone* .................... er-IG-o-nee
*Eris* .................... ER-is
*Eulaira* .................... yoo-LEY-ruh
*Euryopis* .................... yoo-ree-O-pis
*Eustala* .................... YOO-staal-uh

## F

FILISTATIDAE .................... fil-is-TAH-ti-dee
*Floricomus* .................... flor-ee-CO-muhs
*Frontinella* .................... frahn-tin-EL-uh

## G

*Gasteracantha* .................... gaas-ter-i-CAAN-thuh
*Geolycosa* .................... jee-o-ley-CO-suh
*Gladicosa* .................... glaad-i-CO-suh
*Gnaphosa* .................... naa-FO-suh
GNAPHOSIDAE .................... naa-FO-si-dee
*Grammonota* .................... graam-o-NO-tuh

## H

*Habrocestum* .................... haa-bro-SES-tuhm
*Habronattus* .................... haa-bro-NAA-tuhs
*Hahnia* .................... HAW-nee-uh
HAHNIIDAE .................... haw-NEE-i-dee
*Haplodrassus* .................... haap-lo-DRAA-suhs
*Hentzia* .................... HENT-zee-uh
*Herpyllus* .................... hur-PIL-uhs
HERSILIIDAE .................... her-sil-EE-i-dee
*Heteropoda* .................... het-er-O-poda
*Hibana* .................... hee-BAW-nuh
*Hilaira* .................... hi-LEY-ra
*Hogna* .................... HAWG-nuh
*Hololena* .................... ho-lo-LEE-nuh
HOMALONYCHIDAE .................... ho-maa-lo-NEY-ki-dee
HYPOCHILIDAE .................... hey-po-KEY-li-dee
*Hypochiloidea* .................... hey-po-key-LOY-dee-uh
*Hypochilus* .................... hey-po-KEY-luhs
*Hypsosinga* .................... hip-so-SING-uh

## I

*Islandiana* .................... iz-lahn-dee-AHN-uh

## K

*Kibramoa* .................... kib-rah-MO-uh
*Kukulcania* .................... koo-kool-KA-nee-uh

## L

*Lathys* .................... LA-thuhs
*Latrodectus* .................... laa-tro-DEK-tuhs
*Lepthyphantes* .................... lep-thuh-FAHN-tees
LEPTONETIDAE .................... lep-to-NET-i-dee
*Linyphantes* .................... lin-i-FAHN-tees
*Linyphia* .................... lin-I-fee-uh
LINYPHIIDAE .................... lin-i-FEE-i-dee
LIOCRANIDAE .................... ley-o-CRAAN-i-dee
*Liocranoides* .................... ley-o-craan-OY-dees
*Lophomma* .................... lo-FO-muh
*Loxosceles* .................... lawx-AW-sel-ees
*Lycosa* .................... ley-CO-suh
LYCOSIDAE .................... ley-KO-si-dee

## M

*Maevia* .................... MEE-vee-uh
*Mallos* .................... MAAL-uhs
*Mangora* .................... mang-OR-uh
*Marpissa* .................... mahr-PIS-uh
*Mastophora* .................... maas-TAWF-or-uh
*Meioneta* .................... mee-o-NEE-tuh
*Metaphidippus* .................... met-uh-fid-I-puhs
*Metepeira* .................... met-i-PEY-rah
*Micaria* .................... mi-KAH-ree-uh
*Micrathena* .................... meyk-ruh-THEE-nuh
MIMETIDAE .................... mi-MET-i-dee
*Mimetus* .................... mi-MEET-uhs
*Misumenops* .................... mis-OO-men-ahps
MITURGIDAE .................... mi-TUR-ji-dee
*Mygalomorphae* .................... mey-gal-o-MOR-fee

*Myrmekiaphila* .................. meyr-mee-kee-AAF-i-lah
MYSMENIDAE .................. mis-MEN-i-dee

# N
NEMESIIDAE .................. nem-es-EE-i-dee
*Neoantistea* .................. nee-o-aan-TIS-tee-uh
*Neoleptoneta* .................. nee-o-lep-to-NEET-uh
*Neon* .................. NEE-awn
*Neoscona* .................. nee-o-SKO-nuh
*Nephila* .................. NE-fil-uh
*Neriene* .................. ner-ee-EE-nee
NESTICIDAE .................. nes-TIS-i-dee
*Nesticus* .................. NES-ti-kuhs
*Nodocion* .................. no-DO-see-uhn
*Novalena* .................. no-vau-LEE-nuh

# O
OCHYROCERATIDAE .................. o-kee-ro-ser-AAT-i-dee
OECOBIIDAE .................. ek-o-BEE-i-dee
*Oecobius* .................. ek-O-bee-uhs
*Oedothorax* .................. ed-o-THAWR-aax
*Olios* .................. OL-ee-uhs
OONOPIDAE .................. o-o-NO-pi-dee
*Opopaea* .................. op-o-PEE-uh
*Orchestina* .................. or-kes-TEE-nuh
*Oreonetides* .................. or-ee-o-NET-i-deez
*Orthonops* .................. OR-tho-nawpz
*Oxyopes* .................. awx-ee-OP-eez
OXYOPIDAE .................. awx-ee-O-pi-dee
*Ozyptila* .................. o-ZIP-til-uh

# P
*Pachygnatha* .................. paa-KIG-nuh-thuh *or* paak-ig-NA-thuh
*Pardosa* .................. pahr-DO-suh
*Pelecopsis* .................. pel-i-KOP-sis
*Pelegrina* .................. pel-i-GREE-nuh
*Pellenes* .................. pel-EE-neez
*Peucetia* .................. pyoo-SET-ee-uh
*Phanias* .................. FAN-ee-uhs
*Phantyna* .................. fan-TEY-nuh
*Phidippus* .................. fid-I-puhs *or* FID-i-puhs
PHILODROMIDAE .................. fil-o-DRO-mi-dee
*Philodromus* .................. fil-i-DRO-muhs
PHOLCIDAE .................. FOL-si-dee
*Pholcus* .................. FOL-kuhs
*Phrurolithus* .................. fru-RO-ligh-uhs *or* fru-ro-LITH-uhs
*Phruronellus* .................. fru-ro-NEL-uhs
*Phrurotimpus* .................. fru-ro-TIM-puhs
*Piabuna* .................. pey-au-BOO-nuh
*Pimoa* .................. pim-O-uh
PIMOIDAE .................. pim-O-i-dee
*Pimus* .................. PIM-uhs
*Pirata* .................. pey-RA-tuh
PISAURIDAE .................. py-SAWR-i-dee
*Pityohyphantes* .................. pit-ee-o-hey-FAN-teez
PLECTREURIDAE .................. plek-TRUR-i-dee
*Plectreurys* .................. plek-TRUR-is
*Poeciloneta* .................. poy-sil-o-NEET-uh
*Porrhomma* .................. por-HO-muh
PRODIDOMIDAE .................. pro-di-DO-mi-dee
*Psilochorus* .................. sil-o-KOR-uhs

# R
*Rabidosa* .................. raab-i-DO-suh
*Robertus* .................. ro-BER-tuhs
*Rualena* .................. roo-ah-LEE-nuh

# S
SALTICIDAE .................. sahl-TIS-i-dee
*Satilatlas* .................. saa-til-AAT-luhs
*Schizocosa* .................. skiz-o-KO-suh
*Sciastes* .................. skee-AAS-tes
*Scopoides* .................. sko-POY-dees
*Scotinella* .................. sko-ti-NEL-uh
*Scotinotylus* .................. sko-tin-o-TEY-luhs
*Scytodes* .................. skey-TO-dees
SCYTODIDAE .................. skey-TO-di-dee
SEGESTRIIDAE .................. seg-es-TREE-i-dee
*Selenops* .................. SEL-en-awps
*Sergiolus* .................. ser-jee-OL-uhs
SICARIIDAE .................. sik-uh-REE-uh-dee
*Sisicottus* .................. sis-i-KO-tuhs
*Sitticus* .................. SIT-i-kuhs
*Socalchemmis* .................. so-kaal-KEM-uhs
*Sosippus* .................. so-SIP-uhs *or* SO-si-puhs
SPARASSIDAE .................. spahr-AAS-i-dee
*Sphodros* .................. SFO-druhs
*Spirembolus* .................. spir-EM-bo-luhs
*Steatoda* .................. stee-uh-TO-duh
SYMPHYTOGNATHIDAE .................. sim-fey-tog-NAATH-i-dee
*Synageles* .................. sin-AA-jel-eez

# T
*Tachygyna* .................. taak-ee-JEY-na
*Tapinocyba* .................. taap-in-o-SEE-buh
*Tegenaria* .................. tej-in-ER-ee-uh
TELEMIDAE .................. tel-EM-i-dee
TENGELLIDAE .................. ten-GEL-i-dee
*Terralonus* .................. ter-ah-LON-uhs
*Tetragnatha* .................. te-TRAAG-nuh-thuh *or* te-truhg-NA-thuh
TETRAGNATHIDAE .................. tet-raag-NAATH-i-dee
*Thanatus* .................. THAA-nuh-tuhs
THERAPHOSIDAE .................. ther-ah-FO-si-dee
THERIDIIDAE .................. ther-i-DEE-i-dee
*Theridion* .................. ther-I-dee-awn
THERIDIOSOMATIDAE .................. ther-i-dee-o-so-MAAT-i-dee
THOMISIDAE .................. to-MIS-i-dee
*Thomisus* .................. to-MIS-uhs
*Thymoites* .................. they-mo-EE-tes
*Tibellus* .................. tib-EL-uhs
TITANOECIDAE .................. ti-tah-NEES-i-dee
*Tmarus* .................. tim-AHR-uhs
*Trachelas* .................. tra-KEE-luhs
TRECHALEIDAE .................. trey-kel-EE-i-dee
*Tricholathys* .................. trik-o-LA-thuhs

# U
ULOBORIDAE .................. yoo-lo-BOR-i-dee
*Ummidia* .................. yoo-MI-dee-uh

# W
*Wadotes* .................. wah-DO-tees
*Walckenaeria* .................. wahl-ken-AAR-ee-uh
*Wubana* .................. woo-BAH-nah
*Wulfila* .................. WUL-fil-uh

# X
*Xysticus* .................. ZIS-ti-kuhs

# Y
*Yorima* .................. YOR-i-ma

# Z
*Zanomys* .................. zen-O-meez
*Zelotes* .................. zel-O-teez
ZODARIIDAE .................. zo-dahr-EY-i-dee
ZORIDAE .................. ZOR-i-dee
*Zorocrates* .................. zo-RAH-kruh-teez
ZOROCRATIDAE .................. zor-o-CRAAT-i-dee
ZOROPSIDAE .................. zor-AHP-si-dee

# Bibliography

The bibliography is organized into two sections: the first covers the references pertaining strictly to the etymological chapter of H.D. Cameron (chapter 73), and the second covers arachnological references.

Arachnological references follow Platnick's (2005) on-line bibliography, including his designations of letter suffixes with dates containing multiple publications by the same author. For instance, Banks 1904b here corresponds to Platnick's Banks 1904b, even if Banks 1904a is not cited in this book.

When a given paper is not listed in Platnick (2005) our citations continue where his leave off. For example, he lists Banks 1892a, 1892b, 1892c, but does not list the other papers produced by Banks in 1892, we adopted 1892d, 1892e, etc. to maintain consistency.

We did not follow Platnick in cases where more detailed information is available. This is usually true for old references that were serially published. Specifically, Biologia Centrali-Americana Volume 1 by O. Pickard-Cambridge (1889-1902a); Biologia Centrali-Americana Volume 2 by F.O. Pickard-Cambridge (1897-1905); Die Arachniden by C.W. Hahn and then by C.L. Koch (1831-1849); Die Arachniden-Familie der Drassiden (1966-1867) by C.L. Koch and a few other classic works.

The controversy about the authorship and publication date of "Description de l'Égypte..." by Savigny and Audouin 1825 etc... is resolved by using Audouin 1826 as the correct date and author (see Paquin *et al.* 2001: 50-51 for a review of the case and also ICZN 1975, opinion 1038). In other cases where the publication date is uncertain, we use the protocol recommended by the Code (ICZN 1999). The presumed correct date is given in the bibliography, followed by the erroneous date in quotation marks and square brackets (e.g., ["1889"]).

Finally, the publication date for Clerck's *Aranei Svecici...* is a unique case in the Zoological nomenclature as it predates the official standard of 1758 (see Paquin *et al.* 2001: 35-36 for a review of the case), but we follow Platnick (2005) in using 1757 for that particular publication.

**Etymological references —**

Beauchamp, W. 1907. Aboriginal Place Names of New York. 1907. New York State Museum Bulletin No. 108.

Bloomfield, L. 1956. Eastern Ojibwa. Ann Arbor.

Buck, C.D. and W. Peterson. 1948. A Reverse Index of Greek Nouns and Adjectives. Chicago; University of Chicago Press.

Cassiro, S.E. 1885. The International Scientist's Directory. Boston.

Cassiro, S.E. 1890. The Naturalist's Directory. Boston Part II.

Cattell, J. (ed). 1944. American men of Science: Biographical Dictionary, 7th edition. The Science Press, Lancaster Pennsylvania.

Chamberlin, R.V. 1908b. Animal names and anatomical terms of the Gosiute Indians. Proceedings of the Academy of Natural Sciences of Philadelphia, 60: 75-103.

Chamberlin, R.V. 1911. Ethnobotany of the Gosiute Indians. Proceedings of the Academy of Natural Sciences of Philadelphia, 24-99.

Chamberlin, R.V. 1913. Place and personal names of the Gosiute Indians of Utah. Proceedings of the American Philosophical Society, 52(208): 1-20.

Chantraine, P. 1933. La Formation des noms en grec ancien. Paris.

Förstemann, E. 1900. Altdeutsches Namenbuch. Erster Band. Personennamen. Second edition, Bonn.

Freeman, R.B. 1980. British Natural History Books. Folkstone.

Gil Fernandez, L. 1959. Nombres de los insectos en griego antiguo. Madrid.

Heller, J. 1945. Classical mythology in the Systema Naturae of Linnaeus. Transactions and Proceedings of the American Philological Association, 76:333-357.

Hodge, F.W. 1906. Handbook of American Indians north of Mexico. Bulletin Bureau of American Ethnology no. 30, pt. 1-2. s.v. Geneseo.

Jones, H.L. (ed.) 1930. Strabo, Geography. 8 volumes. Loeb Classical Library. Cambridge.

Jones, W.H.S. (ed.) 1963. Pliny, Natural History. 10 volumes. Loeb classical Library. Cambridge.

Keller, O. 1913. Die antike Tierwelt. 2 volumes. Leipzig.

Lampe, 1961 G.W.H. ed. 1961. A Patristic Greek Lexicon. Oxford.

MacCulloch & Gray (eds.) 1916-1932. The mythology of all races in thirteen volumes. Archaeological Institute of America. Marshal Jones Co., Boston.

Mallis, A. 1971. American Entomologists. Rutgers University Press. New Brunswick, New Jersey. 549 pages.

Müller, C. (ed.) 1841-70. Agatharchides, Fragmenta Historica. 5 volumes. Paris.

Peck, A.L. 1970. Aristotle, Historia Animalium, 2 volumes. Loeb Classical Library.

Scholefield, A.F. (ed.) 1972. Aelian, On Animals, 3 volumes. Loeb Classical Library.

Schönfeld, M. 1911. Wörterbuch der altgermanischen Personen-und Völkernamen. Heidelberg.

Simeon, R. 1885. Dictionnaire de la langue Nahuatl oú Mexicaine. Paris. Rpd. Graz 1963. s.v.

**Arachnological references —**

Abbot, J.T. 1792. Spiders of Georgia. Manuscript. British Museum of Natural History. 116 pages.

Adis, J. and M.S. Harvey. 2000. How many Arachnida and Myriapoda are there worldwide and in Amazonia? Studies in Neotropical Fauna and Environment, 35: 139-141.

Agassiz, L. 1848. Nomenclatoris Zoologici Index Universalis, Continens Nomina systematica Classium, Ordinum, Familiarum et Generum Animalium Omnium, tam Viventium Quam Fossilium. Soloduri. 1135 pages.

Agnarsson, I. 2004. Morphological phylogeny of cobweb spiders and their relatives (Araneae, Araneoidea, Theridiidae). Zoological Journal of the Linnean Society, 141: 447-626.

Agnew, C.W., D.A. Dean and J.W. Smith. 1985. Spiders collected from peanuts and non-agricultural habitats in the Texas West Cross-Timbers. Southwestern Naturalist, 30(1): 1-12.

Akre, R.D. and E.A. Myhre. 1991. Biology and medical importance of the aggressive house spider, *Tegenaria agrestis*, in the Pacific Northwest (Arachnida: Araneae: Agelenidae). Melanderia, 47: 1-30.

Alberti, G. and F.A. Coyle. 1991. Ultrastructure of the primary male genital system, sprematozoa and spermiogenesis of *Hypochilus pococki* (Araneae: Hypochilidae). Journal of Arachnology, 19: 136-149.

Albin, E. 1736. A Natural History of Spiders and Other Curious Insects. London. 1-56 + pl. I-XXXVI pages.

Aldrovandrus, U. 1792. De Animalibus Insectis. Bonon. 766 pages.

Alexander, R.D., D.C. Marshall and J.R. Cooley. 1997. Evolutionary perspectives on insect mating. Pages 4-31 *in* J. Choe & B. Crespi (editors): The Evolution of Mating Systems in Insects and Arachnids. Cambridge University Press, Cambridge.

Anonymous. 1842. Strickland Report. Proceedings of the British Association for the Advancement of Science, 1842: 105-212.

Archer, A.F. 1941a. Alabama spiders of the family Mimetidae. Papers of the Michigan Academy of Sciences, Arts, and Letters, 26: 183-193.

Archer, A.F. 1941b. Supplement to the Argiopidae of Alabama. Alabama Museum of Natural History, Museum Paper 18: 5-47.

Archer, A.F. 1946. The Theridiidae or comb-footed spiders of Alabama. Alabama Museum of Natural History, Museum Paper 22: 1-67.

Archer, A.F. 1950. A study of theridiids and mimetid spiders with descriptions of new genera and species. Alabama Museum of Natural History, Museum Paper 30: 7-40.

Archer, A.F. 1953. Studies in the orbweaving spiders (Argiopidae). American Museum Novitates, No. 1622: 1-27.

Arnedo, M.A., J. Coddington, I. Agnarsson and R.G. Gillespie. 2004. From a comb to a tree: phylogenetic relationships of the comb-footed spiders (Araneae, Theridiidae) inferred from nuclear and mitochondrial genes. Molecular Phylogenetics and Evolution, 31: 225-245.

Atkinson, G.F. 1886. Descriptions of some new trap-door spiders, their nests and food habits. II. Entomologica Americana, 2: 109-117, 128-137 + pl. V.

Audouin, V. 1826. Explication sommaire des planches d'Arachnides de l'Égypte et de la Syrie, publiées par Jules-César Savigny, membre de l'Institut; offrant un exposé des caractères naturels des genres, avec la distinction des espèces. Pages 99-186 in Description de l'Égypte ou recueil des observations et des recherches qui ont été faites en Égypte pendant l'expédition de l'armée française, publiée par les ordres de sa Majesté l'Empereur Napoléon Le Grand. Histoire naturelle. Tome premier, 4ème partie.

Ausserer, A. 1871a. Beiträge zur Kenntniss der Arachniden-Familie der Territelariae Thorell (Mygalidae Autor.). Verhandlungen der kaiserlich-königlichen zoologisch-botanischen Gesellschaft in Wien, 21: 117-224.

Ausserer, A. 1871b. Neue Radspinnen. Verhandlungen der kaiserlich-königlichen zoologisch-botanischen Gesellschaft in Wien, 21: 815-832 + pl. V.

Ausserer, A. 1875. Zweiter Beitrag zur Kenntnis der Arachniden-familie der Territelariae. Verhandlungen der zoologisch-botanischen Gesellschaft in Wien, 25: 125-206.

Austad, S.N. 1984. Evolution of sperm priority patterns in spiders. Pages 223-249 in R.L. Smith (editor): Sperm Competition and the Evolution of Animal Mating Systems. Academic Press. New York.

Ayoub, N.A., S.E. Riechert and R.L.Small. 2005. Speciation history of the North American funnel web spiders, *Agelenopsis* (Araneae: Agelenidae): phylogenetic inferences at the population-species interface. Molecular Phylogenetics and Evolution, 36: 42-57.

Baehr, M. and B. Baehr. 1993. The Hersiliidae of the Oriental region including New Guinea. Taxonomy, phylogeny, zoogeography (Arachnida: Araneae: Hersiliidae). Spixiana (Supplement), 19: 1-96.

Baert, L. 1984a. Mysmenidae and Hadrotarsidae from the Neotropical Guaraní zoogeographical province (Paraguay and south Brasil) (Araneae). Revue suisse de zoologie, 91: 603-616.

Baert, L. and R. Jocqué. 1993. *Anapistula caecula* n. sp., the smallest known female spider (Araneae, Symphytognathidae). Journal of African Zoology, 107: 187-189.

Banks, N. 1891c. Notes on the Dysderidae of the United States. The Canadian Entomologist, 23: 207-209.

Banks, N. 1892a. The spider fauna of the Upper Cayuga Lake Basin. Proceedings of the Academy of Natural Sciences of Philadelphia, 1: 11-81 + pl. I-V.

Banks, N. 1892d. A classification of North American spiders. The Canadian Entomologist, 24: 88-97.

Banks, N. 1893a. Notes on spiders. Journal of the New York Entomological Society, 1(3): 123-134.

Banks, N. 1894b. Two families of spiders new to the United States. Entomological News, 5: 298-300.

Banks, N. 1895d. A list of spiders of Long Island, with a description of new species. Journal of the New York Entomological Society, 3(2): 76-93.

Banks, N. 1895e. Some new Attidae. The Canadian Entomologist, 27(4): 96-102.

Banks, N. 1896a. A few new spiders. The Canadian Entomologist, 28(3): 62-65.

Banks, N. 1896b. New californian spiders. Journal of the New York Entomological Society, 4: 88-91.

Banks, N. 1896e. New North American spiders and mites. Transactions of the American Entomological Society, 23: 57-77.

Banks, N. 1897. Descriptions of new spiders. The Canadian Entomologist, 29(8): 193-197.

Banks, N. 1898a. Some new spiders. The Canadian Entomologist, 30: 185-188.

Banks, N. 1898b. Arachnida from Baja California and other parts of Mexico. Proceedings of the California Academy of Sciences. Third Series. Zoology, 1(7): 205-309.

Banks, N. 1899b. Some spiders from Northern Louisiana. Proceedings of the Entomological Society of Washington, 4(3): 188-195.

Banks, N. 1900a. Some new north American spiders. The Canadian Entomologist, 32: 96-102.

Banks, N. 1900b. Papers from the Harriman Alaska Expedition. XI. Entomological results (5): Arachnida. Proceedings of the Washington Academy of Sciences, 2(23): 477-486 + pl. XXIX.

Banks, N. 1900c. Some Arachnida from Alabama. Proceedings of the Academy of Natural Sciences of Philadelphia, 52: 529-543.

Banks, N. 1901a. Some spiders and other Arachnida from southern Arizona. Proceedings of the United States National Museum, 23, no. 1223: 581-590 + pl. XXII.

Banks, N. 1901b. Some Arachnida from New Mexico. Proceedings of the Academy of Natural Sciences of Philadelphia, 53: 568-597 + pl. XXXIII.

Banks, N. 1902d. Some spiders and mites from the Bermuda Islands. Transactions of the Connecticut Academy of Arts and Sciences, 2(11): 267-275.

Banks, N. 1904a. New genera and species of Nearctic spiders. Journal of the New York Entomological Society, 12(2): 109-119 + pl. V-VI.

Banks, N. 1904b. Some Arachnids from California. Proceedings of the California Academy of Sciences, Third Series, 3(13): 331-376 + pl. XXXVIII-XLI.

Banks, N. 1904c. The Arachnida of Florida. Proceedings of the Academy of Natural Sciences of Philadelphia, 120-147.

Banks, N. 1905a. Synopses of North American invertebrates. XX. Families and genera of Araneida. The American Naturalist, 39: 293-323.

Banks, N. 1906b. Descriptions of new American spiders. Proceedings of the Entomological Society of Washington, 7: 94-101.

Banks, N. 1909a. Arachnida from Costa Rica. Proceedings of the Academy of Natural Sciences of Philadelphia, 61: 194-234 + pl. V-VI.

Banks, N. 1909b. Arachnida of Cuba. Estacion Central Agronomica de Cuba, Second report. II, 150-174.

Banks, N. 1910. Catalogue of Nearctic spiders. Bulletin of the United States National Museum, 72: 1-80.

Banks, N. 1911. Some Arachnida from North Carolina. Proceedings of the Academy of Natural Sciences of Philadelphia, 63: 440-456 + pl. XXXIV-XXXV.

Banks, N. 1913. Notes on the types of some American spiders in European collections. Proceedings of the Academy of Natural Sciences of Philadelphia, 65: 177-188.

Banks, N. 1914a. Notes on some Costa Rican Arachnida. Proceedings of the Academy of Natural Sciences of Philadelphia, 65: 676-687 + pl. XXVIII-XXX.

Banks, N. 1914b. Arachnida from South Georgia. Bulletin Brooklyn Institute of Arts and Sciences, 2: 78-79.

Banks, N. 1921. New Californian spiders. Proceedings of the California Academy of Sciences, Fourth Series, 11(7): 99-102.

Banks, N. 1929. Spiders from Panama. Bulletin of the Museum of Comparative Zoology, 69(3): 53-96 + pl. I-IV.

Banks, N., N.M. Newport and R.D. Bird. 1932. II. An annotated list of Oklahoma spiders. In Oklahoma Spiders. Publications of the University of Oklahoma Biological survey, IV(1-2): 12-34. Norman, University of Oklahoma Press.

Barnes, R.D. 1955. North American jumping spiders of the genus *Maevia*. American Museum Novitates, No. 1746: 1-13.

Barnes, R.D. 1958. North American jumping spiders of the subfamily Marpissinae (Araneae, Salticidae). American Museum Novitates, No. 1867: 1-50.

Barnes, R.D. 1959. The *lapidicica* group of the wolf spider genus *Pardosa* (Araneae, Lycosidae). American Museum Novitates, No. 1960: 1-20.

de Barros Machado, A. 1951. Ochyroceratidae (Araneae) de l'Angola. Companhia de diamantes de Angola. Serviços Culturalis. Museu do Dundo. Subsidios para o Estudo da Biologia na Lunda: 1-88.

de Barros Machado, A. 1964. Ochyroceratidae nouveaux d'Afrique (Araneae). Annals of the Natal Museum, 16: 215-230.

Barrows, W.M. 1919b. New spiders from Ohio. Ohio Journal of Science, 19(6): 355-360 + pl. XV.

Barrows, W.M. 1940. New and rare spiders from the Great Smoky Mountain National Park region. Ohio Journal of Science, 40(3): 130-138.

Barrows, W.M. 1943. A new prairie Spider. Ohio Journal of Science, 43(5): 209.

Barrows, W.M. 1945. New spiders from the Great Smoky Mountain National Park. Annals of the Entomological Society of America, 38(1): 70-76.

Barth, F.G. 1985. Slit sensilla and the measurement of cuticular strains. Pages 162-188 in F.G. Barth (editor): Neurobiology of Arachnids. Springer-Verlag, New York.

Barth, F.G. (editor). 2002. A Spider's World: Senses and Behavior. Springer-Verlag, Berlin. 394 pages.

Beatty, J.A. 1970. The spider genus *Ariadna* in the Americas (Araneae, Dysderidae). Bulletin of the Museum of Comparative Zoology, 139(8): 433-517.

Beatty, J.A. and J.M. Berry. 1988. The spider genus *Paratheuma* (Araneae, Desidae). Journal of Arachnology, 16: 47-54.

Beatty, J.A. and J.M. Berry. 1989. Four new species of *Paratheuma* (Araneae, Desidae) from the Pacific. Journal of Arachnology, 16: 339-347.

Becker, L. 1879a. Diagnoses de nouvelles Aranéides américaines. Annales de la Société entomologique de Belgique, 22: 77-86 + pl. I-II.

Beechhold, H.F. 2000. A key to the pronunciation and meaning of scientific names of popular species, part I: pronunciation. The American Tarantula Society, Forum, 6(4): 1.

Benjamin, S.P. and S. Zschokke. 2002. Untangling the tangle-web: web construction behavior of the comb-footed spider *Steatoda triangulosa* and comments on phylogenetic implications (Araneae: Theridiidae). Journal of Insect Behavior, 15: 791-809.

Bennett, R.G. 1985. The natural history and taxonomy of *Cicurina bryantae* Exline (Araneae, Agelenidae). Journal of Arachnology, 13: 87-96.

Bennett, R.G. 1987. Systematics and natural history of *Wadotes* (Araneae, Agelenidae). Journal of Arachnology, 15: 91-128.

Bennett, R.G. 1988. The spider genus *Cybaeota* (Araneae, Agelenidae). Journal of Arachnology, 16: 103-119.

Bennett, R.G. 1989. "Emerit's glands" in *Cybaeota* (Araneae, Agelenidae). Journal of Arachnology, 17(2): 225-235.

Bennett, R.G. 1991. The Systematics of the North American Cybaeid Spiders (Araneae, Dictynoidea, Cybaeidae). Unpublished Ph.D. Thesis, University of Guelph. 308 pages.

Bennett, R.G. 1992. The spermathecal pores of spiders with special reference to Dictynoids and Amaurobioids (Araneae, Araneomorphae, Araneoclada). Proceedings of the Entomological Society of Ontario, 123: 1-21.

Bennett, R.G. in press. Ontogeny, variation, and synonymy in North American *Cybaeus* spiders (Araneae: Cybaeidae). The Canadian Entomologist.

Bennett, R.G. and L.J. Brumwell. 1996. *Zora hespera* in British Columbia: a new family record for Canada (Araneae: Zoridae). Journal of the Entomological Society of British Columbia, 93: 105-109.

Bentzien, M.M. 1973. Biology of the spider *Diguetia imperiosa* (Araneida: Diguetidae). The Pan-Pacific Entomologist, 49: 110-123.

Bentzien, M.M. 1976. Biosystematics of the spider genus *Brachythele* Ausserer (Araneida: Dipluridae). Unpublished Ph.D. thesis, University of California, Berleley, 136 pages.

Berland, L. 1914. Araneae (1ère partie). Pages 37-94 in Voyage de Ch. Alluaud et R. Jeannel en Afrique oriental (1911-1912). Résultats scientifiques, Paris, 3.

Berland, L. 1919b. Note sur le peigne metatarsal que possèdent certaines araignées de la famille des Drassidae. Bulletin du Muséum d'histoire naturelle. Paris, 1919: 458-463.

Berland, L. 1929c. Araignées. Pages 23-71 in R. Perrier (editor): La faune de France en tableaux synoptiques illustrés. Fascicule 2: Arachnides et Crustacés. Delagrave, Paris.

Berland, L. 1932a. Voyage de mm. L. Chopard et A. Méquignon aux Açores (aout-septembre 1930). II. Araignées. Annales de la Société entomologique de France, 101: 69-84 + pl. V-VI.

Berland, L. 1932d. Rectification. Bulletin de la Société entomologique de France, 37: 119.

Berman, J.D. and H.W. Levi. 1971. The orb weaver genus *Neoscona* in North America (Araneae: Araneidae). Bulletin of the Museum of Comparative Zoology, 141(8): 465-500.

Bertkau, P. 1872. Über die Respirationorgane der Aranen. Archiv für Naturgeschichte, 38: 208-233.

Bertkau, P. 1878. Versuch einer natürlichen Anordnung der Spinnen nebst Bemerkungen zu einzelnen Gattungen. Archiv für Naturgeschichte, 44: 351-410.

Bertkau, P. 1880a. Verzeichniss der von Prof. Ed. van Beneden auf seiner im Auftrage der Belgischen Regierung unternommen wissenschaftlichen Reise nach Brasilien und La Plata im Jahren 1872-1873 gensammelten Aarchniden. 43: 1-120.

Bertkau, P. 1880c. [Legt vor und bespricht sein Verzeichniss der Spinnen Bonns]. Verhandlungen des naturhistorischen Vereines der preussischen Rheinlande und Westfalens, 37(7): Sitzungsbericht, 282-285.

Bertkau, P. 1882. Ueber das Cribellum und Calamistrum. Ein Beitrag zur Histologie, Biologie und Systematik der Spinnen. Archiv für Naturgeschichte, 48: 316-362 + pl. XVIII.

Bertkau, P. 1889. Interessante Tiere aus der Umgebung von Bonn. Verhandlungen des naturhistorischen Vereines der preussischen Rheinlande und Westfalens, 46: 69-82.

Bishop, S.C. 1924b. A revision of the Pisauridae of the United States. New York State Museum Bulletin, No. 252: 1-140.

Bishop, S.C. 1950. A new cave spider from North Carolina. Proceedings of the Biological Society of Washington, 63: 9-12.

Bishop, S.C. and C.R. Crosby. 1926. Notes on the spiders of the southeastern United States with decriptions of new species. Journal of the Elisha Mitchell Scientific Society, 41(3/4): 165-212 + pl. XX-XXV.

Bishop, S.C. and C.R. Crosby. 1930. Studies in American spiders: genera *Ceratinopsis*, *Ceratinopsidis* and *Tutaibo*. Journal of the New York Entomological Society, 38: 15-37.

Bishop, S.C. and C.R. Crosby. 1933 ["1932"]. Studies in American spiders: the genus *Grammonota*. Journal of the New York Entomological Society, 40(4): 393-421.

Bishop, S.C. and C.R. Crosby. 1935a. A new genus and two new species of Dictynidae (Araneae). Proceedings of the Biological Society of Washington, 48: 45-48.

Bishop, S.C. and C.R. Crosby. 1935b. American Erigoninae: the spider genera *Pelecopsidis* and *Floricomus*. Journal of the New York Entomological Society, 43(1): 31-45.

Bishop, S.C. and C.R. Crosby. 1935c. Studies in American spiders: miscellaneous genera of Erigoneae, part I. Journal of the New York Entomological Society, 43(2):217-241, 43(3):255-281

Bishop, S.C. and C.R. Crosby. 1936b. A new genus of spiders in the Erigoninae. Proceedings of the Biological Society of Washington, 49: 39-42.

Bishop, S.C. and C.R. Crosby. 1938. Studies in American spiders: miscellaneous genera of Erigoninae, part II. Journal of the New York Entomological Society, 46(1): 55-107.

Blackledge, T.A., J.A. Coddington and R.G. Gillespie. 2003. Are three-dimensional spider webs defensive adaptations? Ecology Letters, 6: 13-18.

Blackwall, J. 1833a. Characters of some undescribed genera and species of Araneidae. The London, Edinburgh and Dublin Philosophical Magazine and Journal of Science, Series 3, Volume 3(14): 104-112.

Blackwall, J. 1833c. Characters of some undescribed genera and species of Araneidae. The London, Edinburgh and Dublin Philosophical Magazine and Journal of Science, Series 3, Volume 3(17): 344-352.

Blackwall, J. 1833d. Characters of some undescribed genera and species of Araneidae. The London, Edinburgh and Dublin Philosophical Magazine and Journal of Science, Series 3, Volume 3(18): 436-443.

Blackwall, J. 1833e. Notice of several recent discoveries in the structure and economy of spiders. Transactions of the Linnean Society of London, 16: 471-485, 767-770 + pl I.

Blackwall, J. 1834a. Characters of some undescribed species of Araneidae. The London, Edinburgh and Dublin Philosophical Magazine and Journal of Science, Series 3, Volume 5(25): 50-53.

Blackwall, J. 1834b. Researches in Zoology: Illustrative of the Manners and Economy of Animals; with Descriptions of Numerous Species New to Naturalists; Accompanied by Plates. Simpkin & Marshall, London. 434 pages.

Blackwall, J. 1841a. The difference in the number of eyes with which spiders are provided proposed as the basis of their distribution into tribes; with descriptions of newly discovered species and the characters of a new family and three new genera of spiders. Transactions of the Linnean Society of London, 18: 601-670.

Blackwall, J. 1841b. On the number and structure of the mammulae employed by spiders in the process of spinning. Transactions of the Linnean Society of London, 18: 219-224.

Blackwall, J. 1844. Descriptions of some newly discovered species of Araneida. The Annals and Magazine of Natural History, 1(13): 179-188.

Blackwall, J. 1846a. Notice of spiders captured by Professor Potter in Canada, with descriptions of such species as appear to be new to science. The Annals and Magazine of Natural History. Series 1, 17: 30-43, 76-82.

Blackwall, J. 1852a. A catalogue of British spiders, including remarks on their structure, functions, œconomy and systematics arrangement. The Annals and Magazine of Natural History, 2(9): 15-22, 268-275, 464-471, 2(10): 182-189, 248-253.

Blackwall, J. 1853. Descriptions of some newly discovered species of Araneidea. The Annals and Magazine of Natural History, 2(11): 14-25.

Blackwall, J. 1858a. Descriptions of six newly discovered species and characters of a new genus of Araneida. The Annals and Magazine of Natural History, 3(1): 426-434.

Blackwall, J. 1859b. Descriptions of newly discovered spiders captured by James Yate Johnson, Esq., in the island of Madeira. The Annals and Magazine of Natural History, 3(4): 255-267.

Blackwall, J. 1862a. Descriptions of newly discovered spiders from the Island of Madeira. The Annals and Magazine of Natural History, 3(9): 370-382.

Blackwall, J. 1863a. Description of newly discovered spiders captured in Rio de Janeiro, by John Gray and the Rev. Hamlet Clarck (continued). The Annals and Magazine of Natural History, 3(11): 29-45.

Blackwall, J. 1864. History of the Spiders of Great Britain and Ireland. Part II. London, published for the Ray Society, 175-384 + pl. XIII-XXIX.

Blackwall, J. 1867a. Descriptions of several species of East Indian spiders, apparently to be new or little known to arachnologists. The Annals and Magazine of Natural History, 3(19): 387-394.

Blauvelt, H.H. 1936. The comparative morphology of the secondary sexual organs of *Linyphia* and some related genera, including a revision of the group. Pages 81-171 + pl. VI-XXIII in Festschrift zum 60. Geburtstag von Professor Dr. Embrik Strand. Riga. Volume II.

Blest, A.D. 1985. The fine structure of photoreceptors in relation to function. Pages 79-102 in F.G. Barth (editor): Neurobiology of Arachnids. Springer-Verlag, New York.

Blest, A.D., L.C. Kao and K. Powell. 1978a. Photoreceptor membrane breakdown in the spider *Dinopis*: the fate of rhabdomere products. Cell Tissue Research, 195: 425-444.

Blest, A.D. and M.F. Land. 1977. The physiological optics of *Dinopis subrufus* L. Koch: a fish lens in a spider. Proceedings of the Royal Society of London (B) 196: 197-222.

Blest, A.D., K. Powell and L.C. Kao. 1978b. Photoreceptor membrane breakdown in the spider *Dinopis*: GERL differentiation in the receptors. Cell Tissue Research, 195: 277-297.

Bode, F., F. Sachs and M.R. Franz. 2001. Tarantula peptide inhibits atrial fibrillation: a peptide from spider venom can prevent the heartbeat from losing its rhythm. Nature, 409: 35-36.

Bonaldo, A.B. 2000. Taxonomia da subfamília Corinninae (Araneae, Corinnidae) nas regiões Neotropical e Neárctica. Iheringia (Serie Zoologia) 89: 3-148.

Bonaldo, A.B. and A.D. Brescovit. 1992. As aranhas do gênero *Cheiracanthium* C.L. Koch, 1839 na Regiao Neotropical (Araneae, Clubionidae). Revista brasileira de Entomologia, 36: 731-740.

Bond, J.E. 2004. Systematics of the Californian euctenizine spider genus *Apomastus* (Araneae: Mygalomorphae: Cyrtaucheniidae): the relationship between molecular and morphological taxonomy. Invertebrate Systematics, 18: 361-376.

Bond, J.E. and B.D. Opell. 1997. Systematics of the spider genera *Mallos* and *Mexitlia* (Araneae, Dictynidae). Zoological Journal of the Linnean Society, 119: 389-445.

Bond, J.E. and B.D. Opell. 2002. Phylogeny and taxonomy of the genera of south-western North American Euctenizinae tarpdoor spiders and their relatives (Araneae: Mygalomorphaeae, Cyrtaucheniidae). Zoological Journal of the Linnean Society, 136: 487-534.

Bonnet, P. 1945. Bibliographia Araneorum. Analyse méthodique de toute la littérature aranéologique jusqu'en 1939. Tome I. Toulouse, les frères Douladoure. vi-xvii + 1-832 pages.

Bonnet, P. 1949. Un point compliqué de nomenclature dans le nom spécifique d'une araignée: *cinerea*, *cicurea* ou *cicur*? Bulletin de la société d'histoire naturelle de Toulouse, 84: 65-72 + pl. I.

Bonnet, P. 1950. La question *Araneus-Epeira*. Bulletin de la société d'histoire naturelle de Toulouse, 85: 176-184.

Bonnet, P. 1953a. Différence de nomenclature chez les Araneides. VII. Le problème *Agelena-Agalena*. 78e Congrès des Sociétés Savantes, 329-332.

Bonnet, P. 1953b. Sur la valeur des tubercules huméraux dans le genre *Araneus*. Transactions of the 9th International Congress of Entomology, volume 2, 355-356.

Bonnet, P. 1955. Bibliographia Araneorum. Analyse méthodique de toute la littérature aranéologique jusqu'en 1939. Tome II, 1re partie: A-B. Toulouse, Les Artisans de l'Imprimerie Douladoure. 1-918 pages.

Bonnet, P. 1956a. Bibliographia Araneorum. Analyse méthodique de toute la littérature aranéologique jusqu'en 1939. Tome II, 2e partie: C-F. Toulouse, Les Artisans de l'Imprimerie Douladoure. Pages 919-1926.

Bonnet, P. 1957a. Bibliographia Araneorum. Analyse méthodique de toute la littérature aranéologique jusqu'en 1939. Tome II, 3e partie: G-M. Toulouse, Les Artisans de l'Imprimerie Douladoure. Pages 1927-3026.

Bonnet, P. 1958. Bibliographia Araneorum. Analyse méthodique de toute la littérature aranéologique jusqu'en 1939. Tome II, 4e partie: N-S. Toulouse, Les Artisans de l'Imprimerie Douladoure. Pages 3027-4230.

Bonnet, P. 1959. Bibliographia Araneorum. Analyse méthodique de toute la littérature aranéologique jusqu'en 1939. Tome II, 5e partie: T-Z. Toulouse, Les Artisans de l'Imprimerie Douladoure. Pages 4231-5058.

Bösenberg, W. 1903. Die Spinnen Deutschlands. V, VI. Zoologica. Stuttgart. 14: 385-465.

Bösenberg, W. and E. Strand. 1906. Japanische Spinnen. Abhandlungen herausgegeben von der senckenbergischen Naturforschenden Gesellschaft, 30: 92-422 + pl. III-XVI.

Bosmans, R. 1994b. Revision of the genus *Zodarion* Walckenaer, 1833 in the Iberian Peninsula and Balearic Islands (Araneae, Zodariidae). Eos, 69: 115-142.

Bosmans, R. 1997. Revision of the genus *Zodarion* Walckenaer, 1833, part II. Western and Central Europe, including Italy (Araneae, Zodariidae). Bulletin of the British Arachnological Society, 10(8): 265-294.

Bosselaers, J. 2002. A cladistic analysis of Zoropsidae (Araneae), with the description of a new genus. Belgian Journal of Zoology, 132: 141-154.

Bosselaers, J. and R. Jocqué. 2002. Studies in Corinnidae: cladistic analysis of 38 corinnid and liocranid genera, and transfer of Phrurolithinae. Zoologica Scripta, 31(3): 241-270.

Bowling, T.A. and R.J. Sauer. 1975. A taxonomic revision of the crab spider genus *Coriarchne* (Araneida, Thomisidae) for North America north of Mexico. Journal of Arachnology, 2: 183-193.

Bradley, R. 2002. Kaston.transy.edu/spiderlist/index.html.

Bradley, R. 2003. A newly introduced jumping spider (*Myrmarachne formicaria*) in North America. Abstract, page 29. American Arachnological Society, 27th Annual Meeting, Denver, CO.

Brady, A.R. 1962. The spider genus *Sosippus* in North America, Mexico, and Central America (Araneae, Lycosidae). Psyche, 69: 129-164.

Brady, A.R. 1964. The lynx spiders of North America, north of Mexico (Araneae; Oxyopidae). Bulletin of the Museum of Comparative Zoology, 131: 429-518.

Brady, A.R. 1969. A reconsideration of the *Oxyopes apollo* species group with the descriptions of two new species (Araneae: Oxyopidae). Psyche, 76(4): 426-438.

Brady, A.R. 1980. Nearctic species of the wolf spider genus *Trochosa* (Araneae: Lycosidae). Psyche, 86(2/3): 167-212.

Brady, A.R. 1987 ["1986"]. Nearctic species of the wolf spider genus *Gladicosa* (Araneae: Lycosidae). Psyche, 93(3/4): 285-319.

Brady, A.R. and K.S. McKinley. 1994. Nearctic species of the wolf spider genus *Rabidosa* (Araneae: Lycosidae). Journal of Arachnology, 22: 138-160.

Braendegaard, J. 1932. Araneae. Göteborgs Kongliga Vetenskaps och Vitterhets Samhälles Handlingar, Serie B, Volume 2(7): 8-36.

Braun, R. 1964. Üeber einige Spinnen aus Tirol, Österreich (Arach., Araneae). Senckenbergiana Biologica, 45: 151-160.

Breene, R.G. 2003. Common names of arachnids. 5th edition. http://www.americanarachnology.org/acn5.pdf. pages: 1-42.

Brescovit, A.D. 1991b. *Hibana*, novo gênero de aranhas da família Anyphaenidae (Arachnida, Araneae). Revista brasileira de Entomologia, 35(4): 729-744.

Brescovit, A.D. 1997a. Revisao de Anyphenidae Bertkau a nivel de generos na regiao Neotropical (Araneae, Anyphaenidae). Revista brasileira de Zoologia, 13 (Supplement 1): 1-187.

Brescovit, A.D., A.B. Bonaldo and K.G. Mikhailov. 1994. Revalidation of the spider genus *Elaver* O. Pickard-Cambridge, 1898 (Aranei Clubionidae). Arthropoda Selecta, 3(1/2): 35-38.

Brescovit, A.D. and C.A. Rheims. 2000. On the synanthropic species of the genus *Scytodes* Latreille (Araneae, Scytodidae) of Brazil, with synonymies and records of these species in other Neotropical countries. Bulletin of the British Arachnological Society, 11: 320-330.

Brescovit, A.D. and C.A. Rheims. 2001. Notes on the genus *Scytodes* (Araneae, Scytodidae) in Central and South America. Journal of Arachnology, 29: 312-329.

Brignoli, P.M. 1970b. Contribution à la connaissance des Symphytognathidae paléarctiques (Arachnida, Araneae). Bulletin du Muséum d'histoire naturelle de Paris, 41: 1403-1420.

Brignoli, P.M. 1971d. Note su ragni cavernicoli italiani (Araneae). Fragmenta entomologica, 7: 121-229.

Brignoli, P.M. 1972i. Some cavernicolous spiders from Mexico (Araneae). Accademia Nazionale dei Lincei, 171: 129-155.

Brignoli, P.M. 1973a. Telemidae, una famiglia di ragni nuova per il continente americano. Fragmenta Entomologica, 8: 247-263.

Brignoli, P.M. 1976a. Beiträge zur kenntnis der Scytodidae (Araneae). Revue suisse de Zoologie, 83(1): 125-191.

Brignoli, P.M. 1976g. Ragni d'Italia XXIV. Note sulla morfologia dei genitalia interni dei Segestriidae e cenni sulle specie italiane. Fragmenta entomologica, 12: 19-62.

Brignoli, P.M. 1977a. Ragni del Brasile III. Note su *Bruchnops melloi* Biraben e sulla posizione sistematica dei Caponiidae (Arachnida, Araneae). Revue suisse de Zoologie, 84: 609-616.

Brignoli, P.M. 1977g. Spiders from Mexico, III. A new leptonetid from Oaxaca (Araneae, Leptonetidae). Accademia Nazionale dei Lincei, 171(3): 213-218.

Brignoli, P.M. 1979c. On some cave spiders from Guatemala and United States (Araneae). Revue suisse de Zoologie, 86(2): 435-443.

Brignoli, P.M. 1979k. Recherches en Afrique de l'Institut de Zoologie de l'Aquila (Italie) II. *Reo latro* nov. gen., nov. sp. du Kenya (Araneae: Mimetidae). Revue de Zoologie africaine, 93: 919-928.

Brignoli, P.M. 1979q. The morphology and the relationships of the Leptonetidae (Arachnida: Araneae). Journal of Arachnology, 7: 231-236.

Brignoli, P.M. 1980j. On few Mysmenidae from the Oriental and Australian regions (Araneae). Revue suisse de Zoologie, 87: 727-738.

Brignoli, P.M. 1981c. Studies on the Pholcidae, I. Notes on the genera *Artema* and *Physocyclus*. Bulletin of the American Museum of Natural History, 170(1): 90-100.

Brignoli, P.M. 1983c. A catalogue of the Araneae described between 1940 and 1981 (edited by P. Merrett). Manchester University Press in association with the British Arachnological Society. vii-xi + 1-755 pages.

Brignoli, P.M. 1985a. On some generic homonymies in spiders (Araneae). Bulletin of the British Arachnological Society, 6: 380.

Bristowe, W. 1938a. The classification of spiders. Proceedings of the Zoological Society of London (B) 108: 285-322.

Bristowe, W. 1939. The Comity of Spiders. Volume I. London. 1-228 + pl. I-XIX pages.

Bristowe, W. 1948. Notes on the structure and systematic position of oonopid spiders based on an examination of the British species. Proceedings of the Zoological Society of London, 118: 878-891.

Bristowe, W. 1958. The World of Spiders. Collins Press. 304 pages.

Brown, K.M. 1973. A preliminary checklist of the spiders of Nacogdoches, Texas. Journal of Arachnology, 1: 229-240.

Bruce, J.A. and J.E. Carico. 1988. Silk use during mating in *Pisaurina mira* (Walckenaer) (Araneae, Pisauridae). Journal of Arachnology, 16: 1-4.

Bruins, E. 1999. Encyclopaedia of Terrarium. Rebo Productions Ltd., The Netherlands. 320 pages.

Bryant, E.B. 1931. Notes on North American Anyphaenidae in the Museum of Comparative Zoology. Psyche, 38: 102-126.

Bryant, E.B. 1933a. New and little known spiders from the United States. Bulletin of the Museum of Comparative Zoology, 74: 171-193 + pl. I-IV.

Bryant, E.B. 1935. A rare spider. Psyche, 42: 163-166.

Bryant, E.B. 1940. Cuban spiders in the Museum of Comparative Zoology. Bulletin of the Museum of Comparative Zoology, 86: 247-554.

Bryant, E. 1944. Three species of *Coleosoma* from Florida (Araneae; Theridiidae). Psyche, 51: 51-58.

Bryant, E.B. 1945b. Some new or little known southern spiders. Psyche, 52: 178-192.

Bryant, E.B. 1949. The male of *Prodidomus rufus* Hentz (Prodidomidae, Araneae). Psyche, 56: 22-25.

Bryant, E.B. 1950. The salticid spiders of Jamaica. Bulletin of the Museum of Comparative Zoology, 103: 163-209.

Buckle, D.J. 1972. Sound production in the courtships of two lycosid spiders, *Schizocosa avida* (Walckenaer) and *Tarentula aculeata* (Clerck). Blue-Jay, 30: 110-113.

Buckle, D.J., D. Carroll, R.L. Crawford and V.D. Roth. 2001. Linyphiidae and Pimoidae of America north of Mexico: Checklists, synonymy, and literature. Part 2. Pages 89-191 in P. Paquin and D.J. Buckle (editors): Contributions à la Connaissance des Araignées (Araneae) d'Amérique du Nord. Fabreries, Supplément 10.

Buddle, C.M. 2000. Life history of *Pardosa moesta* and *Pardosa mackenziana* (Araneae, Lycosidae) in central Alberta, Canada. Journal of Arachnology, 28: 319-328.

Buschman, L.L., W.H. Whitcomb Jr., R.C. Hemenway, D.L. Mays, N. Ru, N.C. Leppla and B.J. Smittle. 1977. Predators of velvetbean caterpillar eggs in Florida soybeans. Environmental Entomology, 6: 403-403.

di Caporiacco, L. 1934b. Aracnidi. Missione zoologica del Dott. E. Festa in Cirenaica. Bollettino dei Musei di zoologia e di anatomia comparata della R. Università di Torino, 44, Serie 3(47): 121-146.

di Caporiacco, L. 1938f. Il sistema degli Araneidi. Archives Zoologiche Italia: 25, Supplemento, Attualita di Zoologiche: 4, 35-155.

di Caporiacco, L. 1940c. Arachnidi raccolte nella Reg. dei Laghi Etiopici della Fossa Galla. Atti della Reale Accademia d'Italia. Rendiconti, 11: 767-873.

Carico, J.E. 1972a. The Nearctic spider genus *Pisaurina* (Pisauridae). Psyche, 79(4): 295-310.

Carico, J.E. 1973. The Nearctic species of the genus *Dolomedes* (Araneae: Pisauridae). Bulletin of the Museum of Comparative Zoology, 114: 435-488.

Carico, J.E. 1976. The spider genus *Tinus* (Pisauridae). Psyche, 83(1): 63-78.

Carico, J.E. 1985. Description and significance of the juvenile web of *Pisaurina mira* (Walck.) (Araneae: Pisauridae). Bulletin of the British Arachnological Society, 6(7): 295-296.

Carico, J.E. 1986. Trechaleidae: a "new" American spider family. Page 305 in W.G. Eberhard, Y.D. Lubin and B.C. Robinson (editors): Proceedings of the 9th International Congress of Arachnology. Smithsonian Institution Press.

Carico, J.E. 1993. Revision of the genus *Trechalea* Thorell (Araneae, Trechaleidae) with a review of the taxonomy of the Trechaleidae and Pisauridae of the Western Hemisphere. Journal of Arachnology, 21: 226-257.

Carico, J.E. and E.W. Minch. 1981. A new species of *Trechalea* (Pisauridae) from North America. Bulletin of the American Museum of Natural History, 170: 154-156.

Carmichael, L.D. 1973. Correlation between segment length and spine counts in two spider species of *Araneus* (Araneae: Araneidae). Psyche, 80: 62-69.

Catley, K.M. 1993. Courtship, mating and post-oviposition behaviour of *Hypochilus pococki* (Araneae, Hypochilidae). Memoirs of the Queensland Museum, 33(2): 469-474.

Catley, K.M. 1994. Descriptions of new *Hypochilus* species from New Mexico and California with a cladistic analysis of the Hypochilidae (Araneae). American Museum Novitates, No. 3088: 1-27.

Cazier, M.A. and M.A. Mortense. 1962. Analysis of the habitat, web design, cocoon and egg sacs of the tube weaving spider *Diguetia canities* (McCook) (Aranea, Diguetidae). Bulletin of the Southern California Academy of Sciences, 61: 65-88.

Chamberlin, R.V. 1904a. Notes on generic characters in the Lycosidae. The Canadian Entomologist, 36(6): 145-148.

Chamberlin, R.V. 1908a. Revision of the North American spiders of the family Lycosidae. Proceedings of the Academy of Natural Sciences of Philadelphia, 60: 158-318.

Chamberlin, R.V. 1909a. The American *Drapestica*. The Canadian Entomologist, 41(10): 368.

Chamberlin, R.V. 1916. Results of the Yale Peruvian Expedition of 1911. Bulletin of the Museum of Comparative Zoology, 60: 177-299.

Chamberlin, R.V. 1917. New spiders of the family Aviculariidae. Bulletin of the Museum of Comparative Zoology, 61(3): 25-75 + pl. I-V.

Chamberlin, R.V. 1919a. New western spiders. Annals of the Entomological Society of America, 12(3): 239-260 + pl. XIV-IXX.

Chamberlin, R.V. 1919b. New Californian spiders. Pomona Collection. Journal of Entomology and Zoology (Claremont), 12: 1-17 + pl. I-IV.

Chamberlin, R.V. 1920a. New spiders from Utah. The Canadian Entomologist, 52(7): 193-201.

Chamberlin, R.V. 1921a. A new genus and a new species of spiders in the group Phrurolitheae. The Canadian Entomologist, 53: 69-70.

Chamberlin, R.V. 1921b. On some Arachnids from Southern Utah. The Canadian Entomologist, 53: 245-247 + pl. X.

Chamberlin, R.V. 1921c. Linyphiidae of St. Paul Island, Alaska. Journal of the New York Entomological Society, 29: 35-44.

Chamberlin, R.V. 1922. The North American spiders of the Family Gnaphosidae. Proceedings of the Biological Society of Washington, 35: 145-172.

Chamberlin, R.V. 1923. The North American species of *Mimetus*. Pomona Collection. Journal of Entomology and Zoology (Claremont), 15: 3-7 + pl. I-II.

Chamberlin, R.V. 1924a. Descriptions of new American and Chinese spiders, with notes on other Chinese species. Proceedings of the United States National Museum, 63(13): 1-38.

Chamberlin, R.V. 1924b. The spider fauna of the shores and islands of the gulf of California. Proceedings of the California Academy of Sciences, Fourth Series, 12(28): 561-694.

Chamberlin, R.V. 1925b. Notes on North American spiders heretofore referred to *Coelotes*. Proceedings of the Biological Society of Washington, 38: 119-124.

Chamberlin, R.V. 1925c. New North American spiders. Proceedings of the California Academy of Sciences, 14: 105-142.

Chamberlin, R.V. 1928a. Notes on the spiders from the La Sal Mountains of Utah. The Canadian Entomologist, 60(4): 93-95.

Chamberlin, R.V. 1929. On three new spiders of the genus *Oxyopes* (Araneina). Entomological News, 40: 17-20.

Chamberlin, R.V. 1933b. On a new eyeless spider of the family Linyphiidae from Potter Creek Cave, California. The Pan-Pacific Entomologist, 9(3): 122-124.

Chamberlin, R.V. 1936a. Records of North American Gnaphosidae with descriptions of new species. American Museum Novitates, No. 841: 1-30.

Chamberlin, R.V. 1936b. Further records and descriptions of North American Gnaphosidae. American Museum Novitates, No. 853: 1-25.

Chamberlin, R.V. 1937b. Two genera of trap-door spiders from California. Bulletin of the University of Utah: 28(3). Biological Series: 3(7), 1-11.

Chamberlin, R.V. 1940a. New American tarantulas of the family Aviculariidae. Bulletin of the University of Utah: 30(13). Biological Series: 5(8), 1-39.

Chamberlin, R.V. 1940b. A new trap-door spider from Texas. Proceedings of the Biological Society of Washington, 53: 5-6.

Chamberlin, R.V. 1947. A summary of the known North American Amaurobiidae. Bulletin of the University of Utah: 38(8). Biological Series: 10(5), 1-31.

Chamberlin, R.V. 1948b. The genera of North American Dictynidae. Bulletin of the University of Utah: 38(15). Biological Series: 10(6), 1-31.

Chamberlin, R.V. 1949 ["1948a"]. On some American spiders of the family Erigonidae. Annals of the Entomological Society of America, 41(4): 483-562.

Chamberlin, R.V. and W.J. Gertsch. 1928. Notes on spiders from Southeastern Utah. Proceedings of the Biological Society of Washington, 41: 175-187.

Chamberlin, R.V. and W.J. Gertsch. 1929. New spiders from Utah and California. Journal of Entomology and Zoology (Claremont), 21: 101-112 + pl. I-V.

Chamberlin, R.V. and W.J. Gertsch. 1930. On fifteen new North American spiders. Proceedings of the Biological Society of Washington, 43: 137-144.

Chamberlin, R.V. and W.J. Gertsch. 1940. Descriptions of new Gnaphosidae from the United States. American Museum Novitates, No. 1068: 1-19.

Chamberlin, R.V. and W.J. Gertsch. 1958. The spider family Dictynidae in America north of Mexico. Bulletin of the American Museum of Natural History, 116: 1-152 + pl. I-ILVII.

Chamberlin, R.V. and W. Ivie. 1932. North American spiders of the genera *Cybaeus* and *Cybaeina*. Bulletin of the University of Utah: 23(2). Biological Series: 3(1): 3-43.

Chamberlin, R.V. and W. Ivie. 1933a. Spiders of the Raft River Mountains of Utah. Bulletin of the University of Utah: 23(4). Biological Series: 2(2): 1-79.

Chamberlin, R.V. and W. Ivie. 1933b. A new genus in the family Agelenidae. Bulletin of the University of Utah: 24(5). Biological Series: 2(3): 1-7.

Chamberlin, R.V. and W. Ivie. 1934. A new genus of Theridiid spiders in which the male develops only one palpus. Bulletin of the University of Utah: 24(4). Biological Series: 2(4): 1-18.

Chamberlin, R.V. and W. Ivie. 1935b. Miscellaneous new American spiders. Bulletin of the University of Utah: 26(4). Biological Series: 2(8): 1-79.

Chamberlin, R.V. and W. Ivie. 1935c. Nearctic spiders of the family Urocteidae. Annals of the Entomological Society of America, 28: 265-279.

Chamberlin, R.V. and W. Ivie. 1936a. New spiders from Mexico and Panama. Bulletin of the University of Utah: 27(5). Biological Series: 3(5): 1-103.

Chamberlin, R.V. and W. Ivie. 1936b. Nearctic spiders of the genus *Wubana*. Annals of the Entomological Society of America, 29(1): 85-98.

Chamberlin, R.V. and W. Ivie. 1937. New spiders of the family Agelenidae from western North America. Annals of the Entomological Society of America, 30: 211-241.

Chamberlin, R.V. and W. Ivie. 1938. Araneida from Yucatan. Publications of the Carnegie Institution of Washington, 491: 123-136.

Chamberlin, R.V. and W. Ivie. 1939a. Studies on North American spiders of the family Micryphantidae. Verhandlungen, VII. International Kongress für Entomologie (Berlin), 1: 56-73 + pl. I-VI.

Chamberlin, R.V. and W. Ivie. 1939b. New tarantulas from the southwestern states. Bulletin of the University of Utah: 29(11). Biological Series: 5(1): 1-17.

Chamberlin, R.V. and W. Ivie. 1940. Agelenid spiders of the genus *Cicurina*. Bulletin of the University of Utah: 30(13). Biological Series: 5(9): 1-108.

Chamberlin, R.V. and W. Ivie. 1941a. North American Agelenidae of the genera *Agelenopsis*, *Calilena*, *Ritalena* and *Tortolena*. Annals of the Entomological Society of America, 34: 585-628.

Chamberlin, R.V. and W. Ivie. 1941b. Spiders collected by L.W. Saylor and others, mostly in California. Bulletin of the University of Utah: 31(8). Biological Series: 6(3): 1-49.

Chamberlin, R.V. and W. Ivie. 1942a. A hundred new species of American spiders. Bulletin of the University of Utah: 32(13). Biological Series: 7(1): 1-117.

Chamberlin, R.V. and W. Ivie. 1942b. Agelenidae of the genera *Hololena*, *Novalena*, *Ruralena*, and *Melpomene*. Annals of the Entomological Society of America, 35: 203-241.

Chamberlin, R.V. and W. Ivie. 1943. New genera and species of North American linyphiid spiders. Bulletin of the University of Utah: 33(10). Biological Series: 7(6): 1-39.

Chamberlin, R.V. and W. Ivie. 1944. Spiders of the Georgia region of North America. Bulletin of the University of Utah: 35(9). Biological Series: 8(5): 1-267.

Chamberlin, R.V. and W. Ivie. 1945a. On some Nearctic mygalomorph spiders. Annals of the Entomological Society of America, 38(4): 549-558.

Chamberlin, R.V. and W. Ivie. 1945b. Some erigonid spiders of the genera *Eulaira* and *Diplocentria*. Bulletin of the University of Utah: 36(2). Biological Series: 9(4), 1-19.

Chamberlin, R.V. and W. Ivie. 1945c. Erigonid spiders of the genera *Spirembolus*, *Disembolus*, and *Bactroceps*. Transactions of the Connecticut Academy of Arts and Sciences, 36: 215-235.

Chamberlin, R.V. and W. Ivie. 1947a. North American dictynid spiders: the *bennetti* group of *Amaurobius*. Annals of the Entomological Society of America, 40: 29-55.

Chamberlin, R.V. and W. Ivie. 1947b. The spiders of Alaska. Bulletin of the University of Utah: 37(10). Biological Series: 10(3): 5-103.

Chapin, E.A. 1922. On *Simonella*, a genus of salticid spiders new to North America. Proceedings of the Biological Society of Washington, 35: 129-132.

Chapman, T., G. Arnqvist, J. Bangham and L. Rowe. 2003. Sexual conflict. Trends in Ecology and Evolution, 18(1): 41-47.

Chickering, A.M. 1968e. The genus *Triaeris* Simon (Araneae, Oonopidae) in Central America and the West Indies. Psyche, 75: 351-359.

Chickering, A.M. 1969c. The family Oonopidae (Araneae) in Florida. Psyche, 76: 144-162.

Chyzer, C. and W. Kulczyński. 1894. Araneae Hungariae. Tomus II, pars prior : Therididoidea. Budapestini. 1-151 + pl. I-V.

Clerck, C. 1757 ["1758"]. Svenska Spindlar, Uti Sina Hufvud - Slågter Indelte Smat under Några Och Sextio Särskildte Arter Beskrefne Och Med Illuminerade Figurer Uplyste, På Kongl. Vetensk. Societ. i Uppsala. Befallning Utgifne. [Aranei Svecici, descriptionibus et figuris oeneis illustrati, ad genera subalterna redacti speciebus ultra LX determinati, Auspiciis Regiae Societatis Scientiarum Uppsaliensis]. Stockholmiae. 1-154 + pl. I-VI pages.

Cockerell, T.D.A. 1894b. The hunting spider of the vine. The Entomologist, 27, No. 374: 207-208.

Cockerell, T.D.A. 1897. A new attid spider. The Canadian Entomologist, 29: 223-224.

Coddington, J.A. 1986a. The genera of the spider family Theridiosomatidae. Smithsonian Contributions to Zoology, No. 422: 1-96.

Coddington, J.A. 1986b. The monophyletic origin of the orb web. Pages 319-363 *in* W.A. Shear (editor): Spiders. Webs Behavior and Evolution . Stanford University Press, Stanford CA.

Coddington, J.A. 1986c. Orb webs in 'non orb weaving' ogre faced spiders (Araneae: Dinopidae): a question of genealogy. Cladistics, 2: 53-67.

Coddington, J.A. 1989. Spinneret silk spigot morphology. Evidence for the monophyly of orb-weaving spiders, Cyrtophorinae (Araneidae), and the group Theridiidae-Nesticidae. Journal of Arachnology, 17: 71-95.

Coddington, J.A. 1990. Ontogeny and homology in the male palpus of orb-weaving spiders and their relatives, with comments on phylogeny (Araneoclada: Araneoidea, Deinopoidea). Smithsonian Contributions to Zoology, No. 496: 1-52.

Coddington, J.A. and R.K. Colwell. 2001. Arachnida. Pages 199-218 *in* S.C. Levin (editor): Encyclopedia of Biodiversity. Volume 1. Academic Press, New York.

Coddington, J.A., G. Giribet, M.S. Harvey, L. Prendini and D.E. Walter. 2004. Arachnida. Pages 296-318 *in* J. Cracraft & M.J. Donoghue (editors): Assembling the Tree of Life. Oxford University Press, New York.

Coddington, J.A., C.E. Griswold, D. Silva Dávila, E. Peñaranda and S.F. Larcher. 1991. Designing and testing sampling protocols to estimate biodiversity in tropical ecosystems. Pages 44-60 *in* E.C. Dudley (editor): The Unity of Evolutionary Biology: Proceedings of the Fourth International Congress of Systematic and Evolutionary Biology, Volume 1. Critical issues in biodiversity. Symposium. Dioscorides Press, Portland, Oregon.

Coddington, J.A., S.F. Larcher and J.C. Cokendolpher. 1990. The systematic status of Arachnida, exclusive of Acari, in North America north of Mexico. Pages 5-20 *in* M. Kosztarab & C.W. Schafer (editors): Systematics of the North American Insects and Arachnids: status and needs. Virginia Agricultural Experimental Station Information. Series 90-1. Virginia Polytechnic Institute and State University, Blacksburg Virginia.

Coddington, J.A. and H.W. Levi. 1991. Systematics and evolution of spiders (Araneae). Annual Review of Ecology and Systematics, 22: 565-592.

Coddington, J.A. and C. Sobrevila. 1987. Web manipulation and two stereotyped attack behaviors in the ogre-faced spider *Deinopis spinosus* Marx (Araneae, Deinopidae). Journal of Arachnology, 15: 213-225.

Cokendolpher, J.C. 1978b. A new species of *Mazax* from Texas (Aranea: Clubionidae). Journal of Arachnology, 6(3): 230-232.

Cokendolpher, J.C. 2004a. *Cicurina* spiders from caves in Bexar County, Texas. Texas Memorial Museum Speleological Monographs, 6. Studies on the Cave and Endogean Fauna of North America. IV: 1-46.

Cokendolpher, J.C. 2004b. Another new species of troglobitic *Cicurina* (Araneae: Dictynidae) from Fort Hood, Texas. Texas Memorial Museum Speleological Monographs, 6. Studies on the Cave and Endogean Fauna of North America. IV: 59-62.

Cokendolpher, J.C. 2004c. A new *Neoleptoneta* spider from a cave at Camp Bullis, Bexar County, Texas (Araneae: Leptonetidae). Texas Memorial Museum Speleological Monographs, 6. Studies on the Cave and Endogean Fauna of North America. IV: 63-69.

Cokendolpher, J.C. and N.V. Horner. 1978. The spider genus *Poultonella* (Araneae: Salticidae). Journal of Arachnology, 6: 133-139.

Cokendolpher, J.C. and J.R. Reddell. 2001a. New and rare nesticid spiders from Texas caves (Araneae: Nesticidae). Texas Memorial Museum, Speleological Monographs, 5. Studies on the Cave and Endogean Fauna of North America. III: 25-34.

Cokendolpher, J.C. and J.R. Reddell. 2001b. Cave spiders (Araneae) of Fort Hood, Texas, with descriptions of new species of *Cicurina* (Dictynidae) and *Neoleptoneta* (Leptonetidae). Texas Memorial Museum Speleological Monographs, 5. Studies on the Cave and Endogean Fauna of North America. III: 35-56.

Comstock, J.H. 1912. The Spider Book. A manual for the study of the spiders and their near relatives, the scorpions, pseudoscorpions, whipscorpions, harvestmen and other members of the class Archnida, found in America north of Mexico, with analytical keys for their classification and popular accounts of their habits. Garden City, New York. 721 pages.

Comstock, J.H. 1940. (revised and edited by W.J. Gertsch). The Spider Book. A manual for the study of the spiders and their near relatives, the Scorpions, Pseudoscorpions, Whip-Scorpions, Harvestmen, and other members of the class Arachnida, found in America north of Mexico, with analytical keys for their classification and popular accounts of their habits. Doubleday, Doran & Company, Inc., New York. v-xi + 729 pages.

Cooke, J.A.L. 1964. A revisionary study of some spiders of the rare family Prodidomidae. Proceedings of the Zoological Society of London, 142: 257-305.

Cooke, J.A.L. 1965a. Spider genus *Dysdera* (Araneae, Dysderidae). Nature, 205: 1027-1028.

Cooke, J.A.L. 1965b. A contribution to the biology of the British spiders belonging to the genus *Dysdera*. Oikos, 16: 20-25.

Corey, D.T. and D.J. Mott. 1991. A revision of the genus *Zora* (Araneae, Zoridae) in North America. Journal of Arachnology, 19: 55-61.

Coyle, F.A. 1968. The mygalomorph spider genus *Atypoides* (Araneae: Antrodiaetidae). Psyche, 75(2): 157-194.

Coyle, F.A. 1971. Systematics and natural history of the mygalomorph spider genus *Antrodiaetus* and related genera (Araneae: Antrodiaetidae). Bulletin of the Museum of Comparative Zoology, 141(6): 269-402.

Coyle, F.A. 1975. Systematics of the trapdoor spider genus *Aliatypus* (Araneae: Antrodiaetidae). Psyche, 81(3/4): 431-500.

Coyle, F.A. 1981. The mygalomorph spider genus *Microhexura* (Araneae: Dipluridae). Bulletin of the American Museum of Natural History, 170(1): 64-75.

Coyle, F.A. 1985a. Observations on the mating behaviour of the tiny mygalomorph spider, *Microhexura montivaga* Crosby & Bishop (Araneae, Dipluridae). Bulletin of the British Arachnological Society, 6(8): 328-330.

Coyle, F.A. 1985b. Two year life cycle and low palpal character variance in a Great Smoky Mountain population of the lamp-shade spider (Araneae: Hypochilidae, *Hypochilus*). Journal of Arachnology, 13: 211-218.

Coyle, F.A. 1986. The role of silk in prey capture by non-araneomorph spiders. Pages 269-305 in W.A. Shear (editor): Spiders. Webs, Behavior, and Evolution. Stanford University Press, Stanford CA.

Coyle, F.A. 1988. A revision of the American funnel-web mygalomorph spider genus *Euagrus* (Araneae, Dipluridae). Bulletin of the American Museum of Natural History, 187(3): 203-292.

Coyle, F.A. and W.R. Icenogle. 1994. Natural history of the Californian trapdoor spider genus *Aliatypus* (Araneae, Antrodiaetidae). Journal of Arachnology, 22: 225-255.

Coyle, F.A. and A.C. McGarity. 1991. Two new species of *Nesticus* spiders from the southern Appalachians (Araneae, Nesticidae). Journal of Arachnology, 19: 161-162.

Coyle, F.A. and W.A. Shear. 1981. Observation on the natural history of *Sphodros abboti* and *Sphodros rufipes* (Araneae, Atypidae), with evidence for a contact sex pheromone. Journal of Arachnology, 9: 317-326.

Craig, C.L. 2003. Spiderwebs and Silk. Tracing evolution from molecules to genes to phenotypes. Oxford University Press. 256 pages.

Crawford, R.L. 1985. Mt. St. Helens and spider biogeography. Proceedings of the Washington State Entomological Society, 46: 700-702.

Crawford, R.L. 1988. An annotated checklist of the spiders of Washington. Burke Museum Contributions in Anthropology and Natural History, No. 5: 1-48.

Crosby, C.R. 1905a. A catalogue of Erigoneae of North America, with notes and description of new species. Proceedings of the Academy of Natural Sciences of Philadelphia, 57: 301-343 + pl. XXVIII-XXIX.

Crosby, C.R. 1906. Two new species of Theridiidae. The Canadian Entomologist, 38(9): 308-310.

Crosby, C.R. 1926. Some arachnids from the Carlsbad Cave of New Mexico. Proceedings of the entomological Society of Washington, 28(1): 1-5.

Crosby, C.R. 1929. Studies in North American spiders: the genus *Cochlembolus* (Araneina). Entomological News, 40(3): 79-83 + pl. IV.

Crosby, C.R. and S.C. Bishop. 1925a. Two new spiders from the Blue Ridge Mountains of North Carolina (Araneina). Entomological News, 36(5): 142-146.

Crosby, C.R. and S.C. Bishop. 1925b. A new genus and two new species of Spiders collected by *Bufo quercius* (Holdbrook). Florida Entomologist, 9(3): 33-36.

Crosby, C.R. and S.C. Bishop. 1925c. Studies in New York spiders. Genera: *Ceratinella* and *Ceraticelus*. New York State Museum Bulletin, No. 264: 1-71.

Crosby, C.R. and S.C. Bishop. 1927. New species of Erigoneae and Theridiidae. Journal of the New York Entomological Society, 35(2): 147-153.

Crosby, C.R. and S.C. Bishop. 1928a. Revision of the spider genera *Erigone*, *Eperigone*, and *Catabrithorax* (Erigonidae). New York State Museum Bulletin, No. 278: 3-74.

Crosby, C.R. and S.C. Bishop. 1931. Studies in American spiders: genera, *Cornicularia*, *Paracornicularia*, *Tigellinus*, *Walckenaera*, *Epiceraticelus* and *Pelecopsis*, with descriptions of new genera and species. Journal of the New York Entomological Society, 39(3): 359-403.

Crosby, C.R. and S.C. Bishop. 1933b. American spiders: Erigoneae, males with cephalic pits. Annals of the Entomological Society of America, 26(1): 105-182.

Crosby, C.R. and S.C. Bishop. 1936b. Studies in American spiders: miscellaneous genera of Erigoninae. Pages 52-64 + pl. IV-V in Festschrift zum 60. Geburtstag von Professor Dr. Embrik Strand. Volume II, Riga.

Cushing, P.E. 1989. Possible eggsac defense behaviors in the spider *Uloborus glomosus* (Araneae, Uloboridae). Psyche, 96(3/4): 269-277.

Cushing, P.E. 1995. Description of the spider *Masoncus pogonophilus* (Araneae, Linyphiidae), a harvester ant myrmecophile. Journal of Arachnology, 23: 55-59.

Cushing, P.E. and B.D. Opell. 1990a. The effect of time and temperature on disturbance behaviors shown by the orb-weaving spider *Uloborus glomosus* (Uloboridae). Journal of Arachnology, 18: 87-93.

Cushing, P.E. and B.D. Opell. 1990b. Disturbance behaviors in the spider *Uloborus glomosus* (Araneae, Uloboridae): possible predator avoidance strategies. Canadian Journal of Zoology, 68(6): 1090-1097.

Cushing, P.E. and R.G. Santangelo. 2002. Notes on the natural history and hunting behavior of an ant eating zodariid spider (Arachnida, Araneae) in Colorado. Journal of Arachnology, 30: 618-621.

Cutler, B. 1979. Jumping spiders of the United States and Canada: changes in the key and list (2). Peckhamia, 1(6): 125.

Cutler, B. 1981b. A revision of the spider genus *Paradamoetas* (Araneae: Salticidae). Bulletin of the American Museum of Natural History, 170: 207-215.

Cutler, B. 1982. *Euophrys erratica* (Walckenaer), new to North America. Peckhamia, 2(3): 36-37.

Cutler, B. 1988a. A revision of the American species of the antlike jumping spider genus *Synageles* (Araneae, Salticidae). Journal of Arachnology, 15(3): 321-348.

Cutler, B. 1990. A revision of the western hemisphere *Chalcoscirtus* (Araneae: Salticidae). Bulletin of the British Arachnological Society, 8(4): 105-108.

Dacke, M., T.-A. Doan and D.C. O'Corroll. 2001a. Polarized light detection in spiders. Journal of Experimental Biology, 204: 2481-2490.

Dacke, M., D.-E Nilsson, E.J. Warrent, A.D. Blest, M.F. Land and D.C. O'Carroll. 2001b. Built-in polarizers form part of a compass organ in spiders. Nature, 401: 470-473.

Dahl, F. 1886. Monographie der Erigonen-arten im Thorell'schen Sinne, nebst andern Beiträge zur Spinnenfauna Schleswig-Holstein. Schriften des naturwissenschaftlichen Vereins für Schleswig-Holstein, 6(2): 65-102.

Dahl, F. 1901b. Nachtrag zur Uebersicht der Zorospiden. Sitzungberichte der Gesellschaft Naturforeshender Fruende zu Berlin, 1901: 244-255.

Dahl, F. 1901c. Über die Seltenheit gewisser Spinnenarten. Sitzungberichte der Gesellschaft Naturforeshender Fruende zu Berlin, 1901: 257-266.

Dahl, F. 1904. Ueber das System der Spinnen (Araneae). Sitzungberichte der Gesellschaft Naturforeshender Fruende zu Berlin, 1904:93-120.

Dahl, F. 1908. Die Lycosiden oder Wolfsspinnen Deutschlands und ihre Stellung im Haushalt der Natur. Nach statischen Untersuchungen dargestellt. Nova Acta Academiae Caesareae Leopoldino-Carolinae Germanicae Naturae Curiosorum, 88: 175-678.

Dahl, F. 1912b. Über die Fauna des Plagefenn gebietes. Spinnen, Araneida. Beiträge zur Naturdenkmalpflege, 3: 575-622.

Dahl, F. 1912c. Arachnoidea. In Handwörterbuch der Naturwissenschaften 1, pages 485-514.

Dahl, F. 1913. Vergleichende Physiologie und Morphologie der Spinnentiere unter besonderer Berücksichtung der Lebensweise I. Jena. 113 pages.

Dahl, F. 1926. Spinnentiere oder Arachnoidea. Springspinnen (Salticidae). Die Tierwelt Deutschlands, 3: 1-55.

Dahl, F. and M. Dahl. 1927. Spinnentiere oder Arachnoidea. II. Lycosidae s. lat. Die Tierwelt Deutschlands, 5: 1-80.

Dalmas, R. 1916. Révision du genre *Orchestina* E.S., suivie de la description de nouvelles espèces du genre *Oonops* et d'un étude sur les Dictynidae du genre *Scololathys*. Annales de la Société entomologique de France, 85: 203-258.

Dalmas, R. 1917a. Araignées de Nouvelle Zélande. Annales de la Société entomologique de France, 86:

Darwin, C. 1909. The Voyage of the Beagle. Harvard Classics Edition, P.F. Collier & Son Company, New York. 524 pages.

Davies, V.T. 1988a. Three new species of the spider genus *Stiphidion* (Araneae: Amaurobioidea: Stiphidiidae) from Australia. Memoirs of the Queensland Museum, 25: 265-271.

Davies, V.T. 1998a. A revision of the Australian metaltellines (Aranaea: Amaurobioidea: Amphinectidae: Metaltellinae). Invertebrate Taxonomy, 12: 211-243.

Davies, V.T. 1999. *Carbinea*, a new spider genus from north Queensland, Australia (Araneae, Amaurobioidea, Kababininae). Journal of Arachnology, 27: 25-36.

Davies, V.T. and C. Lambkin. 2000. *Wabua*, a new spider genus (Araneae: Amaurobioidea: Kababininae) from North Queensland, Australia. Memoirs of the Queensland Museum, 46: 129-147.

De Geer, C. 1778b. Mémoires pour servir à l'histoire des Insectes. Tome 7, 4$^{\text{ème}}$ mémoire. Stockholm. Pages 241-324.

Decae, A.E. 1986. Dispersal: ballooning and other mechanisms. Pages 348-356 in W. Nentwig (editor): Ecophysiology of Spiders. Springer-Verlag, New York.

Deeleman-Reinhold, C.L. 1995d. The Ochyroceratidae of the Indo-Pacific region (Araneae). Raffles Bulletin of Zoology (Supplement) 2: 1-103.

Deeleman-Reinhold, C.L. 2001. Forest spiders of South East Asia: with a revision of the sac and ground spiders (Araneae: Clubionidae, Corinnidae, Liocranidae, Gnaphosidae, Prodidomidae and Trochanterriidae). Brill, Leiden. 591 pages.

Deeleman-Reinhold, C.L. and P.R. Deeleman. 1988. Révision des Dysderinae (Araneae, Dysderidae), les espèces méditerranéennes occidentales exceptées. Tijdschrift voor Entomologie, 131: 141-269.

Deeleman-Reinhold, C.L. and J.D. Prinsen. 1987. *Micropholcus fauroti* (Simon) n. comb., a pantropical, synanthropic spider (Araneae: Pholcidae). Entomologische Berichten, Amsterdam, 47(5): 73-77.

Denis, J. 1948a. Notes sur les érigonides. VII. Remarques sur le genre *Araeoncus* Simon et quelques genres voisins. Bulletin de la Société entomologique de France, 53: 17-32.

Dippenaar-Schoeman, A.S. and R. Jocqué. 1997. African Spiders. An identification manual. Plant Protection Research Institute Handbook, No. 9. Agricultural Research Council, South Africa, Pretoria. 392 pages.

Dondale, C.D. 1959. Definition of the genus *Grammonota* (Araneae: Erigonidae) with descriptions of seven new species. The Canadian Entomologist, 91(4): 232-242.

Dondale, C.D. 1961. Life histories of some common spiders from trees and shrubs in Nova Scotia. Canadian Journal of Zoology, 39: 777-787.

Dondale, C.D. 1964. Sexual behaviour and its application to a species problem in the spider genus *Philodromus* (Araneae: Thomisidae). Canadian Journal of Zoology, 42(5): 817-827.

Dondale, C.D. 1967. Sexual behavior and the classification of the *Philodromus rufus* complex in North America (Araneae: Thomisidae). Canadian Journal of Zoology, 45(4): 453-459.

Dondale, C.D. 1977. Life histories and distribution patterns of hunting spiders (Araneida) in an Ontario meadow. Journal of Arachnology, 4: 73-93.

Dondale, C.D. 1986. The subfamilies of wolf spiders (Araneae: Lycosidae). Pages 327-332 in J.A. Barrientos (editor): Actas X Congreso Internacional de Arachnologia, Volume 1. Barcelona, Spain.

Dondale, C.D. 1995. A new synonym in the genus *Meta* (Araneae, Tetragnathidae). Journal of Arachnology, 23(2): 125-126.

Dondale, C.D. 1999. Revision of the *groenlandicola* subgroup of the genus *Pardosa* (Araneae, Lycosidae). Journal of Arachnology, 27: 435-448.

Dondale, C.D. and D.J. Buckle. 2001. The genus *Maro* in North America (Araneae: Linyphiidae). Fabreries, 26(1): 9-15.

Dondale, C.D. and J.H. Redner. 1968. The *imbecillus* and *rufus* groups of the spider genus *Philodromus* in North America (Araneida: Thomisidae). Memoirs of the Entomological Society of Canada, No. 55, 78 pages.

Dondale, C.D. and J.H. Redner. 1969. The *infuscatus* and *dispar* groups of the spider genus *Philodromus* in North and Central America and the West Indies. The Canadian Entomologist, 101(9): 921-954.

Dondale, C.D. and J.H. Redner. 1975a. The genus *Ozyptila* in North America (Araneae, Thomisidae). Journal of Arachnology, 2(3): 129-181.

Dondale, C.D. and J.H. Redner. 1975b. The *fuscomarginatus* and *histrio* groups of the spider genus *Philodromus* in North America (Araneae: Thomisidae). The Canadian Entomologist, 107(4): 369-384.

Dondale, C.D. and J.H. Redner. 1975c. Revision of the spider genus *Apollophanes* (Araneida: Thomisidae). The Canadian Entomologist, 107: 1175-1192.

Dondale, C.D. and J.H. Redner. 1976a. A review of the spider genus *Philodromus* in the Americas (Araneida: Philodromidae). The Canadian Entomologist, 108: 127-157.

Dondale, C.D. and J.H. Redner. 1976c. A rearrangement of the North American species of *Clubiona*, with descriptions of two new species (Araneida: Clubionidae). The Canadian Entomologist, 108(11): 1155-1165.

Dondale, C.D. and J.H. Redner. 1978a. Revision of the Nearctic wolf spider genus *Schizocosa* (Araneida: Lycosidae). The Canadian Entomologist, 110: 143-181.

Dondale, C.D. and J.H. Redner. 1978b. The Crab Spiders of Canada and Alaska (Araneae: Philodromidae and Thomisidae). The Insects and Arachnids of Canada. Part 5. Agriculture Canada, Ottawa. Publication 1663. 255 pages.

Dondale, C.D. and J.H. Redner. 1979. Revision of the wolf spider genus *Alopecosa* Simon in North America (Araneae: Lycosidae). The Canadian Entomologist, 111(9): 1033-1055.

Dondale, C.D. and J.H. Redner. 1981a. Description of a new wolf spider in the genus *Pirata* (Araneae: Lycosidae). Psyche, 87(3/4): 193-197.

Dondale, C.D. and J.H. Redner. 1981b. Classification of two North American species of *Pirata*, with a description of a new genus (Araneae, Lycosidae). Bulletin of the American Museum of Natural History, 170: 106-110.

Dondale, C.D. and J.H. Redner. 1982. The Sac Spiders of Canada and Alaska (Araneae: Clubionidae and Anyphaenidae). The Insects and Arachnids of Canada. Part 9. Agriculture Canada, Ottawa. Publication 1724. 194 pages.

Dondale, C.D. and J.H. Redner. 1983a. Revision of the wolf spiders of the genus *Arctosa* C.L. Koch in North America (Araneae: Lycosidae). Journal of Arachnology, 11(1): 1-30.

Dondale, C.D. and J.H. Redner. 1983b. The wolf spider genus *Allocosa* in North and Central America (Araneae: Lycosidae). Canadian Entomologist, 115: 933-964.

Dondale, C.D. and J.H. Redner. 1984. Revision of the *milvina* group of the wolf spider genus *Pardosa* (Araneae: Lycosidae). Psyche, 91(1/2): 67-117.

Dondale, C.D. and J.H. Redner. 1986. The *coloradensis*, *xerampelina*, *lapponica*, and *tesquorum* groups of the genus *Pardosa* (Araneae: Lycosidae) in North America. The Canadian Entomologist, 118: 815-835.

Dondale, C.D. and J.H. Redner. 1987. The *atrata*, *cubana*, *ferruginea*, *moesta*, *monticola*, *saltuaria*, and *solituda* groups of the spider genus *Pardosa* in North America (Araneae: Lycosidae). The Canadian Entomologist, 119(1): 1-19.

Dondale, C.D. and J.H. Redner. 1990. The Wolf Spiders, Nurseryweb Spiders and Lynx Spiders of Canada and Alaska (Araneae: Lycosidae, Pisauridae, Oxyopidae). The Insects and Arachnids of Canada. Part 17. Agriculture Canada, Ottawa. Publication 1856. 383 pages.

Dondale, C.D., J.H. Redner and Y.M. Marusik. 1997. Spiders (Araneae) of the Yukon Territory. Pages 73-113 in H.V. Danks and J.A. Downes (editors): Insects of the Yukon. Biological Survey of Canada Monograph series. No. 2. Biological Survey of Canada (Terrestrial Arthropods). Ottawa.

Dondale, C.D., J.H. Redner, P. Paquin and H.W. Levi. 2003. The Orb-weaving Spiders of Canada and Alaska. Uloboridae, Tetragnathidae, Araneidae and Theridiosomatidae (Araneae). The Insects and Arachnids of Canada. Part 23. Agriculture Canada, Ottawa, National Research Council publications, NRC 44466. 371 pages.

Dondale, C.D., A.L. Turnbull and J.H. Redner. 1964. Revision of the Nearctic species of *Thanatus* C.L. Koch (Araneae: Thomisidae). The Canadian Entomologist, 96: 636-656.

Duffey, E. and P. Merrett. 1964. *Carorita limnaea* (Crosby and Bishop), a linyphiid spider new to Britain, from Wybunbury Moss, Cheshire. The Annals and Magazine of Natural History, 13th series, 69(6): 573-576.

Dufour, L. 1820c. Descriptions de cinq Arachnides nouvelles. Annales générales de science Physique, 5: 198-209.

Dufour, L. 1831. Descriptions et figures de quelques Arachnides nouvelles ou mal connues et procédé pour conserver à sec ces invertébrés dans les collections. Annales des Sciences Naturelles. Zoologie. Paris, 22: 355-371.

Dugès, A. 1836a. Observations sur les aranéides. Annales des Sciences Naturelles. Zoologie. Paris, 2(6): 159-219.

Dumitrescu, M. and M. Georgescu. 1983. Sur les Oonopidae (Araneae) de Cuba. Résultats des Expéditions Biospéologiques Cubano-Roumaines à Cuba, 4: 65-114.

Dumitrescu, M. and M. Georgescu. 1992. Ochyroceratides de Cuba (Araneae). Mémoires de Biospéléologie, 19: 143-153.

Dunlop, J.A. and P.A. Selden. 1998. The early history and phylogeny of the chelicerates. Pages 221-235 in R.A. Fortey & R.H. Thomas (editors): Arthropod Relationships. Chapman & Hall, London.

Dunlop, J.A. and M. Webster. 1999. Fossil evidence, terrestrialization and arachnid phylogeny. Journal of Arachnology, 27: 86-93.

Eason, R.R. 1969. Life history and behavior of *Pardosa lapidicina* Emerton (Araneae: Lycosidae). Journal of the Kansas Entomological Society, 42: 339-360.

Eason, R.R. and W.H. Whitcomb. 1965. Life history of the dotted wolf spider, *Lycosa punctulata* Hentz (Araneida: Lycosidae). Proceedings of the Arkansas Academy of Science, 19: 11-20.

Eberhard, W.G. 1967. Attack behavior of diguetid spiders and the origin of prey wrapping in spiders. Psyche, 74: 173-181.

Eberhard, W.G. 1971. The ecology of the web of *Uloborus diversus* (Araneae: Uloboridae). Oecologia, 6: 328-342.

Eberhard, W.G. 1972. The web of *Uloborus diversus* (Araneae: Uloboridae). Journal of the Zoological Society of London, 166: 417-465.

Eberhard, W.G. 1973. Stabilimenta on the webs of *Uloborus diversus* (Araneae: Uloboridae) and other spiders. Journal of the Zoological Society of London, 171: 367-384.

Eberhard, W.G. 1982. Behavioural characters for the higher classification of orb-weaving spiders. Evolution, 36: 1067-1095.

Eberhard, W.G. 1985. Sexual Selection and Animal Genitalia. Harvard University Press, Cambridge Massachussetts. 244 pages.

Eberhard, W.G. 1986. Web-building behavior of anapid, symphytognathid, and mysmenid spiders. Journal of Arachnology, 14: 339-358.

Eberhard, W.G. 1987a. Orb webs and construction behavior in Anapidae, Symphytognathidae, and Mysmenidae. Journal of Arachnology, 14(3): 339-356.

Eberhard, W.G. 1987b. Construction behaviour of non-orb weaving cribellate spiders and the evolutionary origin of orb webs. Bulletin of the British Arachnological Society, 7: 175-178.

Eberhard, W.G. 1990. Function and phylogeny of spider webs. Annual Review of Ecology and Systematics, 21: 341-372.

Eberhard, W.G. 1992. Web construction by *Modisimus* sp. (Araneae, Pholcidae). Journal of Arachnology, 20: 25-34.

Eberhard, W.G. 1996. Female Control: Sexual Selection by Cryptic Female Choice. Princeton University Press, Princeton. 472 pages.

Eberhard, W.G. 2004. Why study spider sex: special traits of spiders facilitate studies of sperm competition and cryptic female choice. Journal of Arachnology, 32: 545-556.

Eberhard, W.G., S. Guzman-Gomes and K.M. Catley. 1993. Correlation between spermathecal morphology and mating systems in spiders. Journal of the Linnean Society, 50: 197-209.

Eberhard, W.G. and F. Pereira. 1993. Ultrastructure of cribellate silk of nine species in eight families and possible taxonomic implications (Araneae: Amaurobiidae, Deinopidae, Desidae, Dictynidae, Filistatidae, Hypochilidae, Stiphidiidae, Tengellidae). Journal of Arachnology, 21: 161-174.

Edwards, G.B. 1979a. Pantropical jumping spiders occurring in Florida. Florida Department of Agriculture and Consumer Services, Division of Plant Industry, Entomology Circular, No. 199: 1-2.

Edwards, G.B. 1979b. The giant crab spider, *Heteropoda venatoria* (Linn.) (Araneae: Sparassidae). Florida Department of Agriculture and Consumer Services, Division of Plant Industry, Entomology Circular, No. 205: 1-2.

Edwards, G.B. 1980. Jumping spiders of the United States and Canada: changes in the key and list (4). Peckhamia, 2(1): 11-14.

Edwards, G.B. 1993c. *Crossopriza lyoni* and *Smeringopus pallidus*: cellar spiders new to Florida (Araneae: Pholcidae). Florida Department of Agriculture and Consumer Services, Division of Plant Industry, Entomology Circular, No. 361: 1-2.

Edwards, G.B. 1999a. The genus *Attidops* (Araneae, Salticidae). Journal of Arachnology, 27(1): 7-15.

Edwards, G.B. 1999b. *Corythalia canosa* (Araneae: Salticidae) reassigned to *Anasaitis*. Insecta Mundi, 12(3/4): 10.

Edwards, G.B. 2003a. A review of the Nearctic jumping spiders (Araneae: Salticidae) of the subfamily Euophryinae north of Mexico. Insecta Mundi, 16(1/3): 65-75.

Edwards, G.B. 2003b. A new species of *Neonella* (Araneae: Salticidae) from southeast Florida. Insecta Mundi, 16(1/3): 157-159.

Edwards, G.B. 2004. Revision of the jumping spiders of the genus *Phidippus* (Araneae: Salticidae). Occasional papers of the Florida State Collection of Arthropods, Volume 11: i -viii + 1-156.

Edwards, G.B. in press. A review of described *Metacyrba*, the status of *Parkella*, and notes on *Platycryptus* and *Balmaceda*, with a comparison of the genera (Araneae: Salticidae: Marpissinae). Insecta Mundi.

Edwards, G.B. and K.L. Hibbard. 1999. The Mexican Redrump, *Brachypelma vagans* (Araneae: Theraphosidae), an exotic tarantula established in Florida. Florida Department of Agriculture and Consumer Services, Division of Plant Industry, Entomology Circular No. 392: 1-2.

Edwards, G.B. and D.E. Hill. 1978. Representatives of the North American salticid fauna. Peckhamia, 1(5): 110-117.

Edwards, J.S., R.L. Crawford, P.M. Sugg and M.A. Peterson. 1987. Arthropod recolonization in the blast zone of Mount St. Helens. Pages 329-333 *in* S.A.C. Keller (editor): Mount St. Helens: five years later. Eastern Washington University Press, Cheney, WA.

Edwards, R.J. 1958. The spider subfamily Clubioninae of the United States, Canada and Alaska (Araneae: Clubionidae). Bulletin of the Museum of Comparative Zoology, 118(6): 363-436.

Edwards, R.L., E.H. Edwards and A.D. Edwards. 2003. Observations of *Theotima minutissimus* [sic] (Araneae, Ochyroceratidae), a parthenogenetic spider. Journal of Arachnology, 31(2): 274-277.

Eichwald, E. 1830. Araneae. Pages 63-73 *in* Zoologia specialis, quam expositis animalibus tum vivis, tum fossilibus et Poloniae in specie, in usum lectionum publicarum in universitate Caesarea Vilnensi habendarum edidit. pars prior, pars altera, pars posterior. Vilna.

Emerit, M. 1981. Sur quelques formations tégumentaires de la patte de *Telema tenella* (Araignée, Telemidae), observées au microscope électronique a balayage. Atti della Società Toscana di Scienze Naturali, 88: 45-52.

Emerit, M. 1984. Les glandes tibiales des Telemidae: Une structure inédite localisée sur l'appendice locomoteur des araignées. Comptes Rendus de l'Académie des Sciences, Paris, 299: 1-4.

Emerit, M. 1985. L'appareil glandulaire tégumentaire de la patte des Telemidae (Araneae). Un critère phylogénique? Mémoires de Biospéléologie, 12: 91-96.

Emerit, M. and C. Juberthie. 1983. Mise en évidence d'un équipement glandulaire tégumentaire sur la patte d'une araignée troglobie : *Telema tenella* (Telemidae). Mémoires de Biospéléologie, 10: 407-411 + pl. I-III.

Emerton, J.H. 1875. Notes on spiders in caves of Kentucky, Virginia and Indiana. The American Naturalist, 9(5): 278-281.

Emerton, J.H. 1882. New England spiders of the family Theridiidae. Transactions of the Connecticut Academy of Arts and Sciences, 6: 1-86 + pl. I-XXIV.

Emerton, J.H. 1885. New England Lycosidae. Transactions of the Connecticut Academy of Arts and Sciences, 6: 481-505 + pl. XLVI-XLIX.

Emerton, J.H. 1888a. New England spiders of the family Ciniflonidae. Transactions of the Connecticut Academy of Arts and Sciences, 7: 443-458 + pl. IX-XI.

Emerton, J.H. 1890 ["1889"]. New England spiders of the families Drassidae, Agalenidae, and Dysderidae. Transactions of the Connecticut Academy of Arts and Sciences, 8: 166-206 + pl. III-VIII.

Emerton, J.H. 1891. New England spiders of the family Attidae. Transactions of the Connecticut Academy of Arts and Sciences, 8: 220-252 + pl. XVI-XXI.

Emerton, J.H. 1894. Canadian spiders. Transactions of the Connecticut Academy of Arts and Sciences, 9: 400-429 + pl. I-IV.

Emerton, J.H. 1909. Supplement to the New England spiders. Transactions of the Connecticut Academy of Arts and Sciences, 14: 173-236 + pl. I-XII.

Emerton, J.H. 1911. New spiders from New England. Transactions of the Connecticut Academy of Arts and Sciences, 16: 385-407 + pl. I-VI.

Emerton, J.H. 1913a. New England spiders identified since 1910. Transactions of the Connecticut Academy of Arts and Sciences, 18: 210-224 + pl. I-II.

Emerton, J.H. 1913b. New and rare spiders from within fifty miles of New York City. Bulletin of the American Museum of Natural History, 32: 255-260 + pl. XLVIII.

Emerton, J.H. 1914a. New spiders from the neighbourhood of Ithaca, N.Y. Journal of the New York Entomological Society, 22(3): 262-264 + pl. VIII.

Emerton, J.H. 1915a. New spiders from New England, XI. Transactions of the Connecticut Academy of Arts and Sciences, 20: 135-142 + pl. I.

Emerton, J.H. 1915b. Canadian spiders , II. Transactions of the Connecticut Academy of Arts and Sciences, 20: 147-156 + pl. II-III.

Emerton, J.H. 1917a. New spiders from Canada and the adjoining states. The Canadian Entomologist, 49(8): 261-272.

Emerton, J.H. 1919a. New spiders from Canada and the adjoining states, No. 2. The Canadian Entomologist, 51(5): 105-108 + pl. VII.

Emerton, J.H. 1923. New spiders from Canada and the adjoining states, No. 3. The Canadian Entomologist, 55(10): 238-243.

Emerton, J.H. 1924b. New California spiders. The Pan-Pacific Entomologist, 1(1): 29-31.

Emerton, J.H. 1925a. New spiders from Canada and the adjoining states, No. 4. The Canadian Entomologist, 57(3): 65-69.

Ennik, F. 1971. Laboratory observations on the biology of *Loxosceles unicolor* Keyserling (Araneae, Loxoscelidae). Santa Barbara Museum of Natural History, Contributions in Science , 3: 1-16.

Erichson, G.F. 1845. Nomina systematica Generum Arachnidarum. *In* Agazziz (L.) Nomenclator Zoologicus. Soloduri. 14 pages.

Eskov, K.Y. 1979. Three new species of spiders of the family Linyphiidae from Siberia (Aranei) [in Russian]. Trudy Zoologicheskii Institut. Akademiia Nauk SSSR. [Proceedings of the Zoological Institute of Leningrad. USSR Academy of Sciences], Vol. 85: 65-72.

Eskov, K.Y. 1984a. Taxonomy of spiders of the genus *Oreonetides* (Aranei, Linyphiidae). Zoologicheskii Zhurnal, 63(5): 662-670.

Eskov, K.Y. 1986b. On spiders of the genus *Pero* Tanasevitch, 1985 (Aranei, Linyphiidae) [in Russian]. Zoologicheskii Zhurnal, 65(11): 1738-1742.

Eskov, K.Y. 1987d. Spiders of Nearctic genera *Ceraticelus* and *Islandiana* (Aranei, Linyphiidae) in the fauna of Siberia and Far East [in Russian]. Zoologicheskii Zhurnal, 66(11): 1748-1752.

Eskov, K.Y. 1988b. Seven new monotypic genera of spiders of the family Linyphiidae (Aranei) from Siberia [in Russian]. Zoologicheskii Zhurnal, 67(5): 678-690.

Eskov, K.Y. 1989a. New Siberian species of erigonine spiders (Arachnida, Aranei, Linyphiidae). Spixiana, 11(2): 97-109.

Eskov, K.Y. 1989b. New monotypic genera of the spider family Linyphiidae (Aranei) from Siberia: Communication 1. Zoologicheskii Zhurnal, 68(9): 68-78.

Eskov, K.Y. 1990a. New monotypic genera of the spider family Linyphiidae (Aranei) from Siberia: Communication 2. Zoologicheskii Zhurnal, 69(1): 43-53.

Eskov, K.Y. 1993. Several new linyphiid spider genera (Araneida, Linyphiidae) from the Russian Far East. Arthropoda Selecta, 2(3): 43-60.

Eskov, K.Y. and Y.M. Marusik. 1992a ["1991"]. On *Tunagyna* and *Thaleria*, two closely related Siberio-Nearctic spider genera (Araneidae: Linyphiidae). Entomologica Scandinavica, 22(4): 405-416.

Eskov, K.Y. and Y.M. Marusik. 1994. New data on the taxonomy and faunistics of North Asian linyphiid spiders. Arthropoda Selecta, 2(4): 41-79.

Exline, H. 1935a. Three new species of *Cybaeus*. The Pan-Pacific Entomologist, 11(3): 129-132.

Exline, H. 1936a. Nearctic spiders of the genus *Cicurina* Menge. American Museum Novitates, No. 850: 1-25.

Exline, H. 1936c ["1935b"]. A new species of *Cybaeus* (Araneae: Agelenidae). Entomological News, 46: 285-286.

Exline, H. 1948. Morphology, habits, and systematics position of *Allepeira lemniscata* (Walckenaer) (Araneida: Argiopidae, Allepeirinae). Annals of the Entomological Society of America, 41(3): 309-325.

Exline, H. 1950a. Conopisthine spiders (Theridiidae) from Peru and Ecaduor. Pages 108-124 in Studies Honoring Trevor Kincaid. University of Washington Press.

Exline, H. and H.W. Levi. 1962. American spiders of the genus *Argyrodes* (Araneae: Theridiidae). Bulletin of the Museum of Comparative Zoology, 127(2): 75-204.

Fabricius, J.C. 1775. Araneae, pages 431-441 In Systema Entomologiae, Sistens Insectorum Classes, Ordines, Genera, Species, Adiectis, Synonymis, Locis Descriptionibus Observationibus. Flensburg and Lipsiae.

Fabricius, J.C. 1787. Araneae, pages 342-346 In Mantissa Insectorum Sistens Eorum Species Nuper Detectas Adiectis Characteribus Genericis, Differentiis Specificis, Emendationibus, Observationibus, Tome 1, Hafniae.

Fabricius, J.C. 1793. Entomologia Systematica Emendata et Aucta, Secundum Classes, Ordines, Genera, Species Adjectis Nominis, Locis Observationibus Descriptionibus. Araneae tomus 2 (1793): 407-428. Hafniae 1792-1796, 4 tomes + index.

Fage, L. 1912. Études sur les araignées cavernicoles. I. Révision des Ochyroceratidae (n. fam.). Biospeleologica 29, Archives de Zoologie expérimentale et générale, 5ème Série, Tome X: 97-162.

Fage, L. 1913. Études sur les araignées cavernicoles. II. Révision de Leptonetidae. Biospeleologica 71, Archives de Zoologie expérimentale et générale, 5ème Série, Tome X: 479-576.

Fage, L. 1929c ["1928"]. Sur quelques Araignées des grottes de l'Amérique du Nord et de Cuba. Bolletino del Laboratorio di Zoologia Generale e Agraria del R. Istituto Superiore Agrario di Portici, 22: 181-187.

Fage, L. 1931a. Araneae, 5ème Série, précédée d'un essai sur l'évolution souterraine et son déterminisme. Biospeleologica LV, Archives de Zoologie expérimentale et générale, Tome 71: 91-291.

Fage, L. and A. de Barros Machado. 1951. Quelques particularités remarquables de l'anatomie des Ochyroceratides (Araneae). Archives de Zoologie expérimentale et générale, 87(3): 95-103.

Falconer, W. 1910. A new genus and species of spider (*Eboria caliginosa*). The Naturalist, 35: 83-88, 253-254.

Fergusson, I.C. 1972. Natural history of the spider *Hypochilus thorelli* Marx (Hypochilidae). Psyche, 79: 179-199.

Foelix, R.F. 1996. Biology of Spiders. Second edition. Oxford University press, New York. 330 pages.

Foelix, R.F. and H. Jung. 1978. Some anatomical aspects of *Hypochilus thorelli* with special reference to the calamistrum and cribellum. Symposium of the Zoological Society of London, No. 42: 417-422.

Forskål, P. 1775. Descriptiones Animalium Avium, Amphibiorium, Piscium, Insectorum, Vermium; quae in itinere orientali observavit Petrus Forskal. Post mortem auctoris edidit Carsten Niebuhr. Adjuncta est materia medica kahirina atque tabula maris Rubri geographica. Hauniae. Ex officina Molleri, 1775. xxxiv + 1-164 pages.

Förster, A. and P. Bertkau. 1883. Beiträge zur Kenntniss der Spinnenfauna der Rheinprovinz. Verhandlungen des naturhistorischen Vereines der preussischen Rheinlande und Westfalens, 40(10): 205-278 + pl. III.

Forster, L.M. 1985. Target discrimination in jumping spiders (Araneae: Salticidae). Pages 249-274 in F.G. Barth (editor): Neurobiology of Arachnids. Springer-Verlag, New York.

Forster, R.R. 1951. New Zealand spiders of the family Symphytognathidae. Records of the Canterbury Museum, 5: 231-244.

Forster, R.R. 1958. Spiders of the family Symphytognathidae from North and South America. American Midland Naturalist, 1885: 1-14.

Forster, R.R. 1959. The spiders of the family Symphytognathidae. Transactions Royal Society New Zealand, 86: 269-329.

Forster, R.R. 1967. The Spiders of New Zealand. Part I. Otago Museum Bulletin. No. 1. Dunedin.

Forster, R.R. 1970b. Desidae, Dictynidae, Hahniidae, Amaurobioididae, Nicodamidae. Pages 11-184 in The Spiders of New Zealand. Part III. Otago Museum Bulletin. No. 3. Dunedin.

Forster, R.R. and L.M. Forster. 1999. Spiders of New Zealand and Their Worldwide Kin. University of Otago Press. 270 pages.

Forster, R.R. and N.I. Platnick. 1977. A review of the spider family Symphytognathidae (Arachnida, Araneae). American Museum Novitates, No. 2619: 1-29.

Forster, R.R. and N.I. Platnick. 1984. A review of the archaeid spiders and their relatives, with notes on the limits of the superfamily Palpimanoidea (Arachnidae, Araneae). Bulletin of the American Museum of Natural History, 178: 1-106.

Forster, R.R. and N.I. Platnick. 1985. A review of the austral spider family Orsolobidae (Arachnida, Araneae), with notes on the superfamily Dysderoidea. Bulletin of the American Museum of Natural History, 181: 1-230.

Forster, R.R., N.I. Platnick and J. Coddington. 1990. A proposal and review of the spider family Synotaxidae (Araneae, Araneoidea), with notes on theridiid interrelationships. Bulletin of the American Museum of Natural History, 193: 1-116.

Forster, R.R., N.I. Platnick and M.R. Gray. 1987. A review of the spider superfamilies Hypochiloidea and Austrochiloidea (Araneae, Araneomorphae). Bulletin of the American Museum of Natural History, 185(1): 1-116.

Forster, R.R. and C.I. Wilton. 1973. Agelenidae, Stiphidiidae, Amphinectidae, Amaurobiidae, Neolanidae, Ctenidae, Psechridae. Pages 1-309 in The Spiders of New Zealand. Part IV. Otago Museum Bulletin No. 4. Dunedin.

Fox, I. 1937c. Notes on North American Agelenid spiders. The Canadian Entomologist, 69: 174-177.

Fox, I. 1937e. The Nearctic spiders of the family Heteropodidae. Journal of the Washington Academy of Sciences, 27(11): 461-474.

Franganillo Balboa, P. 1930. Aracnidos de Cuba. Mas aracnidos nuevos de la isla de Cuba. Instituto Nacional de Investigaciones Cientificos, Havana, 1: 47-99.

Franganillo Balboa, P. 1936b. Aracnidos recogidos durante el verano de 1934 (Prosigue). Estudios de "Belen", 1936 (57-58): 75-82.

Fuesslin, J.C. 1775. Verzeichniss der ihm bekannten schweizerischen Insekten, mit einer ausgemahlten Kupfertafel: nebst der Ankündigung eines neuen Insecten Werkes. Zurich und Winterthur. Pages 1-62 + pl. I.

Galiano, M.E. 1965b. Salticidae (Araneae) formiciformes IV. Revision del genero *Sarinda* Peckham 1892. Revista Museo Argentina Ciencias Naturales Entomología, 1: 267-312.

Galiano, M.E. 1966a. Salticidae (Araneae) formiciformes V. Revision del genero *Synemosyna* Hentz, 1846. Revista Museo Argentina Ciencias Naturales Entomología, 1: 339-380.

Galiano, M.E. 1969a. Salticidae (Araneae) formiciformes IX. Adicion a las revisiones de los generos *Martella* y *Sarinda*. Physis (Buenos Aires), 28: 247-255.

Galiano, M.E. 1972a. Salticidae (Araneae) formiciformes XIII. Revision del genero *Bellota* Peckham, 1892. Physis (Buenos Aires), 31: 463-484.

Galiano, M.E. 1980a. Revision del genero *Lyssomanes* Hentz, 1845 (Araneae, Salticidae). Opera Lilloana, 30: 1-104.

Galiano, M.E. 1988b. New species of *Neonella* Gertsch, 1936 (Araneae, Salticidae). Revue suisse de Zoologie, 95(2): 439-448.

Galiano, M.E. 1989a. Note on the genera *Admestina* and *Akela* (Araneae, Salticidae). Bulletin British Arachnological Society, 8(2): 49-50.

Garb, J.E. 1999. An adaptive radiation of Hawaiian Thomisidae: Biogeographic and genetic evidence. Journal of Arachnology, 27: 71-78.

Gay, C. 1849. Historia Fisica y Politica de Chile. Zoologia (3): 319-543. Paris. Aracnidos by H. Nicolet.

Gerhardt, U. 1928. Biologische Studien an griechischen, corsischen und deutchen Spinnen. Zeitschrift für Morphologie und Ökologie der Tiere, 10(4): 576-675.

Gerhardt, W. and A. Kästner. 1938-1941. Araneae-Echte Spinnen -. In W. Kükenthal and T. Krumbach (editors). Handbuch der Zoologie, 3(2): 497-656.

Gering, R.L. 1953. Structure and function of the genitalia in some American agelenid spiders. Smithsonian Miscellaneous Collections, 121(4): 1-84.

Gerstäcker, A. 1873. Arachnoidea. In C. von der Decken (editor): Reisen in Ostafrica. Leipzig. 3(2): 461-503 pages

Gertsch, W.J. 1932. A new generic name for *Coriarachne versicolor* Keyserling, with new species. American Museum Novitates, No. 563: 1-7.

Gertsch, W.J. 1933c. Diagnoses of new American spiders. American Museum Novitates, No. 637: 1-14.

Gertsch, W.J. 1934a. Notes on American Lycosidae. American Museum Novitates, No. 693: 1-25.

Gertsch, W.J. 1934b. Notes on American crab spiders (Thomisidae). American Museum Novitates, No. 707: 1-25.

Gertsch, W.J. 1934c. Some American spiders of the family Hahniidae. American Museum Novitates, No. 712: 1-32.

Gertsch, W.J. 1934d. Further notes on American spiders. American Museum Novitates, No. 726: 1-26.

Gertsch, W.J. 1935a. Spiders from the Southwestern United States, with description of new species. American Museum Novitates, No. 792: 1-31.

Gertsch, W.J. 1935b. New American spiders with notes on other species. American Museum Novitates, No. 805: 1-24.

Gertsch, W.J. 1936a. Further diagnosis of new American spiders. American Museum Novitates, No. 852: 1-27.

Gertsch, W.J. 1937. New American spiders. American Museum Novitates, No. 936: 1-7.

Gertsch, W.J. 1939b. A revision of the typical crab-spiders (Misumeninae) of America north of Mexico. Bulletin of the American Museum of Natural History, 76(7): 377-442.

Gertsch, W.J. 1941a. Report on some Arachnids from Barro Colorado Island, Canal Zone. American Museum Novitates, No. 1146: 1-14.

Gertsch, W.J. 1941b. New American spiders of the family Clubionidae. I. American Museum Novitates, No. 1147: 1-20.

Gertsch, W.J. 1942. New American spiders of the family Clubionidae. III. American Museum Novitates, No. 1195: 1-18.

Gertsch, W.J. 1945. Two interesting species of *Thallumetus*. Transactions of the Connecticut Academy of Arts and Sciences, 36: 191-197.

Gertsch, W.J. 1946a. Note on American spiders of the family Dictynidae. American Museum Novitates, No. 1319: 1-21.

Gertsch, W.J. 1946b. Five new spiders of the genus *Neoantistea*. Journal of the New York Entomological Society, 54: 31-36.

Gertsch, W.J. 1949. American Spiders. First edition. D. van Nostrand Company, New York. 285 pages.

Gertsch, W.J. 1951. New American linyphiid spiders. American Museum Novitates, No. 1514: 1-11.

Gertsch, W.J. 1953. The spider genera *Xysticus*, *Coriarachne*, and *Ozyptila* (Thomisidae, Misumeninae) in North America. Bulletin of the American Museum of Natural History, 102(4): 413-482.

Gertsch, W.J. 1955. The North American bolas-spiders of the genera *Mastophora* and *Agathostictus*. Bulletin of the American Museum of Natural History, 106(4): 221-254.

Gertsch, W.J. 1958a. The spider family Diguetidae. American Museum Novitates, No. 1904: 1-24.

Gertsch, W.J. 1958b. The spider genus *Loxosceles* in North America, Central America, and the West Indies. American Museum Novitates, No. 1907: 1-2.

Gertsch, W.J. 1958c. The spider family Hypochilidae. American Museum Novitates, No. 1912: 1-28.

Gertsch, W.J. 1958d. The spider family Plectreuridae. American Museum Novitates, No. 1920: 1-53.

Gertsch, W.J. 1960a. Description of American spiders of the family Symphytognathidae. American Museum Novitates, No. 1981: 1-40.

Gertsch, W.J. 1960b. The fulva group of the spider genus *Steatoda* (Araneae, Theridiidae). American Museum Novitates, No. 1982: 1-48.

Gertsch, W.J. 1961. The spider genus *Lutica*. Senckenbergiana Biologica, 42(4): 365-374.

Gertsch, W.J. 1964b. A review of the genus *Hypochilus* and a description of a new species from Colorado (Araneae, Hypochilidae). American Museum Novitates, No. 2203: 1-14.

Gertsch, W.J. 1971a. A report on some Mexican cave spiders. Association for Mexican Cave Studies, Bulletin 4, 47-111.

Gertsch, W.J. 1973a. A report on cave spiders from Mexico and Central America. Association for Mexican Cave Studies, Bulletin 5, 141-163.

Gertsch, W.J. 1974. The spider family Leptonetidae in North America. Journal of Arachnology, 1: 145-203.

Gertsch, W.J. 1979. American Spiders. Second edition. Van Nostrand Reinhold Company, Toronto. 274 pages.

Gertsch, W.J. 1982b. The spider genera *Pholcophora* and *Anopsicus* (Araneae, Pholcidae) in North America, Central America and the West Indies. Bulletin of the Texas Memorial Museum (8). Association for Mexican Cave Studies, Bulletin 28, 95-144.

Gertsch, W.J. 1984. The spider family Nesticidae (Araneae) in North America, Central America, and the West Indies. Bulletin of the Texas Memorial Museum. No. 31. 91 pages.

Gertsch, W.J. 1986. The spider genus *Metagonia* (Araneae: Pholcidae) in North America, Central America, and the West Indies. Texas Memorial Museum, Speleological Monographs, 1. Studies on the cave and endogean fauna of North America. I, 39-62.

Gertsch, W.J. 1992. Distribution patterns and speciation in North American cave spiders with a list of the troglobites and revision of the cicurinas of the subgenus *Cicurella*. Texas Memorial Museum Speleological Monographs, 3. Studies on the Endogean Fauna of North America. II, 75-122.

Gertsch, W.J. and A.F. Archer. 1942. Description of new American Theridiidae. American Museum Novitates, No. 1171: 1-16.

Gertsch, W.J. and L.I. Davis. 1936. New spiders from Texas. American Museum Novitates, No. 881: 1-21.

Gertsch, W.J. and L.I. Davis. 1937. Report of a collection of spiders from Mexico. I. American Museum Novitates, No. 961: 1-29.

Gertsch, W.J. and L.I. Davis. 1946. Report on a collection of spiders from Mexico. V. American Museum Novitates, No. 1313: 1-11.

Gertsch, W.J. and F. Ennik. 1983. The spider genus *Loxosceles* in North America, Central America and the West Indies (Araneae, Loxoscelidae). Bulletin of the American Museum of Natural History, 175: 263-360.

Gertsch, W.J. and W. Ivie. 1936. Descriptions of new American spiders. American Museum Novitates, No. 858: 1-25.

Gertsch, W.J. and W. Ivie. 1955. The spider genus *Neon* in North America. American Museum Novitates, No. 1743: 1-17.

Gertsch, W.J. and S. Mulaik. 1936a. Diagnoses of new southern spiders. American Museum Novitates, No. 851: 1-21.

Gertsch, W.J. and S. Mulaik. 1940. The spiders of Texas. I. Bulletin of the American Museum of Natural History, 77(6): 307-340.

Gertsch, W.J. and S.B. Peck. 1992. The pholcid spiders of the Galápagos Islands, Ecuador (Araneae: Pholcidae). Canadian Journal of Zoology, 70: 1185-1199.

Gertsch, W.J. and N.I. Platnick. 1975. A revision of the trapdoor spider genus *Cyclocosmia* (Araneae, Ctenizidae). American Museum Novitates, No. 2580: 1-20.

Gertsch, W.J. and N.I. Platnick. 1979. A revision of the spider family Mecicobothriidae (Araneae, Mygalomorphae). American Museum Novitates, No. 2687: 1-32.

Gertsch, W.J. and N.I. Platnick. 1980. A revision of the American spiders of the family Atypidae (Araneae, Mygalomorphae). American Museum Novitates, No. 2704: 1-39.

Gertsch, W.J. and S.E. Riechert. 1976. The spatial and temporal partitioning of a desert spider community, with descriptions of new species. American Museum Novitates, No. 2604: 1-25.

Gertsch, W.J. and H.K. Wallace. 1935. Further notes on American Lycosidae. American Museum Novitates, No. 794: 1-22.

Gertsch, W.J. and H.K. Wallace. 1936. Notes on new and rare American mygalomorph spiders. American Museum Novitates, No. 884: 1-25.

Gertsch, W.J. and H.K. Wallace. 1937. New American Lycosidae with notes on other species. American Museum Novitates, No. 919: 1-22.

Getty, R.M. and F.A. Coyle. 1996. Observations on prey capture and antipredator behaviors of ogre-faced spiders (*Deinopis*) in southern Costa Rica (Araneae, Deinopidae). Journal of Arachnology, 24: 93-100.

Giebel, C.G. 1869a. Ueber einige Spinnen aus Illinois. Zeitschrift für die gesammten Naturwissenschaften, 33: 248-253.

Gillespie, R.G., H.B. Croom and S.R. Palumbi. 1994. Multiple origins of a spider radiation in Hawaii. Proceedings of the National Academy of Sciences of the United States of America, 91: 2290-2294.

von Gistel, J.N.F.X. 1848. Naturgeschichte des Thierreichs für höhere Schulen. Hoffman, Stuttgart. i-xvi + 1-216 + pl. I-XXXII.

von Gistel, J.N.F.X. and T. Bromme. 1850. Spinnen, pages 510-514 in Handbuch der Naturgeschichte aller drei Reiche, für Lehrer und Lernende, für Schule und Haus. Stuttgart. 1-1037 + pl. I-XLVIII pages.

Glatz, L. 1967. Zur biologie und morphologie von *Oecobius annulipes* Lucas. Zeitschrift für Morphologie und Ökologie der Tiere, 61: 185-214.

Goloboff, P.A. 1993. A reanalysis of mygalomorph spider families (Araneae). American Museum Novitates, No. 3056: 1-32.

Goloboff, P.A. 1995. A revision of the South American spiders of the family Nemesiidae (Araneae, Mygalomorphae). Part I: Species from Peru, Chile, Argentina, and Uruguay. Bulletin of the American Museum of Natural History, 224: 1-189.

Gosline, J.M., M.E. Demont and M.W. Denny. 1986. The structure and properties of spider silk. Endeavour, 10(1): 37-43.

Gosline, J.M., P.A. Guerette, C.S. Ortlepp and K.N. Savage. 1999. The mechanical design of spider silks: from fibroin sequence to mechanical function. Journal of Experimental Biology, 202: 3295-3033.

Grasshoff, M. 1983. *Larinioides* Caporiacco 1934, der korrekte name für die sogenannte Araneus cornutus-gruppe (Arachnida: Araneae). Senckenbergiana Biologica, 64 (1/3): 225-229.

Gray, M.R. 1983a. The taxonomy of the semi-communal spiders commonly referred to the species *Ixeuticus candidus* (L. Koch) with notes on the genera *Phryganoporus*, *Ixeuticus* and *Badumna* (Araneae, Amaurobioidea). Proceedings of the Linnean Society of New South Wales, 106: 247-261.

Gray, M.R. 1995. Morphology and relationships within the spider family Filistatidae (Araneae: Araneomorphae). Records of the Western Australian Museum. Supplement, 52: 79-89.

Grimm, R. 1986. Die Clubionidae Mitteleuropas: Corinninae und Liocraninae (Arachnida, Araneae). Abhandlungen des Naturwissenschaftlichen Vereins in Hamburg, 27: 1-91.

Griswold, C.E. 1977. Araneae (spiders). Pages 136-14 in W. Berry & E. Schlinger (editors): Inglenook Fen: A Study and Plan. State of California Department of Parks and Recreation.

Griswold, C.E. 1987a. A revision of the jumping spider genus *Habronattus* F.O.P.-Cambridge (Araneae; Salticidae), with phenetic and cladistic analyses. University of California Publications in Entomology, 107: 1-344.

Griswold, C.E. 1993. Investigations into the phylogeny of the lycosoid spiders and their kin (Arachnida: Araneae: Lycosoidea). Smithsonian Contributions to Zoology, No. 539: 1-39.

Griswold, C.E., J.A. Coddington, G. Hormiga and N. Scharff. 1998. Phylogeny of the orb-web building spiders (Araneae, Orbiculariae: Deinopoidea, Araneoidea). Zoological Journal of the Linnean Society, 123: 1-99.

Griswold, C.E., J.A. Coddington, N.I. Platnick and R.R. Forster. 1999a. Towards a phylogeny of entelegyne spiders (Araneae, Araneomorphae, Entelegynae). Journal of Arachnology, 27: 53-63.

Griswold, C.E., C. Long and G. Hormiga. 1999b. A new spider of the genus *Pimoa* from the Gaoligong Shan Mts., Yunnan, China (Araneae, Pimoidae). Acta Botanica Yunnanica (supplement) 11: 91-97.

Griswold, C.E. and D. Ubick. 2001. Zoropsidae: a spider family newly introduced to the USA (Araneae, Entelegynae, Lycosoidea). Journal of Arachnology, 29: 111-113.

Griswold, C.E. and H.M. Yang. 2003. On the egg-guarding behavior of a Chinese symphytognathid spider of the genus *Patu* Marples, 1951 (Araneae, Araneoidea, Symphytognathidae). Proceedings of the California Academy of Sciences, 54: 356-360.

Grothendieck, K. and O. Kraus. 1994. Die Wasserspinne *Argyroneta aquatica*: Verwandtschaft und Spezialisation (Arachnida, Araneae, Agelenidae). Verhanlungden des naturwissenschaftlichen Vereins in Hamburg (Neue Folge), 34: 259-273.

Guarisco, H. 2001. The spider family Zoridae (Araneae) in Kansas. Transactions of the Kansas Academy of Science, 104: 239-240.

Guarisco, H. 2003. *Scytodes globula* Nicolet, a synanthropic spitting spider (Araneae: Scytodidae) newly discovered in North America. Journal of the Kansas Entomological Society, 76: 638-639.

Guarisco, H. and B. Cutler. 2003. *Crossopriza lyoni* (Araneae: Pholcidae), a synanthropic cellar spider newly discovered in Kansas. Transactions of the Kansas Academy of Science, 106(1/2): 105-106.

Hackman, W. 1952b ["1951"]. Contributions to the knowledge of Finnish spiders. Memoranda Societatis pro Fauna et Flora Fennica, 27: 69-79.

Hackman, W. 1954. The spiders of Newfoundland. Acta Zoologica Fennica, 79: 1-99.

Hahn, C.W. 1826. Monographie der Spinnen. Heft 4: 1-2 + pl. I-IV.

Hardman, J.M. and A.L. Turnbull. 1974. The interaction of spatial heterogeneity, predator competition and the functional response to prey density in a laboratory system of wolf spiders (Araneae: Lycosidae) and fruit flies (Diptera: Drosophilidae). Journal of Animal Ecology, 43: 155-171.

Hardman, J.M. and A.L. Turnbull. 1980. Functional response of the wolf spider, *Pardosa vancouveri*, to changes in the density of vestigial-winged fruit flies. Researches in Population Ecology, Kyoto University, 21: 233-259.

Harvey, M.S. 1995. The systematics of the spider family Nicodamidae (Araneae: Amaurobioidea). Invertebrate Taxonomy, 9: 279-386.

van Hasselt, A.W.M. 1887a. Études sur le genre *Nops*. Tijdschrift vor Entomologie 30: 67-86

Hausdorf, B. 1999. Molecular phylogeny of araneomorph spiders. Journal of Evolutionary Biology, 12: 980-985.

Hawthorn, A.C. and B.D. Opell. 2002. Evolution of adhesive mechanisms in cribellar spider prey capture thread: evidence for van der Waals and hygroscopic forces. Biological Journal of the Linnean Society, 77: 1-8.

Haynes, D.L. and P. Sisojevic. 1966. Predatory behavior of *Philodromus rufus* Walckenear (Araneae: Thomisidae). The Canadian Entomologist, 98(2): 113-133.

Hebets, E.A., G.E. Stratton and G.L. Miller. 1996. Habitat and courtship behavior of the wolf spider *Schizocosa retrorsa* (Banks) (Araneae, Lycosidae). Journal of Arachnology, 24: 141-147.

Hedin, M.C. 1997a. Molecular phylogenetics at the population/species interface in cave spiders of the Southern Appalachians (Araneae: Nesticidae: *Nesticus*). Molecular Biology and Evolution, 14(3): 309-324.

Hedin, M.C. 1997b. Speciational history in a diverse clade of habitat-specialized spiders (Araneae: Nesticidae: *Nesticus*): inferences from geographic-based sampling. Evolution, 51(6): 1929-1945.

Hedin, M.C. 2001. Molecular insights in species phylogeny, biogeography, and morphological stasis in the ancient spider genus *Hypochilus* (Araneae: Hypochilidae). Molecular Phylogenetics and Evolution, 18(2): 238-251.

Hedin, M.C. and B. Dellinger. 2005. Descriptions of a new species and previously unknown males of *Nesticus* (Araneae: Nesticidae) from caves in eastern North America, with comments on species rarity. Zootaxa, 904: 1-19.

Hedin, M.C. and W.P. Maddison. 2001. A combined molecular approach to phylogeny of the jumping spider subfamily Dendryphantinae (Araneae: Salticidae). Molecular Phylogenetics and Evolution, 18(3): 386-403.

Hedin, M.C. and D.A. Wood. 2002. Genealogical exclusivity in geographically proximate populations of *Hypochilus thorelli* Marx (Araneae, Hypochilidae) on the Cumberland Plateau of North America. Molecular Ecology, 11: 1975-1988.

Heimer, S. and W. Nentwig. 1991. Spinnen Mitteleuropas, Ein Bestimmungsbuch. Verlag Paul Parcy, Berlin and Hamburg. 543 pages.

Heiss, J.S. A systematic study of the spider genus *Calymmaria* (Araneae, Agelenidae) Unpublished Ph.D. thesis, University of Arkansas, 148 pages.

Heiss, J.S. and M.L. Draney. 2004. Revision of the Nearctic spider genus *Calymmaria* (Araneae, Hahniidae). Journal of Arachnology, 32: 457-525.

Heller, G. 1976. Zum Beutefangverhalten der ameisenfressenden Spinne *Callilepis nocturna* (Aranchnida: Araneae: Drassodidae). Entomologica Germanica, 3: 100-103.

van Helsdingen, P.J. 1968. Comparative notes on the species of the Holarctic genus *Stemonyphantes* Menge (Araneida, Linyphiidae). Zoologische Medededingen. Uitgegeven door het. Rijksmuseum van Natuurlijke Historie te Leiden, 43(10): 117-139.

van Helsdingen, P.J. 1969. A reclassification of the species *Linyphia* Latreille based on the functioning of the genitalia (Araneae, Linyphiidae). I. Zoologische Verhandelingen Uitgegeven Door Het Rijksmuseum van Natuurlijke Historie te Leiden, 105: 1-306.

van Helsdingen, P.J. 1970. A reclassification of the species of *Linyphia* based on the functioning of the genitalia (Araneida, Linyphiidae), II. Zoologische Verhandelingen Rijksmuseum van Natuurlijke Historie te Leiden, 111: 1-86.

van Helsdingen, P.J. 1973b. Annotations on two species of Linyphiid spiders described by the late Wilton Ivie. Psyche, 80(1/2): 48-61.

van Helsdingen, P.J. 1973c. A recapitulation of the Nearctic species of *Centromerus* Dahl (Araneida, Linyphiidae) with remarks on *Tunagyna debilis* (Banks). Zoologische Verhandelingen Rijksmuseum van Natuurlijke Historie te Leiden, 124: 1-45.

van Helsdingen, P.J. 1974. The affinities of *Wubana* and *Allomengea* with some notes on the latter genus (Araneae, Linyphiidae). Zoologische Mededelingen Rijksmuseum van Natuurlijke Historie te Leiden, 46(22): 295-321.

van Helsdingen, P.J. 1981a. The Nearctic species of *Oreonetides* (Araneae, Linyphiidae). Bulletin of the American Museum of Natural History, 170: 229-241.

Hendrixson, B. and J. Bond. 2005. Two sympatric species of *Antrodiaetus* from southwestern North Carolina (Araneae, Mygalomorphae, Antrodiaetidae). Zootaxa, 872: 1-19.

Hentz, N.M. 1832. On North American spiders. The American Journal of Science and Arts, 21: 99-109 + appendix 109-122.

Hentz, N.M. 1835. VII. Araneids Latr. The spiders. Pages 550-552 in A catalogue of the animals and plants in Massachusetts. Report on the Geology, Mineralogy, Botany and Zoology of Massachussets. Second edition. Amherst.

Hentz, N.M. 1841a. Description of an American spider, constituting a new subgenus of the tribe Inaequitelae of Latreille. Silliman's American Journal of Science and Arts, 41: 115-117.

Hentz, N.M. 1841b. Species of mygale of the United States. Proceedings of the Boston Society of Natural History, 1: 41-42.

Hentz, N.M. 1842a. Descriptions and figures of the araneides of the United States. Boston Journal of Natural History, 4: 54-57, + pl. VII.

Hentz, N.M. 1842b. Descriptions and figures of the araneides of the United States. Boston Journal of Natural History, 4: 223-231+ pl. VIII.

Hentz, N.M. 1844. Descriptions and figures of the araneides of the United States. Boston Journal of Natural History, 4: 386-396 + pl. XVII-XIX.

Hentz, N.M. 1845. Descriptions and figures of the araneides of the United States. Boston Journal of Natural History, 5: 189-202 + pl. XVI-XVII.

Hentz, N.M. 1846. Descriptions and figures of the araneides of the United States. Boston Journal of Natural History, 5: 352-370 + pl. XXI-XXII.

Hentz, N.M. 1847. Descriptions and figures of the araneides of the United States. Boston Journal of Natural History, 5: 443-478, + pl. XXIII-XXIV, XXX-XXXI.

Hentz, N.M. 1850a. Descriptions and figures of the araneides of the United States. Boston Journal of Natural History, 6: 18-35 + pl. III-IV.

Hentz, N.M. 1850b. Descriptions and figures of the araneides of the United States. Boston Journal of Natural History, 6: 271-295 + pl. IX-X.

Hentz, N.M. 1875. The spiders of the United States. A collection of the arachnological writings of Nicholas Marcellus Hentz, M.D. Edited by Edward Burgess with notes and descriptions by James H. Emerton. Occasional Papers Boston Society of Natural History, 2: 1-171 + pl. I-XXI.

Herrich-Schäffer, G.W. 1833. Faunae Insectorum Germaniae initia. Regensburg. Arachniden [spiders by C.L. Koch] 119 folio 1-4; 120 folio 1-3, 8-11, 17-24.

Herrich-Schäffer, G.W. 1834. Faunae Insectorum Germaniae initia. Regensburg. Arachniden [spiders by C.L. Koch], 121 folio 13-24; 122 folio 9-24; 123 folio 1-24; 124 folio 16-24; 125 folio 11-24; 127 folio 16-24.

Herrich-Schäffer, G.W. 1835. Faunae Insectorum Germaniae initia. Regensburg. Arachniden [spiders by C.L. Koch] 128 folio 8-16; 23-24; 129 folio 12-24; 130 folio 13-24; 131 folio 1-24; 134 folio 1-24.

Herrich-Schäffer, G.W. 1836b. Faunae Insectorum Germaniae initia. Regensburg. Arachniden [spiders by C.L. Koch], 137 folio 3-6; 138 folio 3-6; 139 folio 3-6; 141 folio 3-6.

Hickman, V.V. 1931. A new family of spiders. Proceedings of the Zoological Society of London, (B) 1931(4): 1321-1328.

Hill, D.E. 1979. The scales of salticid spiders. Zoological Journal of the Linnean Society, 65(3): 193-218.

Hippa, H. and I. Oksala. 1982. Definition and revision of the *Enoplognatha ovata* (Clerck) group (Araneae: Theridiidae). Entomologica Scandinavica, 13(2): 213-222.

Hite, J.M., W.J. Gladney, J.L. Lancaster Jr. and W.H. Whitcomb. 1966. Biology of the brown recluse spider. Agricultural experiment station, Division of Agriculture, University of Kansas, Fayetteville, Bulletin 711: 1-26.

Hlivko, J.T. and A.L. Rypstra. 2003. Spiders reduce herbivory: non-lethal effects of spiders on the consumption of soybean leaves by beetle pests. Annals of the Entomological Society of America, 96: 914-919.

Hodge, M.A. and S.D. Marshall. 1996. An experimental analysis of intraguild predation among three genera of web-building spiders: *Hypochilus, Coras* and *Achaearanea* (Araneae: Hypochilidae, Amaurobiidae and Theridiidae). Journal of Arachnology, 24: 101-110.

Hoffman, A. 1976. Relacion bibliografica preliminar de las Arañas de Mexico (Arachnida: Araneae). Instituto Biologia Universidad Nacional Autonoma Mexico, Publicacion Especial, 3: 1-117.

Hoffman, R.L. 1963. A second species of the spider genus *Hypochilus* from eastern North America. American Museum Novitates, No. 2148: 1-8.

Hogg, H.R. 1915a. Report on the spiders collected by the British Ornithologists' Union Expedition and the Wollaston Expedition in Dutch New Guinea. Transactions of the Zoological Society of London, 1915: 501-528.

Holm, Å. 1939b. Neue Spinnen aus Schweden. Beschreibung neuer Arten der Familien Drassidae, Theridiidae, Linyphiidae und Microphantidae. Arkiv för Zoologi, 31(8): 1-38.

Holm, Å. 1943. Zur Kenntnis der Taxonomie, Ökologie und Verbreitung der swedischen Arten der Spinnengattungen *Rhaebothorax* Sim., *Typhochraestus* Sim., und *Latithorax* n. gen. Arkiv för Zoologi, 34 (A19): 1-32.

Holm, Å. 1944. Revision einiger norwegischer Spinnenarten und Bemerkungen über deren vorkommen in Schweden. Entomologisk Tidskrift, 65: 122-134.

Holm, Å. 1945b. Zur Kenntnis der Spinnenfauna des Tornetträsk-Gebietes. Arkiv för Zoologi, 36(15): 1-80.

Holm, Å. 1960a. Notes on Arctic spiders. Arkiv för Zoologi, 12(32): 511-514.

Holm, Å. 1960b. On a collection of spiders from Alaska. Zoologiska Bidrag fran Uppsala, 33: 109-134.

Holm, Å. 1967. Spiders (Araneae) from west Greenland. Meddelelser om Grønland, 184(1): 1-99.

Holm, Å. 1973. On the spiders collected during the Swedish expeditions to Novaya Zemlya and Yenisey in 1875 and 1876. Zoologica Scripta, 2(2/3): 71-110.

Holmberg, E.L. 1876. Arachnidos argentinos. Anales de Agricultura de la Republica Argentina. Buenos aires, 4: 1-30.

Holmberg, E.L. 1882. Observations à propos du sous-ordre des Araignées territelaires (Territelariae) spécialement du genre Nord-Americain *Catadysas*, Hentz et de la nouvelle famille Mecicobothrioidae, Holm. Boletin de la Academia Nacional de ciencias en Cordoba (Republica Argentina), 4(2): 153-174 + pl. I.

Holmberg, R.G. and A.L. Turnbull. 1982. Selective predation in a euryphagous invertebrate predator, *Pardosa vancouveri* (Arachnida: Araneae). The Canadian Entomologist, 114: 243-257.

Homann, H. 1971. Die Augen der Araneae: anatomie, ontogenie un der bedeutung fur die systematik (Chelicerata, Arachnida). Zeitschrift fur Morphologie und Ökologie der Tiere, 69: 201-272.

Homann, H. 1975. Die Stellung der Thomisidae und der Philodromidae im System der Araneae (Chelicerata, Arachnida). Zeitschrift fur Morphologie und Ökologie der Tiere, 80: 181-202.

Hormiga, G. 1993. Implications of the phylogeny of Pimoidae for the systematics of linyphiid spiders (Araneae, Araneoidea, Linyphiidae). Memoirs of the Queensland Museum, 33: 533-542.

Hormiga, G. 1994a. A revision and cladistic analysis of the spider family Pimoidea (Araneoidea: Araneae). Smithsonian Contributions to Zoology, No. 549: 1-104.

Hormiga, G. 1994b. Cladistics and the comparative morphology of linyphiid spiders and their relatives (Araneae, Araneoidea, Linyphiidae). Zoological Journal of the Linnean Society, 111: 1-71.

Hormiga, G. 2000. Higher level phylogenetics of erigonine spiders (Aranea, Linyphiidae, Erigoninae). Smithsonian Contributions to Zoology, No. 609: 1-160.

Hormiga, G. 2003. *Weintrauboa*, a new genus of pimoid spiders from Japan and adjacent islands, with comments on the monophyly and diagnosis of the family Pimoidae and the genus *Pimoa* (Araneoidea, Araneae). Zoological Journal of the Linnean Society, 139: 261-281.

Hormiga, G. and H.G. Döbel. 1990. A new *Glenognatha* (Araneae: Tetragnathidae) from New Jersey. Journal of Arachnology, 18: 195-204.

Hormiga, G., W.G. Eberhard and J.A. Coddington. 1995. Web-construction behaviour in Australian *Phonognatha* and the phylogeny of nephiline and tetragnathid spiders (Araneae: Tetragnathidae). Australian Journal of Zoology, 43: 313-364.

Hormiga, G., J.A. Miller and F. Alvarez-Padilla. 2005. LinyGen, Linyphioid genera of the world (Pimoidae and Linyphiidae) an illustrated catalog. www.gwu.edu/-clade/spiders/.

Howard, L.O. 1883. A list of the invertebrate fauna of South Carolina. Charleston. 47 pages.

Huber, B.A. 1994. Copulatory mechanics in the funnel-web spiders *Histopona torpida* and *Textrix denticulata* (Agelenidae, Araneae). Acta Zoologica, 75: 379-384.

Huber, B.A. 1995. The retrolateral tibial apophysis in spiders-shaped by sexual selection? Zoological Journal of The Linnean Society, 113: 151-163.

Huber, B.A. 1997a. On the distinction between *Modisimus* and *Hedypsilus* (Pholcidae; Araneae), with notes on behavior and natural history. Zoologica Scripta, 25(3): 233-240.

Huber, B.A. 1998e. Notes on the neotropical spider genus *Modisimus* (Pholcidae, Araneae), with descriptions of thirteen new species from Costa Rica and neighboring countries. Journal of Arachnology, 26: 19-60.

Huber, B.A. 2000. New world pholcid spiders (Araneae: Pholcidae): a revision at generic level. Bulletin of the American Museum of Natural History, 254: 1-348.

Huber, B.A. 2002. Functional morphology of the genitalia in the spider *Spermophora senoculata* (Pholcidae, Araneae). Zoologischer Anzeiger, 241: 105-116.

Huber, B.A. 2004. Evolutionary transformation from muscular to hydraulic movements in spider (Arachnida, Araneae) genitalia: a study based on histological serial sections. Journal of Morphology, 261: 364-376.

Huber, B.A., C.L. Deeleman-Reinhold and A. Pérez González. 1999. The spider genus *Crossopriza* (Araneae, Pholcidae) in the New World. American Museum Novitates, No. 3262: 1-10.

Huber, K.C., T.S. Haider, M.W. Muller, B.A. Huber, R.J. Schweyen and F.G. Barth. 1993. DNA sequence data indicates the polyphyly of the family Ctenidae (Araneae). Journal of Arachnology, 21: 194-201.

Huff, R.P. and F.A. Coyle. 1992. Systematics of *Hypochilus sheari* and *Hypochilus coylei*, two southern Appalachian lampshade spiders (Araneae, Hypochilidae). Journal of Arachnology, 20: 40-46.

Hull, J.E. 1911a. Papers on spiders, I. The genus *Tmeticus* (Simon, 1884; Cambridge, 1900) and some allied genera. Transactions of the Natural History Society of Northumberland, Durham, and Newcastle-upon-Tyne (New Series), 3(3): 573-587 + pl. XV.

Hull, J.E. 1916. Notes and records. Spiders. The Vasculum, 2: 90-91.

Hull, J.E. 1920. The spider family Linyphiidae: an essay in taxonomy. The Vasculum, 6(1): 7-11.

Hull, J.E. 1932. Nomenclature of British linyphiid spiders: a brief examination of Simon's French catalogue. Transactions of the Northern Naturalists' Union, 1: 104-110.

Hutchinson, R. and R. Limoges. 1998. Première mention de *Synageles venator* (Lucas) (Araneae: Salticidae) pour l'Amérique du Nord. Fabreries, 23(1): 10-16.

ICZN (International Commission on Zoological Nomenclature). 1948. Clerck, 1757, "Aranei Svecici" : proposed validation for nomenclatorial purposes of the names published in: discussion on, concluded. Bulletin of Zoological Nomenclature, 4(10\12): 315-319.

ICZN (International Commission on Zoological Nomenclature). 1958. Opinion 517. Rejection of the proposal for the emendation to "*Theridium*" of the generic name "*Theridion*" Walckenaer, 1805; use of the above powers (i) to validate the specific name "*picta*" Walckenaer, 1802, as published in the combination "*Aranea picta*"; (ii) to designate the species so named to be the type species of "*Theridion*" Walckenaer, 1805, (iii) to suppress the generic name "*Phyllonethis*" Thorell, 1869, (Class Arachnida). Opinions and declarations rendered by the International Commission on Zoological Nomenclature, 19(3): 45-110.

ICZN (International Commission on Zoological Nomenclature). 1975. Opinion 1038. *Argiope* Audouin, 1826 (Arachnida, Aranea): placed on the official list of generic names in zoology. Bulletin of Zoological Nomenclature, 32(2): 105-109.

ICZN (International Commission on Zoological Nomenclature). 1987b. Opinion 1426. *Argyrodes* Simon, 1864, and *Robertus* O. Pickard-Cambridge, 1879 (Arachnida, Araneae): conserved. Bulletin of Zoological Nomenclature, 44(1): 59-60.

ICZN (International Commission on Zoological Nomenclature). 1991. Opinion 1637. *Aphonopelma* Pocock, 1901 (Arachnida, Araneae): given precedence over *Rhechostica* Simon, 1892. Bulletin of Zoological Nomenclature, 48(2): 166-167.

ICZN (International Commission on Zoological Nomenclature). 1999. International Code of Zoological Nomenclature. Fourth Edition. International Trust for Zoological Nomenclature. v-xxix + 1-305 pages.

Ivie, W. 1965. The spiders of the genus *Islandiana* (Linyphiidae, Erigoninae). American Museum Novitates, No. 2221: 1-25.

Ivie, W. 1967. Some synonyms in American spiders. Journal of the New York Entomological Society, 75: 126-131.

Ivie, W. 1969. North American spiders of the genus *Bathyphantes* (Araneae, Linyphiidae). American Museum Novitates, No. 2364: 1-70.

Ivie, W. and W.M. Barrows. 1935. Some new spiders from Florida. Bulletin of the University of Utah: 26(6). Biological Series: 3(2), 3-24.

Jackman, J.A. 1997. A Field Guide to Spiders & Scorpions of Texas. Gulf Publishing Company, Houston, Texas. 202 pages.

Jackson, A.R. 1908. On some rare arachnids captured during 1907. Transactions of the Natural History Society of Northumberland, Durham, and Newcastle-upon-Tyne (New Series), 3: 49-78.

Jackson, A.R. 1932a. Results of the Oxford University Expedition to Lapland, 1930. Araneae and Opiliones. Proceedings of the Zoological Society of London, 1: 97-112 + pl. I.

Jackson, A.R. 1932b. On new and rare British spiders. Proceedings of the Dorset Natural History and Antiquarian Field Club, 53: 200-214.

Jackson, R.R. 1978. Life history of *Phidippus johnsoni* (Araneae, Salticidae). Journal of Arachnology, 6: 1-29.

Jäger, P. 1999a. Sparassidae-the valid scientific name for the huntsman spiders (Arachnida: Araneae). Arachnologische Mitteilungen, 17: 1-10.

Jennings, D.T., K.M. Catley and F. Graham Jr. 2003. *Linyphia triangularis*, a palearctic spider (Araneae: Linyphiidae) new to North America. Journal of Arachnology, 455-460.

Jiménez, M.-L. and C.D. Dondale. 1988 ["1987"]. Descripcion de una nueva especie del genero *Varacosa* de Mexico (Araneae, Lycosidae). Journal of Arachnology, 15(2): 171-175.

Jocqué, R. 1991a. A generic revision of the spider family Zodariidae (Araneae). Bulletin of the American Museum of Natural History, 201: 1-160.

Karsch, F. 1878d. Diagnoses Attoidarum aliquot novarum Novae Hollandiae collectionis Musei Zoologici Beroliniensis. Mitteilungen des münchener entomologischen Vereins, 2: 22-32.

Karsch, F. 1879g. Baustoffe zu einer Spinnenfauna von Japan. Verhandlungen des naturhistorischen Vereines der preussischen Rheinlande und Westfalen, 36: 57-105.

Karsch, F. 1880c. Arachnologische Blätter (Decas I). Zeitschrifft für der gesammten Naturwissenschaften, 53: 373-409.

Karsch, F. 1881a. Verzeichniss der während der Rohlfs'schen africanischen Expedition erbeuteten Myriopoden und Arachniden. Archiv für Naturgeschichte, 47(1): 1-14 + pl. I.

Karsch, F. 1881c. Diagnoses Arachnoidarum Japoniae. Berliner entomologische Zeitschrift, herausgegeben von dem entomologischen Verein in Berlin, 25: 15-16.

Kaston, B.J. 1938b. North American spiders of the genus *Agroeca*. American Midland Naturalist, 20(3): 562-570.

Kaston, B.J. 1938d. Checklist of the spiders of Connecticut. State of Connecticut Geological and Natural History Survey, Bulletin 60: 175-201.

Kaston, B.J. 1938f. A note on synonymy in spiders (Araneae: Salticidae and Argiopidae). Entomological News, 49: 258-259.

Kaston, B.J. 1945a. New spiders in the group Dionycha with notes on other species. American Museum Novitates, No. 1290: 1-25.

Kaston, B.J. 1945b. New Micryphantidae and Dictynidae with notes on other spiders. American Museum Novitates, No. 1291: 1-14.

Kaston, B.J. 1946. North American spiders of the genus *Ctenium*. American Museum Novitates, No. 1306: 1-19.

Kaston, B.J. 1948. Spiders of Connecticut. State Geological and Natural History Survey of Connecticut. Bulletin 70. First edition. 1-874 pages.

Kaston, B.J. 1972a. How to Know the Spiders. Second edition. Dubuque, Iowa. 289 pages.

Kaston, B.J. 1973. Four new species of *Metaphidippus*, with notes on related jumping spiders from the Eastern and Central United States. Transactions of the American Microscopical Society, 92(1): 106-122.

Kaston, B.J. 1974. Remarks on the names of families and higher taxa in spiders. Journal of Arachnology, 2: 47-51.

Kaston, B.J. 1977 ["1976"]. Supplement to the Spiders of Connecticut. Journal of Arachnology, 4: 1-72.

Kaston, B.J. 1978. How to Know the Spiders. Third edition. Wm. C. Brown. Co., Dubuque, Iowa. 272 pages.

Kaston, B.J. 1981 (revised edition). Spiders of Connecticut. State Geological and Natural History Survey of Connecticut. Bulletin 70. 1020 pages.

Kesel, A.B., A. Martin and T. Seidl. 2003. Adhesion measurements on the attachment devices of the jumping spider *Evarcha arcuata*. The Journal of Experimental Biology, 206: 2733-2738.

Keyserling, E. 1877a. Ueber amerikanische Spinnenarten der Unterordnung Citigradae. Verhandlungen der kaiserlich-königlichen zoologisch-botanischen Gesellschaft in Wien, 26: 609-708 + pl. VII-VIII.

Keyserling, E. 1878a. Spinnen aus Uruguay und einigen anderen Gerenden Amerikas. Verhandlungen der kaiserlich-königlichen zoologisch-botanischen Gesellschaft in Wien, 27: 571-624 + pl. XIV.

Keyserling, E. 1878c. Amerikanische Spinnenarten aus den Familien der Pholcidae, Scytodidae und Dysderidae. Verhandlungen der kaiserlich-königlichen zoologisch-botanischen Gesellschaft in Wien, 28: 205-234 + pl. vii.

Keyserling, E. 1880a ["1879a"]. Neue Spinnen aus Amerika. Verhandlungen der kaiserlich-königlichen zoologisch-botanischen Gesellschaft in Wien, 29: 293-350 + pl. IV.

Keyserling, E. 1880b. Die Spinnen Amerikas. Laterigradae. Nürnberg, 1: 1-283 + pl. I-VIII.

Keyserling, E. 1882b ["1881"]. Neue Spinnen aus Amerika. III. Verhandlungen der kaiserlich-königlichen zoologisch-botanischen Gesellschaft in Wien, 31: 269-314 + pl. XI.

Keyserling, E. 1883a. Neue Spinnen aus Amerika. IV. Verhandlungen der kaiserlich-königlichen zoologisch-botanischen Gesellschaft in Wien, 32: 195-226 + pl. XV.

Keyserling, E. 1884a. Die Spinnen Amerikas. Theridiidae. Nürnberg, 2, part I: 1-222 + pl. I-X.

Keyserling, E. 1884b. Neue Spinnen aus Amerika. V. Verhandlungen der kaiserlich-königlichen zoologisch-botanischen Gesellschaft in Wien, 33: 649-684 + pl. XXI.

Keyserling, E. 1885. Neue Spinnen aus Amerika. VI. Verhandlungen der kaiserlich-königlichen zoologisch-botanischen Gesellschaft in Wien, 34: 489-534 + pl. XIII.

Keyserling, E. 1886b. Die Spinnen Amerikas. Theridiidae. Nürnberg, 2, part II: 1-295 + pl. XI-XXI.

Keyserling, E. 1887b. Neue Spinnen aus Amerika. VII. Verhandlungen der kaiserlich-königlichen zoologisch-botanischen Gesellschaft in Wien, 37: 421-490 + pl. VI.

Keyserling, E. 1891. Die Spinnen Amerikas. Brasilianische Spinnen. Nürnberg, 3: 1-278 + pl. I-X.

Keyserling, E. 1892. Die Spinnen Amerikas. Epeiridae. Nürnberg, 4, part I: 1-208 + pl. I-IX.

King, G.F., H.W. Tedford and F. Maggio. 2002. Structure and function of insecticidal neurotoxins from Australian funnel-web spiders. Journal of Toxicology, Toxin Reviews, 21(4): 361-389.

Knoflach, B. 1999. The comb-footed genera *Neottiura* and *Coleosoma* in Europe (Araneae, Theridiidae). Bulletin de la Société entomologique Suisse, 72: 341-371.

Knoflach, B. 2004. Diversity in the copulatory behaviour of comb-footed spiders (Araneae, Theridiidae). Pages 161-256 in K. Thaler (editor): Diversität und Biologie vin Webspinnen, Skorpionen und Anderen Spinnentieren, Biologiezentrum des Oberösterreichischen. Landesmuseums, Linz, Austria.

Knoflach, B. and A. van Harten. 2000b. Palpal loss, single palp copulation and obligatory mate consumption in *Tidarren cuneolatum* (Araneae, Theridiidae). Journal of Natural History, 34: 1639-1659.

Knoflach, B. and K. Pfaller. 2004. Kugelspinnen-eine Einführung (Araneae, Theridiidae). Pages 111-160 in K. Thaler (editor): Diversität und Biologie vin Webspinnen, Skorpionen und Anderen Spinnentieren, Biologiezentrum des Oberösterreichischen. Landesmuseums, Linz, Austria.

Knoflach, B. and K. Thaler. 2000. Notes on Mediterranean Theridiidae (Araneae)-I. Memorie della Società entomologica Italiana, 78: 411-442.

Koch, C.L. 1836a-1837a. Die Arachniden. Getreu nach der Natur abgebildet und beschrieben. 3. Nürnberg, 1-120 + pl. LXXIII-CVIII. [**1836a**, no.1: p.1-16; **1836b**, no.2: p.17-32; **1836c**, no.3: p.33-50; **1836d**, no.4: p.51-72; **1836e**, no.5: p.73-104; **1837a**, no.6: p.105-120].

Koch, C.L. 1837b-1838a. Die Arachniden. Getreu nach der Natur abgebildet und beschrieben. 4. Nürnberg, 1-144 + pl. CIX-CXLIV. [**1837b**, no.1: p.1-24; **1837c**, no.2: p.25-44; **1837d**, no.3: p.45-68; **1837e**, no.4: p.69-88; **1837f**, no.5: p.89-108; **1838a**, no.6: p.109-144].

Koch, C.L. 1837g. Übersicht des Arachnidensystems. No. 1. Nürnberg, 1-39 + pl. VI.

Koch, C.L. 1838b-1839a. Die Arachniden. Getreu nach der Natur abgebildet und beschrieben. 5. Nürnberg, 1-158 + pl. CXLV-CLXXX. [**1838b**, no.1: p.1-24; **1838c**, no.2: p.25-44; **1838d**, no.3: p.45-60; **1838e**, no.4: p.61-92; **1838f**, no.5: p.93-124; **1839a**, no.6: p.125-158].

Koch, C.L. 1839b-g. Die Arachniden. Getreu nach der Natur abgebildet und beschrieben. 6. Nürnberg, 1-156 + pl. CLXXXI-CCXVI. [**1839b**, no.1: p.1-24; **1839c**, no.2: p.25-50; **1839d**, no.3: p.51-74; **1839e**, no.4: p.75-102; **1839f**, no.5: p.103-132; **1839g**, no.6: p.133-156].

Koch, C.L. 1840b-1841d. Die Arachniden. Getreu nach der Natur abgebildet und beschrieben. 8. Nürnberg, 1-131 + pl. CCLIII-CCLXXXVIII. [**1840b**, no.1: p.1-20; **1840c**, no.2: p.21-40; **1841a**, no.3: p.41-64; **1841b**, no.4: p.65-86; **1841c**, no.5: p.87-110; **1841d**, no.6: p.111-131].

Koch, C.L. 1841e-1842c. Die Arachniden. Getreu nach der Natur abgebildet und beschrieben. 9. Nürnberg, 1-108 + pl. CCLXXXIX-CCCXXIV. [**1841e**, no.1: p.1-24; **1841f**, no.2: p.25-40; **1841g**, no.3: p.41-56; **1842a**, no.4: p.57-72; **1842b**, no.5: p.73-88; **1842c**, no.6: p.89-108].

Koch, C.L. 1842d-1843d. Die Arachniden. Getreu nach der Natur abgebildet und beschrieben. 10. Nürnberg, 1-142 + pl. CCCXXV-CCCLX. [**1842d**, no.1: p.1-20; **1842e**, no.2: p.21-36; **1843a**, no.3: p.37-60; **1843b**, no.4: p.61-84; **1843c**, no.5: p.85-112; **1843d**, no.6: p.113-142].

Koch, C.L. 1844a-f. Die Arachniden. Getreu nach der Natur abgebildet und beschrieben. 11. Nürnberg, 1-174 + pl. CCCLXI-CCCXCVI. [**1844a**, no.1: p.1-24; **1844b**, no.2: p.25-48; **1844c**, no.3: p.49-76; **1844d**, no.4: p.77-108; **1844e**, no.5: p.109-136; **1844f**, no.6: p.137-142].

Koch, C.L. 1846a-f. Die Arachniden. Getreu nach der Natur abgebildet und beschrieben. 13. Nürnberg, 1-234 + pl. CCCXXXIII-CCCLXVIII. [**1846a**, no.1: p.1-36; **1846b**, no.2: p.37-76; **1846c**, no.3: p.77-112; **1846d**, no.4: p.113-148; **1846e**, no.5: p.149-188; **1846f**, no.6: p.189-234].

Koch, C.L. 1846g-1847d. Die Arachniden. Getreu nach der Natur abgebildet und beschrieben. 14. Nürnberg, 1-210 + pl. CCCLXIX-DIV. [**1846g**, no.1: p.1-48; **1846h**, no.2: p.49-88; **1847a**, no.3: p.89-120; **1847b**, no.4: p.121-144; **1847c**, no.5: p.145-172; **1847d**, no.6: p.173-210].

Koch, C.L. 1847e-j. Die Arachniden. Getreu nach der Natur abgebildet und beschrieben. 15. Nürnberg, 1-136 + pl. DV-DXL. [**1847e**, no.1: p.1-24; **1847f**, no.2: p.25-50; **1847g**, no.3: p.51-66; **1847h**, no.4: p.67-86; **1847i**, no.5: p.87-110; **1847j**, no.6: p.111-136].

Koch, C.L. 1850. Übersicht des Arachnidensystems. No. 5. Nürnberg, 1-104.

Koch, L. 1864. Die europäischen Arten der Arachnidengattung *Cheiracanthium*. Abhandlungen der Naturhistorischen Gesellschaft zu Nürnberg, 14: 3(1): 137-162.

Koch, L. 1866a. Die Arachniden-Familie der Drassiden. No. 1. Nürnberg, 1-64 + pl. I-II.

Koch, L. 1866b. Die Arachniden-Familie der Drassiden. No. 2. Nürnberg, 65-112 + pl. III-IV.

Koch, L. 1866c. Die Arachniden-Familie der Drassiden. No. 3. Nürnberg, 113-160 + pl. V-VI.

Koch, L. 1866f. Die Arachniden-Familie der Drassiden. No. 6. Nürnberg, 257-304 + pl. XI-XII.

Koch, L. 1867b. Beschreibungen neuer Arachniden und Myriapoden. Verhandlungen der kaiserlich-königlichen zoologisch-botanischen Gesellschaft in Wien, 17: 173-250.

Koch, L. 1868. Die Arachnidengattungen *Amaurobius*, *Caelotes* und *Cybaeus*. Abhandlungen der Naturhistorischen Gesellschaft zu Nürnberg, 4: 1-52 + pl. I-II.

Koch, L. 1871. Die Arachniden Australiens, nach der Natur beschrieben und abgebildet. No. 1. Nürnberg, 1-104.

Koch, L. 1872a. Die Arachnidens Australiens. Nünberg, 1, 105-368.

Koch, L. 1872b. Beitrag zur Kenntniss der Arachnidenfauna Tirols. Zweite Abhandlung. Zeitschrift des Ferdinandeums für Tirol und Voralberg, 3(17): 239-328.

Koch, L. 1876a. Die Arachniden Australiens. No. 1. Nürnberg, 741-888.

Koch, L. 1877b. Verzeichniss der bei Nürnberg jetzt beobachteten Arachniden (mit Ausschluss der Ixodiden und Acariden) und Beschreibungen von neuen, hier vorkommenden Arten. Abhandlungen der Naturhistorischen Gesellschaft zu Nürnberg, 6: 113-198 + pl. I.

Koch, L. 1878a. Die Arachniden Australiens. Nürnberg, 1: 969-1044.

Koch, L. 1879b. Uebersicht der von Dr. Finsch in Westsibirien gesammelten Arachniden. Verhandlungen der kaiserlich-königlichen zoologisch-botanischen Gesellschaft in Wien, 28: 481-490.

Koch, L. 1879c. Arachniden aus Sibirien und Novaja Semlja, eingesammelt von der schwedischen Expedition im Jahre 1875. Kongliga Svenska Vetenskaps-Akdemiens Handlingar, 16(5): 1-136 + pl. I-VII.

Komatsu, T. 1961a. Cave spiders of Japan, their taxonomy, chorology, and ecology. Arachnological Society of East Asia, Osaka. 91 pages.

Komatsu, T. 1968b. Cave spiders of Japan, II, and *Heterocybaeus*. Arachnological Society of East Asia, Osaka. 38 pages.

Kovoor, J. 1987. Comparative structure and histochemistry of silk-producing organs in arachnids. Pages 160-186 *in* W. Nentwig (editor): Ecophysiology of Spiders. Springer-Verlag, New York.

Kraus, O. 1967a. Zur Spinnenfauna Deutschlands, I. *Tapinesthis inermis*, eine für Deutschland neue Oonopide (Arachnida: Araneae: Oonopidae). Senckenbergiana Biologica, 48: 381-385.

Krinsky, W.L. 1987. Envenomation by the sac spider *Cheiracanthium mildei*. Cutis, 40: 127-129.

Kronestedt, T. 1975a. Studies on species of Holarctic *Pardosa* groups (Araneae, Lycosidae). Redescription of *Pardosa albomaculata* Emerton and description of two new species from North America with comments on some taxonomic characters. Zoologica Scripta, 4: 217-228.

Kronestedt, T. 1981. Studies on species of Holarctic *Pardosa* groups (Araneae, Lycosidae), II. Redescriptions of *Pardosa modica* (Blackwall), *Pardosa labradorensis* (Thorell), and *Pardosa sinistra* (Thorell). Bulletin of the American Museum of Natural History, 170: 111-124.

Kronestedt, T. 1986a. Studies on species of Holarctic *Pardosa* groups (Araneae, Lycosidae). III Redescriptions of *Pardosa algens* (Kulczynski), *P. septentrionalis* (Westring), and *P. sodalis* Holm. Entomologica Scandinavia, 17: 215-234.

Kronestedt, T. 1988. Studies on species of Holarctic *Pardosa* groups (Araneae, Lycosidae). IV. Redescription of *Pardosa tetonensis* Gertsch and descriptions of two new species from the western United States. Entomologica Scandinavia, 18: 409-4244.

Kronestedt, T. and Y.M. Marusik. 2002. On *Acantholycosa solituda* (Levi & Levi) and *A. sterneri* (Marusik) (Araneae: Lycosidae), a pair of geographically distant allied species. Acta Arachnologica, 51: 63-71.

Kulczyński, W. 1882b. Opisy nowych Gatunkow Pajakow, z Tatra, Babiej gory i Karpat szlazkich przez. Pamietnik Akademji umiejetnosci w Krawkow wydzial matematyczno-przyrodniczy Krawkow, 8: 1-42.

Kulczyński, W. 1885a. Pajaki zebranae na Kamczatce przcz Dra B. Dybowskiego. [Araneae in Camtshchadalia a Dre B. Dybowski collectae]. Pamietnik Akademji umiejetnosci w Krakow wydzial matematyczno-przyrodniczy, 11: 1-60 + pl. IX-XI.

Kulczyński, W. 1894. Ueber die Theridioiden der Spinnenfauna Ungarns. Mathematische und naturwissenschaftliche Berichte aus Ungarn, 12: 321-338.

Kulczyński, W. 1907a. Fragmenta arachnologica V. VIII-Arachnoidea nonnulla in insularis Diomedeis (Isole di Tremiti) a cel. Prof. Dre G. Cecconi lecta. Bulletin international de l'Académie des Sciences de Cracovie, 570-584, 595-596 + pl. XXI, in part.

Kulczyński, W. 1908b. Araneae et Oribatidae expeditionum rosicarum in insulas Novo-Sibiricas annis 1885-1886 & 1900-1903 susceptarum. Zapiski imperatorskoi Akademy naouk. [Mémoires de l'Académie impériale des Sciences de Saint-Pétersbourg], 8, 18(7): i-ii + 1-97 + pl. I-III.

Kulczyński, W. 1916. Araneae Siberiae occidentalis arcticae. Zapinski imperatorskoï Akademy naouk. [Mémoires de l'Académie impériale des sciences de St-Pétersbourg], 8, 28(11): 1-44 + pl. I-II.

Kullmann, E. and W. Zimmermann. 1976a. Ein neuer Beitrag zum Cribellaten-Ecribellaten-Problem: Beschreibung von *Uroecobius ecribellatus* n. gen. n. sp. und Diskussion seiner phylogenetischen Stellung (Arachnida: Araneae: Oecobiidae). Entomologia Germanica, 3: 29-40.

Kurata, T.B. 1944. Two new species of Ontario Spiders. Occasional Papers of the Royal Ontario Museum of Zoology, 8: 1-6.

Labandeira, C.C. 1999. Insects and other hexapods. Pages 604-624 *in* R. Singer (editor): Encyclopedia of paleontology. Fitzroy and Dearborn, Chicago.

Lamoral, B.H. 1981. *Paroecobius wilmotae*, a new genus and species of spider from the Okavango Delta, Botswana (Araneae: Oecobiidae: Oecobiinae). Annals of the Natal Museum, 24: 507-512.

Land, M.F. 1985. The morphology and optics of spider eyes. Pages 53-78 *in* F.G. Barth (editor): Neurobiology of Arachnids. Springer-Verlag, New York.

Latreille, P.A. 1799b. Description d'une nouvelle espèce d'araignée. Bulletin de la Société philomatique de Paris, 1(22): 170.

Latreille, P.A. 1802a. Histoire naturelle, générale et particulière des Crustacés et des Insectes. Ouvrage faisant suite à l'histoire naturelle générale et particulière, composée par Leclerc de Buffon, et rédigée par C.S. Sonnini, membre de plusieurs sociétés savantes. Familles naturelles des genres. Tome troisième. An X. [Arachnides : pp. 48-59). Dufart, Paris. Pages xii + 13-467 (+ 1 p. errata, unnumbered).

Latreille, P.A. 1802b. Mémoire sur une nouvelle distribution méthodique des Araignées; lu à la société philomatique, In Histoire naturelle des Fourmis et recueil de mémoires et d'observations sur les Abeilles, les Araignées, les Faucheurs, et d'autres Insectes, pages 332-353. an X. Paris.

Latreille, P.A. 1804b. Nouveau dictionnaire d'histoire naturelle, appliquée aux arts, principalement à l'agriculture et à l'économie rurale et domestique: par une société de naturalistes et d'agriculteurs: avec des figures tirées des trois règnes de la nature. Tome XXIV. Tableaux méthodiques d'histoire naturelle. Caractères des premières divisions des corps terrestres nommées règnes de la nature. Tableau méthodique des Insectes. Imprimerie de Crapelet, Paris. 1-238 + tables alphabétiques des figures: 1-34 pages.

Latreille, P.A. 1806. Genera crustaceorum et insectorum secundum ordinem naturalem in familias disposita, iconibus exemplisque plurimis explicata. Tomus primus [Aranéides : pp. 82-127 + planches III-V]. Koenig, Parisiis et Argentorati. i-xviii + 1-302 + pl. I-XVI + emendata.

Latreille, P.A. 1809. Genera crustaceorum et insectorum. Paris. 4: 370-371.

Latreille, P.A. 1810. Aranéides pages 119-129 *In* Considérations générales sur l'ordre naturel des Animaux composant la classe des Crustacés, des Arachnides et des Insectes. Paris. 444 pages.

Latreille, P.A. 1816. [Articles sur les Araignées]. Nouveau dictionnaire d'histoire naturelle, appliquée aux arts, à l'agriculture, à l'économie rurale et domestique, à la médecine, etc. par une société de naturalistes et d'agriculteurs. Nouvelle édition presqu'entièrement refondue et considérablement augmentée; avec des figures tirées des trois règnes de la nature. Tomes 1-2-3.

Latreille, P.A. 1817. [Articles sur les Araignées]. Nouveau dictionnaire d'histoire naturelle, appliquée aux arts, à l'agriculture, à l'économie rurale et domestique, à la médecine, etc. par une société de naturalistes et d'agriculteurs. Nouvelle édition presqu'entièrement refondue et considérablement augmentée; avec des figures tirées des trois règnes de la nature. Tomes 7-8-9-10-11-13-17-18.

Latreille, P.A. 1819. Nouveau dictionnaire d'histoire naturelle, appliquée aux arts, à l'agriculture, à l'économie rurale et domestique, à la médecine, etc. par une société de naturalistes et d'agriculteurs. Nouvelle édition presqu'entièrement refondue et considérablement augmentée; avec des figures tirées des trois règnes de la nature. Tome XXXIV. Deterville, Paris, France. v-viii + 1-264 pages.

Latreille, P.A. 1824. Notice sur un nouveau genre d'Aranéides. Annales des sciences naturelles. Zoologie, 3: 23-27.

Latreille, P.A. 1829. Les Arachnides. In G. Cuvier (editor), Le règne animal, nouvelle édition, 4, pages 206-291. Paris.

Laughlin, S., A.D. Blest and S. Stowe. 1980. The sensitivity of receptors in the posterior median eye of the nocturnal spider *Dinopis*. Journal of Comparative Physiology A: Sensory, Neural, and Behavioral Physiology, 141: 53-65.

Lazaris, A., S. Arcidiacono, Y. Huang, J.-F. Zhou, F. Duguay, N. Chretien, E.A. Welsh, J.W. Soares and Karatzas C.N. 2002. Fibers spun from soluble recombinant silk produced in mammalian cells. Science, 295: 472-476.

Leach, W.E. 1815. The zoological miscellany; being descriptions of new, or interesting animals. Volume II [Araneae: 131-134, + pl. CIX-CX in separate volume]. London. 254 pages.

Leach, W.E. 1816. Araneidae. Pages 434-443 in Annulosa in Encyclopoedia Britannia.

Leach, W.E. 1824. Araneidae. Page 438 in Encyclopaedia Brittanica, supplement 4th - 6th Editions. Volume 1(2).

Ledford, J.M. 2004. A revision of the spider genus *Calileptoneta* Platnick (Araneae, Leptonetidae) with notes on morphology, natural history, and biogeography. Journal of Arachnology, 32: 231-269.

Leech, R.E. 1969b. Transfer of the genus *Melocosa* from Pisauridae to Lycosidae, and a key to the Nearctic genera of Lycosidae (Arachnida: Araneida). The Canadian Entomologist, 101(7): 890-894.

Leech, R.E. 1971. The introduced Amaurobiidae of North America, and *Callobius hokkaido* n.sp. from Japan (Arachnida: Araneida). The Canadian Entomologist, 103(1): 23-32.

Leech, R.E. 1972. A revision of the Nearctic Amaurobiidae (Arachnida: Araneida). Memoirs of the Entomological Society of Canada. No. 84, 1-182.

Leech, R.E. and J.K. Ryan. 1972. Notes on Canadian Arctic spiders (Araneida), mainly from Devon Island, N.W.T. The Canadian Entomologist, 104: 1787-1791.

Leech, R.E. and M. Steiner. 1992. *Metaltella simoni* (Keyserling, 1878) (Amaurobiidae) new to Canada, and records for *Tegenaria gigantea* Chamberlin and Ivie, 1935, Agelenidae in Alberta and British Columbia (Arachnida: Araneida). The Canadian Entomologist, 124: 419-420.

Lehtinen, P.T. 1967. Classification of the cribellate spiders and some allied families, with notes on the evolution of the suborder Araneomorpha. Annales Zoologici Fennici, 4: 199-468.

Lehtinen, P.T. 1986. Evolution of the Scytodoidea. Pages 149-157 in W.G. Eberhard, Y.D. Lubin, and B.C. Robinson (editors): Proceedings of the 9th International Congress of Arachnology, 1983, . Panama. Smithsonian Institution Press, Washington.

Lehtinen, P.T. and M. Saaristo. 1980. Spiders of the Oriental-Australian region. II. Nesticidae. Acta Zoologica Fennica, 17: 47-66.

Levi, H.W. 1951. New and rare spiders from Wisconsin and adjacent states. American Museum Novitates, No. 1501: 1-41.

Levi, H.W. 1953b. Spiders of the genus *Dipoena* from America north of Mexico (Araneae, Theridiidae). American Museum Novitates, No. 1647: 1-39.

Levi, H.W. 1954a. Spiders of the new genus *Theridiotis* (Araneae: Theridiidae). Transactions of the American Microscopical Society, 73(2): 177-189.

Levi, H.W. 1954b. Spiders of the genus *Euryopis* from North and Central America (Araneae, Theridiidae). American Museum Novitates, No. 1666: 1-48.

Levi, H.W. 1954c. The spider genus *Theridula* in North and Central America and the West Indies (Araneae: Theridiidae). Transactions of the American Microscopical Society, 73(4): 331-343.

Levi, H.W. 1955a. The spider genera *Coressa* and *Achaearanea* in America north of Mexico (Araneae, Theridiidae). American Museum Novitates, No. 1718: 1-33.

Levi, H.W. 1955b. The spider genera *Episinus* and *Spintharus* from North America, Central America and the West Indies (Araneae: Theridiidae). Journal of the New York Entomological Society, 62: 65-90.

Levi, H.W. 1955c. The spider genera *Oronata* and *Stemmops* in North America, Central America and the West Indies (Araneae: Theridiidae). Annals of the Entomological Society of America, 48(5): 333-342.

Levi, H.W. 1956a. The spider genus *Mysmena* in the Americas (Araneae, Theridiidae). American Museum Novitates, No. 1801: 1-13.

Levi, H.W. 1956b. The spider genera *Neottiura* and *Anelosimus* in America (Araneae: Theridiidae). Transactions of the American Microscopical Society, 75(4): 407-422.

Levi, H.W. 1957a. The spider genera *Enoplognatha*, *Theridion*, and *Paidisca* (Araneae: Theridiidae). Bulletin of the American Museum of Natural History, 112(1): 1-123.

Levi, H.W. 1957b. The spider genera *Crustulina* and *Steatoda* in North America, Central America, and the West Indies (Araneae, Theridiidae). Bulletin of the Museum of Comparative Zoology, 117(3): 365-424.

Levi, H.W. 1957c. The spider genera *Chrysso* and *Tidarren* in America (Araneae: Theridiidae). Journal of the New York Entomological Society, 63: 59-81.

Levi, H.W. 1957d. The North American spider genera *Paratheridula*, *Tekellina*, *Pholcomma* and *Archerius* (Araneae: Theridiidae). Transactions of the American Microscopical Society, 64(2): 105-115.

Levi, H.W. 1957e. Spiders of the new genus *Arctachaea* (Araneae, Theridiidae). Psyche, 64(3): 102-106.

Levi, H.W. 1959a. The spider *Coleosoma* (Araneae, Theridiidae). Brevoria, 110: 1-11.

Levi, H.W. 1959b. The spider genera *Achaearanea*, *Theridion* and *Sphyrotinus* from Mexico, Central America and the West Indies (Araneae, Theridiidae). Bulletin of the Museum of Comparative Zoology, 112(3): 57-163.

Levi, H.W. 1959c. The spider genus *Latrodectus* (Araneae, Theridiidae). Transactions of the American Microscopical Society, 78(1): 7-43.

Levi, H.W. 1960. The spider genus *Styposis* (Araneae: Theridiidae). Psyche, 66(1/2): 13-19.

Levi, H.W. 1962a. The spider genera *Steatoda* and *Enoplognatha* in America (Araneae, Theridiidae). Psyche, 69(1): 11-36.

Levi, H.W. 1962b. More American spiders of the genus *Chrysso* (Araneae, Theridiidae). Psyche, 64(9): 209-237.

Levi, H.W. 1963a. American spiders of the genera *Audifia*, *Euryopis* and *Dipoena* (Araneae: Theridiidae). Bulletin of the Museum of Comparative Zoology, 129(2): 121-185.

Levi, H.W. 1963b. American spiders of the genus *Achaearanea* and the new genus *Echinotheridion* (Aranae, Theridiidae). Bulletin of the Museum of Comparative Zoology, 129(3): 187-240.

Levi, H.W. 1963c. American spiders of the genus *Theridion* (Araneae, Theridiidae). Bulletin of the Museum of Comparative Zoology, 129(10): 481-589.

Levi, H.W. 1963e. The American spider genera *Spintharus* and *Thwaitesia* (Araneae: Theridiidae). Psyche, 70(4): 223-234.

Levi, H.W. 1964b. The spider genus *Thymoites* in America (Araneae: Theridiidae). Bulletin of the Museum of Comparative Zoology, 130(7): 447-479.

Levi, H.W. 1964d. American spiders of the genus *Phoroncidia* (Araneae: Theridiidae). Bulletin of the Museum of Comparative Zoology, 131(3): 65-86.

Levi, H.W. 1964f. The spider genera *Stemmops*, *Chrosiothes* and the new genus *Cabello* from America. Psyche, 71(2): 73-92.

Levi, H.W. 1966b. American spider genera *Theridula* and *Paratheridula* (Araneae: Theridiidae). Psyche, 73(2): 123-130.

Levi, H.W. 1967c. Cosmopolitan and pantropical species of theridiid spiders (Araneae: Theridiidae). Pacific Insects, 9(2): 175-186.

Levi, H.W. 1968a. The spider genera *Gea* and *Argiope* in America (Araneae: Araneidae). Bulletin of the Museum of Comparative Zoology, 136(9): 319-352.

Levi, H.W. 1969. Notes on American theridiid spiders. Psyche, 76(1): 68-73.

Levi, H.W. 1971a. The *diadematus* group of the orb-weaver genus *Araneus* north of Mexico (Araneae: Araneidae). Bulletin of the Museum of Comparative Zoology, 141(4): 131-179.

Levi, H.W. 1971b [1970]. The *ravilla* group of the orbweaver genus *Eriophora* in North America (Araneae: Araneidae). Psyche, 77(3): 280-301.

Levi, H.W. 1972a. The orb-weaver genera *Singa* and *Hypsosinga* in America (Araneae: Araneidae). Psyche, 78(4): 229-256.

Levi, H.W. 1973. Small orb-weavers of the genus *Araneus* north of Mexico (Araneae: Araneidae). Bulletin of the Museum of Comparative Zoology, 145(9): 473-552.

Levi, H.W. 1974a. The orb-weaver genus *Zygiella* (Araneae: Araneidae). Bulletin of the Museum of Comparative Zoology, 146(5): 267-290.

Levi, H.W. 1974b. The orb-weaver genera *Araniella* and *Nuctenea* (Araneae: Araneidae). Bulletin of the Museum of Comparative Zoology, 146(6): 291-316.

Levi, H.W. 1975a. Description of the female of the spider *Pholcomma carota* (Arachnida: Araneae, Theridiidae). Transactions of the American Microscopical Society, 94(2): 281-282.

Levi, H.W. 1975b. The American orb-weaver genera *Larinia*, *Cercidia*, and *Mangora* north of Mexico (Araneae, Araneidae). Bulletin of the Museum of Comparative Zoology, 147(3): 100-135.

Levi, H.W. 1975c. Additinal notes on the orb-weaver genera *Araneus*, *Hypsosinga*, and *Singa* north of Mexico (Araneae, Araneidae). Psyche, 82(2): 265-274.

Levi, H.W. 1976. The orb-weaver genera *Verrucosa*, *Acanthepeira*, *Wagneriana*, *Acacesia*, *Wixia*, *Scoloderus* and *Alpaida* north of Mexico. Bulletin of the Museum of Comparative Zoology, 147(8): 351-391.

Levi, H.W. 1977a. The American orb-weaver genera *Cyclosa, Metazygia* and *Eustala* north of Mexico (Araneae, Araneidae). Bulletin of the Museum of Comparative Zoology, 148(3): 61-127.

Levi, H.W. 1977b. The orb-weaver genera *Metepeira, Kaira* and *Aculepeira* in America north of Mexico (Araneae, Araneidae). Bulletin of the Museum of Comparative Zoology, 148(5): 185-238.

Levi, H.W. 1978a. The American orb-weaver genera *Colphepeira, Micrathena*, and *Gasteracantha* north of Mexico (Araneae, Araneidae). Bulletin of the Museum of Comparative Zoology, 148(9): 417-442.

Levi, H.W. 1980a. The orb-weaver genus *Mecynogea*, the subfamily Metinae and the genera *Pachygnatha, Glenognatha* and *Azilia* of the subfamily Tetragnathinae north of Mexico (Araneae: Araneidae). Bulletin of the Museum of Comparative Zoology, 149(1): 1-74.

Levi, H.W. 1980b. Two new spiders of the genera *Theridion* and *Achaearanea* from North America (Araneae: Theridiidae). Transactions of the American Microscopical Society, 99(3): 334-337.

Levi, H.W. 1981a. The American orb-weaver genera *Dolichognatha* and *Tetragnatha* north of Mexico (Araneae: Araneidae, Tetragnathidae). Bulletin of the Museum of Comparative Zoology, 149(5): 271-318.

Levi, H.W. 1981b. More on the genus *Araneus* from North America. Bulletin of the American Museum of Natural History, 170: 254-256.

Levi, H.W. 1983b. The orb-weaver genera *Argiope, Gea*, and *Neogea* from the Western Pacific region (Araneae: Araneidae: Argiopinae). Bulletin of the Museum of Comparative Zoology, 150(5): 247-338.

Levi, H.W. 1984b. On the value of genitalic structures and coloration in separating species of widow spiders (*Latrodectus* sp.). Mitteilungen des Hamburger Naturforschenden Vereins, 26: 195-200.

Levi, H.W. 1993a. American *Neoscona* and corrections to previous revisions of neotropical orb-weavers (Araneae: Araneidae). Psyche, 99: 221-239.

Levi, H.W. 1993b. The Neotropical orb-weaving spiders of the genera *Wixia, Pozonia*, and *Ocrepeira* (Araneae: Araneidae). Bulletin of the Museum of Comparative Zoology, 153: 47-141.

Levi, H.W. 1993c. The orb-weaver genus *Kaira* (Araneae: Araneidae). Journal of Arachnology, 21: 209-225.

Levi, H.W. 1997. The American orb weavers of the genera *Mecynogea, Manogea, Kapogea* and *Cyrtophora* (Araneae: Araneidae). Bulletin of the Museum of Comparative Zoology, 215-255.

Levi, H.W. 1999. The Neotropical and Mexican orb weavers of the genera *Cyclosa* and *Allocyclosa* (Araneidae). Bulletin of the Museum of Comparative Zoology, 155: 299-378.

Levi, H.W. 2003. The bolas spiders of the genus *Mastophora* (Araneae: Araneidae). Bulletin of the Museum of Comparative Zoology, 157(5): 307-382.

Levi, H.W. and L.R. Levi. 1951. Report on a collection of spiders and harvestmen from Wyoming and neighboring states. Zoologica, 36(4): 219-237.

Levi, H.W. and L.R. Levi. 1962. The genera of the spider family Theridiidae. Bulletin of the Museum of Comparative Zoology, 127(1): 1-99.

Levi, H.W. and L.R. Levi. 2002. Spiders and Their Kin. A Golden Guide. St. Martin's Press, New York. 160 pages.

Levi, H.W., L.R. Levi and H.S. Zim. 1968. Spiders and Their Kin. A Golden Guide. Golden Press, New York. 160 pages.

Levi, H.W. and D.E. Randolph. 1975. A key and checklist of the family Theridiidae north of Mexico (Araneae). Journal of Arachnology, 3: 31-51.

Levi, L.R. and H.W. Levi. 1955. Spiders and Harvestmen from the Waterton and Glacier National Parks. The Canadian Field-Naturalist, 69(2): 32-40.

Li, D., R.R. Jackson and M.L.M. Lim. 2003. Influence of background and prey orientation on an ambushing predator's decisions. Behaviour, 140: 739-764.

Linnaeus, C. von. 1758. Systema naturae per regna tria naturae, secundum classes, ordines, genera, species cum characteribus, differentiis, synonymis, locis. Editio decima, reformata Tomus I. Laurentii Salvii, Holmiae. 823 pages (+ 1 unnumbered page Emendata, Addenda).

Linnaeus, C. von. 1767. Systema naturae per regna tria naturae, secundum classes, ordines, genera, species cum characteribus, differentiis, synonymis, locis. Editio duodecima, reformata. Tom. I. Pars II. Laurentii Salvii, Holmiae. Pages 533-1327 (+ 37 unnumbered pages; Prayer, Index, Errata, Addenda, Synonymy).

Locht, A., M. Yanez and I. Vasquez. 1999. Distribution and natural history of Mexican species of *Brachypelma* and *Brachypelmides* (Theraphosidae, Theraphosinae) with morphological evidence for their synonymy. Journal of Arachnology, 27: 196-200.

Locket, G.H. 1965. A new British species of linyphiid spider. Entomologist's Monthly Magazine, 101(1208): 48-50.

Locket, G.H. and A.F. Millidge. 1953. British Spiders. Volume II. Ray Society (London). Publication 137. iv-viii + 449 pages.

Locket, G.H., A.F. Millidge and P. Merrett. 1974. British Spiders. Volume III. Ray Society (London). Publication 149. v-viii + 315 pages

Logunov, D.V. 1999a. Redefinition of the genus *Habrocestoides* Prószynski, 1992, with establishment of a new genus, *Chinattus* gen n. (Araneae: Salticidae). Bulletin of the British arachnological Society, 11: 139-149.

Logunov, D.V. 2001a ["2000"]. A redefinition of the genera *Bianor* Peckham & Peckham, 1885 and *Harmochirus* Simon, 1885, with the establishment of a new genus *Sibianor* gen. n. (Aranei: Salticidae). Arthropoda Selecta, 9(4): 221-286.

Logunov, D.V. and B. Cutler. 1999. Revision of the genus *Paramarpissa* F.O. P.-Cambridge 1901 (Araneae, Salticidae). Journal of Natural History, 33: 1217-1236.

Lohmander, H. 1944. Vorläufige Spinnennotizen. Arkiv för Zoologi, 35A(16): 1-21.

Lowe, R.T. 1835. Descriptions of two species of Araneidae, natives of Madeira. The Zoological Journal (London), 5: 320-323 + pl. XLVIII.

Lowrie, D.C. and C.D. Dondale. 1981. A revision of the *nigra* group of the genus *Pardosa* in North America (Araneae, Lycosidae). Bulletin of the American Museum of Natural History, 170: 125-139.

Lowrie, D.C. and W.J. Gertsch. 1955. A list of the spiders of the Grand Teton Park area, with description of some new North American spiders. American Museum Novitates, No. 1736: 1-29.

Lubin, Y.D., W.G. Eberhard and G. Montgomery. 1978. The single-line web of *Miagrammopes* (Uloboridae). Psyche, 85: 1-23.

Lubin, Y.D., B.D. Opell, W.G. Eberhard and H.W. Levi. 1982. Orb plus cone-webs in Uloboridae (Araneae), with a description of a new genus and four new species. Psyche, 89: 29-64.

Lucas, H. 1835b. Genre *Pachyloscelis* Lucas. Magasin de zoologie, 6(8): 1-6.

Lucas, H. 1843. Note sur une nouvelle espèce d'Aranéide appartenant au genre *Actinopus* de M. Perty. Revue zoologique par la Société cuviérienne, 318.

Lucas, H. 1844a. Notice sur une nouvelle espèce d'aranéide appartenant au genre *Scytodes* de M. Walckenaer. Revue et Magazin de Zoologie, 1844: 72.

Lucas, H. 1846. Deuxième classe. Arachnides. Pages 89-271. Tome I (i-xxxv + pp. 1-403) + Tome IV (pl. I-XVII) *in* Histoire naturelle des animaux articulés, Exploration scientifique de l'Algérie pendant les années 1840, 1841, 1842, publiée par ordre du gouvernement et avec le concours d'une commission académique. Sciences physiques, Zoologie. Paris.

Lucas, H. 1857a. Arachnides. *In* R. de la Sagra, Historica fisica, politica y natural de la Isla de Cuba. Secunda parte, tomo VII, pages 69-84.

Lucas, H. 1884. [Note on a Theraphosid]. Annales de la Société entomologique de France, Bulletin, 4(6): cxlii-cxliii.

MacLeay, W.S. 1839. On some new forms of Arachnida. The Annals and Magazine of Natural History, 2: 1-14.

Maddison, W.P. 1978. *Bianor aemulus* (Gertsch), new combination (Araneae: Salticidae). Peckhamia, 1(5): 76-77.

Maddison, W.P. 1986. Distinguishing the jumping spiders *Eris militaris* and *Eris flava* in North America (Araneae: Salticidae). Psyche, 93: 141-149.

Maddison, W.P. 1987. *Marchena* and other jumping spiders with an apparent leg-carapace stridulatory mechanism (Araneae: Salticidae: Heliophaninae and Thiodininae). Bulletin of the British Arachnological Society, 7(4): 101-106.

Maddison, W.P. 1996. *Pelegrina* Franganillo and other jumping spiders formerly placed in the genus *Metaphidippus* (Aranaeae: Salticidae). Bulletin of the Museum of Comparative Zoology, 154(4): 215-368.

Maddison, W. and M. Hedin. 2003a. Phylogeny of *Habronattus* jumping spiders (Araneae: Salticidae), with consideration of genitalic and courtship evolution. Systematic Entomology, 28: 1-21.

Maddison, W.P. and M. Hedin. 2003b. Jumping spider phylogeny (Araneae: Salticidae). Invertebrate Systematics, 17: 529-549.

Maloney, D., F.A. Drummond and R. Alford. 2003. Spider predation in agroecosystems: can spiders effectively control pest populations? Maine Agricultural and Experiment Station, Technical Bulletin, 190: 1-32.

Maretić, Z. 1987. Spider venoms and their effect. Pages 142-159 *in* W. Nentwig (editor): Ecophysiology of Spiders. Springer-Verlag, New York.

Marples, B.J. 1959. Spiders from some Pacific islands, III. The Kingdom of Tonga. Pacific Science, 13: 362-367.

Marples, B.J. 1968. The hypochilomorph spiders. Proceedings of the Linnean Society of London, 179: 11-31.

Marshall, S.D. 1995. Natural history, activity patterns, and relocation rates of a burrowing wolf spider, *Geolycosa xera archboldi* McCrone (Araneae, Lycosidae). Journal of Arachnology, 23: 65-70.

Marshall, S.D. 1999. Neighborhood effects within territorial aggregations of a burrowing wolf spider. Canadian Journal of Zoology, 77: 1006-1012.

Marshall, S.D., W.R. Hoeh and M.A. Deyrup. 2000. Biogeography and conservation biology of Florida's *Geolycosa* wolf spiders: threatened spiders in endangered ecosystems. Journal of Insect Conservation, 4: 11-21.

Martyn, T. 1793. Aranei, or a Natural History of Spiders, including the principal parts of the well known work of English spiders by Eleazar Albin as also the whole of the celebrated publication on Swedish spiders by Charles Clerck. London, 1-70 + 1-31 + pl. I-XI, + I-XVII.

Marusik, Y.M. 1989a. Two new species of the spider genus *Xysticus* and synonymy of crab spiders (Aranei, Thomisidae, Philodromidae) from Siberia. Zoologicheskii Zhurnal, 68(4): 140-145.

Marusik, Y.M. and S. Koponen. 1992. A review of *Meta* (Araneae, Tetragnathidae), with description of two new species. Journal of Arachnology, 20(2): 137-143.

Marusik, Y.M. and S. Koponen. 2001c. Revision of the Holarctic spider genus *Procerocymbium* Eskov 1989 (Araneae: Linyphiidae). Acta Arachnologica, 50(2): 145-156.

Marusik, Y.M. and P.T. Lehtinen. 2003. Synaphridae Wunderlich, 1986 (Aranei: Araneoidea), a new family status, with a description of a new species from Turkmenistan. Arthropoda Selecta, 11: 143-152.

Marusik, Y.M. and D.V. Logunov. 1998 ["1997"]. Taxonomic notes on the *Evarcha falcata* species complex (Aranei Salticidae). Arthropoda Selecta, 6(3/4): 95-104.

Marusik, Y.M. and M.I. Saaristo. 1999. Review of east Palearctic species of the genus *Minicia* Thorell, 1875 with descriptions of two new genera (Aranei: Linyphiidae: Erigoninae). Arthropoda Selecta, 8(2): 125-130.

Marx, G. 1881. On some new tube-constructing spiders. The American Naturalist, 15: 396-400.

Marx, G. 1888a. On a new and interesting spider. Entomologica Americana, 4: 160-162.

Marx, G. 1889b. On the new spider of the genus *Dinopis*, from the Southern United States. Proceedings of the Academy of Natural Sciences of Philadelphia, 1889: 341-343.

Marx, G. 1890 ["1889"]. Catalogue of the described Araneae of temperate North America. Proceedings of the United States National Museum, 12, no. 782: 497-594.

Marx, G. 1891. A contribution to the knowledge of North American spiders. Proceedings of the Entomological Society of Washington, 2(1): 28-37 + pl. I.

Maughan, O.E. and H.S. Fitch. 1976. A new pholcid spider from northeastern Kansas. Journal of the Kansas Entomological Society, 49(2): 304-312.

McCook, H.C. 1878a. The Basilica spider and her snare. Proceedings of the Academy of Natural Sciences of Philadelphia, 1878: 124-135.

McCook, H.C. 1887. A note on *Cyrtophora bifurca* (n. sp.) and her cocoons, a new orb-weaving spider. Proceedings of the Academy of Natural Sciences of Philadelphia, 1887: 342-343.

McCook, H.C. 1888a. Necessity for revising the nomenclature of American spiders. Proceedings of the Academy of Natural Sciences of Philadelphia, 1888: 74-79.

McCook, H.C. 1888b. Descriptive notes on new American species of orbweaving spiders. Proceedings of the Academy of Natural Sciences of Philadelphia, 193-199.

McCook, H.C. 1889. American Spiders and Their Spinningwork. A Natural History of the Orbweaving Spiders of the United States with Special Regard to Their Industry and Habits. Volume 1. 1-372 pages.

McCook, H.C. 1890. American Spiders and Their Spinningwork. A Natural History of the Orbweaving Spiders of the United States with Special Regard to Their Industry and Habits. Volume 2. 1-479 pages.

McCook, H.C. 1892. *Drexelia*, a new genus of spiders. Proceedings of the Academy of Natural Sciences of Philadelphia, 1892: 127.

McCook, H.C. 1894 ["1893"]. American Spiders and Their Spinningwork. A Natural History of the Orbweaving Spiders of the United States with Special Regard to Their Industry and Habits. Volume 3. 1-407 pages.

McCrone, J.D. 1963. Taxonomic status and evolutionary history of the *Geolycosa pikei* complex in the southeastern United States (Araneae, Lycosidae). American Midland Naturalist, 70: 47-73.

McCrone, J.D. and H.W. Levi. 1964. North American widow spiders of the *Latrodectus curacaviensis* group (Araneae: Theridiidae). Psyche, 71(1): 12-27.

McQueen, D.J. 1978. Field studies of growth, reproduction, and mortality in the burrowing wolf spider *Geolycosa domifex* (Hancock). Canadian Journal of Zoology, 56: 2037-2049.

de Mello-Leitão, C.F. 1917a. Notas arachnologicas. 5. Especies novas ou pouco conhecidas do Brasil. Broteria, 15: 74-102.

de Mello-Leitão, C.F. 1917b. Generos e especies novas de araneidos. Archivos da Escola Superior de Agricultura e Medicina Veterinaria, Rio-de-Janerio, 1: 3-19.

de Mello-Leitão, C.F. 1931b. Contribuição ao estudo da tribu Mastophoreas. Annaes da Academia brasileira de Sciencias, 3(2): 65-74 + pl. I.

de Mello-Leitão, C.F. 1931c. Notas sobre arachnidos argentinos. Annaes da Academia brasileira de Sciencias, 3: 83-97.

de Mello-Leitão, C.F. 1938a. Algunas arañas nuevas de la Argentina. Revista del Museo de La Plata (Nueva Serie), 1, Zoologia: 89-118.

de Mello-Leitão, C.F. 1941a. Notas sobre a sistematica das Aranhas, com descrição de Algunas novas especies Sul Americana. Anaes de Academia Brasileira de Ciencias, 13(2): 103-127.

de Mello-Leitão, C.F. 1945b. Arañas de Misiones, Corrientes y Entre Rios. Revista del Museo de La Plata (Nueva Serie), 4 (Zoologia 29): 213-302.

Menge, A. 1866. Preussische Spinnen. Schriften der naturforschenden Gesellschaft in Danzig. Neue folge, 1: 1-152 + pl. I-XXVIII.

Menge, A. 1868. Preussische Spinnen. II. Schriften der naturforschenden Gesellschaft in Danzig. Neue folge, 2(1): 153-218 + pl. XXIX-XLIII.

Menge, A. 1869. Preussische Spinnen. III. Schriften der naturforschenden Gesellschaft in Danzig. Neue folge, 2(2): 219-264 + pl. XLIV-XLIX.

Menge, A. 1871. Preussische Spinnen. IV. Schriften der naturforschenden Gesellschaft in Danzig. Neue folge, 2(3/4): 265-296 + pl. L-LIII.

Menge, A. 1872. Preussische Spinnen. V. Schriften der naturforschenden Gesellschaft in Danzig. Neue folge, 3: 297-326 + pl. LIV-LVII.

Menge, A. 1873. Preussische Spinnen. VI. Schriften der naturforschenden Gesellschaft in Danzig. Neue folge, 3: 327-374 + pl. LVIII-LXIII.

Merrett, P. 1963b. The palpus of male spiders of the family Linyphiidae. Proceedings of the Zoological Society of London, 140: 347-467.

Merrett, P., G.H. Locket and A.F. Millidge. 1985. A check list of British spiders. Bulletin of the British Arachnological Society, 6(9): 381-403.

Miller, J.A. 1999. Revision and cladistic analysis of the erigonine spider genus *Sisicottus* (Araneae, Linyphiidae, Erigoninae). Journal of Arachnology, 27(3): 553-603.

Millidge, A.F. 1951. Key to the British genera of subfamily Erigoninae (Family Linyphiidae: Araneae): including the description of a new genus (*Jacksonella*). The Annals and Magazine of Natural History, 12(4): 545-562.

Millidge, A.F. 1976. Re-examination of the erigonine spiders "*Micrargus herbigradus*" and "*Pocadicnemis pumila*" (Araneae: Linyphiidae). Bulletin of the British Arachnological Society, 3(6): 145-155.

Millidge, A.F. 1977. The conformation of the male palpal organs of linyphiid spiders, and its application to the taxonomic and phylogenetic analysis of the family (Araneae: Linyphiidae). Bulletin of the British Arachnological Society, 4(1): 1-60.

Millidge, A.F. 1978b. The genera *Saaristoa* n.gen. and *Metapanamomops* Millidge (Araneae: Linyphiidae). Bulletin of the British Arachnological Society, 4: 113-123.

Millidge, A.F. 1980b. The erigonine spiders of North America. Part 2. The genus *Spirembolus* Chamberlin (Araneae: Linyphiidae). Journal of Arachnology, 8: 109-158.

Millidge, A.F. 1981a. The erigonine spiders of North America. Part 3. The genus *Scotinotylus* Simon (Araneae, Linyphiidae). Journal of Arachnology, 9: 167-213.

Millidge, A.F. 1981b. The erigonine spiders of North America. Part 4. The genus *Disembolus* Chamberlin and Ivie (Araneae, Linyphiidae). Journal of Arachnology, 9: 259-284.

Millidge, A.F. 1981c. The erigonine spiders of North America. Part 5. The genus *Satilatlas* (Araneae, Linyphiidae). Bulletin of the American Museum of Natural History, 170: 242-253.

Millidge, A.F. 1981d. A revision of the genus *Gonatium* (Araneae: Linyphiidae). Bulletin of the British Arachnological Society, 5(6): 253-277.

Millidge, A.F. 1983. The erigonine spiders of North America. Part 6. The genus *Walckenaeria* Blackwall (Araneae, Linyphiidae). Journal of Arachnology, 11(3): 105-200.

Millidge, A.F. 1984a. The erigonine spiders of North America. Part 7. Miscellaneous genera (Araneae, Linyphiidae). Journal of Arachnology, 12(2): 121-169.

Millidge, A.F. 1984b. The taxonomy of the Linyphiidae, based chiefly on the epigynal and tracheal characters (Araneae: Linyphiidae). Bulletin of the British Arachnological Society, 6(6): 229-267.

Millidge, A.F. 1985. Some linyphiid spiders from South America (Araneae, Linyphiidae). American Museum Novitates, No. 2836: 1-78.

Millidge, A.F. 1987. The erigonine spiders of North America. Part 8. The genus *Eperigone* Crosby and Bishop (Araneae, Linyphiidae). American Museum Novitates, No. 2885: 1-75.

Millidge, A.F. 1991. Further linyphiid spiders (Araneae) from South America. 205: 1-199.

Millidge, A.F. 1993a. *Blestia*, a new genus of erigonine spider with clypeal sulci (Araneae: Linyphiidae). Bulletin of the British Arachnological Society, 9: 126-128.

Millidge, A.F. 1993d. A North American species of the genus *Sphecozone* O.P.-Cambridge, 1870 (Araneae: Linyphiidae). Bulletin of the British Arachnological Society, 9: 168.

Millot, J. 1931. Le tubercule anal des Uroctéides et des Oecobiides (Araneidae). Bulletin de la Société zoologique de France, 56: 199-205.

Moggridge, J.T. 1874. Supplement to harvesting ants and trap-door spiders. 157-304 + pl. XIII-XX.

Montgomery, T.H. Jr. 1904. Description of North American Araneae of the family Lycosidae and Pisauridae. Proceedings of the Academy of Natural Sciences of Philadelphia, 56: 261-323 + pl. XVIII-XX.

Morse, D.H. 1983. Foraging patterns and time budgets of the crab spiders *Xysticus emertoni* Keyserling and *Misumena vatia* (Clerck) (Araneae: Thomisidae) on flowers. Journal of Arachnology, 11: 87-94.

Morse, D.H. 1986. Predatory risk to insects foraging at flowers. Oikos, 46: 223-228.

Morse, D.H. 1987. Attendance patterns, prey capture, changes in mass, and survival of crab spiders *Misumena vatia* (Araneae, Thomisidae) guarding their nests. Journal of Arachnology, 15: 193-204.

Morse, D.H. and R.S. Fritz. 1982. Experimental and observational studies of patch-choice at different scales by the crab spider *Misumena vatia*. Ecology, 63: 172-182.

Mott, D.A. 1989. Revision of the genus *Mimetus* in North America (Araneae, Mimetidae) unpublished Ph.D. thesis, Southern Illinois University at Carbondale. i-xv + 182 pages.

Muma, M.H. 1945. New and interesting spiders from Maryland. Proceedings of the Biological Society of Washington, 58: 91-104.

Muma, M.H. 1946. North American Agelenidae of the genus *Coras* Simon. American Museum Novitates, No. 1329: 1-20.

Muma, M.H. 1947. North American Agelenidae of the genus *Wadotes* Chamberlin. American Museum Novitates, No. 1334: 1-12.

Muma, M.H. 1953. A study of the spider family Selenopidae in North America, Central America and the West Indies. American Museum Novitates, No. 1619: 1-55.

Muma, M.H. and W.J. Gertsch. 1964. The spider family Uloboridae in North America north of Mexico. American Museum Novitates, No. 2196: 1-43.

Murphy, F. and J. Murphy. 2000. An Introduction to the Spiders of South East Asia with Notes on All Genera. Malaysian Nature Society, Kuala Lumpur. 624 + pl. I-XXXII pages.

Nentwig, W. 1993. Spiders of Panama. The Sandhill Crane Press, Inc. Gainesville, Florida. vi + 274 pages.

Nyffeler, M., D.A. Dean and W.L. Sterling. 1992. Diets, feeding specialization, and predatory role of two lynx spiders, *Oxyopes salticus* and *Peucetia viridans* (Araneae: Oxyopidae), in a Texas cotton agroecosystem. Enviromental Entomology, 212(6): 1457-1465.

Odenwall, E. 1901. Araneae nonnullae Sibiriae Transbaicalensis. Öfversigt af Finska Vetenskaps societetens Förhandlingar, 43(4): 255-273 + pl. I.

Oi, R. 1960a. Linyphiid spiders of Japan. Journal of the Institutes of Polytechnics, Osaka City University, 11 (Series D): 137-244.

Olivier, G.A. 1789. Araignée, Aranea. In Encyclopédie Méthodique. Histoire naturelle. Tome quatrième. Insectes. A-Bom, pages 173-240. Paris.

Olmstead J.V. 1975. A revision of the spider genus *Syspira* (Araneida: Clubionidae). Unpublished M.Sc. thesis. California State University, Long Beach. 89 pages.

Ono, H. 1988. A revisional study of the spider family Thomisidae (Arachnida, Araneae) of Japan. National Science Museum, Tokyo. ii + 1-252 pages.

Opell, B.D. 1979. Revision of the genera and tropical American species of the spider family Uloboridae. Bulletin of the Museum of Comparative Zoology, 148(10): 443-549.

Opell, B.D. 1982. Post-hatching development and web production of *Hyptiotes cavatus* (Hentz) (Araneae: Uloboridae). Journal of Arachnology, 10: 185-191.

Opell, B.D. 1983. Checklist of American Uloboridae (Arachnida: Araneae). The Great Lakes Entomologist, 16: 61-66.

Opell, B.D. 1984a ["1983"]. *Lubinella*, a new genus of Uloboridae (Arachnida: Araneae). Journal of Arachnology, 11: 441-446.

Opell, B.D. 1984b. Phylogenetic review of the genus *Miagrammopes sensu lato* (Araneae: Uloboridae). Journal of Arachnology, 12: 229-240.

Opell, B.D. 1984c. Eggsac differences in the spider family Uloboridae (Arachnida: Araneae). Transactions of the American Microscopical Society, 103: 122-129.

Opell, B.D. 1987. The influence of web monitoring tactics upon the tracheal systems of spiders of the family Uloboridae. Zoomorphology, 107: 255-259.

Opell, B.D. 1988. Ocular changes accompanying eye loss in the spider family Uloboridae. Journal of Morphology, 196: 119-126.

Opell, B.D. 1989a. Centers of mass and weight distribution in spiders of the family Uloboridae. Journal of Morphology, 202: 351-359.

Opell, B.D. 1989b. Do female *Miagrammopes animotus* (Araneae: Uloboridae) spin color-coordinated egg sacs? Journal of Arachnology, 17: 108-111.

Opell, B.D. 1990. The relationship of book lung and tracheal systems in the spider family Uloboridae (Arachnida, Araneida). Journal of Morphology, 206: 211-216.

Opell, B.D. 1994a. Increased stickiness of prey capture threads accompanying web reduction in the spider family Uloboridae. Functional Ecology, 8: 85-90.

Opell, B.D. 1994b. Factors governing the stickiness of cribellar prey capture threads in the spider family Uloboridae. Journal of Morphology, 221: 111-119.

Opell, B.D. 1996. Functional similarities of spider webs with diverse architectures. American Naturalist, 148: 630-648.

Opell, B.D. 1997. A comparison of capture thread and architectural features of deinopoid and araneoid orb-webs. Journal of Arachnology, 25: 295-306.

Opell, B.D. 1998. The respiratory complementarity of spider book lung and tracheal systems. Journal of Morphology, 236: 57-64.

Opell, B.D. 1999. Changes in spinning anatomy and thread stickiness associated with the origin of orb-weaving spiders. Biological Journal of the Linnean Society, 68: 593-612.

Opell, B.D. 2001a. Cribellum and calamistrum ontogeny in the spider family Uloboridae: linking functionally related but separate silk spinning features. Journal of Arachnology, 29: 220-226.

Opell, B.D. 2001b. Egg sac recognition by female *Miagrammopes animotus* (Araneae, Uloboridae). Journal of Arachnology, 29: 244-248.

Opell, B.D. and J.A. Beatty. 1976. The Nearctic Hahniidae (Arachnida: Araneae). Bulletin of the Museum of Comparative Zoology, 147: 393-433.

Opell, B.D. and P.E. Cushing. 1986. Visual fields of orb web and single line web spiders of the family Uloboridae (Arachnida, Araneida). Zoomorphology, 106: 199-204.

Opell, B.D. and W.G. Eberhard. 1983. Resting postures of orb-weaving uloborid spiders. Journal of Arachnology, 11: 369-376.

Opell, B.D. and A.D. Ware. 1987. Changes in visual fields associated with web reduction in the spider family Uloboridae. Journal of Morphology, 192: 87-100.

Ovtsharenko, V.I., G. Levy and N.I. Platnick. 1994. A review of the ground spider genus *Synaphosus* (Araneae, Gnaphosidae). American Museum Novitates, 3095: 1-27.

Ovtsharenko, V.I. and Y.M. Marusik. 1988. Spiders of the family Gnaphosidae (Aranei) of the north-east USSR (the Magadan Province). Entomologiceskoe Obozrenie, 67:204-217.

Ovtsharenko, V.I., N.I. Platnick and Y.M. Marusik. 1995. A review of the Holarctic ground spider genus *Parasyrisca* (Araneae, Gnaphosidae). American Museum Novitates, No. 3147: 1-55.

Paquin, P. and N. Dupérré. 2003. Guide d'identification des Araignées (Araneae) du Québec. Fabreries, Supplément 11. 251 pages.

Paquin, P., N. Dupérré and R. Hutchinson. 2001. Liste révisée des Araignées (Araneae) du Québec. Partie 1. Pages 5-87 in P. Paquin and D.J. Buckle (editors), Contributions à la Connaissance des Araignées (Araneae) d'Amérique du Nord. Fabreries, Supplément 10.

Paquin, P. and M. Hedin. 2004. The power and perils of "molecular taxonomy": A case study of eyeless and endangered *Cicurina* (Araneae: Dictynidae) from Texas caves. Molecular Ecology, 13: 3239-3255.

Patel, B.H. and T.S. Reddy. 1991. A rare new species of *Homalonychus* Marx (Araneae: Homalonychidae) from coastal Andhra Pradesh, India. Records of the Zoological Survey of India, 89(1/4): 205-207.

Pavesi, P. 1880. Studi sugli Aracnidi africani. Introduzione + I. Aracnidi di Tunisia. Annali del Museo civico di Storia naturale di Genova, 15: 283-309, 310-388.

Peck, W.B. 1981. The Ctenidae of temperate zone North America. Bulletin of the American Museum of Natural History, 170(1): 157-169.

Peck, W.B. and W.H. Whitcomb. 1970. Studies on the biology of a spider, *Chiracanthium inclusum* (Hentz). University of Arkansas, Agricultural Experiment Station Bulletin, 753: 1-76.

Peckham, G.W. and E.G. Peckham. 1883. Descriptions of new or little known spiders of the family Attidae from various parts of the United States of North America. Milwaukee, 1-35 + pl. I-III.

Peckham, G.W. and E.G. Peckham. 1885a. On some new genera and species of the Attidae, from Madagascar and Central America. Proceedings of the Natural History Society of Wisconsin, 23-42 + pl. I.

Peckham, G.W. and E.G. Peckham. 1885b. On some new genera and species of Attidae from the eastern part of Guatemala. Proceedings of the Natural History Society of Wisconsin, 62-86 + pl. II.

Peckham, G.W. and E.G. Peckham. 1885c. On the genera of the family Attidae. Transactions of the Wisconsin Academy of Sciences, Arts and Letters, 6: 257-342 + tables I-IV.

Peckham, G.W. and E.G. Peckham. 1888. Attidae of North America. Transactions of the Wisconsin Academy of Sciences, Arts and Letters, 7: 1-105 + pl. I-VI.

Peckham, G.W. and E.G. Peckham. 1892. Ant-like spiders of the family Attidae. Occasional Papers of the Natural History Society of Wisconsin, 2(1): 1-83 + pl. I-VII.

Peckham, G.W. and E.G. Peckham. 1894. Spiders of the *Marptusa* group. Occasional Papers of the Natural History Society of Wisconsin, 2(2): 85-156 + pl. VIII-XIV.

Peckham, G.W. and E.G. Peckham. 1895. Spiders of the *Homalattus* group of the family Attidae. Occasional Papers of the Natural History Society of Wisconsin, 2(3): 159-178 + pl. XV-XVI + index.

Peckham, G.W. and E.G. Peckham. 1896. Spiders of the family Attidae from Central America and Mexico. Occasional Papers of the Natural History Society of Wisconsin, 3: 1-101.

Peckham, G.W. and E.G. Peckham. 1901a. Spiders of the *Phidippus* group of the family Attidae. Transactions of the Wisconsin Academy of Sciences, Arts and Letters, 13(1): 282-358 + pl. XXIII-XXVIII.

Peckham, G.W. and E.G. Peckham. 1901b. On spiders of the family Attidae found in Jamaica. Proceedings of the Zoological Society of London, 1901(2): 6-16.

Peckham, G.W. and E.G. Peckham. 1909. Revision of the Attidae of North America. Transactions of the Wisconsin Academy of Sciences, Arts and Letters, 16, Part I, (5): 355-646.

Penney, D. 2001. Advances in the taxonomy of spiders in Miocene amber from the Dominican Republic (Arthropoda: Araneae). Paleontology, 44(5): 987-1010.

Penney, D. 2003. A new deinopoid spider from Cretaceous Lebanese amber. Acta Palaeontologica Polonica, 48: 569-574.

Penniman, A.J. 1985. Revision of the britcheri and pugnata groups of *Scotinella* (Araneae, Corinnidae, Phrurolithinae) with a reclassification of Phrurolithine spiders. Unpublished Ph.D. Dissertation, Ohio State University, ii-xx + 1-240 pages.

Pérez-Miles, F., S.M. Lucas, P.I. da Silva Jr. and R. Bertani. 1996. Systematic revision and cladistic analysis of Theraphosidae (Aranee: Theraphosidae). Mygalomorph, 1: 33-68.

Perty, M. 1833. Arachnides Brasilienses. Pages 191-209 + pl. XXXVIII-XXXIX in J.B. de Spix and F.P. Martius (editors), Delectus animalium articulatorum quae in itinere per Braziliam ann. 1817 et 1820 colligerunt. Monachii.

Peters, H.M. 1992. On the spinning apparatus and the structure of the capture threads of *Deinopsis subrufus* (Araneae, Deinopidae). Zoomorphology, 112: 27-37.

Petrunkevitch, A. 1910. Some new or little known spiders. Annals of the New York Academy of Sciences, 19(9): 205-224.

Petrunkevitch, A. 1920. Description of *Orchestina saltitans* Banks (Arachnida). Journal of the New York Entomological Society, 28: 157-160.

Petrunkevitch, A. 1923. On families of spiders. Annals of the New York Academy of Sciences, 29: 145-180.

Petrunkevitch, A. 1925a. Arachnida from Panama. Transactions of the Connecticut Academy of Arts and Sciences, 27: 51-248.

Petrunkevitch, A. 1925b. Descriptions of new or inadequately-known American spiders. Annals of the Entomological Society of America, 18: 313-322.

Petrunkevitch, A. 1925c. New Erigoninae from Tennessee. Journal of the New York Entomological Society, 3: 170-176 + pl. VIII.

Petrunkevitch, A. 1926b. Spiders from the Virgin Islands. Transactions of the Connecticut Academy of Arts and Sciences, 28: 21-78.

Petrunkevitch, A. 1928b. Systema Araneareum. Transactions of the Connecticut Academy of Arts and Sciences, 29: 1-270.

Petrunkevitch, A. 1929b. The spiders of Porto Rico. Part I. Transactions of the Connecticut Academy of Arts and Sciences, 30: 1-158.

Petrunkevitch, A. 1930b. The spiders of Porto Rico, Part III. Transactions of the Connecticut Academy of Arts and Sciences, 31: 1-191.

Petrunkevitch, A. 1933. An inquiry into the natural classification of spiders, based on a study of their internal anatomy. Transactions of the Connecticut Academy of Arts and Siences, 31: 303-389.

Petrunkevitch, A. 1939a. Catalogue of American spiders. Part I. Transactions of the Connecticut Academy of Arts and Sciences, 33: 133-138.

Petrunkevitch, A. 1939b. The status of the genus *Eurypelma* (Order Araneae, family Theraphosidae). The Annals and Magazine of Natural History, (11)4 : 561-568.

Petrunkevitch, A. 1958. Amber spiders in European collections. Transactions of the Connecticut Academy of Arts and Sciences, 41: 97-400.

Pickard-Cambridge, F.O. 1893c. Handbook to the study of British spiders (Drassidae and Agelenidae). British Naturalist, supplement 3. 123-170 + pl. I-VII pages.

Pickard-Cambridge, F.O. 1894b. New genera and species of British spiders. The Annals and Magazine of Natural History, 6(13): 87-111 + pl.I-II.

Pickard-Cambridge, F.O. 1894c. Part II. Synonymic list of the genera of the British Araneida. The British Naturalist (New Series), 165-172, 191-194, 211-215, 240-241.

Pickard-Cambridge, F.O. 1897-1905. Arachnida. Araneida and Opiliones. Biologia Centrali-Americana. Zoologia, Volume 2: 1-610 + pl. I-LIV. [**1897a**: p.1-40; ; **1899a**: p.41-88; **1900**: p.89-192; **1901a**: p.193-312; **1902a**: p.313-424; **1903a**: p.425-464; **1904**: p.465-545; **1905**: p.546-610].

Pickard-Cambridge, F.O. 1901c. Arachnida in The Victoria History of the County of Worcester, pages 125.

Pickard-Cambridge, F.O. 1902d. A revision of the genera of Araneae or spiders with reference to their type species. The Annals and Magazine of Natural History, 7(9): 5-20.

Pickard-Cambridge, O. 1861a. Descriptions of ten new species of spiders lately discovered in England. The Annals and Magazine of Natural History, 3(7): 428-441.

Pickard-Cambridge, O. 1862a. Descriptions of ten new species of British spiders. Zoologist, 20: 7951-7969.

Pickard-Cambridge, O. 1863. Description of twenty-four new species of spiders lately discovered in Dorsetshire and Hampshire; together with a list of rare and some other hitherto unrecorded British spiders. The Zoologist, 21: 8561-8599.

Pickard-Cambridge, O. 1869b. Catalogue of a collection of Ceylon Araneida lately received from Mr. J. Nietner, with descriptions of new species and characters of a new genus. I. Journal of the Linneaen Society of London (Zoology), 10: 373-397.

Pickard-Cambridge, O. 1869d. Notes on some spiders and scorpions from St. Helena, with descriptions of new species. Proceedings of the Zoological Society of London, 1869: 531-544.

Pickard-Cambridge, O. 1870a. Descriptions and sketches of two new species of Araneida with characters of a new genus. Zoological Journal of the Linnean Society, 10: 398-405.

Pickard-Cambridge, O. 1870d. On some new genera and species of Araneida. Proceedings of the Zoological Society of London, 1870: 728-747.

Pickard-Cambridge, O. 1871a. Descriptions of some British spiders new to science, with a notice of others, of which some are now for the first time recorded as British species. Transactions of the Linnean Society of London, 27: 393-464.

Pickard-Cambridge, O. 1871c. Arachnida. Zoological Record, 7: 207-224.

Pickard-Cambridge, O. 1872a. General list of the spiders of Palestine and Syria, with descriptions of numerous new species and characters of two new genera. Proceedings of the Zoological Society of London, 212-254 + pl. XIII-XVI.

Pickard-Cambridge, O. 1872b. Description of twenty-four new species of *Erigone*. Proceedings of the Zoological Society of London, 747-769 + pl. LXV-LXVI.

Pickard-Cambridge, O. 1873b. On new and rare British spiders (being a Second Supplement to "British Spiders new to science," Linn. Trans. vol. xxvii. p. 393). Transactions of the Linnean Society of London, 28: 523-555.

Pickard-Cambridge, O. 1874b. On some new species of *Erigone* from North America. Proceedings of the Zoological Society of London, 428-442 + pl. LV.

Pickard-Cambridge, O. 1874c. On some new genera and species of Araneida. The Annals and Magazine of Natural History, 14: 169-183.

Pickard-Cambridge, O. 1875b. On some new species of *Erigone*. Part I. Proceedings of the Zoological Society of London, 190-224 + pl. XXVII-XXIX.

Pickard-Cambridge, O. 1875c. On some new species of *Erigone* from North America. Proceedings of the Zoological Society of London, 393-405 + pl. XLVI.

Pickard-Cambridge, O. 1875e. Notes and descrptions of some new and rare British spiders. The Annals and Magazine of Natural History, (4)16: 237-260.

Pickard-Cambridge, O. 1875g. On some new species of *Erigone*. Part II. Proceedings of the Zoological Society of London, 323-335 + pl. XLIV.

Pickard-Cambridge, O. 1876b. Catalogue of a Collection of spiders made in Egypt, with descriptions of new species and characters of a new genus. Proceedings of the Zoological Society of London, 541-630 + pl. LVIII-LX.

Pickard-Cambridge, O. 1877b. On some new species of Araneidea, with characters of two new genera and some remarks on the families Podophthalmides and Dinopides. Proceedings of the Zoological Society of London, 1877: 557-578.

Pickard-Cambridge, O. 1878a. Notes on British spiders with description of some new species. The Annals and Magazine of Natural History, 5(1): 105-128 + pl. XI.

Pickard-Cambridge, O. 1879d. On some new and rare spiders from New Zealand, with characters of four new genera. Proceedings of the Zoological Society of London, 681-703 + pl. LII-LIII.

Pickard-Cambridge, O. 1879e. On some new and rare British spiders, with characters of a new genus. The Annals and Magazine of Natural History, 5(4): 190-215 + pl. XII.

Pickard-Cambridge, O. 1879g. The spiders of Dorset, with an appendix containing short descriptions of those British species not yet found in Dorsetshire. Proceedings of the Dorset Natural History and Antiquarian Field Club. Part 1, 1-235.

Pickard-Cambridge, O. 1882b. Notes on British spiders, with descriptions of three new species and characters of a new genus. The Annals and Magazine of Natural History, 5(9): 1-13 + pl. I.

Pickard-Cambridge, O. 1882c. On new genera and species of Araneidea. Proceedings of the Zoological Society of London, 423-442 + pl. XXIX-XXI.

Pickard-Cambridge, O. 1883. On some new genera and species of spiders. Proceedings of the Zoological Society of London, 352-365 + pl. XXXVI-XXXVII.

Pickard-Cambridge, O. 1889-1902a. Class Arachnida. Order Araneida. Biologia Centrali-Americana. Zoologia, Volume 1: v-xv + 1-317 + pl. I-XXXIX. [**1889d**: p.1-56; **1890a**: p.57-72; **1891b**: p.73-88; **1892b**: p.89-104; **1893b**: p.105-120; **1894a**: p.121-144; **1895a**: p.145-160; **1896a**: p.161-224; **1897a**: p.225-232; **1898a**: p.233-288; **1899a**: p.289-304; **1902a**: p.305-317].

Pickard-Cambridge, O. 1889a. On some new species and a new genus of Araneida. Proceedings of the Zoological Society of London, 1889: 34-46.

Pickard-Cambridge, O. 1900a. Some notes on British spiders observed in 1899. Proceedings of the Dorset Natural History and Antiquarian Field Club, 21: 18-394.

Pickard-Cambridge, O. 1905. On new and rare British Arachnida. Proceedings of the Dorset Natural History and Antiquarian Field Club, 26: 40-74 +pl. A-B.

Pickard-Cambridge, O. 1906a. On some new and rare British Arachnida. Proceedings of the Dorset Natural History and Antiquarian Field Club, 27: 72-92 + pl. A.

Pickard-Cambridge, O. 1913. On new and rare British Arachnida noted and observed in 1912. Proceedings of the Dorset Natural History and Antiquarian Field Club, 34: 107-136 + pl. A.

Pickavance, J.R. 2001. Life-cycles of four species of *Pardosa* (Araneae, Lycosidae) from the Island of Newfoundland, Canada. Journal of Arachnology, 29: 367-377.

Piel, W.H. 1992 ["1991"]. The Nearctic jumping spiders of the genus *Admestina* (Araneae: Salticidae). Psyche, 98(4): 265-282.

Piel, W.H. 1995. A new *Chrosiothes* spider from West Virginia (Araneae, Theridiidae). Journal of Arachnology, 22: 181-184.

Piel, W.H. 2001. The systematics of neotropical orb-weaving spiders in the genus *Metepeira* (Araneae: Araneida). Bulletin of the Museum of Comparative Zoology, 157(1): 1-92.

Piel, W.H. and K.J. Nutt. 1997. *Kaira* is a likely sister group to *Metepeira*, and *Zygiella* is an araneid (Araneae, Araneidae): evidence from mitochondrial DNA. Journal of Arachnology, 25: 262-268.

Platnick, N.I. 1974. The spider family Anyphaenidae in America north of Mexico. Bulletin of the Museum of Comparative Zoology, 146(4): 205-266.

Platnick, N.I. 1975b. A revision of the Holarctic spider genus *Callilepis* (Araneae, Gnaphosidae). American Museum Novitates, No. 2573: 1-32.

Platnick, N.I. 1975c. A revision of the spider genus *Eilica* (Araneae, Gnaphosidae). American Museum Novitates, No. 2578: 1-19.

Platnick, N.I. 1976f. On Asian *Prodidomus* (Araneae, Gnaphosidae). Acta Arachnologica Tokyo, 27: 37-42.

Platnick, N.I. 1977c. Notes on the spider genus *Paratheuma* Bryant (Arachnida, Araneae). Journal of Arachnology, 3: 199-201.

Platnick, N.I. 1977g. The Hypochiloid spiders: a cladistic analysis, with notes on the Atypoidea (Arachnida: Araneae). American Museum Novitates, No. 2627: 1-23.

Platnick, N.I. 1984a. On the pseudoscorpion-mimicking spider *Cheliferoides* (Araneae: Salticidae). Journal of the New York Entomological Society, 92: 169-173.

Platnick, N.I. 1984d. On the Gnaphosidae (Arachnida, Araneaea) of the California Channel Islands. Journal of Arachnology, 11: 453-455.

Platnick, N.I. 1986d. On the tibial and patellar glands, relationships, and American genera of the spider family Leptonetidae (Arachnida, Araneae). American Museum Novitates, No. 2855: 1-16.

Platnick, N.I. 1989a. A revision of the spider genus *Segestrioides* (Araneae, Diguetidae). American Museum Novitates, No. 2940: 1-9.

Platnick, N.I. 1989b. Advances in Spider Taxonomy 1981-1987. A supplement to Brignoli's A Catalogue of the Araneae described between 1940 and 1981 (edited by P. Merrett). Manchester University Press, New York. vi-vii + 1-673 pages.

Platnick, N.I. 1990a. Spinneret morphology and phylogeny of ground spiders (Araneae, Gnaphosoidea). American Museum Novitates, No. 2978: 1-42.

Platnick, N.I. 1993a. A new genus of the spider family Caponiidae (Araneae, Haplogynae) from California. American Museum Novitates, No3063: 1-8.

Platnick, N.I. 1993c. Advances in Spider Taxonomy 1988-1991, with synonymies and transfers 1940-1980 (edited by P. Merrett). New York Entomological Society in association with the American Museum of Natural History. 840 pages.

Platnick, N.I. 1995b. A revision of the spider genus *Orthonops* (Araneae, Caponiidae). American Museum Novitates, No. 3150: 1-18.

Platnick, N.I. 1999. A revision of the Appalachian spider genus *Liocranoides* (Aranea: Tengellidae). American Museum Novitates, No. 3285: 1-13.

Platnick, N.I. 2000a. A relimitation and revision of the Australian ground spider family Lamponidae (Araneae: Gnaphosoidea). Bulletin of the American Museum of Natural History, 245: 1-330.

Platnick, N.I. 2002. A revision of the Australasian ground spiders of the families Ammoxenidae, Cithaeronidae, Gallieniellidae, and Trochanteriidae (Araneae: Gnaphosoidea). Bulletin of the American Museum of Natural History, 271: 1-243.

Platnick, N.I. 2005. The world spider catalog. Version 5.5. American Museum of Natural History. http://research.amnh.org/entomology/spiders/catalog/index.html.

Platnick, N.I., J.A. Coddington, R.R. Forster and C.E. Griswold. 1991. Spinneret morphology and the phylogeny of haplogyne spiders (Araneae, Araneomorphae). American Museum Novitates, No. 3016: 1-73.

Platnick, N.I. and D.T. Corey. 1989. On a new spider of the genus *Drassyllus* (Araneae, Gnaphosidae) from Florida. Journal of Arachnology, 17: 114-115.

Platnick, N.I. and C.D. Dondale. 1992. The Ground Spiders of Canada and Alaska (Araneae: Gnaphosidae). The Insects and Arachnids of Canada. Part 19. Agriculture Canada, Ottawa. Publication 1875. 297 pages.

Platnick, N.I. and C. Ewing. 1995. A revision of the tracheline spiders (Araneae, Corinnidae) of southern South America. American Museum Novitates, 3128: 1-41.

Platnick, N.I. and R.R. Forster 1990. On the spider family Anapidae (Araneae, Araneoidea) in the United States. Journal of the New York Entomological Society, 98(1): 108-112.

Platnick, N.I. and W.J. Gertsch. 1976. The suborder of spiders: a cladistic analysis (Arachnida, Araneae). American Museum Novitates, No. 2607: 1-15.

Platnick, N.I. and J.A. Murphy. 1984. A revision of the spider genera *Trachyzelotes* and *Urozelotes* (Araneae, Gnaphosidae). American Museum Novitates, No. 2792: 1-30.

Platnick, N.I. and J.A. Murphy. 1998. On the widespread species *Zelotes schmitzi* (Araneae: Gnaphosidae). Bulletin of the British Arachnological Society, 11: 118-119.

Platnick, N.I. and V.I. Ovtsharenko. 1991. On Eurasian and American *Talanites* (Araneae, Gnaphosidae). Journal of Arachnology, 19: 115-121.

Platnick, N.I. and T.R. Prentice. 1999. A new species of the spider genus *Zelotes* (Araneae, Gnaphosidae) from California. Journal of Arachnology, 27: 672-674.

Platnick, N.I. and M.U. Shadab. 1974a. A revision of the *tranquillus* and *speciosus* groups of the spider genus *Trachelas* (Araneae, Clubionidae) in North and Central America. American Museum Novitates, No. 2553: 1-34.

Platnick, N.I. and M.U. Shadab. 1974c. A revision of the *bispinosus* and *bicolor* groups of the spider genus *Trachelas* (Araneae, Clubionidae) in North and Central America and the West Indies. American Museum Novitates, No. 2560: 1-34.

Platnick, N.I. and M.U. Shadab. 1975a. A revision of the spider genus *Gnaphosa* (Araneae, Gnaphosidae) in America. Bulletin of the American Museum of Natural History, 155: 1-66.

Platnick, N.I. and M.U. Shadab. 1975b. A revision of the spider genera *Haplodrassus* and *Orodrassus* (Araneae, Gnaphosidae) in North America. American Museum Novitates, No. 2583: 1-40.

Platnick, N.I. and M.U. Shadab. 1976a. A revision of the spider genera *Drassodes* and *Tivodrassus* (Araneae, Gnaphosidae) in North America. American Museum Novitates, No. 2593: 1-29.

Platnick, N.I. and M.U. Shadab. 1976b. A revision of the spider genera *Rachodrassus*, *Sosticus*, and *Scopodes* (Araneae, Gnaphosidae) in North America. American Museum Novitates, No. 2594: 1-33.

Platnick, N.I. and M.U. Shadab. 1976c. A revision of the spider genera *Lygromma* and *Neozimiris* (Araneae, Gnaphosidae). American Museum Novitates, No. 2598: 1-23.

Platnick, N.I. and M.U. Shadab. 1977b. A revision of the spider genera *Herpyllus* and *Scotophaeus* (Araneae, Gnaphosidae) in North America. Bulletin of the American Museum of Natural History, 159: 1-44.

Platnick, N.I. and M.U. Shadab. 1978a. A review of the spider genus *Mysmenopsis* (Araneae, Mysmenidae). American Museum Novitates, No. 2661: 1-22.

Platnick, N.I. and M.U. Shadab. 1980a. A revision of the North American spider genera *Nodocion*, *Litopyllus*, and *Synaphosus* (Araneae, Gnaphosidae). American Museum Novitates, No. 2691: 1-26.

Platnick, N.I. and M.U. Shadab. 1980b. A revision of the spider genus *Cesonia* (Araneae, Gnaphosidae). Bulletin of the American Museum of Natural History, 165: 335-386.

Platnick, N.I. and M.U. Shadab. 1981b. A new genus of the spider family Gnaphosidae (Arachnida, Araneae). Bulletin of the American Museum of Natural History, 170: 176-182.

Platnick, N.I. and M.U. Shadab. 1981e. A revision of the spider genera *Sergiolus* (Araneae, Gnaphosidae). American Museum Novitates, No. 2717: 1-41.

Platnick, N.I. and M.U. Shadab. 1982a. A revision of the American spiders of the genus *Drassyllus* (Araneae, Gnaphosidae). Bulletin of the American Museum of Natural History, 173: 1-97.

Platnick, N.I. and M.U. Shadab. 1982b. A revision of the American spiders of the genus *Camillina* (Araneae, Gnaphosidae). American Museum Novitates, No. 2748: 1-38.

Platnick, N.I. and M.U. Shadab. 1983a. A revision of the American spiders of the genus *Zelotes* (Araneae, Gnaphosidae). Bulletin of the American Museum of Natural History, 174: 97-192.

Platnick, N.I. and M.U. Shadab. 1988. A revision of the American spiders of the genus *Micaria* (Araneae, Gnaphosidae). American Museum Novitates, No. 2916: 1-64.

Platnick, N.I. and M.U. Shadab. 1989. A revision of the genus *Teminus* (Araneae, Miturgidae). American Museum Novitates, No. 2963: 1-12.

Platnick, N.I. and D. Ubick. 1989. A revision of the spider genus *Drassinella* (Araneae, Liocranidae). American Museum Novitates, No. 2937: 1-12.

Platnick, N.I. and D. Ubick. 2001. A revision of the North American spiders of the new genus *Socalchemmis* (Araneae, Tengellidae). American Museum Novitates, No. 3339: 1-25.

Platnick, N.I. and D. Ubick. 2005. A revision of the North American spider genus *Anachemmis* Chamberlin (Araneae, Tengellidae). American Museum Novitates, No. 3477: 1-20.

Pocock, R.I. 1899d. Diagnoses of some new Indian Arachnida. Journal of the Bombay Natural History Society, 12: 744-753.

Pocock, R.I. 1900a. The fauna of British India, including Ceylon and Burma. Arachnida. London. 279 pages.

Pocock, R.I. 1901c. Some old and new genera of S. American Aviculariidae. The Annals and Magazine of Natural History, 7(8): 540-555.

Porter, A.H. and E.M. Jakob. 1990. Allozyme variation in the introduced spider *Holocnemus pluchei* (Araneae, Pholcidae) in California. Journal of Arachnology, 18: 313-319.

Prendini, L., R. Hanner and R. DeSalle. 2002. Obtaining, storing and archiving specimens and tissue samples for use in molecular studies. Pages 176-248 in R. DeSalle, G. Giribet & W. Wheeler (editors): Techniques in Molecular Systematics and Evolution. Basel, Birkhaeuser, Verlag.

Prentice, T.R. 1993. A new species of North American tarantula, *Aphonopelma paloma* (Araneae, Mygalomorphae, Theraphosidae). Journal of Arachnology, 20: 189-199.

Prentice, T.R. 1997. Theraphosidae of the Mojave Desert west and north of the Colorado River (Araneae, Mygalomorphae, Theraphosidae). Journal of Arachnology, 25: 137-176.

Prentice, T.R. 2001. Distinguishing the females of *Trochosa terricola* and *Trochosa ruricola* (Araneae, Lycosidae) from populations in Illinois, USA. Journal of Arachnology, 29: 427-430.

Prószyński, J. 1968c. Revision of the spider genus *Sitticus* Simon 1901 (Araneida, Salticidae), I. The *terebratus* group. Annales Zoologici. Warszawa, 26: 391-407.

Prószyński, J. 1971b. Revision of the spider genus *Sitticus* Simon 1901 (Aranei, Salticidae), II. *Sitticus saxicola* (C.L. Koch, 1846) and related forms. Annales Zoologici. Warszawa, 28: 183-204.

Prószyński, J. 1973a. Revision of the spider genus *Sitticus* Simon 1901 (Aranei, Salticidae), III. *Sitticus penicillatus* (Simon, 1875) and related forms. Annales Zoologici. Warszawa, 30: 71-95.

Prószyński, J. 1980. Revision of the spider genus *Sitticus* Simon 1901 (Aranei, Salticidae), IV. *Sitticus floricola* (C.L. Koch) group. Annales Zoologici. Warszawa, 36: 1-35.

Prószyński, J. 1992a. Salticidae (Araneae) of the Old World and Pacific Islands in several US collections. Annales Zoologici. Warszawa, 44: 87-163.

Prószyński, J. 2003. Salticidae (Araneae) of the World. Online document at http://salticidae.org/salticid/main.htm.

Purcell, W.F. 1904. Descriptions of new genera and species of South African spiders. Transactions of the South African Philosophical Society, 15: 115-174.

Putman, W.L. 1967. Life histories and habits of two species of *Philodromus* (Araneida: Thomisidae) in Ontario. The Canadian Entomologist, 99: 622-631.

Ramírez, M.J. 1995a. A phylogenetic analysis of the subfamilies of Anyphaenidae (Arachnida, Araneae). Entomologica Scandinavica, 26: 361-384.

Ramírez, M.J. 1995c. Natural history of the spider genus *Lutica* (Araneae, Zodariidae). Journal of Arachnology, 23: 111-117.

Ramírez, M.J. 2000. Respiratory system morphology and the phylogeny of haplogyne spiders (Araneae, Araneomorphae). Journal of Arachnology, 28: 149-157.

Ramírez, M.J. 2003. The spider subfamily Amaurobioidinae (Araneae, Anyphaenidae): a phylogenetic revision at the generic level. Bulletin of the American Museum of Natural History, 277: 1-262.

Ramírez, M.J. and R.D. Beckwitt. 1995. Phylogeny and historical biogeography of the spider genus *Lutica* (Araneae, Zodariidae). Journal of Arachnology, 23: 177-193.

Ramírez, M.J., A.B. Bonaldo and A.D. Brescovit. 1997. Revisión del género *Macerio* y comentarios sobre la ubicación de *Cheirancanthium*, *Tecution* y *Helebiona* (Araneae, Miturgidae, Eutichurinae). [Revision of the genus *Macerio* and comments on the placement of *Cheirancanthium*, *Tecution* and *Helebiona* (Araneae, Miturgidae, Eutichurinae)]. Iheringia Zoologia, 82: 43-66.

Ramírez, M.J. and C.J. Grismado. 1997. A review of the spider family Filistatidae in Argentina (Arachnida, Araneae), with a cladistic reanalysis of filistatid genera. Entomologica Scandinavia, 28: 319-349.

Raven, R.J. 1979. Systematics of the mygalomorph spider genus *Masteria*. Australian Journal of Zoology, 27: 623-636.

Raven, R.J. 1985a. The spider infraorder mygalomorphae (Araneae): cladistics and systematics. Bulletin of the American Museum of Natural History, 182(1): 1-180.

Raven, R.J. and K. Stumkat. 2003. Problem solving in the spider families Miturgidae, Ctenidae and Psechridae (Araneae) in Australia and New Zealand. Journal of Arachnology, 31: 105-121.

Reddell, J.R. 1981. A review of the cavernicole fauna of Mexico, Guatemala, and Belize. Bulletin 27 of the Texas Memorial Museum. 327 pages.

Reeves, E.M., J.J. Hadam, J.H. Redner and C.D. Dondale. 1984. Identity of the North American crab spiders *Xysticus durus* and *X. keyserlingi* (Araneae: Thomisidae). The Canadian Entomologist, 116: 221-225.

Reeves, R.M. 2000. Invertebrate cavernicoles of the Great Smoky Mountains National Park, USA. Journal of the Elisha Mitchell Scientific Society, 116(4): 334-343.

Reiskind, J. 1969. The spider subfamily Castianeirinae of the North and Central America (Araneae, Clubionidae). Bulletin of the Museum of Comparative Zoology, 138(5): 163-325 + pl. (unnumbered).

Reiskind, J. and P.E. Cushing. 1996. Study of a narrow hybrid zone between two wolf spiders, *Lycosa ammophila* and *Lycosa ericeticola*, in north Florida (Araneae, Lycosidae). Revue suisse de Zoologie, volume hors série, 543-554.

Reissland, A. and P. Görner. 1985. Trichobothria. Pages 138-161 in F.G. Barth (editor), Neorobiology of Arachnids. Springer-Verlag, New York.

Reuss, A. 1834. Beschreibung der Arachniden. Zoologischen Miscellen. Museum Senckenbergianum, 1. Arachniden [spiders by F. Wider], pages 195-282 + pl. XIV-XVIII.

Rheims, C.A. and A.D. Brescovit. 2004a. Revision and cladistics of the spider family Hersiliidae (Arachnida, Araneae) with emphasis on Neotropical and Nearctic species. Insect Systematics and Evolution, 35: 189-239.

Richman, D.B. 1965. Jumping spiders (Salticidae) from Yuma County, Arizona, with a description of a new species and distributional records. Southwestern Naturalist, 10: 132-135.

Richman, D.B. 1973. Field studies on the biology of *Loxosceles arizonica* Gertsch & Mulaik (Araneae, Scytodidae). Journal of the Arizona Academy of Sciences, 8: 124-126.

Richman, D.B. 1978. Key to the jumping spider (Salticidae) genera of North America. Peckhamia, 1(5): 77-81.

Richman, D.B. 1979. Jumping spiders of the United States and Canada: changes in the key and list (1). Peckhamia, 1(6): 125.

Richman, D.B. 1980. Jumping spiders of the United States and Canada: changes in the key and list (3). Peckhamia, 2(1): 11.

Richman, D.B. 1981. A revision of the genus *Habrocestum* (Araneae, Salticidae) in North America. Bulletin of the American Museum of Natural History, 170(1): 197-206.

Richman, D.B. 1989. A revision of the genus *Hentzia* (Araneae, Salticidae). Journal of Arachnology, 17(3): 285-344.

Richman, D.B. 2003. Spiders (Araneae) of pecan orchards in the southwestern United States and their role in pest suppression. Southwestern Entomologist Supplement, 27: 115-122.

Richman, D.B., W.F. Buren and W.H. Whitcomb. 1983. Predatory arthropods attacking the eggs of *Diaprepes abbreviatus* (L.) (Coleoptera: Curculionidae) in Puerto Rico and Florida. Journal of the Georgia Entomological Society, 18: 335-342.

Richman, D.B. and B. Cutler. 1978. A list of the jumping spiders (Araneae: Salticidae) of the United States and Canada. Peckhamia, 1(5): 82-110.

Richman, D.B. and B. Cutler. 1988. A list of the jumping spiders of Mexico. Peckhamia, 2(5): 63-68.

Richman, D.B., R.C. Hemenway jr. and W.H. Whitcomb. 1980. Field cage evaluation of predators of the soybean looper, *Pseudoplusia includens* (Lepidoptera: Noctuidae). Environmental Entomology, 9: 315-317.

Richman, D.B. and R.R. Jackson. 1992. A review of the ethology of jumping spiders (Araneae, Salticidae). Bulletin of the British Arachnological Society, 9: 33-37.

Richman, D.B. and R.S. Vetter. 2004. A review of the spider genus *Thiodina* (Araneae, Salticidae) in the United States. Journal of Arachnology, 32(3): 418-431.

Richman, D.B. and W.H. Whitcomb. 1981. The ontogeny of *Lyssomanes viridis* (Walckenaer) (Araneae: Salticidae) on *Magnolia grandiflora* L. Psyche, 88: 127-133.

Riechert, S.E. and A.B. Cady. 1983. Pattern and resource use and test for competitive release in a spider community. Ecology, 64: 899-913.

Riechert, S.E. and T. Lockley. 1984. Spiders as biological control agents. Annual Review of Entomology, 29: 299-320.

Roberts, M.J. 1985a. The Spiders of Great Britain and Ireland. Volume I. Introduction, Classification and Nomenclature, Key to Families, Description of Species-Atypidae to Theridiosomatidae. Harley Books, England. 229 pages.

Roberts, M.J. 1985b. The Spiders of Great Britain and Ireland. Volume III. The Colour Plates: Atypidae to Linyphiidae. Harley Books, England. 256 pages.

Roberts, M.J. 1987. The Spiders of Great Britain and Ireland. Volume II. Description of Species -Linyphiidae, Check list of the British Spiders. Harley Books, England. 204 pages.

Roberts, M.J. 1995. Spiders of Britain and northern Europe. Collins field guide. Harper Collins Publisher, New York. 383 pages.

Roddy, L.R. 1966. New species, records, of clubionid spiders. Transactions of the American Microscopical Society, 85: 399-407.

Roewer, C.F. 1935. Zwei myrmecophile Spinnen -Arten Brasiliens. Veröffentlichungen aus dem Deutschen Kolonial -und Uebersee- Museum in Bremen, 1(2): 193-197 + pl. X.

Roewer, C.F. 1942a. Katalog der Araneae von 1758 bis 1940. I. Bremen. 1-1040 pages.

Roewer, C.F. 1954a. Katalog der Araneae von 1758 bis 1940, bzw. 1954. IIa. (Lycosaeformia, Dionycha [excl Salticiformia]). Institut royal des Sciences naturelles de Belgique, Bruxelles. Pages 1-923.

Roewer, C.F. 1954b. Katalog der Araneae von 1758 bis 1940, bzw. 1954. IIb. (Salticiformia, Cribellata) (Synonyma -Verzeichnis, Gesamtindex). Institut royal des Sciences naturelles de Belgique, Bruxelles. Pages 927-1751.

Roewer, C.F. 1959b. Exploration du Parc National de l'Upemba. Araneae, Lycosiformia II (Lycosidae), no. 1. Institut des Parcs nationaux du Congo Belge. Fascicule 55. Brussels, Belgium. Pages 1-518.

Roewer, C.F. 1960d. Araneae Lycosaeformia II (Lycosidae) (Fortsetzung und Schluss). Exploration du Parc National de l'Upemba. Miss G.F. de White. 55: 519-1040.

Roth, V.D. 1952b. The genus *Cybaeus* (Arachnida: Agelenidae) in Oregon. Annals of the Entomological Society of America, 45: 205-219.

Roth, V.D. 1952c. Notes and a new species in Cybaeinae (Arachnida: Agelenidae). The Pan-Pacific Entomologist, 28(4): 195-219.

Roth, V.D. 1954. Review of the spider subgenus *Barronopsis* (Archnidae, Agelenidae). American Museum Novitates, No. 1768: 1-7.

Roth, V.D. 1956a. Revision of the genus *Yorima* Chamberlin and Ivie (Arachnida, Agelenidae). American Museum Novitates, No. 1773: 1-10.

Roth, V.D. 1956b. Taxonomic changes in the Agelenidae (Arachnida). The Pan-Pacific Entomologist, 32(4): 175-180.

Roth, V.D. 1968. The spider genus *Tegenaria* in the western hemisphere (Agelenidae). American Museum Novitates, No. 2323: 1-33.

Roth, V.D. 1970. A new genus of spiders (Agelenidae) from the Santa Catalina Mountains. Journal of the Arizona Academy of Science, 6(2): 114-116.

Roth, V.D. 1972a. Proposal for a Nearctic spider catalog. American Arachnology (Newsletter of the American Arachnological Society), 8: 6-9.

Roth, V.D. 1972b. Diguetidae. American Arachnology (Newsletter of the American Arachnological Society), 8: 22.

Roth, V.D. 1972c. Hahniidae. American Arachnology (Newsletter of the American Arachnological Society), 8: 24-26.

Roth, V.D. 1981. A new genus of spider (Agelenidae) from California exhibiting a third type of leg autospasy. Bulletin of the American Museum of Natural History, 170: 101-105.

Roth, V.D. 1982. Handbook for Spider Identification. Published by author on behalf of the American Arachnological Society. Portal, Arizona. i + 1-62 pages.

Roth, V.D. 1984. The spider family Homalonychidae. American Museum Novitates, No. 2790: 1-11.

Roth, V.D. 1985. Spider Genera of North America. Second edition. Published by author on behalf of the American Arachnological Society. Portal, Arizona. 171 pages.

Roth, V.D. 1988a. American Agelenidae and some misidentified spiders (Clubionidae, Oonopidae and Sparassidae) of E. Simon in the Muséum National d'Histoire Naturelle. Bulletin du Muséum National d'Histoire Naturelle. 4ème Série, 10, section A, No. 1: 25-37.

Roth, V.D. 1988b. Linyphiidae of America north of Mexico. American Arachnological Society. Gainsville, Florida. 127 pages.

Roth, V.D. 1994 ["1993"]. Spider Genera of North America, with Keys to Families and Genera, and a Guide to Literature. Third edition. American Arachnological Society. Gainesville, Florida. 203 pages. **Note from the editors:** *book first available only in June 1994.*

Roth, V.D. and P.L. Brame. 1972. Nearctic genera of the spider family Agelenidae (Arachnida, Araneida). American Museum Novitates, No. 2505: 1-52.

Roth, V.D. and W.L. Brown. 1975a. A new genus of the Mexican intertidal zone spider (Desidae) with biological and behavioral notes. American Museum Novitates, No. 2568: 1-7.

Roth, V.D. and W.L. Brown. 1975b. Comments on the spider *Saltonia incerta* Banks (Agelenidae?). Journal of Arachnology, 3: 53-56.

Roth, V.D. and W.L. Brown. 1986. Catalog of Nearctic Agelenidae. The Museum Texas Tech University, Occasional papers, 99: 1-21.

Rovner, J.S. 1968. An analysis of display in the spider *Lycosa rabida* Walckenaer. Animal Behaviour, 16: 358-369.

Rovner, J.S. 1974. Copulation in the lycosid spider *Schizocosa saltatrix* (Hentz): an analysis of palpal insertion patterns. Animal Behaviour, 22: 94-99.

Rovner, J.S. 1975. Sound production by Nearctic wolf spiders: a substratum-coupled stridulatory mechanism. Science, 190: 1309-1310.

Rovner, J.S. 1980a. Morphological and ethological adaptation for prey capture in wolf spiders (Araneae, Lycosidae). Journal of Arachnology, 8: 201-215.

Rovner, J.S. 1980b. Adaptations for prey capture in oxyopid spiders: Phylogenetic implications. Eight Internationaler Arachnologen-Kongress, Wien, 233-237.

Rovner, J.S., G.A. Higashi and R.F. Foelix. 1973. Maternal behavior in wolf spiders: the role of abdominal hairs. Science, 182: 1153-1155.

Saaristo, M.I. 1972. Redelimitation of the genus *Oreonetides* Strand, 1901 (Araneae, Linyphiidae) based on an analysis of the genital organs. Annales Zoologici Fennici, 9: 69-74.

Saaristo, M.I. 1973b. Taxonomical analysis of the type-species of *Agyneta*, *Anomalaria*, *Meioneta*, *Aprolagus*, and *Syedrula* (Araneae, Linyphiidae). Annales Zoologici Fennici, 10: 451-466.

Saaristo, M.I. 1974a. Taxonomical analysis of *Microneta viaria* (Blackwall, 1841), the type-species of the genus *Microneta* Menge, 1869 (Araneae, Linyphiidae). Annales Zoologici Fennici, 11: 166-169.

Saaristo, M.I. 1978. Spiders (Arachnida, Araneae) from the Seychelle Islands, with notes on taxonomy. Acta Zoologica Fennica, 15: 99-126.

Saaristo, M.I. 1997c. Scytotids [sic] (Arachnida, Araneae, Scytodidae) of the granitic islands of Seychelles. Phelsuma, 5: 49-57.

Saaristo, M.I. 1998. Ochyroceratid spiders of the granitic islands of Seychelles (Araneae, Ochyroceratidae). Phelsuma, 6: 20-26.

Saaristo, M.I. 2001a. Dwarf hunting spiders or Oonopidae (Arachnida, Araneae) of the Seychelles. Insect Systematics and Evolution, 32: 307-358.

Saaristo, M.I. and K.Y. Eskov. 1996. Taxonomy and zoogeography of the hypoarctic erigonine spider genus *Semljicola* (Araneida, Linyphiidae). Acta Zoologica Fennica, 201: 47-69.

Saaristo, M.I. and Y.M. Marusik. 2004a ["2003"]. Revision of the Holarctic spider genus *Oreoneta* Kulczynski, 1894 (Arachnida: Aranei: Linyphiidae). Arthropoda Selecta, 12(3/4): 207-249.

Saaristo, M.I. and A.V. Tanasevitch. 1996b. Redelimitation of the subfamily Micronetinae Hull, 1920 and the genus *Lepthyphantes* Menge, 1866 with description of some new genera (Aranei, Linyphiidae). Bericht des Naturwissens-Medizinischen Verein Innsbruck, 83: 163-186.

Saaristo, M.I. and A.V. Tanasevitch. 2000. Systematics of the *Bolyphantes-Poeciloneta* genus-group of the subfamily Micronetinae Hull, 1920 (Arachnida: Araneae: Linyphiidae). Reichenbachia, 33(32): 255-265.

Saito, S. 1939. On the spiders from Tohoku (northernmost part of the main island), Japan. Saito Ho-on Kai Museum Research Bulletin, 18 (zool. 6): 1-91.

Santos, A.J. and A.D. Brescovit. 2003. A revision of the Neotropical species of the lynx spider genus *Peucetia* Thorell 1869 (Araneae: Oxyopidae). Insect Systematics and Evolution, 34: 95-116.

Sarno, P.A. 1973. A new species of *Atypus* (Araneae: Atypidae) from Pennsylvania. Entomological News, 84: 37-51.

Sauer, R.J. and N.I. Platnick. 1972. The crab spider genus *Ebo* (Araneida: Thomisidae) in the United States and Canada. The Canadian Entomologist, 104(1): 35-60.

Sauer, R.J. and J. Wunderlich. 1991. Die schönsten Spinnen Europas. Fauna-Verlag Dr. Frieder Sauer, Karlsfeld. 192 pages.

Savigny, J.C. 1817. Arachnides. pl. I-IX, Description de l'Égypte. Atlas. Histoire naturelle. Tome II. Paris.

Scharff, N. and J.A. Coddington. 1997. A phylogenetic analysis of the orb-weaving spider family Araneidae (Araneae). Zoological Journal of the Linnean Society, 120: 355-434.

Scheibel, T. 2004. Spider silks: recombinant synthesis, assembly, spinning, and engineering of synthetic proteins. Microbial Cell Factories, 3: 14.

Schenkel, E. 1925d. Beitrag zur Kenntniss der schweizerischen Spinnenfauna. Revue suisse de Zoologie, 32: 253-318.

Schenkel, E. 1930b. Die Araneiden der Schwedischen Kamtchatka-Expedition 1920-1922. Arkiv för Zoologi, 21A(15): 1-33.

Schenkel, E. 1950b. Spinnentiere aus dem westlichen Nordamerika, gesammelt von Dr. Hans Schenkel-Rudin. Verhandlungen der Naturforschenden Gesellsschaft, Basel, 61: 28-92.

Schenkel, E. 1963. Ostasiatsche Spinnen aus dem Muséum d'Histoire naturelle de Paris. Mémoires du Musée national d'Histoire naturelle. Nouvelle Série, Série A, Zoologie, Tome XXV, Fascicule 1: 1-288.

Schiapelli, R.D. and Gerschman de Pikelin. 1958a. Arañas argentinas III. Arañas de Misiones. Revista del Museo Argentino de ciencas Naturales

"Bernandino Rivadavia" e Instituto Nacionale de Investigacion de las Ciencas Naturales (Argentina). Zoologia, 3: 187-231.

Schick, R.X. 1965. The crab spiders of California (Araneida, Thomisidae). Bulletin of the American Museum of Natural History, 129(1): 1-180.

Schick, R.X. A Revision of Caponiidae of North America. Unpublished manuscript.

Schmalhofer, V.R. 1999. Thermal tolerances and preferences of the crab spiders *Misumenops asperatus* and *Misumenoides formosipes* (Araneae, Thomisidae). Journal of Arachnology, 27: 470-480.

Schmitz, A. and S.F. Perry. 2001. Bimodal breathing in jumping spiders: morphometric partitioning of the lungs and tracheae in *Salticus scenicus* (Arachnida, Araneae, Salticidae). Journal of Experimental Biology, 204: 4321-4334.

Schrank, F. von P. 1803. Spinne. In Fauna Boica. Durchg dachte Geschichte der in Baiern einheimischen und Zahmen Tiere, 3(1): 229-244.

Schütt, K. 2000. The limits of the Araneoidea (Arachnida: Araneae). Australian Journal of Zoology, 48: 135-153.

Schütt, K. 2003. Phylogeny of Symphytognathidae s.l. (Araneae, Araneoidea). Zoologica Scripta, 32: 129-151.

Scopoli, J.A. 1763. Entomologia carniolica exhibens insecta carnioliae indigena et distributa in ordines, genera, species, varietates. Methodo Linnaeana. [Araneae p. 392-404]. Vindobonae. 420 + errata unnumbered pages.

Scudder, S.H. 1877. The tube constructing ground-spider of Nantucket. Psyche, 2: 2-9.

Selden, P.A. 1996. Fossil mesothele spiders. Nature, 379: 498-499.

Senglet, A. 2001. Copulatory mechanisms in *Hoplopholcus*, *Stygopholcus* (revalidated), *Pholcus*, *Spermophora* and *Spermophorides* (Araneae, Pholcidae), with additional faunistic and taxonomic data. Bulletin de la Société entomologique de Suisse, 74: 43-67.

Shear, W.A. 1969. Observations on the predatory behavior of the spider *Hypochilus gertschi* Hoffman (Hypochilidae). Psyche, 76(4): 407-417.

Shear, W.A. 1970. The spider family Oecobiidae in North America, Mexico and the West Indies. Bulletin of the Museum of Comparative Zoology, 140(4): 129-164.

Shear, W.A. 1976. The spider family Ochyroceratidae new to the United States. Bulletin of the British Arachnological Society, 3(9): 249-250.

Shear, W.A. 1981. Structure of the male palpal organ in *Mimetus*, *Ero*, and *Gelanor* (Araneoidea, Mimetidae). Bulletin of the American Museum of Natural History, 170: 257-262.

Shultz, J.W. 1990. Evolutionary morphology and phylogeny of Arachnida. Cladistics, 6: 1-38.

Sierwald, P. 1989. Morphology and ontogeny of female copulatory organs in American Pisauridae, with special reference to homologous features (Arachnida: Araneae). Smithsonian Contributions to Zoology, No. 484: 1-24.

Sierwald, P. 1990. Morphology and homologous features in the male palpal organ in Pisauridae and other spider families, with notes on the taxonomy of Pisauridae (Arachnida: Araneae). Nemouria, Occasional Papers of the Delaware Museum of Natural History, No. 35: 1-59.

Sierwald, P. 2000. Description of the male of *Sosippus placidus*, with notes on the subfamily Sosippinae (Araneae, Lycosidae. Journal of Arachnology, 28: 133-140.

Silva Dávila, D. 2003. Higher-level relationships of the spider family Ctenidae. Bulletin of the American Museum of Natural History, 274: 1-86.

Simione, F.P. 1995. Storage in fluid preservative. Pages 161-186 in C.L. Rose, C.A. Hawks & H.H. Genoways (editors): Storage of Natural History Collections: A Preventive Conservation Approach. Published by the Society for the Preservation of Natural History Collections, Iowa City, Idaho.

Simon, E. 1864. Histoire naturelle des Araignées (Aranéides). Paris. 1-540 pages

Simon, E. 1868b. Monographie des espèces Européennes de la famille des Attides (Attidae Sundewall. -Saltigradae Latreille. Annales de la Société entomologique de France, 4(8): 11-72, 529-726 + pl. V-VII.

Simon, E. 1870b ["1873"]. Aranéides nouveaux ou peu connus du midi de l'Europe. Mémoires de la Société royale des sciences de Liège, 2(3): 271-358.

Simon, E. 1871a. Révision des Attidae Européens. Supplément à la révision de Attides (Attidae Sund.). Annales de la Société entomologique de France, 5(1): 125-230, 329-360.

Simon, E. 1871b. [Notes sur plusieurs araignées de Corse et d'Espagne]. Annales de la Société entomologique de France, Bulletin, 5(1): vi-viii.

Simon, E. 1872a. Notice sur les arachnides cavernicoles et hypogés. Annales de la Société entomologique de France, 215-255.

Simon, E. 1873a. Aranéides nouveaux ou peu connu du midi de l'Europe (2ème mémoire). Mémoires de la Société royale des sciences de Liège, 2(5): 187-351 + pl. 1-3.

Simon, E. 1874a. Les Arachnides de France. Tome 1, contenant les familles des Epeiridae, Uloboridae, Dictynidae, Enyoidae et Pholcidae, Paris, Librairie encyclopédique de Roret: 1-273 + pl. I-III.

Simon, E. 1874b. Études arachnologiques. 3e mémoire. V. Révision des espèces européennes de la famille des Sparassidae. Annales de la Société entomologique de France, (5)4: 243-279.

Simon, E. 1875a. Les Arachnides de France. Tome 2, contenant les familles des Urocteidae, Agelenidae, Thomisidae et Sparassidae. Paris, Librairie encyclopédique de Roret. 1-358 + pl. IV-VII pages.

Simon, E. 1875e. Liste d'Arachnides de Constantinople et description de deux Opilionides. Annales de la Société entomologique de France, Bulletin, (5)5: CXCVI-CXCVIII.

Simon, E. 1876a. Les Arachnides de France. Tome 3, contenant les familles des Attidae, Oxyopidae et Lycosidae. Paris, Librairie encyclopédique de Roret. 1-370 + pl. VIII-XI pages.

Simon, E. 1876e. Descriptions d'araignées nouvelles de France. Annales de la Société entomologique de France, Bulletin, 5(6): clxxx-clxxxiii.

Simon, E. 1876f. Étude sur les Arachnides du Congo. Bulletin de la Société zoologique de France, 1: 12-15, 215-224.

Simon, E. 1877f. Description of two species of the genus *Agroeca*. Annales de la Société entomologique de France, Bulletin, 7(5): clxxxix.

Simon, E. 1878a. Les Arachnides de France. Paris. Tome 4. 1-334 + pl. XII-XVI pages.

Simon, E. 1879a. Arachnides nouveau de France, d'Espagne et d'Algérie. Premier mémoire. Bulletin de la Société zoologique de France, 4: 251-263.

Simon, E. 1880a. Révision de la famille des Sparassidae (Arachnides). Actes de la Société Linnéenne de Bordeaux, 34: 223-351.

Simon, E. 1880b. Études arachnologiques. 11e Mémoire. XVII. Arachnides recueillis aux environs de Pékin par M.V. Collin de Plancy. Annales de la Société entomologique de France, 5(10): 97-128 + pl. III.

Simon, E. 1881a. Les Arachnides de France. Tome 5. 1 ère partie contenant les familles des Epeiridae (supplément) et des Theridionidae (commencement). Paris, Librairie encyclopédique de Roret. 1-180 + pl. XXV.

Simon, E. 1881g. Descriptions d'arachnides nouveaux du genre *Erigone*. Bulletin de la Société Zoologique de France, 6: 233-257.

Simon, E. 1882b. II. Étude sur les Arachnides du Yemen méridional. In Viaggio ad Assab nel Mar Rosso dei signori C. Doria ed O. Beccari con il r. Aviso exploratore del 16 nov. 1879 ad 26 feb. 1881. Annali del Museo civico di Storia naturale di Genova, 18: 207-260.

Simon, E. 1882c. Description d'espèces et de genres nouveaux de la famille des Dysderidae. Annales de la Société entomologique de France, 6(2): 201-240.

Simon, E. 1884a. Les Arachnides de France. Tome 5. 2 ème partie contenant la famille des Theridionidae (suite). Paris, Librairie encyclopédique de Roret. 181-420 + pl. XXVI pages.

Simon, E. 1884b. Descriptions de quelques arachnides des genres *Miltia* E.S. et *Zimiris* E.S. Annales de la Société entomologique de Belgique, 28: 139-142.

Simon, E. 1884j. Études arachnologiques. 16 e mémoire. XXIII. Matériaux pour servir à la faune des Arachnides de Grèce. Annales de la Société entomologique de France, 6(4): 305-356.

Simon, E. 1884m. Note sur le groupe des Mecicobothria. Bulletin de la Société zoologique de France, 9: 313-317.

Simon, E. 1884o. Arachnides nouveaux d'Algérie. Bulletin de la Société zoologique de France, 9: 321-327.

Simon, E. 1884q. Arachnides recueillis en Birmanie par M. le chevalier J.B. Comotto et appartenant au Musée civique d'histoire naturelle de Gènes. Annali del Museo civico di Storia naturale di Genova, 20: 325-372.

Simon, E. 1884r. Les Arachnides de France. Tome 5. 3 ème partie contenant la famille des Theridionidae (fin), Paris, Librairie encyclopédique de Roret. 421-885 + pl. XXVII.

Simon, E. 1885d. Études arachnologiques. 18e Mémoire XXXVI. Matériaux pour servir à la faune des Arachnides du Sénégal. (Suivi d'une appendice intitulé: Descriptions de plusieurs espèces africaines nouvelles). Annales de la Société entomologique de France, (6)5: 345-396.

Simon, E. 1885e. Matériaux pour servir à la faune arachnologique de l'Asie méridionale. I. Arachnides recueillis à Wagra-Karoor près de Gundacul, district de Bellary par M. M. Chaper. II. Arachnides recueillis à Ramnad, district de Madura par M. l'abbé Fabre. Bulletin de la Société zoologique de France, 10: 1-39.

Simon, E. 1885g. Étude sur les Arachnides recueillis en Tunisie en 1883 et 1884 par MM. A. Letourneux, M. Sédillot et Valery Mayet, membres de la mission d'exploration scientifique de la Tunisie, publiée sous les auspices du Ministère de l'instruction publique. Zoologie.-Arachnides, Paris, Imprimerie Nationale: i-iv + 1-55.

Simon, E. 1886d. Descriptions de quelques espèces nouvelles de la famille des Agelenidae. Annales de la Société entomologique de Belgique. Comptes Rendus, 30: lvi-lviii.

Simon, E. 1886f. Note sur le mico (*Dendryphantes noxiosus*), espèce venimeuse de Bolivie. Annales de la Société entomologique de Belgique. Comptes Rendus, 30: 168-172.

Simon, E. 1887d. Quelques observations sur les Arachnides. Bulletin des séances et bulletin bibliographique de la Société entomologique de France, Bulletin 10: clviii-clix, Bulletin 11: clxvii, Bulletin 11: clxxv-clxxvi, Bulletin 12: clxxxvi-clxxxvii, Bulletin 13: cxciii-cxcv.

Simon, E. 1887e. Arachnides recueillis à Obock en 1886, par M. le Dr L. Faurot. Bulletin de la Société zoologique de France, 12: 452-455.

Simon, E. 1887g. Espèces et genres nouveaux de la famille des Sparassidae. Bulletin de la Société zoologique de France, 12: 466-474.

Simon, E. 1888b. Études arachnologiques. 21e Mémoire. XXIX. Descriptions d'espèces et de genres nouveaux de l'Amérique centrale et des Antilles. Annales de la Société entomologique de France, 6(8): 203-216.

Simon, E. 1889a. Études arachnologiques. 21e Mémoire. XXX. Descriptions de quelques arachnides du Chili et remarques synonymiques sur quelques unes des espèces décrites par Nicolet. Annales de la Société entomologique de France, (6)8: 217-222.

Simon, E. 1889b. Études arachnologiques. 21e Mémoire. XXXI. Descriptions d'espèces et de genres nouveaux de Madagascar et de Mayotte. Annales de la Société entomologique de France, (6)8: 223-236.

Simon, E. 1889e. Description de *Hypochilus Davidi* sp. nov. Annales de la Société entomologique de France, Bulletin, 6(8): ccvii-ccix.

Simon, E. 1889l. Arachnidae transcaspicae ab ill. Dr G. Radde, Dr A. Walter et A. Conchin inventae (annis 1886-1887). Verhandlungen der kaiserlich-königlichen zoologisch-botanischen Gesellschaft in Wien, 39: 373-386.

Simon, E. 1889m. Voyages de M. E. Simon au Venezuela (Décembre 1887-Avril 1888). 4 e Mémoire (1). Arachnides. Annales de la Société entomologique de France, 6(9): 169-220 + pl. I-III.

Simon, E. 1890a. Études arachnologiques, 22e Mémoire. XXXIV. Étude sur les Arachnides de l'Yemen. Annales de la Société entomologique de France, 10: 77-124.

Simon, E. 1891a. Descriptions de quelques arachnides du Costa Rica communiqués par M. A. Getaz (de Genève). Bulletin de la Société zoologique de France, 16: 109-112.

Simon, E. 1891c. On the spiders of the Islands of St. Vincent. Part 1. Proceedings of the Zoological Society of London, 549-575 + pl. XLII.

Simon, E. 1891d. 11e Mémoire. Observations biologiques sur les arachnides. I. Araignées sociables. *In* Voyage de M. E. Simon au Venezuela (Décembre 1887-avril 1888). Annales de la Société entomologique de France, 60: 5-14.

Simon, E. 1891f. Études arachnologiques. 23e Mémoire. XXXVIII. Descriptions d'espèces et de genres nouveaux de la famille des Aviculariidae. Annales de la Société entomologique de France, 60: 300-312.

Simon, E. 1891g. Liste des espèces de la famille des Aviculariidae qui habitent le Mexique et l'Amérique du Nord. Actes de la Société linnéenne de Bordeaux, 44: 307-326.

Simon, E. 1891h. Liste des Aviculariides qui habitent le Mexique et l'Amérique centrale. Actes de la Société linnéenne de Bordeaux, 44: 327-339.

Simon, E. 1892a. Histoire naturelle des Araignées. Tome 1, Fascicule 1. Seconde édition. Paris, Librairie encyclopédique de Roret. 1-256 pages.

Simon, E. 1893a. Histoire naturelle des Araignées. Tome 1, Fascicule 2. Seconde édition. Paris, Librairie encyclopédique de Roret. 257-488 pages.

Simon, E. 1893b. Arachnides. Voyage de M. E. Simon aux îles Philipines (décembre 1887-avril 1888). 21e Mémoire. Annales de la Société entomologique de France, 61: 423-462.

Simon, E. 1893d. Études arachnologiques. 25e Mémoire. XL. Descriptions d'espèces et de genre nouveaux de l'ordre des Araneae. Annales de la Société entomologique de France, 62: 299-330.

Simon, E. 1894a. Histoire naturelle des Araignées. Tome 1, Fascicule 3. Seconde édition. Paris, Librairie encyclopédique de Roret. 489-760 pages.

Simon, E. 1894c. On the spiders of the Islands of St. Vincent. Part II. Proceedings of the Zoological Society of London, 1894: 519-526.

Simon, E. 1895a. Histoire naturelle des Araignées. Tome 1, Fascicule 4. Seconde édition. Paris, Librairie encyclopédique de Roret. 761-1084 pages.

Simon, E. 1895e. Description de quelques Arachnides de Basse-Californie faisant partie des collections du Dr. Geo Marx. Bulletin de la Société zoologique de France, 20: 134-137.

Simon, E. 1895f. Sur les Arachnides recueillis en Basse-Californie par M. Diguet. Bulletin du Muséum national d'Histoire naturelle, 1: 105-107.

Simon, E. 1895g. Études arachnologiques, 26e mémoire. XLI. Descriptions d'espèces et de genres nouveaux de l'ordre des Araneae. Annales de la Société entomologique de France, 63: 131-160.

Simon, E. 1896b. Descriptions d'arachnides nouveaux de la famille des Clubionidae. Annales de la Société entomologique de Belgique, 40: 400-422.

Simon, E. 1897a. Histoire naturelle des Araignées. Tome 2, Fascicule 1. Seconde édition. Paris, Librairie encyclopédique de Roret. 1-192 pages.

Simon, E. 1897d. On the spiders of the island of St Vincent. III. Proceedings of the Zoological Society of London, 1897: 860-890.

Simon, E. 1898a. Histoire naturelle des Araignées. Tome 2, Fascicule 2. Seconde édition. Paris, Librairie encyclopédique de Roret. 193-380 pages.

Simon, E. 1899a. Contribution à la faune de Sumatra. Arachnides receuillis par M. J. l. Weyers, à Sumatra. (Deuxième édition). Annales de la Société entomologique de Belgique, 43: 78-125.

Simon, E. 1899b. Ergebnisse einer Reine nach dem Pacific (Schauinsland 1896-1897). Arachnoideen. Zoologische Jahrbücher. II Abteilung für Systematik, 12: 411-437.

Simon, E. 1900b. Descriptions d'arachnides nouveaux de la famille des Attidae. Annales de la Société entomologique de Belgique, 44: 381-407.

Simon, E. 1901a. Histoire naturelle des Araignées. Tome 2, Fascicule 3. Seconde édition. Paris, Librairie encyclopédique de Roret. 381-668 pages.

Simon, E. 1902a. Études arachnologiques. 32ème Mémoire. LI. Descriptions d'espèces nouvelles de la famille des Salticidae (suite). Annales de la Société entomologique de France, 71: 389-421.

Simon, E. 1902d. Description d'arachnides nouveaux de la famille des Salticidae (Attidae) (suite). Annales de la Société entomologique de Belgique, 46: 24-56, 363-406.

Simon, E. 1902h. Arachnoideen, exclu. Acariden und Gonyleptiden. Ergebnisse der Hamburger Magalhaensische Sammelreise. Hamburg, 6(4): 1-47.

Simon, E. 1903a. Histoire naturelle des Araignées. Tome 2, Fascicule 4. Seconde édition. Paris, Librairie encyclopédique de Roret. 669-1080 pages.

Simon, E. 1903d. Descriptions d'Arachnides nouveaux. Annales de la Société entomologique de Belgique, 21-39.

Simon, E. 1914a. Les Arachnides de France. Synopsis générale et catalogue des espèces françaises de l'ordre des Araneae. Tome 6, 1e partie. Paris. 1-308 pages.

Simon, E. 1922. Description de deux Arachnides cavernicoles du midi de la France. Bulletin de la Société entomologique de France, 15: 199-200.

Simon, E. 1926. Les Arachnides de France. Synopsis général et catalogue des espèces françaises de l'ordre des Araneae. Tome 6, 2e partie. Paris. 309-532 pages.

Simon, E. 1929. Les Arachnides de France. Synopsis général et catalogue des espèces françaises de l'ordre des Araneae. Tome 6, 3e partie. Paris. 533-772 pages

Sissom, W.D., W.B. Peck and J.C. Cokendolpher. 1999. New records of wandering spiders from Texas, with a description of the male of *Ctenus valverdiensis* (Araneae: Ctenidae). Entomological News, 110: 260-266.

Smith, A.M. 1995. Tarantula Spiders: Tarantulas of the USA and Mexico. Fitzgerald Publishing, London. 196 pages.

Smith, C.P. 1908. A preliminary study of the Araneae Theraphoseae of California. Annals of the Entomological Society of America, 1: 207-236 + pl. XIII-XIX.

Smith, D.R. 1997. Notes on the reproductive biology and social behavior of two sympatric species of *Philoponella* (Araneae, Uloboridae). Journal of Arachnology, 25: 11-19.

Smith, F.P. 1904. The spiders of the *Erigone* group. The Journal of the Queckett Microscopical Club, Serie 2, volume 9, No. 54: 109-116.

Smith, F.P. 1905a. The spiders of the *Walckenaëria* group. Journal of the Queckett Microscopical Club, (2)9(57): 239-246.

Smith, F.P. 1908. Some British spiders taken in 1908. The Journal of the Queckett Microscopical Club, Series 2, volume 10, No. 63: 311-334 + pl. XXV.

Song, D., Zhu, M. and Chen, J. 1999. Spiders of China. Hebei Science and Technology Publishing House, Shijiazhuang. 640 pages.

Sørensen, W. 1898. Arachnida Groenlandica (Acaris exeptis). Videnskabelige Meddelelser fra den naturhistorike Forening i Kjøbenhavn, 176-235.

Spagna, J.C. and R.G. Gillespie. 2003. Molecular phylogenetics and spinneret evolution of RTA-clade spiders. Abstract page 23. Annual Meeting of the American Archnological Society, Denver, CO.

Stern, H. and E. Kullmann. 1981. Leben am Seidenen Faden. Kindler Verlag GmbH, München. 300 pages.

Stocks, I.C. 1999. Systematics and natural history of *Barronopsis* (Araneae, Agelenidae). M.Sc. Thesis, Western Carolina University. 113 pages.

Stone, W. 1890. Pennsylvania and New Jersey spiders of the family Lycosidae. Proceedings of the Academy of Natural Sciences of Philadelphia, 3: 420-434.

Strand, E. 1901b. Theridiiden aus dem Nördlichen Norwegen. Archiv for Mathematik og Naturvidenskab, 24(2): 3-66.

Strand, E. 1905. Coleoptera, Hymenoptera, Lepidoptera und Arachnida. Pages 1-30 *in* Report on the Second Norwegian Arctic Expedition in the "Fram". 1898-1902. Volume 1, No 3. Videnskabs-Selskabet i Kristiniana.

Strand, E. 1906a. Die arktischen Araneae, Opiliones und Chernetes. Fauna Arctica, 4: 431-478.

Strand, E. 1907b. Vorläufige diagnosen süd -und ostasiatischer Clubioniden, Ageleniden, Pisauriden, Lycosiden, Oxyopiden und Salticiden. Zoologischer Anzeiger, 31: 558-570.

Strand, E. 1912b. Drei neue Gattungsnamen in Arachnida. Internationale Entomologische Zeitschrift, 346.

Strand, E. 1914a. Neue Namen verschiedener Tiere. Archiv für Naturgeschichte 80A(1): 163.

Strand, E. 1928b. Miscellanea nomenclatorica zoologica et palaeontologica, I-II. [Arachnida: 42-45]. Archiv für Naturgeschichte, 92(A)(8): 30-75.

Strand, E. 1929. Zoological and paleontological nomenclatorical notes. Acta Universitalis Latviensis, 20: 1-29.

Strand, E. 1932. Miscellanea nomenklatorica zoologica et paleontologica, III, IV. Folia Zoologica et Hydrobiologica, 4: 133-147, 193-196.

Strand, E. 1934. Miscellanea nomenclatorica zoologica et palaeontolgica, VI. Folia Zoologica et Hydrobiologica, 6(2): 271-277.

Stratton, G.E. 1985. Behavioral studies in wolf spiders: a review of recent research. Revue Arachnologique, 6: 57-70.

Stratton, G.E. 1991. A new species of wolf spider, Schizocosa stridulans (Araneae: Lycosidae). Journal of Arachnology, 19: 29-39.

Stratton, G.E., E.A. Hebets, P.R. Miller and G.L. Miller. 1996. Pattern and duration of copulation in wolf spiders (Araneae, Lycosidae). Journal of Arachnology, 24: 186-200.

Stratton, G.E. and D.C. Lowrie. 1984. Courtship behavior and life cycle of Schizocosa mccooki from New Mexico. Journal of Arachnology, 12: 223-228.

Stratton, G.E. and G.W. Uetz. 1986. The inheritance of courtship behavior and its role as a reproductive isolating mechanism in two species of Schizocosa wolf spiders (Araneae; Lycosidae). Evolution, 40: 129-141.

Strazny, F. and S.F. Perry. 1987. Respiratory system: structure and function. Pages 78-94 in W. Nentwig (editor), Ecophysiology of spiders. Springer-Verlag, New York.

Sunderland, K. 1999. Mechanisms underlying the effects of spiders on pest populations. Journal of Arachnology, 27: 308-316.

Sundevall, C.J. 1823. Specimen Academicum Genera Araneidum Sveciae Exhibens. Lundae, 1-22.

Sundevall, C.J. 1830 ["1829"]. Svenska Spindlarmes beskrifning. Konglige Svenska Vetenskaps-Academiens Handlingar, 199-219.

Sundevall, C.J. 1833a. Svenska Spindlarmes beskrifning. Fortsättning och slut. Konglige Svenska Vetenskaps-Academiens Handlingar, 1832: 171-272.

Sundevall, J.C. 1833b. Conspectus Arachnidum. Londini Gothorum, 1-39.

Taczanowski, L. 1873 ["1872"]. Les aranéides de la Guyane française. Horae Societatis entomologicae Rossicae, 9: 66-150, 261-286 + pl. III-VI.

Taczanowski, L. 1874. Les aranéides de la Guyane française. Horae Societatis entomologicae Rossicae, 10: 56-115 + pl. II.

Tanasevitch, A.V. 1982. New genus and species of spiders of the family Linyphiidae (Aranei) from the Bolshezemelskaya tundra. Zoologicheskii Zhurnal, 61(10): 1501-1508.

Tanasevitch, A.V. 1984a. New and little known spiders of the family Linyphiidae (Aranei) from Bolshezemelskaya tundra [in Russian]. Zoologicheskii Zhurnal, 63(3): 382-391.

Tanasevitch, A.V. 1985a. A study of spiders (Aranei) of the polar urals [In Russian]. Trudy Zoologicheskii Institut. Akademiia Nauk SSSR. [Proceedings of the Zoological Institute of Leningrad. USSR Academy of Sciences], 139: 52-62.

Tanasevitch, A.V. 1992. New genera and species of the tribe Lepthyphantini (Aranei Linyphiidae Micronetinae) from Asia (with some nomenclatorial notes on Linyphiids). Arthropoda Selecta, 1(1): 39-50.

Taylor, R.M. and W.A. Foster. 1996. Spider nectarivory. American Entomologist, 42: 82-86.

Tellkampf, T. 1844. Beschreibung einiger neuer in der Mammuth-Höhle in Kentucky Aufgeundener Gattungen von Gliedertieren. Archiv für Naturgeschichte: 318-322.

Templeton, R. 1835. On the spiders of the genus Dysdera Latr. with the description of a new allied genus. The Zoological Journal. London, 5: 400-408.

Thaler, K. 1968. Zum Vorkommen von Porrhomma -Arten in Tirol und anderen Alpenländern (Arach., Araneae, Linyphiidae). Bericht des Naturwissens-Medizinischen Verein Innsbruck, 56: 361-388.

Theuer, B. 1954. Contributions to the life history of Deinopis spinosus. Unpubl. M.S. thesis, University of Florida, Gainesville, FL, 75 pages.

Thorell, T. 1856. Recensio critica Aranearum Suecicarum quas descripserunt Clerckius, Linnaeus, de Geerus. Nova Acta regiae Societatis scientiarum Upsaliensis, Serie 3, volume 2(1): 61-176.

Thorell, T. 1869. On European Spiders. Part I. Review of the European genera of spiders, preceded by some observations on zoological nomenclature. Nova Acta Regiae Societatis Scientiarum Upsaliensis, Series 3, volume 7: 1-108 + pl. I.

Thorell, T. 1870b. On European Spiders. Part I. Review of the European genera of spiders, preceded by some observations on zoological nomenclature. Nova Acta Regiae Societatis Scientiarum Upsaliensis, Series 3, volume 7: 109-242 + pl. I-XXIV.

Thorell, T. 1870c. Araneae nonnullae Novae Hollandiae, descriptae. Öfversigt af Kongl. Vetenkaps-Akademiens Förhandlingar, 27(4): 367-389.

Thorell, T. 1871a. Remarks on Synonyms of European spiders, No. 2. Upsala. 97-228 pages.

Thorell, T. 1872a. Remarks on Synonyms of European spiders, No. 3. Upsala. 229-374 pages.

Thorell, T. 1872b. Om några Arachnider från Grönland. Öfversigt af Kongl. Vetenkaps-Akademiens Förhandlingar, 29: 147-166.

Thorell, T. 1873. Remarks on Synonyms of European spiders, No. 4. Upsala, 375-645.

Thorell, T. 1875a. Diagnoses Aranearum Europaearum aliquot novarum. Tijdschrift voor Entomologie, 18: 81-108.

Thorell, T. 1875d. Notice of some spiders from Labrador. Proceedings of the Boston Society of Natural History, 17: 490-504.

Thorell, T. 1877b. Studi sui Ragni Malesi e Papuani. I. Ragni di Selebes raccolti nel 1874 dal Dott. O. Beccari. Annali del Museo civico di Storia naturale di Genova, 10: 341-634.

Thorell, T. 1877c. Descriptions of the Araneae collected in Colorado in 1875 by A.S. Packard Jr., M.D. Bulletin of the United States Geological Survey of the Territories, Volume 3, Bulletin No. 2: 477-529.

Thorell, T. 1878b. Studi sui ragni Malesi e Papuani. II. Ragni di Amboina raccolti Prof. O. Beccari. Annali del Museo civico di Storia naturale di Genova, 13: 1-317.

Thorell, T. 1887. Viaggio di L. Fea in Birmania e regioni vicine. II. Primo saggio sui ragni birmani. Annali del Museo civico di Storia naturale di Genova, 25: 5-417.

Thorell, T. 1890a. Studi sui ragni Malesi e Papuanti. IV, 1. Annali del Museo civico di Storia naturale di Genova, 28: 1-419.

Townsend, V.R. Jr. and B.E. Felgenhauer. 2001. Phylogenetic significance of the morphology of the cuticular scales of the lynx spiders (Araneae: Oxyopidae). Journal of Zoology, 253: 309-332.

Traw, B. 1996 ["1995"]. A revision of the Neotropical orb-weaving spider genus Scoloderus (Araneae: Araneidae). Psyche, 102: 49-72.

Tullgren, A. 1910. Araneae. In Wissenschaftliche Ergebnisse der Schwedischen Zoologischen Expedition nach dem Kilimandjaro, dem Meru und dem Umgebenden Massaisteppen Deutsch-Ostafrikas 1905-1906 unter Leitung von Prof. Dr. Yngve Sjöstedt, 20(6):85-172 + pl. I-IV. Stockholm.

Tullgren, A. 1948. Zwei bemerkenswerte Vertreter der Familie Dictynidae (Araneae). Entomologisk Tidskrift, 155-160.

Ubick, D. 1999. Obituary: Vincent D. Roth (1924-1997). The Pan-Pacific Entomologist, 75(3): 121-129.

Ubick, D. 2005. New genera and species of cribellate coelotine spiders from California (Araneae: Amaurobiidae). Proceedings of the California Academy of Sciences, 56(24): 303-334.

Ubick, D. and M.J. Moody. 1995. On males of Californian Talanites (Araneae, Gnaphosidae). Journal of Arachnology, 23: 209-211.

Ubick, D. and N.I. Platnick. 1991. On Hesperocranum, a new spider genus from western North America, (Araneae, Liocranidae). American Museum Novitates, No. 3019: 1-12.

Ubick, D. and V.D. Roth. 1973a. Nearctic Gnaphosidae including species from adjacent Mexican states. American Arachnology Newsletter, 9, Supplement 2: 1-12.

Ubick, D. and V.D. Roth. 1973b. Index to synonymy and invalid names in Nearctic Gnaphosidae, including Mexican states. American Arachnology Newsletter, 9, supplement 3: 1-6.

Ubick, D. and V.D. Roth. 1973c. Index to synonymy and invalid names in Nearctic Gnaphosidae, including Mexico. American Arachnology Newsletter, 9, Supplement 2: 1-6.

Ubick, D. and R.S. Vetter. 2005. A new species of Apostenus from California, with notes on the genus (Araneae, Liocranidae). Journal of Arachnology, 33: 63-75.

Uetz, G.W. and C.D. Dondale. 1979. A new wolf spider in the genus Schizocosa (Araneae: Lycosidae) from Illinois. Journal of Arachnology, 7: 86-88.

Urquhart, A.T. 1894. Descriptions of new species of Araneae. Papers Proceedings Royal Society of Tasmania, 26: 204-218.

Valerio, C.E. 1981. Spitting spiders (Araneae, Scytodidae, Scytodes) from Central America. Bulletin of the American Museum of Natural History, 170: 80-89.

Vellard, J. 1924a. Études de zoologie. I. Araneidae. Archivos do Instituto Vital Brazil, 1: 1-32.

Vetter, R.S. 1996. The extremely rare Prodidomus rufus Hentz (Araneae, Prodidomidae) in California. Journal of Arachnology, 24: 72-73.

Vetter, R.S. 2000a. A South American spider, Metaltella simoni (Keyserling), (Araneae: Amphinectidae) in southern California. The Pan-Pacific Entomologist, 76: 134-135.

Vetter, R.S. 2000b. Myth: idiopathic wounds are often due to brown recluse or other spider bites throughout the United States. Western Journal of Medicine, 173: 357-358.

Vetter, R.S. 2001. Revision of the spider genus *Neoanagraphis* (Araneae, Liocranidae). Journal of Arachnology, 29: 1-10.

Vetter, R.S., P.E. Cushing, R.L. Crawford and L.A. Royce. 2003. Diagnoses of brown recluse spider bites (loxoscelism) greatly outnumber actual verifications of the spider in four western American states. Toxicon, 42: 413-418.

Vetter, R.S. and G.K. Isbister. 2004. Do hobo spider bites cause dermonecrotic injuries? Annals of Emergency Medicine, 44: 605-607.

Vetter, R.S. and T.R. Prentice. 1997. *Callobius guachama* (Araneae, Amaurobiidae): habitat, distribution and description of the female. Journal of Arachnology, 25: 177-181.

Vetter, R.S. and P.K. Visscher. 1994. A non-native spider, *Metaltella simoni*, found in California. Journal of Arachnology, 22: 256.

Vink, C.J., A.D. Mitchell and A.M. Patterson. 2002. A preliminary molecular analysis of phylogenetic relationships of Australian wolf spider genera (Araneae, Lycosidae). Journal of Arachnology, 30: 227-237.

Vink, C.J., S.M. Thomas, P. Paquin, C.Y. Hayashi and M.C. Hedin. 2005. The effects of preservatives and temperatures on arachnid DNA. Invertebrate Systematics, 19: 99-104.

Vinson, A. 1863. Aranéides des îles de la Réunion, Maurice et Madagascar. Paris. i-cxx + 1-337 + pl. I-XIV.

Vogel, B.R. 1964. A taxonomic revision of the *distincta* group of the wolf-spider genus *Pardosa* in America north of Mexico. Postilla, 82: 1-30.

Vogel, B.R. 1968a. A zodariid spider from Pennsylvania (Araneae: Zodariidae). Journal of the New York Entomological Society, 76(2): 96-100.

Vogel, B.R. 1970a. Taxonomy and morphology of the *sternalis* and *falcifera* species groups of *Pardosa* (Araneida: Lycosidae). The Armadillo Papers, 3: 31.

Vogel, B.R. 1970b. Bibliography of Texas spiders. The Armadillo Papers, 2: 1-37.

Vollrath, F. 1978. A close relationship between two spiders (Arachnida, Araneidae): *Curimagua bayano* synecious on a *Diplura* species. Psyche, 85: 347-353.

Walckenaer, C.A. 1802. Faune parisienne, insectes. ou histoire abrégée des Insectes des environs de Paris, classés d'après le système de Fabricius; précédée d'un discours sur les insectes en général, pour servir d'introduction à l'étude de l'entomologie; accompagnée de sept planches gravées. [Araignée. (Araneae) : 187-250]. Tome second. An XI (1802). Paris. i-xxii + 1-438 + errata i-ii pages.

Walckenaer, C.A. 1805. Tableau des Aranéides ou caractères essentiels des tribus, genres, familles et races que renferme le genre Aranea de Linné, avec la désignation des espèces comprises dans chacune de ces divisions. Paris, de l'Imprimerie de Dentu. i-vii + 1-88 + pl. I-IX pages.

Walckenaer, C.A. 1826. Aranéides. *In* Faune française ou histoire naturelle générale et particulière des animaux qui se trouvent en France, constamment ou passagèrement, à la surface du sol, dans les eaux qui le baignent et dans le littoral des mers qui le bornent par Viellot, Desmarrey, Ducrotoy, Audinet, Lepelletier et Walckenaer. Paris, 11-12. 96 pages.

Walckenaer, C.A. 1830. Aranéides. *In* L.P. Vieillot, A.G. Desmarest, H. Ducrotay de Blainville, Audinet-Serville, Lepeletier de Saint-Fargeau, C.A Walckenaer, Faune française ou histoire naturelle générale et particulière des animaux qui se trouvent en France, constamment ou passagèrement à la surface du sol, dans les eaux qui baignent et dans le littoral des mers qui le bornent. Livre 27, pages 97-175. Paris.

Walckenaer, C.A. 1833. Mémoire sur une nouvelle classification des Aranéides. Annales de la Société entomologique de France, 2: 414-446.

Walckenaer, C.A. 1835. Mémoire sur une nouvelle espèce de Mygale, sur les Théraphoses et les divers genres dont se compose cette tribu d'Araneides. Annales de la Société entomologique de France, 4: 637-651 + pl. XIX.

Walckenaer, C.A. 1837. Histoire Naturelle des Insectes Aptères. Tome I. Paris. 1-682.

Walckenaer, C.A. 1841. Histoire Naturelle des Insectes Aptères. Tome II. Paris. 1-549 + 1-19 + pl. I-XXII pages.

Walckenaer, C.A. 1847a. Histoire Naturelle des Insectes. Aptères. (Araneae, pp. 365-564). Paris. 623 pages.

Wallace, H.K. 1942a. A revision of the burrowing spiders of the genus *Geolycosa* (Araneae, Lycosidae). American Midland Naturalist, 27: 1-62.

Wallace, H.K. 1942b. A study of the lenta group of the genus *Lycosa*, with descriptions of new species (Araneae, Lycosidae). American Museum Novitates, No. 1185: 1-21.

Wallace, H.K. and H. Exline. 1978. Spiders of the genus *Pirata* in North America, Central America and the West Indies (Araneae: Lycosidae). Journal of Arachnology, 5(1): 1-112.

Wang, X.-P. 2002. A generic-level revision of the spider subfamily Coelotinae (Araneae, Amaurobiidae). Bulletin of the American Museum of Natural History, 269: 1-150.

Waterhouse, C.O. 1902. Index Zoologicus. An Alphabetical List of Names and Genera and Subgenera Proposed for Use in Zoology as Recorded in the "Zoological Record" 1880-1900, Together with Other Names Not Included in the "Nomenclator Zoologicus" of S.H. Scudder. London. i-xii + 1-421 pages.

Westring, N. 1851. Förteckning öfver de till närvarande tid Kände, i Sverige förekommande spindlarter, utgörande ett antal af 253, deraf 132 äro nya för svenska Faunan. Göteborgs Kongl. Vetenskaps-och Vitterhets-Samhälles Handlingar, Serie 3, volume 2: 25-62.

Westring, N. 1861. Araneae Svecicae Descriptae. 1-615 pages.

Westring, N. 1874. Bemerkungen über die Arachnologischen Abhandlungen von Dr. T. Thorell unter dem Titel: 1°, On European spiders etc. part 1 & 2, Upsala, 1869-70. 2°, Remarks on Synonyms of Eur. Spid., Upsala, 1870-73. Göteborgs Kongl. Vetenskaps och Vitterhets Samhälles Handlingar, 14: 1-68.

Westwood, J.O. 1835. Insectorum Arachnoidumque novorum Decades duo. The Zoological Journal, 5: 440-453 + pl. XXII.

White, A. 1841. Descriptions of new or little known Arachnida. The Annals and Magazine of Natural History, 1(7): 471-477.

Wiehle, H. 1956. Spinnentiere oder Arachnoidea (Araneae). 28. Familie Linyphiidae-Baldachinspinnen. Die Tierwelt Deutschlands, 44: iii-viii + 1-337.

Williams, D.S. and P.D. McIntyre. 1980. The principal eyes of a jumping spider have a telephoto component. Nature, 288: 578-580.

Winchester, S. 2003. Krakatoa: the Day the World Exploded: August 27, 1883. Harper-Collins Publishers Inc., New York, New York. 416 pages.

Wise, D.H. 1993. Spiders in Ecological Webs. Cambridge University Press, New York. 328 pages.

Wolff, R.J. 1978. The cribellate genus *Tengella*. Journal of Arachnology, 5: 139-144.

Worley, L.G. 1928. New Nebraska spiders. Annals of the Entomological Society of America, 21: 619-622.

Wu, C., D. Song and M. Zhu. 2002. On the phylogeny of some important groups of spiders by using the third domain of 12S rRNA gene sequence analyses. Acta Arachnologica Sinica, 11: 65-73.

Wunderlich, J. 1970. Zur synonymie einiger Spinnen-Gattungen und -Arten aus Europa und Nordamerika. Senckenbergiana Biologica, 51: 403-408.

Wunderlich, J. 1972a. Zur Kenntnis der Gattung *Walckenaeria* Blackwall 1833 unter besonderer Berücksichtigung der europäischen subgenera und arten (Arachnida: Araneae: Linyphiidae). Zoologische Beiträge (Neue Folge) 18: 371-427.

Wunderlich, J. 1978c. Die synonymie der Gattungen *Notiomaso* Banks 1914, *Drepanotylus* Holm 1945, *Perimaso* Tambs-Lyche 1954 und *Micromaso* Tambs-Lyche 1954 (Arachnida: Araneae: Linyphiidae). Senckenbergiana Biologica, 58: 253-255.

Wunderlich, J. 1979b. Revision der europäischen Arten der Gattung *Micaria* Westring 1851, mit Anmerkungen zu den übrigen paläarktischen Arten (Arachnida: Araneida: Gnaphosidae). Zoologische Beiträge (Neue Folge), 25: 233-341.

Wunderlich, J. 1980h. Sternal-Organe der Theridisomatidae-eine bisher übersehene Autapomorphie (Arachnida: Araneae). Verhanlungden des naturwissen schaftlichen Vereins in Hamburg (Neue Folge), 23: 255-257.

Wunderlich, J. 1986. Spinnenfauna gestern und heute: Fossile Spinnen in Bernstein und ihre heute lebenden Verwandten. Quelle and Meyer, Wiesbaden. 283 pages.

Wunderlich, J. 1987a. Die Spinnen der Kanarischen Inseln und Madeiras: Adaptive Radiation, Biogeographie, Revisionen und Neubeschreibungen. Triops Verlag, Langen, West Germany. 435 pages.

Wunderlich, J. 1988. Die Fossilen Spinnen im Dominikanischen Bernstein. Straubenhardt, West Germany. 378 pages.

Wunderlich, J. 1992a ["1991"]. Die Spinnen-Fauna der Makaronesischen Inseln. Taxonomie, Ökologie, Biogeographie und Evolution. [The Spider Fauna of the Macronesian Islands. Taxonomy, Ecology, Biogeography and Evolution]. Beiträge zur Araneologie. I. Jörg Wunderlich Publishing House, Straubenhardt. i-iv + 1-619 pages.

Wunderlich, J. 1994b. Beschreibung der neuen Spinnen-Gattung *Megalepthyphantes* aus der Familie Baldachinspinnen und einer bisher unbekannten Art aus Griechenland (Arachnida: Araneae: Linyphiidae). Entomologische Zeitchrift, 104(9): 168-170.

Wunderlich, J. 1995n. Revision und Neubeschreibung einiger Gattungen der Familie Theridiidae aus der Neartkis und Neotropis (Arachnida: Araneae). Beiträge zur Araneologie, 4 (1994): 609-615.

Wunderlich, J. 1999a. *Liocranoeca*-eine bisher unbekannte Gattung der Feldspinnen aus Europa und Nordamerika (Arachnida: Araneae: Liocranidae). Entomologische Zeitschrift, Frankfurt am Main, 109: 67-70.

Yaginuma, T. 1960. Spiders of Japan in Colour. 1-Chome Uehonmachi, Higashi-ku, Osaka, Japan. 186 pages.

Yaginuma, T. 1974a. The spider fauna of Japan (IV). Faculty of Letters Review, Otemon Gakuin University, 8: 169-173.

Yaginuma, T. 1987a. On amaurobiid spiders of Japan. Pages 451-465 *in* Essays and studies published in commemoration of the twentieth anniversary of Otemon-Gakuin University. Otemon-Gakuin University, Ibaraki.

Yamashita, S. 1985. Photoreceptor cells in the spider eye: spectral sensitivity and efferent control. Pages 103-117 *in* F.G. Barth (editor): Neurobiology of Arachnids. Springer-Verlag, New York.

Yoshida, H. 2001b. The spider genera *Robertus, Enoplognatha, Steatoda* and *Crustulina* (Araneae: Theridiidae) from Japan. Acta Arachnologica, 50: 31-48.

Yoshida, H. 2001c. A revision of the Japanese genera and species of the subfamily Theridiinae (Araneae: Theridiidae). Acta Arachnologica, 50: 157-181.

Yoshida, H. 2001d. The genus *Rhomphaea* (Araneae: Theridiidae) from Japan, with notes on the subfamily Argyrodinae. Acta Arachnologica, 50: 183-192.

Yoshida, H. 2002. A revision of the Japanese genera and species of the subfamily Hadrotarsinae (Aranae: Theridiidae). Acta Arachnologica, 51: 7-18.

Yoshida, H. 2003. The spider family Theridiidae (Arachnida: Araneae) from Japan. Arachnological Society of Japan, Osaka. 223 pages.

Young, O.P. and T.C. Lockley. 1985. The striped lynx spider, *Oxyopes salticus* [Araneae: Oxyopidae], in agroecosystems. Entomophaga, 30(4): 329-346.

Zhang, D., W.B. Cook and N.V. Horner. 2004. ITS2 rDNA variation of two black widow species, *Latrodectus mactans* and *Latrodectus hesperus* (Araneae, Theridiidae). Journal of Arachnology, 32(2): 349-352.

Zhu, M.S. 1998. Fauna Sinica: Arachnida: Araneae: Theridiidae. Science Press, Beijing. xi + 436 pages.

Zorsch, H.M. 1937. The spider genus *Lepthyphantes* in the United States. American Midland Naturalist, 18(5): 856-898.

Zujko-Miller, J. 1999. On the phylogenetic relationships of *Sisicottus hibernus* (Araneae, Linyphiidae, Erigoninae). Journal of Arachnology, 27: 44-52.

Zyuzin, A.A. and D.V. Logunov. 2000. New and little-known species of the Lycosidae from Azerbaijan, the Caucasus (Araneae, Lycosidae). Bulletin of the British Arachnological Society, 11: 305-319.

# Index

## A

*abbotii, Sphodros* 41
*abnormis, Neriene* 318
*Acacesia* 69, 72, 276
*Acanthepeira* 8, 68, 69, 71, 276, 304
ACANTHOCTENIDAE 259
ACANTHOCTENINAE 83
*Acanthoctenus* 28, 83, 84, 276
*Acantholycosa* 165, 166, 274, 276
*Acanthoscurria* 279
Acari 11, 18
*Acartauchenius* 276
*Achaea* 280
*Achaearanea* 236, 241, 243, 276, 280, 315, 332
*Acrosoma* 306
ACTINOPODIDAE 20
*Actinopus* 286
*Actinoxia* 46, 276, 277, 278, 279
*aculeata, Alopecosa* 170
*aculeata, Epeira* 276
*aculeata, Phoroncidia* 315
*Aculepeira* 69, 72, 276
*acuminata, Entelecara* 292
*acuminata, Walckenaeria* 150
*acuminatum, Theridium* 292
*acuminatus, Sciastes* 155
*acutiventer, Coleosoma* 241
*adansoni, Hasarius* 211
*Adelosimus* 278
*Admestina* 206, 207, 210, 276
*Aduva* 274, 276
AELURILLINAE 206
*aemulus, Bianor* 206
*aemulus, Sibianor* 213
*aestivus, Tiso* 150
*affine, Baryphyma trifrons* 139
*affinis, Azilia* 234
*affinis, Temenius* 174
*Agalena* 276
*agalena, Aranea* 276
*agalena, Epeira* 276
*agalenoides, Leptoctenus* 302
*Agalenopsis* 277
*Agassa* 206, 207, 210, 276
*Agathostichus* 276, 317
*Agatostichus* 276, 317
*Agelena* 276, 277, 284, 299, 303, 305, 317, 321
AGELENIDAE 14, 15, 20, 24, 35, **56**, 57, 60, 61, 85, 86, 95, 96, 112, 113, 190, 276, 277, 283, 284, 287, 289, 299, 305, 310, 317, 319, 323, 325, 328, 332
AGELENINAE 57
*agelenoides, Amaurobius* 280
*agelenoides, Arctobius* 62
*Agelenopsis* 56, 57, 58, 277, 283, 317, 332
*Agenelopsis* 277
*Agnyphantes* 127, 277
*Agobardus* 298
*Agraeca* 277
*agrestis, Tegenaria* 15
*Agroeca* 66, 162, 163, 277, 302, 332
*Agyneta* 125, 127, 132, 134, 135, 160, 161, 277, 296
*Aigola* 277
*ajo, Styposis* 238, 240
*alarius, Phrurolithus* 315
*albupilis, Agenelopsis* 277
*alboimmaculata, Poultonella* 210
*albolineata, Diguetia* 102
*albomaculata, Chrysso* 241, 243
*Alcmena* 330
*algens, Pardosa* 169
*Aliatypus* 39, 40, 277
*Allepeira* 277, 304
*Allocosa* 164, 165, 166, 274, 277, 318
ALLOCOSINAE 165

*Allocyclosa* 69, 70, 277
*Allomengea* 125, 127, 128, 157, 277
*Alopecosa* 164, 165, 168, 169, 170, 274, 277, 280, 313, 318
*Alpaida* 277
*alpha, Disembolus* 142
*alpinus, Cochlembolus* 287
*Altella* 305
*altera, Crustulina* 237
*alteranda, Drapetisca* 130, 157
*alternata, Bredana* 210
*alticeps, Maso* 147
AMAUROBIIDAE 20, 24, 27, 28, 35, 56, 57, **60**, 61, 62, 63, 85, 93, 95, 96, 104, 248, 254, 258, 277, 278, 280, 284, 286, 287, 288, 289, 313, 315, 328, 329, 332
AMAUROBIINAE 61
*Amaurobioidea* 24, 56, 86, 113, 173
AMAUROBIOIDINAE 66
*Amaurobius* 57, 60, 61, 62, 86, 277, 278, 280, 284, 287, 328, 332
*Amazula* 278
*Ambika* 274, 297
*ambita, Maymena* 175, 177
Amblypygi 11, 18
*americana, Pholcophora* 195
*americana, Phoroncidia* 237
*americana, Storena* 321
*americana, Titanoeca* 248
*americanus, Araeoncus* 287
AMMOXENIDAE 20
*amoenus, Episinus* 239
*amphiloga, Gertschosa* 110
*Amphinecta* 278
AMPHINECTIDAE 20, 24, 28, 60, 61, **63**, 248, 258, 278, 305, 332
*Amphissa* 315
*amplus, Maro* 133, 161
*Amycoida* 206
*Anachemmis* 230, 231, 278
*Anacornia* 127, 146, 278
*Anagraphis* 309
*Anahita* 83, 84, 274, 278
*analyticus, Anelosimus* 240
ANAPIDAE 20, 31, 33, **64**, 175, 226, 227, 245, 266, 278, 279, 285, 288, 296, 332
*Anapis* 278
*Anapistula* 30, 64, 226, 227, 278
*Anasaitis* 206, 211, 278
*anastera, Epeira* 294
*anastera, Eustala* 294
*anax, Aphonopelma* 54, 55
*anceps, Hyptiotes* 300
*ancilla, Euophrys* 294
ANELOSIMINAE 236
*Anelosimus* 236, 240, 278
*Anerigone* 293
*anglicana, Ceratinopsis* 327
*anglicanus, Tutaibo* 150
*angustinus, Gaucelmus* 178, 179, 180
*Anibontes* 127, 134, 278
*Annapolis* 127, 155, 278
*annulipes, Ceratinella* 285
*annulipes, Ceratinops* 142
*annulipes, Oecobius* 184
*anomala, Brachythele* 52, 53
*anopica, Neoleptoneta* 123
*antennatus, Scotinotylus* 319
*Anthrobia* 127, 147, 155, 279
*Anthrodiaetus* 279
*Antistea* 112, 113, 114, 278, 297, 309
*Antrobia* 279
ANTRODIAETIDAE 20, 21, 25, **39**, 41, 50, 277, 279, 282, 283, 310
*Antrodiaetus* 39, 40, 279, 283, 310, 332
*Anyphaena* 66, 67, 279
*Anyphaenella* 279
ANYPHAENIDAE 5, 14, 20, 24, 37, **66**, 77, 224, 279, 282, 296, 299, 303, 312, 315, 324, 328, 332

ANYPHAENINAE 66
*anyphenoides, Dirksia* 291
*anyphenoides, Ethobuella* 291
*apacheca, Tosyna* 325
*apertus, Sisicus* 134
*Aphantopelma* 294
*Aphileta* 127, 134, 279
*Aphonopelma* 54, 55, 279, 285, 287, 290, 292, 294, 297, 317
*Apollophanes* 192, 193, 279
*Apomastus* 45, 46, 47, 279
*Apostenus* 38, 162, 163, 279
*Appaleptoneta* 122, 123, 279, 332
*approximatus, Bathyphantes* 159
*approximatus, Spirembolus* 125, 137
*approximatus, Tortembolus* 325
*Apralogus* 307
*Aptostichus* 45, 46, 47, 276, 279, 317
*aquatica, Argyroneta* 86, 278, 293
*aquilonaris, Erigone* 299
Arachnida 18
*Arachosia* 66, 67, 279
*Araeoncus* 287, 296, 304
Araneae 18, 19, 20, 25, 276, 284, 286, 290, 293, 296, 307, 308, 319, 327, 329, 332
ARANEIDAE 14, 20, 23, 34, **68**, 69, 171, 232, 233, 244, 245, 250, 276, 277, 279, 280, 285, 288, 289, 292, 293, 294, 295, 296, 298, 299, 300, 301, 302, 303, 304, 305, 306, 309, 310, 318, 320, 328, 329, 332
ARANEINAE 306
Araneoclada 20
Araneoidea 18, 19, 20, 23, 124
*araneoides, Galeodes* 305
Araneomorphae 6, 11, 12, 18, 20, 21, 23, 25, 27, 29, 32, 36, 120, 121, 260, 261, 263, 264, 269, 270, 271
*Araneus* 69, 73, 74, 279, 301, 307, 310, 315, 332
*Araniella* 69, 72, 279
ARCHAEIDAE 20, 171
*archboldi, Tekellina* 240
*Archerius* 64, 65, 279, 288
*Archoleptoneta* 122, 123, 228, 280
ARCHOLEPTONETINAE 123
*arcifera, Meriola* 80
*Arctachaea* 236, 241, 243, 280
*Arctella* 96, 99, 280
*Arcterigone* 127, 145, 280
*Arctilaira* 280
ARCTOBIINAE
*Arctobius* 60, 61, 62, 280
*Arctosa* 164, 165, 166, 168, 169, 170, 274, 277, 280, 303, 313, 318, 332
*arctosaeformis, Trochosina* 313
*arcuata, Linyphia* 280
*arcuatulus, Arcuphantes* 280
*Arcuphantes* 127, 130, 158, 280, 332
*Arcys* 301
*arenarium, Soucron* 153
*arenata, Epeira* 328
*arenata, Verrucosa* 71
*arenicola, Lycosa* 296
*argelasius, Sparassus* 311
*Argenna* 96, 100, 280, 297
*Argennina* 96, 99, 280
*Argiope* 68, 69, 70, 280, 304, 323, 332
ARGIOPIDAE 179, 233, 245
*Argystes* 299
*Argus* 296, 303, 304, 306
*argyra, Plesiometa* 233
*argyrobapta, Leucauge* 302
*Argyrodes* 33, 236, 238, 280, 281, 282, 283, 285, 286, 287, 288, 291, 295, 296, 300, 301, 302, 307, 310, 317, 318, 319, 320, 332
*argyrodes, Argyrodes* 281
*argyrodes, Linyphia* 280
*Argyrodina* 288
ARGYRODINAE 236
*Argyroepeira* 302
*Argyroneta* 86, 278, 281, 293, 302, 305, 306, 311, 316, 326, 329
*Ariadna* 219, 220, 281, 287
*Arietina, Ochyrocera* 310
*Ariston* 251, 282, 320
*arizonensis, Metacyrba* 206
*arizonica, Kukulcania* 104
*arizonica, Loxosceles* 222
*arizonicus, Entychides* 46, 47
*armipotens, Callioplus* 284
*Artema* 194, 196, 282, 289
*artemis, Olios* 311
*arthuri, Ariadna* 219

*Asagena* 282
*Ashtabula* 282, 283
*Astrosoga* 46, 274, 277, 282, 284, 315, 325
*Athena* 277
*atlanta, Artema* 196
*atomarius, Aptostichus* 45
*Atopogyna* 282
*atra, Masikia* 304
*atrata, Pardosa* 170
*atrica, Tegenaria* 58
*atrotibialis, Walckenaeria* 139
*atrox, Clubiona* 287
*Atta* 318
*Attidops* 206, 210, 282
*Attus* 274, 282, 286, 291, 297, 318, 321, 324
ATYPIDAE 20, 21, 25, **39**, 41, 282, 299, 321
Atypoidea 20
*Atypoides* 39, 40, 277, 282
*Atypus* 41, 42, 277, 282, 321
*auberti, Laseola* 301
*audouinii, Pachyloscelis* 327
*augustinus, Gaucelmus* 178
*Aulacocyba* 282
*aurantiaca Trabeops* 325
*aurantiacus, Trabeops* 165
*aureoceps, Brachypelma* 55
*aurigera, Eris* 312
*australis, Notiomaso* 310
Austrochilidae 20
Austrochilioidea 20
*avara, Varacosa* 327, 328
*Avella* 92
*Avellopsis* 92
*Avicosa* 274, 280, 282
*Avicularia* 294, 307, 323
Avicularioidea 20, 21
*avida, Avicosa* 282
*Aysha* 66, 67, 274, 282
*Azilia* 233, 234, 282

# B

baboon spider 54
*Bactroceps* 282
*Badumna* 28, 60, 93, 94, 283, 300
*Bagheera* 206, 207, 216, 283, 305
*bajulus, Creugas* 82
*Ballus* 210, 282, 283, 330
*bandina, Opopaea* 187, 188
*barbipes, Sassacus* 212, 213
*Barronopsis* 56, 57, 59, 283
*barrowsi, Agelenopsis* 283
*barrowsi, Nesticus* 179
BARYCHELIDAE 20
*Baryphyma* 127, 139, 144, 283, 307
*basilica, Allepeira* 304
*basilica, Epeira* 277
*basilica, Hentzia* 304
*basilica, Mecynogea* 298
*Bassania* 283
*Bassaniana* 246, 247, 283
*Batesiella* 329
*Bathyphantes* 127, 128, 131, 156, 159, 278, 280, 283, 284, 288, 296, 300, 302, 310, 311, 321, 332
*Bathyphantoides* 283
*Beata* 206, 207, 213, 283
*beattyi, Neocryphoeca* 115
*Bedriacum* 315
*behlei, Aphonopelma* 54
*bella, Segestria* 219
*Bellota* 206, 207, 208, 283, 289
*bennetti, Amaurobius* 284
*bennetti, Callobius* 60
*beringiana, Pardosa* 165
*bernardino, Hypochilus* 121
*Bianor* 206, 213, 283, 319
*bicalcarata, Septentrinna* 82, 319
*bicapillata, Paracornicularia* 142
*bicolor, Ariadna* 219
*bicolor, Centromerita* 130, 158
*bicolor, Eilica* 109
*bicolor, Stemmops* 240, 321
*bicornis, Delorrhipis* 288

*bidentata, Scoutoussa* 319
*bidentatus, Tmeticus* 291
*bifrons, Dismodicus* 291
*bifurca, Allocyclosa* 70
*bilineata, Tapinopa* 129, 157
*bimaculata, Neottiura* 241, 243, 309
*biovatus, Thyreosthenius* 324
*birostra, Savignia* 136
*birostrum, Cephalethus* 285
*bispinosus, Traematosisis* 138
*bivittatus, Menemerus* 215
*Blabomma* 56, 96, 97, 98, 283, 286, 329, 332
*blackwallii, Scotophaeus* 106, 111
*blauveltae, Stemonyphantes* 129, 157
*Blestia* 127, 144, 145, 283
*Boethus* 285
*Bolyphantes* 127, 283
*bonneti, Hypochilus* 120, 121
*bostoniensis, Sougambus* 131
*Bothriocyrtum* 43, 44, 283
*braccata, Laseola* 301
*Brachybothrium* 283, 310
*Brachypelma* 54, 55, 283
*Brachythele* 52, 53, 283
*Breda* 283
*Bredana* 206, 207, 210, 283
*brevicauda, Miagrammopes* 306
*brevipes, Pisaurina* 199
*brevispina, Piabuna* 81
*brevitarsus, Dictyna* 95
*Brommella* 96, 99, 284, 312
brown recluse spider 222
brown spider 222
*Bruchnops* 310
*brunnea, Antistea* 114
*brunneipes, Coriarachne* 246
*bryantae, Cicurina* 86, 96
*bryantae, Pirata* 166
*bungei, Thanatus* 193
*byrrhus, Leptoctenus* 84

# C

*cacuminatus, Wabasso* 149
*caerulescens, Tapinauchenius* 323
*Caira* 300
*calcar, Pippuhana* 67
*calcaratum, Liocranum* 86
*calcaratus, Wadotes* 328
*calcarifera, Frontina* 295
*Caledonia* 315
*californica, Agelena* 284
*californica, Marptusa* 315
*californica, Meriola* 80
*californica, Metacyrba* 206
*californicum, Bothriocyrtum* 44
*californicum, Cteniza* 283
*californicum, Megamyrmaekion* 305
*californicum, Megamyrmecion* 310
*californicum, Theridion* 242
*californicus, Apostenus* 163
*californicus, Sosticus* 110
*californicus, Titiotus* 231
*caloginosa, Eboria* 292
*Calilena* 57, 58, 275, 284, 299, 317, 332
*Calileptoneta* 122, 123, 284, 332
*Calisoga* 26, 52, 53, 54, 282, 283, 284, 288
*Callilepis* 37, 106, 107, 109, 284, 332
*Callioplus* 60, 62, 284, 289
*Callobius* 60, 61, 62, 284, 332
*Calodipoena* 175, 176, 177, 284
*calona, Opopaea* 187, 188
*Calopelma* 294
*Calponia* 32, 75, 76, 201, 284
*calycina, Aranea* 307
*Calymmaria* 35, 56, 112, 113, 115, 197, 284, 332
*Camilla* 284
*camillae, Neonella* 213
*Camillina* 106, 107, 108, 284
*canadensis, Clubiona* 77
*cancriformis, Gasteracantha* 71
*cancroides, Chelifer* 286
*canionis, Ero* 171

*canities, Diguetia* 102
*canities, Segestria* 290
cannibal spider 171
*canosa, Anasaitis* 211
*capillifolium, Eupatorium* 189
*Caponia* 75, 284
CAPONIIDAE 20, 30, 32, **75**, 103, 201, 203, 284, 310, 311, 323, 332
CAPONIINAE 75
*Caponina* 284
*caraiba, Zosis* 329
*carbonarioides, Aculepeira* 72
*carolina, Ceratinops* 285
*Carorita* 127, 154, 284
*cascadeus, Araeoncus* 288
*castanea, Plectreurys* 202
*castanea, Walckenaeria* 139
*castaneus, Metaphidippus* 296
*Castianeira* 79, 80, 81, 284, 332
CASTIANEIRINAE 38, 79, 80
*castrata, Dipoena* 329
*Catabrithorax* 284, 293, 294, 301
*catalina, Linyphia* 125, 129, 157
*catawba, Colphepeira* 70
*catawba, Epeira* 288
*cavatica, Scotolathys* 326
*Cavator* 287
*cavatus, Hyptiotes* 250
*cavernicola, Porrhmomma* 316
*Cavernocymbium* 27, 61, 284
*Caviphantes* 127, 148, 284
cellar spider 194
*cellariorum, Oecobius* 183
*cellicola, Mygale* 309
*cellicola, Nemesia* 309
*cellulanus, Nesticus* 178, 179, 180
*Centromeria* 285
*Centromerita* 127, 130, 158, 284, 285
*Centromeroides* 285
*Centromerus* 125, 127 130, 158, 282, 284, 285, 332
*centronethus, Hesperopholis* 52
*Cephalethus* 285
*Cepheia* 176
*Ceraticelus* 124, 127, 143, 148, 285, 293, 326, 332
*Ceratina* 285, 310
*Ceratinella* 127, 148, 285, 291, 300, 310, 315, 332
*Ceratinops* 127, 140, 142, 285, 332
*Ceratinopsidis* 127, 149, 285, 332
*Ceratinopsis* 125, 127, 138, 146, 150, 285, 310, 321, 327
*Cercidia* 69, 71, 285
*Cesonia* 107, 111, 285, 332
*cespitum, Philodromus* 192
*chalcodes, Aphonopelma* 54
*Chalcoscirtus* 206, 207, 211, 285
*chamberlini, Neoanagraphis* 163
*championi, Scytodes* 218
*Chasmocephalon* 64, 285, 296
*Chaunopelma* 285
*Cheiracanthium* 15, 77, 173, 174, 285, 286, 290
*Chelifer* 286
*Cheliferoides* 206, 207, 208, 286
*Chemmis* 80, 278, 286, 320
*Cheniseo* 127, 147, 286
*Chinattus* 206, 207, 211, 286
*Chindellum* 240, 243
*Chiracanthium* 174, 285, 286, 305
*Chisosa* 194, 195, 286
*Chocorua* 286
*Chorizomma* 286, 329
*Chorizommoides* 286
*Chorizops* 286
*Chrosioderma* 286
*Chrosiothes* 236, 239, 286, 332
*Chrysso* 236, 241, 242, 243, 280, 286
CHUMMIDAE 20
*cicurea, Aranea* 286
*Cicurella* 286
*Cicurina* 56, 86, 96, 97, 98, 286, 295, 332
*Cicurininae* 56, 96, 97
*Cimex* 288
*cinctipes, Dirksia* 115
*cinerea, Hahnia* 112
*Ciniflo* 277, 278, 286, 287
CINIFLONIDAE 96
CITHAERONIDAE *20*, 119
*Citharacanthus* 279

*Citharoceps* 220, 287
*citigrada, Loxosceles* 303
*citricola, Cyrtophora* 70
*clathrata, Neriene* 309
*claveata, Ozyoptila* 312
*clavipes, Nephila* 233, 309
*Clavopelma* 287
*Clitaetra* 287
*Clitistes* 287, 299
*Clitolyna* 125, 287
*Clubiona* 77, 78, 287, 332
CLUBIONIDAE 20, 24, 38, 66, 77, 79, 80, 83, 94, 162, 173, 224, 230, 257, 286, 287, 292, 297, 315, 325, 332
CLUBIONINAE 230
*Clubionoides* 78, 287
*clypeata, Epeira* 306
*Cnephalocotes* 127, 139, 287
cobweb weaver 235
*Cocalus* 330
*coccinea, Florinda* 128, 156
*coccinea, Linyphia* 302
*Cochlembolus* 287
*Coelotes* 57, 278, 287, 328
COELOTINAE 61, 85
*cognatus, Episinus* 239
*Coleosoma* 236, 241, 242, 287, 332
collardoor spider 39
*Collinsia* 125, 278, 284, 287
*Coloncus* 150, 287, 127
*Colophon* 284
*coloradensis, Drassus* 311
*coloradensis, Pardosa* 169
*Colphepeira* 69, 70, 288
*Comaroma* 31, 64, 65, 279, 288
*communis, Frontinella* 132, 159
*completa, Dictyna* 292
*complicata, Bredana* 210
*concolor, Stemmops* 321
*confusa, Cybaeina* 88
*congener, Wamba* 240, 243
*Conifaber* 251
*Conigerella* 288
*Coniphantes* 288
*Conopistha* 288
*conspecta, Corythalia* 211
*convexa, Lasaeola* 301
*Coras* 57, 60, 61, 288, 332
*Coreorgonal* 127, 145, 288
*Coressa* 288, 237
*Coriarachne* 246, 247, 288, 316
*Corinna* 288, 319, 330
CORINNIDAE 20, 24, 38, 77, 79, 80, 85, 103, 106, 162, 205, 254, 256, 284, 286, 288, 289, 291, 295, 297, 304, 305, 308, 315, 319, 329, 332
CORINNINAE 38, 79, 80, 81
*Cornicularia* 278, 288, 312
*cornigera, Deinopis* 290
*cornigera, Epeira* 304
*cornigera, Mastophora* 304
*cornupalpis, Centromerus* 282
*cornutus, Araneus* 301, 310
*cornutus, Larinioides* 301, 310
*coroniger, Pocobletus* 132, 159
*Corteza* 94
*corticarius, Soulgas* 151, 320
*corticeus, Miagrammopes* 307
*corvallis, Zygottus* 151
*Coryphaeolana* 288, 310
*Coryphaeolanus* 288
*Coryphaeus* 288
*Corythalia* 206, 211, 283, 288, 289, 295, 313
*coylei, Hypochilus* 121
crab spider 246
*crandalli, Pellenes* 212
*crassimana, Aigola* 277
*crassipalpis, Filistatinella* 105
*crassipalpum, Gonatium* 146
Crassitarsae 20
*Crematogaster* 209
*crenata, Ceratinops* 285
*Creugas* 80, 82, 289
Cribellatae 23, 24
*crinita, Dugesiella* 292
*crinitum, Aphonopelma* 55
*crispulus, Wamba* 240, 243
*cristatus, Diplocephalus* 144

*crocata, Dysdera* 103
*Crocodilosa* 274, 280, 313
*Crossopriza* 194, 196, 289
*Crustulina* 235, 236, 237, 289
*cruzana, Segestria* 219
*Cryphoeca* 113, 115, 289, 309
CRYPHOECINAE 56, 85, 95, 112, 113, 114
*crypticola, Lepthyphantes* 302
CRYPTOTHELIDAE 20
*cryptodon, Zornella* 141
*Ctenia* 289
CTENIDAE 20, 24, 28, 37, 77, 83, 164, 230, 256, 257, 259, 276, 278, 289, 302, 332
CTENINAE 230
*Ctenium* 238, 289, 295
*Cteniza* 283, 289, 294
CTENIZIDAE 20, 21, 26, 39, 41, 43, 44, 45, 46, 52, 279, 283, 286, 289, 298, 312, 316, 327, 332
Ctenoidea 83, 119
*Ctenus* 83, 84, 276, 289, 302
*cubana, Arachosia* 67
*cubana, Pseudosparianthis* 224, 225
*cubanus, Litopyllus* 111
*cuernavaca, Jalapyphantes* 132, 160
*culicinus, Hedypsilus* 298
*culicinus, Modisimus* 195
*cultrigera, Zornella* 141
*cuneata, Chocorua* 286
*Cupiennius* 83, 288
*Curimagua* 226
*curta, Hololena* 59
*curvata, Pimoa* 197
*cutleri, Sarinda* 209
*cyanea, Agassa* 210
*cyanea, Singa* 73
CYATHOLIPIDAE 20
CYBAEIDAE 20, 29, 31, 35, 57, 85, 86, 95, 112, 256, 264, 281, 289, 332
*Cybaeina* 85, 86, 87, 88, 289
CYBAEINAE 86, 87
*Cybaeopsis* 60, 61, 62, 284, 289, 332
*Cybaeota* 85, 86, 87, 289
*Cybaeozyga* 31, 85, 86, 87, 89, 90, 289, 330
*Cybaeus* 85, 86, 87, 89, 90, 289, 325, 332
*Cyclocosmia* 26, 43, 44, 286, 289, 298
CYCLOCTENIDAE 20
*Cyclosa* 69, 70, 71, 277, 289, 332
*cylindraceus, Deinopis* 290
*Cyllodania* 206, 207, 209, 289
*Cyllopodia* 316
*cymbia, Mysmenopsis* 176, 177
*Cyrba* 305
CYRTAUCHENIIDAE 21, 26, 39, 41, 43, 45, 46, 52, 276, 279, 282, 289, 292, 294, 308, 309, 316, 332
*Cyrtauchenius* 289
*Cyrtocarenum* 283
*Cyrtocephalus* 289
*Cyrtonota* 289
*Cyrtophora* 69, 70, 289

# D

*Dactylopisthes* 127, 143, 290
daddylonglegged spider 194
*Damoetas* 312
*Daramuliana* 251
*davidi, Hypochilus* 121
*debilipes, Tutaibo* 327
*debilis, Tunagyna* 153
*decepta, Meriola* 80
*decollatus, Hybocoptus* 299
*deformis, Perregrinus* 144
DEINOPIDAE 20, 23, 27, 91, 92, 251, 290, 332
*Deinopis* 27, 91, 92, 290
Deinopoidea 20, 23
*dela, Tortolena* 59
*Delena* 321
*Delopelma* 290
*Delorrhipis* 288
*Dendryphantes* 206, 216, 290, 305, 312
DENDRYPHANTINAE 206
*dentichelis, Lessertia* 153
*depressus, Xysticus* 288
*deserta, Loxosceles* 222

*deserticola, Opopaea* 187, 188
DESIDAE 20, 24, 28, 34, 63, **93**, 94, 190, 283, 290, 300, 312, 313, 332
DESINAE 60
*Desis* 290
*Diaea* 246, 247, 290
*Diana* 290
*Dicrostichus* 317
*Dictyna* 95, 96, 100, 101, 274, 290, 292, 298, 314, 325, 328, 332
DICTYNIDAE 20, 24, 27, 28, 29, 31, 34, 35, 56, 57, 60, 63, 85, 86, 93, **95**, 96, 112, 124, 248, 251, 256, 263, 280, 283, 284, 286, 290, 292, 297, 298, 300, 301, 303, 306, 310, 312, 314, 318, 319, 324, 325, 326, 328, 329, 332
*Dictynina* 298
DICTYNINAE 96, 97
*Dictynoidea* 57, 85, 86, 95, 113
*Dicymbium* 127, 144, 290
*Dicyphus* 291, 292
*Dietrichia* 127, 138, 290
*digiticeps, Erigone* 290
*Diguetia* 30, 102, 290, 332
DIGUETIDAE 20, 22, 30, **102**, 201, 290, 332
*diluta, Chisosa* 195
*Dingosa* 274, 280, 313
*Dinopis* 92, 290
*Dionycha* 20, 24, 83, 162, 205, 257
*Diplocentria* 127, 155, 291, 306, 319, 320
*Diplocephalus* 127, 137, 144, 286, 291, 332
*Diplostyla* 291, 322
*Diplostylus* 291
*Diplura* 291, 294, 299, 316
DIPLURIDAE 21, 26, **48**, 50, 52, 291, 294, 299, 306, 332
*Dipoena* 235, 236, 239, 284, 291, 301, 306, 308, 316, 329, 332
*Dirksia* 112, 113, 115, 291
*Disembolus* 127, 137, 139, 142, 146, 291, 332
*Dismodicus* 127, 139, 291
*displicata, Araniella* 72
*displicata, Epeira* 279
*distincta, Pardosa* 169
*Dolichognatha* 233, 234, 291
*dolichopus, Anachemmis* 278
*dolloff, Meta* 234
*Dolomedes* 199, 200, 249, 291, 329, 332
DOLOMEDIDAE 199
*domestica, Tegenaria* 58
Domiothelina 20
*dondalei, Procerocymbium* 147
*doriai, Zimiris* 309
*Drapetisca* 127, 130, 157, 291
DRASSIDAE 257
*Drassinella* 79, 80, 81, 291, 297, 332
*Drassodes* 24, 107, 109, 291, 319, 332
DRASSODINAE 107
*Drassus* 274, 282, 289, 291, 292, 296, 298, 302, 311, 317, 318, 321, 324, 329
*Drassyllus* 107, 108, 292, 332
*Drepanotylus* 292
*Drexelia* 292
*Drosophila* 122, 228
DRYMUSIDAE 20, 218
*Dryptopelma* 294
*dubia, Pisaurina* 199
*dubius, Ctenus* 289
*dubius, Sciastes* 151
*duellica, Tegenaria* 58
*Dugesiella* 55, 292, 294, 317, 329
dwarf spider 124
*Dysdera* 30, 103, 292, 327, 332
DYSDERIDAE 20, 22, 30, 75, **103**, 219, 269, 292, 332
DYSDERINAE 103
Dysderoidea 75

# E

*Ebo* 192, 193, 292, 298, 328, 332
*Eboria* 292
*ecclesiasticus, Herpyllus* 106, 111
ECHEMINAE 107
*echinus, Rachodrassus* 317
*Ectatosticta* 121
*ectypa, Wixia* 310
*edax, Eriophora* 72
*edenticulata, Pimoa* 197, 198
*edwardsi, Tama* 322
*Egilona* 295

*Eidmannella* 178, 179, 180, 292, 332
*Eilica* 106, 107, 109, 292, 328
*Elaver* 77, 78, 287, 292, 315, 332
*eldorado, Metleucauge* 234
*electa, Clitolyna* 125
*elegans, Camillina* 108
*elegans, Xysticus* 246
*elongata, Pelecopsis* 313
*elongata, Pholcus* 320
*elongatum, Dicymbium* 144
*elongatum, Lophocarenum* 303
*Emblyna* 96, 100, 101, 275, 292, 332
*Emertonella* 236, 239, 292
*emertoni, Agroeca* 163
*emertoni, Euryopis* 292
*emertoni, Liocranoeca* 163
*Enidia* 292
*Enoplognatha* 235, 236, 238, 292, 303, 317, 318, 331, 332
*Entelecara* 127, 138, 292, 308
Entelegynae 20, 22, 29, 30, 32, 23, 24
*Entychides* 46, 47, 292
*Enyo* 329
*Epecthina* 176
*Epecthinula* 176
*Epeïra* 276, 293
*Epeira* 8, 275, 276, 277, 288, 293, 294, 302, 304, 305, 306, 307, 309, 310, 328
EPEIRIDAE 171, 233, 251
*epeirides, Argyrodes* 280
*epeirides, Linyphia* 280
EPEIROTYPINAE 245
*Eperigone* 125, 127, 148, 152, 154, 293, 313, 332
*ephedrus, Linyphia* 302
*Epiblemum* 298
*Epiceraticelus* 127, 146, 293
*Episinus* 236, 239, 293
*erectus, Mythoplastoides* 141
ERESIDAE 20, 23, 116, 184
Eresoidea 20, 116, 184
*Eridantes* 127, 139, 141, 293
*Erigone* 124, 126, 127, 131, 143, 148, 149, 280, 288, 290, 292, 293, 297, 299, 313, 326, 332
*Erigonella* 127, 293
ERIGONIDAE 307
ERIGONINAE 124, 125, 127, 128, 136, 288, 310
*erigonoides, Eridantes* 139, 141
*erigonoides, Lophocarenum* 293
*Eriophora* 69, 72, 293
*Eris* 206, 207, 214, 216, 293, 312, 332
*Ero* 171, 172, 274, 293, 317
*erratica, Pseudeuophrys* 206, 215
*Ervig* 295
*erythropus, Lasaeola* 301
*eschatologica, Eperigone* 148
*Eskovia* 127, 147, 294
*Estrandia* 127, 130, 158, 294
*Ethobuella* 112, 113, 115, 285, 291, 294
*Euagrus* 48, 49, 294, 295
*Eucalyptus* 219
*Eucteniza* 26, 45, 46, 47, 282, 294
EUCTENIZINAE 46
*eugenii, Singa* 73
*Eulaira* 127, 154, 294, 332
EUOPHRYINAE 206, 207
*Euophrys* 206, 209, 215, 289, 294, 309
*euoplus, Callioplus* 284
*europea, Mutilla* 305
*Europhrys* 313, 317
*Euryopis* 183, 235, 236, 238, 239, 292, 294, 332
*Eurypelma* 279, 285, 287, 294, 297
EUSPARASSIDAE 224
*Eusparassus* 224, 225
*Eustala* 68, 69, 72, 73, 74, 294, 332
*eustala, Epeira* 294
EUTICHURINAE 173, 174
*Eutichurus* 294
*eutypus, Cybaeus* 85
*eutypus, Mimetus* 317
*eutypus, Reo* 171, 172
*eutypus, Scotinotylus* 142
*Evagrus* 294
*Evarcha* 206, 212, 294, 295
*exarmata, Eskovia* 147
*exarmata, Minicia* 294
*excepta, Clubiona* 286
*excisa, Hilaira* 299

*exiguus, Mythoplastoides* 141
*exlinae, Cryphoeca* 115
*exlinae, Ctenus* 83
*extensa, Tetragnatha* 232
*extremus, Sciastes* 151

# F

*Faiditus* 236, 238, 295
*falcata, Evarcha* 206
*falcifera, Pardosa* 169
*falconeri, Maro* 300
*falconerii, Jacksonella* 155
*Falconia* 295
*Falconina* 79, 80, 82, 295
*famelicum, Gnathonarium* 10
*fasciata, Phlegra* 215
*fasciculatus, Olios* 224
*fasiculatum, Erigonium* 189
*fauroti, Microphorus* 195
*fauroti, Pholcus* 307
*faustus, Sintula* 301
*Favila* 295
*felix, Messua* 216
*felix, Metaphidippus* 216
*fervidus, Talanites* 322
*festivus, Phrurolithus* 82
*fidelis, Misumena* 247
*fidicina, Ariadna* 219, 220
*Filistata* 104, 105, 295, 300
FILISTATIDAE 20, 22, 28, **104**, 105, 201, 219, 295, 300, 332
FILISTATINAE 105
*Filistatinella* 104, 105, 295
*Filistatoides* 104, 105, 295
*fissiceps, Ceraticelus* 124
*fitchi, Sphodros* 41
*flava, Eris* 206
*flava, Usofila* 229
*flavescens, Heterodictyna* 298
*flavescens, Nigma* 298
*flavescens, Styposis* 322
*flavescens, Titiotus* 231
*flavidus, Spintharus* 239, 321
*flavomaculata, Euryopis* 294
*florens, Hypselistes* 299
*Floricomus* 127, 143, 295, 332
*floricomus, Floricomus* 295
*floridana, Coriarachne* 246
*floridana, Masonetta* 151, 153
*floridana, Thalamia* 315
*floridana, Tinyna* 325
*floridanus, Platoecobius* 183, 184
*Florinda* 127, 128, 156, 274, 295, 302, 328
*Floronia* 295
*fluvialis, Epiceraticelus* 146
*fluviatilis, Mygale* 308
foldingdoor spider 39
*foliata, Myrmekiaphila* 308
*folium, Larinioides* 301, 310
*fontanus, Paradamoetas* 209, 212
*fordum, Theridion* 325
*formacina, Paradamoetas* 312
*formacinus, Paradamoetas* 212
*formica, Phrurolithus* 315
*formica, Synemosyna* 208
*formicaria, Aranea* 308
*formicaria, Myrmarachne* 207, 209
*formicaria, Tutelina* 207
*formicarius, Salticus* 305
*formicinus, Paradamoetas* 209, 212
*formicum, Tennesseellum* 134, 161
*formosa, Ceratinella* 300
*formosa, Ceratinopsidis* 149
*formosipes, Misumenoides* 247
*foxi, Theridion* 307
*fradeorum, Eperigone* 148
*fragilis, Arcuphantes* 280
*frederici, Chemmis* 278
*frondicola, Hogna* 167, 168
*frontalis, Pelecopsidis* 143
*Frontina* 295
*Frontinella* 127, 132, 159, 295, 332
*Frontinus* 295
*Fuentes* 283, 295, 311, 322

*fulgens, Micaria* 306
*fulva, Hexura* 304
*fulva, Megahexura* 50, 51
*fulvonigrum, Zodarion* 254
*fumosa, Lycosa* 305
*fumosa, Melocosa* 167
*funerea, Lycosa* 277
funnel web spider 48
funnel web weaver 56
*fusca, Melanophora* 319
*fusca, Scytodes* 217

# G

*Galeodes* 305
GALLIENIELLIDAE 20
*Gamasomorpha* 185, 187, 188, 295
GAMASOMORPHINAE 185
*Gamasus* 295
*Garricola* 295
*garrina, Cicurina* 295
*garrina, Hilaira* 295
*Garritus* 295
*garza, Archoleptoneta* 123
*Gasteracantha* 68, 69, 71, 295, 296, 306, 332
*Gaucelmus* 178, 179, 180, 296
*Gayenna* 296
*Gayennina* 296
*Gea* 68, 69, 70, 296, 304
*gemmosum, Theridiosoma* 244, 245
*geniculata, Zosis* 251, 252, 253
*Geodrassus* 320
*Geolycosa* 164, 165, 296, 332
*georgia, Neoleptoneta* 123
*Gertschanapis* 33, 64, 65, 285, 296
*gertschi, Atypoides* 40
*gertschi, Dolomedes* 199
*gertschi, Hypochilus* 120, 121
*gertschi, Neocryphoeca* 115, 309
*gertschi, Trechalea* 249
*gertschi, Willisus* 115, 328
*Gertschia* 296
*Gertschosa* 106, 107, 110, 296
*Ghelna* 206, 216, 296
giant crab spider 224
*gibbosus, Oedothorax* 311
*gigas, Grammonota* 145
*glacialis, Pardosa* 168
*Gladicosa* 164, 165, 166, 296, 332
*glauca, Nops* 310
*glaucopis, Tortolena* 59
*Glenognatha* 124, 232, 233, 234, 296, 307
*globosa, Synema* 322
*globosum, Synaema* 322
*globosus, Physocyclus* 315
*globula, Scytodes* 217
*Glyphesis* 127, 138, 296
*Glyptocranium* 276, 304
*Glyptogona* 327
*Gnaphosa* 106, 107, 109, 291, 296, 303, 318, 322, 324, 332
GNAPHOSIDAE 20, 24, 37, 77, 79, 94, **106**, 107, 119, 162, 203, 224, 257, 284, 285, 291, 292, 296, 297, 298, 302, 303, 304, 306, 310, 311, 312, 313, 315, 316, 317, 319, 320, 322, 326, 327, 329, 332
GNAPHOSINAE 107
Gnaphosoidea 20, 24
*Gnathantes* 296
*Gnathonagrus* 296
*Gnathonargus* 127, 144, 296
*Gnathonarioides* 297
*Gnathonarium* 127, 148, 296, 297
*Gnathonaroides* 127, 151, 296
*Gonatium* 127, 146, 297
*Goneatara* 127, 145, 297
*Gonglydium* 297
*Gongylidiellum* 127, 297
*Gongylidium* 297
*Gosipelma* 297
*Gosiphrurus* 297
*gowerense, Baryphyma* 139
*gracilis, Falconina* 79, 82
*gracilis, Usofila* 229, 327
GRADUNGULIDAE 20
*Grammonota* 127, 145, 146, 150, 289, 291, 297, 321, 332
GRAMMOSTOLINAE 55

*granadensis, Sparianthis* 317
*grandaeva, Estrandia* 130, 158
*Graphomoa* 127, 132, 159, 277, 297
*grisea, Hamataliwa* 191, 298
*grisea, Marxia* 304
*groenlandicum, Baryphyma* 144
ground spider 106, 203
*gulosa, Lycosa* 296
*gulosus, Creugas* 82
*guttata, Microdipoena* 175, 177

# H

*habanensis, Tama* 116, 329
*habanensis, Yabisi* 116, 117
*Habrocestoides* 286
*Habrocestum* 206, 211, 297, 332
*Habronattus* 206, 207, 212, 297, 332
hackled band orb-weaver 250
*Hackmania* 96, 100, 297
HADROTARSINAE 236
*haemorrhoidale, Tidarren* 241, 243
*Hahnia* 112, 113, 114, 297, 309, 332
HAHNIIDAE 20, 35, 56, 57, 85, 95, **112**, 113, 278, 284, 289, 291, 294, 297, 309, 319, 328, 332
HAHNIINAE 113, 114
*Hahnistea* 297
*Halorates* 125, 127, 152, 154, 278, 288, 297, 307, 313
HALIDAE 20
*hamata, Acacesia* 72
*hamata, Euophrys* 294
*Hamataliva* 298
*Hamataliwa* 189, 190, 191, 297, 298
*hansii, Eusparassus* 225
*Haplodrassus* 107, 110, 298, 332
Haplogynae 20, 22, 24, 29, 30, 32, 265
*harrisonfordi, Calponia* 76, 284
*hartlandianus, Tusukuru* 140
*Hasarius* 207, 211, 298
*hastatus, Sciastes* 151
*Hebestatis* 43, 44, 298
*Hedypsilus* 195, 298
*helia, Hamataliwa* 191
HELIOPHANINAE 207, 213
*Helophora* 127, 129, 157, 298, 318, 327
*hemicloeinus, Lauricius* 231
*hentzi, Phlegra* 215
*hentzi, Sarinda* 209
*hentzi, Tapinopa* 129, 157
*Hentzia* 206, 207, 214, 277, 298, 304, 328, 332
*heptagon, Gea* 70, 296
*herbigradus, Micrargus* 140
HERPYLLINAE 107
*Herpyllus* 106, 107, 111, 292, 298, 302, 303, 332
*Hersilia* 298
HERSILIIDAE 20, 23, 32, **116**, 184, 298, 309, 322, 329, 332
*hespera, Ethobuella* 115
*hespera, Labulla* 301
*hespera, Scaphiella* 185
*hespera, Steatoda* 235
*hespera, Zora* 256, 257
*Hesperauximus* 93
*hesperia, Dietrichia* 138
*Hesperocosa* 165, 167, 169, 298
*Hesperocranum* 38, 162, 163, 298
*Hesperopholis* 52
*Heterodictyna* 100, 298, 310
*Heteroonops* 185, 186, 298
*heterophthalmus, Oxyopes* 311
*Heteropoda* 224, 225, 298, 299, 321, 332
HETEROPODIDAE 224
HETEROPODINAE 224
*heterops, Cybaeozyga* 89, 90
*Hexathele* 299
HEXATHELIDAE 20
*Hexura* 50, 51, 299, 304, 306
*Hexurella* 50, 51, 299
HEXURINAE 299
*Hibana* 66, 67, 282, 299, 332
*hibernalis, Carorita* 154
*hibernalis, Filistata* 300
*hibernalis, Kukulcania* 105
*hiemalis, Drassus* 298

*Hilaira* 127, 128, 131, 280, 294, 295, 299, 307, 332
*Hillhousia* 299
*hirtipes, Alopecosa* 165
*Hogna* 165, 167, 168, 169, 299, 300, 317, 332
HOLARCHAEIDAE 20
HOLOCNEMINAE 194
*Holocnemus* 194, 196, 299
*Hololena* 57, 59, 275, 284, 299, 310, 317, 325, 332
HOMALONYCHIDAE 20, 37, **118**, 119, 224, 299, 332
*Homalonychus* 37, 118, 119, 299
*hooki, Lauricius* 231
*hoplites, Callioplus* 284
*haplomachus, Callioplus* 284
*Horcotes* 127, 138, 274, 299
*Hormathion* 299
*Hormathium* 299
*hortorum, Epeira* 302
*hoyi, Evarcha* 206, 212
*huachuca, Frontinella* 132, 159
huntsman spider 224
HUTTONIIDAE 20
*Hyaenosa* 280, 313
*Hybauchenidium* 127, 137, 299
*Hybocoptus* 299
*Hyctia* 299
HYETUSSINAE 206
*hyperborea, Pardosa* 164
*Hyphantes* 296
HYPOCHILIDAE 20, 22, 27, **120**, 121, 194, 261, 270, 271, 299, 332
Hypochiloidea 265, 332
*Hypochilus* 27, 120, 121, 299, 332
*Hypomma* 127, 137, 292, 299
*Hypophthalma* 318
*Hypselistes* 127, 141, 299
*Hypselos* 299
*hypsigaster, Parogulnius* 175, 176, 177, 313
*Hypsosinga* 34, 69, 73, 74, 299, 332
*Hyptiotes* 244, 245, 250, 251, 253, 274, 300, 316

# I

*Iardinis* 176
*Icelus* 300
*Icius* 300, 312, 317
*idahoana, Microhexura* 48, 49
*idahoanus, Glyphesis* 138
*Idastrandia* 294
*Idionella* 300
IDIOPIDAE 20
*imperialis, Saltonia* 318
*imperiosa, Diguetia* 102
*improbulus, Lepthyphantes* 300
*Improphantes* 127, 300
*inaequalis, Micryphantes* 303, 313
*inaurata, Argyrodes* 280
*inaurata, Linyphia* 280
*incerta, Saltonia* 97
*Incestophantes* 127, 300
*incestus, Lepthyphantes* 300
*inclusum, Cheiracanthium* 173, 174
*incompta, Euryophrya* 294
*incredula, Calodipoena* 175, 177
*incursa, Hibana* 66
*indica, Zimiris* 309
*Indonops* 310
*inermis, Tapinesthis* 186, 323
*infimus, Chalcoscirtus* 285
*inflata, Ceratinops* 285
*inflata, Scyletria* 155
*infuscata, Leptoneta* 302
*inornata, Lasaeola* 301
*insidatrix, Filistata* 295
*insignis, Filistatoides* 105
*insolita, Paratrochosina* 165, 168, 170
*insulana, Paratheuma* 93, 94
*insulanus, Eutichurus* 94
*insularis, Sosticus* 110
*insularis, Temenius* 174
*interaesta, Paratheuma* 93, 94
*interpellator, Oecobius* 183, 184
*interpres, Notionella* 310
*intricata, Scytodes* 218
*inuncans, Styloctetor* 322

*invicta, Solenopsis* 79, 82
*iodius, Aphonopelma* 54
*Isala* 300
*Isaloides* 246, 300
*Ischnaspis* 300
*Ischnothyreus* 185, 188, 300
ISCHNOCOLINAE 20
*Ischnyphantes* 307
*Islandiana* 127, 142, 147, 149, 276, 300, 332
*Isohogna* 300
*istokpoga, Theridion* 240
*Ivesia* 300
*Iviella* 96, 99, 300
*Ivielum* 127, 151, 300
*Ixeuticus* 60, 93, 283, 300

## J

*Jacksonella* 127, 155, 300
*Jalapyphantes* 127, 132, 160, 300
*Janus* 330
*jemez, Hypochilus* 121
*johnsoni, Phidippus* 206
*jona, Scyletria* 153
jumping spider 205

## K

*Kaestneria* 127, 131, 159, 288, 300
*Kaira* 69, 71, 300
*kaspari, Mazax* 80
*kastoni, Hypochilus* 121
*Katadysas* 257
*keeni, Bathyphantes* 128, 156
*Keijia* 236, 242, 243, 300
*Keyserlinigiella* 329
*Kibramoa* 201, 202, 274, 300, 332
*kiplingi, Bagheera* 283
*kochii, Alopecosa* 170
*kristenae, Apomastus* 47
*Kukulcania* 104, 105, 219, 274, 295, 300, 332
*kulczynskii, Baryphyma* 144

## L

*Labidognatha* 19
*Labuella* 301
*Labulla* 301
*labyrinthea, Epeira* 307
*labyrintheca, Agelena* 276
LACHESANINAE 254
*laeta, Frontinella* 295
*laeta, Loxosceles* 222
*lamina, Deinopis* 290
LAMPONIDAE 20
lampshade weaver 120
*lapidaria, Drassodes* 291
*lapidicina, Pardosa* 169
*lapidicola, Drassodes* 291
*lapidicolens, Drassodes* 291
*lapidosa, Drassodes* 291
*lapidosus, Drassodes* 291
*lapponica, Arctella* 99, 280
*lapponica, Pardosa* 169
*Larinia* 69, 72, 274, 292, 301
*Larinioides* 69, 73, 74, 301, 310
LARONIINAE 107
*Lasaeola* 236, 239, 301
*Laseola* 236, 301
*Lasiodora* 279
*Lasiopelma* 294
*Lasoeola* 301
*lata, Ceratinops* 285
*Lathys* 27, 96, 98, 124, 301, 319, 326, 332
*Latithorax* 301
*latithorax, Typhochrestus* 145
*latro, Reo* 317
LATRODECTINAE 236
*Latrodectus* 11, 15, 235, 236, 237, 280, 301, 323, 332

*laurae, Pimoa* 197
*Lauricius* 36, 230, 231, 301
*lectularius, Cimex* 288
*leechi, Thaleria* 134, 324
*leechi, Theridion* 236
*lemniscata, Linyphia* 304
*lemniscata, Mecynogea* 69
*Lephthyphantes* 302
*Lepthyphantes* 125, 127, 129, 131, 157, 158, 277, 280, 284, 296, 300, 301, 302, 304, 312, 318, 323, 332
*Leptoctenus* 83, 84, 302
*Leptoneta* 122, 123, 279, 280, 281, 284, 302, 305, 309
LEPTONETIDAE 20, 29, 31, 86, **122**, 123, 124, 181, 228, 229, 264, 279, 280, 284, 302, 309, 332
LEPTONETINAE 123
*Leptorhoptrum* 125, 302, 308
*Leptyphantes* 302
*Lessertia* 127, 153, 302
*Lethia* 301
*Leucauge* 232, 233, 302, 305, 316
*leucoplagiata, Dipoena* 308
*Liger* 292, 315
*limbata, Messua* 216
*limitaneus, Pityohyphantes* 156
*limnaea, Carorita* 154
*limnatum, Hormathion* 141
*lineata, Fuentes* 295
*lineatus, Miagrammopes* 306
*lineatus, Myrmecotypus* 81
*linsdalei, Nigma* 100
*Linyphantes* 127, 132, 160, 278, 284, 285, 302, 332
*Linyphia* 125, 127, 128, 129, 156, 157, 280, 295, 302, 304, 306, 309, 312, 317, 332
*Linyphiella* 302
LINYPHIIDAE 20, 23, 29, 31, 34, **124**, 178, 181, 185, 197, 198, 232, 244, 263, 264, 266, 268, 269, 272, 276, 277, 278, 279, 280, 282, 283, 284, 285, 286, 287, 288, 290, 291, 292, 293, 294, 295, 296, 297, 298, 299, 300, 301, 302, 303, 304, 305, 306, 307, 308, 309, 310, 311, 312, 313, 314, 315, 316, 317, 318, 319, 320, 321, 322, 323, 324, 325, 326, 327, 328, 329, 330, 332
LINYPHIINAE 124, 125, 127, 128, 156, 277, 307, 310, 316
LIOCRANIDAE 20, 24, 38, 66, 77, 79, **162**, 224, 277, 279, 286, 297, 298, 302, 309, 315, 332
LIOCRANINAE 80, 162, 163, 173, 230
*Liocranoeca* 162, 163, 302
*Liocranoides* 77, 230, 231, 302, 332
*Liocranum* 86, 298, 302, 317
*Liodrassus* 302
LIPHISTIIDAE 11, 20, 21
*Lipocrea* 301
*Lithyphantes* 302
*Litopyllus* 106, 107, 111, 303
*littoralis, Scolopembolus* 154
*livens, Diaea* 247
*livida, Neriene* 289
long legged water spider 249
longjawed orb weaver 232
*longimanus, Cheliferoides* 208
*longipalpa, Castianeira* 79
*longipalpis, Peucetia* 191
*longipes, Anibontes* 134, 161
*longipes, Hahnistea* 297
*longipes, Lyssomanes* 303
*longipes, Scytodes* 217
*longiqua, Badumna* 93, 94
*longitarsis, Brachythele* 52, 53
*longitarsis, Calisoga* 52, 53
*longitarsum, Baryphyma* 140
*longitarsus, Micrargus* 140, 142
LOPHOCARENINAE 303
*Lophocarenum* 293, 299, 303, 313
*Lophomma* 127, 140, 303, 332
*loricatus, Chorizops* 286
*loricatus, Sosticus* 111
*lorna, Argenna* 100
*Louisfagea* 198
*Loxosceles* 15, 30, 222, 223, 303, 332
LOXOSCELINAE 222
*Loxoscelis* 303
*Lubinella* 251
*lucifuga, Aranea* 296
*lucifugus, Drassus* 296
*Lupettiana* 66, 67, 303
*Lutica* 254, 255, 303
*lutzi, Gamasomorpha* 187, 188

*Lycaena* 329
*Lycodes* 329
*Lycodia* 329
*Lycoides* 329
*Lycosa* 164, 165, 167, 169, 274, 276, 277, 280, 282, 296, 298, 299, 303, 305, 313, 315, 317, 318, 325, 326, 327, 329, 332
LYCOSIDAE 6, 12, 14, 15, 18, 19, 20, 24, 35, 54, 56, 83, 118, **164**, 165, 166, 167, 168, 169, 170, 173, 190, 199, 249, 256, 257, 259, 265, 267, 268, 269, 276, 277, 280, 282, 296, 298, 299, 300, 303, 305, 313, 315, 317, 318, 320, 325, 326, 327, 332
LYCOSINAE 165
Lycosoidea 24, 113, 164, 173, 190, 257, 259
lynx spider 189
*Lynxosa* 274, 280, 313
*lyoni, Crossopriza* 196, 289
*Lyssomanes* 206, 207, 208, 303
LYSSOMANINAE 207

# M

*Macaria* 306
*macellus, Tibellus* 325
*Macrargus* 127, 133, 160, 303, 304, 306
*Macrothele* 299, 316
*macrura, Mygale* 291
*mactans, Latrodectus* 11
*maculata, Eperigone* 293
*maculata, Lutica* 254, 303
*Madognatha* 321
*Maevia* 206, 207, 210, 303, 312, 332
*magnificus, Pacifiphantes* 125, 159
*magnipalpis, Sphecozone* 146, 152, 321
*Majella* 303
*Majellula* 246, 303
*majesticum, Yukon* 329
MALKARIDAE 20
*Malloneta* 281
*Mallos* 95, 96, 98, 100, 298, 303, 306, 332
*mammouthia, Anthrobia* 147, 279
*mandibulare, Theridion* 292
*mandibulata, Linyphia* 317
*Mangora* 69, 70, 303, 332
*manni, Metaphidippus* 216
*manueli, Pholcus* 195
*Marchena* 206, 207, 209, 303
*Marfisa* 304
*Marmatha* 303
*marmorata, Marmatha* 303
*Maro* 127, 133, 135, 161, 300, 303
*Marpessa* 304
*Marpissa* 205, 206, 207, 210, 299, 304, 311, 312, 332
MARPISSINAE 206, 283
*Marpissus* 304
*Marptusa* 304, 315
*martius, Ixeuticus* 93
*marxi, Trebacosa* 166
*Marxia* 276, 304
*marxii, Satiatlas* 318
*Masikia* 125, 127, 304
*Maso* 127, 147, 153, 304, 310
*Masoncus* 127, 142, 304
*Masonetta* 127, 151, 153, 304
*Mastophora* 34, 68, 69, 70, 276, 304, 332
*Maturna* 295
*mauriciana, Artema* 282
*maxillosa, Aranea* 290
*mayanus, Nesticus* 304
*Maymena* 64, 175, 176, 177, 304
*Mazax* 79, 80, 304
MECICOBOTHRIIDAE 20, 21, 25, 39, 41, 48, **50**, 299, 304
*Mecicobothrium* 50
*Mecynargus* 127, 147, 288, 304, 317
*Mecynogea* 68, 69, 277, 298, 304
*Mecynometa* 316
MECYSMAUCHENIIDAE 20
*Megahexura* 50, 51, 304
*Megalepthyphantes* 127, 304
*Megalostrata* 80, 286, 304
*Megamyrmaecium* 321
*Megamyrmaekion* 274, 304, 305, 319, 326
*Megamyrmecion* 305, 310
*Meioneta* 125, 132, 135, 160, 161, 296, 305, 307, 332
*Melanophora* 319, 329

*melanopygius, Ostearius* 133, 311
*Melocosa* 165, 167, 305
*Melpomene* 57, 58, 305, 317
*menardi, Meta* 233
*mendocino, Archerius* 279, 288
*mendocino, Comaroma* 64, 65, 288
*Menemerus* 207, 215, 305, 327
*Mengea* 277
*Menneus* 92
*mentasta, Erigone* 131
*Meotipa* 236, 241, 242
*meridionalis, Spermophora* 195
*Meriola* 80, 305
Mesothelae 11, 13, 19, 20, 120, 260, 268, 269, 272
*Messua* 206, 207, 216, 274, 283, 305
*Meta* 34, 197, 232, 233, 234, 305, 306, 316
*Metacyrba* 206, 305, 315
*Metagonia* 195, 305
*Metaltella* 28, 60, 63, 278, 305
METALTELLINAE 20, 60, 63
*Metaphidippus* 206, 216, 296, 305, 323, 332
*Metazygia* 69, 73, 74, 305
*Metella* 305
*Metellina* 232, 233, 234, 305
*Metepeira* 69, 72, 306, 332
METINAE 233, 306
*Metleucauge* 233, 234, 305
*Metopobactrus* 127, 147, 306
*mexicana, Neotama* 116, 117
*mexicana, Tama* 116
*mexicanus, Miagrammopes* 251, 253, 306
*mexicanus, Synageles* 209
*Mexigonus* 206, 207, 215, 306
*Mexitlia* 96, 100, 306
*Miagrammopes* 27, 250, 251, 253, 306, 307
*Miagrammopides* 251
*micans, Bellota* 208
*Micaria* 37, 77, 79, 106, 107, 108, 162, 306, 332
MICARIINAE 80, 107, 230, 297
*Micrargus* 127, 140, 142, 143, 306
*Micrathena* 68, 69, 70, 306, 332
*Microcentria* 306
*Microctenonyx* 127, 141, 282, 306
*Microdipoena* 124, 175, 176, 177, 306
*Microhexura* 48, 49, 306
*Microlinyphia* 125, 127, 129, 157, 306, 317
*Micrommata* 224
*Microneta* 127, 132, 160, 277, 278, 281, 305, 306, 316
*Microphantus* 328
MICROPHOLCOMMATIDAE 20, 227
*Micropholcus* 194, 195, 307
*microps, Anacornia* 146
*microps, Neoleptoneta* 123
MICROSTIGMATIDAE 20
*Micryphantes* 303, 305, 307, 313, 328
MICRYPHANTIDAE 307
MIGIDAE 20
Migoidea 20
*mildei, Cheiracanthium* 174
*militaris, Eris* 206
*milvina, Pardosa* 169
MIMETIDAE 20, 23, 33, **171**, 293, 307, 317, 332
*Mimetus* 171, 172, 307, 317, 332
*Mimognatha* 307
*mimus, Anibontes* 134, 161
*Minicia* 294
*minimum, Symmigma* 141
*minor, Tidarren* 325
*minor, Trachelas* 326
*minus, Tidarren* 325
*minuta, Cybaeina* 88
*minuta, Marchena* 209
*minuta, Talavera* 213
*minutissima, Theotima* 181, 182
*minutus, Cybaeus* 86
*minutus, Stenoonops* 186
*Minyrioloides* 307
*Minyriolus* 146, 307
*Minyrus* 307
*mira, Pisaurina* 199
*mirabilis, Ocyale* 315
*mirus, Sprirembolus* 145
*misera, Aphileta* 134
*Missulena* 321
*Misumena* 246, 247, 307, 308

*Misumenoides* 246, 247, 307
*Misumenops* 246, 247, 307, 332
*Mithras* 300
*mitrata, Hentzia* 214
*Miturga* 307
Miturgeae 173
MITURGIDAE 15, 20, 24, 37, 66, 77, 83, **173**, 224, 230, 257, 258, 285, 286, 301, 302, 307, 321, 322, 323, 329, 332
MITURGINAE 173, 174
*mixtum, Tidarren* 325
*modica, Pardosa* 170
*Modisimus* 194, 195, 298, 307
*moesta, Pardosa* 168, 169, 170
*monadnock, Euophrys* 209, 215
money spider 124
*monisticus, Anachemmis* 278
*monmouthia, Anthrobia* 147, 279
*monoceros, Cornicularia* 288
*monoceros, Erigone* 288
*monocerus, Ochyrocera* 310
*monocerus, Theridion* 288
*montana, Cryphoeca* 113, 115
*monticola, Nanavia* 125, 135, 308
*monticola, Pardosa* 170
*Montilaira* 294, 307
*montivaga, Microhexura* 48, 49, 306
*montivagus, Microhexura* 306
mouse spider 106
*mordax, Lupettiana* 67
*mossi, Annapolis* 155
*mossi, Sciastes* 278
*mulegensis, Varyna* 328
*multesimus, Macrargus* 133, 160
*Mumaia* 307
*muralicola, Pholcus* 195
*muscicola, Lepthyphantes* 302
*Mustelicosa* 274, 280, 313
*Mutilla* 305
*Mygale* 291, 308, 324
Mygalomorphae 11, 12, 14, 19, 20, 21, 25, 120, 260, 261, 263, 265, 267, 268, 269, 270, 272, 332
*mylothrus, Metaphidippus* 323
*Myrmarachne* 207, 209, 308
MYRMARACHNINAE 207
*Myrmecarachne* 308
*Myrmecotypus* 79, 80, 81, 308
*Myrmekiaphila* 45, 46, 47, 305, 308, 316, 333
*Mysmena* 176, 284, 304, 306, 308
Mysmeneae 176
MYSMENIDAE 20, 33, 64, 68, 124, **175**, 176, 226, 227, 245, 257, 262, 284, 304, 306, 308, 326, 333
MYSMENINAE 176, 306
*Mysmenopsis* 175, 176, 177, 308
*Mythoplastis* 308
*Mythoplastoides* 127, 141, 287, 308

# N

*Nagaina* 283, 305
*Nanavia* 125, 127, 128, 135, 308
*Naphrys* 206, 207, 211, 309
*natalensis, Caponia* 284
*navus, Oecobius* 183, 184
*nearcticus, Maro* 135, 161
*nearcticus, Tibioplus* 133
*nebulosus, Megalepthyphantes* 304
*nelli, Neon* 309
*nelsoni, Mimetus* 172
*Nemesia* 309
NEMESIIDAE 21, 26, **52**, 54, 283, 284, 333
*Nemesoides* 46, 309
*Neoanagraphis* 162, 163, 309
*Neoantistea* 112, 113, 114, 309, 333
*Neoapachella* 45, 46, 47, 309
Neocribellatae *20*, 22
*Neocryphoeca* 112, 113, 115, 309
NEOLANIDAE 20
*Neoleptoneta* 122, 123, 309, 333
*neomexicanum, Theridion* 236, 242
*Neon* 207, 212, 274, 309, 333
*Neonella* 206, 207, 213, 309
*Neoschoena* 309
*Neoscona* 69, 73, 74, 293, 309, 333

*Neospintharus* 236, 238, 309
*Neotama* 116, 117, 309
*Neotoma* 75, 176
*Neottiura* 235, 236, 241, 243, 309
*Neozimiris* 203, 204, 274, 309
*Nephila* 34, 233, 309, 333
NEPHILINAE 233
*Neriene* 125, 127, 129, 157, 289, 309, 316, 318, 333
*nesophilus, Selenops* 221
NESTICIDAE 20, 23, 29, 30, 31, 32, 33, 171, **178**, 179, 235, 262, 292, 296, 300, 310, 326, 333
*Nesticodes* 235, 236, 242, 310
*Nesticus* 30, 178, 179, 180, 296, 300, 304, 310, 326, 333
net casting spider 91
NICODAMIDAE 20, 23
*Nidivalvata* 310
*nidulans, Aranea* 327
*niger, Sphodros* 41
*Nigma* 96, 100, 298, 310
*nigra, Pardosa* 169
*nigriceps, Scoloderus* 70
*nigromaculatus, Dendryphantes* 216, 290
*nigrum, Dicymbium* 144
*Nilakantha* 318
NINETINAE 194
*nitidus, Damoetas* 312
*nobilis, Marxia* 304
*nocturna, Aranea* 284
*noda, Wala* 310
*Nodocion* 107, 111, 302, 310, 322, 333
NOPINAE 75
*Nops* 284, 310, 311
*Nopsides* 75, 310, 323
*nordica, Achaea* 280
*nordica, Arctachaea* 241
*norvegica sudetica, Acantholycosa* 276
*Notiomaso* 292, 310
*Notionella* 310
*Novalena* 57, 59, 299, 310, 333
*noxiosa, Synemosyna* 296
*Nuctenea* 73, 74, 293, 301, 310
*Nuctobia* 310
*nuecesensis, Poultonella* 210
nursery web spider 199
*Nyctimus* 314

# O

*Oaphantes* 127, 133, 161, 310
*oatimpa, Corypheaolana* 310
*obesa, Argenna* 100
*oblivium, Tidarren* 325
*oblongiusculus, Tibellus* 325
*oblongus, Tibellus* 325
*obscura, Pholcophora* 195
*Obscuriphantes* 127
*obscurus, Cnephalocotes* 139, 287
*Ochyrocera* 181, 229, 310
OCHYROCERATIDAE 20, 22, 31, 122, **181**, 185, 228, 229, 268, 310, 324, 333
OCHYROCERATINAE 181
*Ocrepeira* 68, 69, 71, 310
*octonarius, Uloborus* 251
*Octonoba* 250, 251, 252, 253, 310, 320
*Ocyale* 315
*Odo* 298, 328
OECOBIIDAE 20, 23, 28, 116, **183**, 184, 278, 311, 315, 324, 333
OECOBIINAE 184
*Oecobius* 28, 183, 184, 278, 311, 315, 324, 333
*Oedothorax* 127, 137, 284, 297, 311, 316, 326, 333
Ogre-faced spider 91
OGULNINAE 245
*Ogulnius* 313
*ohioensis, Iviella* 99, 300
*ohioensis, Tricholathys* 300
*Olios* 224, 225, 311, 333
*Onondaga* 311
OONOPIDAE 20, 30, 31, 75, 103, **185**, 220, 228, 261, 267, 268, 295, 298, 300, 311, 313, 314, 318, 321, 323, 326, 333
OONOPINAE 185
*Oonops* 185, 186, 298, 311, 321
*Oophora* 321
Opiliones 11, 12, 18
*opima, Corythalia* 211

*Opisthothelae* 11, 19, 20
*Opisthotheles* 21
*Opopaea* 185, 187, 188, 311, 313, 333
*Opuntia* 102
orb weaver 68
Orbiculariae 20, 23
*Orchestina* 185, 186, 311, 333
*oregona, Usofila* 228, 229
*oregonensis, Habronattus* 207
*oregonus, Zygottus* 151
*Oreoneta* 131, 281, 311
*Oreonetides* 125, 127, 133, 135, 160, 161, 277, 301, 311, 333
*Oreophantes* 127, 130, 158, 311
*Origanates* 127, 138, 311
*Orinomana* 251
*orites, Orodrassus* 312
*orites, Parasyrisca* 110
*ornata, Agroeca* 162
*ornatus, Stemmops* 240
*ornatus, Tmeticus* 148
*ornithes, Taranucnus* 131, 158
*Orodrassus* 37, 106, 107, 110, 311, 312
*Oronota* 237, 311, 327
*Oroodes* 311, 327
ORSOLOBIDAE 20, 103, 220
*Orthognatha* 18, 19, 24
*Orthonops* 75, 76, 310, 311, 323, 333
*Ostearius* 127, 133, 311
*oteroana, Drapetisca* 130, 157
*ovalis, Meta* 233, 234
*Oxyopa* 311
*Oxyopes* 189, 190, 191, 311, 333
OXYOPIDAE 20, 24, 35, 83, 119, **189**, 190, 205, 297, 311, 314, 333
*Oxyptila* 311
*Oxysoma* 66, 67, 296, 312
*Ozyptila* 246, 247, 274, 311, 312, 333

# P

*Pachygnatha* 232, 233, 234, 312, 331, 333
*Pachylomerides* 44, 312, 327
PACHYLOMERINAE 312
*Pachylomerus* 312, 327
*Pachyloscelis* 327
*pacifica, Ochyrocera* 229
*pacifica, Segestria* 219, 220
*pacificus, Micrargus* 143
*Pacifiphantes* 125, 127, 159, 312
*packardii, Aculepeira* 72
*paetulus, Rhaebothorax* 317
*pagana, Tegenaria* 58
*Pagomys* 99, 284, 312, 329
*Paidisca* 242, 243, 312
Paleocribellatae 20, 22
*Pales* 313
*pallida, Dictynina* 298
*pallida, Eidmannella* 178, 179, 180
*pallidula, Bathyphantes* 310
*pallidulus, Bathyphantes* 310
*pallidulus, Oaphantes* 133, 161
*pallidus, Scylaceus* 151
*pallidus, Smeringopus* 196
*palmarum, Epiblemum* 298
*paloma, Aphonopelma* 54
*palomara, Ceratinopsis* 138
PALPIMANIDAE 20
Palpimanoidea 20, 22, 23, 171
*paludis, Synaphosus* 111
*palustris, Pardosa* 168
*pampia, Minyriolus* 146, 320
*pampia, Silometopoides* 146
*pantoplus, Callioplus* 284
*papenhoei, Sassacus* 213
*Paracornicularia* 127, 142, 312
*Paradamoetas* 206, 207, 208, 209, 212, 312
*paradoxa, Epeira* 311, 327
*paradoxa, Trogloneta* 175, 176, 177
*paradoxus, Hyptiotes* 300
*paradoxus, Mithras* 300
*Parahedrus* 118
*parallelum, Lophocarenum* 303
*Paramaevia* 206, 207, 210, 312
*Paramarpissa* 206, 207, 210, 312

*Paraphidippus* 206, 207, 214, 305, 312
PARARCHAEIDAE 20
*Parasaitis* 278
*parasita, Argyrodes* 280
*parasita, Linyphia* 280
*parasiticus, Thyreosthenius* 141
*Parasyrisca* 106, 107, 110, 312
*Paratheridula* 236, 242, 312
*Paratheuma* 34, 93, 94, 312, 313
*Paratrochosina* 165, 313
PARATROPIDIDAE 20
*Parazanomys* 61, 313
*Pardosa* 164, 165, 166, 168, 169, 170, 274, 277, 280, 313, 317, 318, 326
PARDOSINAE 165
*Parerigone* 293, 313
*parietalis, Thalamia* 324
*parisiensis, Tapinocyba* 282
*Parogulnius* 175, 176, 177, 313
*parva, Souessoula* 152
*parvulum, Habrocestum* 211
*parvulus, Chinattus* 211
*Pasithea* 314
*passivum, Tidarren* 325
*pauper, Dactylophistes* 290
*pax, Mazax* 80
*paykullii, Plexippus* 211
*pearcei, Neoanagraphis* 163
*peckhamae, Salticus* 207
*Peckhamia* 206, 207, 209, 296, 313
*pectinatus, Scirites* 149
*pedalis, Araeoncus* 296
*pedalis, Gnathonaroides* 151
Pedipalpi 18
*Pelayo* 295
*Pelecinus* 313
*Pelecopsides* 313
*Pelecopsidis* 127, 143, 313
PELECOPSINAE 303
*Pelecopsis* 127, 136, 303, 313, 326 333
*Pelegrina* 206, 216, 313, 333
*Pelicinus* 185, 187, 188, 313, 314
*Pellenes* 206, 212, 313, 333
PELLENINAE 206
*Pelopatis* 313
*peltifer, Ischnothyreus* 188
*pelyx, Arctachaea* 241
*penifusifer, Sisicus* 134
*pentagona, Dolichognatha* 234
*perditus, Cybaeus* 89, 90
*Peregrinus* 313
*peregrinus, Tinus* 199, 200
PERIEGOPIDAE 20
PERIEGOPINAE 102
*Perimones* 313
*perita, Arctosa* 169
*perniciosa, Paratheridula* 242
*Pero* 313
*Perregrinus* 127, 144, 313
*Perro* 127, 152, 313
*persica, Calymmaria* 113
*personata, Ulesanis* 327
*pertinax, Fuentes* 295
*petrunkevitchi, Hypochilus* 121
*petrunkevitchi, Synemosyna* 208
*Peucetia* 189, 190, 191, 314, 333
*phalangioides, Pholcus* 194, 195
*phana, Tapinocyba* 314
*Phanetta* 127, 148, 314
*Phanias* 206, 212, 215, 216, 314, 333
*Phantyna* 96, 100, 101, 275, 292, 314, 325, 328, 333
*Phidippus* 205, 206, 207, 214, 300, 305, 312, 314, 316, 331, 333
*Philesius* 314, 313
PHILODROMIDAE 20, 36, 37, 116, **192**, 221, 224, 273, 279, 292, 314, 324, 325, 333
*Philodromus* 192, 193, 246, 314, 333
*Philoponella* 250, 251, 252, 253, 314
*Philoponellus* 314
*Philoponus* 314
*Phlattothrata* 274, 314
*Phlegra* 206, 215, 314
PHOLCIDAE 20, 22, 30, 32, 102, 120, 122, **194**, 201, 217, 228, 232, 262, 282, 286, 289, 298, 299, 305, 307, 314, 315, 317, 320, 321, 333
PHOLCINAE 194
*Pholcomma* 124, 235, 236, 238, 240, 295, 314
PHOLCOMMATINAE 236

*Pholcophora* 194, 195, 314
*Pholcus* 194, 195, 274, 307, 314, 320, 333
*Phoneutria* 83
*Phormictopus* 55, 279, 314
*Phoroncidia* 236, 237, 311, 314, 327
PHRUROLITHINAE 38, 79, 80, 81, 85, 106, 162, 256, 297, 315
*Phrurolithus* 80, 82, 297, 302, 315, 331, 333
*Phruronellus* 80, 82, 315, 333
*Phrurotimpus* 79, 80, 81, 274, 315, 333
*Phryganoporus* 283
*phrygiana*, *Pityohyphantes* 315
*phthisica*, *Lipocrea* 301
*Phycosoma* 236, 239, 276, 315
*Phycus* 315
*Physocyclus* 194, 196, 315
PHYXELIDIDAE 20
*Piabuna* 79, 80, 81, 274, 315, 333
*picea*, *Hexura* 50, 51
*pictilis*, *Erigone* 297
*pictipes*, *Theridion* 236
*pikei*, *Geolycosa* 296
*pikes*, *Enoplognatha* 317
*pilifera*, *Ariadna* 219
*pilifrons*, *Arcterigone* 145
*Pimoa* 34, 197, 198, 274, 301, 315, 333
PIMOIDAE 20, 23, 34, 124, **197**, 198, 232, 315, 333
*Pimus* 60, 61, 62, 315, 325, 333
*pineus*, *Thallumetus* 100
*pingue*, *Ctenium* 289
*pinocchio*, *Cornicularia* 288
*pior*, *Tosyna* 315, 325
*Pippuhana* 66, 67, 315
pirate spider 171
*Pirata* 164, 165, 166, 315, 320, 333
*piratica*, *Lycosa* 315
*piraticus*, *Araneus* 315
*piraticus*, *Icius* 312
*piraticus*, *Pseudicius* 206
*Pisaura* 315
*Pisauria* 315
PISAURIDAE 13, 20, 24, 35, 54, 83, 118, 119, 164, 190, **199**, 249, 291, 305, 313, 315, 325, 333
*Pisaurina* 199, 200, 313, 315
*Pisauroidea* 119
*piscatorius*, *Strotarchus* 173, 174
*pitus*, *Pimus* 315, 325
*Pityohyphantes* 127, 128, 156, 277, 278, 300, 312, 315, 333
*planeticus*, *Strotarchus* 174
*Platoecobius* 183, 184, 315
PLATONINAE 245
*platus*, *Phormictopus* 55
*Platycryptus* 206, 214, 315
*platyrhinus*, *Goneatara* 297
*Platyxysticus* 288, 316
*Plectana* 276, 304, 306
PLECTREURIDAE 20, 32, 102, **201**, 300, 316, 333
*Plectreurys* 201, 202, 300, 301, 316, 333
*Plesiometa* 34, 232, 233, 316
*plesius*, *Sisis* 142
PLEXIPPINAE 206
*Plexippus* 206, 211, 316
*pluchei*, *Holocnemus* 196, 299
*Pocadicnemis* 127, 140, 316
*Pocobletus* 132, 159
*pococki*, *Hypohilus* 120, 121
*Poecilarcys* 301
*Poecilochroa* 110, 316
*Poeciloneta* 127, 131, 159, 281, 316, 333
*polaris*, *Perro* 152
*Polenecia* 251
*Ponella* 251
*Porrhomma* 127, 131, 159, 316, 328, 333
*Porrhothele* 316
*Porropis* 316
*positivum*, *Theridion* 240
*potteri*, *Agelenopsis* 56, 277
*potteria*, *Tuganobia* 326
*Poultonella* 206, 207, 210, 316
*praecursa*, *Erigone* 148
*prentoglei*, *Cavernocymbium* 61
*princeps*, *Islandiana* 300
*pristina*, *Artema* 289
PRITHINAE 105
*probata*, *Erigone* 293, 313
*probatus*, *Catabrithorax* 293

*proceps*, *Anacornia* 146
*Procerocymbium* 127, 147, 316
PRODIDOMIDAE 20, 24, 37, 75, 103, 106, **203**, 309, 316, 333
PRODIDOMINAE 203
*Prodidomus* 203, 204, 316
*projectum*, *Theridion* 314
*prominens*, *Cercidia* 71
*prominula*, *Hackmania* 100
*prominulus*, *Metopobactrus* 147
*Promyrmekiaphila* 45, 46, 47, 276, 316
*prona*, *Lasaeola* 301
*propinquus*, *Herpyllus* 111
*propinquus*, *Tibellus* 325
*proserpina*, *Porrhomma* 316
*Prosopotheca* 316
*prosper*, *Bagheera* 216
*Prosthesima* 329
*proszynskii*, *Evarcha* 206, 212
*proxima*, *Araniella* 72
PSECHRIDAE 20, 190
*Pselothorax* 316
*Pseudeuophrys* 206, 207, 215, 317
*Pseudicius* 206, 213, 312, 317
*Pseudometa* 316
*Pseudomyrmex* 208
*Pseudoscorpionida* 11, 18, 286
*Pseudosparianthis* 36, 224, 225, 317, 323
*Psilochorus* 194, 196, 317, 333
PSILODERCINAE 181
*pubescens*, *Neozimiris* 204
*pubescens*, *Zimiris* 309
*pulcher*, *Oonops* 186
*pulcherrima*, *Meotipa* 241, 242
*pulchra*, *Camillina* 108
*pullatus*, *Bathyphantes* 288
*pullulus*, *Psilochorus* 196
*pulverulenta*, *Alopecosa* 169
*pumila*, *Zora* 256, 257
*punctatus*, *Fuentes* 295
*puncigera*, *Chemmis* 278
*punctipes*, *Theridion* 236
*punctulata*, *Anahita* 84
*purpurescens*, *Styloctetor* 150
purseweb spider 41
*Purumitra* 251
*pusilla*, *Hahnia* 297
*pusilla*, *Linyphia* 206, 317
*Pusillia* 317
*pusillum*, *Theridion* 307
*pusillus*, *Minyriolus* 307
*putus*, *Oecobius* 184
*pygmaeus*, *Typhochrestus* 145, 152
*pyramidalis*, *Lasaeola* 301
*pyramitela*, *Frontinella* 132, 159
*Pythochrestus* 327
*Pythonissa* 316

#

*quaestio*, *Wabasso* 149
*quercicus*, *Bufo* 295

#

*rabida*, *Lycosa* 317
*Rabidosa* 165, 167, 169, 317, 333
*Rachodrassus* 110, 317, 322
*radiata*, *Hogna* 165
*radiata*, *Lycosa* 299
*raptor*, *Arctosa* 168
Rastelloidina 20
*ravilla*, *Eriophora* 72
ray spider 244
*reclusa*, *Iviella* 99
*reclusa*, *Loxosceles* 222
*reclusa*, *Tricholathys* 300
*recurvatus*, *Bathyphantes* 311
*recurvatus*, *Oreophantes* 130, 158
*referena*, *Siratoba* 320
*referens*, *Ariston* 282, 320
*referens*, *Siratoba* 251, 252, 253

*regius, Phidippus* 205
*Reo* 171, 172, 317
*reprobus, Bathyphantes* 128, 156
*republicanus, Uloborus* 314
*reticulata, Smodix* 147, 320
*reticulatus, Cybaeus* 325
*reticulatus, Neon* 309
*reticulatus, Smodix* 320
*reticulatus, Tmeticus* 320
*retionax, Smodigoides* 320
*rex, Eucteniza* 46, 47
*Rhaeboctesis* 317
*Rhaebothorax* 317
*rhagavi, Homalonychus* 119
*Rhechostica* 317
*Rhechosticha* 317
*Rhecostica* 276, 279, 292, 301, 309, 317
*Rhecosticha* 290
*Rhetenor* 206, 207, 213, 274, 317
*Rhomphaea* 236, 238, 281, 317
*rita, Linyphia* 125, 129, 157
*rita, Melpomene* 58
*Ritalena* 57, 317
*rivalis, Diplocentria* 306
*rivalis, Tmeticus* 291
*riversi Atyoides* 282
*Robertus* 124, 235, 236, 238, 289, 295, 317, 333
*robustum, Leptorhoptrum* 125
*riscidus, Coelotes* 278
*rostratum, Pholcomma* 295
*rostratus, Floricomus* 295
*rostratus, Origanates* 138
*rothi, Hesperocranum* 163, 298
*rothi, Hexura* 50, 51
*rothi, Neoapachella* 46, 47
*rotundata, Aranea* 322
*rotundata, Synema* 322
*rotundus, Sisis* 142, 151
*rua, Agelena* 317
*Rualena* 57, 59, 275, 299, 310, 317, 333
*rubidum, Zodarion* 254, 255
*rufescens, Loxosceles* 222, 303
*rufipes, Nesticodes* 242
*rufipes, Sphodros* 41
*rufipes, Theridion* 310
*rufum, Theridion* 303
*rufus, Argus* 303
*rufus, Prodidomus* 203, 204
*Rugatha* 317, 318
*Rugathodes* 236, 242, 243, 318
*rugosa, Ceratinops* 285
*rugosa, Enoplognatha* 317
*rugosus, Ceratinops* 142
running crab spider 192
*ruralis, Sitalcas* 153
*rurestris, Meioneta* 305
*ruricola, Trochosa* 167, 168
*rustica, Sisyrbe* 153
*rusticus, Urozelotes* 106, 108

# S

*Saaristoa* 127, 160, 318, 133
*sacra, Calisoga* 52
sac spider 77
*saeva, Tegenaria* 58
*Saitis* 278
*saltabunda, Anyphaena* 279
*saltabundus, Wulfila* 66
SALTICIDAE 6, 10, 12, 14, 15, 18, 19, 20, 36, 66, 86, 189, **205**, 206, 207, 208, 209, 210, 211, 212, 213, 214, 215, 216, 270, 276, 278, 282, 283, 285, 286, 288, 289, 290, 293, 294, 295, 296, 297, 298, 299, 300, 303, 304, 305, 306, 308, 309, 311, 312, 313, 314, 315, 316, 317, 318, 319, 320, 322, 323, 324, 327, 328, 330, 338, 333
SALTICINAE 207
*Salticus* 207, 210, 282, 283, 298, 305, 318, 320
*salticus, Oxyopes* 189, 190
*Saltonia* 34, 56, 93, 95, 96, 97, 318
*saltuaria, Pardosa* 170
*samensis, Caviphantes* 284
*sammamish, Lepthyphantes* 318
*sammamish, Saaristoa* 133, 160
*sandra, Leptoneta* 122, 123

*saphes, Argenna* 100
*saphes, Hackmania* 100
*sarcocuon, Blestia* 145
*Sarinda* 206, 207, 209, 318
*Sassacus* 206, 207, 212, 213, 215, 216, 318
*Satilatlas* 127, 140, 313, 318, 333
*savannum, Theridiosoma* 244, 245
*Savignia* 127, 136, 285, 318
*saxatilis, Clubiona* 287
*saxetorum, Caviphantes* 148
*scabricula, Erigone* 326
*scabriculus, Troxochrus* 326
*scalaris, Oxyopes* 189
*Scaphiella* 185, 187, 318
*Scaptophorus* 298
*Scatophorus* 318
*scenicus, Salticus* 210
*schantzi, Gertschanapis* 64, 65
*Schizocosa* 164, 165, 166, 167, 169, 274, 282, 298, 303, 318, 333
*Schizogyna* 318
*Schizomida* 18
*schlingeri, Apomastus* 47
*Schoenobates* 311
*schusteri, Archoleptoneta* 123
*Sciastes* 127, 147, 151, 155, 278, 318, 333
*Scirites* 127, 149, 318
*Scironis* 127, 152, 318
*Scoloderus* 69, 70, 318
*Scolopembolus* 127, 154, 319
*Scopodes* 109, 319
*Scopoides* 37, 106, 107, 109, 203, 305, 319, 333
*scopulifer, Glyphesis* 138
*scopus, Tetragnatha* 319
Scorpiones 11, 18
*scorpionia, Synageles* 286
*Scotina* 319
*Scotinella* 79, 80, 82, 315, 319, 333
*Scotinotylus* 127, 136, 137, 142, 287, 288, 319, 329, 333
*Scotolathys* 319, 326
*Scotophaeus* 106, 107, 111, 319
*Scotoussa* 291, 319
*Scotussa* 319
*Scotynotylus* 287, 288, 319, 329
*scutulatus, Scotophaeus* 319
*Scylaceus* 127, 151, 319
*Scyletria* 127, 153, 155, 319
*Scytodes* 30, 217, 218, 222, 319, 333
SCYTODIDAE 20, 22, 30, 201, **217**, 218, 222, 319, 333
Scytodoidea 75, 201
*secreta, Anapistula* 226
*seemanni, Aphonopelma* 55
*Segestria* 219, 220, 290, 319
SEGESTRIIDAE 20, 22, 30, 103, **219**, 281, 287, 319, 333
*Segestrioides* 102
*segmentatus, Cheliferoides* 208
SELENOPIDAE 20, 36, 118, 192, **221**, 224, 319
*selenopoides, Homalonychus* 118
*Selenops* 36, 118, 221, 319, 333
*seminola, Diaea* 247
*Semljicola* 127, 148, 292, 301, 319
*senoculata, Spermophora* 195
SENOCULIDAE 20, 190
*Septentrinna* 80, 82, 319
*sequoia, Cybaeina* 88
*sequoiae, Chorizomma* 329
*Sergia* 319
*Sergiolus* 107, 110, 316, 319, 333
*sericata, Laseola* 301
*sexpunctata, Fuentes* 295
*sexpunctatum, Rugathodes* 318
*shantzi, Chasmocephalon* 285, 296
*shantzi, Gertschanapis* 64
*sheari, Hypochilus* 121
sheet web spider 124
*Sibianor* 206, 212, 213, 319
*sibiricum, Ivielum* 151, 300
SICARIIDAE 15, 20, 30, 102, 201, 217, 218, **222**, 223, 290, 333
*Sicarius* 223
*Sidusa* 319
*signifer, Cybaeus* 85, 86
*Silometopoides* 127, 146, 320
*Silometopus* 320
*silvaticus, Chrosiothes* 286
*silvestrii, Nesticus* 178
*sima, Scironis* 152

*simile, Theridion* 320
*Simitidion* 236, 242, 243, 320
*Simonella* 329
*simoni, Comaroma* 288
*simoni, Metaltella* 63
*Simonida* 319
*sinensis, Octonoba* 251, 252, 253
*Singa* 69, 73, 74, 299, 302, 320
*Sintula* 278, 301
*sira, Siratoba* 320
*Siratoba* 250, 251, 252, 253, 282, 320
*Sisicottus* 127, 152, 154, 320, 330, 333
*Sisicus* 127, 128, 134, 320
*Sisis* 127, 142, 151, 274, 318, 319, 320, 322, 326
*sisyphoides, Tidarren* 241, 243
*Sisyrbe* 127, 153, 320
*Sitalcas* 127, 153, 320
*siticulosus, Pseudicius* 213
SITTICINAE 206
*Sitticus* 206, 211, 320, 333
*Smeringopus* 194, 196, 320
*Smodigoides* 320
*Smodingoides* 320
*Smodix* 127, 147, 320
*snetsingeri, Atypus* 42
*sober, Anachemmis* 230, 278
*Socalchemmis* 230, 231, 278, 320, 333
*Solenopsis* 79
Solifugae 11, 18
*solituda, Acantholycosa* 165
*solituda, Pardosa* 166
*Sosilaus* 320
SOSIPPINAE 165
*Sosippus* 56, 164, 165, 167, 168, 320, 333
*Sosticus* 107, 109, 110, 111, 320
*Sostogeus* 111, 320
*Soucron* 127, 153, 274, 293, 320
*Soudinus* 320
*Souessa* 127, 145, 320
*Souessoula* 127, 152, 320
*Sougambus* 127, 128, 131, 320
*Souidas* 127, 151, 320
*Soulgas* 127, 151, 320
SPARASSIDAE 20, 36, 77, 118, 192, 221, **224**, 225, 298, 311, 317, 321, 323, 333
*Sparassus* 224, 274, 282, 291, 299, 311, 317, 321, 324
SPARIANTHINAE 224
*Sparianthis* 317
*Spariolenus* 317
*spathula, Tivyna* 325
*spatulata, Epeira* 294
*Spencerella* 329
*Speocera* 182
*Spermophora* 194, 195, 321
*Sphasus* 282, 291, 321, 324
*Sphecozone* 127, 146, 152, 321
*Sphodros* 41, 42, 321, 333
*Sphyrotinus* 242, 243, 321
*spinifera, Souessa* 145
*spiniger, Acanthoctenus* 84, 276
*spiniger, Pirata* 166
*spinimana, Zora* 329
*spinimana, Zoropsis* 259
*spinimanus, Heteroonops* 185, 186
*spinipes, Acanthoctenus* 84
*spinosa, Deinopis* 91, 92
*spinosa, Mazax* 304
SPINTHARINAE 236
*Spintharus* 236, 239, 321
*spiralis, Spiropalpus* 316, 321
*Spirembolus* 125, 127, 137, 139, 145, 146, 149, 283, 291, 321, 325, 333
*Spiropalpus* 316, 321
spitting spider 217
*Staphylococcus* 15
*stativus, Styloctetor* 150
*Steatoda* 23, 235, 236, 238, 282, 302, 321, 324, 333
*steckleri, Septentrinna* 82
*stellata, Acanthepeira* 8, 68, 276, 304
*stellata, Epeira* 8
*stellata, Marxia* 304
*stellata, Plectana* 276
*Stemmops* 236, 240, 321
*Stemonyphantes* 125, 127, 129, 157, 321
*stenaspis, Triaeris* 188
STENOCHILIDAE 20

*Stenoonops* 185, 186, 321
*sternalis, Pardosa* 169
*sternitzkii, Hesperauximus* 93
*Sterrhochrotus* 286
*sticta, Crustulina* 237
STIPHIDIIDAE 20, 24, 190
*stolida, Eucteniza* 46, 47
*Storena* 303, 321
*Storenomorpha* 119
*Streptococcus* 15
*striata, Agroeca* 302
*stridula, Theonoe* 237
*Strotarchus* 173, 174, 321
*studiosus, Anelosimus* 240
*Styloctetor* 125, 127, 150, 322
*Stylophora* 291, 322
*Styposis* 236, 238, 240, 322
*Subbekasha* 127, 322
*subitaneus, Microctenonyx* 141
*subterranea, Phanetta* 148, 314
*subterraneum, Chorizomma* 286
*sudetica, Acantholycosa norvegica* 276
*suprenans, Kibramoa* 201, 202
*suprenans, Plectreurys* 300
*swainsonii, Micrathena* 306
*Sybota* 251
*syllepsicus, Mimetus* 172
*Symmigma* 127, 141, 322
*Symphytognatha* 278, 322
SYMPHYTOGNATHIDAE 20, 30, 64, 175, 176, **226**, 227, 245, 278, 284, 322, 333
*Synaema* 322
*Synageles* 206, 207, 209, 296, 322, 333
SYNAGELINAE 206
*Synaphosus* 106, 107, 110, 111, 322
SYNAPHRIDAE 20
SYNAPHRINAE 176
*Synaphris* 176
*Synema* 246, 247, 322
*Synemosyna* 206, 207, 208, 296, 322
SYNEMOSYNINAE 206
SYNOTAXIDAE 20, 235
*syntheticus, Nodocion* 322
*syntheticus, Synaphosus* 111
*Syrisca* 174, 312, 322
*Syriscus* 323
*Syspira* 173, 174, 322
*systematicus, Tarsonops* 75, 76
*szechwanensis, Habrocestoides* 286

# T

*Tachygyna* 127, 149, 322, 333
*taeniola, Attus* 305
*taeniola, Fuentes* 295
*takayense, Theridion* 322
*Takayus* 236, 242, 243, 322
*Talanites* 106, 107, 110, 317, 322
*Talavera* 207, 213, 274, 295, 322
*Tama* 116, 309, 322, 329
*Tangaroa* 251
*Tapinauchenius* 323
*Tapinesthis* 185, 186, 323
*Tapinillus* 190
*Tapinocyba* 127, 137, 140, 141, 282, 314, 323, 333
*Tapinopa* 127, 129, 157, 323
*Taranucnus* 127, 131, 158, 274, 323
Tarentula 54
*Tarentula* 278
*tarsalis, Scironis* 152, 318
*Tarsonops* 75, 76, 310, 311, 323
*Tarsonopsis* 323
*tauphora, Linyphia* 125, 129, 157
*tauricornis, Epeira* 328
*tauricornis, Wagneriana* 71
*Tegenaria* 15, 35, 56, 57, 58, 284, 286, 323, 333
*Tekella* 275, 323
*Tekellina* 236, 240, 323
*Telema* 229, 323, 327
TELEMIDAE 20, 29, 31, 86, 122, 123, 181, 185, 264, **228**, 229, 323, 327
TELEMINAE 229
*Teminius* 173, 174, 323
*temporarius, Litopyllus* 111

*tenebrosus, Dolomedes* 199
*Tengella* 230, 258, 323
TENGELLIDAE 20, 24, 36, 38, 77, 83, 225, **230**, 248, 256, 258, 259, 278, 301, 302, 320, 323, 329, 333
*Tengellinae* 230, 258, 302
*Tennesseellum* 124, 127, 134, 161, 323
*tennesseensis, Ivesia* 300
*Tentabuna* 323
*Tentabunda* 224, 323
*Tenuiphantes* 127, 323
*tenuis, Lepthyphantes* 323
*tepidariorum, Theridion* 276
*tepidariorum, Theridium* 307
*Teraphosa* 324
*Terralonus* 206, 216, 323, 333
*terricola, Trochosa* 167, 168, 313
*tesquorum, Pardosa* 168, 170
*testacea, Filistata* 295
*testaceomarginata, Lasaeola* 301
TETRABLEMMIDAE 20
*Tetragnatha* 232, 233, 291, 292, 296, 307, 312, 319, 322, 323, 331, 333
TETRAGNATHIDAE 6, 14, 20, 32, 34, 68, 91, 124, 197, 198, **232**, 233, 251, 264, 282, 291, 296, 305, 307, 309, 312, 316, 323, 333
Tᴇᴛʀᴀɢɴᴀᴛʜɪɴᴀᴇ 179, 233
*Tetrapneumonae* 24
*Teudis* 66, 67, 274, 324
*Teutana* 274, 324
*texana, Pholcophora* 195
*texanus, Modisimus* 195
*texanus, Rhetenor* 213
*texense, Aphonopelma* 55
*texensis, Tapinauchenius* 323
*Thalamia* 315, 324
*Thaleria* 127, 128, 324
*Thallumetus* 95, 96, 100, 324
*Thanatus* 192, 193, 224, 324, 333
*Theoclia* 324
*theologus, Homalonychus* 118
*Theonoe* 124, 176, 236, 237, 288, 324
*Theonoeae* 176
*Theotima* 31, 181, 182, 324
Tʜᴇᴏᴛɪᴍɪɴᴀᴇ 181, 182
*Theraphosa* 324
THERAPHOSIDAE 20, 21, 26, 52, **54**, 279, 283, 285, 287, 290, 292, 294, 297, 308, 314, 317, 323, 324, 333
Tʜᴇʀᴀᴘʜᴏsɪɴᴀᴇ 20
*Theraphosodina* 20
*Theraphosoidea* 20
THERIDIIDAE 4, 11, 15, 20, 23, 29, 33, 64, 124, 171, 175, 176, 178, 179, 183, 232, **235**, 244, 262, 272, 276, 278, 279, 280, 282, 284, 286, 287, 288, 289, 291, 292, 293, 294, 295, 300, 301, 302, 303, 306, 309, 310, 311, 312, 314, 316, 317, 318, 320, 321, 322, 323, 324, 325, 327, 328, 329, 333
Tʜᴇʀɪᴅɪɪɴᴀᴇ 236
*theridioides, Graphomoa* 132, 159
*Theridion* 235, 236, 240, 241, 242, 243, 276, 278, 288, 292, 303, 307, 310, 314, 318, 320, 321, 322, 324, 325, 333
*Theridionidae* 171
*Theridiosoma* 33, 244, 245, 324
*Theridiosomateae* 245
THERIDIOSOMATIDAE 20, 33, 64, 68, 124, 175, 176, 226, 227, **244**, 245, 288, 313, 324, 333
Tʜᴇʀɪᴅɪᴏsᴏᴍᴀᴛɪɴᴀᴇ 245
*Theridiosomatini* 245
*Theridiotis* 239
*Theridium* 292, 305, 307, 321, 324
*Theridula* 236, 240, 242, 312, 324
*Theuma* 312
*theveneti, Brachythele* 52
*theveneti, Calisoga* 53
*theveneti, Hebestatis* 43, 44, 298
*Thiodina* 206, 207, 209, 324
Tʜɪᴏᴅɪɴɪɴᴀᴇ 206
*Thomisa* 322
THOMISIDAE 4, 14, 15, 20, 36, 192, 205, 221, 224, **246**, 266, 283, 288, 290, 300, 303, 307, 311, 312, 316, 322, 324, 325, 329, 333
*Thomisus* 246, 274, 291, 312, 321, 324
*thoracica, Aranea* 319, 333
*thoracica, Scytodes* 217, 319
*thoracica, Vermontia* 154
*thoracicus, Tmeticus* 328
*thorelli, Hypochilus* 120, 121, 299
*thwaitesi, Miagrammopes* 306
*thyasionnes, Parazanomys* 61
*Thymoites* 29, 124, 235, 236, 242, 243, 312, 321, 324, 333

*Thyreosthenius* 127, 141, 299, 324
*Tibellus* 37, 192, 193, 325, 333
*tibialis, Souidas* 151
*Tibioplus* 127, 128, 133, 325
*Tidarren* 235, 236, 241, 243, 274, 325
*Tigellinus* 274, 325
*tigrina, Euophrys* 294
*tigrinus, Coelotes* 278
*Tinus* 199, 200, 325
*Tiso* 127, 150, 325
*Titanoeca* 28, 60, 248, 325
TITANOECIDAE 20, 24, 28, 60, 63, **248**, 258, 325, 333
Tɪᴛᴀɴᴏᴇᴄɪɴᴀᴇ 60, 248
*Titidius* 325
*Titiotus* 77, 230, 231, 325
*Titurius* 325
*tius, Cybaeus* 325
*tivior, Lycosa* 325
*Tivyna* 95, 96, 100, 101, 314, 325, 328
*Tmarus* 246, 247, 325, 333
*Tmeticus* 127, 148, 288, 291, 293, 313, 320, 325, 328
*Tomopisthes* 290
*torreya, Cyclocosmia* 44
*Tortembolus* 287, 325
*Tortolena* 57, 59, 275, 299, 317, 325
*Tortula* 278
*tosior, Ceratinella* 315
*Tosnyna* 292, 314, 325, 328
*Tosyna* 275, 315, 325
*Trabaea* 325
*Trabea* 165, 325
*Trabeops* 165, 325
*Trachelas* 79, 80, 305, 325, 326, 333
Tʀᴀᴄʜᴇʟɪɴᴀᴇ 38, 79, 80, 103, 263, 333
*Trachyzelotes* 107, 108, 326
*Traematosis* 274, 326
*Traematosisis* 127, 138, 326
trapdoor spider 39
*Trebacosa* 165, 166, 326
*Trechalea* 35, 199, 249, 326
TRECHALEIDAE 20, 24, 35, 164, 190, 199, **249**, 326, 333
*Trematosisis* 326
*Triaeris* 185, 188, 326
*triangularis, Linyphia* 128, 156
*Tricholathys* 96, 99, 300, 326, 333
Tʀɪᴄʜᴏʟᴀᴛʜʏsɪɴᴀᴇ 56, 96, 97
*Trichopterna* 326
*Triclaria* 326
*tricornis, Walckenaeria* 145
*tridentiger, Xeropigo* 82
*trifrons affine, Baryphyma* 139
*Trigonobothrys* 236, 239, 301
*Trionycha* 24
*tristis, Plectreurys* 202
*trivittata, Mexitlia* 96, 100
*trivittatus, Mallos* 306
TROCHANTERIIDAE 20
*Trochosa* 164, 165, 167, 168, 274, 313, 326, 327
*Trochosina* 313
*Troglonata* 326
*Trogloneta* 175, 176, 177, 326
*Troxochrus* 293, 326
*truncata, Cyclocosmia* 44
*truncatus, Sciastes* 155
*tuberculata, Aranea* 293
*tuberculifer, Scoloderus* 70
*tuberella, Aigola* 277
*tuberosum, Gongylidium* 277
*tuckeri, Eidmannella* 179
*Tugana* 326
*Tuganobia* 326
*tuganus, Ceraticelus* 326
*Tunabo* 326
*Tunagyna* 127, 153, 322, 326
*tunagyna, Helophora* 327
*tuoba, Tunagyna* 322
*tuonops, Ethobuella* 115
turret spider 39
*Tusukuru* 127, 140, 327
*Tutaibo* 127, 150, 327
*Tutelina* 206, 207, 214, 327
*Tutilina* 327
*Tylogonus* 206, 215, 319, 327
*Typhistes* 299
*Typhochraestus* 274, 327

*Typhochrestus* 127, 145, 152, 327
*typicus, Cybeopsis* 289

# U

*uintanus, Typhochrestus* 152
*Ulesanis* 275, 311, 327
ULOBORIDAE 14, 20, 23, 27, 28, 92, **250**, 251, 282, 300, 306, 307, 310, 314, 320, 327, 329, 333
ULOBORINAE 251
*Uloborus* 250, 251, 252, 253, 310, 314, 327, 329
*umbraticus, Nuctenea* 301, 310
*Umidia* 327
*Ummidia* 43, 44, 312, 327, 333
*unca, Hamataliwa* 191
*uncata, Erigone* 292
*undata, Metacyrba* 206, 315
*undulata, Pisaurina* 199
*unica, Argennina* 99, 280
*unica, Hesperocosa* 167, 169
*unica, Schizocosa* 298
*unicolor, Antrodiaetus* 39, 40
*unicolor, Chemmis* 80, 278
*unicolor, Liocranoides* 302
*unicolor, Loxosceles* 222
*unicolor, Megalostrata* 80
*unicolor, Pholcus* 195, 307
*unicolor, Zorocrates* 258
*unicorn, Gnathonargus* 144
*unicornis, Cornicularia* 288
*unicornis, Islandiana* 142
*unicornis, Walckenaeria* 288
*Uroctea* 183
UROCTEIDAE 183
*Uropygi* 11, 18
*Urozelotes* 106, 107, 108, 327
*Usofila* 29, 31, 228, 229, 327
*uta, Hilaira* 294, 307
*utahensis, Coriarachne* 246
*utibilis, Eridantes* 139, 141

# V

*Vaejovis* 305
*vagans, Brachypelma* 54, 55
*vagans, Tiso* 150
*vaginatus, Oreonetides* 277
*Varacosa* 165, 168, 274, 280, 327
*variabilis, Nops* 310
*variana, Pythonissa* 316
*variata, Tama* 309
*variegata, Neriene* 316
*Varyna* 292, 314, 325, 328
*vasingtonus, Spirembolus* 125, 149
*vatia, Misumena* 246, 247
*vatius, Araneus* 307
*velox, Hibana* 66
*venatoria, Heteropoda* 224, 225
VENONIINAE 165
*venusta, Leucauge* 233, 302
*Vermontia* 127, 154, 328
*vernalis, Pelecinus* 187, 188
*Verrucosa* 69, 71, 328
*verrucosa, Epeira* 328
*versicolor, Coriarachne* 246, 288
*Verucosa* 328
*vetteri, Cavernocymbium* 61
*viaria, Microneta* 132, 160
*video, Dactylopisthes* 143
*Viderius,* 328
*vinnula, Neonella* 213
violin spider 222
*virginica Calymmaria* 113
*viridans, Peucetia* 189, 190, 191
*viridis, Argyrodes* 280
*viridis, Linyphia* 280
*viridis, Lyssomanes* 206, 208, 303
*vitis, Icius* 215
*vitis, Sassacus* 215
*Vixia* 328
*Vulfila* 329
*vulgaris, Thanatus* 192, 193

# W

*Wabasso* 127, 149, 328
*Wadotes* 57, 60, 61, 274, 328, 333
*Wagneria* 328
*Wagneriana* 69, 71, 328
*Waitkera* 251
*Wala* 277, 298, 310, 328
*Walckenaera* 328
*Walckenaeria* 125, 127, 139, 143, 145, 150, 288, 316, 321, 325, 328, 333
*Walmus* 60, 62, 328
*Wamba* 236, 240, 243, 295, 328
*Weintrauboa* 198
*Wendilgarda* 298
*wheeleri, Bellota* 208
*wickhami, Beata* 213
*Wideria* 328
*Widerius* 328
*wilburi, Collinsia* 125, 149
*Willibaldia* 328
*Willisus* 113, 115, 328
*Witica* 295, 328
*wittfeldae, Epeira* 305
*Wixia* 71, 310, 328
wolf spider 164
*Wubana* 127, 134, 161, 328, 333
*Wulfila* 66, 67, 279, 328, 329, 333

# X

*xantha, Cybaeina* 88
*Xenesthis* 323
*xerampelina, Pardosa* 170
*Xeropigo* 80, 82, 329
*Xystica* 329
*Xysticus* 246, 247, 288, 329, 333

# Y

*Yabisi* 116, 117, 329
*Yaginumena* 236, 239, 329
*yakima, Argenna* 100
*Yanigumena* 239
*Yorima* 56, 96, 97, 286, 329, 333
*Yucca* 104
*Yukon* 329

# Z

*Zanomys* 60, 61, 312, 313, 329, 333
*zebra, Orthonops* 75
*Zelotes* 107, 108, 326, 327, 329, 333
ZELOTINAE 107
*Zilla* 282, 329
*Zimiris* 309
ZODARIIDAE 20, 23, 34, 118, **254**, 303, 321, 329, 333
ZODARIINAE 254
*Zodarion* 254, 255, 329
*Zodariones* 329
*Zodarium* 305, 329
*zonata, Argyrodes* 280
*zonata, Linyphia* 280
*Zora* 37, 256, 257, 329
ZORIDAE 20, 37, 77, 83, 85, 164, **256**, 257, 259, 329, 333
ZORINAE 257
*Zornella* 124, 127, 141, 329
*Zorocrates* 28, 258, 329
ZOROCRATIDAE 20, 28, 63, 83, 230, 248, **258**, 259, 329, 333
ZOROPSIDAE 20, 24, 28, 83, 164, 230, 258, **259**, 329, 333
*Zoropsis* 28, 259, 329
*Zosis* 250, 251, 252, 253, 329
*zosis, Uloborus* 329
*Zygia* 305, 329
*Zygiella* 34, 68, 69, 73, 74, 305, 329
*Zygoballus* 206, 213, 274, 330
*Zygottus* 127, 151, 330

family: llll llll llll llll l
genus: llll llll llll llll
species: llll llll ll

+6

21 × 3 = 63
+ 6
  69
  24
  93
 100